Downloaded Music

Digital music players such as the iPod have made it possible for people to buy music on the Internet. Sales of music in downloadable form have increased dramatically in recent years. Internet music sales information is often represented graphically in magazines or online publications. To estimate such sales for a particular year, we may need to read a line graph. See the Real-World Connection and Example 10 on pages 61 and 62.

Video Games

In order to achieve high scores, video game players often need to complete levels of a game quickly. The amount of time that it takes to complete a single task such as avoiding a collision can be found by adding *reaction time* and *performance time*. Polynomials can be used to describe reaction and performance times, so the total time needed to complete a single task can be found by adding polynomials. See A Look into Math for Section 8.3 on page 512 and Exercises 91 and 92 on page 521.

Wilderness Maps

When people hike or camp in remote areas, they often rely on a map to guide the way. To be useful, wilderness maps need to be made so that it is possible to accurately identify the locations of relatively small landmarks. Once a hiker finds his or her position on a map, proportions can be used to estimate distances to surrounding landmarks. See A Look into Math for Section 6.2 on page 389 and Exercise 55 on page 399.

Social Networks

From 2006 to 2009, the number of people who joined Facebook increased dramatically. The number of Facebook users for each of these years can be represented numerically in a table of values and visually in a scatterplot or line graph. See Example 7 on page 545.

Setting Yourself Up for SUCCESS in MATH

Make the Most of Class

- Attend class regularly.
- Bring your textbook and a notebook to every class.
- Take notes in class, and write down everything that is on the board.
- Ask questions. (You won't be the only one who wants to know the answer!)
- Study with others in your class.

Get Help When You Need It

- Find your instructor's office, and write down the number and office hours.
- Find the location and hours of the math tutor resources on campus.
- Try to solve the problems and write down your questions before you ask for help.
- Use the online resources, such as videos, that accompany this book.

Master the Exam

- Schedule a time and place to study every day instead of "cramming" before the exam.
- Solve the problems in the Chapter Test and check your answers in the back of the book.
- Read all your notes and go over your homework assignments to review important concepts.
- Pick a few problems from your assignments at random and try to solve them without help.
- Carefully write out all your work on the exam whenever possible.

Relax and Visualize doing well.
You CAN Learn Math.

Prealgebra

Gary K. Rockswold
Minnesota State University, Mankato

Terry A. Krieger
Rochester Community and Technical College

Boston Columbus Indianapolis New York San Francisco Upper Saddle River
Amsterdam Cape Town Dubai London Madrid Milan Munich Paris Montréal Toronto
Delhi Mexico City São Paulo Sydney Hong Kong Seoul Singapore Taipei Tokyo

Editorial Director Chris Hoag
Executive Editor Cathy Cantin
Executive Content Editor Kari Heen
Senior Content Editor Lauren Morse
Editorial Assistant Kerianne Okie
Executive Director of Development Carol Trueheart
Senior Development Editor Elaine Page
Senior Managing Editor Karen Wernholm
Associate Managing Editor Tamela Ambush
Senior Production Project Manager Sheila Spinney
Digital Assets Manager Marianne Groth
Supplements Production Coordinator Kerri Consalvo
Associate Media Producer Jonathan Wooding
Content Development Manager Rebecca E. Williams
QA Manager Assessment Content Marty Wright
Marketing Manager Rachel Ross
Marketing Assistant Ashley Bryan
Senior Author Support/Technology Specialist Joe Vetere
Rights and Permissions Advisor Michael Joyce
Image Manager Rachel Youdelman
Procurement Manager Evelyn Beaton
Procurement Specialist Debbie Rossi
Media Procurement Specialist Ginny Michaud
Associate Director of Design, USHE North and West Andrea Nix
Senior Designer Barbara T. Atkinson
Text Design, Production Coordination, Composition, and Illustrations Nesbitt Graphics, Inc.
Cover Image Water rushing by stones ©Yuriy Kulyk / Shutterstock
Cover Design Studio Montage
Art Director Barbara T. Atkinson

For permission to use copyrighted material, grateful acknowledgment is made to the copyright holders on page G-6, which is hereby made part of this copyright page.

Many of the designations used by manufacturers and sellers to distinguish their products are claimed as trademarks. Where those designations appear in this book, and Addison-Wesley was aware of a trademark claim, the designations have been printed in initial caps or all caps.

Library of Congress Cataloging-in-Publication Data
Rockswold, Gary K.
    Prealgebra / Gary K. Rockswold,
    Terry A. Krieger. –1st ed.
       p. cm.
  ISBN 978-0-321-56799-4
    1. Mathematics–Textbooks. I. Krieger, Terry A. II. Title.
QA39.3.R635 2012

513–dc22   2011012262

1 2 3 4 5 6 7 8 9 10—QGD—14 13 12 11

ISBN 13: 978-0-321-56799-4
ISBN 10: 0-321-56799-4

To my wife Carrie Krieger

who is a kindred spirit and my wonderful, darling friend.

– TAK

Terry Krieger has taught mathematics for over 20 years at the middle school, high school, vocational, community college, and university levels. His undergraduate degree in secondary education is from Bemidji State University in Minnesota, where he graduated summa cum laude. He received his MA in mathematics from Minnesota State University–Mankato. In addition to his teaching experience in the United States, Terry has taught mathematics in Tasmania, Australia, and in a rural school in Swaziland, Africa, where he served as a Peace Corps volunteer. Terry is currently teaching at Rochester Community and Technical College in Rochester, Minnesota. He has been involved with various aspects of mathematics textbook publication for more than 15 years and has joined his friend Gary Rockswold as co-author of a developmental math series published by Pearson Education. In his free time, Terry enjoys spending time with his wife and two boys, physical fitness, wilderness camping, and trout fishing.

Gary Rockswold has been a professor and teacher of mathematics, computer science, astronomy, and physical science for over 35 years. Not only has he taught at the undergraduate and graduate college levels, but he has also taught middle school, high school, vocational school, and adult education. He received his BA degree with majors in mathematics and physics from St. Olaf College and his PhD in applied mathematics from Iowa State University. He has been a principal investigator at the Minnesota Supercomputer Institute, publishing research articles in numerical analysis and parallel processing. He is currently an emeritus professor of mathematics at Minnesota State University–Mankato. He is an author for Pearson Education and has over 10 current textbooks at the developmental and precalculus levels. His developmental co-author and friend is Terry Krieger. They have been working together for over a decade. Making mathematics meaningful for students and professing the power of mathematics are special passions for Gary. In his spare time he enjoys sailing, doing yoga, and spending time with his family.

Contents

1 Whole Numbers 1

5   Decimals   299

6   Ratios, Proportions, and Measurement   380

7 Percents 441

Preface

Prealgebra connects the real world to mathematics in ways that are both meaningful and motivational. Students using this textbook will have no shortage of realistic and convincing answers to the question, "Why is math important in my life?" We introduce mathematical concepts using relevant applications from students' lives to support their conceptual understanding. Math is no longer a sequence of disconnected procedures to be memorized; rather, we show that math does make sense.

In addition to presenting algebraic techniques, we present concepts in a visual way whenever possible. Tables of values, line graphs, and bar charts are included to provide students with an opportunity to understand math in more than one way. In future algebra courses, students will learn that equations can be solved algebraically with symbols, numerically with tables of values, and visually with graphs. This text prepares students for these important concepts by teaching them both reasoning and problem-solving skills.

As mathematicians, we understand the profound impact that mathematics has on the world around us; as teachers with a combined experience of over 50 years, we know that our students often have difficulty learning mathematics and recognizing its relevance. We believe that applications and visualization are essential pathways to empowering students in mathematics. Our goal in *Prealgebra* is to help students succeed.

This textbook is one of four textbooks in our series:

- *Prealgebra*
- *Beginning Algebra with Applications and Visualization*, Third Edition
- *Beginning & Intermediate Algebra with Applications and Visualization*, Third Edition
- *Intermediate Algebra with Applications and Visualization*, Fourth Edition

Features

Real Math for the Real World

The Rockswold/Krieger series places an emphasis on teaching mathematics in context. Students typically understand best when concepts are tied to the real world or presented visually. We believe that meaningful applications and visualization of important concepts are pathways to success in mathematics.

- **Chapter Openers** Each chapter opens with an application that motivates students by offering insight into the relevance of that chapter's mathematical concepts.

- **A Look into Math** Each section is introduced with a practical application of the math topic students are about to learn.

- **Real World Connection** Where appropriate, we expand on specific math topics and their connections to the everyday world.

- **Applications and Models** We integrate applications and models into both the discussions and the exercises to help students become more effective problem solvers.

Understanding the Concepts

Conceptual understanding is critical to success in mathematics.

- **Math from Multiple Perspectives** Throughout the text, we present concepts by means of verbal, graphical, numerical, and symbolic representations to support multiple learning styles and problem-solving methods.

- **New Vocabulary** At the beginning of each section, we direct the students' attention to important terms before they are discussed in context.

- **Reading Check** Reading Check questions appear along with important concepts, ensuring that students understand the material they have just read.

- **Study Tips** Study tips offer just-in-time suggestions to help students stay organized and focused on the material at hand.

- **Making Connections** Throughout the text, we help students understand the relationship between previously learned concepts and new concepts.

- **Putting It All Together** At the end of each section, we summarize the techniques and reinforce the mathematical concepts presented in the section.

Practice

Multiple types of exercises in this series support the application-based and conceptual nature of the content. These exercise types are designed to reinforce the skills students need to move on to the next concept.

- **Extensive Exercise Sets** The exercise sets cover basic concepts, skill-building, writing, applications, and conceptual mastery. The exercise sets are further enhanced by several special types of exercises that appear throughout the text:

  - **Now Try Exercises** Suggested exercises follow every example for immediate reinforcement of skills and concepts.

  - **Checking Basic Concepts Exercises** These mixed review exercises appear after every other section and can be used for individual or group review. These exercises require 10–20 minutes to complete and are also appropriate for in-class work.

  - **Writing About Mathematics** This exercise type is at the end of most sections. Students are asked to explain the concepts behind the mathematics procedures they have just learned.

- **Group Activities: Working with Real Data** This feature occurs once or twice per chapter, and provides an opportunity for students to work collaboratively on a problem that involves real-world data. Most activities can be completed with limited use of class time.

Mastery

By reviewing the material and putting their abilities to the test, students will be able to assess their level of mastery as they complete each chapter.

- **Chapter Summary** In a quick and thorough review, we combine key terms, topics, and procedures with illuminating examples to assist in test preparation.

- **Chapter Review Exercises** Students can work these exercises to gain confidence that they have mastered the material.

- **Chapter Test** Students can reduce math anxiety by using these tests as a rehearsal for the real thing.

- **Cumulative Reviews** Starting with Chapter 2 and appearing in all subsequent chapters, Cumulative Reviews help students see the big picture of math by reviewing topics and skills they've already learned.

Supplements

STUDENT RESOURCES

Student's Solutions Manual

Contains solutions for the odd-numbered section-level exercises (excluding Writing About Mathematics and Group Activity exercises), and solutions to all Concepts and Vocabulary exercises, Checking Basic Concepts exercises, Chapter Review exercises, Chapter Test exercises, and Cumulative Review exercises.

ISBNs: 0-321-77249-0, 978-0-321-77249-7

Worksheets

These lab- and classroom-friendly worksheets provide

- Learning objectives and key vocabulary terms for every text section, along with vocabulary practice problems.
- Extra practice exercises for every section of the text with ample space for students to show their work.

- Additional opportunity to explore multiple representation solutions.

  ISBNs: 0-321-75700-9, 978-0-321-75700-5

Video Resources

- A series of lectures correlated directly with the content of each section of the text.
- Material presented in a format that stresses student interaction, often using examples from the text.
- Ideal for distance learning and supplemental instruction.
- Video lectures include English captions.
- Available in MyMathLab®.

INSTRUCTOR RESOURCES

Annotated Instructor's Edition

- Contains Teaching Tips and provides answers to every exercise in the textbook excluding the Writing About Mathematics exercises.
- Answers that do not fit on the same page as the exercises themselves are supplied in the Instructor Answer Appendix at the back of the textbook.

  ISBNs: 0-321-78284-4, 978-0-321-78284-7

Instructor's Resource Manual with Tests (Download Only)

Includes resources designed to help both new and adjunct faculty with course preparation and classroom management:

- Teaching tips and additional exercises for selected content.
- Three sets of Cumulative Review Exercises that cover Chapters 1–3, 1–6, and 1–10.
- Notes for presenting graphing calculator topics as well as supplemental activities.
- Three free-response alternate test forms and one multiple-choice test form per chapter; one free-response and one multiple-choice final exam.
- Available in MyMathLab and on the Instructor's Resource Center.

Instructor's Solutions Manual (Download Only)

- Provides solutions to all section-level exercises (excluding Writing About Mathematics and Group Activity exercises), and solutions to all Checking Basic Concepts exercises, Chapter Review exercises, Chapter Test exercises, and Cumulative Review exercises.
- Available in MyMathLab and on the Instructor's Resource Center.

PowerPoint Slides

- Key concepts and definitions from the text.
- Available in MyMathLab or can be downloaded from the Instructor's Resource Center.

TestGen®

TestGen (www.pearsoned.com/testgen) enables instructors to build, edit, print, and administer tests using a computerized bank of questions developed to cover all the objectives of the text. TestGen is algorithmically based, allowing instructors to create multiple equivalent versions of the same question or test with the click of a button. Instructors can also modify test bank questions or add new questions. The software and testbank are available for download from Pearson Education's online catalog.

MathXL Online Course (access code required)

MathXL is the homework and assessment engine that runs MyMathLab. (MyMathLab is MathXL plus a learning management system.) With MathXL, instructors can:

- Create, edit, and assign online homework and tests using algorithmically generated exercises correlated at the objective level with the textbook.
- Create and assign their own online exercises and import TestGen tests for added flexibility.
- Maintain records of all student work tracked in MathXL's online gradebook.

With MathXL, students can:

- Take chapter tests in MathXL and receive personalized study plans and/or personalized homework assignments based on their test results.
- Use the study plan and/or the homework to link directly to tutorial exercises for the objectives they need to study.
- Access supplemental animations and video clips directly from selected exercises.

MathXL is available to qualified adopters. For more information, visit www.mathxl.com or contact your Pearson representative.

MyMathLab Online Course (access code required)

MyMathLab delivers **proven results** in helping individual students succeed.

- MyMathLab has a consistently positive impact on the quality of learning in higher education math instruction. MyMathLab can be successfully implemented in any environment—lab-based, hybrid, fully online, traditional—and demonstrates the quantifiable difference that integrated usage has on student retention, subsequent success, and overall achievement.
- MyMathLab's comprehensive online gradebook automatically tracks students' results on tests, quizzes, and homework in the study plan. Instructors can use the gradebook to intervene quickly if students have trouble or to provide positive feedback on a job well done. The data within MyMathLab are easily exported to a variety of spreadsheet programs, such as Microsoft Excel®. Instructors can determine which points of data they want to export, and then analyze the results to determine success.

MyMathLab provides **engaging experiences** that personalize, stimulate, and measure learning for each student.

- **Tutorial Exercises:** The homework and practice exercises in MyMathLab and MyStatLab® are correlated with the exercises in the textbook, and they regenerate algorithmically to give students unlimited opportunity for practice and mastery. The software offers immediate, helpful feedback when students enter incorrect answers.
- **Multimedia Learning Aids:** Exercises include guided solutions, sample problems, animations, videos, and eText clips for extra help at point-of-use.
- **Expert Tutoring:** Although many students describe the whole of MyMathLab as "like having your own personal tutor," students using MyMathLab and MyStatLab do have access to live tutoring from Pearson, from qualified math and statistics instructors who provide tutoring sessions for students via MyMathLab and MyStatLab.

And, MyMathLab comes from a **trusted partner** with educational expertise and an eye on the future.

Knowing that you are using a Pearson product means knowing that you are using quality content. That means that our eTexts are accurate, that our assessment tools work, and that our questions are error-free. And whether you are just getting started with MyMathLab or have a question along the way, we're here to help you learn about our technologies and how to incorporate them into your course.

To learn more about how MyMathLab combines proven learning applications with powerful assessment, visit **www.mymathlab.com** or contact your Pearson representative.

Acknowledgments

Many individuals contributed to the development of this textbook. We thank the following reviewers, whose comments and suggestions were invaluable in preparing *Prealgebra*.

Ali Ahmad, *Dona Ana Community College*
Christy Babu, *Laredo Community College*
Dean Barchers, *Red Rocks Community College*
Susan Beane, *University of Houston—Downtown*
David Behrman, *Somerset Community College*
Marion G. Ben-Jacob, *Mercy College*
Annette Benbow, *Tarrant County College—Northwest Campus*
Rosanne B. Benn, *Prince George's Community College*
Carole Bergen, *Mercy College*
Patricia Blus, *National—Louis University*
Harriet "Gale" Brewer, *Amarillo College*
Mary Sarvis Brown, *Harrisburg Area Community College*
Sharon Brown, *San Juan College*
Rebecca Burkala, *Rose State College*
Edythe "Edie" Carter, *Amarillo College*
Danny Cowan, *Tarrant County College—Northwest Campus*
Ann Darke, *Bowling Green State University*
Monette Elizalde, *Palo Alto College*
Mona Ellis, *Palomar College*
Mary Ann Elzerman, *Lansing Community College*
Jennifer Feenstra, *Gainesville State College*
Richard Fielding, *Southwestern College*
Tammy Ford, *North Carolina A & T State University*
Barbara Gardner, *Carroll Community College*
Jane Go, *Hillsborough Community College – Brandon Campus*
Rene Gottwig, *Sierra College*
Shannon Martin Gracey, *Southwestern College*
Felicia Graves, *Cuyahoga Community College—Western Campus*
Edna Greenwood, *Tarrant County College*
Anne Grice, *Fontbonne University*
Susan Grody, *Broward College—North*
Jin Ha, *Northeast Lakeview College*
Richard Halkyard, *Gateway Community College*
Christine Harding, *Saddleback College*
Lynn Hargrove, *Sierra College*
Alan Hayashi, *Oxnard College*
Tonja Hester, *Amarillo College*
Amy Hobbs, *Blinn College*
Stephanie Houdek, *St. Cloud State University*
Michelle Ingram, *Pasadena City College*
Nancy R. Johnson, *Manatee Community College*

Philip Kaatz, *Mesalands Community College*
Fred Katiraie, *Montgomery College*
Mary Ann Klicka, *Bucks County Community College*
Jan LaTurno, *Rio Hondo College*
David N. Magallanes, *Oxnard College*
Gayathri Manikandan, *East Los Angeles College*
Dorothy Marshall, *Edison State College*
Stacy Martig, *St. Cloud State University*
Aimee Martin, *Amarillo College*
Anna Maria Mendiola, *Laredo Community College*
Karen Mifflin, *Palomar College*
Jeff Milner, *The Art Institute of California—Orange County*
Christine Mirbaha, *Community College of Baltimore County, Dundalk*
Michael Murphy, *Texas State Technical College*
Joanna Oberthur, *Ivy Tech Community College*
Maria Olivas, *Southwestern College*
Barbara Pearl, *Bucks County Community College*
CaLandra Pervis, *Lone Star College—Kingwood*
Sara Pries, *Sierra College*
Don Reichman, *Mercer County Community College*
Heather Roth, *Nova Southeastern University*
Bonnie Rountree, *Northwest Arkansas Community College*
Mario Scribner, *Tidewater Community College*
Julia Simms, *Southern Illinois University—Edwardsville*
Joseph A. Spadaro, *Gateway Community College*
Marie St. James, *St. Clair County Community College*
Donna Statler, *Black River Technical College*
Janet E. Teeguarden, *Ivy Tech Community College*
Maria Usher, *Lansing Community College*
Diane Veneziale, *Burlington Community College*
Paul Vroman, *St. Louis Community College—Florissant Valley Campus*
Sandra Williams, *Front Range Community College*
Catalina Yang, *Oxnard College*
Bashar Zogheib, *Nova Southeastern University*

We would like to welcome Jessica Rockswold to our team. She has been instrumental in developing, writing, and proofing new applications, graphs, and visualizations for this text.

Paul Lorczak, Gary Williams, Lynn Baker, Sarah Sponholz, and David Atwood deserve special credit for their help with accuracy checking. Without the excellent cooperation from the professional staff at Pearson Education, this project would have been impossible. Thanks go to Greg Tobin and Maureen O'Connor for giving their support. Particular recognition is due Elaine Page and Cathy Cantin, who gave essential advice and assistance. The outstanding contributions of Lauren Morse, Sheila Spinney, Joe Vetere, Michelle Renda, Rachel Ross, Tracy Rabinowitz, Ashley Bryan, and Jonathan Wooding are greatly appreciated. Special thanks go to Kathy Diamond, who was instrumental in the success of this project.

Thanks go to Wendy Rockswold and Carrie Krieger, whose unwavering encouragement and support made this project possible. We also thank the many students and instructors with whom we have been associated over the years. Their insights have helped shape the presentation methods used in this text.

Please feel free to send us your comments and questions at either of the following e-mail addresses: gary.rockswold@mnsu.edu or terry.krieger@roch.edu. Your opinion is important to us.

Gary K. Rockswold
Terry A. Krieger

Index of Applications

Work

1 Whole Numbers

Success is the sum of small efforts, repeated day in and day out.

— ROBERT COLLIER

The amount of information on the Internet has grown enormously in recent years. Web sites such as Facebook, Twitter, and YouTube offer ordinary people the opportunity to add content to the World Wide Web.

A 2002 estimate indicated that the public side of the Internet (known as the *Surface Web*) contained about the same amount of information as the Library of Congress. By 2010, more than 250 libraries the size of the Library of Congress were necessary to store all of the information on the Web! The extraordinary growth of the Internet and the development of related technologies such as cell phones and iPads are evidence that society is moving quickly into the information age.

Math plays an important role in generating, retrieving, and storing all forms of information. Using math, we can analyze numbers from everyday life and make informed decisions about the future. As we move into the information age, the ability to reason mathematically is becoming increasingly important. In fact, people with a strong math background will be better prepared for careers and life in the 21st century.

In this chapter, we discuss whole numbers. Understanding whole numbers is essential to understanding all types of numbers used in modern society.

Source: Lyman, Peter, and Hal R. Varian, "How Much Information?" 2003.

1.1 Introduction to Whole Numbers

Natural Numbers and Whole Numbers • Place Value • Word Form • Expanded Form • The Number Line • Bar Graphs, Line Graphs, and Tables

A LOOK INTO MATH ▶

The need for numbers first arose when people wanted to count and measure things such as time, possessions, and money. Originally, there was no need for the number zero because people thought that "none" was not a quantity. As societies developed, more types of numbers were needed. In today's society, we need many kinds of numbers that would have been considered useless in the past. In this section, we discuss the set of whole numbers— a fundamental building block of mathematics. (*Source: Historical Topics for the Mathematics Classroom, Thirty-first Yearbook, NCTM.*)

NEW VOCABULARY

- ☐ Natural numbers
- ☐ Whole numbers
- ☐ Period
- ☐ Standard form
- ☐ Place value
- ☐ Word form
- ☐ Expanded form
- ☐ Number line
- ☐ Graph of a whole number
- ☐ Bar graph
- ☐ Line graph
- ☐ Table

Natural Numbers and Whole Numbers

When children learn to count, they begin with 1 and follow the *counting numbers* in a way that seems *natural* to us. This thought may help you remember that **natural numbers** are the same as counting numbers and can be expressed as follows:

$$1, 2, 3, 4, 5, 6, \ldots$$

Because there are infinitely many natural numbers, three dots called an *ellipsis* are used to show that the list continues in the same pattern without end. A second set of numbers is called the **whole numbers** and can be expressed as follows:

$$0, 1, 2, 3, 4, 5, \ldots$$

Whole numbers include the natural numbers and the number 0.

STUDY TIP

Bring your book, notebook, and a pen or pencil to every class. Write down major concepts presented by your instructor. Your notes should also include the meaning of words written in bold type in the text. Be sure that you understand the meaning of these important words.

▶ **REAL-WORLD CONNECTION** Natural numbers and whole numbers can be used when data are not broken into fractional parts. For example, Table 1.1 lists the number of residents on selected Hawaiian islands. The island of Hawaii, for example, has a population of 201,109 people, and the island of Kahoolawe has a population of 0 people. Note that both natural numbers and whole numbers are appropriate to describe these data because a fraction of a person is not possible.

READING CHECK

- What numbers are the same as the counting numbers?
- What is the difference between the natural numbers and the whole numbers?

TABLE 1.1 Population of Hawaiian Islands

Island	Hawaii	Kahoolawe	Lanai	Oahu
Population	201,109	0	3484	1,012,000

Source: Hawaii Visitor and Convention Bureau—2007 Estimates.

Place Value

Numbers are written using the *digits* 0, 1, 2, 3, 4, 5, 6, 7, 8, and 9. When a number with more than four digits is written, commas are used to separate the digits of the number into groups called **periods**. For example, the number 18,376,403 has **eight** digits and **three** periods. Numbers written this way are said to be in **standard form**.

eight digits

18,376,403

three periods

READING CHECK

- How many digits are there in any period that is located to the right of the first (left-most) comma?

NOTE: The period to the left of the first comma in a whole number may contain one, two, or three digits. However, all periods to the right of the first comma must contain three digits.

The **place value** of a digit in a number written in standard form is determined by the position that the digit takes in the number. In Figure 1.1, the number 18,376,403 is shown in a place value chart that gives the first 15 place values.

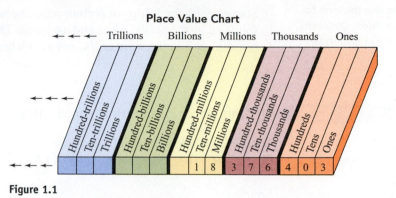

Figure 1.1

Arrows to the left of the place value chart in Figure 1.1 indicate that the periods continue indefinitely. As we move to the left from trillions, the next three periods are quadrillions, quintillions, and sextillions.

NOTE: When a four-digit number is written in standard form, the comma that separates the ones period and the thousands period is optional. For example, the number 3,971 can also be written as 3971. In this text, four-digit numbers are written without a comma.

EXAMPLE 1

Finding the place value of a digit

For the whole number 3,928,107,465,
(a) Determine the place value of the digit 2.
(b) Name the digit that is in the hundred-thousands place.

Solution
(a) The digit 2 is the eighth digit from the right. It is in the ten-millions place.

ten-millions place
3,928,107,465

(b) The hundred-thousands place is the sixth place from the right. That digit is 1.

hundred-thousands place
3,928,107,465

Now Try Exercises 11, 21

Word Form

A whole number written in standard form can also be written in words. For example, the whole number 17,024,863 can be written in **word form** as

seventeen million, twenty-four thousand, eight hundred sixty-three.

This number has three periods—millions, thousands, and ones. The names of the millions and thousands periods are used in writing the number in words, but the name of the ones period is not. Regardless of whether a whole number is written in standard form or in word form, commas are used to separate the periods. To write a whole number in word form, use the following procedure:

STUDY TIP

Ideas and procedures written in boxes like the one to the right are major concepts. Be sure that you understand them.

WRITING A WHOLE NUMBER IN WORDS

Starting with the left-most period, write the word form of the number in each period, followed by the period name and a comma. The name of the ones period is commonly not written, and the word *and* is not used when writing a whole number in words.

EXAMPLE 2 **Writing whole numbers in word form**

Write each whole number in word form.
(a) 62,407   (b) 15,075,410   (c) 2011

Solution
(a) The number in the thousands period is sixty-two, and the number in the ones period is four hundred seven. The word form is sixty-two thousand, four hundred seven.
(b) The number in the millions period is fifteen, the number in the thousands period is seventy-five, and the number in the ones period is four hundred ten. The word form is fifteen million, seventy-five thousand, four hundred ten.
(c) Although the number 2011 does not contain a comma, there are two periods. The number in the thousands period is two, and the number in the ones period is eleven. The word form is two thousand, eleven.

Now Try Exercises 31, 33, 35

▶ **REAL-WORLD CONNECTION** When writing a check, the dollar amount (without the cents) is a whole number written in word form. This is illustrated in the next example.

EXAMPLE 3 **Writing a whole number in standard form**

Write the standard form of the dollar amount written on the check in Figure 1.2.

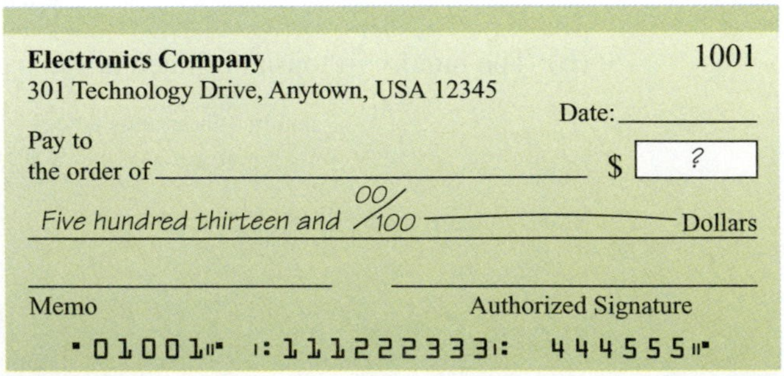

Electronics Company   1001
301 Technology Drive, Anytown, USA 12345
                                                     Date: _____
Pay to
the order of _____ $ [?   ]

Five hundred thirteen and 00/100 _____ Dollars

Memo _____   _____
                               Authorized Signature
   ⑆01001⑈ ⑆111222333⑆ 444555⑈

Figure 1.2

Solution
Five hundred thirteen is written in standard form as 513.

Now Try Exercise 37

Expanded Form

▶ **REAL-WORLD CONNECTION** A student who cashes a paycheck for $784 receives seven hundred-dollar bills, eight ten-dollar bills, and four one-dollar bills. The total received is

700   and   80   and   4

dollars. By replacing each "and" with a *plus sign* (+), we can write the number 784 in **expanded form** as

700 + 80 + 4.

Note that **700** is the standard form of the number given by the digit **7** in the **hundreds** place, **80** is the standard form of the number given by the digit **8** in the **tens** place, and **4** is the digit in the **ones** place. To write a whole number in expanded form, use the following procedure:

READING CHECK

- How can you tell how many zeros are needed to write the standard form for a digit?

WRITING A WHOLE NUMBER IN EXPANDED FORM

Starting with the left-most digit, write the standard form of the number given by each digit and its corresponding place value. Place a plus sign between each of these results.

EXAMPLE 4 ▶ **Writing a whole number in expanded form**

Write each whole number in expanded form.
(a) 45,923   **(b)** 709,416

Solution
(a) The digit **4** in the **ten-thousands** place represents **40,000**. The digit **5** in the **thousands** place represents **5000**. Likewise, the digit 9 represents 900, and the digit 2 represents 20. Finally, the digit 3 is in the ones place. The number can be written in expanded form as 40,000 + 5000 + 900 + 20 + 3.
(b) The expanded form is 700,000 + 9000 + 400 + 10 + 6.

NOTE: The digit **0** in the ten-thousands place is not used in the expanded form because we do not write 0 ten-thousands as **0**0,000. Whenever 0 appears in the standard form of a whole number, it will not be included as part of the expanded form.

Now Try Exercises 41, 45

The Number Line

Sometimes it is helpful to visualize whole numbers using a **number line**. Starting with 0, whole numbers are written below equally spaced *tick marks* as shown in Figure 1.3. The arrow on the right end of the number line means that the whole numbers continue without end. Since we are working only with 0 and numbers greater than 0, we will not place an arrow on the left end of the number line.

Visualizing Whole Numbers

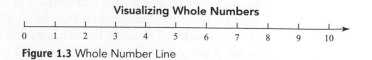

Figure 1.3 Whole Number Line

The **graph of a whole number** shows a dot placed on the number line at the whole number's position. For example, the whole numbers 3 and 7 are graphed in Figure 1.4.

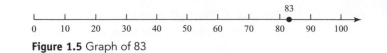

Figure 1.4 Graph of 3 and 7

READING CHECK

• How is a number graphed on a number line?
• Why would we need to adjust the scale of a number line?

When graphing larger whole numbers, it is often necessary to adjust the *scale* of the number line so that the distance between consecutive tick marks represents a more convenient value. A graph of the whole number 83 is shown in Figure 1.5. Tick marks are labeled by tens, and the position of the dot is approximated between the tick marks for 80 and 90.

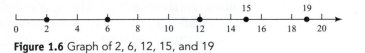

Figure 1.5 Graph of 83

When two numbers are graphed on the same number line, the number to the left is always *less than* the number to the right. Similarly, the number to the right is always *greater than* the number to the left. To show that one number is less than another, we use the symbol $<$. For example, $3 < 7$ because 3 is located to the left of 7 on the number line. Likewise, to show that one number is greater than another, we use the symbol $>$. For example, $7 > 3$ because 7 is located to the right of 3 on the number line.

EXAMPLE 5   **Graphing and comparing whole numbers**

Graph the whole numbers 12, 6, 19, 2, and 15 on the same number line. Use your graph to compare the numbers in parts (a)–(d) and write the correct symbol, $<$ or $>$, in the blank.

(a) 6 _____ 15   **(b)** 19 _____ 12   **(c)** 6 _____ 2   **(d)** 12 _____ 15

Solution

The graph is shown in Figure 1.6. Note that the scale has been adjusted so that the numbers may be more easily represented on the number line.

Figure 1.6 Graph of 2, 6, 12, 15, and 19

(a) On the number line, the number 6 is located to the **left** of the number 15, so we know that 6 is **less than** 15. We write $6 < 15$ and estimate the graph of 15 to be halfway between the tick marks for 14 and 16.

(b) Because 19 is located to the **right** of 12 on the number line, we know that 19 is **greater than** 12, and we write $19 > 12$. The graph of 19 is plotted halfway between the tick marks for 18 and 20.

(c) The number 6 is **greater than** the number 2 because it is located to the **right** of 2 on the number line. We write $6 > 2$.

(d) We write $12 < 15$ because 12 is located to the **left** of 15, or 12 is **less than** 15.

Now Try Exercises 55, 57

Bar Graphs, Line Graphs, and Tables

▶ **REAL-WORLD CONNECTION** To make it easier to read, compare, and analyze data, numbers are often displayed in graphs and tables. For example, Figure 1.7 shows a **bar graph** displaying a seven-day temperature forecast. A glance at this bar graph can provide information very quickly. The tallest bar indicates that Tuesday will have the warmest temperature, and the shortest bar indicates that Friday will be the coolest day of the week.

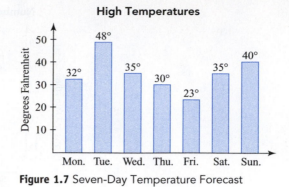

Figure 1.7 Seven-Day Temperature Forecast

READING CHECK

• Why is it sometimes helpful to display data in graphs and tables?

EXAMPLE 6   **Reading a bar graph**

The bar graph in Figure 1.8 shows the approximate U.S. Asian-American population in millions for selected years. Double hash marks // are used on the vertical axis to indicate that there is a break in the scale. The vertical scale starts at 0 and then jumps to 12 (million) before it proceeds in units of 1 million people. (*Source:* U.S. Census Bureau.)

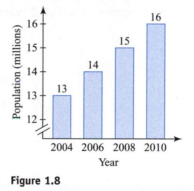

Figure 1.8

(a) What was the population in 2008?
(b) In which of these years was the population the smallest?
(c) In which year was the population 16 million?

Solution
(a) The bar representing 2008 has a height of 15. Since the units are in millions, we say that the U.S. Asian-American population was 15 million in 2008.
(b) The shortest bar corresponds to the smallest population. This occurred in 2004.
(c) The population was 16 million in 2010, as shown by the bar with height 16.

Now Try Exercises 71, 73

Another type of graph for displaying data is the **line graph**. With this type of graph, we can quickly identify any trends in the data being displayed. The line graph in Figure 1.9 on the next page suggests a downward trend in the number of farms in Ohio. In this graph, double hash marks // are used on each axis to indicate a break in each scale. The horizontal scale starts at 0 and jumps to 2001 before it shows every 2 years up through 2009. The vertical scale starts at 0 and jumps to 75 (thousand) before it shows every 1 thousand farms.

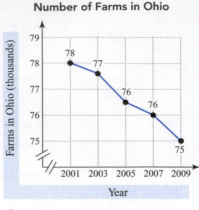

Figure 1.9

NOTE: Since the label on the vertical axis states that the data are shown in "thousands," each number on the vertical scale is a number of thousands. For example, the number 77 on the vertical scale represents 77,000 farms.

EXAMPLE 7   **Reading a line graph**

The line graph in Figure 1.10 shows actual and projected federal income tax receipts in billions of dollars for selected years. (*Source:* Office of Management and Budget.)

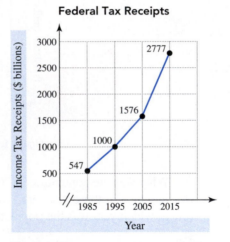

Figure 1.10

(a) How much income tax did the federal government collect in 1995?
(b) In which of these years was $547 billion collected?
(c) Comment on the general trend of income tax receipts.

READING CHECK

• What type of graph is useful in identifying trends in data?

Solution
(a) In 1995, the line graph reaches $1000 billion, or $1 trillion.
(b) The line graph indicates that $547 billion was collected in 1985.
(c) Income tax receipts increased over this time period.

Now Try Exercises 83, 85

Information can also be displayed visually in a **table**. The farm data shown in Figure 1.9 are also given in Table 1.2. Even though the data in the line graph are identical to the data in the table, the line graph shows the downward trend in the number of Ohio farms more easily than does the table. However, the number of Ohio farms in a given year can be seen at a glance within the table.

TABLE 1.2 **Number of Ohio Farms**

Year	2001	2003	2005	2007	2009
Farms	78,000	77,600	76,500	76,000	75,000

Source: U.S. Department of Agriculture.

EXAMPLE 8 | **Reading a table**

Table 1.3 shows the number of music albums sold by genre (category) in 2007 and 2008.

TABLE 1.3 **Albums Sold by Genre**

	Alternative	Country	R&B	Rap
2007	88,604,000	62,696,000	95,555,000	41,662,000
2008	80,919,000	47,657,000	77,014,000	33,410,000

Source: Nielsen SoundScan.

(a) How many R&B albums were sold in 2008?
(b) Which of these genres sold the fewest albums in 2007?

Solution
(a) Move downward in the "R&B" column until you find the number listed in the row labeled "2008." There were 77,014,000 R&B albums sold in 2008.
(b) The smallest number in the "2007" row is 41,662,000. Moving upward, we find that the corresponding type of music is Rap.

Now Try Exercises 87, 89

STUDY TIP

Putting It All Together gives a summary of important concepts in each section. Be sure that you have a good understanding of these concepts.

1.1   Putting It All Together

CONCEPT	COMMENTS	EXAMPLES
Natural Numbers	Sometimes referred to as the *counting numbers*	1, 2, 3, 4, 5, …
Whole Numbers	Includes the natural numbers and 0	0, 1, 2, 3, 4, …
Standard Form	A whole number written in digits with commas separating the periods is in standard form.	345,690,274
Place Value	The place value of a digit in a number written in standard form is determined by the position that the digit takes in the number.	In the number 83,451,276, the digit 7 is in the tens place, and the digit 3 is in the millions place.

continued on next page

continued from previous page

CONCEPT	COMMENTS	EXAMPLES
Word Form	Starting with the left-most period, write the word form of the number in each period followed by the period name and a comma.	In word form, 34,506 is written as thirty-four thousand, five hundred six.
Expanded Form	Starting with the left-most digit, write the standard form of the number given by each digit and its corresponding place value. Place a plus sign between each of these results.	In expanded form, 500,349 is written as $500,000 + 300 + 40 + 9$.
Number Line	Whole numbers can be visualized on a number line. The graph of a whole number is a dot placed on a number line at the whole number's position.	The numbers 3 and 5 are graphed.
Comparing Whole Numbers	To compare two whole numbers, determine their positions on a number line. Then use the symbols $>$ or $<$ to write an appropriate comparison.	$7 > 3$ and $2 < 11$
Bar Graph	A bar graph can be used to represent data visually and is helpful when analyzing data.	
Line Graph	A line graph can be used to represent data visually and is helpful when looking for trends in data.	
Table	A table can be used to display data in an at-a-glance way.	

Year	2010	2011	2012
Price	$53	$67	$78

1.1   Exercises

MyMathLab | Math XL PRACTICE | WATCH | DOWNLOAD | READ | REVIEW

CONCEPTS AND VOCABULARY

1. The natural numbers are also referred to as the _____ numbers.

2. The whole numbers include the natural numbers and the number _____.

3. Commas are used to separate the digits of a whole number into groups called _____.

4. A whole number expressed in digits with commas separating the periods is in _____ form.

5. The position of a digit in a whole number determines the digit's _____.

6. In a place value chart, the period immediately to the left of billions is _____.

7. The number "six thousand, four hundred seventeen" is written in (word/expanded) form.

8. The number 30,000 + 2000 + 40 + 3 is written in (word/expanded) form.

9. The _____ of a whole number is a dot placed on a number line at the whole number's position.

10. Name two kinds of graphs that can be used to display data visually.

PLACE VALUE

Exercises 11–20: For the given whole number, determine the place value of the digit 8.

11. 18,450

12. 456,981

13. 89,104,765

14. 6,830,142

15. 310,842

16. 4,982,017

17. 3008

18. 48,362,710

19. 890,247,135

20. 38,907,142,516

Exercises 21–30: Name the digit with the given place value in the whole number 3,409,816,725.

21. hundreds

22. ten-thousands

23. hundred-thousands

24. ones

25. billions

26. millions

27. tens

28. thousands

29. hundred-millions

30. ten-millions

WRITING WHOLE NUMBERS

Exercises 31–36: Write the whole number in word form.

31. 472,500

32. 79

33. 93,206

34. 10,000,015

35. 1651

36. 632

Exercises 37 and 38: Write the standard form of the dollar amount written on the given check.

37.

Electronics Company   1001
301 Technology Drive, Anytown, USA 12345

Date: _____

Pay to the order of _____ $ [?]

Two thousand fifty-five and 00/100 —————— Dollars

Memo _____   Authorized Signature _____

⑈01001⑈ ⑆111222333⑆ 444555⑈⑈

38.

Electronics Company   1001
301 Technology Drive, Anytown, USA 12345

Date: _____

Pay to the order of _____ $ [?]

Four hundred seventy-one and 00/100 —————— Dollars

Memo _____   Authorized Signature _____

⑈01001⑈ ⑆111222333⑆ 444555⑈⑈

Exercises 39 and 40: Write the standard form of the whole number that is expressed in word form in the sentence.

39. A typical student entering college immediately after high school will be five hundred ninety-nine million, six hundred sixteen thousand, four hundred twenty-three seconds old at some point during his or her freshman year.

40. The Nile River is the longest river in the world, at four thousand, one hundred thirty-five miles.

Exercises 41–46: Write the whole number in expanded form.

41. 2,510,036

42. 8004

43. 629

44. 63,907

45. 603,138

46. 17

Exercises 47–54: Write the whole number in standard form.

47. Thirty-nine million, four hundred ten thousand

48. Fifty-two thousand, three hundred sixty-seven

49. Eighty-three billion, six hundred thousand, twelve

50. One million, four hundred two thousand, eighty-one

51. $300{,}000 + 40{,}000 + 2000 + 500 + 60 + 3$

52. $5000 + 500 + 50 + 1$

53. $7{,}000{,}000 + 900{,}000 + 5000 + 300 + 70 + 7$

54. $4{,}000{,}000 + 500{,}000 + 7000 + 200 + 9$

WHOLE NUMBERS AND THE NUMBER LINE

Exercises 55–60: Graph the given whole numbers on the same number line. Place the correct symbol, $<$ or $>$, in the blank between the whole numbers in parts (a)–(c).

55. 3, 5, 4, 1
   (a) 3 _____ 5   **(b)** 4 _____ 3   **(c)** 5 _____ 1

56. 2, 5, 8, 3
   (a) 3 _____ 2   **(b)** 5 _____ 8   **(c)** 2 _____ 5

57. 11, 22, 4, 8
   (a) 22 _____ 4   **(b)** 8 _____ 11   **(c)** 11 _____ 22

58. 23, 31, 12, 40
   (a) 12 _____ 31   **(b)** 23 _____ 40   **(c)** 31 _____ 40

59. 86, 24, 64, 10
   (a) 86 _____ 64   **(b)** 64 _____ 24   **(c)** 24 _____ 10

60. 98, 27, 73, 15
   (a) 27 _____ 73   **(b)** 98 _____ 15   **(c)** 15 _____ 73

Exercises 61–70: Place the correct symbol, $<$ or $>$, in the blank between the whole numbers.

61. 34 _____ 0   **62.** 0 _____ 56

63. 45 _____ 54   **64.** 72 _____ 27

65. 300 _____ 299   **66.** 175 _____ 155

67. 30,000 _____ 300,000 **68.** 2100 _____ 2001

69. 50,101 _____ 51,010   **70.** 630,020 _____ 632,202

READING GRAPHS

Exercises 71–74: The bar graph at the top of the next column shows the 10 countries that had the largest number of Internet users in 2009. (*Source:* Internet World Stats.)

Top Ten Countries in Internet Usage

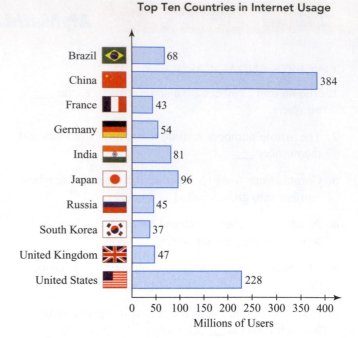

71. Which of the countries shown had 81,000,000 Internet users in 2009?

72. How many Internet users were there in Japan in 2009?

73. Which of these countries had the fewest number of Internet users in 2009?

74. Which country had more Internet users in 2009, Brazil or Russia?

Exercises 75–78: The following bar graph shows the four longest rivers in Canada. (*Source:* Statistics Canada.)

Rivers in Canada

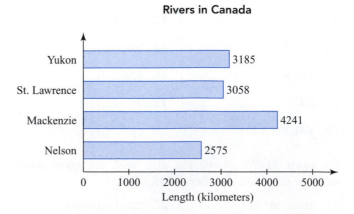

75. What is the longest river in Canada?

76. Which river is 3058 kilometers long?

77. How long is the Nelson River?

78. Which river is longer, the Nelson or the Yukon?

Exercises 79–82: The following line graph shows the box office receipts for top-grossing movies of selected years. These values have not been adjusted for inflation.
(*Source:* MovieWeb.)

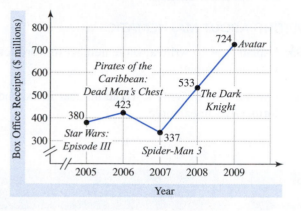

Top-Grossing Movies

79. In what year did the top-grossing movie take in the smallest amount in box office receipts?

80. Which movie had box office receipts of $533,000,000?

81. What were the box office receipts of the top-grossing movie of 2006?

82. Which movie grossed more, *Spider-Man 3* or *Avatar*?

Exercises 83–86: The following line graph shows the federal minimum wage in cents for selected years.
(*Source:* Bureau of Labor Statistics.)

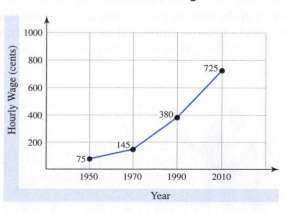

Minimum Wage

83. Comment on the general trend of the minimum wage over this time period.

84. What was the minimum wage in 1990?

85. What 20-year period had the largest increase in the minimum wage?

86. In what year was the minimum wage 75¢?

Exercises 87–90: The following table shows average SAT math scores for males and females during selected years.

	2000	2002	2004	2006	2008
Female	498	500	501	502	500
Male	533	534	537	536	533

Source: The College Board.

87. In what year did females score 502?

88. In what year did males score the highest?

89. Which gender had a higher average score in 2008?

90. In what two years did males record identical average scores?

Exercises 91–94: The following table shows the cost in dollars of tuition and fees at public and private four-year colleges for selected years.

	2005	2006	2007	2008	2009
Public	6053	6142	6401	6453	7020
Private	23,122	23,607	24,549	24,649	26,273

Source: The College Board.

91. In what year did public colleges cost $6142?

92. In what year did private colleges cost $24,649?

93. Which type of college had a cost of $6401 in 2007?

94. Did tuition and fees at public colleges ever decrease from one year to the next?

APPLICATIONS

95. *Speed of Sound* In dry air at a temperature of 65° Fahrenheit, the speed of sound is about 1124 feet per second. Write this whole number in expanded form.

96. *Speed of Light* The speed of light in a vacuum is about 186,282 miles per second. Write this whole number in expanded form.

97. *iPod Memory* Digital information is stored in units called bytes. A 32-gigabyte iPod Touch holds thirty-four billion, three hundred fifty-nine million, seven hundred thirty-eight thousand, three hundred seventy-eight bytes. Write this whole number in standard form.

98. *iPod Memory* A 1-gigabyte iPod Shuffle holds one billion, seventy-three million, seven hundred forty-one thousand, eight hundred twenty-four bytes. Write this whole number in standard form.

99. *Text Messaging* In June 2009, cell phone users in the U.S. sent over 135,000,000,000 text messages. Write this whole number in word form. (*Source:* CTIA— The Wireless Association.)

100. *Text Messaging* In June 2005, U.S. cell phone users sent about 7,200,000,000 text messages. Write this whole number in word form. (*Source:* CTIA—The Wireless Association.)

101. *Annual Income* Who has a greater yearly income, an electrician earning $41,627 per year or a truck driver earning $41,804 per year?

102. *Bacteria* There are 12,678,453 bacteria in a white dish and 12,687,435 bacteria in a black dish. Which dish has fewer bacteria?

WRITING ABOUT MATHEMATICS

103. Explain how to write a whole number in word form. Give an example.

104. Explain how to write a whole number in expanded form. Give an example.

1.2 Adding and Subtracting Whole Numbers; Perimeter

Adding Whole Numbers • Subtracting Whole Numbers • Solving Equations • Perimeter and Other Applications Involving Addition and Subtraction

A LOOK INTO MATH ▶

A person planning a Super Bowl party could send out electronic invitations using an online service such as Evite (www.evite.com). The RSVP feature of this service can be used to determine the number of people who will be coming to the party. The total number of guests is found by *adding* the numbers on the electronic RSVP cards. However, at the last minute, some people may call the host to say that they cannot attend. When this happens, a new attendance total is found by *subtracting* the number of cancellations. The need to add and subtract whole numbers can come up in any part of everyday life.

Adding Whole Numbers

NEW VOCABULARY

☐ Sum
☐ Addends
☐ Commutative property for addition
☐ Associative property for addition
☐ Identity property for addition
☐ Difference
☐ Minuend
☐ Subtrahend
☐ Equation
☐ Solution
☐ Solving an equation
☐ Perimeter

▶ **REAL-WORLD CONNECTION** Two single moms can use an eight-passenger van to take their children to reading time at the local library if the total number of people is not more than 8. Including the mother for each family, one family has 3 members, and the other has 4. The total number of people is found by *adding* 3 and 4. By simply counting the people in Figure 1.11, we see that the total number of people is 7.

Figure 1.11 The total of 3 people and 4 people is 7 people.

When whole numbers are added, the result is called the **sum**, and the numbers being added are called the **addends**. For Figure 1.11, the **sum** is **7** and the **addends** are **3** and **4**.

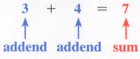

ADDING WHOLE NUMBERS WITHOUT CARRYING To add whole numbers with more than one digit, it is often convenient to stack the numbers vertically with corresponding place values aligned. Then we add the digits in each place value, starting with the ones place. For example, the numbers 521 and 6374 are added as follows:

$$
\begin{array}{r}
5\ 2\ 1 \quad \leftarrow \text{Addend} \\
+\ 6\ 3\ 7\ 4 \quad \leftarrow \text{Addend} \\
\hline
6\ 8\ 9\ 5 \quad \leftarrow \text{Sum}
\end{array}
$$

NOTE: The digits in corresponding place values must be aligned vertically when adding numbers. For example, the digit **5** in the top number in the previous example represents 500 and the digit **3** in the second number represents 300. When added, the total is **8**, which represents 800 in the sum.

EXAMPLE 1 **Adding whole numbers without carrying**

Add.
(a) 2416 + 332   **(b)** 11 + 314 + 5473

Solution
Start by stacking the addends vertically so that the place values are aligned.

(a)
$$
\begin{array}{r}
2416 \\
+\ 332 \\
\hline
2748
\end{array}
$$

(b)
$$
\begin{array}{r}
11 \\
314 \\
+\ 5473 \\
\hline
5798
\end{array}
$$

Now Try Exercises 21, 31

ADDING WHOLE NUMBERS WITH CARRYING Sometimes adding digits within a particular place value results in a sum larger than 9. For example, adding digits in the ones place of the numbers 378 and 607 results in 15. Because **15** is written in expanded form as **10** + **5**, there are **5** ones and **1** ten. We write the **5** in the ones column, and the ten is *carried* as the digit **1** (because it is 1 ten) to the tens column as shown here.

$$
\begin{array}{r}
1 \quad \leftarrow \text{Carry the ten as a 1.} \\
3\ 7\ 8 \\
+\ 6\ 0\ 7 \\
\hline
9\ 8\ 5
\end{array}
$$

EXAMPLE 2 **Adding whole numbers with carrying**

Add.
(a) 328 + 4169   **(b)** 38 + 367 + 2276

Solution
(a) The digits in the ones column add to 17, so 7 is written as the result in the ones column and **1** ten is carried to the tens column. The remaining digits of the sum are found by adding the digits in each column.

$$
\begin{array}{r}
1 \\
328 \\
+\ 4169 \\
\hline
4497
\end{array}
$$

CALCULATOR HELP

To add whole numbers with a calculator, see the Appendix (page AP-1).

READING CHECK

- In adding whole numbers, when is it necessary to carry?

(b) The sum of the digits in the ones column is 21. In this case, a 1 is written as the result in the ones column, and **2** tens are carried to the tens column. Adding the digits in the tens column results in 18, meaning that there are 18 tens. An 8 is written as the result in the tens column, and the 10 tens or **1** hundred is carried to the hundreds column.

$$
\begin{array}{r}
\mathbf{12} \\
38 \\
367 \\
+\ 2276 \\
\hline
2681
\end{array}
$$

■ **Now Try Exercises 23, 33**

STUDY TIP

The information in Making Connections ties the current concepts to those studied earlier. By reviewing your notes often, you can gain a better understanding of mathematics.

MAKING CONNECTIONS

Carrying and Expanded Form

To see how carrying works, we write the addends in expanded form before performing addition. For example, the sum $357 + 876$ is found as follows:

300	$+$	50	$+$	7		Write 357 in expanded form.
$+\ 800$	$+$	70	$+$	6		Write 876 in expanded form.
1100	$+$	**120**	$+$	**13**		Add columns.

Carry 100   Carry 10

$\mathbf{1000 + 100}\quad +\quad \mathbf{100 + 20}\quad +\quad \mathbf{10 + 3}$ Write each result in expanded form.

$1000\ +\ 200\quad +\quad 30 + 3$ Add hundreds; add tens.

1233 Write the sum in standard form.

When adding whole numbers, the following procedure can be used.

ADDING WHOLE NUMBERS

To add whole numbers,

1. Stack the numbers vertically with corresponding place values aligned.
2. Add the digits in each place value. Use carrying when necessary.

PROPERTIES OF ADDITION Adding a long list of whole numbers by hand can be time-consuming. However, there are three properties of addition that often make the process easier. The first property, the **commutative property for addition**, states that changing the *order* of the addends does not change the resulting sum. For example,

$$3 + 2 = 5 \quad \text{and} \quad 2 + 3 = 5.$$

COMMUTATIVE PROPERTY FOR ADDITION

The *order* in which two addends are written does not affect the sum.

The commutative property for addition can be visualized as shown in Figure 1.12.

Figure 1.12 A Visual Representation of the Commutative Property: 3 + 2 = 2 + 3

The second property is called the **associative property for addition**. This property allows for the *regrouping* of addends when more than two numbers are being added. Parentheses are used to show which two addends should be grouped (added) first.

$$(3 + 2) + 1 = 5 + 1 = 6 \quad \text{and} \quad 3 + (2 + 1) = 3 + 3 = 6$$

ASSOCIATIVE PROPERTY FOR ADDITION

The way in which three or more addends are *grouped* does not affect the sum.

The associative property for addition can be visualized as shown in Figure 1.13.

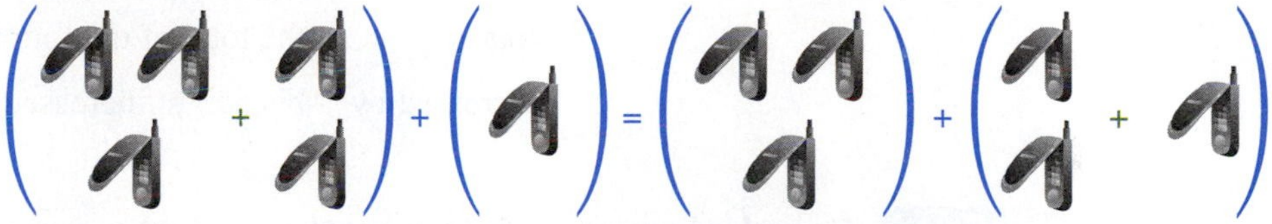

Figure 1.13 A Visual Representation of the Associative Property: (3 + 2) + 1 = 3 + (2 + 1)

The **identity property for addition** is the third property. It states that adding 0 to a number does not change the number. If you have $12 and a friend gives you $0, you still have $12.

$$12 + 0 = 12 \quad \text{and} \quad 0 + 12 = 12$$

IDENTITY PROPERTY FOR ADDITION

When 0 is added to any number, the result is that number.

Sometimes the three addition properties can be used to add mentally. Consider the sum $14 + 9 + 0 + 6 + 8 + 1 + 12 + 7$. Note that the addend 0 has no effect on the sum, so it can be ignored. The remaining numbers can be arranged and grouped so that the sum can be computed mentally.

READING CHECK

- Which addition property allows us to add in any order?
- Which addition property allows us to regroup the addends?

$$14 + 9 + 6 + 8 + 1 + 12 + 7$$
$$(14 + 6) + (9 + 1) + (8 + 12) + 7$$
$$20 + 10 + 20 + 7$$
$$57$$

The sum of 57 is found by adding the numbers 20, 10, 20, and 7.

EXAMPLE 3   **Adding whole numbers mentally**

Add mentally.
16 + 23 + 12 + 8 + 5 + 9 + 7 + 0 + 15 + 4

Solution
Ignore 0 and regroup to get a sum of 99.

$$16 + 23 + 12 + 8 + 5 + 9 + 7 + 15 + 4 =$$
$$20 + 30 + 20 + 20 + 9 = 99$$

Now Try Exercise 35

WORDS ASSOCIATED WITH ADDITION Many times information is given in words rather than in symbols or numbers. To find a required result, it may be necessary to translate words into a mathematical expression. Table 1.4 shows some words commonly associated with addition, along with a sample phrase using each word.

TABLE 1.4 **Words Associated with Addition**

Words	Sample Phrase
add	add the two temperatures
plus	her age plus his age
more than	10 miles more than the distance
sum	the sum of the length and the width
total	the total of the four prices
increased by	his height increased by 3 inches

EXAMPLE 4   **Translating words into a mathematical expression**

Translate each phrase into a mathematical expression. Find the result.
(a) the total of 5 inches and 63 inches   **(b)** 17 medals more than 12 medals already won

Solution
(a) The word *total* suggests that we add 5 and 63. The corresponding mathematical expression is 5 + 63, which results in 68 inches.
(b) The words *more than* suggest that we add 17 to 12. The corresponding mathematical expression is 12 + 17, which results in 29 medals.

Now Try Exercises 61, 67

Subtracting Whole Numbers

▶ **REAL-WORLD CONNECTION** On February 1, a student's blog had a cumulative total of 1454 hits, and on March 1, the blog had a cumulative total of 1878 hits. The number of hits that the blog had during the month of February was 424, which can be found by *subtracting* the number 1454 from the number 1878. The result of subtracting one whole number from another is called the **difference**. The number we are subtracting from is called the **minuend**, and the number being subtracted is called the **subtrahend**. In this example, the **minuend** is **1878**, the **subtrahend** is **1454**, and the **difference** is **424**.

$$1878 \quad - \quad 1454 \quad = \quad 424$$

minuend   subtrahend   difference

SUBTRACTING WHOLE NUMBERS WITHOUT BORROWING To subtract one whole number from another, stack the minuend vertically above the subtrahend with corresponding place values aligned. Then subtract the digits in each place value, starting with the ones place. For example, 431 is subtracted from 7573 as follows:

$$
\begin{array}{r}
7\ 5\ 7\ 3 \quad \leftarrow \text{Minuend} \\
-\ 4\ 3\ 1 \quad \leftarrow \text{Subtrahend} \\
\hline
7\ 1\ 4\ 2 \quad \leftarrow \text{Difference}
\end{array}
$$

NOTE: As with addition, it is important to remember that digits in corresponding place values must be aligned vertically when subtracting one number from another.

EXAMPLE 5 **Subtracting whole numbers without borrowing**

Subtract.
(a) $1688 - 437$ (b) $12{,}877 - 10{,}641$

Solution
Stack the minuend vertically above the subtrahend so that the place values are aligned.

(a)
$$
\begin{array}{r}
1688 \\
-\ 437 \\
\hline
1251
\end{array}
$$
(b)
$$
\begin{array}{r}
12{,}877 \\
-\ 10{,}641 \\
\hline
2236
\end{array}
$$

NOTE: In part (b), subtracting the digits in the ten-thousands place results in 0, which is not written as the leading digit in the resulting difference.

Now Try Exercises 43, 45

SUBTRACTING WHOLE NUMBERS WITH BORROWING Sometimes the digit in a particular place value of the minuend (top number) is smaller than the corresponding digit in the subtrahend (bottom number). When this happens, *borrowing* is necessary. For example, the ones digit in 753 is smaller than the ones digit in 318 (because $3 < 8$). To subtract $753 - 318$, we must borrow. The number **753** has **5** tens and **3** ones. We borrow **1** ten from the tens place and "lend" it to the ones place. After doing this, the number **753** has **4** tens and **13** ones.

$$
\begin{array}{r}
4\ 13 \quad \leftarrow \text{Borrow 1 ten as a } \mathbf{1}. \\
7\ \cancel{5}\ \cancel{3} \\
-\ 3\ 1\ 8 \\
\hline
4\ 3\ 5
\end{array}
$$

EXAMPLE 6 **Subtracting whole numbers with borrowing**

Subtract.
(a) $3653 - 1481$ (b) $4039 - 372$

Solution
(a) Borrow **1** (hundred) from the hundreds place and lend it to the tens place, resulting in **15** tens. Then perform the subtraction.

$$
\begin{array}{r}
5\ 15 \\
3\ \cancel{6}\ \cancel{5}\ 3 \\
-\ 1\ 4\ 8\ 1 \\
\hline
2\ 1\ 7\ 2
\end{array}
$$

READING CHECK

• In subtracting whole numbers, when is it necessary to borrow?

(b) Before we can borrow **1** (hundred) from the hundreds place, which contains a 0, we must first borrow **1** (thousand) from the thousands place and lend it to the hundreds place, resulting in **10** hundreds. We may then borrow **1** (hundred) from the hundreds place and lend it to the tens place, resulting in **13** tens.

$$\begin{array}{r} \overset{9}{3}\,\overset{10}{\cancel{10}}\,\overset{13}{\cancel{13}} \\ \cancel{4}\,\cancel{0}\,\cancel{3}\,9 \\ -\ 3\ 7\ 2 \\ \hline 3\ 6\ 6\ 7 \end{array}$$

Now Try Exercises 47, 53

CALCULATOR HELP

To subtract whole numbers with a calculator, see the Appendix (page AP-1).

MAKING CONNECTIONS

Borrowing and Expanded Form

To see how borrowing works, we can write the minuend in *modified* expanded form before performing subtraction. For example, the difference $847 - 372$ is found as follows:

800	+	40	+	7	Write 847 in expanded form.
700 + 100	+	40	+	7	Write 847 in modified expanded form.

$$\text{Borrow 100}$$

700	+	**100** + 40 +		7	Lend 100 to the tens column.
700	+	140	+	7	Write $100 + 40$ as 140.
− (300	+	70	+	2)	Write 372 in expanded form.
400	+	70	+	5	Subtract columns.
		475			Write the difference in standard form.

When subtracting one whole number from another, we can use the following procedure.

SUBTRACTING WHOLE NUMBERS

To subtract one whole number from another,

1. Stack the numbers vertically with corresponding place values aligned.
2. Subtract the digits in each place value. Use borrowing when necessary.

PROPERTIES OF SUBTRACTION The commutative property for addition does **not** hold true for subtraction. For example, $7 - 3$ results in 4, while $3 - 7$ does not give a whole number result. Similarly, the associative property for addition does **not** hold true for subtraction, as shown.

$$(10 - 8) - 1 = 2 - 1 = 1, \quad \text{but} \quad 10 - (8 - 1) = 10 - 7 = 3.$$

Subtraction does, however, have two properties that are related to the identity property for addition. The first property states that subtracting 0 from a number does not change the number. If you have $15 and you give away $0, you still have $15. The second property says that subtracting a number from itself results in 0. If you have $7 and you give away $7, you have no money left. These results can be illustrated as follows.

$$15 - 0 = 15 \quad \text{and} \quad 7 - 7 = 0$$

IDENTITY PROPERTIES FOR SUBTRACTION

1. When 0 is subtracted from any number, the result is that number.
2. When a number is subtracted from itself, the result is 0.

WORDS ASSOCIATED WITH SUBTRACTION Just as some words suggest that addition should be used to write a mathematical expression, other words suggest that subtraction is appropriate. Table 1.5 shows some words associated with subtraction, together with a sample phrase using each word.

TABLE 1.5 Words Associated with Subtraction

Words	Sample Phrase
subtract	subtract the cost from the revenue
minus	his income minus his taxes
fewer than	18 fewer flowers than shrubs
difference	the difference between their heights
less than	his age is 4 years less than hers
decreased by	the number of boxes decreased by 7
take away	the subtotal take away the cash back

EXAMPLE 7 **Translating words into a mathematical expression**

Translate each phrase into a mathematical expression. Find the result.
(a) 13 days fewer than 23 days   **(b)** the difference between 18 cards and 12 cards

Solution
(a) The word *fewer* suggests that we should subtract 13 from 23. The corresponding mathematical expression is $23 - 13$, which results in 10 days.
(b) The word *difference* suggests that we should subtract 12 from 18. The corresponding mathematical expression is $18 - 12$, which results in 6 cards.

Now Try Exercises 63, 69

Solving Equations

In mathematics, an **equation** can be written when one quantity is equal to another. Every equation contains an equal sign, $=$. The following words or phrases suggest that an equal sign is needed and an equation can be written.

equals,   *is*,   *gives*,   *results in*,   *is the same as*

An equation can be true or false. For example, the equation $4 + 8 = 12$ is true, and the equation $19 - 4 = 10$ is false. However, an equation such as

$$\square + 4 = 9$$

READING CHECK

• How do we know if a number is a solution to an equation?

contains an *unknown value* and may be either true or false depending on the number that is written in the box. Any such number that makes an equation true is called a **solution** to the equation. In general, **solving an equation** means finding all of its solutions. The whole number 5 is the only solution to the equation above because $5 + 4 = 9$ is a true equation, and writing any other number in the box would result in a false equation.

EXAMPLE 8 **Solving equations**

Solve each equation by finding the unknown value.
(a) $14 - \square = 3$ **(b)** $\square - 17 = 5$ **(c)** $38 + \square = 67$

Solution
(a) The solution is **11** because $14 - \mathbf{11} = 3$ is a true equation.
(b) Because $\mathbf{22} - 17 = 5$, the solution is **22**.
(c) Add 29 to 38 to obtain a sum of 67. The solution is 29.

Now Try Exercises 73, 75, 79

Perimeter and Other Applications Involving Addition and Subtraction

▶ **REAL-WORLD CONNECTION** The in-bounds playing surface of an official doubles tennis court is a rectangular shape that is 36 feet wide and 78 feet long. The *perimeter* of the court is marked by a white line that is the boundary between the in-bounds surface and the out-of-bounds surface. In geometry, the **perimeter** of an enclosed region is the distance around the region. Figure 1.14 shows that the perimeter of a doubles tennis court is the sum of the lengths of its four sides, or $36 + 78 + 36 + 78 = 228$ feet.

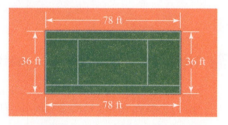

Figure 1.14 Perimeter of a Doubles Tennis Court

READING CHECK

• How is the perimeter of an enclosed region found?

For some enclosed regions, the length of each side is known, and the perimeter is found by adding the given lengths. Other regions may be missing a length. In this case, we must find the missing length before we can find the perimeter.

EXAMPLE 9 **Finding perimeters of shaded regions**

Find the perimeter of each shaded region.
(a)   **(b)**

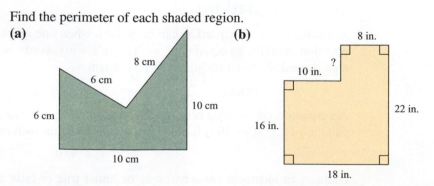

Solution
(a) The length of every side is given. The perimeter is $6 + 6 + 8 + 10 + 10 = 40$ cm.
(b) The missing length can be found by subtraction, as shown in Figure 1.15.

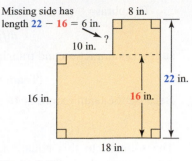

Missing side has
length **22** − **16** = 6 in.

8 in.

10 in.   ?

22 in.

16 in.   **16** in.

18 in.

Figure 1.15

The perimeter is $16 + 10 + 6 + 8 + 22 + 18 = 80$ in.

■ **Now Try Exercises 87, 91**

In the next two examples, addition and subtraction are used to solve applications involving real-world data.

EXAMPLE 10 **Finding an average temperature in Alaska**

The average summertime high temperature in Nome, Alaska is 11°F more than the average summertime high temperature in Barrow, Alaska. If this temperature is 58°F in Nome, find the corresponding temperature in Barrow. (*Source: USA Today.*)

Solution
The words

average summertime high temperature in Nome

is 11°*F more than*

the average summertime high temperature in Barrow

translates to an equation of the form

(Nome temperature) = (Barrow temperature) + 11.

Since the average summertime high temperature in Nome is **58**°F, and the corresponding temperature in Barrow is unknown, then the equation to be solved is

$$58 = \square + 11.$$

The solution to this equation is 47 because $58 = 47 + 11$. The average summertime high temperature in Barrow is 47°F.

■ **Now Try Exercise 93**

EXAMPLE 11 **Finding the number of goals scored**

Joe Malone, circa 1920

In 1920, Joe Malone of the Quebec Bulldogs set the NHL record for the number of goals scored by an *individual* in a single game. That same year, the Montreal Canadiens set the NHL record for the number of goals scored by a *team* in a single game with a 16-3 victory over the Quebec Bulldogs. If the difference between the records is 9, how many goals did Malone score in a single game to set the individual scoring record? (*Source: NHL.com*)

Solution
When the Montreal Canadiens defeated the Quebec Bulldogs, the score was 16 to 3, which means that the team scoring record is 16 goals in a single game. The individual scoring record is unknown.

The phrase

difference *between the records* **is** 9

suggests subtraction and translates to an equation of the form

(Team record) − (Individual record) = 9.

Because the team record is **16** and the individual record is unknown, the equation is

16 − □ = 9.

The solution to this equation is 7 because 16 − 7 = 9. Therefore, Joe Malone scored 7 goals to set the individual scoring record.

Now Try Exercise 99

1.2   Putting It All Together

CONCEPT	COMMENTS	EXAMPLES
Addition	The numbers being added are the *addends*, and the result is the *sum*.	8 + 7 = 15 **addend addend sum**
Adding Whole Numbers	1. Stack the numbers vertically with corresponding place values aligned. 2. Add the digits in each place value. Use carrying if necessary.	$\begin{array}{r} 1 \\ 2452 \\ +\,6374 \\ \hline 8826 \end{array}$
Properties for Addition	1. Commutative property 2. Associative property 3. Identity property	**1.** $3 + 6 = 6 + 3$ **2.** $(2 + 5) + 4 = 2 + (5 + 4)$ **3.** $4 + 0 = 4$ and $0 + 4 = 4$
Translating Words to Addition	Words associated with addition include *add*, *plus*, *more than*, *sum*, *total*, and *increased by*.	7 points more than her score The total of the coins
Subtraction	The number that we subtract from is the *minuend*. The number being subtracted is the *subtrahend*. The result is the *difference*.	24 − 13 = 11 **minuend subtrahend difference**
Subtracting Whole Numbers	1. Stack the numbers vertically with corresponding place values aligned. 2. Subtract the digits in each place value. Use borrowing if necessary.	$\begin{array}{r} {}^{6\ 14} \\ 7\!\!\!/ 4\!\!\!/ 8\,5 \\ -\,2\,8\,3\,1 \\ \hline 4\,6\,5\,4 \end{array}$
Properties for Subtraction	There are two identity properties, each involving the number 0.	**1.** $19 - 0 = 19$ **2.** $43 - 43 = 0$
Translating Words to Subtraction	Words associated with subtraction include *subtract*, *minus*, *fewer than*, *difference*, *less than*, *decreased by*, and *take away*.	The number of bugs decreased by 4 19 days fewer than 10 weeks

CONCEPT	COMMENTS	EXAMPLES
Solving Equations	An *equation* can be written when one quantity is equal to another. A *solution* is any number that makes an equation true when it replaces the unknown value. *Solving an equation* means finding all of its solutions.	The solution to $$\square - 13 = 5$$ is 18 because $18 - 13 = 5$ is a true equation.
Perimeter	The distance around an enclosed region is called its perimeter.	 The perimeter is $3 + 7 + 4 + 6 = 20$ feet.

1.2 Exercises

MyMathLab   Math XL PRACTICE   WATCH   DOWNLOAD   READ   REVIEW

CONCEPTS AND VOCABULARY

1. When adding whole numbers, the numbers being added are called the _____.

2. When whole numbers are being added, the result is called the _____.

3. Is *carrying* necessary when adding $468 + 215$?

4. The equation $4 + 3 = 3 + 4$ illustrates the _____ property for addition.

5. The equation $(1 + 2) + 6 = 1 + (2 + 6)$ illustrates the _____ property for addition.

6. The _____ property for addition is illustrated by the equation $7 + 0 = 7$.

7. The word *increase* suggests that (addition/subtraction) should be used.

8. When subtracting whole numbers, the number we are subtracting from is the _____ and the number being subtracted is the _____.

9. When whole numbers are being subtracted, the result is called the _____.

10. Is *borrowing* needed to subtract $864 - 521$?

11. The equation $8 - 8 = 0$ illustrates one of the _____ properties for subtraction.

12. The operation (addition/subtraction) should be used for the word *fewer*.

13. A(n) _____ is any number that makes an equation true when it replaces the unknown value.

14. Solving an equation means finding all of its _____.

ADDITION OF WHOLE NUMBERS

Exercises 15–34: Add.

15. $11 + 17$

16. $34 + 21$

17. $65 + 534$

18. $742 + 56$

19. $624 + 261$

20. $322 + 516$

21. $\begin{array}{r} 357 \\ + 7511 \end{array}$

22. $\begin{array}{r} 671 \\ + 2128 \end{array}$

23. $\begin{array}{r} 3748 \\ + 4124 \end{array}$

24. $\begin{array}{r} 3352 \\ + 1539 \end{array}$

25. $\begin{array}{r} 16{,}491 \\ + 10{,}573 \end{array}$

26. $\begin{array}{r} 12{,}458 \\ + 23{,}975 \end{array}$

27. $28{,}529 + 53{,}298$

28. $340{,}982 + 72{,}099$

29. $409{,}377 + 654{,}782$

30. $500{,}809 + 499{,}765$

31.
```
    230
   5602
  +3135
```

32.
```
    528
   6377
  +8327
```

33.
```
  10,669
  45,127
 +32,255
```

34.
```
  73,417
  56,830
 +22,804
```

Exercises 35–38: Add mentally.

35. $11 + 8 + 13 + 6 + 0 + 7 + 12 + 9$

36. $6 + 3 + 25 + 8 + 0 + 14 + 22 + 5$

37. $0 + 33 + 11 + 6 + 0 + 7 + 9 + 4$

38. $20 + 0 + 44 + 1 + 0 + 6 + 19 + 2$

SUBTRACTION OF WHOLE NUMBERS

Exercises 39–60: Subtract.

39. $24 - 11$

40. $55 - 31$

41. $468 - 37$

42. $282 - 61$

43. $1769 - 347$

44. $3857 - 554$

45.
```
   3672
  - 3521
```

46.
```
   8175
  - 8042
```

47.
```
   5534
  - 3218
```

48.
```
   6452
  - 3327
```

49.
```
  56,431
 - 23,526
```

50.
```
  81,647
 - 58,329
```

51. $45,832 - 14,399$

52. $184,297 - 98,428$

53. $517,056 - 416,029$

54. $873,870 - 649,335$

55.
```
   3007
  - 389
```

56.
```
   6004
  -576
```

57.
```
  40,063
 - 22,378
```

58.
```
  70,036
 - 67,873
```

59.
```
  100,703
 - 89,827
```

60.
```
  400,102
 - 398,516
```

TRANSLATING WORDS INTO MATH

Exercises 61–72: Translate the phrase into a mathematical expression and then find the result.

61. The sum of $22 and $57

62. 107 songs decreased by 39 songs

63. The difference between 793 photos and 54 photos

64. 873 toothpicks more than 1011 toothpicks

65. Subtract 19 eggs from 62 eggs

66. The total of 13, 89, and 104 cell phone minutes

67. 1200 patients increased by 300 patients

68. 89 degrees fewer than 107 degrees

69. 645 DVDs take away 3 DVDs

70. 58 plates less than 185 plates

71. Add 39 Web pages and 71 Web pages

72. 539 downloads plus 267 downloads

SOLVING EQUATIONS

Exercises 73–84: Solve the given equation by finding the unknown value.

73. $8 - \square = 5$

74. $\square + 7 = 11$

75. $\square + 14 = 33$

76. $24 - \square = 17$

77. $87 - 63 = \square$

78. $31 + 58 = \square$

79. $\square - 10 = 141$

80. $53 + \square = 99$

81. $81 + \square = 107$

82. $84 - 18 = \square$

83. $153 + 379 = \square$

84. $\square - 102 = 14$

APPLICATIONS

Exercises 85–92: Find the perimeter of the shaded region.

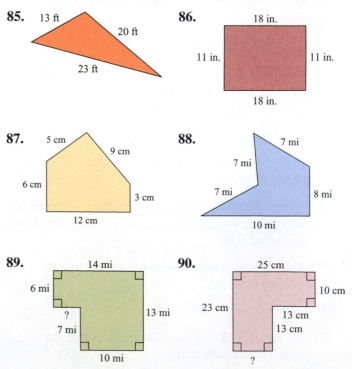

85. 13 ft, 20 ft, 23 ft

86. 18 in., 11 in., 11 in., 18 in.

87. 5 cm, 9 cm, 6 cm, 3 cm, 12 cm

88. 7 mi, 7 mi, 7 mi, 8 mi, 10 mi

89. 14 mi, 6 mi, ?, 7 mi, 10 mi, 13 mi

90. 25 cm, 10 cm, 23 cm, 13 cm, 13 cm, ?

91.   **92.**

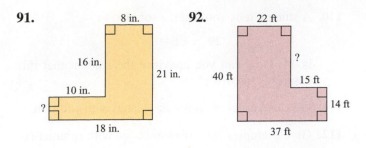

93. *MLB Players* At 83 inches in height, Jon Rauch is the tallest person ever to play major league baseball. The height of the shortest player, Eddie Gaedel, was 40 inches less than Rauch's height. Find Gaedel's height. (*Source: Major League Baseball.*)

94. *Tennis Serves* Andy Roddick has the fastest serve ever recorded in men's tennis. His record serve is 26 miles per hour faster than the fastest serve ever recorded in women's tennis, a 129-mph serve by Venus Williams. How fast is Roddick's record serve? (*Source: United States Tennis Association, 2008.*)

95. *Deadly Tornados* The following table lists the total number of tornado-related fatalities in the U.S. during selected years. Find the sum of the fatalities from the two deadliest years.

Year	2006	2007	2008	2009
Deaths	67	81	126	21

Source: NOAA, Storm Prediction Center.

96. *U.S. Fires* The following table lists the total number of fires reported in the U.S. during selected years. Find the difference between the largest number and the smallest number of fires.

Year	2002	2004	2006	2008
Fires	1,687,500	1,550,500	1,642,500	1,451,500

Source: National Fire Protection Association.

97. *Nuclear Power* Illinois has 4 more nuclear power plants than Nebraska. If there are 2 nuclear power plants in Nebraska, how many are there in Illinois? (*Source: Energy Information Administration.*)

98. *National Parks* There are 22 fewer national parks in West Virginia than there are in California. If West Virginia has 9 national parks, how many are there in California? (*Source: National Parks Service.*)

99. *NHL Players* There is a 35-year difference between Wayne Gretzky's age at the start of his professional hockey career and Gordie Howe's age at the end of his professional career. If Howe retired at age 52, how old was Gretzky when he began his professional career? (*Source: National Hockey League.*)

100. *NFL Scores* In 1929, Ernie Nevers set the NFL individual scoring record for a single game when he scored *every* point for the Chicago Cardinals in a 40-6 victory over the Chicago Bears. The single-game scoring record for an NFL team, set in 1966 by the Washington Redskins, is 32 points more than the record set by Nevers. Find the team scoring record. (*Source: National Football League.*)

101. *Bench Press* In 2008, the world record for an unassisted bench press, known as a *raw* bench press, was 715 pounds. A weight lifter wearing a bench shirt was able to bench 360 pounds more to set an equipment-assisted world record. What was the world record for the weight lifter wearing a bench shirt? (*Source: Powerlifting Watch.*)

102. *Stock Market* One of the largest one-day leaps in the history of the Dow Jones Industrial Average occurred on October 13, 2008. That day, the Dow increased by 936 points to close at 9390. What was the value of the Dow when trading began that day? (*Source: Wall Street Journal.*)

Exercises 103–106: Welfare The following bar graph shows the number of people, in millions, who received temporary assistance during selected years. (*Source: Administration for Children and Families.*)

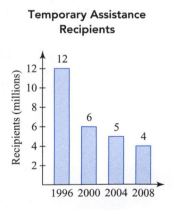

103. Which year had 8,000,000 fewer recipients than there were in 1996?

104. Which year had 1,000,000 more recipients than there were in 2008?

105. To find the number of recipients in 2000, we must decrease the number of recipients in 1996 by what amount?

106. By increasing the number of recipients in 2008 by 2,000,000, we find the number of recipients for which year?

107. Explain what it means to *carry* when adding whole numbers. Give an example.

108. Explain what it means to *borrow* when subtracting whole numbers. Give an example.

109. A student tells you that the solution to

$$\square + 14 = 23$$

is 11. How can you convince the student that this answer is incorrect?

110. A student tells you that the solution to

$$29 - \square = 17$$

is 14. How can you convince the student that this answer is incorrect?

111. Give examples of words associated with addition.

112. Give examples of words associated with subtraction.

STUDY TIP

Checking Basic Concepts provides a review of concepts from recent sections.

SECTIONS 1.1 and 1.2	**Checking Basic Concepts**

1. Give the place value of the digit 3 in each of the given whole numbers.
   (a) 132,458   **(b)** 45,267,309

2. Write the number 74,293 in expanded form.

3. Write forty-eight million, two hundred thirty-nine thousand, six hundred ten in standard form.

4. Graph the whole numbers 2, 5, and 1 on the same number line.

5. Place the correct symbol, < or >, in the blank between the whole numbers.
   (a) 67 _____ 25   **(b)** 15 _____ 51

6. Add.
   (a) 581 + 3736   **(b)** 204,633 + 5897

7. Subtract.
   (a) 8783 − 124   **(b)** 713,448 − 112,564

8. Translate each phrase into a mathematical expression and then find the result.
   (a) 97 minus 45   **(b)** 73 more than 106

9. Solve each equation.
   (a) 3 + $\square$ = 8   **(b)** $\square$ − 22 = 7

10. Find the perimeter of the shaded region.

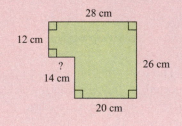

28 cm
12 cm
26 cm
?
14 cm
20 cm

1.3   Multiplying and Dividing Whole Numbers; Area

Multiplying Whole Numbers • Dividing Whole Numbers • Solving Equations • Area and Other Applications Involving Multiplication and Division

A LOOK INTO MATH ▶

To make shopping more appealing for consumers, shelves at retail stores are usually kept well stocked. An employee with several cases of liquid hand soap can use *multiplication* to figure out the total number of individual soap bottles to place on a shelf by multiplying the number of cases by the number of bottles in each case. Similarly, the employee can use *division* to find the numbers of cases needed to stock an empty shelf by dividing the number of bottles needed on the shelf by the number of bottles in each case. In this section we discuss multiplication and division of whole numbers.

continued on page 33

Multiplying Whole Numbers

When addends in a sum are the same, we can use **multiplication** as a fast way to perform *repeated addition*. For example, there are seven addends in the sum

$$4 + 4 + 4 + 4 + 4 + 4 + 4 = 28,$$

and each addend is 4. Rather than adding the 4s, the result can be found much faster by using the *multiplication fact*, **seven 4**s are **28**. In multiplication notation, we write

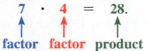

$$7 \quad \cdot \quad 4 \quad = \quad 28.$$

factor   factor   product

The numbers being multiplied are called **factors**, the result is called the **product**, and the symbol · is called the **multiplication sign**.

NOTE: Multiplication can be written in several ways. Each of the following expressions indicates that we are multiplying 7 and 4.

$$7 \cdot 4, \quad 7 \times 4, \quad 7(4), \quad (7)4, \quad \text{and} \quad (7)(4)$$

Before we can multiply larger whole numbers effectively, we must first *memorize* the products that result when multiplying two single-digit numbers. These basic multiplication facts are shown in Table 1.6 where the product $7 \cdot 4 = 28$ has been highlighted to demonstrate how a product can be found in the table.

STUDY TIP

Remember that a positive attitude is important. The first step to success is believing in yourself.

READING CHECK

• Why is it important to memorize the basic multiplication facts?

TABLE 1.6 Basic Multiplication Facts

·	0	1	2	3	4	5	6	7	8	9
0	0	0	0	0	0	0	0	0	0	0
1	0	1	2	3	4	5	6	7	8	9
2	0	2	4	6	8	10	12	14	16	18
3	0	3	6	9	12	15	18	21	24	27
4	0	4	8	12	16	20	24	28	32	36
5	0	5	10	15	20	25	30	35	40	45
6	0	6	12	18	24	30	36	42	48	54
7	0	7	14	21	28	35	42	49	56	63
8	0	8	16	24	32	40	48	56	64	72
9	0	9	18	27	36	45	54	63	72	81

PROPERTIES OF MULTIPLICATION Several properties are helpful when performing computations involving multiplication. Some of these properties are similar to those used for addition. The **commutative property for multiplication** states that changing the *order* of the factors does not change the resulting product. For example,

$$5 \cdot 8 = 40 \quad \text{and} \quad 8 \cdot 5 = 40.$$

COMMUTATIVE PROPERTY FOR MULTIPLICATION

The *order* in which two factors are written does not affect the product.

The **associative property for multiplication** allows for the *regrouping* of factors when more than two numbers are multiplied. Parentheses are used to indicate which two factors should be grouped (multiplied) first. For example,

$$(2 \cdot 3) \cdot 8 = 6 \cdot 8 = 48 \quad \text{and} \quad 2 \cdot (3 \cdot 8) = 2 \cdot 24 = 48.$$

ASSOCIATIVE PROPERTY FOR MULTIPLICATION

The way in which three or more factors are *grouped* does not affect the product.

Two other properties for multiplication can be illustrated as follows. One four-can box of energy drinks contains four cans, and zero four-can boxes of energy drinks contain zero cans. The first example demonstrates the **identity property for multiplication**, which states that multiplying a number by 1 does not change the number. The second example illustrates the **zero property for multiplication**, which states that multiplying a number by 0 results in 0.

IDENTITY PROPERTY FOR MULTIPLICATION

When any number is multiplied by 1, the result is that number.

ZERO PROPERTY FOR MULTIPLICATION

When any number is multiplied by 0, the result is 0.

READING CHECK

- Which multiplication property allows us to multiply in any order?
- Which multiplication property allows us to regroup the factors?

Examples of the identity and zero properties for multiplication include

$$4 \cdot 1 = 4 \quad \text{and} \quad 1 \cdot 4 = 4 \qquad \text{and} \qquad 0 \cdot 4 = 0 \quad \text{and} \quad 4 \cdot 0 = 0.$$

One final property of multiplication is illustrated in Figure 1.16. This property, called the **distributive property**, allows us to multiply a sum (or difference) by a number. Note that 4 rows of 5 faces (3 blue and 2 red) equals 4 rows of 3 faces plus 4 rows of 2 faces.

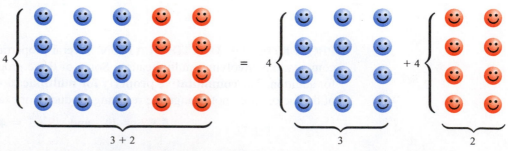

Figure 1.16 A Visual Representation of the Distributive Property: $4(3 + 2) = 4 \cdot 3 + 4 \cdot 2$

Figure 1.16 demonstrates that *multiplication distributes over addition.* The property also holds if the sum is replaced by a difference. That is, *multiplication distributes over subtraction.* These results are summarized as follows.

DISTRIBUTIVE PROPERTIES

When a sum or difference in parentheses is multiplied by a number outside the parentheses, the outside number can be multiplied by each of the inside numbers *before* the sum or difference is computed.

NOTE: Curved arrows are often used to illustrate the distributive properties.

$$3(5 + 7) = 3 \cdot 5 + 3 \cdot 7 \quad \text{and} \quad 4(8 - 5) = 4 \cdot 8 - 4 \cdot 5$$

EXAMPLE 1 **Applying the distributive properties**

Use a distributive property to rewrite each expression. Do not find the product.
(a) $6(3 + 8)$ **(b)** $5(7 - 2)$

Solution

(a) $6(3 + 8) = 6 \cdot 3 + 6 \cdot 8$

(b) $5(7 - 2) = 5 \cdot 7 - 5 \cdot 2$

Now Try Exercises 19, 21

THE DISTRIBUTIVE PROPERTIES AND MULTIPLICATION When multiplying larger whole numbers, we can use the expanded form and a distributive property to find a product. For example, the product 6×47 can be written as $6(40 + 7)$, and a distributive property may be applied as follows:

$$6(40 + 7) = 6 \cdot 40 + 6 \cdot 7$$
$$= 240 + 42$$
$$= 282$$

However, by stacking the numbers vertically and using carrying when appropriate, we can perform multiplication more efficiently without the need to write any of the factors in expanded form. The product 6×47 can be multiplied using the following two steps:

STEP 1: Multiply 6×7 (ones)   **STEP 2:** Multiply 6×4 (tens)

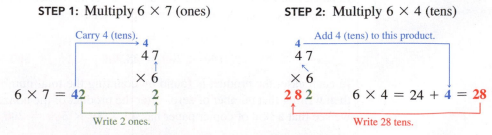

The next example shows how these steps can be combined into a single process in order to multiply larger whole numbers.

EXAMPLE 2 **Multiplying whole numbers**

Multiply.
(a) $83(4)$ **(b)** $92 \cdot 35$ **(c)** 386×73 **(d)** $208(867)$

Solution

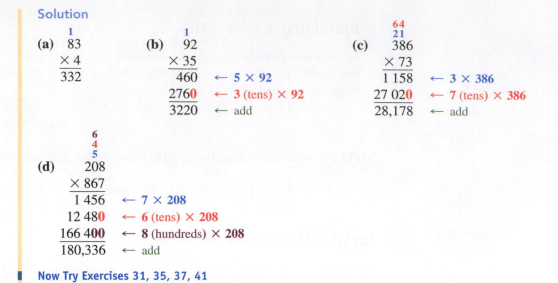

(a)
$$\begin{array}{r} \overset{1}{83} \\ \times\ 4 \\ \hline 332 \end{array}$$

(b)
$$\begin{array}{r} \overset{1}{92} \\ \times\ 35 \\ \hline 460 \\ 2760 \\ \hline 3220 \end{array}$$
$\leftarrow 5 \times 92$
$\leftarrow 3$ (tens) $\times$ **92**
$\leftarrow$ add

(c)
$$\begin{array}{r} \overset{64}{\overset{21}{386}} \\ \times\ 73 \\ \hline 1\ 158 \\ 27\ 020 \\ \hline 28{,}178 \end{array}$$
$\leftarrow 3 \times 386$
$\leftarrow 7$ (tens) $\times$ **386**
$\leftarrow$ add

(d)
$$\begin{array}{r} \overset{6}{\overset{4}{\overset{5}{208}}} \\ \times\ 867 \\ \hline 1\ 456 \\ 12\ 480 \\ 166\ 400 \\ \hline 180{,}336 \end{array}$$
$\leftarrow 7 \times 208$
$\leftarrow 6$ (tens) $\times$ **208**
$\leftarrow 8$ (hundreds) $\times$ **208**
$\leftarrow$ add

Now Try Exercises 31, 35, 37, 41

READING CHECK

• Which multiplication property makes it possible to multiply as shown in Example 2?

CALCULATOR HELP

To multiply whole numbers with a calculator, see the Appendix (page AP-1).

MAKING CONNECTIONS

The Distributive Property and Expanded Form

The multiplication process used in Example 2 provides a short way to apply the distributive property to the expanded form of each factor. The *partial products* 2760 and 460 found in Example 2(b) are also shown in the following multiplication process:

$92 \times 35 = 92(30 + 5)$	Write 35 in expanded form.
$= 92(30) + 92(5)$	Distribute the 92.
$= (90 + 2)(30) + (90 + 2)(5)$	Write 92 in expanded form.
$= (90)(30) + (2)(30) + (90)(5) + (2)(5)$	Distribute the 30 and the 5.
$= \mathbf{2700 + 60 + 450 + 10}$	Find each product.
$= 3220$	Add.

▶ **REAL-WORLD CONNECTION** A *skid* of copier paper contains 40 boxes of paper. Each box contains 10 individually wrapped *reams*, and each ream has 500 sheets of paper. To find the total number of sheets of copier paper in a skid, we multiply

$$40 \times 10 \times 500.$$

Because each of these numbers ends in one or more zeros, we can perform the multiplication mentally. Consider the following products:

$$10 \times 10 = 100 \qquad\qquad 20 \times 300 = 6000$$
$$140 \times 200 = 28{,}000 \qquad\qquad 500 \times 600 = 300{,}000$$

In each case, the product is found by counting the total number of zeros in the factors and then writing that number of zeros after the product of the nonzero digits. Using this process, we see that a skid of copier paper has $40 \times 10 \times 500 = 200{,}000$ sheets.

EXAMPLE 3 **Multiplying whole numbers that end in zeros**

Multiply.
(a) 130×40 **(b)** 700×2500

Solution
(a) $130 \times 40 = 5200$ **(b)** $700 \times 2500 = 1{,}750{,}000$

Now Try Exercises 45, 47

WORDS ASSOCIATED WITH MULTIPLICATION Just as there are words associated with addition and subtraction, there are also words associated with multiplication. Table 1.7 shows some of these words, together with a sample phrase using each word.

TABLE 1.7 Words Associated with Multiplication

Word	Sample Phrase
multiply	multiply the length and the width
times	the number purchased times the price
product	the product of the measurements
double	double the recipe
triple	triple the score

EXAMPLE 4

Translating words into a mathematical expression

Translate each phrase into a mathematical expression. Find the result.
(a) 15 times 20 cars   **(b)** the product of 7 and 38 coffee drinks

Solution
(a) The word *times* suggests that we multiply 15 and 20. The corresponding mathematical expression is 15×20, which results in 300 cars.
(b) The word *product* suggests that we multiply 7 and 38. The corresponding mathematical expression is 7×38, which results in 266 coffee drinks.

Now Try Exercises 79, 81

Music for your ears!

NEW VOCABULARY

continued from page 29
☐ Division
☐ Quotient
☐ Dividend
☐ Divisor
☐ Division sign
☐ Identity properties for division
☐ Zero properties for division
☐ Undefined
☐ Remainder
☐ Partial dividend
☐ 1 square unit
☐ Area

Dividing Whole Numbers

▶ **REAL-WORLD CONNECTION** Some gas stations and convenience stores sell music CDs at discounted prices. A customer can use *repeated subtraction* to compute the number of CDs costing $6 each that can be purchased for $24. Each time 6 is subtracted, another CD can be purchased.

$24 - 6 = 18$	$18 - 6 = 12$	$12 - 6 = 6$	$6 - 6 = 0$
1st Subtraction	**2nd Subtraction**	**3rd Subtraction**	**4th Subtraction**

Because 6 can be subtracted a total of 4 times, the customer can buy 4 CDs for $24. Just as multiplication is a fast way to perform repeated addition, **division** is a fast way to perform repeated subtraction. We say that **24** divided by **6** is **4** and write $24 \div 6 = 4$. The result of dividing one whole number by another is called the **quotient**. The number we are dividing *into* is called the **dividend**, the number we are dividing *by* is called the **divisor**, and symbol $\div$ is called the **division sign**. In this example, the **dividend** is **24**, the **divisor** is **6**, and the **quotient** is **4**.

$$24 \div 6 = 4$$

dividend divisor quotient

NOTE: Like multiplication, division can be written in several ways. Each of the following expressions represents dividing 24 by 6.

$$24 \div 6, \quad \frac{24}{6}, \quad 24/6, \quad \text{and} \quad 6\overline{)24}$$

Division can be checked by multiplying as follows.

Quotient $\times$ Divisor = Dividend

EXAMPLE 5 **Dividing whole numbers**

Find each quotient. Check your answers by multiplying.

(a) $48 \div 6$ **(b)** $\dfrac{63}{9}$

Solution

(a) The quotient $48 \div 6 = 8$ checks by multiplying $8 \times 6 = 48$. ✓

(b) The quotient $\frac{63}{9} = 7$ checks by multiplying $7 \times 9 = 63$. ✓

Now Try Exercises 55, 63

PROPERTIES OF DIVISION The commutative properties for addition and multiplication do *not* hold true for division. For example, $10 \div 5$ results in 2, whereas $5 \div 10$ does not give a whole number result. Similarly, the associative properties for addition and multiplication do *not* hold true for division. For example,

$$(24 \div 6) \div 2 = 4 \div 2 = 2, \quad \text{but} \quad 24 \div (6 \div 2) = 24 \div 3 = 8.$$

READING CHECK

• Which two multiplication properties do not hold true for division?

CALCULATOR HELP

To divide whole numbers with a calculator, see the Appendix (page AP-1).

If we think of division as a way to perform repeated subtraction, then we can find several properties for division. Consider the following four questions where "a number" can be any whole number *except* 0:

1. How many times can a number be subtracted from itself?   (Dividing a number by itself)
2. How many times can 1 be subtracted from a number?   (Dividing a number by 1)
3. How many times can a number be subtracted from 0?   (Dividing 0 by a number)
4. How many times can 0 be subtracted from a number?   (Dividing a number by 0)

Questions 1 and 2 illustrate the **identity properties for division**. The answer to the first question is **1**. For example, a person with 5 dimes can give away all 5 dimes exactly **1** time, or $5 \div 5 = 1$. In the second question, the result is always the **dividend**. For example, a person with **7** nickels can give away 1 nickel **7** times, or $7 \div 1 = 7$.

Questions 3 and 4 illustrate the **zero properties for division**. The third question always results in **0**. For example, a person with 0 pennies can give away 12 pennies **0** times because there are no pennies to give away, or $0 \div 12 = 0$. The last question does not have a single, correct answer. For example, a person with 6 quarters can give away 0 quarters *any* number of times. We say that $6 \div 0$ is **undefined**.

IDENTITY PROPERTIES FOR DIVISION

1. When any number (except 0) is divided by itself, the result is 1.
2. When any number is divided by 1, the result is the number (dividend).

ZERO PROPERTIES FOR DIVISION

1. When 0 is divided by any number (except 0), the result is 0.
2. When any number is divided by 0, the result is undefined.

| EXAMPLE 6 | **Applying the division properties** |

Use division properties to find each quotient, when possible.

(a) $23 \div 1$ **(b)** $\dfrac{0}{14}$ **(c)** $83 \div 83$ **(d)** $\dfrac{62}{0}$

Solution
(a) Dividing by 1 results in the dividend, 23.
(b) Dividing 0 by a number that is not 0 results in 0.
(c) Dividing a nonzero number by itself results in 1.
(d) Division by 0 is undefined.

Now Try Exercises 51, 53, 57, 59

REMAINDER Sometimes a divisor does not divide evenly into a dividend. For example, Figure 1.17 shows that 17 books can be divided into **3** stacks of 5 books each, with 2 books remaining. We say that **2** is the **remainder** and write the quotient as **3 r2**.

5 books   5 books   5 books   2 books left over

Figure 1.17 Remainder 2, when 17 is divided by 5.

READING CHECK

• What is a remainder?

NOTE: When a remainder exists, division can be checked as follows.

Quotient × Divisor + Remainder = Dividend

LONG DIVISION When we need to find a quotient involving a larger dividend, we can use a process called *long division*, which allows us to break a large division problem into several smaller division problems. For example, to divide **2461** by **5** using long division, we set up the problem as follows.

Divisor → $5\overline{)2461}$ ← **Dividend**

Starting at the left end of the dividend, select the fewest digits that give a number that is greater than the divisor. This (highlighted) number is called the **partial dividend**.

$$5\overline{)2461}$$
└── The partial dividend is 24.

The divisor 5 will "go into" the partial dividend 24 at most 4 times. Write **4** above the rightmost digit of the partial dividend.

$$\begin{array}{r} 4 \\ 5\overline{)2461} \end{array}$$

Next, multiply 4 and 5, write the result **20** below the partial dividend, and subtract.

$$\begin{array}{r} 4 \\ 5\overline{)2461} \\ -\,20 \\ \hline 4 \end{array}$$

Now, "bring down" the first digit in the original dividend that is aligned to the right of the partial dividend. The number 46 is the new partial dividend. We are now ready to begin the process again.

$$
\begin{array}{r}
4 \\
5\overline{)2461} \\
-\ 20\downarrow \\
\hline
46
\end{array}
$$

The divisor 5 goes into 46 at most 9 times. Write **9** next to 4 in the quotient. Multiply 9 and 5, write the result **45** below the partial dividend, and subtract. Bring down the 1. The number 11 is the new partial dividend. Go through the process one more time.

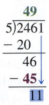

$$
\begin{array}{r}
49 \\
5\overline{)2461} \\
-\ 20 \\
\hline
46 \\
-\ 45\downarrow \\
\hline
11
\end{array}
$$

The divisor 5 goes into 11 at most 2 times. Write **2** next to 9 in the quotient. Multiply 2 and 5, write the result **10** below the partial dividend, and subtract.

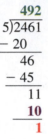

$$
\begin{array}{r}
492 \\
5\overline{)2461} \\
-\ 20 \\
\hline
46 \\
-\ 45 \\
\hline
11 \\
10 \\
\hline
1
\end{array}
$$

When there are no more digits to bring down from the original dividend, the process is done. The final difference **1** is the remainder. We write the quotient as **492 r1**.

PERFORMING LONG DIVISION

STEP 1: Determine the number of times that the divisor will "go into" the partial dividend. Write this digit above the right-most digit of the partial dividend.

STEP 2: Multiply the digit found in Step 1 by the divisor, and write the product below the partial dividend.

STEP 3: Subtract the product found in Step 2 from the partial dividend.

STEP 4: From the original dividend, "bring down" the first digit aligned to the right of the partial dividend. The number formed becomes the new partial dividend. If there is no digit to bring down, you are done.

READING CHECK

• When doing long division, where is the quotient written?

EXAMPLE 7   **Performing long division**

Divide: $2875 \div 3$. Check your answer.

Solution

The partial dividend is highlighted each time through the steps.

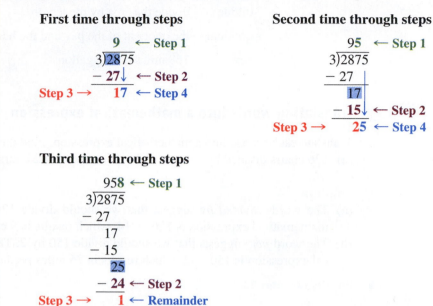

First time through steps

$$
\begin{array}{r}
9 \quad \leftarrow \textbf{Step 1} \\
3\overline{)28\!75} \\
-27\!\downarrow \quad \leftarrow \textbf{Step 2} \\
\textbf{Step 3} \rightarrow \quad 17 \quad \leftarrow \textbf{Step 4}
\end{array}
$$

Second time through steps

$$
\begin{array}{r}
95 \quad \leftarrow \textbf{Step 1} \\
3\overline{)2875} \\
-27\ \ \\
17\!\downarrow \quad \leftarrow \textbf{Step 2} \\
\textbf{Step 3} \rightarrow \quad 25 \quad \leftarrow \textbf{Step 4}
\end{array}
$$

Third time through steps

$$
\begin{array}{r}
958 \quad \leftarrow \textbf{Step 1} \\
3\overline{)2875} \\
-27\ \ \\
17\ \ \\
-15\ \ \\
25 \\
-24 \quad \leftarrow \textbf{Step 2} \\
\textbf{Step 3} \rightarrow \quad 1 \quad \leftarrow \textbf{Remainder}
\end{array}
$$

The quotient is 958 with remainder 1, or 958 r1. This result can be checked as follows.

$$958 \times 3 + 1 = 2875$$

Now Try Exercise 65

We do not need to rewrite a division problem every time we go through the steps. Typically, all of the steps used to perform long division are stacked vertically. The next example illustrates this process for divisors with more than 1 digit.

EXAMPLE 8   **Performing long division**

Divide. Check your answer.
(a) $2511 \div 23$ **(b)** $89,285 \div 258$

Solution

In part (a), note that 23 (the divisor) does not divide into the partial dividend 21. As a result, 0 is written in the quotient, and the steps are continued as usual.

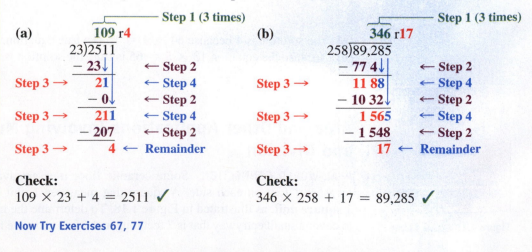

(a)

$$
\begin{array}{r}
\overbrace{\qquad}^{\textbf{Step 1 (3 times)}} \\
109\ \text{r}\mathbf{4} \\
23\overline{)2511} \\
-23\!\downarrow \quad \leftarrow \textbf{Step 2} \\
\textbf{Step 3} \rightarrow \quad 21\ \ \quad \leftarrow \textbf{Step 4} \\
-0\!\downarrow \quad \leftarrow \textbf{Step 2} \\
\textbf{Step 3} \rightarrow \quad 211 \quad \leftarrow \textbf{Step 4} \\
-207 \quad \leftarrow \textbf{Step 2} \\
\textbf{Step 3} \rightarrow \quad 4 \quad \leftarrow \textbf{Remainder}
\end{array}
$$

(b)

$$
\begin{array}{r}
\overbrace{\qquad}^{\textbf{Step 1 (3 times)}} \\
346\ \text{r}\mathbf{17} \\
258\overline{)89,285} \\
-77\ 4\!\downarrow \quad \leftarrow \textbf{Step 2} \\
\textbf{Step 3} \rightarrow \quad 11\ 88\ \ \quad \leftarrow \textbf{Step 4} \\
-10\ 32\!\downarrow \quad \leftarrow \textbf{Step 2} \\
\textbf{Step 3} \rightarrow \quad 1\ 565 \quad \leftarrow \textbf{Step 4} \\
-1\ 548 \quad \leftarrow \textbf{Step 2} \\
\textbf{Step 3} \rightarrow \quad 17 \quad \leftarrow \textbf{Remainder}
\end{array}
$$

Check:
$109 \times 23 + 4 = 2511$ ✓

Check:
$346 \times 258 + 17 = 89,285$ ✓

Now Try Exercises 67, 77

WORDS ASSOCIATED WITH DIVISION Table 1.8 shows some words associated with division, together with a sample phrase using each word.

TABLE 1.8 Words Associated with Division

Word	Sample Phrase
divide	divide the area by the length
quotient	the quotient of the pay and the hours worked
per	168 miles per 12 gallons

EXAMPLE 9 | **Translating words into a mathematical expression**

Translate each phrase into a mathematical expression. Find the result.
(a) 126 chairs divided by 14 rows   **(b)** 150 miles per 2 hours

Solution
(a) The words *divided by* suggest that we should divide 126 by 14. The corresponding mathematical expression is $126 \div 14$, which results in 9 chairs per row.
(b) The word *per* suggests that we should divide 150 by 2. The corresponding mathematical expression is $150 \div 2$, which results in 75 miles per hour.

▌ **Now Try Exercises 83, 85**

Solving Equations

In the previous section, we solved addition and subtraction equations containing unknown values by finding the value that, when written in the empty box, makes the equation true. Equations involving multiplication or division can be solved in this way also. For example, the solution to the equation

$$\square \times 4 = 36$$

is 9 because $9 \times 4 = 36$ is a true equation and writing any other number in the box would result in a false equation.

EXAMPLE 10 | **Solving equations**

Solve each equation by finding the unknown value.
(a) $64 \div \square = 16$ **(b)** $13 \times \square = 65$

Solution
(a) The solution is **4** because $64 \div \mathbf{4} = 16$ is a true equation.
(b) Because the equation $13 \times \mathbf{5} = 65$ is true, the solution is **5**.

▌ **Now Try Exercises 87, 93**

Area and Other Applications Involving Multiplication and Division

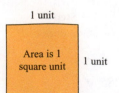

1 unit

Area is 1 square unit   1 unit

Figure 1.18 Square Unit

▶ **REAL-WORLD CONNECTION** Some ceramic floor tiles are available in square pieces that measure 1 foot on each side. A square that measures 1 unit on each side has an *area* of **1 square unit**, as illustrated in Figure 1.18. To determine the number of square tiles needed to cover a small entryway that is 7 feet long and 5 feet wide, we must find the area of the floor.

Area is computed by finding the number of square units that are needed to cover a region. Figure 1.19 shows that a 7-feet by 5-feet entryway has an area of 35 square feet.

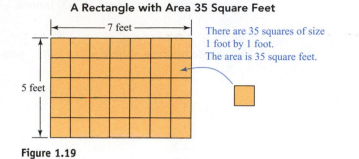

A Rectangle with Area 35 Square Feet

7 feet

5 feet

There are 35 squares of size 1 foot by 1 foot. The area is 35 square feet.

Figure 1.19

The area of the rectangle in Figure 1.19 can be found by counting the number of 1-foot by 1-foot tiles that cover the entire region. However, we can also find the area by multiplying the length **7** and the width **5**.

$$7 \text{ feet} \times 5 \text{ feet} = 35 \text{ square feet}$$

To find the area of a rectangle, we use the following formula.

READING CHECK

• What kind of units are used to describe area?

AREA OF A RECTANGLE

The area of a rectangle is found by multiplying its length and width.

$$\text{Area} = \text{Length} \times \text{Width}$$

EXAMPLE 11   **Finding the area of an Olympic swimming pool's surface**

An Olympic swimming pool is 50 meters long and 25 meters wide. (A meter is a unit of length in the metric system.) Find the area of the pool's surface in square meters.

Solution

The area of the pool's surface is found by multiplying the length **50** and the width **25**.

$$50 \times 25 = 1250 \text{ square meters}$$

Now Try Exercise 99

▶ **REAL-WORLD CONNECTION** The E-126 wind turbine is among the largest in the world. Each rotor blade is approximately 413 feet long. It is estimated that just twenty of these turbines can generate enough electricity to power 35,500 American homes. In the next example, division is used to find the number of homes that can be powered by a single E-126 wind turbine. (*Source: WindBlatt, Enercon Magazine for Wind Energy, 2007.*)

EXAMPLE 12   **Finding the number of homes powered by a wind turbine**

Find the number of American homes that can be powered by a single E-126 wind turbine if 20 such turbines can power 35,500 homes.

Solution
To find the number of homes that can be powered by one E-126 wind turbine, divide 35,500 by 20.

$$
\begin{array}{r}
1\,775 \\
20\overline{)35{,}500} \\
-\,20 \\
\hline
15\,5 \\
-\,14\,0 \\
\hline
1\,50 \\
-\,1\,40 \\
\hline
100 \\
-\,100 \\
\hline
0
\end{array}
$$

A single E-126 wind turbine can power 1775 American homes.

Now Try Exercise 103

1.3 Putting It All Together

CONCEPT	COMMENTS	EXAMPLES
Multiplication	The numbers being multiplied are the *factors,* and the result is the *product.*	$8 \times 4 = 32$ factor factor product
Properties for Multiplication	**1.** Commutative property **2.** Associative property **3.** Identity property **4.** Zero property **5.** Distributive properties	**1.** $4 \times 3 = 3 \times 4$ **2.** $(2 \cdot 4) \cdot 5 = 2 \cdot (4 \cdot 5)$ **3.** $5 \times 1 = 5$ and $1 \times 5 = 5$ **4.** $3 \times 0 = 0$ and $0 \times 3 = 0$ **5.** $3(6 + 5) = 3 \cdot 6 + 3 \cdot 5$ $2(7 - 4) = 2 \cdot 7 - 2 \cdot 4$
Words Associated with Multiplication	Words associated with multiplication include *multiply, times, product, double,* and *triple.*	3 times her age Double the number of cups
Division	The number divided into is the *dividend*; the number divided by is the *divisor,* and the result is the *quotient.*	$48 \div 6 = 8$ dividend divisor quotient
Properties for Division	**1.** Identity properties **2.** Zero properties	**1.** $7 \div 7 = 1$ $9 \div 1 = 9$ **2.** $0 \div 3 = 0$ $8 \div 0$ is undefined.

CONCEPT	COMMENTS	EXAMPLES
Long Division	When the dividend is large, we use a process called *long division*. When the divisor does not divide perfectly into the dividend, the *remainder* is the amount left over.	$$\begin{array}{r} 36\ r6 \\ 7\overline{)258} \\ -21 \\ \hline 48 \\ -42 \\ \hline 6 \end{array}$$ **Check:** $36 \times 7 + 6 = 258$ ✓
Words Associated with Division	Words associated with division include *divide*, *quotient*, and *per*.	Divide the area by the width Student athletes per van
Area of a Rectangle	Area = Length × Width	6 inches 3 inches Area is $3 \times 6 = 18$ square inches.

1.3 Exercises

CONCEPTS AND VOCABULARY

1. Multiplication is a fast way to perform repeated _____.

2. When multiplying, the two numbers being multiplied are called _____.

3. The result when multiplying is called the _____.

4. The equation $13 \cdot 26 = 26 \cdot 13$ is an example of the _____ property for multiplication.

5. The equation $(2 \cdot 5) \cdot 4 = 2 \cdot (5 \cdot 4)$ is an example of the _____ property for multiplication.

6. The _____ property for multiplication states that multiplying any number by 1 results in that number.

7. The _____ property for multiplication states that multiplying any number by 0 results in 0.

8. The equation $3(7 + 5) = 3 \cdot 7 + 3 \cdot 5$ illustrates a(n) _____ property for multiplication.

9. The word *times* indicates that (multiplication/division) should be used.

10. Division is a fast way to perform repeated _____.

11. In a division problem, the number we are dividing into is called the _____ and the number we are dividing by is called the _____.

12. The result when dividing is called the _____.

13. The equation $7 \div 7 = 1$ illustrates one of the _____ properties for division.

14. $0 \div 5 =$ _____, but $5 \div 0$ is _____.

15. A process for finding a quotient involving a larger dividend is called _____.

16. The operation (multiplication/division) should be used when we see the word *per*.

17. A square that measures 1 unit on each side has an area of _____.

18. We compute _____ by finding the number of square units that cover a region.

MULTIPLYING WHOLE NUMBERS

Exercises 19–24: Use a distributive property to rewrite the expression. Do not find the product.

19. $5(6 + 9)$

20. $7(2 + 5)$

21. $4(8 - 1)$

22. $6(9 - 3)$

23. $(6 - 2)3$

24. $(5 + 7)4$

Exercises 25–44: Multiply.

25. 7×1

26. $0 \cdot 9$

27. $0 \cdot 5$

28. 1×12

29. $6(9)$

30. $(4)(8)$

31. $7 \cdot 48$

32. 5×83

33. $(302)6$

34. $(479)(8)$

35. $71(24)$

36. $94 \cdot 18$

37. 172×14

38. $(23)(492)$

39. $35 \cdot 1475$

40. 56×9012

41. $376(754)$

42. $126 \cdot 533$

43. $109 \cdot 1074$

44. $2348(342)$

Exercises 45–50: Multiply mentally.

45. 70×300

46. $30 \cdot 2000$

47. $340 \cdot 2000$

48. 4000×800

49. $30 \cdot 100 \cdot 500$

50. $40 \times 80 \times 20$

DIVIDING WHOLE NUMBERS

Exercises 51–78: Divide, when possible.

51. $9 \div 1$

52. $\dfrac{0}{4}$

53. $\dfrac{17}{17}$

54. $12 \div 0$

55. $88 \div 8$

56. $81 \div 9$

57. $\dfrac{25}{0}$

58. $\dfrac{72}{24}$

59. $0 \div 13$

60. $34 \div 1$

61. $391 \div 391$

62. $\dfrac{354}{6}$

63. $\dfrac{72}{6}$

64. $\dfrac{1026}{1026}$

65. $\dfrac{6729}{7}$

66. $\dfrac{5812}{9}$

67. $2487 \div 31$

68. $4679 \div 53$

69. $6000 \div 30$

70. $8000 \div 20$

71. $9874 \div 0$

72. $\dfrac{0}{5430}$

73. $\dfrac{10,651}{84}$

74. $24,682 \div 99$

75. $36,855 \div 567$

76. $76,383 \div 943$

77. $49,777 \div 791$

78. $31,896 \div 665$

TRANSLATING WORDS INTO MATH

Exercises 79–86: Translate the phrase into a mathematical expression and then find the result.

79. The product of 14 feet and 3 feet

80. Double 50 pounds

81. Multiply $5 by 15

82. 31 text messages times 65

83. 126 miles per 7 gallons

84. Divide 1200 boxes by 20

85. The quotient of 75 days and 15

86. 516 people per 43 tables

SOLVING EQUATIONS

Exercises 87–94: Solve the given equation by finding the unknown value.

87. $24 \div \square = 8$

88. $\square \times 7 = 56$

89. $\square \times 9 = 72$

90. $124 \div \square = 31$

91. $\square \div 13 = 5$

92. $5 \times \square = 105$

93. $16 \times \square = 128$

94. $\square \div 11 = 9$

APPLICATIONS

*Exercises 95–98: **Area** Find the area of the rectangle.*

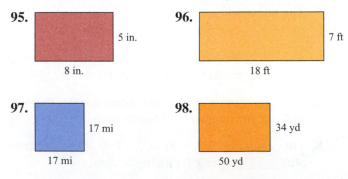

95. 5 in. / 8 in.

96. 7 ft / 18 ft

97. 17 mi / 17 mi

98. 34 yd / 50 yd

99. *Baseball Diamonds* A baseball diamond has the shape of a square measuring 90 feet on each side. Find the area of a baseball diamond.

100. *Tennis Courts* A doubles tennis court is 78 feet long and 36 feet wide. What is the area of a doubles tennis court?

101. *Panda Births* The following table lists the number of pandas born in captivity during selected years. What number should be multiplied by the number of births in 2002 to find the number of births in 2006?

Year	2002	2004	2006	2008
Pandas	10	9	30	31

Source: Environmental News Network.

102. *Endangered Species* The following table shows the number of species listed as threatened or endangered in selected states. Which state has 5 times as many species listed as Alaska, which lists 13 species?

State	Delaware	Maine	Missouri	Virginia
Species	22	16	30	65

Source: U.S. Fish and Wildlife Service.

103. *Wind Power* Twenty-five E-126 wind turbines can power 125,000 European homes. Find the number of European homes that can be powered by a single E-126 wind turbine. (*Source: WindBlatt*, Enercon Magazine for Wind Energy, 2007.)

104. *National Forests* The number of national forests in California is double the number in Montana. If there are 18 national forests in California, how many are in Montana? (*Source:* U.S. Forest Service.)

105. *Counting Calories* One bottle of drinking water has 0 calories. How many calories are in 73 bottles of drinking water?

106. *Counting Calories* One can of grape soda has 190 calories. How many calories are in 30 cans of grape soda?

107. *Digital Photos* The following digital image of Saturn was created using a rectangular pattern of small image points called *pixels*. Find the total number of pixels in a photo with a length of 600 pixels and a width of 400 pixels.

108. *Digital Photos* (Refer to Exercise 107.) Find the total number of pixels in a photo with a length of 400 pixels and a width of 300 pixels.

109. *iPod Music* A 160-gigabyte iPod holds 40,000 songs. How many songs per gigabyte is this?

110. *iPod Photos* A 16-gigabyte iPod Touch can hold 20,000 iPod-viewable photos. How many photos per gigabyte is this?

111. *Fencing Property* A homeowner wants to put a fence around a rectangular property with an area of 7200 square feet. If one side of the property is 60 feet long, find each of the following.
(a) The length of the other side
(b) The total amount of fencing needed

112. *Area of a Rectangle* A rectangle with an area of 168 square inches has a length of 14 inches. What is the width of the rectangle?

113. *A Number Puzzle* When a particular number between 2 and 20 is divided by 2, 3, or 4, the remainder is always 1. Find the number.

114. *Numbers* The product of two numbers is 132. If one of the numbers is 6, find the other number.

115. *CDs* What is the maximum number of CDs costing $8 each that a person can buy with $75? How much change will the person receive?

116. *DVDs* What is the maximum number of DVDs costing $16 each that a person can buy with $80? How much change will the person receive?

WRITING ABOUT MATHEMATICS

117. Explain what a *remainder* is. Give an example.

118. Noting that division can be checked by multiplying

quotient $\times$ divisor $=$ dividend,

explain why dividing by zero is undefined.

119. Give examples of words associated with multiplication.

120. Give examples of words associated with division.

121. Explain how repeated subtraction can be used to show that $79 \div 11 = 7\,r\,2$.

122. Explain how repeated addition can be used to show that $15 \cdot 7 = 105$.

Group Activity   Working with Real Data

Directions: Form a group of 2 to 4 people. Select someone to record the group's responses for this activity. All members of the group should work cooperatively to answer the questions. If your instructor asks for your results, each member of the group should be prepared to respond.

Winning the Lottery In the multistate lottery game Powerball, there are 120,526,770 possible number combinations, only one of which is the grand prize winner. The cost of a single ticket (one number combination) is $1. (*Source:* Powerball.com)

Suppose that a wealthy person decides to buy tickets for every possible number combination, to be assured of winning a $150 million grand prize.

(a) If this strategy is carried out, what would be the total profit?

(b) Suppose this person could purchase one ticket every second. Divide 120,526,770 by 60 to find the number of minutes required to purchase all of the tickets.

(c) Ignoring the remainder in part (b), divide your result by 60 to find the number of hours needed to purchase all of the tickets.

(d) Using the quotient in part (c) without the remainder, divide your result by 24 to find the number of days needed to purchase all of the tickets.

(e) Leaving off the remainder in part (d), divide your result by 365 to find the number of years required to purchase all of the tickets.

(f) If there were a way for this person to buy all possible number combinations quickly, discuss reasons why this strategy would probably lose money.

1.4   Exponents, Variables, and Algebraic Expressions

Exponential Notation • Variables • Algebraic Expressions • Formulas • Translating Words to Expressions and Formulas • Solving Equations

A LOOK INTO MATH ▶

In the previous two sections, we discussed addition, subtraction, multiplication, and division of whole numbers. The branch of mathematics that uses these four *mathematical operations* is known as **arithmetic**. In this section, we begin to study the fundamentals of **algebra**—a generalization of arithmetic in which letters representing numbers are combined according to the rules of arithmetic. Algebra is useful for solving problems in many areas, including electronics, carpentry, surveying, law enforcement, astronomy, landscaping, and business.

NEW VOCABULARY

☐ Arithmetic
☐ Algebra
☐ Exponential notation
☐ Base
☐ Exponent
☐ Base-10 number system
☐ Variable
☐ Algebraic expression
☐ Evaluate
☐ Formula

Exponential Notation

Before we begin to study algebra, let's consider one idea from arithmetic. Just as multiplication is a fast way to perform repeated addition and division is a fast way to perform repeated subtraction, there is a mathematical concept called **exponential notation** that is used to perform repeated multiplication. The following equation shows how repeated multiplication can be written in exponential notation.

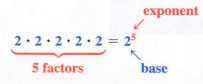

$$\underbrace{2 \cdot 2 \cdot 2 \cdot 2 \cdot 2}_{\text{5 factors}} = 2\overset{\text{exponent}}{^5}$$

The expression on the right side of the equation is an *exponential expression* with **base 2** and **exponent 5**. It is read as "two to the fifth power." For an exponential expression with a natural number exponent, the base is used as a repeated factor, and the exponent indicates *how many times* the base should be multiplied by itself.

NOTE: An exponent of 1 is usually not written. For example, $9^1 = 9$ and $7^1 = 7$.

READING CHECK

• Why is exponential notation used?

STUDY TIP

Do you know your instructor's name? Do you know the location of his or her office and the hours when he or she is available for help? Make sure that you have the answers to these important questions so that you can get help when needed.

EXAMPLE 1 | **Writing repeated multiplication in exponential notation**

Write each of the following in exponential notation.
(a) $4 \cdot 4 \cdot 4 \cdot 4 \cdot 4 \cdot 4 \cdot 4$ **(b)** $2 \cdot 2 \cdot 2 \cdot 5 \cdot 5 \cdot 5 \cdot 5$

Solution
(a) The factor **4** is repeated **7** times. The exponential notation is 4^7.
(b) The factor **2** is repeated **3** times and the factor **5** is repeated **4** times. This product has two bases, each with its own exponent. The exponential notation is $2^3 \cdot 5^4$.

Now Try Exercises 17, 21

CALCULATOR HELP

To evaluate exponents with a calculator, see the Appendix (page AP-2).

The value of an exponential expression with a whole number base and a natural number exponent can be found by first writing the expression as repeated multiplication and then finding the product.

EXAMPLE 2 | **Finding the value of an exponential expression**

Find the value of each exponential expression.
(a) 8^2 **(b)** 2^4 **(c)** 4^5

Solution
(a) $8^2 = 8 \cdot 8 = 64$
(b) $2^4 = 2 \cdot 2 \cdot 2 \cdot 2 = 16$
(c) $4^5 = 4 \cdot 4 \cdot 4 \cdot 4 \cdot 4 = 1024$

Now Try Exercises 33, 35, 39

Figure 1.20 3 Squared

SQUARING AND CUBING The word *squared* is commonly used when reading an exponential expression with exponent 2, and the word *cubed* is used when the exponent is 3. For example, 6^2 is read as "six squared," and 7^3 is read as "seven cubed." The terms *squared* and *cubed* come from geometry. If the length of each side of a square is 3 units, then its area is

$$3 \cdot 3 = 3^2 = 9 \text{ square units,}$$

as shown in Figure 1.20. Similarly, if the length of each edge of a *cube* is 3 units, then the *volume* of the cube is

$$3 \cdot 3 \cdot 3 = 3^3 = 27 \text{ cubic units,}$$

as shown in Figure 1.21.

NOTE: The expressions 3^2 and 3^3 can also be read as "three to the second power" and "three to the third power," respectively.

Figure 1.21 3 Cubed

POWERS OF 10 A pattern emerges when we look at base 10 raised to natural number powers. Table 1.9 shows that for each power of 10, the **exponent** equals the number of **0**s in the standard form for that power of 10. The table also displays the period name that corresponds to each power of 10. Is it any wonder that our number system is called the **base-10 number system**?

TABLE 1.9 Powers of 10

Power of 10	Repeated Multiplication	Standard Form	Period Name
10^1	10	**10**	tens
10^2	$10 \cdot 10$	**100**	hundreds
10^3	$10 \cdot 10 \cdot 10$	**1000**	thousands
10^4	$10 \cdot 10 \cdot 10 \cdot 10$	**10,000**	ten-thousands
10^5	$10 \cdot 10 \cdot 10 \cdot 10 \cdot 10$	**100,000**	hundred-thousands
10^6	$10 \cdot 10 \cdot 10 \cdot 10 \cdot 10 \cdot 10$	**1,000,000**	millions

NOTE: From Table 1.9, it may seem reasonable that the standard form of 10^0 should be a 1 followed by **zero** 0s, or simply 1. We will see later that $10^0 = 1$.

EXAMPLE 3   **Finding the value of a power of 10**

Find the value of the expression $4 \cdot 10^5$.

Solution

$$4 \cdot 10^5 = 4 \cdot \mathbf{100{,}000} = 400{,}000$$

Now Try Exercise 43

Variables

▶ **REAL-WORLD CONNECTION** Because there are 3 feet in 1 yard, 4 yards are equal to $4 \cdot 3 = 12$ feet and 7 yards are equal to $7 \cdot 3 = 21$ feet. Table 1.10 might be useful when converting from yards to feet frequently.

TABLE 1.10 Converting Yards to Feet

Yards	1	2	3	4	5	6	7
Feet	3	6	9	12	15	18	21

However, this table is not helpful if we need to convert more than 7 yards. To find the number of feet in 10 yards, for example, we would need to expand Table 1.10 to include $10 \cdot 3 = 30$ feet. However, expanding the table to include every possible value for yards would not be possible.

One of the most important ideas in mathematics is the notion of a *variable*. A **variable** is a symbol, typically an italic letter such as x, y, or z, used to represent an unknown quantity. (Uppercase letters such as F, P, and Y can also be variables.) Variables can be used when tables of numbers are inadequate. In the previous example, the number of feet is found by multiplying the corresponding number of **yards** by **3**. This is represented by the product

$$\textbf{(Yards)} \cdot \textbf{3}.$$

Any letter can be used as a variable. However, choosing the first letter of the quantity being represented may help give meaning to the variable. For example, if the variable Y is used to represent the number of **y**ards, then the product becomes

$$Y \cdot 3.$$

READING CHECK

- Why are variables used?
- How can a letter be chosen for a variable so that the variable has meaning?

EXAMPLE 4   **Using a variable to express an unknown quantity**

Rewrite the given expression using an appropriate variable.
(a) (Inches) ÷ 12   **(b)** 17 − (Weeks)

Solution
(a) Using the variable I to represent **i**nches, we can write the expression as $I \div 12$.
(b) Using the variable W to represent **w**eeks, we can write the expression as $17 - W$.

▌ Now Try Exercises 47, 53

Algebraic Expressions

When finding the number of feet in a given number of yards, we used the variable Y to represent the number of yards. The product $Y \cdot 3$ is used to find the number of feet in Y yards. This product, $Y \cdot 3$, is an example of an *algebraic expression*. An **algebraic expression** may contain numbers; variables; exponents; operation symbols such as $+$, $-$, $\times$, and $\div$; and grouping symbols, such as parentheses.

If we replace the variable Y in the algebraic expression $Y \cdot 3$ with the number **12**, then we can find the number of feet in 12 yards. That is, we **evaluate** the expression for a given value of the variable. In the next two examples, we evaluate algebraic expressions for given values of variables.

READING CHECK

- How is an algebraic expression evaluated for a given value of the variable?

EXAMPLE 5   **Evaluating algebraic expressions with one variable**

Evaluate each algebraic expression for $x = 3$.

(a) $5 + x$ **(b)** $\dfrac{18}{x}$ **(c)** $10x$ **(d)** x^2

Solution
(a) Replace x with **3** in the expression $5 + x$ to get $5 + 3 = 8$.

(b) Replace x with **3** in the expression $\dfrac{18}{x}$ to get $\dfrac{18}{3} = 6$.

(c) The expression $10x$ indicates multiplication of 10 and x. So, $10x = 10(3) = 30$.

(d) $x^2 = 3^2 = 9$

▌ Now Try Exercises 57, 59, 61, 63

▶ **REAL-WORLD CONNECTION** Some algebraic expressions contain more than one variable. For example, if a car travels 160 miles on 5 gallons of gasoline, then the car's *mileage* is $\frac{160}{5} = 32$ miles per gallon. Generally, if a car travels M miles on G gallons of gasoline, then its mileage is given by the expression $\frac{M}{G}$. Note that the expression $\frac{M}{G}$ contains two variables, M and G.

EXAMPLE 6 **Evaluating algebraic expressions with two variables**

Evaluate each algebraic expression for $y = 3$ and $z = 12$.
(a) $5yz$ **(b)** $z - y$ **(c)** $z \div y$

Solution
(a) Replace y with 3 and z with 12 to get $5yz = 5 \cdot 3 \cdot 12 = 180$.
(b) $z - y = 12 - 3 = 9$
(c) $z \div y = 12 \div 3 = 4$

Now Try Exercises 65, 67, 69

Formulas

In Section 1.2, we learned about equations. Recall that an equation is a mathematical statement that two algebraic expressions are equal. Equations *always* contain an equal sign. A **formula** is a special type of equation that expresses a relationship between two or more quantities. The formula $F = Y \cdot 3$ means that to find the number of feet F in Y yards, we multiply Y by 3.

The dot ($\cdot$) is often used to indicate multiplication because the symbol ($\times$) can be confused with the variable x. Sometimes the multiplication sign is omitted altogether. The four formulas

$$F = 3 \times Y, \quad F = 3 \cdot Y, \quad F = 3Y, \quad \text{and} \quad F = 3(Y)$$

represent the same relationship between yards and feet. For example, to find the number of feet in 12 yards, we can replace Y in any of these formulas with the number 12. If we let $Y = 12$ in the second formula, we have

$$F = 3 \cdot 12 = 36.$$

This means that there are 36 feet in 12 yards.

Earlier in this chapter we used the formula

$$\textbf{Area} = \textbf{length} \times \textbf{width}$$

to find the area of a rectangle. By using the variables A, l, and w to represent area, length, and width, respectively, this formula can be written in the form

$$A = lw.$$

Figure 1.22 shows three common figures from geometry and their associated formulas.

READING CHECK

• How is a formula related to an equation?

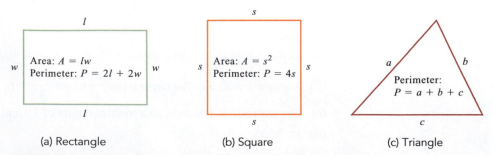

(a) Rectangle   (b) Square   (c) Triangle

Figure 1.22 Formulas Associated with Three Common Geometric Figures

In Figure 1.22, the variable A represents area, and P represents perimeter. For a rectangle, l and w represent length and width, respectively. The variable s represents the measure of one side of a square, and the variables a, b, and c represent the measures of the three sides of a triangle. (The *area* of a triangle will be discussed in Chapter 4.)

EXAMPLE 7   **Working with geometric formulas**

Use the geometric formulas in Figure 1.22 to find each of the following.
(a) Find the area of a square with a side length of 6 inches.
(b) Find the perimeter of a triangle with side lengths of 47, 32, and 55 feet.
(c) Find the perimeter of a rectangle with length 7 miles and width 4 miles.

Solution
(a) Substituting **6** for s in the formula $A = s^2$ gives $A = \mathbf{6}^2 = 6 \cdot 6 = 36$ square inches.
(b) Substitute **47**, **32**, and **55** for a, b, and c in the formula $P = a + b + c$ to find the perimeter of the triangle. $P = \mathbf{47} + \mathbf{32} + \mathbf{55} = 134$ feet
(c) $P = 2l + 2w = 2(\mathbf{7}) + 2(\mathbf{4}) = 14 + 8 = 22$ miles

Now Try Exercises 71, 73, 75

Translating Words to Expressions and Formulas

Many times in mathematics we need to write our own algebraic expressions. To do this, we translate words to symbols. Recall that the symbols $+$, $-$, $\times$, and $\div$ have special mathematical words associated with them. Table 1.11 summarizes many of the words commonly associated with these operations.

READING CHECK

• What words are associated with each of the arithmetic symbols?

TABLE 1.11 **Words Associated with Arithmetic Symbols**

Symbol	Associated Words
$+$	add, plus, more, sum, total, increase
$-$	subtract, minus, less, difference, fewer, decrease
$\times$	multiply, times, twice, double, triple, product
$\div$	divide, divided by, quotient, per

EXAMPLE 8   **Translating words into expressions**

Translate each phrase into an algebraic expression. Explain what each variable represents.
(a) Double the cost of an MP3 player
(b) Eight less than a friend's age
(c) The number of students divided by the number of books
(d) The sum of two different numbers

Solution
(a) If the cost is $79, then twice this cost would be $2 \cdot \mathbf{79} = \$158$. Generally, if we let C be the cost of an MP3 player, then twice the cost would be $2 \cdot C$, or $2C$.
(b) If a friend's age is 17, then eight less than the age would be $\mathbf{17} - 8 = 9$. If we let A represent the friend's age, then eight less than the age would be $A - 8$.
(c) Let S represent the number of students, and let B represent the number of books. The number of students divided by the number of books is $S \div B$ or $\frac{S}{B}$.
(d) Let x be one of the numbers, and let y be the other number. The sum is $x + y$.

Now Try Exercises 77, 79, 81

▶ **REAL-WORLD CONNECTION** When we perform physical activity, the energy used is measured in *calories*. A 175-pound person burns about 8 calories per minute while stacking firewood and about 7 calories per minute while whitewater rafting. We can write a formula

that expresses the relationship between the number of minutes spent performing an activity and the number of calories burned. For example, whitewater rafting for **1** minute burns $7 \cdot 1 = 7$ calories, and whitewater rafting for **5** minutes burns $7 \cdot 5 = 35$ calories. If we let **C** represent the number of calories burned during **w** minutes of whitewater rafting, then the formula $C = 7w$ expresses the relationship between the number of minutes spent whitewater rafting and the number of calories burned. (*Source:* The Calorie Control Council.)

EXAMPLE 9   **Translating words to a formula**

A 250-pound person burns 2 calories per minute while watching television and 9 calories per minute while gardening.
(a) Write a formula that gives the calories C burned while watching television for t minutes.
(b) Write a formula that gives the calories C burned while gardening for g minutes.
(c) Use your formulas to find the number of calories burned while performing each of these activities for 3 hours.

Solution
(a) Watching television for **1** minute burns $2 \cdot 1 = 2$ calories, and watching television for **10** minutes burns $2 \cdot 10 = 20$ calories. The formula $C = 2t$ gives the number of calories burned while watching television for **t** minutes.
(b) The formula $C = 9g$ gives the number of calories burned while gardening for **g** minutes.
(c) Because the formulas require that time is given in minutes, we must convert 3 hours to $3 \cdot 60 = 180$ minutes. For a 250-pound person, watching television for 180 minutes burns $C = 2(180) = 360$ calories, and gardening for that same amount of time burns $C = 9(180) = 1620$ calories.

▌ Now Try Exercise 101

Solving Equations

Because a variable can be used to represent an unknown quantity, we can use a variable rather than an empty box when solving equations that contain unknown values. For example, solving an equation of the form

$$\Box + 7 = 13$$

means that we look for the number that can be written in the box to make the equation true. Similarly, solving an equation of the form

$$n + 7 = 13$$

means that we look for the value of the variable n that makes the equation true. For either equation, the solution is 6 because writing 6 in the box or replacing the variable **n** with **6** results in $6 + 7 = 13$, which is a true equation.

Recall that a number is a *solution* to an equation if it makes the equation true. In the next example, we check a given number to determine if it is a solution to a given equation.

EXAMPLE 10   **Checking solutions**

Determine if the given number is a solution to the given equation.
(a) Is 3 a solution to $5x = 15$?
(b) Is 11 a solution to $23 - m = 9$?
(c) Is 6 a solution to $y^2 = 12$?

Solution

(a) Replace the variable x in the equation with **3**. Because $5(3) = 15$ is a **true** equation, 3 is a solution.

(b) Replacing m in the equation with **11** results in $23 - 11 \stackrel{?}{=} 9$, which is a **false** equation. Subtracting 11 from 23 results in $23 - 11 = 12$, so 11 is not a solution.

(c) Because $6^2 = 36$, replacing the variable y with 6 results in a **false** equation. So, 6 is not a solution.

Now Try Exercises 83, 85, 87

READING CHECK

• What math symbol means "not equal to"?

NOTE: In Example 10(b), we could have written $23 - 11 \neq 9$ instead of saying that $23 - 11 \stackrel{?}{=} 9$ is **false**. The symbol $\neq$ means "not equal to." This notation will be used throughout the remainder of this text.

EXAMPLE 11 **Solving equations**

Solve each equation.

(a) $114 + y = 125$ (b) $48 \div m = 4$ (c) $x^3 = 8$

Solution

(a) Because $114 + 11 = 125$, the solution is **11**.

(b) The solution is **12** because $48 \div 12 = 4$ is true.

(c) Cube **2** to get 8. The solution is **2**.

Now Try Exercises 89, 91, 93

EXAMPLE 12 **Finding the number of panda births**

The number of pandas born in captivity in 2008 is 6 more than the number born in 2007. If 31 pandas were born in 2008, how many were born in 2007? (*Source:* Environmental News Network.)

Solution

The question "how many were born in 2007?" indicates that the unknown value is the number of pandas born in 2007. We will let the variable p represent the number of pandas born in captivity in 2007. Since the number of births in 2008 is **31**, the phrase

the number born in 2008 *is* 6 *more than the number born in* 2007

translates to the equation

$$31 = p + 6.$$

The solution to this equation is 25 because $31 = 25 + 6$ is true. There were 25 pandas born in captivity in 2007.

Now Try Exercise 103

MAKING CONNECTIONS

Equations and Expressions

The words "equation" and "expression" are *not* interchangeable. An equation is similar to a sentence, while an expression is more like a phrase. An equation *always* contains an equal sign, but an expression *never* contains an equal sign. We *solve* an equation for an unknown value, but we *evaluate* an expression for a given value of the variable or variables. Also, the equal sign in an equation separates two expressions. For example, $4 + x = 2x$ is an equation, while $4 + x$ and $2x$ are both expressions.

READING CHECK

- How do the words "equation" and "expression" differ in meaning?

1.4 Putting It All Together

CONCEPT	COMMENTS	EXAMPLES
Exponential Notation	A natural number exponent indicates the number of times that the base is used as a repeated factor.	Exponent $\searrow$ $3^4 = 3 \cdot 3 \cdot 3 \cdot 3$ $\nearrow$ Base
Powers of 10	A natural number exponent on 10 equals the number of 0s that follow a 1 in the standard form of the power of 10.	$10^3 = 1000$ $10^5 = 100,000$ $10^{12} = 1,000,000,000,000$
Variable	Used to represent an unknown quantity	P represents the number of pets. n represents an unknown number.
Algebraic Expression	May contain numbers, variables, exponents, operation symbols, and grouping symbols	$3x + 7$, $2(4 - x) + 9$ $2y^2 - 5y + 1$, $2l + 2w$
Formula	A special type of equation that expresses a relationship between two or more quantities	The formula $C = 10D$ gives the number of cents C in D dimes.
Translating Words to Expressions and Formulas	We can translate words into expressions and formulas using math symbols commonly associated with the words.	"More than" means add, while "double" means multiply.
Solving Equations	We can use variables to represent unknown values when solving an equation.	$\square + 5 = 12$ can be written as $x + 5 = 12$. The solution is 7 because $7 + 5 = 12$.

1.4 Exercises

MyMathLab

CONCEPTS AND VOCABULARY

1. The branch of mathematics that involves the four mathematical operations is _____.

2. A generalization of arithmetic in which letters representing numbers are combined according to the rules of arithmetic is called _____.

3. A mathematical concept called _____ can be used to denote repeated multiplication.

4. In the exponential expression 4^7, the base is _____ and the exponent is _____.

5. The word "squared" means that the exponent is _____.

6. The word "cubed" means that the exponent is _____.

7. When writing 10^9 in standard form, how many zeros should be written after a 1?

8. Write 10,000,000 as an exponential expression.

9. A(n) _____ is a symbol or italic letter such as x, y, or z, used to represent an unknown quantity.

10. A(n) _____ may contain numbers; variables; exponents; operation symbols, such as $+$, $-$, $\times$, and $\div$; and grouping symbols, such as parentheses.

11. A(n) _____ is a mathematical statement that two algebraic expressions are equal.

12. A(n) _____ is a special type of equation that expresses a relationship between two or more quantities.

13. To _____ an expression, replace the variable with a given value and find the result.

14. To represent an unknown quantity in algebra, we use a _____ rather than an empty box.

15. Is $7x + 21$ an equation or an expression?

16. Is $7x = 21$ an equation or an expression?

EXPONENTIAL NOTATION

Exercises 17–24: Use exponential notation to write each repeated multiplication.

17. $8 \cdot 8 \cdot 8$
18. $4 \cdot 4 \cdot 4 \cdot 4 \cdot 4 \cdot 4$
19. $2 \cdot 2 \cdot 2 \cdot 2 \cdot 2$
20. $9 \cdot 9$
21. $2 \cdot 2 \cdot 2 \cdot 5 \cdot 5$
22. $4 \cdot 4 \cdot 6 \cdot 6 \cdot 6 \cdot 6$
23. $5 \cdot 5 \cdot 5 \cdot 7 \cdot 7 \cdot 7$
24. $3 \cdot 9 \cdot 9 \cdot 9$

Exercises 25–32: Write the phrase in exponential notation.

25. Seven squared
26. Five cubed
27. Four to the ninth
28. One to the third
29. Two cubed
30. Ten to the sixth
31. Three to the fifth
32. Eight squared

Exercises 33–46: Evaluate the exponential expression.

33. 9^2
34. 2^3
35. 2^5
36. 3^4
37. 4^4
38. 7^3
39. 6^3
40. 5^3
41. 10^3
42. 10^7
43. $8 \cdot 10^6$
44. $3 \cdot 10^4$
45. $10^2 \cdot 30$
46. $10^4 \cdot 2000$

VARIABLES, EXPRESSIONS, AND FORMULAS

Exercises 47–54: Rewrite the given expression using an appropriate variable.

47. (Age) $- 5$
48. $60 \div$ (Rate)
49. $6 \cdot$ (Goals)
50. (Time) $+ 6$
51. (Score) $+ 10$
52. (Laps) $\cdot 4$
53. (Pieces) $\div 2$
54. $14 -$ (Days)

Exercises 55–64: Evaluate the expression for $x = 5$.

55. $25 - x$
56. $x + 13$
57. $\dfrac{0}{x}$
58. $7x$
59. $13x$
60. $x \div 5$
61. $41 + x$
62. $x - 1$
63. x^3
64. x^4

Exercises 65–70: Evaluate the algebraic expression for $x = 8$ and $y = 2$.

65. $y + x$
66. $3xy$
67. $x \div y$
68. $x - y$
69. x^y
70. y^x

Exercises 71–76: Use the appropriate geometric formula from the following list to find the requested measure.

Rectangle: $A = lw$, $P = 2l + 2w$

Square: $A = s^2$, $P = 4s$

Triangle: $P = a + b + c$

71. The perimeter of a square with a 4-inch side

72. The area of a square with a side of length 15 feet

73. The area of a rectangle with a 50-inch length and a 30-inch width

74. The perimeter of a rectangle with width 3 miles and length 8 miles

75. The perimeter of a triangle with sides of length 14, 32, and 23 yards

76. The area of two squares, each with a 5-foot side

TRANSLATING WORDS INTO MATH

Exercises 77–82: Translate the word phrase into an algebraic expression. Explain what each variable represents.

77. Six times an individual's monthly income

78. Seven fewer than a person's age

79. The quotient of the number of pizza slices and 3

80. A child's weight increased by 9

81. The sum of a person's age and heart rate

82. Triple the cost

SOLVING EQUATIONS

Exercises 83–88: Determine if the given number is a solution to the given equation.

83. Is 7 a solution to $21 \div m = 3$?

84. Is 3 a solution to $x^3 = 9$?

85. Is 48 a solution to $2y = 24$?

86. Is 17 a solution to $34 - n = 17$?

87. Is 9 a solution to $a^2 = 18$?

88. Is 56 a solution to $x + 8 = 64$?

Exercises 89–96: Solve the equation.

89. $b + 7 = 15$

90. $9x = 36$

91. $y^3 = 27$

92. $17 - z = 3$

93. $72 \div d = 6$

94. $n \div 9 = 5$

95. $56 = 7x$

96. $64 = a^3$

APPLICATIONS

Exercises 97–100: Use the given information to find the unknown length represented by the variable x.

97. Square with area 100 square feet

98. Rectangle with area 48 square yards

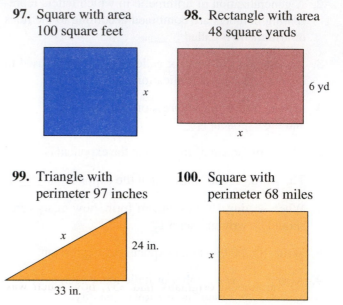

99. Triangle with perimeter 97 inches

100. Square with perimeter 68 miles

101. *Burning Calories* A 160-pound person burns 6 calories per minute while painting over graffiti and 8 calories per minute while clearing out an illegal dump site. (*Source:* The Calorie Control Council.)
(a) Write a formula that gives the number of calories C burned while painting over graffiti for p minutes.
(b) Write a formula that gives the number of calories C burned while clearing out an illegal dump site for d minutes.
(c) Use your formulas to find the number of calories burned while performing each of these activities for 4 hours.

102. *Counting Calories* Each ounce of a sweetened soft drink contains 15 calories, while each ounce of a "light" version of the same drink contains 2 calories.
(a) Write a formula that gives the number of calories C in x ounces of sweetened drink.
(b) Write a formula that gives the number of calories C in y ounces of "light" drink.
(c) If a person drinks a 12-ounce can of "light" drink rather than a 12-ounce can of sweetened drink, what is the calorie difference?

103. *Starbucks Stores* From 2008 to 2009, the number of Starbucks stores increased by 879, bringing the worldwide total to 16,635. How many Starbucks stores were there in 2008? (*Source:* Starbucks.)

104. *Large Mammals* The average birth weight of a blue whale is about 30 times the average birth weight of an African elephant. If the average newborn African elephant weighs 232 pounds, find the weight of the average newborn blue whale.

105. *Movie Downloads* A person has $28 in her online movie account. Find the maximum number of movie downloads costing $9 each that this person can purchase. How much is left in the account?

106. *iPhone Downloads* A person has $32 in his online iTunes account. Find the maximum number of iPhone apps costing $3 each that this person can purchase. How much is left in the account?

107. *Flea Market* A person bought 7 items at a flea market and received $6 in change. If each item was priced at $2, how much did the person give the vendor when paying?

108. *Lending Money* A student gave the same amount of money to each of 4 friends and had $9 left over. If the student originally had $57, how much was given to each friend?

109. *Large Cities* Of the 25 largest U.S. cities, 6 are located in Texas. If Texas has twice as many large cities as California, how many large cities does California have?

110. *Canadian Lakes* The province of Saskatchewan in Canada has 10 times as many lakes as Minnesota. Minnesota is commonly known as "the land of 10,000 lakes." How many lakes are located in Saskatchewan?

111. *Numbers* The quotient of two numbers is 5. If the dividend is 65, what is the divisor?

112. *Numbers* The difference of two numbers is 17. If the subtrahend is 9, what is the minuend?

WRITING ABOUT MATHEMATICS

113. Explain the meaning of a *variable*. Give an example.

114. Explain the difference between an expression and an equation. Give an example of each.

115. Give an example of a *formula*. State what each variable in the formula represents.

116. Explain how to write the standard form of 10 raised to a natural number power.

Sections 1.3 and 1.4

Checking Basic Concepts

1. Multiply.
   (a) 13×22　　　　(b) $0 \cdot 11$
   (c) $1(207)$　　　　　　(d) $317 \cdot 204$

2. Divide, when possible.
   (a) $35 \div 1$　　　　　(b) $1125 \div 8$
   (c) $\dfrac{34}{0}$　　　　　　(d) $\dfrac{12,312}{72}$

3. Solve each equation.
   (a) $3 \cdot \square = 36$　　(b) $\square \div 9 = 8$

4. Multiply 4000×30 mentally.

5. Use exponential notation to write each repeated multiplication.
   (a) $7 \cdot 7 \cdot 7 \cdot 7$　　(b) $2 \cdot 2 \cdot 8 \cdot 8$

6. Evaluate each exponential expression.
   (a) $5 \cdot 2^3$　　　　　(b) $3 \cdot 10^5$

7. Evaluate the expression for $x = 2$ and $y = 3$.
   (a) $5y - x$　　　　　(b) $5x^2y$

8. Is 9 a solution to $17 - n = 8$?

9. Solve each equation.
   (a) $48 \div d = 6$　　(b) $19 - z = 7$

10. Find the perimeter of a rectangle with width 8 inches and length 14 inches.

11. *Numbers* The quotient of two numbers is 7. If the dividend is 49, what is the divisor?

12. *Area* A rectangle is 40 feet long and 20 feet wide. Find the area of the rectangle.

1.5   Rounding and Estimating; Square Roots

Rounding Whole Numbers • **Estimating and Approximating** • **Square Roots** •
Solving Problems Using Estimation • **Estimating Visually**

A LOOK INTO MATH ▶

In May 2008, the movie *Iron Man* had opening weekend, domestic box office receipts of $100 million. The motion picture industry used both *rounding* and *estimating* to compute this amount. Calculating the *exact* total of the opening weekend box office receipts for any movie would be very costly and time-consuming. The process used to determine the $100 million figure involves rounding the average ticket price and estimating the number of moviegoers. In this section, we find that rounding and estimating can be useful in every-day life. (*Source:* Motion Picture Association of America.)

NEW VOCABULARY

☐ Round
☐ Estimation
☐ Approximation
☐ Perfect square
☐ Square root
☐ Radical
☐ Radicand

Rounding Whole Numbers

We **round** a whole number when we approximate it to a given level of accuracy. For example, a person who buys a new car for $28,912 could say that the purchase price was about $29,000 or that the price was about $28,900. The number 29,000 is the result of rounding 28,912 to the nearest *thousand*, and the number 28,900 is the result of rounding 28,912 to the nearest *hundred*. Both results represent an approximation of the price of the car. The difference between these numbers is simply a matter of accuracy.

Number lines can be helpful when rounding whole numbers. To round the number 278 to the nearest **hundred**, plot the number 278 on a number line with tick marks at every **100**, as shown in Figure 1.23. Because 278 is **closer to 300** than to 200, we round **up to 300**.

Figure 1.23 Rounding 278 to the Nearest Hundred Results in 300.

Similarly, when rounding 3198 to the nearest thousand, we round **down to 3000** because 3198 is **closer to 3000** than to 4000 on the number line. See Figure 1.24.

Figure 1.24 Rounding 3198 to the Nearest Thousand Results in 3000.

NOTE: A number that is exactly halfway between two tick marks on the number line will always be rounded up. For example, rounding 55 to the nearest ten results in 60.

READING CHECK

• How is a number line used to round a number?

Although number lines are convenient for visualizing the rounding process, it is not necessary to draw a number line every time we wish to round a whole number. The following procedure can be used to round whole numbers to a given place value without using a number line.

ROUNDING WHOLE NUMBERS

STEP 1: Identify the first digit to the *right* of the given place value.

STEP 2: If this digit is:

    (a) less than 5, do not change the digit in the given place value.

    (b) 5 or more, add 1 to the digit in the given place value.

STEP 3: Replace each digit to the right of the given place value with 0.

EXAMPLE 1 | **Rounding whole numbers**

Round 35,147,289 to the given place value.
(a) ten-thousands   **(b)** millions   **(c)** hundreds

Solution

(a) **STEP 1:** The ten-thousands digit is 4. The first digit to the right of **4** is **7**.

$$35,1\mathbf{4}\mathbf{7},289$$

STEP 2: Because 7 is greater than 5, we add 1 to the 4 in the ten-thousands place.

$$35,1\mathbf{5}\mathbf{7},289$$

STEP 3: Replace each digit to the right of the ten-thousands place with **0**.

$$35,1\mathbf{50,000}$$

Rounding 35,147,289 to the ten-thousands place results in 35,150,000.

(b) The millions digit is **5**, and the first digit to the right of the millions place is **1** (Step 1). Because 1 is less than 5, we do nothing to the millions place (Step 2), and we replace each digit to the right of the millions place with **0** (Step 3). The result of rounding 35,147,289 to the millions place is 3**5,000,000**.

(c) The first digit to the right of the hundreds place is 8, which is greater than 5. We add 1 to the hundreds place and replace each digit to the right of the hundreds place with 0. The result is 35,147,300.

▌ **Now Try Exercises 17, 19, 25**

EXAMPLE 2 | **Rounding whole numbers**

Round each whole number to the given place value.
(a) 10,517, thousands
(b) 89, hundreds

Solution

(a) The thousands digit is 0. The first digit to the right of the 0 is 5. We round *up* to 11,000.
(b) There is no digit in the hundreds place, which means there are 0 hundreds. The number 89 can be written as 089. The first digit to the right of the hundreds place is 8, so we round *up* to 100.

▌ **Now Try Exercises 21, 23**

READING CHECK

• Why is it useful to estimate or approximate before finding an exact answer?

Estimating and Approximating

▶ **REAL-WORLD CONNECTION** At a local thrift store, a student buys a pair of jeans for $8, a leather jacket for $29, a pair of dress shoes for $11, and a lava lamp for $9. At checkout, the clerk claims that the total bill is $112. Do you agree with this number? Even without calculating the *exact* total, it should be clear that the clerk made a mistake. Some people can "sense" when a computation is incorrect without performing complicated mental calculations. They *estimate* or *approximate*. Can you see that the total should be about $60?

An **estimation** is a rough calculation used to find a reasonably accurate answer. An estimated answer is usually not exactly accurate and is called an **approximation** of the actual answer. The next example demonstrates how rounding is used in the estimating process.

EXAMPLE 3 | **Estimating a sum and a difference**

Round each number to the nearest thousand to estimate the sum or difference. Then give the actual sum or difference.
(a) $3084 + 10{,}987 + 6905$ **(b)** $13{,}893 - 4019$

Solution
(a) Round each addend to the nearest thousand as follows: **3084** rounds to **3000**, **10,987** rounds to **11,000**, and **6905** rounds to **7000**. Estimate by adding the rounded values, **3000** + **11,000** + **7000** = 21,000. The actual value is 20,976.
(b) Round each number as follows: **13,893** rounds to **14,000** and **4019** rounds to **4000**. Estimate by finding the difference of the rounded values, **14,000** − **4,000** = 10,000. The actual value is 9874.

■ **Now Try Exercises 35, 37**

▶ **REAL-WORLD CONNECTION** When estimating real-world data, a place value for rounding is rarely given. For example, a real estate developer may have an empty plot of land that measures 617 feet by 18,875 feet. If these numbers are rounded to the nearest hundred, we can estimate the area by finding the product $600 \cdot 18{,}900$. Although this product may be easier to find than $617 \cdot 18{,}875$, it is still difficult to do mentally. However, if we round each number to its *highest place value*, then the estimation can be done more easily. Round **6**17 to the nearest **hundred** to get **600** and **1**8,875 to the nearest **ten thousand** to get **20,000**. A reasonable estimate of the area is **600** · **20,000** = 12,000,000 square feet. The actual value is 11,645,875.

EXAMPLE 4 | **Estimating a product and a quotient**

Round each number to its highest place value to estimate the product or quotient.
(a) $83 \cdot 47{,}978$ **(b)** $798 \div 38$

Solution
(a) Round **83** to the nearest **ten** to get **80** and **47,978** to the nearest **ten thousand** to get **50,000**. An estimate of the product is **80** · **50,000** = 4,000,000. The actual value is 3,982,174.
(b) Round 798 to **800** and 38 to **40**. An estimate of the quotient is **800** ÷ **40** = 20. The actual value is 21.

■ **Now Try Exercises 39, 41**

Square Roots

In Section 1.4, we discussed the meaning of the word *squared*. A number is squared when its exponent is 2. For example, 5 squared is written as 5^2 and represents the product $5 \cdot 5 = 25$. Because 25 is the result of a whole number being squared, we say that 25 is a **perfect square**. Other perfect squares include 16, 81, and 289 because $4^2 = 16$, $9^2 = 81$, and $17^2 = 289$, respectively. Note that many numbers, such as 14 and 312, are *not* perfect squares because there is no whole number that can be squared to give 14 or 312. The perfect squares associated with the first 22 whole numbers are shown in Table 1.12.

TABLE 1.12 Perfect Squares Associated with the First 22 Whole Numbers

Whole Number	0	1	2	3	4	5	6	7	8	9	10
Perfect Square	0	1	4	9	16	25	36	49	64	81	100

Whole Number	11	12	13	14	15	16	17	18	19	20	21
Perfect Square	121	144	169	196	225	256	289	324	361	400	441

In Table 1.12, each perfect square is the *square* of a whole number, and each whole number is a *square root* of a perfect square. For example, **7** is a square root of **49**, and **13** is a square root of **169**. A **square root** of a given whole number is a number whose square is the given whole number.

We can use mathematical notation to find or compute a square root. The square root of 121, for example, can be written as the *radical expression*

$$\sqrt{121}.$$

The symbol $\sqrt{}$ is called the **radical sign** (or simply the **radical**), and the number under the radical (in this case, **121**) is called the **radicand**. Because $11^2 = 121$, the square root of 121 is 11, and we write

$$\sqrt{121} = 11.$$

READING CHECK

- How do you know if a whole number is a perfect square?

EXAMPLE 5 **Computing square roots**

Compute each square root.
(a) $\sqrt{36}$ (b) $\sqrt{256}$ (c) $\sqrt{1156}$

Solution
(a) We look for a number whose square is 36. Because $6^2 = 36$, $\sqrt{36} = \mathbf{6}$.
(b) $\sqrt{256} = \mathbf{16}$ because $16^2 = 256$.
(c) To find a number whose square is 1156, first note that $30^2 = 900$ and $40^2 = 1600$. Because 1156 is between 900 and 1600, $\sqrt{1156}$ is between 30 and 40. By trial and error, $34^2 = 1156$, so $\sqrt{1156} = 34$.

Now Try Exercises 43, 45, 49

As noted earlier, only perfect squares have square roots that are whole numbers. However, we can approximate square roots of numbers that are not perfect squares. For example, we approximate $\sqrt{11}$ to the nearest whole number by using a number line to examine the location of the **radicand** between consecutive perfect squares. Figure 1.25 shows that **11** is located between the consecutive perfect squares **9** and **16**. Because the *radicand* 11 is closer to 9 than to 16, we know that $\sqrt{11}$ is closer to $\sqrt{9} = 3$ than to $\sqrt{16} = 4$. We conclude that $\sqrt{11} \approx \sqrt{9}$. To the nearest whole number, $\sqrt{11} \approx \mathbf{3}$.

NOTE: The symbol $\approx$ means "is approximately equal to."

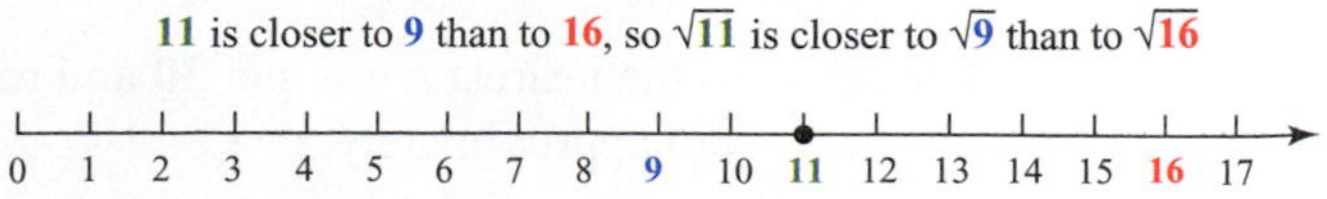

11 is closer to 9 than to 16, so $\sqrt{11}$ is closer to $\sqrt{9}$ than to $\sqrt{16}$

Figure 1.25 Approximating $\sqrt{11}$ to the Nearest Whole Number

EXAMPLE 6   **Approximating square roots**

Approximate each square root to the nearest whole number.
(a) $\sqrt{62}$ **(b)** $\sqrt{230}$

Solution
(a) The radicand 62 is between the consecutive perfect squares 49 and 64. Because it is closer to 64 than to 49, we conclude that $\sqrt{62} \approx \sqrt{64}$, or $\sqrt{62} \approx 8$.
(b) Because 230 is between the consecutive perfect squares 225 and 256 and is closer to 225, we conclude that $\sqrt{230} \approx \sqrt{225}$, or $\sqrt{230} \approx 15$.

Now Try Exercises 51, 53

READING CHECK

• How can a number line be used to approximate the square root of a whole number that is not a perfect square?

Solving Problems Using Estimation

Many real-world problems do not require exact solutions. Sometimes a reasonable estimate will be sufficient. The next three examples illustrate how we can find reasonable estimates of solutions to application problems involving real data.

EXAMPLE 7   **Estimating numbers of icebergs**

Table 1.13 shows the number of icebergs that passed south of 48° North latitude during selected years. Estimate the total number of icebergs recorded south of 48° North latitude for this five-year period by rounding each value to the nearest hundred. How does your estimate compare with the actual value?

TABLE 1.13 Number of Icebergs South of 48° North Latitude

Year	2005	2006	2007	2008	2009
Icebergs	11	0	324	976	1204

Source: International Ice Patrol.

Solution
When rounded to the nearest hundred, the values become 0, 0, 300, 1000, and 1200. The sum of these numbers gives an estimated total of $300 + 1000 + 1200 = 2500$ icebergs. This estimate compares favorably with the actual value of 2515.

Now Try Exercise 67

EXAMPLE 8   **Estimating MySpace page hits**

During the month of January (31 days), a popular MySpace page received an average of 772 daily hits. Estimate the total number of hits that the page received during January. Then, find the actual value.

Solution
Round **31** to the nearest **ten** to get **30** and round **772** to the nearest **hundred** to get **800**. The page received approximately $30 \cdot 800 = 24{,}000$ hits. The actual value is 23,932.

Now Try Exercise 59

EXAMPLE 9   **Estimating distance on a baseball diamond**

A baseball diamond is actually a square with the bases located at the corners. Because of limited space, the planners of a new city-owned baseball diamond have limited the area of the new diamond to 5900 square feet. Estimate the distance between the bases to the nearest foot.

Solution

We must find the length of a side of a square whose area is 5900 square feet. Because the area of a square is found by squaring the length of one of its sides, we look for a number whose square is 5900. By definition, this number is $\sqrt{5900}$. Because $76^2 = 5776$, $77^2 = 5929$, and $78^2 = 6084$, we conclude that $\sqrt{5900} \approx 77$. The distance between bases is about 77 feet.

▌ **Now Try Exercise 69**

Estimating Visually

▶ **REAL-WORLD CONNECTION** Newspapers and magazines often illustrate numerical information in the form of a graph. When reading these graphs, we may need to estimate values visually. For example, the graph shown in Figure 1.26 displays the annual sales of digital music in billions of dollars from 2000 to 2008. In the next example, we use Figure 1.26 to estimate information regarding digital music sales. (*Source:* Recording Industry of America.)

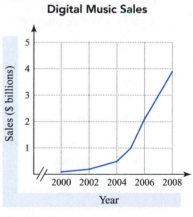

Figure 1.26

EXAMPLE 10   **Using a graph to estimate digital music sales information**

Use the graph in Figure 1.26 to answer each question.
(a) Estimate digital music sales for 2007 to the nearest billion.
(b) Estimate the year when digital music sales hit $1 billion.

Solution

(a) Locate the year 2007 halfway between 2006 and 2008 at the bottom of the graph. Move **vertically** upward to the graphed line. From this position, move **horizontally** to the left to the sales values. See Figure 1.27 on the next page. In 2007, digital music sales were about $3 billion.

(b) Find $1 billion among the sales values at the left edge of the graph. Move **horizontally** to the right to the graphed line. From this position, move **vertically** downward to the years. Note that 2005 is located halfway between 2004 and 2006. See Figure 1.28 on the next page. Digital music sales reached $1 billion in about 2005.

READING CHECK

• How can a graph be used to estimate visually?

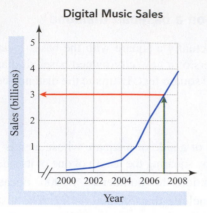

Figure 1.27 In 2007, sales were $3 billion.

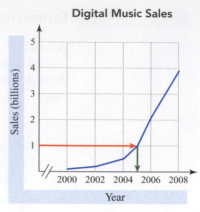

Figure 1.28 Sales reached $1 billion in 2005.

■ **Now Try Exercises 71, 73**

1.5 Putting It All Together

CONCEPT	COMMENTS	EXAMPLES
Rounding	Approximating a number to a given level of accuracy	To the nearest thousand, 52,789 rounds to 53,000.
Estimation	A rough calculation used to find a reasonably accurate answer	When rounded to the nearest ten, $52 + 78 + 13$ can be estimated by $50 + 80 + 10$.
Approximation	• The result when estimating or rounding • Not exactly accurate in most cases • Represented by the symbol $\approx$	To the nearest ten, $147 \approx 150$. $52 + 78 + 13 \approx 140$
Perfect Square	A perfect square results when a whole number is squared.	121 is a perfect square because $11^2 = 121$.
Square Root	• The square root of a whole number is a number whose square is that whole number. • The symbol $\sqrt{}$ is called the radical, and the number under it is called the radicand.	$\sqrt{169} = 13$ because $13^2 = 169$. $\sqrt{83} \approx 9$ because $9^2 = 81 \approx 83$.
Estimating Visually	Values can be estimated visually from a graph.	In 2004, sales were about $25,000.

1.5 Exercises

CONCEPTS AND VOCABULARY

1. Approximating a whole number to a given level of accuracy is called _____.

2. When rounding a whole number, we first identify the digit to the (right/left) of the given place value.

3. A(n) _____ is a rough calculation used to find a reasonably accurate answer.

4. A(n) _____ is the result when estimating or rounding. It is usually not exactly accurate.

5. One way to estimate a product when no place value for rounding is specified is to round each factor to its _____ place value.

6. The number that results when a whole number is squared is called a(n) _____ square.

7. A(n) _____ of a given whole number is a number whose square is the given whole number.

8. In the expression $\sqrt{64}$, the symbol $\sqrt{}$ is the _____, and the number 64 is the _____.

9. Because 39 is closer to 36 than to 49 on the number line, $\sqrt{39} \approx$ _____, to the nearest whole number.

10. The symbol $\approx$ means _____.

ROUNDING WHOLE NUMBERS

Exercises 11–14: Use the given number line to round the whole number to the given place value.

11. 732, hundreds

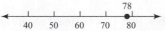

12. 187,654, ten-thousands

170,000 180,000 190,000 200,000 210,000

13. 58,923, thousands

58,000 59,000 60,000 61,000 62,000

14. 78, tens

40 50 60 70 80

Exercises 15–26: Round the whole number to the given place value.

15. 82, tens

16. 43,903, thousands

17. 850, hundreds

18. 4397, thousands

19. 375,803, ten-thousands

20. 6,702,732, millions

21. 54,208, hundreds

22. 509,982, hundred-thousands

23. 783, tens

24. 74,803, ten-thousands

25. 30,092,441, millions

26. 513,783, thousands

Exercises 27–34: Round the whole number to its highest place value.

27. 3409   28. 347

29. 87,430,933   30. 11,908

31. 68   32. 730,982

33. 15,000   34. 350,000

ESTIMATING AND APPROXIMATING

Exercises 35–38: Round each number to the nearest hundred to estimate the sum or difference.

35. $759 + 311 + 406$ 36. $3209 + 287 + 1521$

37. $1739 - 1341$ 38. $5866 - 209$

Exercises 39–42: Round each number to its highest place value to estimate the product or quotient.

39. $72{,}091 \cdot 68$ 40. $311 \cdot 5924$

41. $1007 \div 53$ 42. $27{,}470 \div 510$

SQUARE ROOTS

Exercises 43–50: Compute the square root.

43. $\sqrt{25}$ 44. $\sqrt{9}$ 45. $\sqrt{121}$

46. $\sqrt{100}$ 47. $\sqrt{361}$ 48. $\sqrt{196}$

49. $\sqrt{625}$ 50. $\sqrt{900}$

Exercises 51–58: Approximate the square root to the nearest whole number.

51. $\sqrt{10}$ 52. $\sqrt{47}$ 53. $\sqrt{170}$

54. $\sqrt{97}$ 55. $\sqrt{250}$ 56. $\sqrt{290}$

57. $\sqrt{782}$ 58. $\sqrt{3845}$

APPLICATIONS

59. *Yearly Work Hours* There are 52 full weeks in one year. If a person takes no time off and works 38 hours each week, estimate the total number of hours worked in one year.

60. *Career Work Hours* Use your answer from Exercise 59 to estimate the total number of hours this person works in 29 years at this job.

61. *Bowling Experience* A woman has 37 years of bowling experience. The other four members of the team have 29, 32, 41, and 43 years of experience, respectively. Estimate the combined number of years of bowling experience for the team by rounding each value to the nearest ten.

62. *Estimating Distance* A truck driver makes three deliveries with distances of 189 miles, 57 miles, and 112 miles. Estimate the total distance traveled by rounding each value to the nearest ten.

63. *Population Growth* There are 14,938 births every hour worldwide but only 6459 deaths. By rounding each value to the nearest thousand, estimate the increase in world population each hour. (*Source:* U.S. Census Bureau, International Data Base, 2007.)

64. *Sunny Days* On average, there are 201 cloudless days each year in Bishop, California but only 59 such days in Kodiak, Alaska. Estimate the difference between these values by rounding to the nearest ten. (*Source:* NOAA, National Climate Data Center.)

65. *Highest Elevations* The highest point in Montana is Granite Peak with an elevation of 12,799 feet. The highest point in Mississippi is Woodall Mountain with an elevation of 806 feet. By rounding to the nearest hundred, estimate the difference between these two points. (*Source:* U.S. Geological Survey.)

66. *Lowest Elevations* The lowest point in Wyoming is Belle Fourche River with an elevation of 3099 feet. The lowest point in New Mexico is Red Bluff Reservoir with an elevation of 2842 feet. By rounding to the nearest hundred, estimate the difference between these two points. (*Source:* U.S. Geological Survey.)

67. *Starbucks Stores* The following table shows the number of Starbucks stores in selected states. Estimate the total number of Starbucks stores in these states by rounding each value to the nearest ten.

State	Texas	Ohio	Hawaii	Idaho
Stores	604	203	59	39

Source: Starbucks, 2008.

68. *Wal-Mart Stores* The following table shows the number of Wal-Mart stores in selected states. Estimate the total number of Wal-Mart stores in these states by rounding each value to the nearest ten.

State	Illinois	Iowa	Maine	Utah
Stores	79	16	10	2

Source: Wal-Mart, 2008.

69. *Farm Building* A square shed has a floor area of 1700 square feet. Estimate the length of one side of the shed to the nearest foot.

70. *Baseball Diamond* A baseball diamond has an area of 7400 square feet. Estimate the distance between the bases (corners of the diamond) to the nearest foot.

Exercises 71–74: *U.S. Internet Use* The following graph shows the number of U.S. Internet users in millions for selected years. (*Source:* Department of Commerce.)

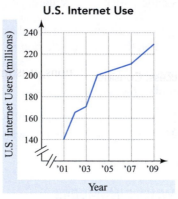

71. Estimate the number of U.S. Internet users in 2008 to the nearest ten-million.

72. Estimate the year when there were 200 million U.S. Internet users.

73. Estimate the year when there were 170 million U.S. Internet users.

74. Estimate the number of U.S. Internet users in 2007 to the nearest ten-million.

WRITING ABOUT MATHEMATICS

75. Explain how to round a whole number to a given place value. Give examples.

76. Explain how a number line can be used as an aid in rounding whole numbers.

77. Create a plan for finding all the whole numbers between 1000 and 1500 that are perfect squares.

78. Explain how a number line can be used when approximating a square root to the nearest whole number.

1.6   Order of Operations

Order of Operations • **Evaluating Algebraic Expressions** • **Translating Words to Symbols**

A LOOK INTO MATH ▶

Retail store employees must occasionally take inventory of items sold in the store. Many times, a particular item is found on the store shelves as well as in boxes in a storage room. Suppose there are 17 light bulbs on the shelf and 4 boxes of 24 light bulbs each in the storage room. The total number of light bulbs can be represented by the expression $17 + 4 \cdot 24$. How would you find the total number of light bulbs? Is the total

$$17 + 4 \cdot 24 = 21 \cdot 24 \overset{?}{=} 504 \quad \text{or} \quad 17 + 4 \cdot 24 = 17 + 96 \overset{?}{=} 113?$$

(add first, then multiply)   (multiply first, then add)

In this section, we discuss rules called the *order of operations*, which are used to compute mathematical expressions involving more than one arithmetic operation.

NEW VOCABULARY

☐ Order of operations agreement

Order of Operations

In the example above, there are $4 \cdot 24 = 96$ light bulbs in the storage room and 17 light bulbs on the shelf. The total number of light bulbs is $17 + 96 = 113$. This result implies that in the expression $17 + 4 \cdot 24$, multiplication should be done before addition. This is just one of the rules stated in the **order of operations agreement**. Because arithmetic expressions may contain parentheses, exponents, radicals, and several arithmetic operations, it is important to evaluate these expressions consistently. To ensure that we all find the same result when evaluating an arithmetic expression, the following rules are used.

STUDY TIP

Studying with other students can greatly improve your chances of success. Consider exchanging phone numbers or email addresses with some of your classmates so that you can find a time and place to study together. If possible, set up a regular meeting time and invite other classmates to join you.

ORDER OF OPERATIONS

1. Do all calculations within grouping symbols, such as parentheses and radicals, or above and below a fraction bar.
2. Evaluate all exponential expressions.
3. Do all multiplication and division from *left to right*.
4. Do all addition and subtraction from *left to right*.

NOTE: If there is more than one arithmetic operation within grouping symbols, the order of Steps 2, 3, and 4 must be followed when performing Step 1.

READING CHECK

• Why is it important to have an order of operations agreement?

EXAMPLE 1   Using the order of operations to evaluate expressions

Evaluate each expression.
(a) $24 - 3 \cdot 6$ **(b)** $12 \div 3 + 4 \cdot 5$ **(c)** $3^2 + 8 \div (3 - 1)$

Solution
(a) Perform multiplication before subtraction.

$$24 - 3 \cdot 6 = 24 - 18 \quad \text{Multiply.}$$
$$= 6 \quad \text{Subtract.}$$

CALCULATOR HELP

To use a calculator to evaluate expressions with parentheses, see the Appendix (page AP-2).

(b) Perform multiplication and division from left to right.

$$12 \div 3 + 4 \cdot 5 = 4 + 4 \cdot 5 \qquad \text{Divide.}$$
$$= 4 + 20 \qquad \text{Multiply.}$$
$$= 24 \qquad \text{Add.}$$

(c) The expression within parentheses is evaluated first.

$$3^2 + 8 \div (3 - 1) = 3^2 + 8 \div 2 \qquad \text{Subtract within parentheses.}$$
$$= 9 + 8 \div 2 \qquad \text{Evaluate } 3^2.$$
$$= 9 + 4 \qquad \text{Divide.}$$
$$= 13 \qquad \text{Add.}$$

> **Now Try Exercises 7, 17, 19**

EXAMPLE 2 **Using the order of operations to evaluate expressions**

Evaluate each expression.

(a) $2\sqrt{11 + 14} + (8 - 5)^3$ **(b)** $\dfrac{10 \cdot (2 + 2)}{\sqrt{64} - 3}$

Solution
(a) The expression under the radical and the expression in parentheses are evaluated first.

$$2\sqrt{11 + 14} + (8 - 5)^3 = 2\sqrt{25} + 3^3 \qquad \text{Evaluate radicand and parentheses.}$$
$$= 2 \cdot 5 + 27 \qquad \text{Evaluate } \sqrt{25} \text{ and } 3^3.$$
$$= 10 + 27 \qquad \text{Multiply.}$$
$$= 37 \qquad \text{Add.}$$

(b) Evaluate the expression under the radical and the expression in parentheses first.

$$\frac{10 \cdot (2 + 2)}{\sqrt{64} - 3} = \frac{10 \cdot 4}{8 - 3} \qquad \text{Evaluate radicand and parentheses.}$$
$$= \frac{40}{5} \qquad \text{Multiply top; subtract bottom.}$$
$$= 8 \qquad \text{Divide.}$$

> **Now Try Exercises 21, 35**

Sometimes expressions contain grouping symbols *within* grouping symbols, or *nested* grouping symbols, that must be evaluated starting with the innermost grouping and working outward. For example, in the expression

$$(3 + (6 - 1)) \cdot 8$$

the grouping $(6 - 1)$ is evaluated first.

EXAMPLE 3 **Evaluating expressions with nested grouping symbols**

Evaluate each expression.
(a) $(7 - (4 + 2)^2 \div 9) + 3$
(b) $(11 - \sqrt{50 - 1} - 3)^3$

Solution

(a) Start by evaluating the innermost grouping, $(4 + 2)$.

$$
\begin{aligned}
(7 - (4 + 2)^2 \div 9) + 3 &= (7 - 6^2 \div 9) + 3 && \text{Evaluate innermost grouping.} \\
&= (7 - 36 \div 9) + 3 && \text{Evaluate } 6^2. \\
&= (7 - 4) + 3 && \text{Divide.} \\
&= 3 + 3 && \text{Evaluate parentheses.} \\
&= 6 && \text{Add.}
\end{aligned}
$$

(b) Start by evaluating the innermost grouping, the radicand $50 - 1$.

$$
\begin{aligned}
(11 - \sqrt{50 - 1} - 3)^3 &= (11 - \sqrt{49} - 3)^3 && \text{Evaluate innermost grouping.} \\
&= (11 - 7 - 3)^3 && \text{Evaluate } \sqrt{49}. \\
&= (4 - 3)^3 && \text{Subtract.} \\
&= 1^3 && \text{Evaluate parentheses.} \\
&= 1 && \text{Evaluate } 1^3.
\end{aligned}
$$

▌ **Now Try Exercises 29, 39**

Evaluating Algebraic Expressions

So far, the order of operations agreement has been used to evaluate expressions without variables. Now we will see that the same rules apply when evaluating algebraic expressions for given values of the variables, as shown in the next example.

EXAMPLE 4 **Evaluating algebraic expressions**

Evaluate each algebraic expression for $x = 3$, $y = 8$, and $z = 4$.

(a) $25 - xy$ **(b)** $(15 - x) \div z + xy$ **(c)** $\dfrac{5xz}{y - 4}$

Solution

(a) Start by replacing x with **3** and y with **8** in the given expression. Recall that writing two variables next to each other implies multiplication. Note that z is not used.

$$
\begin{aligned}
25 - xy &= 25 - 3 \cdot 8 && x = 3 \text{ and } y = 8. \\
&= 25 - 24 && \text{Multiply.} \\
&= 1 && \text{Subtract.}
\end{aligned}
$$

(b) This expression contains three variables. Replace x with **3**, y with **8**, and z with **4**.

$$
\begin{aligned}
(15 - x) \div z + xy &= (15 - 3) \div 4 + 3 \cdot 8 && x = 3, y = 8, \text{ and } z = 4. \\
&= 12 \div 4 + 3 \cdot 8 && \text{Evaluate parentheses.} \\
&= 3 + 24 && \text{Divide; multiply.} \\
&= 27 && \text{Add.}
\end{aligned}
$$

(c) Replace x with **3**, y with **8**, and z with **4**.

$$
\begin{aligned}
\frac{5xz}{y - 4} &= \frac{5 \cdot 3 \cdot 4}{8 - 4} && x = 3, y = 8, \text{ and } z = 4. \\
&= \frac{60}{4} && \text{Multiply top; subtract bottom.} \\
&= 15 && \text{Divide.}
\end{aligned}
$$

▌ **Now Try Exercises 51, 55, 59**

Translating Words to Symbols

Often mathematical expressions are not given in application problems. Instead, words are used to describe the mathematics necessary to find a result. When this occurs, we must translate the words into mathematical symbols. However, some phrases may be difficult to translate. Consider the phrase

$$\text{six plus nine divided by three.}$$

Does this phrase translate to

$$6 + 9 \div 3 \quad \text{or} \quad 6 + \frac{9}{3} \quad \text{or} \quad \frac{6 + 9}{3}?$$

READING CHECK

• How does a comma affect the way in which we translate a phrase into math symbols?

Because of the order of operations agreement, the first two expressions are equivalent, and each evaluates to 9. However, the third expression evaluates to 5. It is correct to translate the phrase to either of the first two expressions. A phrase that translates to the third expression must contain a comma, words such as "the quantity," or both. For example, either of the following phrases translates to the third expression above.

six plus nine, divided by three   or   **the quantity six plus nine, divided by three**

In the next example, we translate words to symbols. Look for the words "the quantity" in each phrase and pay close attention to the placement of any commas.

EXAMPLE 5   **Translating Words to Symbols**

Use symbols to write each expression and then evaluate it.
(a) Five squared plus eight
(b) Two less than the quantity three times six
(c) Four plus five, times seven

Solution
(a) The phrase translates to $5^2 + 8$, which evaluates to $25 + 8 = 33$.
(b) Translate "the quantity three times six" by using parentheses and then subtract two. The phrase translates to $(3 \cdot 6) - 2$ and evaluates to $18 - 2 = 16$.
(c) The phrase translates to $(4 + 5) \cdot 7$, which evaluates to $9 \cdot 7 = 63$.

Now Try Exercises 67, 69, 71

1.6   Putting It All Together

CONCEPT	COMMENTS	EXAMPLES
Order of Operations Agreement	1. Do all calculations within grouping symbols, such as parentheses and radicals, or above and below a fraction bar. 2. Evaluate all exponential expressions. 3. Do all multiplication and division from *left to right*. 4. Do all addition and subtraction from *left to right*.	$14 \div 7 + 3 \cdot 6 = 2 + 18$ $\qquad\qquad\qquad = 20$ $20 \div (3 - 1)^2 = 20 \div 2^2$ $\qquad\qquad\quad = 20 \div 4$ $\qquad\qquad\quad = 5$ $\sqrt{6 + 10} = \sqrt{16}$ $\qquad\qquad = 4$

CONCEPT	COMMENTS	EXAMPLES
Evaluating Algebraic Expressions	Replace each variable in the algebraic expression with its given value, and then use the order of operations agreement to evaluate.	Evaluate $5 + 2x$ for $x = 4$. $$5 + 2x = 5 + 2 \cdot 4$$ $$= 5 + 8$$ $$= 13$$
Translating Words to Symbols	The words "the quantity" and the placement of any commas are important when translating words to symbols.	The phrase "nine minus two, times six" translates to $(9 - 2) \cdot 6$.

1.6 Exercises

MyMathLab · MathXL PRACTICE · WATCH · DOWNLOAD · READ · REVIEW

CONCEPTS AND VOCABULARY

1. The _____ agreement must be followed to ensure that we all find the same result when evaluating an arithmetic expression.

2. A fraction bar is an example of a grouping symbol. Name two other examples of grouping symbols from this section.

3. In the expression $3 + 4 \cdot 7$, multiplication should be done (before/after) addition.

4. In the expression $(3 - 1)^2 \cdot 6$, which is done first, subtraction, squaring, or multiplication?

5. When translating words to symbols, parentheses can be used to express the words "_____".

6. In the phrase "six plus two, times ten," which is done first, addition or multiplication?

ORDER OF OPERATIONS

Exercises 7–44: Evaluate the expression.

7. $6 + 2 \cdot 9$

8. $23 - 3 \cdot 5$

9. $36 \div 4 + 7$

10. $55 \div 11 - 3$

11. $28 \div 4 \cdot 3$

12. $6 \cdot 8 \div 3$

13. $8 + 10 \cdot 3 \div 5$

14. $15 - 12 \div 4 \cdot 3$

15. $80 - 45 \div 5 + 2$

16. $77 - 12 \cdot 5 + 13$

17. $63 \div 7 + 2 \cdot 8$

18. $2 \cdot 100 - 72 \div 8$

19. $32 - 4^2 \div (5 + 3)$

20. $96 \div (5^2 - 13) + 17$

21. $\sqrt{58 - 9} + (14 - 2)$

22. $110 - \sqrt{99 + 1}$

23. $13^2 + \sqrt{130 - 9}$

24. $\sqrt{3^2 + 16} - 2^2$

25. $\sqrt{256} \cdot \sqrt{300 - 44}$

26. $\sqrt{64} - \sqrt{40 + 24}$

27. $99 - (5 + 4^2 \div 2)$

28. $(72 \div 6^2 + 3) \cdot 12$

29. $((3 + 7)^2 \div 5) - 20$

30. $(3 \cdot (4 + 6) \div 10)^3$

31. $\dfrac{18 - 2}{7 - 3}$

32. $\dfrac{69 + 3}{3 \cdot 8}$

33. $\dfrac{(4 + 10) \cdot 5}{37 - 5 \cdot 6}$

34. $\dfrac{2 \cdot (5 + 1)^2}{5 \cdot 4 - 8}$

35. $\dfrac{(6 - 2) \cdot 9}{\sqrt{49} - 3}$

36. $\dfrac{(\sqrt{81} + 2) \cdot 6}{6^2 - 3}$

37. $(2 + 5)^2 + 3\sqrt{4 \cdot 9}$

38. $(8 - 7)^2 \cdot \sqrt{25 - 16}$

39. $4(15 + 5 \cdot 3) - ((50 \div 5)^2 + 12)$

40. $42 + 5^2 \cdot 2 - 64 \div 2^3 + 8$

41. $35 - (8 \cdot 5 - (20 \div 1^4) + 3)$

42. $35 - \sqrt{60 \div 5 - (2 \cdot 3^2 - 10)}$

43. $\sqrt{49 - 3 + (2 \cdot 6^2 \div 4)} + 14$

44. $17 - 3 \cdot 5 + 70 + 2^2 \cdot 3 - 27 \div 3^3 - 9$

Exercises 45–50: Insert parentheses as needed in the expression to make it equal to 0. More than one set of parentheses may be needed.

45. $14 - 12 \cdot 5 - 10$

46. $18 - 12 - 5 + 1$

47. $36 - 6^2 \div 5 - 1$

48. $39 - 7 + 8^2 \div 2$

49. $32 \div 4^2 - 2 \cdot 9$

50. $5 - 5 \cdot 3^2 \div 3$

ALGEBRAIC EXPRESSIONS

Exercises 51–62: Evaluate the algebraic expression for the given values of the variables.

51. $2 \cdot y + x$, for $x = 4, y = 2$

52. $a - 12 + 5b$, for $a = 13, b = 3$

53. $c(d - 8) + 3$, for $c = 2, d = 10$

54. $12 - g(2 + h)$, for $g = 4, h = 1$

55. $m + (9^2 - n) \div 8$, for $m = 31, n = 1$

56. $w + ((8 - v)^2 \div w)$, for $v = 2, w = 4$

57. $\sqrt{p^2 + 9} - q^2$, for $p = 4, q = 2$

58. $\sqrt{r - 7} + (19 + s)$, for $r = 71, s = 13$

59. $\dfrac{3 \cdot (4 + c)}{d - 2 \cdot 7}$, for $c = 10, d = 16$

60. $\dfrac{g(6 - 2)}{(4 + \sqrt{h} - 3)^2}$, for $g = 8, h = 9$

61. $x + (y^2 - x) \div x$, for $x = 4, y = 8$

62. $m + n \cdot m - n \div m$, for $m = 3, n = 15$

TRANSLATING WORDS TO SYMBOLS

Exercises 63–78: Use symbols to write the expression and then evaluate it.

63. Twelve more than five

64. Three less than forty-five

65. Nine fewer than twenty-one

66. Thirty-eight increased by two

67. Four more than seven squared

68. Three squared minus eight

69. Six plus five, times nine

70. Seven plus four times two

71. The quantity six plus two, times five

72. Ten minus the quantity four plus six

73. Two cubed times three squared

74. The square root of the quantity two plus two

75. The square root of sixteen, plus nine

76. Eight divided by the quantity six minus two

77. Seven times three decreased by two

78. Ten divided by two increased by nine

APPLICATIONS

79. *Heart Rate* The average heart rate R in beats per minute (bpm) of an animal weighing W pounds can be approximated by

$$R = \frac{885\sqrt{W}}{W}.$$

Find the heart rate for a 25-pound dog.

80. *Heart Rate* If x is the number of minutes that have passed since exercise has stopped, a person's heart rate R in beats per minute can be approximated by

$$P = \frac{4(10 - x)^2}{5} + 80,$$

where $x < 10$. Find the person's heart rate 5 minutes after exercise has stopped.

81. *Wind Power* Electrical power generated by a particular wind turbine is given by

$$W = \frac{5l^2 s^3}{32},$$

where W is power in watts, l is the length of a turbine blade in feet, and s is the speed of the wind in miles per hour. How many watts are generated by a wind turbine with a blade length of 10 feet if the wind has a speed of 8 miles per hour?

82. *Skid Marks* Vehicles in accidents often leave skid marks. To determine how fast a vehicle was traveling, officials often use a test vehicle to compare skid marks on the same section of road. If a vehicle involved in a crash left skid marks that are D feet long and a test vehicle traveling at v miles per hour leaves skid marks that are d feet long, then the speed of the vehicle in the crash is given by

$$V = \sqrt{\frac{v^2 D}{d}}.$$

Determine V if $v = 40$ miles per hour, $D = 225$ feet, and $d = 100$ feet.

83. *Insect Population* Suppose that an insect population P, in thousands per acre, is given by

$$P = \frac{10x - 6}{x + 1},$$

where x represents time in months. Find the insect population after 7 months.

84. *Stopping Distance* On dry, level pavement, the stopping distance D in feet for a car traveling at x miles per hour can be estimated by

$$D = \frac{x^2}{11} + \frac{11x}{3}.$$

Find the stopping distance for a car traveling on dry, level pavement at 33 miles per hour.

85. *Converting Temperature* To convert a temperature F given in degrees Fahrenheit to an equivalent temperature C in degrees Celsius, use the formula

$$C = \frac{5(F - 32)}{9}.$$

Find the Celsius temperature that is equivalent to a temperature of 104°F.

86. *Converting Temperature* To convert a temperature C given in degrees Celsius to an equivalent temperature F in degrees Fahrenheit, use the formula

$$F = \frac{9C}{5} + 32.$$

Find the Fahrenheit temperature that is equivalent to a temperature of 35°C.

WRITING ABOUT MATHEMATICS

87. Write a paragraph describing how to apply the order of operations to an expression. Give examples.

88. Give an example of a phrase containing the words "the quantity." Translate your phrase into symbols.

SECTIONS 1.5 and 1.6

Checking Basic Concepts

1. Round 45,277 to the nearest thousand.

2. Round each number to the nearest hundred to estimate the sum or difference.
 (a) $789 + 403$ **(b)** $5311 - 694$

3. Compute each square root.
 (a) $\sqrt{81}$ **(b)** $\sqrt{169}$

4. Approximate each given square root to the nearest whole number.
 (a) $\sqrt{26}$ **(b)** $\sqrt{200}$

5. Evaluate each expression.
 (a) $24 \div 6 + 9$ **(b)** $\sqrt{5^2 - 9} + 2^3$

6. Evaluate $12 - (4 \cdot 2 - (27 \div 3^2) + 5)$.

7. Evaluate $x + (3^2 - y) \div 2$ for $x = 7, y = 5$.

8. Use symbols to write the expression

 the quantity four plus two, times three

and then evaluate the expression.

9. *World Population* There are 14,938 births every hour worldwide but only 6459 deaths. Estimate the increase in world population each hour by rounding to the nearest hundred. (*Source:* U.S. Census Bureau, International Data Base, 2008.)

10. *Building Size* A square garage has a floor area of 410 square feet. Estimate the length of one side of the garage to the nearest foot.

1.7 More with Equations and Problem Solving

Simplifying Algebraic Expressions • **Checking a Solution to an Equation** • **A Problem Solving Strategy**

A LOOK INTO MATH ▶

Almost every occupation involves solving problems. Highway engineers need to determine safe speed limits for curves in the road; graphic designers are required to fit text and visual information within specified page dimensions; doctors and pharmacists must correctly calculate medication doses for patients; and store managers often need to figure out both part-time and full-time employee work schedules. Ask someone who is currently working in *your* chosen field about the types of problems that need to be solved. In this section, we discuss a problem solving strategy that can be used to solve many types of real-world problems.

Simplifying Algebraic Expressions

As we learned in Section 1.4, there is a difference between an equation and an algebraic expression. An equation *always* contains an equal sign (=), but an expression *never* contains an equal sign. We can solve an equation, but we cannot solve an expression. An expression can be evaluated, and as we will see in this section, an expression can sometimes be *simplified*.

STUDY TIP

If you are studying with classmates, make sure that they do not "do the work for you." A class-mate with the best intentions may give too many verbal hints while helping you work through a problem. Remember that members of your study group will not be giving hints during an exam.

EXAMPLE 1 **Identifying equations and expressions**

Identify each of the following as an equation or an expression.
(a) $3x + 5 = 17$ **(b)** $24y - 34 + 2y$

Solution
(a) Because $3x + 5 = 17$ has an equal sign, it is an equation.
(b) Because there is no equal sign in $24y - 34 + 2y$, it is an expression.

Now Try Exercises 7, 9

COMBINING LIKE TERMS One way to simplify expressions is to *combine like terms*. A **term** is a number, a variable, or a product of numbers and variables raised to powers. Examples of terms include

$$7, \quad x, \quad 3y, \quad 12xy, \quad \text{and} \quad 9y^2z^3.$$

Terms do not contain addition or subtraction, but the plus and minus signs in an expression separate terms within an expression. For example, the expression $3x + y + 4z^2$ has three terms: $3x$, y, and $4z^2$. The number in a term is called the **coefficient**. If no number appears, the coefficient is 1. The coefficients of $3x$, y, and $4z^2$ are 3, 1, and 4, respectively.

If two terms have the same variables raised to the same powers, then they are called **like terms**. Suppose that we have two boards with lengths $2x$ and $3x$, where the value of x could be any length. See Figure 1.29. These two lengths represent like terms because they have the same variable (in this case, x) raised to the same power (in this case, 1). Because $2x$ and $3x$ are like terms, we can find the total length of the two boards by applying the distributive property and *adding like terms*.

$$2x + 3x = (2 + 3)x = 5x$$

The combined length of the two boards is $5x$ units.

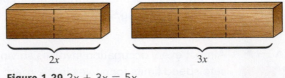

Figure 1.29 $2x + 3x = 5x$.

We can also determine how much longer the second board is than the first board by sub-tracting like terms.

$$3x - 2x = (3 - 2)x = 1x = x$$

The second board is x units longer than the first board.

We can add or subtract like terms but not *unlike* terms. If one board has length $2x$ and the other board has length $3y$, then we cannot determine the total length other than to say it is $2x + 3y$. See Figure 1.30. The terms $2x$ and $3y$ are unlike because the lengths of x and y might not be equal. The unlike terms $2x$ and $3y$ *cannot* be combined.

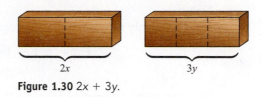

Figure 1.30 $2x + 3y$.

EXAMPLE 2 **Identifying like terms**

Determine whether the terms are like or unlike.
(a) $5w, 9w$ **(b)** $7x^2, 7y^2$ **(c)** $4a^2b, 3b^2a$ **(d)** x, x^2

Solution
(a) The variable in both terms is w (with power 1), so they are like terms.
(b) The variables are different, so they are unlike terms.
(c) Although the variables are the same, the powers on the variables do not match, so they are unlike terms.
(d) The variables are both x, but the powers do not match. They are unlike terms.

Now Try Exercises 15, 17, 19, 21

EXAMPLE 3 **Combining like terms**

Combine like terms in each expression. Write "not possible" if terms cannot be combined.
(a) $4y + 7y$ **(b)** $8xy^2 - 3xy^2$ **(c)** $7m + 3n$

Solution
(a) The variable in both terms is y, so they are like terms and can be combined. Using the distributive property, we have

$$4y + 7y = (4 + 7)y = 11y.$$

We can think of "4 y's plus 7 y's is 11 y's" in the same way as "4 apples plus 7 apples is 11 apples."
(b) The variables and powers match, so the terms are like terms and can be combined. Using the distributive property, we have $8xy^2 - 3xy^2 = (8 - 3)xy^2 = 5xy^2$.
(c) It is not possible to combine $7m$ and $3n$ because the variables are not the same. That is, the terms are unlike.

Now Try Exercises 23, 25, 27

APPLYING ARITHMETIC PROPERTIES Another way to simplify expressions is to apply the commutative, associative, and distributive properties. For example, $(2x + 3) + 5$ can be simplified by using the associative property to group the terms in the expression as $2x + (3 + 5)$. Now the terms in the parentheses are like terms and can be combined to form the *simplified expression* $2x + 8$. Similarly, the commutative property can be used to rewrite the expression $4x + 1 + 3x$ so that it becomes $4x + 3x + 1$. The first two terms can now be combined to form the simplified expression $7x + 1$.

The next example illustrates how algebraic expressions can be simplified by applying the commutative, associative, and distributive properties.

EXAMPLE 4 **Simplifying expressions**

Simplify each expression.
(a) $3 + 5w + 1$ **(b)** $15x + (2x + y)$ **(c)** $4y + 5(y - 3)$

Solution
(a) First apply the commutative property to the first two terms.

$$3 + 5w + 1 = 5w + 3 + 1 \qquad \text{Commutative property}$$
$$= 5w + 4 \qquad \text{Add: } 3 + 1 = 4.$$

(b) Apply the associative property to regroup the terms.

$$15x + (2x + y) = (15x + 2x) + y \qquad \text{Associative property}$$
$$= (15 + 2)x + y \qquad \text{Distributive property}$$
$$= 17x + y \qquad \text{Add: } 15 + 2 = 17.$$

(c) First, use the distributive property to multiply 5 and $(y - 3)$.

$$4y + 5(y - 3) = 4y + 5 \cdot y - 5 \cdot 3 \qquad \text{Distributive property}$$
$$= 4y + 5y - 15 \qquad \text{Multiply.}$$
$$= (4 + 5)y - 15 \qquad \text{Distributive property}$$
$$= 9y - 15 \qquad \text{Add: } 4 + 5 = 9.$$

Now Try Exercises 31, 33, 37

Checking a Solution to an Equation

To see if a number is a solution to an equation, recall that we replace each occurrence of the variable with the given number. For example, to see if 5 is a solution to the equation $4x + 10 = 6x$, replace each x with **5** and determine if the resulting equation is true or false. A question mark above an equal sign means that we are checking a possible solution.

$$4x + 10 = 6x \qquad \text{Given equation}$$
$$4(5) + 10 \stackrel{?}{=} 6(5) \qquad \text{Replace } x \text{ with 5.}$$
$$20 + 10 \stackrel{?}{=} 30 \qquad \text{Multiply}$$
$$30 = 30 \checkmark \qquad \text{Add; the solution checks.}$$

READING CHECK

• How is a solution checked to see if it is correct?

Every equation has an expression on each side of the equal sign. In some equations, it may be possible to simplify one or both of these expressions. In the next example, we simplify the expressions in an equation and then check to see that a given value is a solution to both the given equation and the equation formed by simplifying the expressions.

EXAMPLE 5 **Simplifying expressions in an equation and checking a solution**

For the equation $5x + 3x = 16 + 2(x + 1)$, do the following.
(a) Simplify the expression on each side of the equal sign.
(b) Check to see if 3 is a solution to both the given equation and the one formed in part (a).

Solution
(a) First apply the distributive property on each side of the equation.

$$5x + 3x = 16 + 2(x + 1) \qquad \text{Given equation}$$
$$(5 + 3)x = 16 + 2x + 2 \qquad \text{Distributive property}$$
$$8x = 2x + 16 + 2 \qquad \text{Add; commutative property}$$
$$8x = 2x + 18 \qquad \text{Add.}$$

(b) To see if 3 is a solution to the given equation, replace each occurrence of *x* with **3**.

$$5x + 3x = 16 + 2(x + 1) \quad \text{Given equation}$$
$$5(3) + 3(3) \overset{?}{=} 16 + 2(3 + 1) \quad \text{Replace } x \text{ with 3.}$$
$$15 + 9 \overset{?}{=} 16 + 2(4) \quad \text{Multiply; add.}$$
$$15 + 9 \overset{?}{=} 16 + 8 \quad \text{Multiply.}$$
$$24 = 24 \checkmark \quad \text{Add; the solution checks.}$$

To see if 3 is a solution to the equation formed in part (a), replace each occurrence of *x* with **3**.

$$8x = 2x + 18 \quad \text{Given equation}$$
$$8(3) \overset{?}{=} 2(3) + 18 \quad \text{Replace } x \text{ with 3.}$$
$$24 \overset{?}{=} 6 + 18 \quad \text{Multiply.}$$
$$24 = 24 \checkmark \quad \text{Add; the solution checks.}$$

Now Try Exercise 45

A Problem Solving Strategy

Some application problems can be challenging because formulas and equations are not given. To solve such problems, it is often helpful to follow a strategy. The following strategy is based on George Polya's (1888–1985) four-step process for solving problems.

STUDY TIP

One of the best ways to prepare for class is to read a section *before* it is covered by your instructor. For example, reading ahead about the four steps for solving a problem would give you the chance to consider the process and formulate any questions that you might have about it.

STEPS FOR SOLVING A PROBLEM

STEP 1: Read the problem carefully and be sure you understand it. (You may need to read the problem more than once.) Assign a variable to what you are being asked to find.

STEP 2: Write an equation that relates the quantities described in the problem. You may need to sketch a diagram or refer to known formulas.

STEP 3: Solve the equation. Use the solution to determine the solution(s) to the original problem. Include any necessary units.

STEP 4: Look back and check your solution in the original problem. Does your solution seem reasonable?

READING CHECK

• What are the four steps for solving a problem?

Even when we understand a problem, we may not be able to find a solution if we cannot write an appropriate equation. In the next example, we practice the second step in the four-step process by translating sentences into equations.

EXAMPLE 6   **Translating sentences into equations**

Translate the sentence into an equation using the variable *x*. Do not solve the equation.
(a) Four times the number of feet plus 3 times the same number of feet is 28.
(b) A student's age decreased by 7 is 12.
(c) Sixteen thousand, five hundred is 8000 more than the population.

Solution

(a) If x represents the number of feet, then the phrase "**four times** the number of feet" is written $4x$, and the phrase "**3 times** the same number of feet" is written $3x$. The word "**plus**" indicates that these two quantities should be added to get $4x + 3x$. The word "**is**" suggests an equal sign. The entire sentence translates to $4x + 3x = 28$.

(b) If x represents the student's age, then "**decreased by** 7" indicates that 7 should be subtracted from x to get $x - 7$. Again the word "**is**" suggests an equal sign. The entire sentence translates to $x - 7 = 12$.

(c) If x represents the population, then "8000 **more than** the population" can be written as $x + 8000$. The sentence translates to $16{,}500 = x + 8000$.

▌ **Now Try Exercises 49, 51, 53**

▶ **REAL-WORLD CONNECTION** When a new technology is introduced, it often takes time for the technology to "catch on." For example, thirty-eight years passed between the time that radios were first available to the public and the time when a significant number of people used radios on a regular basis. In the next example, we apply the four-step problem-solving process to a word problem that compares newly introduced technologies.

EXAMPLE 7 ▌ **Comparing newly introduced technologies**

After its introduction to consumers, television took 13 years to catch on in U.S. households. This is eight more years than it took the Internet to catch on. How many years passed between the first availability of the Internet and its widespread use? (*Source: Internet World Stats.*)

Solution

STEP 1: We must find the number of years that it took for the Internet to catch on in the U.S. We assign the variable x to this unknown amount of time.

STEP 2: Reading the paragraph carefully reveals that

> **13** years **is 8 more** years than it took for the Internet to catch on.

Because x represents the time it took for the Internet to catch on, the equation is

$$13 = x + 8.$$

STEP 3: To solve the equation in Step 2 we must find the value of x that makes the equation true. Because $13 = 5 + 8$, the solution is 5 years.

STEP 4: Because 13 is 8 more than 5, the solution checks in the original problem. Based on how quickly new technologies become popular in today's society, it seems reasonable that the Internet caught on faster than television.

▌ **Now Try Exercise 57**

EXAMPLE 8 ▌ **Analyzing doctorate degrees**

In 2008, there were about 48,000 doctorate degrees awarded in the United States. Twice as many of these doctorates were awarded to U.S. citizens than were awarded to foreign citizens. Find the number of doctorate degrees awarded in the United States to foreign citizens in 2008. (*Source: U.S. National Science Foundation.*)

Solution

STEP 1: Let x represent the number of doctorates awarded to foreign citizens. Because there were twice as many doctorates awarded to U.S. citizens, $2x$ represents the number of doctorates awarded to U.S. citizens.

STEP 2: The total number of doctorates is found by adding.

U.S. citizen doctorates + *foreign citizen doctorates* = *total number of doctorates*

Because the total is **48,000**, the equation can be written as

$$2x + x = 48{,}000.$$

STEP 3: To solve the equation in Step 2, first we simplify the expression on the left side of the equation. By combining like terms, the equation becomes

$$3x = 48{,}000.$$

The solution to this equation is 16,000 because $3(16{,}000) = 48{,}000$ is a true equation. There were 16,000 doctorates awarded to foreign citizens in 2008.

STEP 4: If the number of doctorates awarded to foreign citizens was 16,000, and twice this number of doctorates, or 32,000, were awarded to U.S. citizens, then the total number of doctorates awarded was $16{,}000 + 32{,}000 = 48{,}000$. The solution checks in the original problem.

Now Try Exercise 63

1.7	**Putting It All Together**

CONCEPT	COMMENTS	EXAMPLES
Equations and Expressions	Equations always contain an equal sign (=), but expressions never contain an equal sign.	$3x + 7 = 10$ is an equation. $4y - 19$ is an expression.
Like Terms	• Terms that contain the same variables raised to the same powers • Like terms can be combined.	$4x$ and $7x$ are like terms. $9m$ and $9n$ are unlike terms. $5x - 3x = (5 - 3)x = 2x$
Simplifying an Expression	Use arithmetic properties and combine like terms to write an expression more simply.	$3y + (2y + 4) = (3y + 2y) + 4$ $ = (3 + 2)y + 4$ $ = 5y + 4$

1.7	**Exercises**	

CONCEPTS AND VOCABULARY

1. Is $3x - 4$ an equation or an expression?

2. Is $3x = 4$ an equation or an expression?

3. A(n) _____ is a number, a variable, or the product of numbers and variables raised to powers.

4. The number in a term is called the _____ of the term.

5. The terms $3xy$ and $7xy$ are (like/unlike).

6. The terms $4y$ and $4z$ are (like/unlike).

EQUATIONS AND EXPRESSIONS

Exercises 7–14: Identify each of the following as an equation or an expression.

7. $3x + 12$

8. $9 = 3x$

9. $17y + 15 = 49$

10. $38z - 20$

11. $5x = 3x + 10$

12. $2x + (5 - 3x)$

13. $4003 - x$

14. $3m + 2 = 4m - 6$

LIKE TERMS

Exercises 15–22: Determine whether the given terms are like or unlike.

15. $7w, 4w$

16. $2a, 9a$

17. $4bc, 3bc^2$

18. $8x^2y, 17x^2y$

19. $3xy^3, 2xy^3$

20. $9a^2b, 7ab^2$

21. y^2, y

22. $pq, 7p^2q$

Exercises 23–30: Combine like terms in the expression. Write "not possible" if terms cannot be combined.

23. $8x + 3x$

24. $4b + b$

25. $13yz - 6yz$

26. $x - y$

27. $6p - 5q^2$

28. $7z^2 - z^2$

29. $ab + 15ab$

30. $8m^2n - 3m^2n$

SIMPLIFYING EXPRESSIONS

Exercises 31–44: Simplify the expression.

31. $4 + 7x + 5$

32. $2 + 4n + 9$

33. $9y + (2y + 5)$

34. $15m + (3m + 7)$

35. $3a + 4 + 2a$

36. $8x + 7 + 2x$

37. $6z + 2(z - 7)$

38. $10z + 7(z - 3)$

39. $3(x + 2) - 4$

40. $9(q + 1) + 6$

41. $2x + 5 + 3x + 4$

42. $7y + 2 + y + 5$

43. $ab + y + 2ab + 3y$

44. $2x^2 + 3x + 5x^2 + x$

Exercises 45–48: For the given equation, do the following:

(a) Simplify the expression on each side of the equal sign.
(b) See if 5 is a solution to both the given equation and the equation formed in part (a).

45. $3x + 4x = 4(x + 2) + 7$

46. $3x + 2(x + 4) = 8 + 4x + 5$

47. $2(3 + x) + 5 = x + (2x + 6)$

48. $x^2 + 2x^2 = 3x + 12x$

Exercises 49–56: Translate the sentence into an equation using the variable x. Do not solve the equation. State what the variable represents.

49. Six times the number of inches minus two times the same number of inches is 36.

50. Seven pounds less than his weight is 156.

51. Fourteen is 9 less than her score.

52. The total of his age and twice his age is 30.

53. The product of 4 and her shoe size is 28.

54. The total miles divided by 14 is 31.

55. The score is 8 fewer than triple the score.

56. Eight more than 3 times the height is 107.

APPLICATIONS

57. *Education and Pay* A 2008 survey of employed college graduates found that those whose highest degree is a master's earned, on average, $11,000 per year more than graduates whose highest degree is a bachelor's. If those with a master's degree earned $64,000 per year, on average, how much did those with a bachelor's degree earn? (*Source:* National Science Foundation.)

58. *Heart Rate* During exercise, a physically fit male may experience a heart rate that is 3 times his resting heart rate. If this person's heart rate is 186 beats per minute during exercise, what is his resting heart rate?

59. *Roadway Congestion* A recent study found that the average commuter in Atlanta, Georgia burns 24 gallons of gas each year while sitting in traffic. This is 6 times the amount burned by the average commuter in Cleveland, Ohio. How many gallons of gas does the average Cleveland driver burn each year while stuck in traffic? (*Source:* Federal Highway Administration.)

60. *Roadway Congestion* A recent study found that the average commuter in Houston, Texas spends 16 more hours each year stuck in traffic than the average commuter in Portland, Oregon. If Houston drivers spend 36 hours each year stuck in traffic, how many hours do Portland drivers spend each year stuck in traffic? (*Source:* Federal Highway Administration.)

61. *Population Growth* Every hour the population of the world increases by about 8480. If the worldwide death rate is 6460 deaths per hour, what is the worldwide (hourly) birth rate? (*Source:* U.S. Census Bureau, International Data Base, 2008.)

62. *Housing Market* A homeowner recently reduced the price of his home to $218,000. If this represents a $17,000 decrease in price, what was the price before the decrease?

63. *Fireworks* Three friends bought several identical packages of bottle rockets. The first friend bought 2 packages, the second friend bought 4 packages, and the third friend bought 5 packages, giving the three friends a total of 132 bottle rockets. How many bottle rockets are in a single package?

64. *Sale Price* If the sale price of an item is multiplied by 4, then the result is the original price. If the original price is $64, what is the sale price?

65. *A Number Puzzle* If a number is tripled and then added to itself, the result is 12 more than the number. Find the number.

66. *A Number Puzzle* If doubling a natural number has the same result as squaring it, what is the number?

67. *Antique Value* A glass vase purchased at a garage sale has an appraised value of $200. If the appraised value is $179 more than the purchase price, how much did the vase cost at the garage sale?

68. *Counting Coins* A person has 6 coins in his pocket that total 65¢. If 3 of the coins are dimes, what are the other 3 coins?

69. *Perimeter* The rectangle in the following figure has a perimeter of 42 inches. If the length measures $5x$ inches and the width measures $2x$ inches, find x.

70. *Perimeter* A triangle with sides that measure $3x$, $4x$, and $5x$ feet has a perimeter of 60 feet. Find x.

WRITING ABOUT MATHEMATICS

71. Explain what it means to simplify an expression. Give several examples.

72. Describe in your own words how to use the four-step process to solve word problems.

SECTION 1.7

Checking Basic Concepts

1. Identify each of the following as an equation or an expression.
 (a) $4x = 40$ (b) $y + 16$

2. Determine whether the terms are like or unlike.
 (a) $m, 5m$ (b) $3y, 10x$

3. Combine like terms in each expression. Write "not possible" if terms cannot be combined.
 (a) $8pq - 3pq$ (b) $6m - n^2$

4. Simplify each expression.
 (a) $2(x + 3) - 5$ (b) $6x + 3 + 3x$

5. Translate the displayed sentence into an equation using the variable x. Do not solve the equation. State what the variable represents.

 His age decreased by 7 is 23.

6. *Perimeter* The rectangle in the following figure has a perimeter of 32 feet. If the length measures $3x$ feet and the width measures x feet, find x.

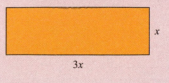

CHAPTER 1 Summary

SECTION 1.1 ■ INTRODUCTION TO WHOLE NUMBERS

Natural Numbers	1, 2, 3, 4, …
Whole Numbers	0, 1, 2, 3, …
Place Value	The place value of a digit in a number written in standard form is determined by the position that the digit takes in the number.
	Example: The 4 in the number 34,879 is in the thousands place.

Word Form	Starting with the left-most period, write the word form of the number in each period followed by the period name and a comma. **Example:** 18,207 is written as eighteen thousand, two hundred seven.
Expanded Form	Starting with the left-most digit, write the standard form of the number given by each digit and its corresponding place value. Place a plus sign (+) between each of these results. **Example:** 184,079 is written as $100{,}000 + 80{,}000 + 4000 + 70 + 9$.
Number Line	A number line can be used to visualize whole numbers. The graph of a whole number is a dot placed at the whole number's position on a number line. **Examples:** 2 and 6 are graphed.
Comparing Whole Numbers	To compare two whole numbers, determine their positions on a number line. Then use the symbols > or < to write an appropriate comparison. **Example:** From the number line above, $2 < 6$.
Bar Graph and Line Graph	A bar graph can be helpful when analyzing data. A line graph can be helpful when looking for trends in data.

Examples:

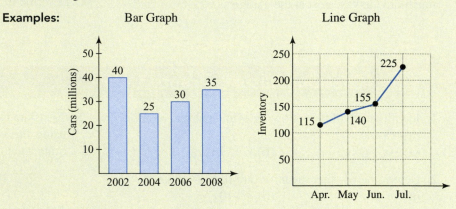

Table	A table can be used to display data in an at-a-glance way.

Example:

Year	2009	2011	2013
Births	207	215	198

SECTION 1.2 ■ ADDING AND SUBTRACTING WHOLE NUMBERS; PERIMETER

Adding Whole Numbers	The numbers being added are the *addends*, and the result is the *sum*. 1. Stack the numbers vertically with corresponding place values aligned. 2. Add the digits in each place value. Use carrying when necessary.

Examples:

$$\begin{array}{r} 6532 \\ +\ 2413 \\ \hline 8945 \end{array} \qquad \begin{array}{r} {}^{1\ 1} \\ 2758 \\ +\ 3617 \\ \hline 6375 \end{array}$$

Addition Properties

Property	**Examples**
1. Commutative property	$4 + 9 = 9 + 4$
2. Associative property	$(3 + 6) + 7 = 3 + (6 + 7)$
3. Identity property	$8 + 0 = 8$ and $0 + 8 = 8$

Translating Words to Addition	Words associated with addition include *add*, *plus*, *more than*, *sum*, *total*, and *increased by*. **Examples:** 7 dimes plus 3 dimes; the number of fish increased by 5.
Subtracting Whole Numbers	The number subtracted from is the *minuend*, the number being subtracted is the *subtrahend*, and the result is the *difference*. **1.** Stack the numbers vertically with corresponding place values aligned. **2.** Subtract the digits in each place value. Use borrowing when necessary.

Examples:

$$\begin{array}{r} 8437 \\ -\ 2216 \\ \hline 6221 \end{array} \qquad \begin{array}{r} {}^{613} \\ 5\overset{\,}{7}\overset{\,}{3}2 \\ -\ 2480 \\ \hline 3252 \end{array}$$

Subtraction Properties	**Property** **Examples** **1.** First identity property $38 - 0 = 38$ **2.** Second identity property $22 - 22 = 0$
Translating Words to Subtraction	Words associated with subtraction include *subtract*, *minus*, *fewer than*, *difference*, *less than*, *decreased by*, and *take away*. **Examples:** 9 fewer than the number of pies; the price decreased by 5.
Solutions	A *solution* is any number that makes an equation true when it replaces the unknown value. *Solving an equation* means finding all of its solutions. **Example:** The solution to $\square + 3 = 15$ is 12 because $12 + 3 = 15$ is a true equation.
Perimeter	The distance around an enclosed region **Example:** The perimeter of the region shown is $5 + 9 + 6 + 8 = 28$ feet.

SECTION 1.3 ■ MULTIPLYING AND DIVIDING WHOLE NUMBERS; AREA

Multiplying Whole Numbers	The numbers being multiplied are the *factors*, and the result is the *product*. **Examples:**

$$\begin{array}{r} {}^{1} \\ 82 \\ \times\ 25 \\ \hline 410 \\ 1640 \\ \hline 2050 \end{array} \begin{array}{l} \\ \\ \\ \leftarrow\ \mathbf{5 \times 82} \\ \leftarrow\ \mathbf{2\ (tens) \times 82} \\ \leftarrow\ \mathbf{add} \end{array} \qquad \begin{array}{r} {}^{42} \\ {}^{52} \\ 274 \\ \times\ 67 \\ \hline 1\,918 \\ 16\,440 \\ \hline 18{,}358 \end{array} \begin{array}{l} \\ \\ \\ \\ \leftarrow\ \mathbf{7 \times 274} \\ \leftarrow\ \mathbf{6\ (tens) \times 274} \\ \leftarrow\ \mathbf{add} \end{array}$$

Multiplication Properties	**Property** **Examples** **1.** Commutative property $5 \cdot 7 = 7 \cdot 5$ **2.** Associative property $(2 \cdot 6) \cdot 5 = 2 \cdot (6 \cdot 5)$ **3.** Identity property $3 \times 1 = 3$ and $1 \times 3 = 3$ **4.** Zero property $8 \times 0 = 0$ and $0 \times 8 = 0$ **5.** Distributive property $3(4 + 2) = 3 \cdot 4 + 3 \cdot 2$ $2(5 - 1) = 2 \cdot 5 - 2 \cdot 1$
Translating Words to Multiplication	Words associated with multiplication include *multiply*, *times*, *product*, and *double*. **Examples:** Double the number of cups; the length times the width.

Dividing Whole Numbers	The number divided into is the *dividend*, the number divided by is the *divisor*, and the result is the *quotient*.
Long Division	A process that can be used when the dividend is a large whole number

Example:

$$\begin{array}{r} 301 \text{ r}52 \\ 263\overline{)79{,}215} \\ -78\ 9 \\ \hline 31 \\ -0 \\ \hline 315 \\ -263 \\ \hline 52 \end{array}$$

Division Properties

Property	Examples
1. Identity properties	$6 \div 6 = 1$ and $8 \div 1 = 8$
2. Zero properties	$0 \div 7 = 0$ and $5 \div 0$ is undefined.

Translating Words to Division

Words associated with division include *divide*, *quotient*, and *per*.

Examples: Days per project;   the area divided by the width

Area of a Rectangle

Area = Length × Width

Example: The area of the region shown is
$7 \times 4 = 28$ square inches.

7 inches

4 inches

SECTION 1.4 ■ EXPONENTS, VARIABLES, AND ALGEBRAIC EXPRESSIONS

Exponential Notation	Used to represent repeated multiplication

Example: $3^5 = 3 \cdot 3 \cdot 3 \cdot 3 \cdot 3$

Powers of Ten

A natural number power on ten equals the number of 0s that follow a 1 in the standard form of the power of 10.

Example: $10^7 = 10{,}000{,}000$

Variable

A symbol or letter used to represent an unknown quantity

Example: *W* represents the weight of an animal.

Algebraic Expression

May contain numbers, variables, operation symbols, and grouping symbols

Examples: $3x + 2$ and $4y$

Formula

An equation that expresses a relationship between two or more quantities

Example: The formula $Q = 4G$ gives the number of quarts *Q* in *G* gallons.

SECTION 1.5 ■ ROUNDING AND ESTIMATING; SQUARE ROOTS

Rounding

Approximating a number to a given level of accuracy

Example: To the nearest hundred, 4588 rounds to 4600.

Estimating

A rough calculation used to find a reasonably accurate answer

Example: When rounded to the nearest thousand, $3967 + 2019$ can be estimated by $4000 + 2000$.

Approximation	An approximation is the result when rounding or estimating. It is usually not exactly accurate. The symbol $\approx$ means "is approximately equal to."
	Example: To the nearest ten, $87 + 72 \approx 160$.
Perfect Square	A perfect square results when a whole number is squared.
	Example: 144 is a perfect square because $12^2 = 144$.
Square Root	The square root of a whole number is a number whose square is the given whole number. The symbol $\sqrt{}$ is called the *radical*, and the number under it is called the *radicand*.
	Examples: $\sqrt{100} = 10$ because $10^2 = 100$, $\sqrt{50} \approx 7$ because $7^2 \approx 50$

SECTION 1.6 ■ ORDER OF OPERATIONS

Order of Operations	1. Do all calculations within grouping symbols, such as parentheses and radicals, or above and below a fraction bar. 2. Evaluate all exponential expressions. 3. Do all multiplication and division from *left to right*. 4. Do all addition and subtraction from *left to right*.
	NOTE: If there is more than one arithmetic operation within grouping symbols, the order of Steps 2, 3, and 4 must be followed when performing Step 1.
	Example: $$\begin{aligned} 5 + 36 \div (4 - 2)^2 &= 5 + 36 \div 2^2 \\ &= 5 + 36 \div 4 \\ &= 5 + 9 \\ &= 14 \end{aligned}$$
Evaluating Expressions	Replace each variable in the expression with its given value and then use the order of operations agreement to evaluate.
	Example: Evaluating $13 - 4x$ for $x = 2$ gives $13 - 4(2) = 13 - 8 = 5$.
Translating Words to Symbols	When translating words to symbols, watch for the words "the quantity" and pay special attention to the placement of any commas.
	Example: "four plus seven, times three" translates to $(4 + 7) \cdot 3$.

SECTION 1.7 ■ MORE WITH EQUATIONS AND PROBLEM SOLVING

Equations and Expressions	Equations always contain an equal sign ($=$), but expressions never contain an equal sign. Equations are often solved, and expressions are often simplified.
	Example: $2x - 5 = 9$ is an equation; $5x + 8$ is an expression.
Like Terms	Terms with the same variables raised to the same powers can be combined.
	Example: $7x^2$ and $3x^2$ are like terms; thus $7x^2 + 3x^2 = 10x^2$.
Simplifying Expressions	Use arithmetic properties and combine like terms.
	Example: $4x + 3 + 6x = 4x + 6x + 3 = 10x + 3$

CHAPTER 1   Review exercises

SECTION 1.1

Exercises 1 and 2: For the given whole number, determine the place value of the digit 3.

1. 25,304

2. 365,719

Exercises 3 and 4: Name the digit with the given place value in the number 2,819,065,347.

3. ten-millions

4. hundred-thousands

Exercises 5 and 6: Write the number in words.

5. 48,309

6. 37

Exercises 7 and 8: Write the number in expanded form.

7. 673

8. 61,004

9. Write *fifty-eight thousand, three hundred forty-five* in standard form.

10. Graph the whole numbers 1 and 4 on a number line.

Exercises 11 and 12: Place the correct symbol, $<$ or $>$, in the blank between the whole numbers.

11. 28 _____ 0

12. 14 _____ 23

SECTION 1.2

Exercises 13–16: Add.

13. $21 + 14$

14. $176 + 949$

15.   378
        $+ 5627$

16.   6952
        $+ 4934$

Exercises 17–20: Subtract.

17. $863 - 97$

18. $2492 - 358$

19.   59,415
        $- 26,588$

20.   41,637
        $- 8,929$

Exercises 21 and 22: Translate the phrase into a mathematical expression and then find the result.

21. The difference between 83 and 21

22. 48 more than 103

Exercises 23 and 24: Solve the given equation by finding the unknown value.

23. $\square + 17 = 39$

24. $99 - \square = 88$

SECTION 1.3

Exercises 25 and 26: Use the distributive property to rewrite the expression. Do not compute the product.

25. $5(4 + 2)$

26. $7(8 - 5)$

Exercises 27–30: Multiply.

27. $0 \cdot 58$

28. 1×99

29. $43 \cdot 1852$

30. 516×712

Exercises 31–34: Divide, when possible.

31. $84 \div 4$

32. $37,721 \div 563$

33. $4239 \div 51$

34. $132 \div 0$

Exercises 35 and 36: Translate the phrase into a mathematical expression and then find the result.

35. The quotient of 66 and 11

36. 26 times 17

Exercises 37 and 38: Solve the given equation by finding the unknown value.

37. $64 \div \square = 16$

38. $\square \times 7 = 63$

SECTION 1.4

Exercises 39 and 40: Use exponential notation to write the repeated multiplication.

39. $8 \cdot 8 \cdot 8 \cdot 8 \cdot 8$

40. $9 \cdot 9 \cdot 9$

Exercises 41–44: Evaluate the exponential expression.

41. 7^2

42. 5^3

43. $4 \cdot 10^2$

44. $9 \cdot 10^5$

Exercises 45 and 46: Evaluate the algebraic expression for $x = 6$ and $y = 3$.

45. $3xy$

46. x^y

Exercises 47 and 48: Use the appropriate geometric formulas to find the measure.

47. The perimeter of a triangle with sides of length 7, 12, and 16 feet

48. The area of a rectangle with a 24-inch length and a 20-inch width

Exercises 49–54: Solve the equation.

49. $b + 9 = 14$ **50.** $12x = 36$

51. $63 = 7x$ **52.** $29 - z = 2$

53. $48 \div d = 8$ **54.** $n \div 7 = 6$

SECTION 1.5

55. Round 162 to the nearest ten.

56. Round 978,423 to the nearest ten-thousand.

Exercises 57 and 58: Round the whole number to its highest place value.

57. 52,809   **58.** 393,001

Exercises 59 and 60: Round each number to the nearest hundred to estimate the sum or difference.

59. $689 + 325 + 286$ **60.** $4739 - 3341$

Exercises 61 and 62: Compute the square root.

61. $\sqrt{256}$ **62.** $\sqrt{121}$

Exercises 63 and 64: Approximate the square root to the nearest whole number.

63. $\sqrt{38}$ **64.** $\sqrt{60}$

SECTION 1.6

Exercises 65–72: Evaluate the expression.

65. $34 - 24 \div 6 + 5$ **66.** $17 - 3 \cdot 5 + 29$

67. $55 \div 11 + 3 \cdot 6$ **68.** $2 \cdot 30 - 72 \div 9$

69. $\dfrac{(6 + 10) \cdot 4}{50 - 7 \cdot 6}$ **70.** $\dfrac{(8 - 2) \cdot 7}{\sqrt{9 - 5} - 1}$

71. $23 - (3 \cdot 5 - (20 \div 2^2) + 3)$

72. $\sqrt{25 - 6 + 4 \cdot 3^2 \div 6} + 8$

Exercises 73 and 74: Evaluate the algebraic expression for the given values of the variables.

73. $7y + 2x$,   for $x = 5, y = 3$

74. $a - 8 + 3b$,   for $a = 17, b = 4$

Exercises 75 and 76: Use symbols to write an expression and then evaluate it.

75. Nine minus the quantity two plus six

76. Four times three decreased by one

SECTION 1.7

Exercises 77 and 78: Identify each of the following as an equation or an expression.

77. $5x - 17$ **78.** $8 = 3x - 1$

Exercises 79 and 80: Determine whether the given terms are like or unlike.

79. $5xy, 2xy^2$ **80.** $a^2b, 7a^2b$

Exercises 81–84: Combine like terms in the expression. Write "not possible" if terms cannot be combined.

81. $7x + 11x$ **82.** $9b + b$

83. $22mn - 9mn$ **84.** $3x - 2y$

Exercises 85–88: Simplify the expression.

85. $3y + (y + 7)$ **86.** $8z + 2(z - 3)$

87. $7a + 4 + 8a$ **88.** $4y + 3 + y + 6$

Exercises 89 and 90: Translate the sentence into an equation using the variable x. Do not solve the equation. State what the variable represents.

89. Seven inches less than his height is 64.

90. Fourteen is 12 more than her score.

APPLICATIONS

*Exercises 91 and 92: **Postage** The following line graph shows the price of a first class postage stamp in cents for selected years. (**Source:** United States Postal Service.)*

91. Comment on the general trend of the price of a stamp over this time period.

92. What was the price of a stamp in 1978?

93. ***National Monuments*** There are 7 fewer national monuments in Nebraska than in New Mexico. If New Mexico has 10 national monuments, how many are in Nebraska? (**Source:** National Parks Service.)

94. *Geometry* Find the area of the rectangle.

8 in.

11 in.

95. *Walgreens* There were 554 more Walgreens pharmacy locations in May 2008 than there were a year earlier. If there were 6252 Walgreens pharmacies in May 2008, how many were there a year earlier? (*Source:* Walgreens.)

96. *Estimating Time* An athlete stretches for 13 minutes, jogs for 48 minutes, and then walks for 19 minutes. Estimate the total time for this workout by rounding to the nearest ten.

97. *Heart Rate* The average heart rate R in beats per minute (bpm) of an animal weighing W pounds can be approximated by

$$R = \frac{885\sqrt{W}}{W}.$$

Find the heart rate for a 225-pound bear.

98. *Heart Rate* During exercise, a physically fit female experiences a heart rate that is twice her resting heart rate. If this person's heart rate is 136 beats per minute during exercise, what is her resting heart rate?

99. *Geometry* Find the perimeter of the figure.

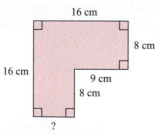

16 cm

8 cm

16 cm

9 cm

8 cm

?

100. *Buying Books* What is the maximum number of books costing $9 that a person can buy with $65? How much change will the person receive?

101. *Yard Sale* A person bought 13 items at a yard sale and received $1 in change. If each item was priced at $3, how much did the person give to the cashier?

102. *Gardening* A garden is being built so that it has the shape of a square with an area of 230 square feet. Estimate the length of one side of the garden to the nearest foot.

103. *Converting Temperature* To convert a temperature F given in degrees Fahrenheit to an equivalent temperature C in degrees Celsius, use the formula

$$C = \frac{5(F - 32)}{9}.$$

Find the Celsius temperature that is equivalent to a temperature of 77°F.

104. *A Number Puzzle* If a number is doubled and then added to itself, the result is 10 more than the number. Find the number.

Exercises 105–108: Music Sales The following line graph shows sales in billions of dollars for recorded music in physical form (not downloaded) during selected years. (*Source:* Recording Industry of America.)

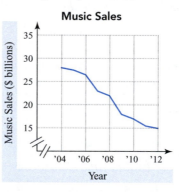

Music Sales

105. Estimate the 2012 sales of music that was purchased in physical form.

106. Estimate the year when sales of music in physical form first fell to $22 billion.

107. Estimate the year when sales of music in physical form first fell to $17 billion.

108. Estimate the 2009 sales of music that was purchased in physical form.

CHAPTER 1 Test

1. Determine the place value of the digit 5 in the whole number 158,902.

2. Write 7341 in expanded form.

3. Round 78,423 to the nearest thousand.

4. Place the symbol, $<$ or $>$, in the blank: 71 _____ 17.

Exercises 5–8: Perform the arithmetic.

5. $3472 + 869$

6. $15{,}902 - 9876$

7. $27 \cdot 4817$

8. $34{,}476 \div 67$

9. Write $3 \cdot 3 \cdot 3 \cdot 3$ using exponential notation.

10. Evaluate $5^2 \cdot 2^3$.

Exercises 11 and 12: Solve the equation.

11. $47 - z = 34$

12. $8x = 96$

Exercises 13–16: Evaluate the expression.

13. $18 - 20 \div 5 + 5$

14. $2 \cdot 10 - 60 \div 4$

15. $\dfrac{(5 + 3) \cdot 3}{20 - 3 \cdot 6}$

16. $29 - (3 \cdot 7 - (4^2 \div 2) + 5)$

17. Evaluate $b - 5 + 6a$ for $a = 2$, $b = 9$.

18. Compute $\sqrt{81}$.

Exercises 19 and 20: Combine like terms in the expression. Write "not possible" if terms cannot be combined.

19. $12p^2 - 7p^2$

20. $4w + 4y$

21. Approximate $\sqrt{35}$ to the nearest whole number.

22. Graph the whole numbers 2 and 5 on a number line.

Exercises 23 and 24: Simplify the algebraic expression.

23. $3x + (5x + 2)$

24. $3y + 9 + y + 4$

25. *Geometry* Find the perimeter of the figure.

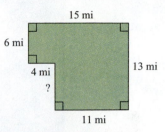

26. *Buying Shirts* What is the maximum number of shirts costing $12 that a person can buy with $80? How much change will the person receive?

27. *College Tuition* A student paid $407 more for tuition this semester than he did last semester. If his tuition bill is $1675 this semester, what was the bill last semester?

28. *Burning Calories* A 180-pound person burns 14 calories each minute while playing handball. Write a formula that gives the total number of calories C burned while playing handball for h minutes. (*Source: The Calorie Control Council.*)

2 Integers

The future is not a gift —it is an achievement.

—HARRY LAUDER

The Republic of Maldives, an island nation in the Indian Ocean about 435 miles southwest of Sri Lanka, is considered the "flattest nation on Earth," with a *maximum* natural elevation of less than 8 feet above sea level. Scientists believe that global warming could cause large portions of the polar ice sheets to melt at an accelerated rate, which would raise world sea levels. A rise in the Indian Ocean of just 2 or 3 feet would have a huge impact on the Maldives. Even a modest sea level rise would be devastating to many coastal locations, including the Maldives.

The following table lists elevations for two locations that are below sea level and two locations that are above sea level.

Elevations for Selected World Locations

Location	Elevation
Amsterdam	13 feet **below** sea level
Mt. Kilimanjaro	19,340 feet **above** sea level
Death Valley	282 feet **below** sea level
The Maldives	7 feet **above** sea level

Source: The World Atlas.

Source: S. Lovgren, *National Geographic News*, April 2004.

In this chapter, we discuss numbers called *integers*, which can be used to describe elevations above sea level (positive elevation) or below sea level (negative elevation).

2.1   Integers and the Number Line

Signed Numbers • **Integers and Their Graphs** • **Comparing Integers** • **Absolute Value** • **Applications**

A LOOK INTO MATH ▶

The idea of negative numbers was a difficult concept for early mathematicians. As late as the 18th century, negative numbers were not readily accepted. After all, if numbers represent quantities, how could a person have −3 apples? However, negative numbers make more sense when someone is working with money, temperatures, or elevations. For example, if you owe someone $200, then this amount is a debt and can be thought of as −200.

Signed Numbers

NEW VOCABULARY

☐ Positive number
☐ Negative number
☐ Signed number
☐ Opposite
☐ Integers
☐ Origin
☐ Absolute Value

In Chapter 1, we discussed whole numbers. All whole numbers other than zero (the natural numbers) are positive numbers. A **positive number** is a number that is greater than zero. Rather than writing the positive whole numbers as

$$+1, +2, +3, +4, \ldots,$$

we usually omit the positive sign (+) and simply write them as follows:

$$1, 2, 3, 4, \ldots$$

READING CHECK

• What is a positive number?
• What is a negative number?

For every positive number, there is a corresponding negative number called its *opposite*. A **negative number** is a number that is less than zero. Together, positive numbers, negative numbers, and zero are the **signed numbers**. We indicate that a number is negative by placing a negative sign (−) immediately in front of the number. For example, the opposite of 3 is −3, and the opposite of 12 is −12. In fact, *every* number has an opposite. If we let a represent any number, then its **opposite** (also known as the *additive inverse*) is represented by $−a$.

NOTE: Zero is neither positive nor negative, so the opposite of 0 is 0.

CALCULATOR HELP

To enter a negative number in a calculator, see the Appendix (page AP-2).

TABLE 2.1 Opposites of Signed Numbers

Number	Opposite	Opposite of Opposite
4	−4	$−(−4) = 4$
−7	7	$−(7) = −7$
5	−5	$−(−5) = 5$

The middle column of Table 2.1 above displays the opposites of the signed numbers in the first column. The third column shows that the opposites of the numbers in the second column are equal to the numbers in the first column. That is, the opposite of the opposite of a number is equal to that number. Table 2.1 suggests the following double negative rule:

DOUBLE NEGATIVE RULE

Let a be any number. Then $−(−a) = a$.

EXAMPLE 1   **Finding opposites (or additive inverses)**

Simplify each of the following.
(a) $-(5)$ **(b)** $-(-(-14))$ **(c)** $-(-9)$

Solution
(a) The opposite of 5 is -5.
(b) By the double negative rule, $-(-14) = 14$. So, $-(-(-14)) = -(14) = -14$.
(c) By the double negative rule, $-(-9) = 9$.

Now Try Exercises 19, 21, 27

NOTE: To find the opposite of an exponential expression, evaluate the exponent first. For example, the opposite of 5^2 is

$$-(5^2) = -(5 \cdot 5) = -25.$$

MAKING CONNECTIONS

Plus, Minus, Positive, and Negative

The symbol $(+)$ is used for addition and also shows that a number is positive. Similarly, the symbol $(-)$ is used for subtraction and also shows that a number is negative. When one of these symbols is used in a printed text to indicate a positive or negative number, there is no space between the symbol and the number to its right. However, when these symbols are used for addition or subtraction, there is a space both before and after the symbol.

$4 + 9$	Four plus nine
$+7$	Positive seven
$13 - 8$	Thirteen minus eight
-12	Negative twelve
$-(-6)$	The opposite of negative six

STUDY TIP

If you have tried to solve a problem but need help, ask a question in class. Other students will likely have the same question.

Integers and Their Graphs

The **integers** are a set of numbers that includes the natural numbers, zero, and the opposites of the natural numbers.

The Integers: $\ldots \; -3, \, -2, \, -1, \, 0, \, 1, \, 2, \, 3, \, \ldots$

Because the negative integers are less than 0, they are located to the left of 0 on the number line. To graph a negative integer, we must extend (to the left) the number line used for graphing whole numbers so that it can be used for numbers less than 0. The number line in Figure 2.1 can be used to graph integers.

READING CHECK

• Where are negative numbers found on the number line?

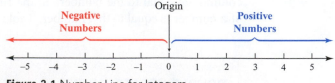

Figure 2.1 Number Line for Integers

Numbers to the left of 0 on the number line are negative, and numbers to the right of 0 are positive. The number 0 is called the **origin** and is neither positive nor negative.

Just as with whole numbers, the graph of an integer includes a dot placed on the number line at the number's position.

EXAMPLE 2   **Graphing integers**

Graph the integers −3, 0, and 4 on the same number line.

Solution
The integers −3, 0, and 4 are graphed as shown in Figure 2.2.

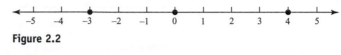

Figure 2.2

■ **Now Try Exercise 31**

Comparing Integers

Recall that when two whole numbers are graphed on the same number line, the number to the left is *less than* the number to the right and the number to the right is *greater than* the number to the left. This method of comparison also holds for integers. As a result, a negative integer is *always less than* a positive integer and a positive integer is *always greater than* a negative integer. For example, −5 < 2 because −5 is **left** of 2 on the number line. Similarly, 4 > −1 because 4 is **right** of −1 on the number line. See Figure 2.3.

READING CHECK

• On a number line, how do we know if a number is greater or less than another number?

Figure 2.3 Comparing Integers

▶ **REAL-WORLD CONNECTION** When comparing two negative integers, it is helpful to think about temperature. On a cold day in Idaho, the temperature might be −8 degrees Fahrenheit. If the temperature later dips to −13 degrees Fahrenheit, we would say that it got colder. In other words, −13°F is colder than (less than) −8°F, and −8°F is warmer than (greater than) −13°F. In math symbols, we write

$$-13 < -8 \quad \text{or} \quad -8 > -13.$$

Note that −13 is located to the left of −8 on the number line and −8 is located to the right of −13 on the number line.

EXAMPLE 3   **Comparing two integers**

Place the correct symbol, < or >, in the blank between each pair of integers.
(a) 5 _____ −9   **(b)** −3 _____ −12   **(c)** −7_____ −6

Solution
(a) Because 5 is located to the **right** of −9 on the number line, 5 **>** −9.
(b) Because −3 is located to the **right** of −12 on the number line, −3 **>** −12.
(c) Because −7 is located to the **left** of −6 on the number line, −7 **<** −6.

■ **Now Try Exercises 37, 39, 43**

CALCULATOR HELP

To find an absolute value with a calculator, see the Appendix (page AP-2).

Absolute Value

The **absolute value** of an integer equals its distance on the number line from 0 (the origin). Because distance is never negative, the absolute value of an integer is *never negative*. If the variable a represents an integer, the absolute value of a is written as $|a|$ and reads as "the absolute value of a." Figure 2.4 shows that $|-3| = 3$ and $|3| = 3$ because both -3 and 3 are located a distance of 3 units from the origin on the number line.

READING CHECK

• How is an absolute value written in math symbols?

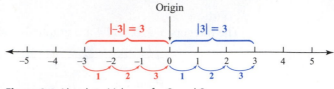

Figure 2.4 Absolute Values of -3 and 3

EXAMPLE 4   **Finding absolute value**

Evaluate each absolute value.
(a) $|-7|$ **(b)** $|5|$ **(c)** $|0|$

Solution
(a) Because -7 is **7** units from the origin, $|-7| = $ **7**.
(b) The integer 5 is **5** units from the origin, so $|5| = $ **5**.
(c) Because 0 is **0** units from the origin, $|0| = $ **0**.

Now Try Exercises 49, 51, 53

MAKING CONNECTIONS

Absolute Value and Opposites

Finding the absolute value of an integer is *not* the same as finding its opposite. An absolute value is *never* negative, but the opposite of any positive integer is *always* negative. The following table shows how the absolute values of some integers compare to their opposites.

READING CHECK

• Explain why the absolute value of a number may not be the opposite of the number.

Integer	Absolute Value	Opposite
4	4	-4
-2	2	2
0	0	0
11	11	-11

NOTE: The vertical lines used to show absolute value are grouping symbols. When evaluating expressions such as $-|21|$ or $-|-16|$, the absolute value should be evaluated first before finding the opposite.

absolute value of 21 is 21   absolute value of -16 is 16

$$-|21| = -(21) = -21 \quad \text{and} \quad -|-16| = -(16) = -16$$

opposite of 21 is -21 opposite of 16 is -16

| EXAMPLE 5 | **Comparing expressions involving absolute value** |

Place the correct symbol, $<$, $>$, or $=$, in each blank between the expressions.
(a) $|-7|$ ____ -7 **(b)** $-|-5|$ ____ $-|5|$ **(c)** $-|3|$ ____ $|-3|$

Solution
(a) Since $|-7| = 7$ and 7 is **right** of -7 on the number line, $|-7| > -7$.
(b) For the expression on the left, $-|-5| = -(5) = -5$, and for the expression on the right, $-|5| = -(5) = -5$. The expressions are **equal**, so $-|-5| = -|5|$.
(c) Evaluating each expression gives $-|3| = -3$ and $|-3| = 3$. Because -3 is **left** of 3 on the number line, $-|3| < |-3|$.

▌ **Now Try Exercises 63, 65, 67**

Applications

▶ **REAL-WORLD CONNECTION**   The U.S. Census Bureau predicts changes in the U.S. population. By analyzing projected (predicted) population growth or decline data, government officials can plan for possible changes in demands on social programs, roadways, and vital resources such as water and electricity. The next example illustrates how both positive and negative integers are used in projecting population changes.

| EXAMPLE 6 | **Analyzing projected population change** |

Table 2.2 lists the projected population change from 2000 to 2030 for selected states.

TABLE 2.2 Projected Population Change: 2000–2030

State	Arkansas	North Dakota	Utah	West Virginia
Change	567,000	$-36,000$	1,252,000	$-88,000$

Source: U.S. Census Bureau.

(a) Which states have a projected decline in population?
(b) Which state has the largest projected growth in population?

Solution
(a) A decline in population is represented by a negative number. The states with negative population change are North Dakota and West Virginia.
(b) Population growth is represented by a positive number. The largest positive number in the table is 1,252,000, which is the projected population change for Utah.

▌ **Now Try Exercise 75**

The next example shows a bar graph of positive and negative temperatures.

| EXAMPLE 7 | **Reading a bar graph involving integers** |

International Falls is a small community in northern Minnesota located on the Canadian border. It is often called "the nation's ice box." Figure 2.5 on the next page shows a bar graph of record low temperatures in International Falls by month. (*Source:* NOAA.)

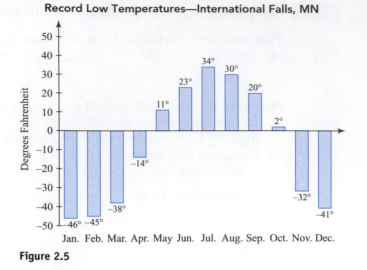

Figure 2.5

(a) What is the record low temperature for April?

(b) Which month has the warmest record low temperature?

(c) Does March have a colder or warmer record low when compared to November?

Solution

(a) The bar for April extends below the horizontal line representing 0°F, so its record low temperature is negative. From the bar graph, the record low for April is −14°F.

(b) The tallest bar reaches above the horizontal line representing 0°F and shows a record low temperature of 34°F in July.

(c) The record low temperature for March is −38°F, as compared to −32°F for November. The March temperature is colder.

▌ **Now Try Exercise 81**

The next example demonstrates how integers can be estimated from a line graph.

EXAMPLE 8 ▶ **Estimating integer values from a line graph**

It is often colder at the top of a mountain than it is at sea level because air temperature decreases as altitude increases. The line graph in Figure 2.6 shows the air temperature in degrees Fahrenheit at various altitudes.

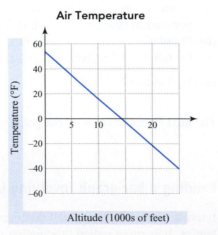

Figure 2.6 Temperatures at Various Altitudes

(a) Estimate the air temperature at an altitude of 25,000 feet.

(b) Estimate the altitude where the air temperature is 0°F.

Solution

(a) Because altitude is displayed in thousands of feet, locate 25 on the horizontal scale. Move **vertically** downward to the graphed line. From this position, move **horizontally** to the left to the temperature values. See Figure 2.7. At an altitude of 25,000 feet, the air temperature is −40°F.

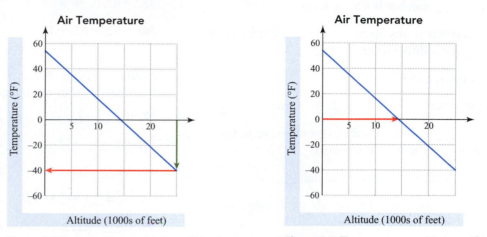

Figure 2.7 Temperatures at Various Altitudes

Figure 2.8 Temperatures at Various Altitudes

(b) Find 0°F at the left edge of the graph. Move **horizontally** to the right to the graphed line. There is no need to move vertically upward or downward. See Figure 2.8. Because altitude is displayed in thousands of feet, the air temperature is 0°F at an altitude of 15,000 feet.

Now Try Exercises 83, 85

2.1 Putting It All Together

CONCEPT	COMMENTS	EXAMPLES
Signed Numbers	A positive number is greater than 0. A negative number is less than 0. Zero is neither positive nor negative.	$3, +14, 137$, and $+900$ $-1, -28, -271$, and -1170
Opposite (Additive Inverse)	The opposite (additive inverse) of a number a is written as $-a$.	The opposite of 7 is -7. The opposite of -5 is $-(-5) = 5$.
Integers	The integers include the natural numbers, zero, and the opposites of the natural numbers.	$\ldots -3, -2, -1, 0, 1, 2, 3 \ldots$
Absolute Value	The absolute value of an integer equals its distance on the number line from 0 (the origin). The absolute value of a is written as $\lvert a \rvert$.	$\lvert 9 \rvert = 9$ $\lvert -12 \rvert = 12$ $\lvert 0 \rvert = 0$

2.1   Exercises

MyMathLab   Math XL PRACTICE   WATCH   DOWNLOAD   READ   REVIEW

CONCEPTS AND VOCABULARY

1. A(n) _____ number is greater than zero.

2. A(n) _____ number is less than zero.

3. If a represents any number, then $-a$ represents the _____ (or additive inverse) of a.

4. The opposite of 0 is _____ .

5. If a represents any number, then $-(-a) =$ _____ .

6. The _____ include the natural numbers, zero, and the opposites of the natural numbers.

7. On the number line, 0 is called the _____ .

8. The _____ of a number equals its distance on the number line from 0.

9. Express the temperature "3 below zero" as an integer.

10. Express an elevation of "17 feet above sea level" as an integer.

OPPOSITES

Exercises 11–18: Find the opposite of the given integer.

11. 7

12. 13

13. -43

14. -21

15. -237

16. 452

17. 93,000

18. -3967

Exercises 19–30: Simplify the expression.

19. $-(8)$

20. $-(11)$

21. $-(-26)$

22. $-(-13)$

23. $-(0)$

24. $-(-0)$

25. $-(-(23))$

26. $-(-(39))$

27. $-(-(-5))$

28. $-(-(-9))$

29. $-(-(-(-1)))$

30. $-(-(-(-6)))$

THE NUMBER LINE

Exercises 31–36: Graph the integers on a number line.

31. $-4, -2, 3$

32. $-2, 0, 4$

33. $-16, -8, 12$

34. $-20, 10, 25$

35. $-87, 5, 76$

36. $-92, -63, -12$

Exercises 37–48: Place the correct symbol, $<$ or $>$, in the blank between the integers.

37. 4 ___ -7

38. -2 ___ 9

39. -8 ___ -12

40. -17 ___ -1

41. 43 ___ 206

42. 99 ___ 34

43. -34 ___ -29

44. -63 ___ -36

45. 0 ___ -293

46. -349 ___ 0

47. 0 ___ 167

48. 682 ___ 0

ABSOLUTE VALUE

Exercises 49–56: Evaluate the absolute value.

49. $|10|$

50. $|-8|$

51. $|0|$

52. $|-0|$

53. $|-18|$

54. $|45|$

55. $|-87|$

56. $|-53|$

Exercises 57–62: Simplify the absolute value expression.

57. $-|2|$

58. $-|-3|$

59. $-|-19|$

60. $-|12|$

61. $-|0|$

62. $-|-0|$

Exercises 63–70: Place the correct symbol, $<$, $>$, or $=$, in the blank between the expressions.

63. 2 ___ $-|2|$

64. $-|8|$ ___ -8

65. $|-12|$ ___ $-|12|$

66. $|-8|$ ___ $-|-8|$

67. $-|-29|$ ___ $-|29|$

68. $-|10|$ ___ $|-10|$

69. $-|25|$ ___ 25

70. $-|-46|$ ___ $|-46|$

APPLICATIONS

Exercises 71–74: Elevation Refer to the following table. Express the elevations of the given locations as positive or negative integers.

Elevations for Selected World Locations

Location	Elevation
Amsterdam	13 feet below sea level
Mt. Kilimanjaro	19,340 feet above sea level
Death Valley	282 feet below sea level
The Maldives	7 feet above sea level

Source: The World Atlas.

71. Death Valley   **72.** The Maldives

73. Mt. Kilimanjaro   **74.** Amsterdam

75. *Population* The following table lists the population change from 1990 to 2010 for selected countries.

Country	Latvia	Malta	Romania	Tonga
Change	−446,000	48,000	−685,000	31,000

Source: U.S. Census Bureau.

  (a) Which of these countries had the largest decline in population from 1990 to 2010?

  (b) List the countries that had a growth in population from 1990 to 2010.

76. *High School Enrollment* The following table lists the change in public high school enrollment from 2000 to 2009 for selected states.

State	Alaska	Florida	Nevada	Ohio
Change	−4,000	336,000	122,000	−33,000

Source: U.S. National Center for Educational Statistics.

  (a) Which of these states had the largest increase in enrollment from 2000 to 2009?

  (b) List the states that had a decrease in enrollment from 2000 to 2009.

Finances Exercises 77–80: Even though −$84 is a negative value (a debt), it is a larger financial quantity than a positive balance of $52 in a checking account (an asset). The debt is larger than the asset because $|-84| > |52|$. Use absolute value to determine which of each pair of financial quantities at the top of the next column is larger.

77. An employee paycheck: $1050
A credit card debt: −$1745

78. Owing the baby sitter: −$32
Cash in your wallet: $44

79. A friend owes you: $160
Club membership dues: −$200

80. Savings account balance: $12,900
Tuition and fees due: −$7800

81. *Ocean Depth* The following bar graph shows the maximum depth of each ocean. *(Source: The World Atlas.)*

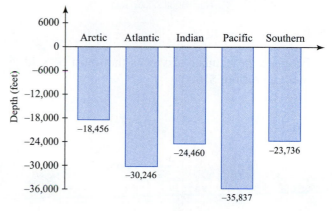

Deepest Points in the Oceans

  (a) Which ocean is the deepest?

  (b) Which ocean is the least deep?

  (c) Which ocean is 23,736 feet deep?

  (d) Which is deeper, the Indian Ocean or the Southern Ocean?

82. *Cold Temperatures* Refer to the following figure.

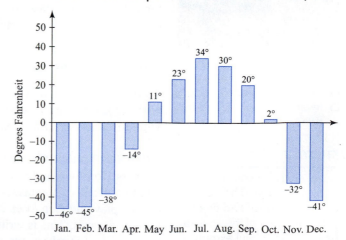

Record Low Temperatures—International Falls, MN

  (a) Which month's record low temperature has the largest absolute value?

  (b) Which month's record low temperature has the smallest absolute value?

Exercises 83–86: Music Videos The following line graph shows the profit made from selling music videos. A negative profit represents a loss.

Music Video Profit

Number of Videos Sold

83. Estimate the profit if no videos are sold.

84. Estimate the number of videos that must be sold to make a profit of $1000.

85. Estimate the number of videos that must be sold to make a profit of $5000.

86. Estimate the profit when 150 videos are sold.

WRITING ABOUT MATHEMATICS

87. Explain how absolute value is computed. Give two examples.

88. Sketch a number line and label the locations of the negative numbers, origin, and positive numbers.

89. Explain why a positive number is always greater than a negative number.

90. Explain how the absolute value of a number compares to the absolute value of its opposite.

2.2 Adding Integers

Adding Integers • Addition Properties • Adding Integers Visually • Applications

A LOOK INTO MATH ▶ One way to measure a football team's performance is to compute total yardage. On some plays, a team may advance the ball toward the opposing team's goal. When this happens, the team has made a *gain* in yardage. On other plays, the ball may end up farther from the opposing team's goal than it was when the play began. In this case, the team has suffered a *loss* in yardage. A yardage gain is recorded as a positive number, and a yardage loss is recorded as a negative number. A team's total yardage is found by adding the gains and losses, or more simply, adding positive and negative numbers.

NEW VOCABULARY

☐ Additive inverse
☐ Inverse property for addition

STUDY TIP

Have you been completing all of the assigned homework on time? Regular and timely practice is one of the keys to having a successful experience in any math class. Don't miss the important opportunity to learn math through doing math.

Adding Integers

To fully understand how to add integers, we must learn how to find the sum of two integers with the same sign (both positive or both negative) and how to find the sum of two integers with different signs (one positive and one negative).

ADDING INTEGERS WITH LIKE SIGNS We already know how to add two positive integers from our study of whole numbers in Chapter 1. To understand what it means to add two negative numbers, consider the total yardage of a football team that loses 4 yards on one play and then loses 2 yards on the next play. For the two plays, the team has lost a total of 6 yards. This is written mathematically as $-4 + (-2) = -6$. To find the sum of two **negative** numbers, we can *add the absolute values of the numbers* and then keep the **negative** sign in the sum. For example, since $|-4| + |-2| = 4 + 2 = 6$, we can compute the related sum as $-4 + (-2) = -6$.

Because the absolute values of positive numbers are positive, a similar rule can be written for adding two positive numbers. To find the sum of two **positive** numbers, we *add the absolute values of the numbers* and then keep the **positive** sign in the sum.

READING CHECK

• How do we add integers with like signs?

ADDING INTEGERS WITH LIKE SIGNS

To add two integers with like signs,

1. Find the sum of the absolute values of the integers.
2. Keep the common sign of the two integers as the sign of the sum.

EXAMPLE 1   **Finding the sum of two integers with like signs**

Find each sum.
(a) $-7 + (-11)$ **(b)** $38 + 9$ **(c)** $-15 + (-26)$

Solution
(a) Start by finding the sum of the absolute values.

$$\left|-7\right| + \left|-11\right| = 7 + 11 = 18$$

Because the common sign is **negative**, the sum is $-7 + (-11) = -\mathbf{18}$.

(b) From our work with whole numbers, we know that $38 + 9 = 47$. However, the rules for adding integers with like signs can be applied to get the same result.

$$\left|38\right| + \left|9\right| = 38 + 9 = 47$$

Because the common sign is **positive**, the sum is $38 + 9 = \mathbf{47}$.

(c) Because $\left|-15\right| + \left|-26\right| = 15 + 26 = 41$ and the common sign is **negative**, the sum is $-15 + (-26) = -\mathbf{41}$.

Now Try Exercises 11, 13, 15

ADDING INTEGERS WITH UNLIKE SIGNS To gain an understanding of how to find the sum of two integers with unlike signs, consider the following pattern. What values should replace the question marks to continue the pattern?

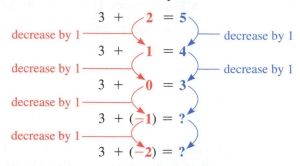

The pattern suggests that $3 + (-1) = \mathbf{2}$ and $3 + (-2) = \mathbf{1}$. If a football team gains 3 yards and then loses 1 yard, the total yardage for the two plays is 2 yards. Similarly, if a team gains 3 yards and then loses 2 yards, the total yardage for the two plays is 1 yard.

Continuing the pattern results in the following:

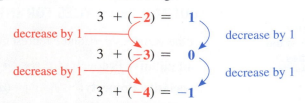

Therefore, if a football team gains 3 yards and then loses 3 yards, the total yardage for the two plays is 0 yards. If a team gains 3 yards and then loses 4 yards, the total yardage for the two plays is -1 yard.

This discussion suggests the following method for adding two integers with unlike signs.

ADDING INTEGERS WITH UNLIKE SIGNS

To add two integers with unlike signs,

1. Find the absolute values of the integers.
2. Subtract the smaller absolute value from the larger absolute value.
3. Keep the sign of the integer with the larger absolute value as the sign of the sum.

NOTE: If the absolute values (Step 1) are equal, the sum is 0. For example, $-8 + 8 = 0$.

EXAMPLE 2 **Finding the sum of two integers with unlike signs**

Find each sum.
(a) $-12 + 4$ **(b)** $14 + (-9)$

Solution
(a) STEP 1: Find the absolute values of -12 and 4.

$$|-12| = 12 \quad \text{and} \quad |4| = 4$$

STEP 2: Subtract the smaller absolute value 4 from the larger absolute value 12.

$$12 - 4 = 8$$

CALCULATOR HELP

To add integers with a calculator, see the Appendix (page AP-2).

STEP 3: Because $|-12| > |4|$, keep the **negative** sign as the sign for the sum.

$$-12 + 4 = \mathbf{-8}$$

(b) STEP 1: The absolute values are $|14| = 14$ and $|-9| = 9$.
STEP 2: Subtract $14 - 9 = 5$.
STEP 3: Because $|14| > |-9|$, keep the **positive** sign in the sum, $14 + (-9) = \mathbf{5}$.

▌**Now Try Exercises 17, 19**

Addition Properties

In Section 2.1, we learned that the opposite of an integer is also called the *additive inverse*. More formally, we say that two numbers are **additive inverses** of each other if their sum is 0. For example, -7 is the additive inverse of 7 because $-7 + 7 = 0$, and 7 is the additive inverse of -7 because $7 + (-7) = 0$. The **inverse property for addition** states that the sum of a number and its additive inverse (opposite) is always 0.

Because the addition properties for whole numbers studied in Section 1.2 also apply to integers, we can create a complete list of addition properties for integers, summarized as follows:

READING CHECK

• Using integers, give an example of each of the addition properties.

ADDITION PROPERTIES FOR INTEGERS

Let the variables a, b, and c represent integers.

Commutative Property:	$a + b = b + a$
Associative Property:	$(a + b) + c = a + (b + c)$
Identity Property:	$a + 0 = a \quad \text{and} \quad 0 + a = a$
Inverse Property:	$a + (-a) = 0 \quad \text{and} \quad -a + a = 0$

<table>
<tr><td>**EXAMPLE 3**</td></tr>
</table>

Identifying addition properties

State the addition property illustrated by each equation.
(a) $-99 + 0 = -99$ **(b)** $4 + (-3) = -3 + 4$
(c) $-65 + 65 = 0$ **(d)** $(-2 + 4) + 1 = -2 + (4 + 1)$

Solution
(a) The sum of an integer and 0 is that integer. This illustrates the identity property.
(b) The commutative property is illustrated by this equation. Changing the order of the addends does not affect the sum.
(c) This is the inverse property. Adding an integer and its opposite results in 0.
(d) The way in which three or more addends are grouped does not affect the sum. This equation illustrates the associative property.

Now Try Exercises 31, 33, 35, 37

Adding Integers Visually

In this subsection, we explore two visual methods that can be used to add integers. The first method involves the use of a number line. The second method uses the symbol ⌒ to represent a positive unit and the symbol ⌣ to represent a negative unit.

ADDING INTEGERS USING A NUMBER LINE On a number line, we can represent a positive integer with an arrow pointing to the right and a negative integer with an arrow pointing to the left. The length of the arrow is equal to the absolute value of the integer. For example, 3 can be represented with an arrow that is 3 units long (absolute value of 3 is 3) and points to the right. Similarly, -4 can be represented with an arrow that is 4 units long (absolute value of -4 is 4) and points to the left. See Figure 2.9.

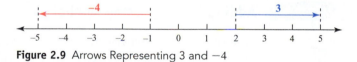

Figure 2.9 Arrows Representing 3 and -4

NOTE: The position of an arrow above the number line is not important. Only the length and direction of an arrow are needed to represent an integer.

To add integers using a number line, start at 0 and draw an arrow that represents the first addend. From the tip of this arrow, draw an arrow that represents the second addend. The sum is located at the tip of the second arrow. This process is illustrated in the next example.

<table>
<tr><td>**EXAMPLE 4**</td></tr>
</table>

Adding integers using a number line

Use a number line to find each sum.
(a) $-2 + 5$ **(b)** $-1 + (-3)$ **(c)** $4 + (-4)$

Solution
(a) To add $-2 + 5$, start at 0 and draw an arrow representing -2. From the tip of this arrow, draw a second arrow representing 5. The sum is **3**, as shown in Figure 2.10.

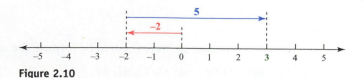

Figure 2.10

(b) To add $-1 + (-3)$, start at 0 and draw an arrow representing -1. From the tip of this arrow, draw a second arrow representing -3. The sum is -4, as shown in Figure 2.11.

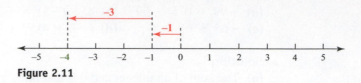

Figure 2.11

(c) The sum $4 + (-4)$ is equal to **0**, as shown in Figure 2.12.

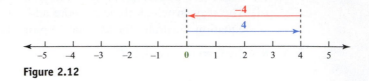

Figure 2.12

▌ **Now Try Exercises 55, 57, 61**

ADDING INTEGERS USING SYMBOLS If we use the symbol ⌢ to represent a positive unit and the symbol ⌣ to represent a negative unit, then adding opposites results visually in "zero" as shown.

$$\frown + \smile = \bigcirc$$

This is a visual representation of the equation $1 + (-1) = 0$. By combining positive units with negative units, it becomes possible to *see* the sum when two integers are added. The next example illustrates this process.

EXAMPLE 5   **Adding integers using symbols**

Perform each addition visually, using the symbols ⌢ and ⌣.
(a) $2 + 4$ **(b)** $-5 + 2$ **(c)** $7 + (-3)$ **(d)** $-2 + (-3)$

Solution
(a) Draw two positive units and then draw four more positive units.

⌢ ⌢   ⌢ ⌢ ⌢ ⌢

Because no "zeros" could be formed, the sum is six positive units, or 6.
(b) Start by drawing five negative units and then draw two positive units. Remember to form "zeros" when possible.

◯ ◯ ⌣ ⌣ ⌣

The "zeros" add no value and can be ignored. The sum is three negative units, or -3.
(c) Draw seven positive units and then draw three negative units. Form "zeros" when possible.

◯ ◯ ◯ ⌢ ⌢ ⌢ ⌢

Ignoring the three "zeros" that were formed, the sum is four positive units, or 4.
(d) Draw two negative units and then draw three negative units.

⌣ ⌣   ⌣ ⌣ ⌣

Because no "zeros" could be formed, the sum is five negative units, or -5.

▌ **Now Try Exercises 63, 65, 67, 69**

READING CHECK

• What number is repre-
sented when a positive
unit and a negative unit
are written together?

Applications

▶ **REAL-WORLD CONNECTION** At the start of this section, we saw how football teams compute total yardage by adding positive and negative integers. The next two examples illustrate other real-world situations that involve adding integers.

| EXAMPLE 6 | **Computing temperature change** |

One of the quickest and most dramatic temperature changes on record in the United States occurred in Great Falls, Montana on January 11, 1980. In just 7 minutes, the temperature increased by 47°F. If the temperature was −32°F before the increase, what was the temperature after the change? (*Source: Montana Almanac.*)

Solution
To find the temperature after a 47°F increase, we must add 47 to −32.

STEP 1: The absolute values are $|-32| = 32$ and $|47| = 47$.
STEP 2: Subtract $47 - 32 = 15$.
STEP 3: Because $|47| > |-32|$, keep the **positive** sign in the sum $-32 + 47 = $ **15**.

The temperature increased to 15°F.

Now Try Exercise 71

| EXAMPLE 7 | **Finding underground depth** |

The activity of exploring caves is known as *spelunking*. If two spelunkers are 135 feet below ground level and they descend an additional 42 feet, what is their new position relative to ground level?

Solution
First, we use the integer 0 to represent ground level. A position of 135 feet **below** ground level can be represented by the integer **− 135**. To find the final position after a **descent** of an additional 42 feet, we must add **− 42**. Start by finding the sum of the absolute values.

$$|-135| + |-42| = 135 + 42 = 177$$

Because the common sign is **negative**, the sum is $-135 + (-42) = $ **− 177**. The spelunkers are located 177 feet below ground level.

Now Try Exercise 75

| 2.2 | **Putting It All Together** |

CONCEPT	COMMENTS	EXAMPLES
Adding Integers with Like Signs	1. Find the sum of the absolute values of the integers. 2. Keep the common sign of the two integers as the sign of the sum.	Because $\|-5\| + \|-6\| = 11$, $-5 + (-6) = -11$. Because $\|3\| + \|12\| = 15$, $3 + 12 = 15$.

continued on next page

continued from previous page

CONCEPT	COMMENTS	EXAMPLES
Adding Integers with Unlike Signs	1. Find the absolute values of the integers. 2. Subtract the smaller absolute value from the larger absolute value. 3. Keep the sign of the integer with the larger absolute value as the sign of the sum.	To add $-7 + 11$, find the absolute values: $\|-7\| = 7$ and $\|11\| = 11$. Subtract $11 - 7 = 4$. Because $\|11\| > \|-7\|$, the sum is positive. So, $-7 + 11 = 4$.
Addition Properties for Integers	1. Commutative Property 2. Associative Property 3. Identity Property 4. Inverse Property	1. $3 + (-2) = -2 + 3$ 2. $(-2 + 1) + 3 = -2 + (1 + 3)$ 3. $-1 + 0 = -1$ and $0 + 2 = 2$ 4. $2 + (-2) = 0$ and $-5 + 5 = 0$
Adding Integers Using a Number Line	Starting at 0, draw an arrow representing the first addend. From the tip of this arrow, draw an arrow representing the second addend. The sum is located at the tip of the second arrow.	To add $-1 + 3$ using a number line, draw arrows as shown. The sum is **2**.
Adding Integers Using Symbols	Using the symbol ⌢ to represent a positive unit and the symbol ⌣ to represent a negative unit, draw the appropriate number of units for each addend. Form "zeros" when possible.	Add $3 + (-5)$ as shown. Ignoring the three "zeros," the sum is -2.

2.2   Exercises

MyMathLab   Math XL PRACTICE   WATCH   DOWNLOAD   READ   REVIEW

CONCEPTS AND VOCABULARY

1. The first step when adding two integers with like signs is to find the sum of the _____ of the integers.

2. When adding two negative integers, what is the sign of the sum?

3. When adding the integers $-1324 + 5678$, what is the sign of the sum?

4. When adding the integers $-32,264 + 11,902$, what is the sign of the sum?

5. Two numbers are _____ of each other if their sum is 0.

6. The sum of a number and its opposite is _____ .

7. When adding integers using a number line, an arrow pointing to the _____ represents a positive number, and an arrow pointing to the _____ represents a negative number.

8. When adding two integers visually using symbols, the symbol _____ represents a positive unit, and the symbol _____ represents a negative unit.

ADDING INTEGERS

Exercises 9–30: Find the sum.

9. $3 + 9$

10. $7 + 12$

11. $-5 + (-7)$

12. $-8 + (-2)$

13. $13 + 28$

14. $33 + 21$

15. $-25 + (-17)$ **16.** $-30 + (-24)$

17. $-28 + 13$ **18.** $-31 + 17$

19. $35 + (-12)$ **20.** $50 + (-30)$

21. $39 + (-39)$ **22.** $47 + (-47)$

23. $-100 + 139$ **24.** $-75 + 150$

25. $61 + (-62)$ **26.** $77 + (-79)$

27. $-33 + (-33)$ **28.** $-41 + (-41)$

29. $-143 + 0$ **30.** $0 + (-78)$

ADDITION PROPERTIES

Exercises 31–38: State the addition property illustrated by the given equation.

31. $13 + (-56) = -56 + 13$

32. $0 + (-1289) = -1289$

33. $347 + (-347) = 0$

34. $13 + (-18 + 47) = (13 + (-18)) + 47$

35. $(-19 + 7) + 43 = -19 + (7 + 43)$

36. $-457 + 457 = 0$

37. $-671 + 0 = -671$

38. $-72 + 561 = 561 + (-72)$

Exercises 39–48: The associative and commutative properties for addition allow for three or more integers to be added in any order. Find the given sum.

39. $-5 + 3 + (-2)$ **40.** $4 + 8 + (-4)$

41. $-1 + (-9) + (-7)$

42. $-4 + (-8) + 12$ **43.** $-7 + (-17) + 24$

44. $-11 + 9 + (-7)$

45. $-18 + 53 + 29$ **46.** $34 + (-51) + 38$

47. $-31 + (-29) + (-47) + 62$

48. $111 + (-15) + (-152) + 68$

Exercises 49–54: Evaluate the expression $x + y$ for the given values of the variables.

49. $x = -12, y = -4$

50. $x = -2, y = 19$

51. $x = 27, y = -14$

52. $x = 32, y = 22$

53. $x = 0, y = -93$

54. $x = -65, y = 1$

ADDING VISUALLY

Exercises 55–62: Use a number line to find the sum.

55. $3 + (-7)$ **56.** $-6 + 8$

57. $-5 + 9$ **58.** $-2 + (-3)$

59. $-4 + (-1)$ **60.** $4 + (-7)$

61. $-5 + 5$ **62.** $9 + (-9)$

Exercises 63–70: Perform the addition visually, using the symbols $\frown$ and $\smile$.

63. $3 + 1$ **64.** $5 + 4$

65. $4 + (-6)$ **66.** $-7 + (-3)$

67. $-4 + (-5)$ **68.** $-4 + 2$

69. $-2 + 7$ **70.** $8 + (-2)$

APPLICATIONS

71. *Temperature Change* In 1972, Loma, Montana set a record for the largest 24-hour temperature swing ever recorded in the United States. On January 14, the temperature was $-54°F$ and increased $103°F$ by the next day. What was the temperature after this change? (*Source: Montana Almanac.*)

72. *Temperature Change* In 1916, the second largest 24-hour temperature swing ever recorded in the United States took place in Browning, Montana. On January 23, the temperature was $44°F$ and dropped $100°F$ by the next day. What was the temperature after this change? (*Source: Montana Almanac.*)

73. *Football Stats* A running back carries the ball four times. Find his total yardage for the four plays if the yardages were $-1, -3, 13,$ and 4 yards.

74. *Game Tokens* A game is played with red and blue tokens. If each red token represents -1 point and each blue token represents $+1$ point, what is the total point value of 7 red tokens and 4 blue tokens?

75. *Underground Depth* If a coal miner is 203 feet below ground level and then descends 816 feet, what is the miner's new position relative to ground level?

76. *Underwater Depth* If a diver who is 97 feet below sea level ascends 56 feet, what is the diver's new position relative to sea level?

77. *Corporate Losses* For the fourth quarter of 2009, DirecTV reported earnings of −$32 million. For the same quarter, Delta Airlines reported earnings that were $7 million higher than those of DirecTV. What were Delta Airline's earnings for the fourth quarter of 2009? (*Source: New York Times.*)

78. *Geography* The elevation of the highest point in Florida, Britton Hill, is 627 feet higher than the elevation of Death Valley, which is −282 feet. What is the elevation of Britton Hill? (*Source: The World Atlas.*)

79. *Finances* A student's savings account starts the month with a balance of $3534. The following positive entries (deposits) and negative entries (withdrawals) are made in the savings register.

$$-282, 445, 390, \text{ and } -1598$$

What is the ending balance?

80. *Finances* A student's checking account starts the month with a balance of $617. The following positive entries (deposits) and negative entries (withdrawals) are made in the checking register.

$$-17, -120, 200, \text{ and } -40$$

What is the ending balance?

WRITING ABOUT MATHEMATICS

81. Give two examples of real-world situations involving the addition of positive and negative integers.

82. Explain how absolute value is used when adding two negative integers.

83. Explain how to find the sum of two integers with like signs. Give two examples.

84. Explain how to find the sum of two integers with unlike signs. Give two examples.

SECTIONS 2.1 and 2.2

Checking Basic Concepts

1. Find the opposite of each number.
   (a) 23   **(b)** −16

2. Simplify each expression.
   (a) −(−52)   **(b)** −(−(−9))

3. Graph the integers −3, 0, and 4 on the same number line.

4. Place the correct symbol, < or >, in each blank between the integers.
   (a) 67 _____ −68   **(b)** 0 _____ −10,003

5. Evaluate each absolute value.
   (a) $|17|$ **(b)** $|-31|$

6. Find each sum.
   (a) −14 + 22   **(b)** −27 + (−8)
   (c) 4 + (−25)   **(d)** 52 + 31

7. Use a number line to find each sum.
   (a) −4 + 8   **(b)** 3 + (−9)

8. Perform the addition visually, using the symbols ⌢ and ⌣.
   (a) −6 + 8   **(b)** 3 + (−8)

9. Use absolute value to determine which of the two given financial quantities is larger.

   A credit card debt: −$420
   Checking account balance: $380

10. *Geography* The elevation of the highest point in Alabama, Cheaha Mountain, is 3783 feet higher than the elevation of the Dead Sea basin, which is −1378 feet. What is the elevation of Cheaha Mountain? (*Source: The World Atlas.*)

2.3 | Subtracting Integers

Subtracting Integers • Adding and Subtracting • Subtracting Integers Visually • Applications

A LOOK INTO MATH ▶ Elevation, temperature, and account balances are all examples of real-world quantities that may involve finding a difference between positive and negative numbers. For example, if a submarine is 235 feet below sea level at a location where the ocean is 640 feet deep, then we can find the distance between the submarine and the bottom of the ocean by subtracting $-235 - (-640)$. In this section, we learn about a process for finding such differences involving integers.

Subtracting Integers

One of the identity properties for subtraction in Chapter 1 states that subtracting a number from itself results in 0. For example, $8 - 8 = 0$ and $32 - 32 = 0$. Similarly, the inverse property for addition from Section 2.2 states that the sum of a number and its additive inverse (opposite) is 0. That is, $8 + (-8) = 0$ and $32 + (-32) = 0$. Together, these properties suggest that we can subtract one number from another by adding the first number to the opposite of the second number.

$$12 - 4 = 8 \qquad 100 - 45 = 55 \qquad 87 - 85 = 2$$
$$12 + (-4) = 8 \qquad 100 + (-45) = 55 \qquad 87 + (-85) = 2$$

This procedure can be used to find a difference regardless of the signs of the two numbers involved. It is common to say that we subtract integers by *adding the opposite*.

> ### SUBTRACTING INTEGERS
>
> If the variables a and b represent two numbers, then
> $$a - b = a + (-b).$$

EXAMPLE 1 | **Subtracting integers by adding the opposite**

Find each difference.
(a) $7 - (-6)$ **(b)** $-13 - 8$ **(c)** $-4 - (-11)$

Solution
(a) Rather than subtracting -6, add its opposite, 6.
$$7 - (-6) = 7 + 6 = 13$$
(b) Instead of subtracting 8, add its opposite, -8.
$$-13 - 8 = -13 + (-8) = -21$$
(c) Rather than subtracting -11, add its opposite, 11.
$$-4 - (-11) = -4 + 11 = 7$$

Now Try Exercises 13, 17, 23

Although the procedure used in Example 1 can always be used to find a difference, it is important to remember that some subtraction problems can be done easily without adding the opposite. The next example illustrates this.

STUDY TIP

If you miss something in class, the **Video Lectures on CD or DVD*** provide a short lecture for each section in this text. These lectures, taught by actual math instructors, offer you the opportunity to review topics that you may not have fully understood before.

*www.mypearsonstore.com

READING CHECK

• How do we subtract integers?

CALCULATOR HELP

To subtract integers with a calculator, see the Appendix (page AP-3).

> **EXAMPLE 2**   **Subtracting integers without adding the opposite**
>
> Subtract $22 - 17$.
>
> **Solution**
> Because both numbers are whole numbers with the second number smaller than the first, this difference can be found using simple subtraction as discussed in Section 1.2. The difference is $22 - 17 = 5$. Adding the opposite is not necessary in this case.
>
> **Now Try Exercise 9**

Adding and Subtracting

Some arithmetic expressions involve a mix of addition and subtraction. When we change each subtraction to the addition of the opposite, the result is an expression that includes only addition. Since both the commutative and associative properties apply to addition, we can rearrange addends. The next example illustrates this process.

> **EXAMPLE 3**   **Adding and subtracting integers**
>
> Simplify each expression.
> **(a)** $10 - (-3) + 5 - 1$ **(b)** $4 - 18 + 6 - 2$
>
> **Solution**
> **(a)** Begin by changing each subtraction to the addition of the opposite.
>
> $$10 - (-3) + 5 - 1 = 10 + 3 + 5 + (-1) \quad \text{Add the opposite.}$$
> $$= 13 + 5 + (-1) \quad \text{Add: } 10 + 3 = 13.$$
> $$= 18 + (-1) \quad \text{Add: } 13 + 5 = 18.$$
> $$= 17 \quad \text{Add.}$$

READING CHECK

• Why is it useful to change subtraction to addition of the opposite?

> **(b)** After changing each subtraction to the addition of the opposite, rearrange the addends and find the sum by applying the commutative and associative properties.
>
> $$4 - 18 + 6 - 2 = 4 + (-18) + 6 + (-2) \quad \text{Add the opposite.}$$
> $$= 4 + 6 + (-18) + (-2) \quad \text{Rearrange addends.}$$
> $$= 10 + (-20) \quad \text{Add.}$$
> $$= -10 \quad \text{Add.}$$
>
> **Now Try Exercises 39, 43**

Subtracting Integers Visually

Integers can be subtracted visually in two ways that are similar to those used for addition. Rather than using arrows on a number line to represent integers, a stepping process can be used to subtract integers using a number line. Also, the symbols $\frown$ and $\smile$ can be used to perform subtraction visually.

SUBTRACTING INTEGERS USING A NUMBER LINE When using a number line to subtract integers, it is important to recognize the difference between a subtraction symbol and a negative sign. In a printed text, a subtraction symbol has a space both before and after it,

READING CHECK

• How can you tell subtraction symbols and negative signs apart in a printed text?

but a negative sign has a number (or variable) immediately to its right. For example, in the expression $-3 - 10$, the first $(-)$ is a negative sign and the second $(-)$ is a subtraction symbol. Calculators have different keys for these operations.

When reading an arithmetic expression, the left-most number indicates the starting position on the number line. After the left-most number, we use the following interpretations for any signs or symbols that occur.

An **addition** symbol means **step to the right** on the number line.

A **subtraction** symbol means **step to the left** on the number line.

A **negative** sign means **stop and change direction**.

To review these symbols, see Making Connections on page 90 in Section 2.1. Examples of these interpretations are as follows:

$-2 - 5$:   Start at -2 on the number line and **step to the left** 5 units.

$4 + 3$:   Start at 4 on the number line and **step to the right** 3 units.

$2 + (-7)$:   Start at 2 on the number line, **step to the right**, **stop and change direction**. **Step to the left** 7 units.

$-5 - (-6)$:   Start at -5 on the number line, **step to the left**, **stop and change direction**. **Step to the right** 6 units.

Reading expressions in this way allows us to read the entire expression from left to right without changing from subtraction to the addition of the opposite. The next example illustrates this process.

EXAMPLE 4   **Subtracting integers using a number line**

Use a number line to find each difference.
(a) $3 - 5$ **(b)** $-2 - 3$ **(c)** $-1 - (-4)$

Solution
(a) To subtract $3 - 5$, start at 3 on the number line and **step to the left** 5 units. The difference is -2, as shown in Figure 2.13.

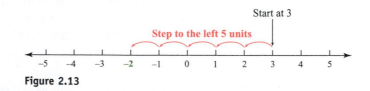

Figure 2.13

(b) To subtract $-2 - 3$, start at -2 on the number line and **step to the left** 3 units. The difference is -5, as shown in Figure 2.14.

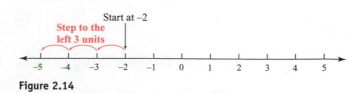

Figure 2.14

(c) To subtract $-1 - (-4)$, start at -1 on the number line, **step to the left**, **stop and change direction**. **Step to the right** 4 units. The difference is **3**, as shown in Figure 2.15.

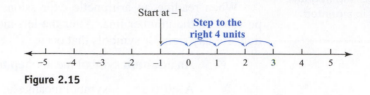

Figure 2.15

Now Try Exercises 45, 47, 49

| EXAMPLE 5 | **Adding and subtracting integers using a number line** |

Use a number line to simplify the expression $-2 + 5 - (-1) + (-7)$.

Solution
To simplify $-2 \; \mathbf{+} \; 5 \; \mathbf{-} \; (-1) \; \mathbf{+} \; (-7)$, start at -2 on the number line. The first addition symbol means that we **step to the right** 5 units to reach 3. From there, we **step to the left**, **stop and change direction**, and then **step to the right** 1 unit to reach 4. Finally, we **step to the right**, **stop and change direction**, and **step to the left** 7 units to reach -3. The expression simplifies to **-3**, as shown in Figure 2.16.

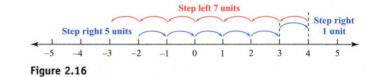

Figure 2.16

Now Try Exercise 53

SUBTRACTING INTEGERS USING SYMBOLS To subtract integers visually using symbols, we again use ⌒ to represent a positive unit and ⌣ to represent a negative unit. Subtraction occurs by "taking away" the appropriate number of positive or negative units from the visual representation. We cross out units to indicate that they have been taken away. For example, to show that $7 - 3 = 4$, we start by drawing seven positive units. Crossing out three positive units results in 4 positive units.

Differences that involve integers with unlike signs such as $3 - (-5)$ pose a special problem. If we start by drawing three positive units, how can we cross out five negative units? To do this, consider that 3 can be written as a sum of 3 and *any number* of zeros. For the difference $3 - (-5)$, we add five zeros to 3 because that is the number of negative symbols we need to cross out. That is, $3 = 3 + 0 + 0 + 0 + 0 + 0$. The difference $3 - (-5)$ is simplified in part (b) of the next example.

| EXAMPLE 6 | **Subtracting integers using symbols** |

Perform each subtraction visually, using the symbols ⌒ and ⌣.
(a) $-6 - (-2)$ **(b)** $3 - (-5)$ **(c)** $-3 - (-7)$

Solution

(a) To subtract $-6 - (-2)$, draw six negative units and then cross out two of them.

The difference is four negative units, or -4.

(b) To subtract $3 - (-5)$, start by drawing three positive units.

Because there are no negative units available to cross out and five are needed, insert five "zeros" in the visual representation of 3. This new representation is also equal to 3.

Now cross out five negative units to result in eight positive units, or 8.

(c) To subtract $-3 - (-7)$, start by drawing three negative units.

Because there are only three negative units available to cross out and seven are needed, insert four "zeros" in the visual representation of -3. This new representation is also equal to -3.

Now cross out seven negative units to result in four positive units, or 4.

▌ **Now Try Exercises 59, 61, 63**

Applications

▶ **REAL-WORLD CONNECTION** At the start of this section, we considered finding the distance between a submarine's location and the ocean floor. We solve this problem in the next example.

EXAMPLE 7 **Analyzing the depth of a submarine**

If a submarine is 235 feet below sea level at a location where the ocean is 640 feet deep, find the distance between the submarine and the bottom of the ocean.

Solution

The distance is found by subtracting: $-235 - (-640)$. Rather than subtracting -640, add its opposite, or 640. The related sum is $-235 + 640$. Find this sum as follows:

STEP 1: The absolute values are $|-235| = 235$ and $|640| = 640$.
STEP 2: Subtract $640 - 235 = 405$.
STEP 3: Because $|640| > |-235|$, keep the **positive** sign in the sum $-235 + 640 = $ **405**.

The submarine is 405 feet from the bottom of the ocean.

▌ **Now Try Exercise 69**

EXAMPLE 8   **Analyzing an account balance**

An overdrawn checking account has a balance of $-\$413$. An amount of money is deposited to bring the balance to 776. How much money has been deposited?

Solution

To determine the amount deposited, we must find the difference $776 - (-413)$. Rather than subtracting -413, we add its opposite, or 413. The related sum is $776 + 413$, which is equal to 1189. The amount deposited is $\$1189$.

Now Try Exercise 73

2.3 Putting It All Together

CONCEPT	COMMENTS	EXAMPLES
Subtracting Integers	If the variables a and b represent two numbers, then $a - b = a + (-b)$.	$3 - (-1) = 3 + 1 = 4$ $-6 - 8 = -6 + (-8) = -14$
Subtracting Integers Using a Number Line	Interpret symbols and signs as follows. **Addition** means **step to the right**. **Subtraction** means **step to the left**. A **negative** sign means **stop and change direction**.	To subtract $-1 - (-2)$, start at -1, **step to the left**, **stop and change direction**, and then **step to the right** 2 units. The difference is **1**. Start at −1 Step to the right 2 units −3 −2 −1 0 1 2 3
Subtracting Integers Using Symbols	Use ◯ to represent a positive unit and ◡ to represent a negative unit.	Subtract: $-2 - (-5)$. The difference is 3.

2.3 Exercises

MyMathLab   Math XL PRACTICE   WATCH   DOWNLOAD   READ   REVIEW

CONCEPTS AND VOCABULARY

1. To subtract $a - b$, add a and the _____ of b.

2. $4 - 7 = 4 +$ _____

3. $-2 - (-9) = -2 +$ _____

4. When doing arithmetic on a number line, an addition symbol means _____.

5. What does a subtraction symbol mean when doing arithmetic on a number line?

6. When doing arithmetic on a number line, a negative sign means _____.

7. In a printed text, which has space both before and after it, a subtraction symbol or a negative sign?

8. (True or False?) $8 = 8 + 0 + 0 + 0 + 0$.

SUBTRACTING INTEGERS

Exercises 9–28: Find the difference.

9. $8 - 2$

10. $12 - 5$

11. $13 - 18$

12. $22 - 25$

13. $-10 - 5$

14. $-20 - 7$

15. $-25 - 17$

16. $-24 - 24$

17. $21 - (-6)$

18. $33 - (-10)$

19. $5 - (-24)$

20. $11 - (-29)$

21. $-14 - (-9)$

22. $-40 - (-12)$

23. $-21 - (-29)$

24. $-17 - (-33)$

25. $34 - 0$

26. $-28 - 0$

27. $0 - (-52)$

28. $0 - 75$

Exercises 29–34: Evaluate the expression $x - y$ for the given values of the variables.

29. $x = -8, y = -17$

30. $x = -3, y = 20$

31. $x = 30, y = -15$

32. $x = 19, y = 43$

33. $x = -70, y = -3$

34. $x = -48, y = 1$

ADDING AND SUBTRACTING INTEGERS

Exercises 35–44: Simplify the expression.

35. $-3 - 3 - (-4)$

36. $5 - 9 - (-3)$

37. $-1 - (-9) - (-5)$

38. $-4 - (-8) - 12$

39. $-2 + (-3) - (-9)$

40. $4 - (-15) + 10$

41. $-17 - 7 + (-7)$

42. $-14 + 2 - (-8)$

43. $-31 + (-16) - (-40) + 21$

44. $-111 - (-99) + (-55) + 68$

SUBTRACTING VISUALLY

Exercises 45–52: Use a number line to find the difference.

45. $5 - 9$

46. $4 - 11$

47. $-4 - (-7)$

48. $-2 - (-8)$

49. $-3 - 3$

50. $5 - 5$

51. $-4 - (-4)$

52. $2 - (-2)$

Exercises 53–56: Use a number line to simplify the given expression.

53. $-3 + (-2) - (-8) + 1$

54. $5 - 9 + (-1) - (-6)$

55. $3 + (-8) + 7 - (-3)$

56. $-2 - (-5) - (-2) - 5$

Exercises 57–64: Perform the subtraction visually, using the symbols ⌢ and ⌣.

57. $2 - 7$

58. $4 - 8$

59. $-9 - (-7)$

60. $-9 - (-2)$

61. $4 - (-6)$

62. $1 - (-7)$

63. $-2 - (-9)$

64. $-5 - (-6)$

APPLICATIONS

65. *Temperature Change* A traveler leaves South Padre Island, where the temperature is 78°F, and flies to Chicago, where the temperature is −9°F. What is the temperature difference for this trip?

66. *Liquid Nitrogen* In some medical procedures, a small amount of liquid nitrogen is applied to the skin. If normal skin temperature is 91°F and the temperature of liquid nitrogen is −321°F, then find the temperature difference.

67. *Elevations in California* The elevation of Mount Whitney is 14,494 feet, and the elevation of Death Valley is −282 feet. Find the elevation difference between these California landmarks. (*Source:* U.S. Geological Survey.)

68. *Elevations in Louisiana* The elevation of Driskill Mountain is 535 feet, and an elevation found in New Orleans is −8 feet. Find the elevation difference between these Louisiana locations. (*Source:* U.S. Geological Survey.)

69. *Underwater Depth* If a diver is 37 feet below sea level at a location where the ocean is 52 feet deep, find the distance between the diver and the bottom of the ocean.

70. *Underwater Depth* If the highest point on an underwater structure is 26 feet below sea level and the ocean is 73 feet deep at the base of the structure, how tall is the structure?

71. *Net Worth* A student with a net worth of $-\$35,600$ inherits an amount of money so that his net worth becomes $\$153,800$. How much money did he inherit?

72. *Game Tokens* In a game, each red token represents -1 point and each blue token represents $+1$ point. How many more points does a player with 4 blue tokens have than a player with 5 red tokens?

73. *Account Balance* An overdrawn checking account has a balance of $-\$47$. An amount of money is deposited to bring the balance to $\$129$. How much money has been deposited?

74. *Account Balance* An overdrawn checking account has a balance of $-\$55$. An amount of money is deposited to bring the balance to $\$217$. How much money has been deposited?

75. *Temperature Change* Over a four-day period in January, a city in Michigan had the following weather changes. The low temperature on Monday was $-6°F$. On Tuesday, the low temperature increased by $8°F$. The low temperature then dropped $5°F$ on Wednesday. Finally, the low temperature decreased by another $7°F$ on Thursday. What was the low temperature recorded on Thursday?

76. *Sea Diving* A diver located 54 feet below sea level swims up 17 feet and then dives down 26 feet. Write the final position of the diver relative to sea level as a negative integer.

WRITING ABOUT MATHEMATICS

77. Give two examples of real-world situations that involve the subtraction of positive and negative integers.

78. Give an example of a temperature or elevation difference that requires a negative number to be subtracted from a positive number. Explain why subtraction can be accomplished by "adding the opposite."

2.4 Multiplying and Dividing Integers

Multiplying Integers • Multiplication Properties • More Than Two Factors • Exponential Expressions • Dividing Integers • Square Roots • Applications

A LOOK INTO MATH ▶

If an all-inclusive vacation package for a Disney World trip costs $4800 and the trip's cost is divided evenly among three friends, then each friend is responsible for $4800 \div 3 = \$1600$ of the total cost. However, if we consider that the *cost* of the trip is actually a *debt* for the friends, written as -4800, then we can divide -4800 by 3 to find each friend's portion of the debt. Since we know that each friend's debt is $1600, written as -1600, we know that $-4800 \div 3 = -1600$. We see that dividing a negative number by a positive number results in a negative number. In this section, we will examine this result more closely and discuss other rules for multiplying and dividing integers.

NEW VOCABULARY

☐ Principal square root

Multiplying Integers

Finding the product of two integers is similar to finding the product of two whole numbers. Since integers may be positive or negative, the product of two integers may be positive or

READING CHECK

• When a positive integer and a negative integer are multiplied, is the product positive or negative?

READING CHECK

• When two negative integers are multiplied, is the product positive or negative?

CALCULATOR HELP

To multiply integers with a calculator, see the Appendix (page AP-3).

negative also. To find a general rule for multiplying integers, we can consider the following pattern. What values should replace the question marks to continue the pattern?

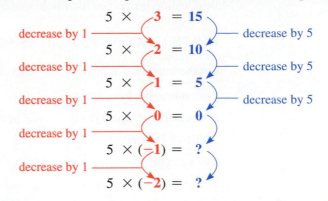

Continuing the pattern, we find that $5 \times (-1) = -5$ and $5 \times (-2) = -10$. This pattern suggests that if we multiply a positive integer and a negative integer, the product is negative. What can be said about the product of two negative integers? To answer this question, consider the following pattern. This time, what values should replace the question marks to continue the pattern?

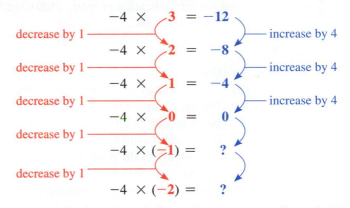

Continuing the pattern results in $-4 \times (-1) = 4$ and $-4 \times (-2) = 8$. This pattern suggests that if we multiply two negative integers, the product is positive.

We now have general rules for multiplying integers. To find the product of two integers, multiply their absolute values and use the following rules to find the sign of the product.

SIGNS OF PRODUCTS

The product of two integers with *like* signs is positive.

The product of two integers with *unlike* signs is negative.

EXAMPLE 1 **Multiplying integers**

Multiply.
(a) $-3 \cdot 8$ **(b)** $-7(-4)$ **(c)** $2 \times (-9)$

Solution
(a) The factors have *unlike* signs, so the product is negative: $-3 \cdot 8 = -24$.
(b) The factors have *like* signs, so the product is positive: $-7(-4) = 28$.
(c) The factors have *unlike* signs, so the product is negative: $2 \times (-9) = -18$.

Now Try Exercises 15, 23, 25

Multiplication Properties

Multiplication properties for whole numbers (Section 1.3) can also be applied to integers.

MULTIPLICATION PROPERTIES FOR INTEGERS

Let the variables a, b, and c represent integers.

Commutative Property:	$a \cdot b = b \cdot a$
Associative Property:	$(a \cdot b) \cdot c = a \cdot (b \cdot c)$
Identity Property:	$a \cdot 1 = a$ and $1 \cdot a = a$
Zero Property:	$a \cdot 0 = 0$ and $0 \cdot a = 0$
Distributive Properties:	$a(b + c) = a \cdot b + a \cdot c$
	$a(b - c) = a \cdot b - a \cdot c$

EXAMPLE 2 **Identifying multiplication properties**

State the multiplication property illustrated by each of the following.
(a) $-47 \cdot 1 = -47$ **(b)** $-3(2 - 8) = -3 \cdot 2 - (-3) \cdot 8$
(c) $13 \times (-9) = -9 \times 13$ **(d)** $(-2 \cdot 7) \cdot 5 = -2 \cdot (7 \cdot 5)$

Solution
(a) The product of an integer and 1 is that integer. This illustrates the identity property.
(b) The distributive property is illustrated by this equation. The integer outside the parentheses is multiplied by each integer inside the parentheses.
(c) This equation illustrates the commutative property. Changing the order of the factors does not affect the product.
(d) The way in which three or more factors are grouped does not affect the product. This equation illustrates the associative property.

Now Try Exercises 41, 43, 45, 47, 49

More Than Two Factors

Because integer multiplication is both commutative and associative, a product involving three or more factors can be rearranged so that all the positive factors are listed first. Doing so will help us discover a rule for finding the sign of a product involving several integers. For example, in the product

$$-2 \cdot 3 \cdot 2 \cdot (-1) \cdot 4 \cdot (-2),$$

we write the **positive factors** first and then group the negative factors into as many *pairs* as possible. Every **pair** of negative factors simplifies to a new positive factor. Any negative factor that **cannot be paired** with another negative factor remains negative.

$$\underbrace{3 \cdot 2 \cdot 4}_{\text{Positive}} \cdot \underbrace{(-2) \cdot (-1)}_{\text{Positive}} \cdot \underbrace{(-2)}_{\text{Negative}}$$

Simplifying further reveals that the final product is negative.

$$\underbrace{24 \cdot 2}_{\text{Positive}} \cdot (-2) = 48 \cdot (-2) = -96$$

READING CHECK

• When more than two integers are multiplied, how can we tell if the product is positive or negative?

This discussion suggests that to find the sign of a product involving more than two factors, we need only to determine if there are any *unpaired* negative factors. This means that we determine if there is an odd or even number of negative factors.

PRODUCTS WITH MORE THAN TWO FACTORS

If there is an *even* number of negative factors, the product is positive.

If there is an *odd* number of negative factors, the product is negative.

EXAMPLE 3   **Multiplying more than two factors**

Multiply.
(a) $-2 \cdot (-1) \cdot 4 \cdot (-5)$ (b) $-2 \cdot (-4) \cdot (-1) \cdot (-3)$

Solution
(a) Because there is an *odd* number of negative factors, the product is negative, and so $-2 \cdot (-1) \cdot 4 \cdot (-5) = -40$.
(b) Because there is an *even* number of negative factors, the product is positive, and so $-2 \cdot (-4) \cdot (-1) \cdot (-3) = 24$.

Now Try Exercises 33, 39

READING CHECK

• If we want to raise a negative integer to a power, what grouping symbol should be used?

Exponential Expressions

Is there a difference in how the expressions -5^2 and $(-5)^2$ are evaluated? To answer this question, it may be helpful to translate each expression into words. The first expression says "the opposite of five squared," which is "the opposite of 25," or -25. However, the second expression says "the square of negative five," which is $(-5)(-5) = 25$. The expressions do not give the same result. The base of the exponential expression -5^2 is **positive**, while the base of $(-5)^2$ is **negative**.

EXAMPLE 4   **Evaluating exponential expressions**

Evaluate each exponential expression.
(a) $(-2)^4$ (b) -3^3 (c) $(-4)^3$

Solution
(a) The base of the expression is negative. Because the exponent is 4, there is an even number of negative factors: $(-2)^4 = (-2) \cdot (-2) \cdot (-2) \cdot (-2) = 16$.
(b) The base of the expression is positive. First we find the cube of 3 and then find the opposite of the result: $-3^3 = -(3 \cdot 3 \cdot 3) = -27$.
(c) The base of the expression is negative. Because the exponent is 3, there is an odd number of negative factors: $(-4)^3 = (-4) \cdot (-4) \cdot (-4) = -64$.

Now Try Exercises 51, 53, 59

CALCULATOR HELP

To use a calculator to evaluate exponential expressions involving integers, see the Appendix (page AP-3).

Dividing Integers

Division and multiplication are closely related. For example, the equation $3 \cdot 4 = 12$ can be used to check the equation $12 \div 4 = 3$. Extending this idea to integers will reveal the sign rules for quotients. Consider the following:

$$3 \cdot 4 = 12 \text{ checks } 12 \div 4 = 3 \quad (\text{\textbf{Positive} divided by \textbf{positive} equals \textbf{positive}.})$$

$$3 \cdot (-4) = -12 \text{ checks } -12 \div (-4) = 3 \quad (\text{\textbf{Negative} divided by \textbf{negative} equals \textbf{positive}.})$$

$$-3 \cdot (-4) = 12 \text{ checks } 12 \div (-4) = -3 \quad (\text{\textbf{Positive} divided by \textbf{negative} equals \textbf{negative}.})$$

$$-3 \cdot 4 = -12 \text{ checks } -12 \div 4 = -3 \quad (\text{\textbf{Negative} divided by \textbf{positive} equals \textbf{negative}.})$$

READING CHECK

• When a positive integer is divided by a negative integer, is the quotient positive or negative?
• When a negative integer is divided by a negative integer, is the quotient positive or negative?

When the dividend and divisor have like signs, the quotient is positive, and when the dividend and divisor have unlike signs, the quotient is negative. This result is similar to the rules for the signs of products and is summarized as follows:

SIGNS OF QUOTIENTS

The quotient of two integers with *like* signs is positive.

The quotient of two integers with *unlike* signs is negative.

EXAMPLE 5 **Dividing integers**

Divide.

(a) $36 \div (-12)$ (b) $\dfrac{-60}{-5}$ (c) $-48 \div 6$

Solution
(a) The numbers have *unlike* signs, so the quotient is negative: $36 \div (-12) = -3$.
(b) The numbers have *like* signs, so the quotient is positive: $\dfrac{-60}{-5} = 12$.
(c) The numbers have *unlike* signs, so the quotient is negative: $-48 \div 6 = -8$.

Now Try Exercises 65, 67, 71

CALCULATOR HELP

To divide integers with a calculator, see the Appendix (page AP-3).

Recall that quotients involving 0 are sometimes undefined. The next example illustrates this and other properties of division.

EXAMPLE 6 **Dividing integers**

Divide, if possible. If a quotient is undefined, state so.

(a) $\dfrac{0}{-7}$ (b) $\dfrac{-12}{-12}$ (c) $-15 \div 0$ (d) $-24 \div 1$

Solution
(a) Dividing 0 by any number (except 0) results in 0: $\dfrac{0}{-7} = 0$.
(b) Dividing a number (except 0) by itself results in 1: $\dfrac{-12}{-12} = 1$.
(c) Any number divided by 0 is undefined: $-15 \div 0$ is undefined.
(d) Dividing a number by 1 results in the number itself: $-24 \div 1 = -24$.

Now Try Exercises 69, 73, 75, 77

READING CHECK

• How many square roots does every positive integer have?

Square Roots

In Section 1.5, we discovered that 7 is a square root of 49 because $7^2 = 49$. However, because the product of two negative integers is positive, we also know that $(-7)^2 = 49$. Does this mean that -7 is also a square root of 49? The answer is yes. The number 49 has two integer square roots, -7 and 7. In fact, *every positive integer has two square roots*. Notice that the square roots of a positive integer are opposites of each other.

EXAMPLE 7 **Finding square roots of integers**

Find all integer square roots of the given integer, if possible.
(a) 36 (b) -9

Solution
(a) Because $6^2 = 36$ and $(-6)^2 = 36$, the integer square roots of 36 are -6 and 6.
(b) The product of two integers with like signs is always positive. As a result, the square of any integer (other than 0) is always positive. Therefore, there is no integer whose square is -9. We say that -9 has *no integer square roots*.

Now Try Exercises 81, 85

Recall that the radical sign $\sqrt{}$ means to find the square root of a number. For example, $\sqrt{100} = 10$. However, this notation does not account for the negative square root of 100, or -10. To avoid confusion, we use the notation $\sqrt{100}$ when finding the positive square root or **principal square root** of 100. To find the negative square root of 100, we write $-\sqrt{100}$. For example, $\sqrt{100} = 10$ and $-\sqrt{100} = -10$.

READING CHECK

• What special name is given to the positive square root of an integer?

SQUARE ROOTS OF INTEGERS

A positive integer a has one positive and one negative square root, as shown below:

positive square root	and	negative square root
$\sqrt{a}$		$-\sqrt{a}.$

A negative integer has no integer square roots.

EXAMPLE 8   **Finding square roots of integers**

Simplify each expression, if possible.
(a) $\sqrt{81}$ (b) $-\sqrt{25}$ (c) $\sqrt{-49}$

Solution
(a) Because $9^2 = 81$ and 9 is positive, $\sqrt{81} = 9$.
(b) The notation $-\sqrt{25}$ means to find the negative square root of 25, or $-\sqrt{25} = -5$.
(c) $\sqrt{-49}$ is not an integer because a negative number has no integer square roots.

CALCULATOR HELP

To find the square roots of an integer with a calculator, see the Appendix (page AP-3).

Now Try Exercises 89, 91, 95

Applications

The next two examples illustrate applications involving products and quotients of integers.

EXAMPLE 9   **Finding the average surface temperature of a planet**

The average surface temperature of Mars is about $-65°C$. Because Uranus is much farther from the sun, its average surface temperature is 3 times as cold as that of Mars. What is the average surface temperature of Uranus? (*Source:* NASA.)

Solution
The average surface temperature of Uranus is $-65 \cdot 3 = -195°C$.

Now Try Exercise 111

EXAMPLE 10 **Calculating investment losses**

An investor lost money on six different investments. If the amount lost was $1500 on each investment, write the total loss as a product of integers and find the total loss.

Solution

We can write a loss as a negative number. The total loss can be written as $-1500 \cdot 6$, which equals -9000. The total loss is $9000.

Now Try Exercise 117

2.4   Putting It All Together

CONCEPT	COMMENTS	EXAMPLES
Multiplying Integers	The product of two numbers with *like* signs is positive.	$3 \cdot 12 = 36$ $-4 \cdot (-6) = 24$
	The product of two numbers with *unlike* signs is negative.	$-5 \cdot 7 = -35$ $8 \cdot (-9) = -72$
Multiplication Properties	1. Commutative Property 2. Associative Property 3. Identity Property 4. Zero Property 5. Distributive Properties	1. $4 \cdot (-2) = -2 \cdot 4$ 2. $(-5 \cdot 1) \cdot 7 = -5 \cdot (1 \cdot 7)$ 3. $-2 \cdot 1 = -2$ and $1 \cdot (-2) = -2$ 4. $-7 \cdot 0 = 0$ and $0 \cdot (-7) = 0$ 5. $-5(2 + 4) = -5 \cdot 2 + (-5) \cdot 4$ $\quad -2(1 - 5) = -2 \cdot 1 - (-2) \cdot 5$
Products with More Than Two Factors	If there is an *even* number of negative factors, the product is positive. If there is an *odd* number of negative factors, the product is negative.	$3 \cdot (-2) \cdot 1 \cdot (-4)$ is positive. (even number of negative factors) $-3 \cdot (-5) \cdot 1 \cdot (-7)$ is negative. (odd number of negative factors)
Dividing Integers	The quotient of two numbers with *like* signs is positive. The quotient of two numbers with *unlike* signs is negative.	$36 \div 9 = 4$ $-45 \div (-5) = 9$ $-63 \div 9 = -7$ $28 \div (-7) = -4$
Square Roots	Every positive integer a has one positive and one negative square root. The positive square root, or *principal square root*, is written as $\sqrt{a}$. The negative square root is written as $-\sqrt{a}$.	$\sqrt{25} = 5$ $-\sqrt{64} = -8$ $\sqrt{-121}$ is not an integer.

2.4 Exercises

MyMathLab | Math XL PRACTICE | WATCH | DOWNLOAD | READ | REVIEW

CONCEPTS AND VOCABULARY

1. The product of two numbers with like signs is _____.

2. When two numbers have unlike signs, their product is _____.

3. The equation $1 \cdot (-17) = -17$ is an example of the _____ property of multiplication.

4. The equation $-3(4 - 5) = -3 \cdot 4 - (-3) \cdot 5$ is an example of the _____ property of multiplication.

5. If there is an even number of negative factors, the product is _____.

6. If there is an odd number of negative factors, the product is _____.

7. (True or False?) $(-11)^2 = -11^2$.

8. (True or False?) $(-2)^3 = -2^3$.

9. The quotient of two numbers with like signs is _____.

10. When two numbers have unlike signs, their quotient is _____.

11. The principal square root of 4 is written as _____.

12. The negative square root of 4 is written as _____.

MULTIPLYING INTEGERS

Exercises 13–40: Multiply.

13. $2(-6)$

14. $-7(4)$

15. $-5(-8)$

16. $-8(-7)$

17. $-1 \cdot 18$

18. $-14 \cdot 0$

19. $-10 \cdot (-17)$

20. $-50 \cdot (-2)$

21. $0 \cdot (-21)$

22. $1 \cdot (-34)$

23. $14 \cdot (-3)$

24. $15 \cdot (-4)$

25. $-25 \cdot 6$

26. $-30 \cdot 4$

27. $-2 \cdot 6 \cdot 3$

28. $4 \cdot (-3) \cdot 2$

29. $-3 \cdot 5 \cdot (-2)$

30. $6 \cdot (-4) \cdot (-2)$

31. $-7(-1)(-3)$

32. $-8(-8)(-1)$

33. $5(-2)(-3)(-3)$

34. $5(5)(-1)(-4)$

35. $12(-1)(0)(2)$

36. $9(0)(-2)(-5)$

37. $2(-1)(5)(-2)(-4)$

38. $-2(5)(3)(-2)(-1)$

39. $2(-1)(5)(-2)(-4)(5)(-1)$

40. $-1(-3)(5)(-2)(-3)(5)(-2)$

MULTIPLICATION PROPERTIES

Exercises 41–50: State the multiplication property illustrated by the given equation.

41. $12 \cdot (-22 \cdot 41) = (12 \cdot (-22)) \cdot 41$

42. $(-11 \cdot 5) \cdot 23 = -11 \cdot (5 \cdot 23)$

43. $-341 \cdot 0 = 0$

44. $1 \times (-30{,}412) = -30{,}412$

45. $-25 \times (-37) = -37 \times (-25)$

46. $0(-64{,}901) = 0$

47. $-2(-3 + 9) = -2(-3) + (-2)(9)$

48. $-10 \cdot (-19) = -19 \cdot (-10)$

49. $-15{,}400 \cdot 1 = -15{,}400$

50. $3(-1 - 8) = 3(-1) - 3(8)$

EXPONENTIAL EXPRESSIONS

Exercises 51–64: Evaluate the exponential expression.

51. -2^3

52. $(-3)^2$

53. $(-4)^2$

54. -5^2

55. -9^2

56. $(-8)^2$

57. -1^4

58. -3^4

59. $(-3)^3$

60. -1^5

61. -10^6

62. $(-10)^3$

63. $(-10)^4$

64. -10^5

DIVIDING INTEGERS

Exercises 65–80: Divide, if possible. If a quotient is undefined, state so.

65. $18 \div (-6)$

66. $-48 \div 8$

67. $\dfrac{-40}{-8}$

68. $\dfrac{24}{-3}$

69. $-12 \div (-1)$

70. $-20 \div 1$

71. $\dfrac{-50}{25}$

72. $\dfrac{-72}{-12}$

73. $-35 \div 0$

74. $0 \div (-3)$

75. $-24 \div (-24)$

76. $-10 \div 10$

77. $0 \div (-9)$

78. $-63 \div 0$

79. $\dfrac{72}{-12}$

80. $\dfrac{-64}{16}$

SQUARE ROOTS

Exercises 81–88: Find all integer square roots of the given number, if possible.

81. 25

82. 9

83. 81

84. 100

85. -36

86. -4

87. 0

88. 1

Exercises 89–98: Simplify the expression, if possible.

89. $\sqrt{16}$

90. $\sqrt{49}$

91. $-\sqrt{36}$

92. $-\sqrt{144}$

93. $\sqrt{100}$

94. $-\sqrt{81}$

95. $\sqrt{-121}$

96. $\sqrt{-25}$

97. $-\sqrt{1}$

98. $\sqrt{0}$

EVALUATING EXPRESSIONS

Exercises 99–110: Evaluate the expression for the given value(s) of the variable(s), if possible.

99. $3x$ $\qquad\qquad x = -7$

100. $-8y$ $\qquad\qquad y = -2$

101. $\dfrac{x}{6}$ $\qquad\qquad x = -60$

102. $4xy$ $\qquad\qquad x = -3, y = 5$

103. $\dfrac{a}{b}$ $\qquad\qquad a = -30, b = 6$

104. $-ab$ $\qquad\qquad a = -6, b = 11$

105. $m \cdot (-n)$ $\qquad\qquad m = -5, n = 5$

106. $2 \cdot (-m) \cdot (-n)$ $\qquad\qquad m = 4, n = -6$

107. $\sqrt{-x}$ $\qquad\qquad x = -100$

108. $-\sqrt{y}$ $\qquad\qquad y = 4$

109. $-\sqrt{a}$ $\qquad\qquad a = -64$

110. $-\sqrt{-a}$ $\qquad\qquad a = 81$

APPLICATIONS

111. *Cold Temperatures* Every state in the U.S. has a record low temperature below 0°C. The warmest of these lows was recorded in Hawaii at Mauna Kea, where the temperature dipped to -11°C. In Iowa, the town of Elkader recorded the state's lowest temperature, which is 4 times as cold as the record in Hawaii. What is Iowa's record low? (*Source: NOAA.*)

112. *Cold Temperatures* The record low temperature for Florida is -19°C. If the record low temperature for Montana is 3 times as cold as that of Florida, what is the record low for Montana? (*Source: NOAA.*)

113. *Spelunking* A cave explorer descends to the floor of a cave in 5 stages, dropping 107 feet each time. Write the total depth of the cave as a product of integers and find the depth of the cave.

114. *Game Tokens* Each red token in a board game represents -5 points. Write the total number of points in a stack of 4 red tokens as a product of integers and find the total number of points.

115. *State Prisoners* According to a recent study, the number of prisoners in Arkansas decreased by 300 inmates over a one-year period. If the decrease was the same for each of the 12 months, write the monthly decrease as a quotient of integers and find the monthly decrease. (*Source: Arkansas News Bureau.*)

116. *Homeless Population* A New York City government study suggested that the number of people living on the streets or in the city's subways dropped by 1089 from the end of 2005 to the beginning of 2008. Assuming that the decrease was the same for each of the 3 years of the study, write the yearly decrease as a quotient of integers and find the yearly decrease. (*Source: New York Times.*)

117. *Late Fees* Each time that a credit card holder fails to make a credit card payment by the due date, a late fee appears on the monthly statement as –$29. If the card holder missed the payment due date five times over the past year, write the total charges for late fees as a product of integers and find the total charges.

118. *ATM Fees* Each time that a credit card holder uses an ATM for a cash advance, a transaction fee appears on the credit card statement. If a year-end summary shows –$129 for 43 ATM transactions, write the transaction fee as a negative integer.

WRITING ABOUT MATHEMATICS

119. Give an example of a product of three or more integers with a negative result. Give an example of a product of three or more integers with a positive result. Explain how you chose your examples.

120. Give a real-world example of when a negative number is divided by a positive number.

121. Give an example of an exponential expression with a negative base and an example of the opposite of an exponential expression with a positive base. Explain the differences between your expressions.

122. Choose a positive integer that is a perfect square and write both of its square roots using radical notation. Which of your expressions represents the principal square root?

SECTIONS 2.3 and 2.4 | **Checking Basic Concepts**

1. Find each difference.
(a) $-11 - 23$ (b) $-21 - (-7)$
(c) $3 - (-30)$ (d) $54 - 39$

2. Use a number line to find each difference.
(a) $-4 - 4$ (b) $3 - (-6)$

3. Perform the subtraction visually, using the symbols $\frown$ and $\smile$.
(a) $-5 - 3$ (b) $-8 - (-3)$

4. Simplify the expression.
$-2 - (-7) + 5 + (-6)$

5. Find each product.
(a) $-11 \cdot 4$ (b) $-3 \cdot (-13)$
(c) $6 \cdot (-8)$ (d) $5 \cdot 10$

6. Simplify the expression.
$2(-1)(4)(-2)(-3)$

7. Find each quotient.
(a) $24 \div (-6)$ (b) $-60 \div 12$
(c) $\dfrac{-36}{-4}$ (d) $\dfrac{25}{-1}$

8. Simplify each expression.
(a) $\sqrt{64}$ (b) $-\sqrt{16}$

9. *Bank Account* An overdrawn checking account has a balance of $-\$57$. Find the amount of a deposit that would bring the balance to $\$108$.

10. *Ocean Depth* A diver descends in 3 stages, dropping 23 feet each time. Write the total depth of the diver as a product of integers and find the diver's final depth.

2.5 Order of Operations; Averages

Order of Operations • Evaluating Algebraic Expressions • Averages

A LOOK INTO MATH ▶

A person getting dressed puts on socks *before* shoes. An airplane can land only *after* it lowers its landing gear. In mathematics, we may need to add, subtract, multiply, divide, take square roots, raise numbers to powers, or find absolute value—all within one problem. When evaluating an expression involving several of these operations, the order in which the operations are performed is important. In Section 1.6, we discussed the *order of operations agreement*, which we now apply to expressions involving integers.

STUDY TIP

Before visiting your instructor or going to a tutor center for help, be sure that you have tried a problem several times in different ways. Organize your questions so that you can be specific about the part of the problem that is giving you difficulty.

NEW VOCABULARY

☐ Average

Order of Operations

In this section, we apply the order of operations agreement to expressions involving integers.

ORDER OF OPERATIONS

1. Do all calculations within grouping symbols, such as parentheses and radicals, or above and below a fraction bar.
2. Evaluate all exponential expressions.
3. Do all multiplication and division from *left to right*.
4. Do all addition and subtraction from *left to right*.

READING CHECK

- Is the order of operations agreement for integers the same as or different from the order of operations agreement for whole numbers?

EXAMPLE 1   **Using the order of operations to evaluate expressions**

Evaluate each expression.
(a) $-16 + 5 \cdot 4$ **(b)** $24 \div 8 - 2 \cdot 6$ **(c)** $17 - 4^2 \div 2$

Solution
(a) Perform multiplication before addition.

$$-16 + \mathbf{5 \cdot 4} = -16 + \mathbf{20} \qquad \text{Multiply.}$$
$$= 4 \qquad \text{Add.}$$

(b) Perform multiplication and division from left to right.

$$\mathbf{24 \div 8} - 2 \cdot 6 = \mathbf{3} - \mathbf{2 \cdot 6} \qquad \text{Divide.}$$
$$= 3 - \mathbf{12} \qquad \text{Multiply.}$$
$$= -9 \qquad \text{Subtract.}$$

(c) The exponential expression is evaluated first.

$$17 - \mathbf{4^2} \div 2 = 17 - \mathbf{16} \div \mathbf{2} \qquad \text{Evaluate } 4^2.$$
$$= 17 - \mathbf{8} \qquad \text{Divide.}$$
$$= 9 \qquad \text{Subtract.}$$

Now Try Exercises 5, 7, 11

Parentheses, radical symbols, fraction bars, and absolute value symbols are all grouping symbols. Remember that expressions within grouping symbols are evaluated first.

EXAMPLE 2   **Evaluating expressions involving grouping symbols**

Evaluate each expression.
(a) $3^2 - (5 + 4)$ **(b)** $20 \div |6 - 11| - 8$

Solution
(a) The expression within parentheses is evaluated first.

$$3^2 - (\mathbf{5 + 4}) = 3^2 - \mathbf{9} \qquad \text{Add within parentheses.}$$
$$= \mathbf{9} - 9 \qquad \text{Evaluate } 3^2.$$
$$= 0 \qquad \text{Subtract.}$$

(b) Evaluate the absolute value expression first.

$$20 \div |6 - 11| - 8 = 20 \div |-5| - 8 \quad \text{Subtract within absolute value.}$$
$$= 20 \div 5 - 8 \quad \text{Evaluate } |-5|.$$
$$= 4 - 8 \quad \text{Divide.}$$
$$= -4 \quad \text{Subtract.}$$

■ **Now Try Exercises 17, 23**

EXAMPLE 3 **Evaluating expressions involving grouping symbols**

Evaluate each expression.

(a) $-32 \div \sqrt{17 - 1}$ **(b)** $\dfrac{-8 + 2 \cdot 3}{10 \div 5 - 1}$

Solution
(a) The expression under the radical is evaluated first.

$$-32 \div \sqrt{17 - 1} = -32 \div \sqrt{16} \quad \text{Subtract under radical.}$$
$$= -32 \div 4 \quad \text{Evaluate } \sqrt{16}.$$
$$= -8 \quad \text{Divide.}$$

(b) Use the order of operations both above and below the fraction bar.

$$\frac{-8 + 2 \cdot 3}{10 \div 5 - 1} = \frac{-8 + 6}{2 - 1} \quad \text{Multiply above; divide below.}$$
$$= \frac{-2}{1} \quad \text{Add above; subtract below.}$$
$$= -2 \quad \text{Divide.}$$

■ **Now Try Exercises 15, 21**

Some expressions contain grouping symbols that are *nested* within other grouping symbols. Expressions of this type are evaluated by starting with the innermost grouping and working outward, as shown in the next example.

EXAMPLE 4 **Evaluating expressions with nested grouping symbols**

Evaluate each expression.
(a) $7 - (4^2 - (2 + 3) \cdot 5)$ **(b)** $21 \div (5 + |3 - 9| - 8)$

Solution
(a) Start by evaluating the innermost grouping, $(2 + 3)$.

$$7 - (4^2 - (2 + 3) \cdot 5) = 7 - (4^2 - 5 \cdot 5) \quad \text{Evaluate innermost grouping.}$$
$$= 7 - (16 - 5 \cdot 5) \quad \text{Evaluate } 4^2.$$
$$= 7 - (16 - 25) \quad \text{Multiply.}$$
$$= 7 - (-9) \quad \text{Evaluate parentheses.}$$
$$= 16 \quad \text{Subtract.}$$

(b) The absolute value, $|3 - 9|$, is the innermost grouping.

$$
\begin{aligned}
21 \div (5 + |3 - 9| - 8) &= 21 \div (5 + |-6| - 8) && \text{Subtract within absolute value.} \\
&= 21 \div (5 + 6 - 8) && \text{Evaluate } |-6|. \\
&= 21 \div (11 - 8) && \text{Add.} \\
&= 21 \div 3 && \text{Evaluate parentheses.} \\
&= 7 && \text{Divide.}
\end{aligned}
$$

Now Try Exercises 27, 35

READING CHECK

• In evaluating an algebraic
expression, when are
parentheses needed?

Evaluating Algebraic Expressions

An algebraic expression can be evaluated when particular values are assigned to the variables in the expression. If a negative value is assigned to a variable, then it is best to place the negative value within parentheses when replacing the variable. For example, if we let $x = -3$ and $y = -2$ in the expression $5x + y$, then replacing each variable with its given value results in the expression $5(-3) + (-2)$. The next example illustrates how parentheses are used in evaluating algebraic expressions.

EXAMPLE 5 **Evaluating algebraic expressions**

Evaluate each expression for $x = -4$, $y = 5$, and $z = -2$.

(a) $17 - 2xz$ **(b)** $6y \div (z + |x| - 7)$ **(c)** $\dfrac{3yz}{x - 2}$

Solution
(a) Start by replacing x with -4 and z with -2 in the given expression.

$$
\begin{aligned}
17 - 2xz &= 17 - 2(-4)(-2) && x = -4 \text{ and } z = -2. \\
&= 17 - (-8)(-2) && \text{Multiply } 2(-4). \\
&= 17 - 16 && \text{Multiply } (-8)(-2). \\
&= 1 && \text{Subtract.}
\end{aligned}
$$

(b) This expression contains all three variables. Replace x with -4, y with 5, and z with -2. Note that the positions of x and z in the expression allow for each variable to be replaced by a negative value *without* the need for additional parentheses.

$$
\begin{aligned}
6y \div (z + |x| - 7) &= 6(5) \div (-2 + |-4| - 7) && x = -4, y = 5, \text{ and } z = -2. \\
&= 6(5) \div (-2 + 4 - 7) && \text{Evaluate } |-4|. \\
&= 6(5) \div (2 - 7) && \text{Add } -2 + 4. \\
&= 6(5) \div (-5) && \text{Subtract within parentheses.} \\
&= 30 \div (-5) && \text{Multiply.} \\
&= -6 && \text{Divide.}
\end{aligned}
$$

(c) Replace x with -4, y with 5, and z with -2.

$$\frac{3yz}{x-2} = \frac{3(5)(-2)}{-4-2} \qquad x = -4, \ y = 5, \text{ and } z = -2.$$

$$= \frac{-30}{-4-2} \qquad \text{Multiply } 3(5)(-2).$$

$$= \frac{-30}{-6} \qquad \text{Subtract } -4 - 2.$$

$$= 5 \qquad \text{Divide.}$$

▌ **Now Try Exercises 49, 55, 59**

READING CHECK

• What is being measured when we find the average of a group of numbers?

Averages

▶ **REAL-WORLD CONNECTION** When graded exams are returned to students, instructors often report the average score for the class. **Average** is a measure of the middle of a group of numbers. The average of a list of numbers is found as follows:

$$\text{Average} = \frac{\text{the sum of the numbers in the list}}{\text{the } \textit{number} \text{ of numbers in the list}}$$

EXAMPLE 6   **Finding average monthly profit for an iPhone application**

The bar graph in Figure 2.17 shows the monthly profit for a small gaming company after the release of a new iPhone application. A negative profit indicates a loss. Find the average monthly profit for the months shown.

Figure 2.17 Monthly Profit

Solution
To find the average of -16, -8, -3, 10, 16, **and** 19, add the numbers together and then divide the result by 6 because there are **6** numbers in the list.

$$\text{Average} = \frac{(-16) + (-8) + (-3) + 10 + 16 + 19}{6} = \frac{18}{6} = 3$$

Since the profit is given in thousands of dollars, the average monthly profit for January through June is $3000.

▌ **Now Try Exercise 67**

2.5   Putting It All Together

CONCEPT	COMMENTS	EXAMPLES
Order of Operations Agreement	1. Do all calculations within grouping symbols, such as parentheses and radicals, or above and below a fraction bar. 2. Evaluate all exponential expressions. 3. Do all multiplication and division from *left to right*. 4. Do all addition and subtraction from *left to right*.	$6 \div (-2) + 3 \cdot 5 = -3 + 3 \cdot 5$ $= -3 + 15$ $= 12$ $18 \div (3 - 6)^2 = 18 \div (-3)^2$ $= 18 \div 9$ $= 2$ $\sqrt{10 - (-6)} = \sqrt{16}$ $= 4$
Evaluating Algebraic Expressions	Replace each variable in the expression with its given value and then evaluate using the order of operations agreement.	Evaluate $7 + 3x$ for $x = -5$. $7 + 3x = 7 + 3(-5)$ $= 7 + (-15)$ $= -8$
Average	Average is a measure of the middle of a group of numbers. It is found by using the formula: $\text{Average} = \dfrac{\text{the sum of a list of numbers}}{\text{number of numbers in the list}}.$	The average of $-5, 0, 4, 7,$ and 9 is $\dfrac{(-5) + 0 + 4 + 7 + 9}{5} = 3.$

2.5   Exercises

CONCEPTS AND VOCABULARY

1. The _____ agreement ensures that algebraic expressions are evaluated in the same way by everyone.

2. When grouping symbols are nested, we evaluate the _____ grouping first.

3. When replacing a variable with a negative number, we often use _____.

4. The _____ is a measure of the middle of a group of numbers.

ORDER OF OPERATIONS

Exercises 5–40: Evaluate the expression.

5. $2 + (-3) \cdot 4$ 6. $-8 - 4(-5)$

7. $-36 \div 3^2 + 7$ 8. $48 \div 12 - 11$

9. $60 + (-35) \div 7 - 28$

10. $-55 + 9 \cdot 7 - 18$

11. $49 \div 7 + (-2) \cdot 4$

12. $-2 \cdot (-50) - 8 \div 2$

13. $\dfrac{3 - 25}{7 + 4}$ **14.** $\dfrac{19 + 16}{5(-1)}$

15. $\dfrac{45 + 9(-10)}{-5}$ **16.** $\left| \dfrac{19 - 59}{2 \cdot 5} \right|$

17. $36 - 3^2 \div (6 - 9)$ **18.** $-28 \div (2 - 3^2) + 3$

19. $35 - |3 + 4^2 \div (-2)|$

20. $(36 \div 6^2 - 4) \cdot |-3|$

21. $\sqrt{41 + 8} + (-7)$ **22.** $-78 - \sqrt{80 + 1}$

23. $-8^2 + |9 \cdot (-8)|$ **24.** $\sqrt{-5^2 + 61} - 3^2$

25. $\sqrt{100} \cdot \sqrt{|0 - 25|}$

26. $|-43| - |20 - 6^2|$

27. $((2 - 7)^2 \div 5) - 18$

28. $(20 \cdot (6 - 7) \div 10)^3$

29. $\dfrac{(13 - 9) \cdot 6}{|37 - 41| \cdot 2}$ **30.** $\dfrac{2 \cdot (4 + 1)^2 + 10}{5 \cdot |4 - 7|}$

31. $\dfrac{(6 - 3) \cdot 5}{2^3 - \sqrt{81}}$ **32.** $\dfrac{(\sqrt{64} + 2) \cdot 5}{|6^2 - 41|}$

33. $(7 - 9)^2 - 3\sqrt{2 \cdot 8}$

34. $(4 - 5)^2 \cdot \sqrt{50 - 25}$

35. $-4|2 - 4 \cdot 3| + ((72 \div 9)^2 + 6)$

36. $50 - 5^2 \cdot 3 - 32 \div 2^3 + 8$

37. $29 - (3 \cdot 9 - (32 \div 2^4) + 3)$

38. $0 - \sqrt{-40 \div 5 - (3 - 3 \cdot 2^2)}$

39. $\sqrt{25 - 3} - (-1 \cdot 6^2 \div 12) + 11$

40. $-34 - 3 \cdot 7 + 2 + 2^2(-3) - 16 \div 2^3 - 9$

Exercises 41–48: Insert parentheses as needed in the expression in order to make it equal to 0. More than one set of parentheses may be needed.

41. $-20 + 10 \cdot 14 - 12$ **42.** $-4 - 3 - 8 - 1$

43. $-5^2 \div 3 + 2 + 5$ **44.** $7 - 10 \cdot 3^2 + 27$

45. $32 \div 4^2 - 2 \cdot 9$ **46.** $5 - 5 \cdot 3^2 \div 3$

47. $16 - 4^2 \div 4 - 9$ **48.** $8 - 5 + 6^2 \div 12$

ALGEBRAIC EXPRESSIONS

Exercises 49–62: Evaluate the expression for the given values of the variables.

49. $4 \cdot y + x$, for $x = -5, y = 1$

50. $3a + 9 - b$, for $a = 5, b = -6$

51. $24v - 6w$, for $v = -2, w = -8$

52. $3c - 5d$, for $c = -5, d = -3$

53. $2m + (4^2 + n) \div 8$, for $m = 7, n = -32$

54. $w + ((3 - v)^2 \div w)$, for $v = 0, w = -3$

55. $2m + |2^3 + n| \div 8$, for $m = 5, n = -16$

56. $w + ((5 - 2v)^2 \div 3w)$, for $v = 4, w = -1$

57. $\sqrt{p^2 - 7} - 2q^2$, for $p = -4, q = 2$

58. $\sqrt{r - 3} - (14 + s)$, for $r = 39, s = -6$

59. $\dfrac{2(7 + c)}{|d - 5| \cdot (-4)}$, for $c = 5, d = 4$

60. $\dfrac{(6 - g) + 7}{(\sqrt{2h} - 3)^2}$, for $g = -5, h = 2$

61. $xy + (y^2 - x) \div y$, for $x = -12, y = -3$

62. $a \cdot b - a \div b + b$, for $a = 0, b = -1$

APPLICATIONS

63. *Converting Temperature* To convert a temperature F given in degrees Fahrenheit to an equivalent temperature C in degrees Celsius, use the formula

$$C = \frac{5(F - 32)}{9}.$$

Find the Celsius temperature that is equivalent to $-4°F$.

64. *Converting Temperature* Use the formula in Exercise 63 to find the Celsius temperature that is equivalent to $-40°F$.

65. *Converting Temperature* To convert a temperature C given in degrees Celsius to an equivalent temperature F in degrees Fahrenheit, use the formula

$$F = \frac{9C}{5} + 32.$$

Find the Fahrenheit temperature that is equivalent to $-15°C$.

66. *Converting Temperature* Use the formula in Exercise 65 on the previous page to find the Fahrenheit temperature that is equivalent to −40°C.

Exercises 67 and 68: Profit and Loss The following bar graph shows the monthly profit for a jewelry-making company. A negative profit indicates a loss.

67. Find the average monthly profit for the months shown.

68. Find the average monthly profit for the months of August, September, and October only.

69. *Average Temperature* Find the average of the list of Fahrenheit temperatures.

$$-16°, -22°, 1°, 11°, -2°, -21°, \text{ and } 7°F$$

70. *Average Temperature* Find the average of the list of Celsius temperatures.

$$-6°, -2°, 15°, 18°, -5°, \text{ and } 4°C$$

WRITING ABOUT MATHEMATICS

71. Why is it sometimes necessary to insert parentheses when evaluating an expression for a negative value of a variable? Give an example of this situation.

72. Give an example of an expression that contains nested grouping symbols. Explain how to evaluate this expression.

Group Activity   Working with Real Data

Directions: Form a group of 2 to 4 people. Select someone to record the group's responses for this activity. All members of the group should work cooperatively to answer the questions. If your instructor asks for your results, each member of the group should be prepared to respond.

Global Temperatures If the Greenland ice sheet melted completely, it would produce approximately 650,000,000,000,000,000 gallons of water. This means that 650 *quadrillion* gallons of water would enter the ocean, raising sea level significantly. The formula

$$F = \left(\frac{6g}{13}\right) \div 12$$

can be used to estimate the increase in sea level F in feet that would result from g quadrillion gallons of water entering the oceans. (*Source:* U.S. Geological Survey.)

Compute each value of F in the table by replacing g in the formula with each value of g given in the table.

g (quadrillion gallons)	26	130	312	468	650
F (feet)					

In the United States, several large cities have low average elevations. Three examples are Boston (14 feet), New Orleans (4 feet), and San Diego (13 feet). Discuss how the melting of large portions of the Greenland ice sheet into the ocean might affect these cities.

2.6 Solving Equations with Integer Solutions

**Checking a Solution • Solving Equations Using Guess-and-Check •
Solving Equations Using Tables of Values • Solving Equations Visually**

A LOOK INTO MATH ▶

Each year billions of dollars are spent to solve equations. If we did not have the ability to solve equations, we would not have cell phones, TiVo, YouTube, digital cameras, satellite TV, or fast Internet search engines. In fact, almost every modern product would not exist without mathematics and the ability to solve equations. In this section, we extend our knowledge of solving equations to include equations with integer solutions.

NEW VOCABULARY

☐ Equation
☐ Solution

Checking a Solution

An **equation** is a mathematical statement that two algebraic expressions are equal. If an equation contains one variable, then any value of the variable that makes the equation true is called a **solution** to the equation. Examples of equations with one variable include

$$4m = 20, \quad 3x - 15 = -12, \quad -80 \div b = 10, \quad \text{and} \quad 1 + y^2 = 37.$$

The solutions to these equations are given by

$$m = 5, \quad x = 1, \quad b = -8, \quad \text{and} \quad y = -6 \text{ or } 6.$$

For example, the solution **1** can be checked in the equation $3x - 15 = -12$ as follows.

$$
\begin{array}{ll}
3x - 15 = -12 & \text{Given equation} \\
3(\mathbf{1}) - 15 \stackrel{?}{=} -12 & \text{Replace } x \text{ with 1.} \\
3 - 15 \stackrel{?}{=} -12 & \text{Simplify.} \\
-12 = -12 \checkmark & \text{The solution checks.}
\end{array}
$$

Because the resulting equation is true, the solution checks.

CHECKING A SOLUTION

To check a solution, replace *every* occurrence of the variable in the equation with the proposed solution and check to see if the resulting equation is true.

READING CHECK

• What is an equation and how do we know if a number is a solution to an equation?

STUDY TIP

If you have trouble keeping your course materials organized or have difficulty managing your time, you may be able to find help at the student support services office on your campus. Students who learn to manage their time and keep organized find the college experience to be less overwhelming and more enjoyable.

EXAMPLE 1 **Checking solutions**

Check each solution.
(a) Is -5 a solution to $3x + 7 = -8$?
(b) Is -1 a solution to $2y^2 = \sqrt{3 - y}$?
(c) Is 2 a solution to $|5w - 11| = -1$?

Solution

(a) Replace the variable x with -5.

$$3x + 7 = -8 \qquad \text{Given equation}$$
$$3(-5) + 7 \stackrel{?}{=} -8 \qquad \text{Replace } x \text{ with } -5.$$
$$-15 + 7 \stackrel{?}{=} -8 \qquad \text{Multiply.}$$
$$-8 = -8 \checkmark \qquad \text{The solution checks.}$$

The solution checks, which means that -5 is a solution to the equation $3x + 7 = -8$.

(b) Replace *every* occurrence of the variable y with -1.

$$2y^2 = \sqrt{3 - y} \qquad \text{Given equation}$$
$$2(-1)^2 \stackrel{?}{=} \sqrt{3 - (-1)} \qquad \text{Replace } y \text{ with } -1.$$
$$2(-1)^2 \stackrel{?}{=} \sqrt{3 + 1} \qquad \text{Add the opposite.}$$
$$2(1) \stackrel{?}{=} \sqrt{4} \qquad \text{Simplify.}$$
$$2 = 2 \checkmark \qquad \text{The solution checks.}$$

The solution checks, which means that -1 is a solution to the equation $2y^2 = \sqrt{3 - y}$.

(c) Replace the variable w with 2.

$$|5w - 11| = -1 \qquad \text{Given equation}$$
$$|5(2) - 11| \stackrel{?}{=} -1 \qquad \text{Replace } w \text{ with } 2.$$
$$|10 - 11| \stackrel{?}{=} -1 \qquad \text{Multiply.}$$
$$|-1| \neq -1 \; \textcolor{red}{✗} \qquad \text{The solution does not check.}$$

Because an absolute value result cannot be negative, the solution does not check and 2 is **not** a solution to the equation $|5w - 11| = -1$.

▌ **Now Try Exercises 7, 11, 15**

Solving Equations Using Guess-and-Check

Some equations can be solved by thinking of the variable as a missing addend, subtrahend, minuend, factor, divisor, dividend, radicand, or base. For example, the equation $4 + x = -13$ has a missing addend. To solve this equation, we must find the number that is added to 4 to get -13. Using a *guess-and-check* strategy, we find that the solution is -17 because $4 + (-17) = -13$.

EXAMPLE 2 **Solving equations**

Solve each equation.

(a) $7x = -63$ (b) $4 - y = -9$ (c) $w \div (-3) = 11$

Solution

(a) This equation has a missing factor. The solution is -9 because $7(-9) = -63$.

(b) Here the subtrahend is missing. To get a result of -9, we must subtract 13 from 4. The solution is **13** because $4 - 13 = -9$.

(c) This equation has a missing dividend. The solution is -33 because it checks in the equation $-33 \div (-3) = 11$.

▌ **Now Try Exercises 23, 25, 27**

EXAMPLE 3   **Solving equations**

Solve each equation.
(a) $m^3 = -27$ (b) $\sqrt{n} = 11$

Solution
(a) To solve this equation, find a base that, when cubed, results in -27. Because the equation $(-3)^3 = -27$ checks, the solution is **−3**.
(b) Find a number whose square root is 11. The solution is **121** because $\sqrt{121} = 11$.

Now Try Exercises 29, 31

Solving Equations Using Tables of Values

▶ **REAL-WORLD CONNECTION** Some equations can be solved by making a table of values and selecting the solution from the table. Suppose that a landscaping crew is able to place 500 cement blocks before noon and 50 blocks each hour afterwards until 5:00 P.M. The total number of blocks B placed after various elapsed times t is shown in Table 2.3. Note that $t = 0$ corresponds to noon (0 hours past noon), $t = 1$ corresponds to 1:00 P.M. (1 hour past noon), and so on until $t = 5$ corresponds to 5:00 P.M.

TABLE 2.3 **Blocks B Placed t Hours Past Noon**

Elapsed Time: t (hours)	0	1	2	3	4	5
Total Blocks Placed: B	500	550	600	650	700	750

A mathematical formula that *describes* or *models* these data is given by

$$B = 500 + 50t.$$

For example, at **2** hours past noon, or 2:00 P.M., a total of

$$B = 500 + 50(2) = 500 + 100 = 600$$

blocks have been placed. By replacing the variable t in the formula with the other values of t in Table 2.3, we can find the corresponding values of B in the table.

Now suppose that the crew supervisor wants to know the time when 700 blocks had been placed. Replacing B in the formula with **700** results in the equation

$$700 = 500 + 50t.$$

Although the techniques for solving this type of equation are not discussed until Chapter 3, we are able to use Table 2.3 to solve it. From the table, the number of blocks B is 700 when t is 4, or at 4:00 P.M. The solution to the equation $700 = 500 + 50t$ is 4.

EXAMPLE 4   **Using a table to solve an equation**

Complete Table 2.4 for the given values of x and then solve the equation $5x - 13 = -3$.

TABLE 2.4

x	−1	0	1	2	3
$5x - 13$					

Solution

Begin by replacing x in $5x - 13$ with -1. Because $5(-1) - 13$ evaluates to -18, the left side of the equation equals -18 when $x = -1$. We write the result -18 below -1, as shown in Table 2.5. Similarly, we write -13 below 0 because the expression $5(0) - 13$ evaluates to -13. The remaining values are found in a similar fashion and are shown in Table 2.5.

TABLE 2.5

x	-1	0	1	2	3
$5x - 13$	-18	-13	-8	-3	2

An equation is true when its left side equals its right side. Table 2.5 reveals that the left side, $5x - 13$, equals the right side, -3, when $x = 2$. Therefore, the solution to the equation $5x - 13 = -3$ is **2**.

▌ **Now Try Exercise 43**

READING CHECK

• Explain how a solution can be found in a table of values.

Example 4 shows that making a table of values is simply an organized way of selecting and checking possible values of the variable to see if any value makes the equation true. If a value makes an equation true, then that value is a solution.

▶ **REAL-WORLD CONNECTION** In the next example, a table of values is used to find the year when average tuition and fees at public four-year colleges reached $6185.

EXAMPLE 5 **Analyzing tuition and fees**

For any year after 2001, the average tuition and fees T paid by students attending public four-year colleges can be approximated using the formula

$$T = 350(x - 2002) + 4085,$$

where x represents the year. By replacing T with 6185 in the formula, we form the equation $6185 = 350(x - 2002) + 4085$. The solution represents the year when average tuition and fees reached $6185. Solve this equation by completing Table 2.6. (*Source:* The College Board.)

TABLE 2.6 **Average Tuition and Fees at Public Four-Year Colleges**

x	2002	2003	2004	2005	2006	2007	2008
$350(x - 2002) + 4085$							

Solution

Begin by replacing x in $350(x - 2002) + 4085$ with **2002** and then evaluate the resulting expression.

$$350(\mathbf{2002} - 2002) + 4085 = 350(0) + 4085 = 0 + 4085 = 4085$$

Write 4085 below **2002**, as shown in Table 2.7. The remaining values are found in a similar fashion and are shown in Table 2.7.

TABLE 2.7 **Average Tuition and Fees at Public Four-Year Colleges**

x	**2002**	2003	2004	2005	2006	2007	**2008**
$350(x - 2002) + 4085$	4085	4435	4785	5135	5485	5835	6185

Table 2.7 shows that the right side of the equation, $350(x - 2002) + 4085$, equals the left side, 6185, when $x = 2008$. Therefore, average tuition and fees reached $6185 in 2008.

▌ **Now Try Exercise 63**

Solving Equations Visually

▶ **REAL-WORLD CONNECTION** If the air temperature at ground level is 80°F, then the air temperature T at an altitude of x miles can be found using the formula

$$T = 80 - 19x.$$

A table of values corresponding to the expression $80 - 19x$ is shown in Table 2.8. (*Source:* Miller, A., and R. Anthes, *Meteorology*, 5th ed.)

TABLE 2.8 Air Temperature at an Altitude of x Miles

x	0	1	2	3	4	5	6
$80 - 19x$	80	61	42	23	4	-15	-34

Note that Table 2.8 does not show that air temperature decreases *gradually* as altitude increases. For example, the temperature is *not* 80°F from the ground up to an altitude of 1 mile, where it suddenly drops to 61°F. Rather, between the ground and 1 mile of altitude, air takes on all temperature values from 61°F to 80°F. The line graph in Figure 2.18 more accurately displays this situation. The line in Figure 2.18 displays the *graph* of the expression $80 - 19x$. It shows *all* values of the expression $80 - 19x$ for altitudes between 0 miles and 6 miles.

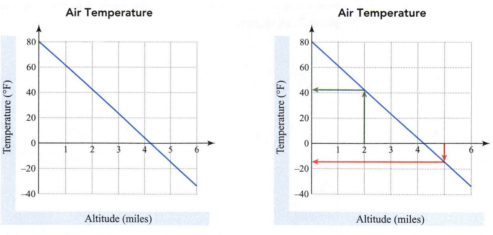

Figure 2.18 Temperatures at Various Altitudes   **Figure 2.19** Temperatures at Various Altitudes

Table 2.8 shows that at an altitude of 2 miles, the temperature is 42°F, and at an altitude of 5 miles, the temperature is -15°F. These values are displayed visually in Figure 2.19. The other values from Table 2.8 can be found visually in a similar way.

When a graph is provided, we can solve some equations visually.

SOLVING AN EQUATION VISUALLY

To solve an equation visually,

1. Use the graph to estimate a solution.
2. Check the estimated solution to be sure it is correct. If it is not, try a new value.

NOTE: A visual solution to an equation is an *estimate* of the actual solution. Be sure to check a visual solution in the given equation to be sure that it is correct.

EXAMPLE 6 **Solving an equation visually**

The graph of the expression $2x - 21$ is shown in Figure 2.20. Use Figure 2.20 to solve the equation $2x - 21 = -7$ visually.

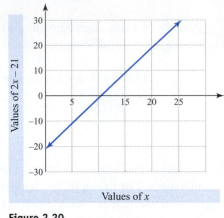

Figure 2.20

Solution
Locate -7 at the left edge of the graph. Move **horizontally** to the right to the graphed line. From this position, move **vertically** upward to the value of x. The solution appears to be about 7, as shown in Figure 2.21.

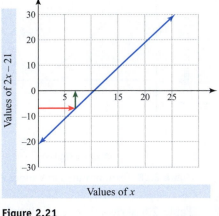

Figure 2.21

READING CHECK

• Why is it a good idea to check a solution that is found visually?

To see that 7 is the correct solution, check it in the given equation.

$$2x - 21 = -7 \qquad \text{Given equation}$$
$$2(7) - 21 \overset{?}{=} -7 \qquad \text{Replace } x \text{ with 7.}$$
$$14 - 21 \overset{?}{=} -7 \qquad \text{Multiply.}$$
$$-7 = -7 \checkmark \qquad \text{The solution checks.}$$

The solution to the equation $2x - 21 = -7$ is 7.

Now Try Exercise 57

2.6   Putting It All Together

CONCEPT	COMMENTS	EXAMPLES
Checking a Solution	To check a solution, replace *every* occurrence of the variable with the proposed solution and see if the resulting equation is true.	The solution to $-4x = 24$ is -6. $-4(-6) \stackrel{?}{=} 24$ Replace x with -6. $24 = 24$ ✓ The solution checks.
Solving Equations Using Guess-and-Check	Determine what portion of the equation is missing: factor, dividend, etc. Test sample values in the equation.	The solution to $x - 4 = 10$ is **14** because $14 - 4 = 10$.
Solving Equations Using Tables of Values	A table of values can be used to solve some equations by completing the table for various values of the variable and then selecting the solution from the table.	The solution to $5x + 3 = 8$ is **1**. | x	-1	0	1
---	---	---	---
$5x + 3$	-2	3	**8**	
Solving Equations Visually	1. Use the graph to estimate a solution. 2. Check the estimated solution to be sure it is correct. If it is not, try a new value.	The solution to $3x - 7 = 20$ is 9.

2.6   Exercises

MyMathLab   Math XL PRACTICE   WATCH   DOWNLOAD   READ   REVIEW

CONCEPTS AND VOCABULARY

1. A(n) _____ is a mathematical statement that two algebraic expressions are equal.

2. Any value of the variable that makes an equation true is a(n) _____ to the equation.

3. To check a solution, replace every occurrence of the _____ in the equation with the proposed solution and see if the resulting equation is true.

4. When solving an equation visually, be sure to _____ the solution in the given equation.

CHECKING SOLUTIONS

Exercises 5–20: Determine if the given value is a solution to the given equation.

5. -7, $2 - x = 9$

6. -12, $3 + n = -9$

7. -3, $4x - 2 = -8$

8. 3, $5a - 12 = -2$

9. -6, $\frac{3x - 2}{5} = 6$

10. 9, $\frac{52}{y + 4} = 4$

11. -3, $\qquad$ $\sqrt{1-y}=3y+10$

12. 1, $\qquad$ $|2w-8|=6$

13. 4, $\qquad$ $6+b\div 2=5$

14. 4, $\qquad$ $\sqrt{6m+1}=m+1$

15. -9, $\qquad$ $|w-8|=8-w$

16. -2, $\qquad$ $(3-x)^2=5x+15$

17. -7, $\qquad$ $x^2+2x=35$

18. -6, $\qquad$ $10-y\div 2=y+19$

19. -2, $\qquad$ $3x^2-x+5=15$

20. -1, $\qquad$ $x^3+5x-2=4$

SOLVING EQUATIONS

Exercises 21–40: Solve the equation.

21. $b+3=-12$ $\qquad$ **22.** $3x=36$

23. $11-z=16$ $\qquad$ **24.** $m-(-4)=3$

25. $-48\div d=-6$ $\qquad$ **26.** $n\div(-5)=-9$

27. $35=-7x$ $\qquad$ **28.** $7+y=-8$

29. $w^3=-1$ $\qquad$ **30.** $8=m^3$

31. $\sqrt{b}=8$ $\qquad$ **32.** $-\sqrt{a}=-3$

33. $(-x)^3=27$ $\qquad$ **34.** $(-y)^3=-1$

35. $-\sqrt{n}=-10$ $\qquad$ **36.** $\sqrt{a}=12$

37. $2x+1=11$ $\qquad$ **38.** $3x-2=10$

39. $18-2x=4$ $\qquad$ **40.** $25-3m=7$

Exercises 41–48: Complete the table. Then solve the given equation.

41. $x+2=1$

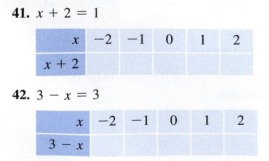

x	-2	-1	0	1	2
$x+2$					

42. $3-x=3$

x	-2	-1	0	1	2
$3-x$					

43. $3x+5=8$

x	-2	-1	0	1	2
$3x+5$					

44. $7x-4=-18$

x	-2	-1	0	1	2
$7x-4$					

45. $5+6x=17$

x	-2	-1	0	1	2
$5+6x$					

46. $3-4x=7$

x	-2	-1	0	1	2
$3-4x$					

47. $\sqrt{3-x}=3$

x	-13	-6	-1	2	3
$\sqrt{3-x}$					

48. $\sqrt{x+7}=2$

x	-7	-6	-3	2	9
$\sqrt{x+7}$					

Exercises 49–56: Solve the equation by making a table of values. Use $-3, -2, -1, 0, 1, 2,$ and 3 for the values of x in your table.

49. $x-2=-3$ $\qquad$ **50.** $6-x=5$

51. $-3x+8=14$ $\qquad$ **52.** $-2x+5=-1$

53. $4-9x=-14$ $\qquad$ **54.** $4-3x=7$

55. $1+3x=1$ $\qquad$ **56.** $8+2x=2$

Exercises 57–62: Solve the given equation visually.

57. $x-4=3$ $\qquad$ **58.** $12-2x=-6$

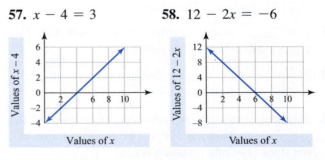

59. $-4x + 27 = 15$ **60.** $5x - 40 = -10$

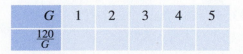

61. $6x - 38 = -20$ **62.** $-9x + 36 = 0$

APPLICATIONS

63. *Worldwide Internet Users* The number of Internet users I in millions during year x, where x is from 2004 to 2010, can be approximated by the formula

$$I = 168x - 335,778.$$

Replace I in the formula with 1566 and use a table of values to solve the resulting equation. In what year were there 1566 million Internet users? (*Source:* Internet World Stats.)

64. *HIV Infections* The cumulative number of HIV infections N in thousands for the United States in year x, where x is from 2002 to 2008, can be approximated by the formula

$$N = 42x - 83,197.$$

Replace N in the formula with 929 and use a table of values to solve the resulting equation. In what year did the number of HIV infections reach 929 thousand? (*Source:* Centers for Disease Control and Prevention.)

65. *Gas Mileage* The gas mileage M of a car using G gallons of gasoline to travel a distance of 120 miles is given by

$$M = \frac{120}{G}.$$

Complete the table and find the number of gallons used by a car traveling 120 miles if the car gets $M = 30$ miles per gallon.

G	1	2	3	4	5
$\frac{120}{G}$					

66. *Altitude and Temperature* If the air temperature on the ground is 75°F, then the air temperature T at an altitude of x miles is given by the formula

$$T = 75 - 19x.$$

Complete the table and find the altitude where the air temperature is -1°F.

x	1	2	3	4	5
$75 - 19x$					

67. *Medicare Costs* Total Medicare costs C in billions of dollars can be found using the formula

$$C = 18x - 35,750,$$

where x is a year from 1996 to 2008. Find the year when Medicare costs were \$250 billion by solving the equation $250 = 18x - 35,750$ visually, using the following graph. (*Source:* Office of Management and Budget.)

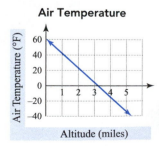

Medicare Costs

68. *Air Temperature* If the air temperature on the ground is 61°F, then the air temperature T at an altitude of x miles is given by the formula

$$T = 61 - 19x.$$

Find the altitude where the air temperature is -15°F by solving the equation $-15 = 61 - 19x$ visually, using the following graph.

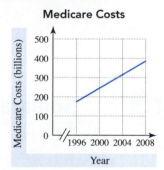

Air Temperature

WRITING ABOUT MATHEMATICS

69. For the equation $3x = 21$, give a value of x that is a solution and a value of x that is not a solution. Explain how to check these values.

70. Explain why the solutions to some equations may be difficult to find using a table of values.

SECTIONS 2.5 and 2.6 · Checking Basic Concepts

1. Evaluate each expression.
(a) $40 + (-28) \div 4 - 10$
(b) $4 - |5 + 3^2 \div (-9)|$
(c) $(5 - 9)^2 - 5\sqrt{2 \cdot 18}$

2. Evaluate the expression for the given values of the variables.
(a) $6c - 5d$, for $c = -2, d = -4$
(b) $((7 - v)^2 \div w) + v$, for $v = 1, w = -3$
(c) $\dfrac{(5 - g) + 14}{(\sqrt{h} - 5)^2}$, for $g = -5, h = 9$

3. Is -2 a solution to $6a - 12 = 0$?

4. Is -3 a solution to $(1 - x)^2 = 1 - 5x$?

5. Solve each equation.
(a) $b + 2 = -3$
(b) $-24 \div d = -4$

6. Solve $5 - 2x = 7$ by completing the table.

x	-2	-1	0	1	2
$5 - 2x$					

7. *Converting Temperature* To convert a temperature C in degrees Celsius to an equivalent temperature F in degrees Fahrenheit, use the formula

$$F = \frac{9C}{5} + 32.$$

Find the Fahrenheit temperature that is equivalent to $-50°C$.

CHAPTER 2 Summary

SECTION 2.1 ■ INTEGERS AND THE NUMBER LINE

Positive Numbers   A positive number is greater than 0.   **Examples:** $7, +17, 152$, and $+10{,}079$.

Negative Numbers   A negative number is less than 0.   **Examples:** $-1, -77, -509$, and -6592.

Signed Numbers   Signed numbers are the negative numbers, positive numbers, and zero.

Opposite (Additive Inverse)   The opposite (or additive inverse) of a number a is written as $-a$.

    Examples: The opposite of 13 is -13, and $13 + (-13) = 0$.
                  The opposite of -32 is 32, and $-32 + 32 = 0$.

Double Negative Rule   Let a be any number. Then $-(-a) = a$.

    Examples: $-(-7) = 7$ and $-(-25) = 25$

Integers $\ldots, -3, -2, -1, 0, 1, 2, 3, \ldots.$

    The integers include the natural numbers, zero, and the opposites of the natural numbers.

Absolute Value   The absolute value of an integer equals its distance from 0 on a number line. The absolute value of a is written as $|a|$.

    Examples: $|4| = 4$, $|-17| = 17$,   and $|0| = 0$

SECTION 2.2 ■ ADDING INTEGERS

Adding Integers with Like Signs   **1.** Find the sum of the absolute value of the integers.
    2. Keep the common sign of the two integers as the sign of the sum.

    Example: Since $|-2| + |-11| = 13$, we know that $-2 + (-11) = -13$.

Adding Integers with Unlike Signs	**1.** Find the absolute value of the integers.
	2. Subtract the smaller absolute value from the larger absolute value.
	3. Keep the sign of the integer with the larger absolute value as the sign of the sum.
	Example: To add $4 + (-9)$, find the absolute values: $\|4\| = 4$ and $\|-9\| = 9$. Subtract $9 - 4 = 5$. Because $\|-9\| > \|4\|$, the sum is negative. $4 + (-9) = -5$

Addition Properties

Property	**Examples**
1. Commutative Property	$-24 + 9 = 9 + (-24)$
2. Associative Property	$(-3 + 5) + 1 = -3 + (5 + 1)$
3. Identity Property	$-2 + 0 = -2$ and $0 + (-2) = -2$
4. Inverse Property	$5 + (-5) = 0$ and $-5 + 5 = 0$

Adding Integers Using a Number Line

Starting at 0, draw an arrow representing the first addend. From the tip of this arrow, draw an arrow representing the second addend. The sum is located at the tip of the second arrow.

Example: To add $-2 + 3$ on a number line, draw arrows as shown. The sum is **1**.

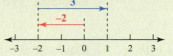

Adding Integers Using Symbols

Using the symbol ⌢ to represent a positive unit and the symbol ⌣ to represent a negative unit, draw the appropriate number of units for each addend. Form "zeros" when possible.

Example: Add $2 + (-5)$ as shown. Ignoring the two "zeros," the sum is -3.

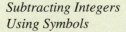

SECTION 2.3 ■ SUBTRACTING INTEGERS

Subtracting Integers by Adding the Opposite

If the variables a and b represent two numbers, then $a - b = a + (-b)$.

Examples: $4 - (-1) = 4 + 1 = 5$ and $-3 - 7 = -3 + (-7) = -10$

Subtracting Integers Using a Number Line

Addition means **step to the right**. **Subtraction** means **step to the left**. A **negative** sign means **stop and change direction**.

Example: Subtract $0 - (-2)$ on a number line as shown. The difference is **2**.

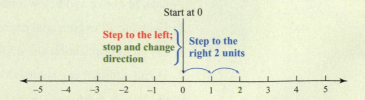

Subtracting Integers Using Symbols

The symbol ⌢ can be used to represent a positive unit and the symbol ⌣ can be used to represent a negative unit. Think of subtraction as "take away."

Example: Subtract $-1 - (-4)$ as shown. The difference is 3.

SECTION 2.4 ■ MULTIPLYING AND DIVIDING INTEGERS

Multiplying Integers with Like Signs

The product of two integers with *like* signs is positive.

Examples: $-3(-4) = 12$ and $5 \times 2 = 10$

Multiplying Integers with Unlike Signs

The product of two integers with *unlike* signs is negative.

Examples: $-4 \cdot 7 = -28$ and $3 \times (-6) = -18$

Multiplication Properties

Property	Examples
1. Commutative Property	$5 \cdot (-6) = -6 \cdot 5$
2. Associative Property	$(-3 \cdot 2) \cdot 7 = -3 \cdot (2 \cdot 7)$
3. Identity Property	$-8 \cdot 1 = -8$ and $1 \cdot (-8) = -8$
4. Zero Property	$0 \cdot (-3) = 0$ and $-3 \cdot 0 = 0$
5. Distributive Properties	$-2(1 + 5) = -2(1) + (-2)(5)$ $-3(4 - 7) = -3(4) - (-3)(7)$

Products with More Than Two Factors

If there is an *even* number of negative factors, the product is positive. If there is an *odd* number of negative factors, the product is negative.

Examples: The product $-4 \cdot 7 \cdot 1 \cdot (-6)$ is positive.

The product $-5 \cdot (-2) \cdot 3 \cdot (-8)$ is negative.

Exponential Expressions with Integers as the Base

Exponential expressions can have integers as the base.

Examples: $(-6)^2 = 36$ and $-6^2 = -36$

Dividing Integers with Like Signs

The quotient of two integers with *like* signs is positive.

Examples: $-12 \div (-4) = 3$ and $10 \div 5 = 2$

Dividing Integers with Unlike Signs

The quotient of two integers with *unlike* signs is negative.

Examples: $-15 \div 5 = -3$ and $20 \div (-4) = -5$

Number of Square Roots

A positive integer a has one positive and one negative square root. The positive square root, or *principal square root*, is written as $\sqrt{a}$, and the negative square root is written as $-\sqrt{a}$.

Examples: $\sqrt{81} = 9$ and $-\sqrt{49} = -7$

SECTION 2.5 ■ ORDER OF OPERATIONS; AVERAGES

Order of Operations

1. Do all calculations within grouping symbols, such as parentheses and radicals, or above and below a fraction bar.

2. Evaluate all exponential expressions.

3. Do all multiplication and division from *left to right*.

4. Do all addition and subtraction from *left to right*.

Example: $5 - 36 \div (2 - 5)^2 = 5 - 36 \div (-3)^2$

$$= 5 - 36 \div 9$$

$$= 5 - 4$$

$$= 1$$

Evaluating Algebraic Expressions	Replace each variable in the expression with its given value and then evaluate using the order of operations agreement.

Example: Evaluating $8 - 3x$ for $x = -7$ gives
$$8 - 3(-7) = 8 - (-21) = 8 + 21 = 29.$$

Average   Average is a measure of the middle of a group of numbers.

$$\text{Average} = \frac{\text{the sum of a list of numbers}}{\text{the number of numbers in the list}}$$

Example: The average of 3, 6, and 15 is

$$\frac{3 + 6 + 15}{3} = \frac{24}{3} = 8.$$

SECTION 2.6 ■ SOLVING EQUATIONS WITH INTEGER SOLUTIONS

Checking a Solution   To check a solution, replace every occurrence of the variable with the proposed solution and check to see if the resulting equation is true.

Example: Check that the solution to $3 + 2x = 17$ is 7.

$$
\begin{array}{ll}
3 + 2x = 17 & \text{Given equation} \\
3 + 2(7) \stackrel{?}{=} 17 & \text{Replace } x \text{ with 7.} \\
17 = 17 \ \checkmark & \text{Simplify; the solution checks.}
\end{array}
$$

Solving Equations Using Guess-and-Check   Guess-and-check is a strategy that can be used to solve some basic equations.

Example: The solution to $5x = -15$ is -3 because $5(-3) = -15$.

Solving Equations Using Tables of Values   We can solve some equations by completing a table of values for the variable and selecting the solution from the table.

Example: The solution to $1 - 3x = -2$ is **1**.

x	-1	0	**1**
$1 - 3x$	4	1	**-2**

Solving Equations Visually   To solve an equation visually,

1. Use the graph to estimate a solution.

2. Check the estimated solution. If it is not correct, try a new value.

Example: The solution to $-4x + 29 = 5$ is 6. Check this solution.

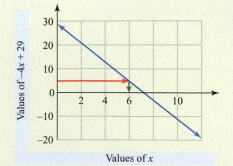

CHAPTER 2   Review Exercises

SECTION 2.1

Exercises 1 and 2: Find the opposite of the given integer.

1. 19

2. −52

Exercises 3 and 4: Simplify the expression.

3. −(−31)

4. −(−(−2))

Exercises 5 and 6: Graph the integers on a number line.

5. −3, −1, 2

6. −4, 0, 3

Exercises 7–10: Place the correct symbol, < or >, in the blank between the integers.

7. −7 _____ 19

8. 2 _____ −5

9. −11 _____ −15

10. −32 _____ −3

Exercises 11–14: Simplify the absolute value expression.

11. −|6|

12. −|−1|

13. |−0|

14. |−12|

Exercises 15 and 16: Place the correct symbol, <, >, or =, in the blank between the expressions.

15. 2 _____ −|2|

16. −|8| _____ −8

Exercises 17 and 18: The following table lists the change in temperature from 6:00 A.M. to 6:00 P.M. for selected days of the week.

Day	Monday	Tuesday	Friday	Sunday
Change	−4°	15°	19°	−8°

17. Which day had the largest decrease in temperature from 6:00 A.M. to 6:00 P.M.?

18. Was there a temperature increase or decrease between 6:00 A.M and 6:00 P.M. on Friday?

SECTION 2.2

Exercises 19–26: Find the sum.

19. −14 + 13

20. −3 + (−12)

21. −21 + (−30)

22. 45 + (−23)

23. −14 + 45 + 22

24. 27 + (−53) + 8

25. −42 + (−21) + (−37) + 54

26. 105 + (−35) + (−64) + 13

Exercises 27 and 28: Evaluate the expression $x + y$ for the given values of the variables.

27. $x = 12, y = -7$

28. $x = -2, y = -3$

Exercises 29–32: State the addition property illustrated by the given equation.

29. 48 + (−48) = 0

30. 1 + (−8 + 7) = (1 + (−8)) + 7

31. 54 + (−59) = −59 + 54

32. −189 + 0 = −189

Exercises 33 and 34: Use a number line to find the sum.

33. 2 + (−5)

34. −5 + 9

Exercises 35 and 36: Perform the addition visually, using the symbols ⌒ and ⌣.

35. 3 + (−7)

36. −6 + 8

SECTION 2.3

Exercises 37–42: Find the difference.

37. 15 − (−4)

38. −16 − 16

39. −23 − 7

40. 11 − 29

41. −17 − 0

42. 0 − 22

Exercises 43 and 44: Evaluate the expression $x - y$ for the given values of the variables.

43. $x = -5, y = -12$

44. $x = -4, y = 18$

Exercises 45–48: Simplify the expression.

45. −7 − 13 + (−1)

46. 8 + (−18) − 2

47. −33 − (−15) + (−40) + 9

48. 101 − (−99) + (−50) + 10

Exercises 49–52: Use a number line to evaluate.

49. −6 − (−9)

50. 5 − 7

51. 1 − 7 + (−2) − (−5)

52. 2 + (−7) + 5 − (−9)

Exercises 53 and 54: Perform the subtraction visually, using the symbols ⌒ and ⌣.

53. $-7 - (-4)$ **54.** $1 - (-8)$

SECTION 2.4

Exercises 55–62: Find the product or quotient.

55. $-42 \div 7$ **56.** $-10 \times (-9)$

57. $-3 \cdot 8$ **58.** $-14 \div (-14)$

59. $\dfrac{-75}{5}$ **60.** $\dfrac{-36}{-12}$

61. $3(-1)(3)(-2)(-5)$

62. $-2(4)(3)(2)(-1)$

Exercises 63–66: State the multiplication property illustrated by the given equation.

63. $2 \cdot (-12 \cdot 31) = (2 \cdot (-12)) \cdot 31$

64. $1 \times (-417) = -417$

65. $-6522 \cdot 0 = 0$

66. $-5(-2 + 9) = -5(-2) + (-5)(9)$

Exercises 67 and 68: Evaluate the exponential expression.

67. -7^2 **68.** $(-7)^2$

Exercises 69 and 70: Simplify the expression if possible.

69. $-\sqrt{16}$ **70.** $\sqrt{-49}$

Exercises 71–74: Evaluate the expression for the given values of the variables, if possible.

71. $\dfrac{2a}{b}$ $a = -20, b = 5$

72. $3 \cdot (-x) \cdot (-y)$ $x = 3, y = -5$

73. $-\sqrt{y}$ $y = 25$

74. $\sqrt{a}$ $a = -36$

SECTION 2.5

Exercises 75–80: Evaluate the expression.

75. $-2 \cdot (10) - 12 \div 6$

76. $5 - 3^2 \div (4 - 7)$

77. $\sqrt{52 + 12} + (-7)$

78. $\sqrt{-5^2 + 50} - 3^2$

79. $\dfrac{-39 + 6(-4)}{-3}$ **80.** $-\left|\dfrac{39 - 49}{2 \cdot (-1)}\right|$

Exercises 81–84: Insert parentheses as needed in the expression in order to make it equal to 0.

81. $-10 + 5 \cdot 8 - 6$ **82.** $14 - 16 - 3 - 1$

83. $7 - 11 \cdot 4^2 + 64$ **84.** $-3^2 \div 5 - 2 + 3$

Exercises 85–88: Evaluate the expression for the given values of the variables.

85. $12v - 3w$, for $v = -1, w = -4$

86. $3m + |2^3 + n| \div 4$, for $m = 2, n = -8$

87. $w + ((3 - v)^2 \div w)$, for $v = 9, w = -2$

88. $\sqrt{13 - p^2} - 2q^2$, for $p = -2, q = 2$

SECTION 2.6

Exercises 89–92: Determine if the given value is a solution to the given equation.

89. 2, $6a - 14 = -2$

90. -3, $|3w - 2| = 11$

91. -2, $\dfrac{5x + 1}{3} = 3$

92. -3, $\sqrt{1 - y} = 4y + 5$

Exercises 93–96: Solve the equation.

93. $b + 9 = -2$ **94.** $-27 = m^3$

95. $\sqrt{x} = 6$ **96.** $n \div (-5) = -4$

Exercises 97 and 98: Complete the table shown. Then solve the given equation.

97. $3x - 5 = 1$

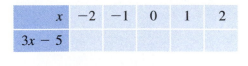

x	-2	-1	0	1	2
$3x - 5$					

98. $7 - 2x = 9$

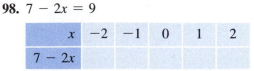

x	-2	-1	0	1	2
$7 - 2x$					

Exercises 99 and 100: Solve the equation visually.

99. $-3x + 28 = 10$ **100.** $4x - 30 = -10$

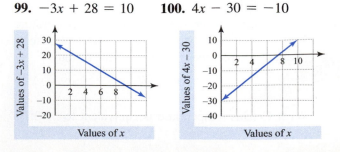

APPLICATIONS

101. *Religion* The following table lists the change from 2001 to 2008 in the number of U.S. citizens who describe themselves as part of selected religious communities.

Religion	Lutheran	Mormon	Jewish	Buddhist
Change	−906,000	461,000	−157,000	107,000

Source: American Religious Identification Survey, 2008.

(a) Which of these religions had the largest increase in membership from 2001 to 2008?

(b) List the religions that experienced a decline in membership from 2001 to 2008.

102. *Finances* A checking account starts the month with a balance of $534. The following positive entries (deposits) and negative entries (withdrawals) are made in the checking register.

$$-72, -125, 300, \text{ and } -45$$

What is the ending balance?

103. *Music Sales* The total music sales *S* in thousands of dollars for a small band during year *x* can be computed using the formula

$$S = 7(x - 2005) + 8.$$

Find the year when the music sales were $15,000 by solving the equation $15 = 7(x - 2005) + 8$ visually, using the following graph.

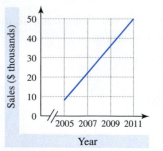

Music Sales

104. *Pittsburgh Population* From January 2000 to January 2008, the population of the Pittsburgh area decreased by about 60,000 people. If the decrease was the same for each of the 8 years, write the yearly decrease as a quotient of integers and find the yearly decrease. (*Source:* New York Times.)

105. *Converting Temperature* To convert a temperature *F* given in degrees Fahrenheit to an equivalent temperature *C* in degrees Celsius, use the formula

$$C = \frac{5(F - 32)}{9}.$$

What Celsius temperature is equivalent to −13°F?

106. *Citrus Imports* The amount of citrus in millions of pounds imported into the United States during year *x* can be found using

$$C = 133(x - 2002) + 707.$$

Replace *C* in the formula with 1239 and use a table of values to solve the resulting equation. For the values of *x* in your table, use 2003 to 2008. In what year were 1239 million pounds of citrus imported? (*Source:* U.S. Department of Agriculture.)

107. *Temperature Change* Over a four-day period in January, a city in Iowa had the following weather changes. The low temperature on Monday was −2°F. On Tuesday, the low temperature increased by 10°F. The low temperature then fell 3°F on Wednesday. Finally, on Thursday, the low temperature decreased by 8°F. What was the low temperature on Thursday?

108. *Late Fees* Each time that a credit card holder fails to make a credit card payment by the due date, a late fee appears on the monthly statement as −$19. If the card holder missed the payment due date eight times over the past year, write the total charges for late fees as a product of integers and find the total charges for late fees.

109. *Corporate Losses* An Internet company reported fourth quarter earnings of −$16 million. During the same quarter, an investment company reported earnings that were $9 million higher than those of the Internet company. What were the investment company's earnings for the fourth quarter?

110. *Average Temperature* Find the average of the list of Celsius temperatures.

$$-8°, -4°, 13°, 16°, -3°, \text{ and } 4°C$$

Exercises 111 and 112: *Finances* *Absolute value can be used to find the "size" of a financial quantity. Even though −$63 is a negative value (a debt), it is a larger financial quantity than a positive balance of $41 in a checking account (an asset). The debt is larger than the asset because $|-63| > |41|$. Use absolute value to determine which of each pair of financial quantities is larger.*

111. Owing money to a friend: −$64
Cash in savings jar: $32

112. An employee paycheck: $850
A credit card debt: −$753

CHAPTER 2   Test

1. Graph the integers -3, 0, and 4 on a number line.

2. Evaluate $x + y$ for $x = -3$ and $y = 8$.

Exercises 3 and 4: Place the correct symbol, $<$, $>$, or $=$, in the blank between the expressions.

3. -19 ____ -25 4. $-|8|$ ____ -8

Exercises 5–8: Perform the arithmetic.

5. $16 + (-21)$ 6. $-38 - 12$

7. $-12 \cdot (-7)$ 8. $-63 \div 9$

9. Evaluate $3 - 5 + (-1) - (-4)$.

10. Evaluate $2(-1)(-3)(-2)(7)$.

11. Evaluate -11^2.

12. Simplify the expression $-\sqrt{100}$.

Exercises 13–18: Evaluate the expression.

13. $-21 + 4 \cdot 7 - 15$

14. $9 - 4^2 \div 8 - 14$

15. $\sqrt{-13 + 38} - (-3)$

16. $5 - |6^2 \div (-4)|$

17. $\dfrac{(13 - 7) \cdot 5}{|33 - 48| \cdot 2}$ 18. $\dfrac{(-5 - 3) \cdot 3}{2^3 - \sqrt{4}}$

Exercises 19 and 20: Check each solution as specified.

19. Is 10 a solution to $\dfrac{-42}{y + 4} = -3$?

20. Is 7 a solution to $\sqrt{5m + 1} = -m + 1$?

Exercises 21 and 22: Solve the equation.

21. $72 = -12x$

22. $-\sqrt{n} = -3$

23. Complete the table and solve $5 - 4x = -3$.

x	-2	-1	0	1	2
$5 - 4x$					

24. Solve the equation $-3x + 11 = -10$ visually.

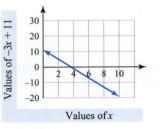

25. ***Net Worth*** A student with a net worth of $-\$25{,}700$ inherits an amount of money so that his net worth becomes $\$107{,}200$. How much money did he inherit?

26. ***Altitude and Temperature*** If the air temperature on the ground is $70°F$, then the air temperature T at an altitude of x miles is given by the formula

$$T = 70 - 19x.$$

Complete the table and determine the altitude where the air temperature is $-6°F$.

x	0	1	2	3	4
$70 - 19x$					

CHAPTERS 1–2   Cumulative Review Exercises

1. Identify the digit in the ten-thousands place in the number 9,145,283,705.

2. Write 32,010 in expanded form.

Exercises 3–6: Perform the arithmetic.

3.   289
   + 5775

4.   19,043
   − 7 938

5. 23 · 279

6. 3672 ÷ 45

7. Write 7 · 7 · 7 using exponential notation.

8. Solve the equation $x - 13 = 6$.

9. Round 32,673,905 to the nearest million.

10. Estimate the sum $789 + 502 + 197$ by rounding each value to the nearest hundred.

11. Approximate $\sqrt{80}$ to the nearest whole number.

12. Evaluate $3 \cdot 15 - 60 \div 10$.

13. Simplify the expression $4x + (x - 5)$.

14. Place the correct symbol, $<, >,$ or $=$, in the blank between the whole numbers: $-|5|$ _____ 5.

Exercises 15–18: Perform the arithmetic.

15. $-14 + (-3)$

16. $-3 - (-8)$

17. $-50 \div 5$

18. $-5 \cdot (-20)$

Exercises 19 and 20: Use a number line to evaluate.

19. $3 + (-4)$

20. $3 + (-4) - (-5)$

21. Multiply $4(-2)(5)(-1)(-2)$.

22. State the multiplication property illustrated by the equation $-41 \times 0 = 0$.

23. Evaluate $6 - 4^2 \div (3 - 11)$.

24. Evaluate $(w + (2 - v)^2) \div 2$ for $v = 3, w = 1$.

25. Is 8 a solution to $4 + b \div 2 = -7$?

26. Complete the table and solve $7x - 12 = -5$.

x	-2	-1	0	1	2
$7x - 12$					

27. *Finances* A checking account starts the month with a balance of $1296. The following positive entries (deposits) and negative entries (withdrawals) are made in the checking register.

$$-504, -81, 700, \text{ and } -432$$

What is the ending balance?

28. *Converting Temperature* To convert a temperature C given in degrees Celsius to an equivalent temperature F in degrees Fahrenheit, use the formula

$$F = \frac{9C}{5} + 32.$$

Find the Fahrenheit temperature that is equivalent to $-20°C$.

29. *Heart Rate* The average heart rate R in beats per minute (bpm) of an animal weighing W pounds can be approximated by

$$R = \frac{885\sqrt{W}}{W}.$$

Find the heart rate for a 25-pound dog.

30. *Geometry* Find the perimeter of the figure.

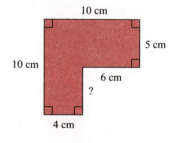

3 Algebraic Expressions and Linear Equations

Whether you think you can or that you can't, you are usually right.

— HENRY FORD

Every November in Bridgeville, Delaware, a world-class competition involving mathematics, physics, and engineering occurs. However, this competition is not held at an elite university, and the competitors are not among the world's top scientists. What started in 1986 as a bet among four friends to see who could toss a pumpkin the farthest has become the World Championship Punkin' Chunkin' competition. The winning toss in 1986 was just over 126 feet. Today, top competitors are creating machines that can hurl an 8- to 10-pound pumpkin over 4000 feet—almost a mile!

In this chapter, we will learn how to solve *linear equations*. The ability to solve mathematical equations makes it possible for people to create airplanes, cell phones, submarines, electric cars, and even machines that can hurl a pumpkin nearly a mile.

3.1   Simplifying Algebraic Expressions

Reviewing Concepts • Combining Like Terms • Adding Expressions •
Subtracting Expressions • Multiplying Expressions • Simplifying Expressions

A LOOK INTO MATH ▶

The U.S. government's tax code involves complicated tax formulas. When working with such formulas, it would be helpful if several formulas could be *combined* into one formula. In other cases, some large formulas might be easier to evaluate if they were more concise and *simple*. In mathematics, we sometimes *combine like terms* when we *simplify* expressions. In this section, we will see how some algebraic expressions can be simplified to make them easier to evaluate for given values of the variables.

Reviewing Concepts

In Chapters 1 and 2, we discussed how to simplify algebraic expressions involving whole numbers and integers. Here we repeat some of the most important concepts and definitions.

A **term** is a number, a variable, or a product of numbers and variables raised to powers. Examples of terms include

$$6, \quad -3x, \quad 5y^2, \quad 17ab, \quad \text{and} \quad -7u^2v.$$

The number within a term is called the **coefficient**. If no number appears in a term, then the coefficient is 1 or -1. For example, the coefficients of the terms $2x$, $-a^2$, and $-14y$ are 2, -1, and -14, respectively.

Two terms with the same variables raised to the same powers are called **like terms**. Examples of like terms include

$$-5 \text{ and } 8, \quad -2x \text{ and } 93x, \quad y^2 \text{ and } -5y^2, \quad \text{and} \quad -ab^2 \text{ and } -2ab^2.$$

Two terms that are not like terms are **unlike terms**. Examples of unlike terms include

$$-7 \text{ and } 2x, \quad -3a \text{ and } 10b, \quad 4y^2 \text{ and } 9y, \quad \text{and} \quad -4x^2y \text{ and } -3xy^2.$$

We can combine (add or subtract) like terms but cannot combine unlike terms.

Several properties of arithmetic are useful when simplifying algebraic expressions. They are listed here with examples corresponding to each property.

Arithmetic Property	**Examples**
Commutative Property	$3x + (-7x) = (-7x) + 3x$
	$(5x)(-2) = (-2)(5x)$
Associative Property	$(-2y + 4) + 3 = -2y + (4 + 3)$
	$(-b \cdot 4) \cdot 2 = -b \cdot (4 \cdot 2)$
Distributive Property	$5(-7y + 1) = 5(-7y) + 5(1)$
	$6(-2a - 3) = 6(-2a) - 6(3)$

NOTE: The commutative and associative properties allow us to add terms in *any* order or multiply terms in *any* order.

Combining Like Terms

When an expression contains like terms, it can often be simplified using the distributive property. To do this, it is helpful to write the distributive property in a more practical form. For example, when we apply the commutative property for multiplication to the left side and to each term on the right side of the distributive property,

$$c(a + b) = ca + cb,$$

the result is an equation that is equivalent to the distributive property,

$$(a + b)c = ac + bc.$$

Switching the expression on the left side of the equal sign with the expression on the right side results in the following form of the distributive property,

$$ac + bc = (a + b)c.$$

This same process works when using the distributive property with subtraction.

READING CHECK

- What does it mean to combine like terms?

DISTRIBUTIVE PROPERTIES FOR COMBINING LIKE TERMS

If the variables a, b, and c represent numbers, then

$$ac + bc = (a + b)c \qquad \text{and} \qquad ac - bc = (a - b)c.$$

The next example demonstrates how these distributive properties can be used to simplify algebraic expressions.

EXAMPLE 1 **Combining like terms**

Simplify each expression by combining like terms.
(a) $3x + 11x$ (b) $7y^2 - 13y^2$ (c) $8a - a + 3$

Solution
(a) $3x + 11x = (3 + 11)x = 14x$
(b) $7y^2 - 13y^2 = (7 - 13)y^2 = -6y^2$
(c) The coefficient of the middle term a is understood to be 1. The first two terms are like terms, so $8a - 1a + 3 = (8 - 1)a + 3 = 7a + 3$.

❚ **Now Try Exercises 9, 11, 13**

Because subtraction is neither commutative nor associative, it is helpful to change each subtraction in an algebraic expression to the addition of its opposite before combining like terms.

EXAMPLE 2 **Combining like terms**

Simplify each expression by combining like terms.
(a) $5x + 9 - 2x - 3$ (b) $3y - 18 + 6y - 2$

Solution
(a) Begin by changing each subtraction to the addition of its opposite.

$$
\begin{aligned}
5x + 9 - 2x - 3 &= 5x + 9 + (-2x) + (-3) && \text{Add the opposite.} \\
&= 5x + (-2x) + 9 + (-3) && \text{Commutative property} \\
&= [5 + (-2)]x + 9 + (-3) && \text{Distributive property} \\
&= 3x + 6 && \text{Simplify.}
\end{aligned}
$$

(b)
$$
\begin{aligned}
3y - 18 + 6y - 2 &= 3y + (-18) + 6y + (-2) && \text{Add the opposite.} \\
&= 3y + 6y + (-18) + (-2) && \text{Commutative property} \\
&= (3 + 6)y + (-18) + (-2) && \text{Distributive property} \\
&= 9y + (-20) && \text{Simplify.} \\
&= 9y - 20 && \text{Change to subtraction.}
\end{aligned}
$$

❚ **Now Try Exercises 17, 19**

Adding Expressions

When two or more expressions are added, we often use parentheses to identify the expressions in the sum. For example, in the sum $(3x - 5) + (2x + 7)$, we are adding the expressions $3x - 5$ and $2x + 7$. However, the parentheses are simply grouping symbols and are not needed to complete the addition.

$$(3x - 5) + (2x + 7) = 3x - 5 + 2x + 7$$

The next example illustrates how to add expressions.

EXAMPLE 3 **Adding expressions**

Simplify each sum.
(a) $(3x - 1) + (4x + 5)$ **(b)** $(7y^2 + 3) + (y^2 + 5y)$

Solution
(a) Begin by removing the parentheses from the sum.

$$
\begin{aligned}
(3x - 1) + (4x + 5) &= 3x - 1 + 4x + 5 & &\text{Remove grouping symbols.}\\
&= 3x + (-1) + 4x + 5 & &\text{Add the opposite.}\\
&= 3x + 4x + (-1) + 5 & &\text{Commutative property}\\
&= (3 + 4)x + (-1) + 5 & &\text{Distributive property}\\
&= 7x + 4 & &\text{Simplify.}
\end{aligned}
$$

(b) The coefficient of the second y^2-term is understood to be **1**.

$$
\begin{aligned}
(7y^2 + 3) + (y^2 + 5y) &= 7y^2 + 3 + y^2 + 5y & &\text{Remove grouping symbols.}\\
&= 7y^2 + 1y^2 + 5y + 3 & &\text{Commutative property}\\
&= (7 + 1)y^2 + 5y + 3 & &\text{Distributive property}\\
&= 8y^2 + 5y + 3 & &\text{Simplify.}
\end{aligned}
$$

▮ **Now Try Exercises 21, 25**

NOTE: In Example 3(b), since the exponent on y in the term $5y$ is 1, we cannot combine $5y$ with the y^2-terms (which each have an exponent of 2).

Subtracting Expressions

In Chapter 2, we found that subtracting one integer from another is the same as adding the first integer to the opposite of the second integer. For example, $3 - 9 = 3 + (-9)$. We can extend this idea to include the subtraction of one expression from another. To do so, we must first understand the meaning of *the opposite of an expression*. The next example shows us how to find the opposite of an expression by using a general rule.

OPPOSITE OF AN EXPRESSION

If the variable a represents a number or an expression, then

$$-a = -1 \cdot a.$$

CHAPTERS 1–3   Cumulative Review Exercises

1. Identify the digit in the hundred-thousands place in the number 4,591,083,276.

2. Write $5 \cdot 5 \cdot 5 \cdot 5$ using exponential notation.

3. Round 79,401 to the nearest ten-thousand.

4. Estimate the sum $2989 + 4002 + 997$ by rounding each value to the nearest thousand.

5. Approximate $\sqrt{120}$ to the nearest whole number.

6. Evaluate $65 \div 5 + 3 \cdot 2$.

7. Simplify the expression $3x - (x - 2)$.

8. Place the correct symbol, $<$, $>$, or $=$, in the blank between the two numbers: $-|7|$ _____ -7.

Exercises 9 and 10: Perform the arithmetic.

9. $-45 \div (-9)$ 10. $-14 - (-6)$

11. Evaluate $4 - 4^2 \div (7 - 3)$.

12. Evaluate $x + (3 - y)^2 \div 2y$ for $x = 7$, $y = -1$.

13. Is -14 a solution to $4 + b \div 2 = -3$?

14. Complete the table and solve $4x - 7 = -3$.

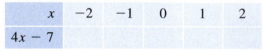

x	-2	-1	0	1	2
$4x - 7$					

Exercises 15 and 16: Simplify the expression.

15. $-4w - (w + 3)$ 16. $(2y - 1) + (5y - 8)$

Exercises 17 and 18: Translate each sentence to an equation using x as the variable. Do not solve the equation.

17. Ten equals a number increased by 7.

18. Triple the sum of a number and 6 equals -18.

Exercises 19–22: Solve the equation.

19. $72 = -8m$ 20. $-5n + 9 = -6n$

21. $9 + 7x - 4 = 2x - 5 - 5x$

22. $3 + (6x - 9) = 7x - (6 + 3x)$

Exercises 23 and 24: Solve the number problem by finding the value of the unknown number.

23. Subtracting 3 from the product of 2 and a number results in 6 more than the number.

24. The product of -2 and the sum of a number and 8 is the same as the number decreased by 1.

25. *Buying DVDs* What is the maximum number of DVDs costing $12 each that a person can buy with $100? How much change will the person receive?

26. *Geometry* Find the perimeter of the figure.

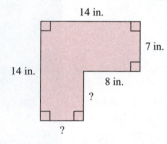

27. *ATM Fees* Each time that a credit card holder uses an ATM machine for a cash advance, a transaction fee appears on the credit card statement. If a year-end summary shows $-\$84$ for 42 ATM transactions, write the fee for a single ATM transaction as a negative integer.

28. *Net Worth* A person with a net worth of $-\$65,200$ wins money in a lottery game so that his net worth becomes $\$134,800$. How much money did the person win?

29. *Salary* Before receiving a raise in salary, a worker earned $1400 per week. Write an equation for which the solution is the amount of the worker's raise if the new salary is $1550 per week.

30. *Farm Land* In 2009, there were 11 million fewer acres of farm land in Arkansas than there were in California. If the sum of acreage for these two states was 39 million acres, find the number of farm land acres in each state in 2009. (*Source:* U.S. Department of Agriculture, 2007.)

4 Fractions

High achievement always takes place in the framework of high expectation.

— JACK KINDER

Throughout history, many calendars have been created to record time accurately. One difficulty with creating an accurate calendar lies in the fact that neither the astronomical month nor year can be measured in a whole number of days. For example, the true length of one year is 365 whole days and part of an additional day. This extra *fraction* of a day is the reason why we have leap years.

Fractions are commonly used in measuring various lengths of time. When a clock reads 5:30 P.M. we say that it is *half* past five. A company's fourth *quarter* earnings refer to the earnings for the last three months of the year. A semester in school is *one-half* of a school year. In addition to their use in measuring and recording time, fractions also occur in retail sales, cooking, carpentry, and a variety of other everyday situations.

4.1 Introduction to Fractions and Mixed Numbers

Fractions • Rational Numbers • Improper Fractions and Mixed Numbers • Graphing Fractions and Mixed Numbers

A LOOK INTO MATH ▶ Although it may not be obvious, the process of preparing and serving a meal almost always involves *fractions*. A recipe may call for one-fourth cup of cooking wine or one-third stick of butter. Four friends can each have two slices of pizza when each slice is one-eighth of the pizza. Even the nutritional information listed on the packaging of some food products may require the use of fractions to determine the total amount of fiber or carbohydrates consumed. Whenever we work with a part of a whole, we are dealing with fractions.

Fractions

NEW VOCABULARY

☐ Fraction
☐ Numerator
☐ Denominator
☐ Rational numbers
☐ Proper fraction
☐ Improper fraction
☐ Mixed number

In Chapters 1 and 2, we used whole numbers and integers to describe data that were not broken into parts. When a situation involves a portion of a whole, we need fractions. A **fraction** is a number that describes a portion of a whole. For example, when a carpenter cuts a board into 4 equal parts, each part is one-fourth of the original board. One-fourth is represented by the fraction $\frac{1}{4}$. See Figure 4.1.

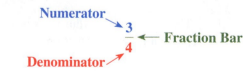

Figure 4.1 A Board Cut into Fourths

STUDY TIP

A new concept is often easier to learn when we find a relationship between the concept and our personal experience. List ten instances from your life when you have encountered fractions.

If **3** out of the **4** pieces of wood in Figure 4.1 are needed to make shelving, then the carpenter will use $\frac{3}{4}$ of the board for the shelving. The top number in a fraction is called the **numerator**, and the bottom number is called the **denominator**. In the fraction $\frac{3}{4}$, the numerator is **3**, and the denominator is **4**.

READING CHECK

• What are the numerator and denominator of a fraction?

The denominator **4** gives the total number of parts in the whole board. The numerator **3** gives the number of parts needed for the shelving.

EXAMPLE 1 **Identifying numerators and denominators**

Write the numerator and denominator of each fraction.

(a) $\frac{5}{7}$ (b) $\frac{4x}{19}$

Solution
(a) In the fraction $\frac{5}{7}$, the numerator is **5**, and the denominator is **7**.
(b) The numerator is $4x$, and the denominator is 19.

Now Try Exercises 7, 13

One way to visualize a fraction is to identify shaded parts of a whole. The diagram shown in Figure 4.2 has a total of 8 equal parts, 5 of which are shaded. Because there are **5** shaded parts out of **8** total parts, the numerator of a fraction representing this shading is **5**, and the denominator is **8**. The fraction for this shading is $\frac{5}{8}$.

Figure 4.2 A Shading Representing $\frac{5}{8}$

EXAMPLE 2   **Identifying a fraction represented by shading**

Write the fraction represented by the shading.

(a)   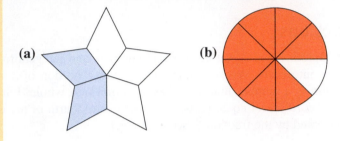   **(b)**

Solution
(a) Because the whole star is divided into 5 equal parts, the denominator of the fraction that represents the shading is **5**. Of these parts, 2 are shaded, so the numerator is **2**. The fraction for this shading is $\frac{2}{5}$.

(b) There are **7** parts shaded out of a total of **8** equal parts. The fraction for this shading is $\frac{7}{8}$.

Now Try Exercises 15, 19

▶ **REAL-WORLD CONNECTION**   Fractions are used to describe many kinds of real data. For example, in 2008, it was estimated that 21 out of every 100 U.S. men who had 8 or fewer years of education were cigarette smokers. This compares to only 7 smokers out of every 100 U.S. men who had graduate degrees. We can express the fraction of men who smoked and had 8 or fewer years of education as $\frac{21}{100}$, and the fraction of men who smoked and had a graduate degree as $\frac{7}{100}$. (*Source:* Centers for Disease Control and Prevention.)

The next example illustrates how fractions can be used when part of a whole is needed to represent given information.

EXAMPLE 3   **Writing a fraction for given information**

The seven continents of the world are Africa, Antarctica, Asia, Australia, Europe, North America, and South America. What fraction of continent names begin with a vowel?

Solution
Because **5** of the **7** continents have names that begin with a vowel, the fraction is $\frac{5}{7}$.

Now Try Exercise 71

Rational Numbers

When a number is or *can be* expressed as a fraction where both the numerator and the denominator are integers (with the denominator not equal to 0), the number belongs to the set of numbers known as the **rational numbers**.

RATIONAL NUMBER

A rational number is a number that can be expressed as

$$\frac{p}{q},$$

where p and q are integers with $q \neq 0$.

Because every integer can be expressed as a fraction by writing the integer in the numerator and 1 in the denominator, we know that *every integer is a rational number*. For example, the integer -92 can be expressed as $\frac{-92}{1}$, and 0 can be expressed as $\frac{0}{1}$. Examples of other rational numbers include

$$\frac{3}{11}, \quad -\frac{1}{8}, \quad 5 = \frac{5}{1}, \quad \text{and} \quad -2 = \frac{-2}{1}.$$

READING CHECK

- How do rational numbers differ from whole numbers and integers?

MAKING CONNECTIONS

Negative Rational Numbers

If the variables p and q represent integers with $q \neq 0$, then an expression of the form $\frac{p}{q}$ can represent both division (as discussed in Chapter 2) and a rational number. Using the rules for signs of quotients from Chapter 2, we see that every negative rational number can be written in three ways, each representing the same fraction. For example, the following three fractions are all equal to "negative five-twelfths."

$$\frac{-5}{12} = \frac{5}{-12} = -\frac{5}{12}$$

Because the fraction bar in a rational number represents division, the identity and zero properties for division discussed in Section 1.3 apply to fractions. These properties can be summarized as follows:

IDENTITY AND ZERO PROPERTIES FOR FRACTIONS

Let the variable n represent any integer.

1. When $n \neq 0$, the expression $\frac{n}{n}$ simplifies to 1.
2. The expression $\frac{n}{1}$ simplifies to n.
3. When $n \neq 0$, the expression $\frac{0}{n}$ simplifies to 0.
4. The expression $\frac{n}{0}$ is undefined.

The next example demonstrates how the identity and zero properties can be used to evaluate some kinds of fractions. Be sure to pay special attention to any fractions containing zeros. A fraction with a zero in the numerator evaluates differently than a fraction with a zero in the denominator.

EXAMPLE 4 **Simplifying fractions**

Simplify each fraction, if possible.

(a) $\dfrac{0}{9}$ **(b)** $\dfrac{-13}{-13}$ **(c)** $\dfrac{5}{0}$ **(d)** $\dfrac{-8}{1}$

Solution

(a) By letting $n = 9$ in Property 3, we see that $\frac{0}{9}$ simplifies to 0.

(b) If $n = -13$ in Property 1, then $\frac{-13}{-13}$ simplifies to 1.

(c) By letting $n = 5$ in Property 4, the fraction $\frac{5}{0}$ is undefined.

(d) If $n = -8$ in Property 2, then $\frac{-8}{1}$ simplifies to -8.

Now Try Exercises 23, 25, 29, 33

Improper Fractions and Mixed Numbers

When the absolute value of the numerator of a fraction is less than the absolute value of the denominator, the fraction is a **proper fraction**. Examples of proper fractions include

$$\frac{3}{7}, \quad \frac{-1}{3}, \quad \frac{5}{9}, \quad \text{and} \quad -\frac{21}{50}.$$

A positive proper fraction represents an amount that is less than 1. For example, $\frac{5}{6}$ is less than 1 and is represented by the shading in Figure 4.3.

When the absolute value of the numerator is greater than or equal to the absolute value of the denominator, the fraction is an **improper fraction**. Examples include

$$\frac{9}{4}, \quad \frac{-7}{2}, \quad \frac{13}{13}, \quad \text{and} \quad -\frac{37}{20}.$$

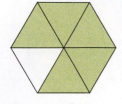

Figure 4.3 The Proper Fraction $\frac{5}{6}$

READING CHECK

• When is a fraction a proper fraction?
• When is a fraction an improper fraction?

> **PROPER AND IMPROPER FRACTIONS**
>
> The fraction $\frac{a}{b}$ with $b \neq 0$ is a proper fraction whenever $|a| < |b|$ and it is an improper fraction whenever $|a| \geq |b|$.

A positive improper fraction represents an amount that is more than 1. For example, $\frac{11}{6}$ is more than 1 and is represented by the shading in Figure 4.4. Each whole is divided into **6** equal parts and a total of **11** parts are shaded. The shading represents $\frac{11}{6}$.

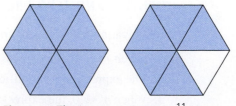

Figure 4.4 The Improper Fraction $\frac{11}{6}$

Figure 4.4 suggests that the improper fraction $\frac{11}{6}$ is equal to 1 whole and $\frac{5}{6}$ more, which can be written as the *mixed number* $1\frac{5}{6}$. A **mixed number** is an integer written with a proper fraction. Examples of mixed numbers include

$$7\frac{1}{3}, \quad 11\frac{3}{8}, \quad -5\frac{21}{22}, \quad \text{and} \quad -10\frac{3}{14}.$$

EXAMPLE 5 | **Identifying an improper fraction and mixed number represented by shading**

Write the improper fraction and mixed number represented by the shading.

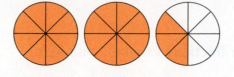

Solution
Each circle is divided into 8 equal parts, giving a denominator of **8**. Because a total of 19 parts are shaded, the numerator is **19**. The improper fraction represented is $\frac{19}{8}$. There are 2 *whole* circles shaded, so the integer part of the mixed number is **2**. Because **3** out of **8** parts of the last circle are shaded, the fraction part is $\frac{3}{8}$. The mixed number is $2\frac{3}{8}$.

Now Try Exercise 35

MAKING CONNECTIONS

Positive and Negative Mixed Numbers

We read the mixed number $2\frac{1}{3}$ as "two and one-third," which means that there are 2 wholes plus one-third of another whole. Therefore $2\frac{1}{3}$ can be written as $2 + \frac{1}{3}$. Similarly, the negative mixed number $-3\frac{2}{5}$ is "the opposite of the mixed number three and two-fifths." This means that $-3\frac{2}{5}$ can be written as $-\left(3 + \frac{2}{5}\right)$ or $-3 - \frac{2}{5}$.

WRITING IMPROPER FRACTIONS AS MIXED NUMBERS Example 5 illustrates that an improper fraction can be written as a mixed number. We can write any improper fraction as a mixed number by remembering that a fraction bar indicates division. For example, the improper fraction $\frac{13}{5}$ indicates that we divide **13** by **5** as follows.

$$\begin{array}{r} 2 \\ 5\overline{)13} \\ \underline{10} \\ 3 \end{array} \begin{array}{l} \leftarrow \textbf{Quotient} \\ \leftarrow \textbf{Dividend} \\ \\ \leftarrow \textbf{Remainder} \end{array} \qquad \frac{13}{5} = 2\frac{3}{5}$$

Divisor →

An improper fraction can be written as a mixed number using the following steps.

WRITING AN IMPROPER FRACTION AS A MIXED NUMBER

To write a positive improper fraction as a mixed number, do the following.

STEP 1: Divide the denominator into the numerator.

STEP 2: The quotient is the integer part of the mixed number. The fraction part of the mixed number has the remainder in its numerator and the divisor in its denominator. Write the mixed number in the form

$$\text{Mixed Number} = \text{Quotient} \frac{\text{Remainder}}{\text{Divisor}}.$$

If the remainder is 0, then the improper fraction is equal to the quotient. For example, the quotient is 2 and the remainder is 0 when 6 is divided by 3, so $\frac{6}{3} = 2$.

NOTE: To write a negative improper fraction as a mixed number, perform the same two steps on the absolute value of the fraction and then place a negative sign in front of the resulting mixed number.

EXAMPLE 6 **Writing improper fractions as mixed numbers**

Write each improper fraction as a mixed number or an integer.

(a) $\dfrac{15}{8}$ (b) $\dfrac{24}{4}$ (c) $-\dfrac{29}{3}$

Solution

(a) Divide 8 into 15.

$$
\begin{array}{r}
1 \leftarrow \textbf{Quotient} \\
\textbf{Divisor} \rightarrow 8)\overline{15} \leftarrow \textbf{Dividend} \\
\underline{8} \\
7 \leftarrow \textbf{Remainder}
\end{array}
$$

As a result, $\frac{15}{8} = 1\frac{7}{8}$.

(b) Divide 4 into 24.

$$
\begin{array}{r}
6 \leftarrow \textbf{Quotient} \\
\textbf{Divisor} \rightarrow 4)\overline{24} \leftarrow \textbf{Dividend} \\
\underline{24} \\
0 \leftarrow \textbf{Remainder}
\end{array}
$$

The remainder is 0, so $\frac{24}{4} = 6$.

(c) Perform the steps on the absolute value of $-\frac{29}{3}$. Because $\left|-\frac{29}{3}\right| = \frac{29}{3}$, first write $\frac{29}{3}$ as a mixed number by dividing 3 into 29 and then place a negative sign in front of the result.

$$
\begin{array}{r}
9 \leftarrow \textbf{Quotient} \\
\textbf{Divisor} \rightarrow 3)\overline{29} \leftarrow \textbf{Dividend} \\
\underline{27} \\
2 \leftarrow \textbf{Remainder}
\end{array}
$$

As a result, $\frac{29}{3} = 9\frac{2}{3}$ and so $-\frac{29}{3} = -9\frac{2}{3}$.

Now Try Exercises 41, 43, 49

WRITING MIXED NUMBERS AS IMPROPER FRACTIONS In Example 6(a), we used *division* to write the improper fraction $\frac{15}{8}$ as the mixed number $1\frac{7}{8}$. If we wish to reverse the process, we can use *multiplication* to write $1\frac{7}{8}$ as $\frac{15}{8}$. First, we note that the numerator of the improper fraction is the dividend and the denominator is the divisor, as follows.

$$
\frac{\textbf{Dividend}}{\textbf{Divisor}}
$$

Next, recall from Chapter 1 that we can check the result of a division problem as follows.

$$
\textbf{Divisor} \cdot \textbf{Quotient} + \textbf{Remainder} = \textbf{Dividend}
$$

As a result, we can replace the numerator **Dividend** in an improper fraction with the expression **Divisor** · **Quotient** + **Remainder** to obtain the expression

$$
\frac{\textbf{Divisor} \cdot \textbf{Quotient} + \textbf{Remainder}}{\textbf{Divisor}}.
$$

Finally, because the mixed number from Example 6(a) has the form

$$
\textbf{Quotient} \rightarrow 1\frac{7}{8}, \begin{array}{l} \leftarrow \textbf{Remainder} \\ \leftarrow \textbf{Divisor} \end{array}
$$

the expression for the improper fraction becomes

$$
\frac{\textbf{Divisor} \cdot \textbf{Quotient} + \textbf{Remainder}}{\textbf{Divisor}} = \frac{8 \cdot 1 + 7}{8} = \frac{8 + 7}{8} = \frac{15}{8}.
$$

To write a mixed number as an improper fraction, we use the following procedure.

WRITING A MIXED NUMBER AS AN IMPROPER FRACTION

To write a positive mixed number as an improper fraction, do the following.

STEP 1: Multiply the denominator of the fraction by the integer part.

STEP 2: Add the numerator of the fraction to the product from Step 1.

STEP 3: The sum found in Step 2 is the numerator of the improper fraction. The denominator of the improper fraction is the denominator from the original mixed number. So a mixed number of the form

$$\text{Quotient } \frac{\text{Remainder}}{\text{Divisor}}$$

can be written as the improper fraction

$$\frac{\text{Divisor} \cdot \text{Quotient} + \text{Remainder}}{\text{Divisor}}.$$

NOTE: To write a negative mixed number as an improper fraction, perform the same three steps on the absolute value of the mixed number and then place a negative sign in front of the resulting improper fraction.

EXAMPLE 7 **Writing mixed numbers as improper fractions**

Write each mixed number as an improper fraction.

(a) $7\dfrac{2}{3}$ **(b)** $-5\dfrac{4}{9}$

Solution

(a) $7\dfrac{2}{3} = \dfrac{3 \cdot 7 + 2}{3} = \dfrac{21 + 2}{3} = \dfrac{23}{3}$

(b) Because $5\dfrac{4}{9} = \dfrac{9 \cdot 5 + 4}{9} = \dfrac{45 + 4}{9} = \dfrac{49}{9}$, we know that $-5\dfrac{4}{9} = -\dfrac{49}{9}$.

Now Try Exercises 51, 53

Graphing Fractions and Mixed Numbers

READING CHECK

• When graphing a fraction, how are the whole distances on the number line divided?

When graphing a fraction, we divide each whole distance on the number line into an equal number of parts determined by the denominator of the fraction to be graphed. For example, to graph the proper fraction $\frac{3}{5}$, divide each whole distance on the number line into **5** equal parts by marking equally spaced tick marks between the whole units on the number line. Then, starting from 0, count **3** tick marks to the right. See Figure 4.5.

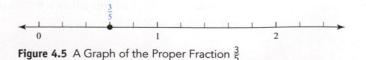

Figure 4.5 A Graph of the Proper Fraction $\frac{3}{5}$

NOTE: To divide a whole unit on a number line into **5** equal parts, mark **4** equally spaced tick marks between the whole units, as shown in green in Figure 4.5.

EXAMPLE 8 **Graphing proper and improper fractions**

Graph each fraction on a number line.

(a) $\dfrac{5}{8}$ (b) $-\dfrac{3}{4}$ (c) $\dfrac{7}{3}$

Solution

(a) To graph $\dfrac{5}{8}$, divide each whole unit on the number line into **8** equal parts. Then count **5** tick marks to the right, starting from 0, and place a dot on the number line as shown.

(b) Divide each whole unit on the number line into **4** equal parts. Locate $-\dfrac{3}{4}$ by counting **3** tick marks to the left, starting from 0. Place a dot on the number line as shown.

(c) To graph $\dfrac{7}{3}$, divide each whole unit on the number line into **3** equal parts. Then count **7** tick marks to the right, starting from 0, and place a dot on the number line as shown.

Now Try Exercises 61, 63, 65

The process for graphing a mixed number on a number line is similar to that for graphing a fraction. Mixed numbers are graphed in the next example.

EXAMPLE 9 **Graphing mixed numbers**

Graph each mixed number on a number line.

(a) $2\dfrac{1}{3}$ (b) $-1\dfrac{3}{4}$

Solution

(a) To graph $2\dfrac{1}{3}$, divide each whole unit on the number line into **3** equal parts. Then count **2** whole units and **1** additional tick mark to the right, starting from 0, and place a dot on the number line as shown.

(b) Divide each whole unit on the number line into **4** equal parts. Locate $-1\dfrac{3}{4}$ by counting **1** whole unit and **3** additional tick marks to the left, starting from 0. Place a dot on the number line as shown.

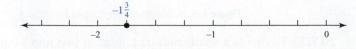

Now Try Exercises 67, 69

4.1 Putting It All Together

CONCEPT	COMMENTS	EXAMPLES
Fraction	A number that can be used to describe a portion of a whole	Numerator → $\dfrac{2}{5}$ ← Fraction Bar Denominator →
Rational Number	A rational number can be expressed as $$\dfrac{p}{q},$$ where p and q are integers with $q \neq 0$.	$\dfrac{3}{8}, \quad -\dfrac{7}{5}, \quad 12 = \dfrac{12}{1}$
Identity and Zero Properties for Fractions	**1.** When $n \neq 0$, $\frac{n}{n}$ simplifies to 1. **2.** $\frac{n}{1}$ simplifies to n. **3.** When $n \neq 0$, $\frac{0}{n}$ simplifies to 0. **4.** $\frac{n}{0}$ is undefined.	**1.** $\dfrac{3}{3} = 1$ **2.** $\dfrac{5}{1} = 5$ **3.** $\dfrac{0}{7} = 0$ **4.** $\dfrac{8}{0}$ is undefined.
Proper and Improper Fractions	The fraction $\frac{a}{b}$ with $b \neq 0$ is a proper fraction whenever $|a| < |b|$. It is an improper fraction whenever $|a| \geq |b|$.	Proper Fractions: $\dfrac{1}{3}, \quad -\dfrac{2}{5}, \quad \dfrac{5}{9}$ Improper Fractions: $\dfrac{4}{3}, \quad -\dfrac{8}{5}, \quad \dfrac{11}{11}$
Mixed Number	An integer written with a proper fraction	$8\dfrac{1}{4}, \quad -2\dfrac{3}{5}, \quad 6\dfrac{5}{9}$
Writing Improper Fractions as Mixed Numbers	**1.** Divide the denominator into the numerator. **2.** The quotient is the integer part of the mixed number. The fraction part of the mixed number has the remainder in its numerator and the divisor in its denominator.	$\dfrac{8}{3} = 2\dfrac{2}{3}$ Divisor → $3\overline{)8}$ ← Dividend, with **2** ← Quotient, $\underline{6}$, **2** ← Remainder
Writing Mixed Numbers as Improper Fractions	**1.** Multiply the denominator of the fraction by the integer part. **2.** Add the numerator of the fraction to the product from Step 1. **3.** The sum found in Step 2 is the numerator of the improper fraction. The denominator of the improper fraction is the denominator from the original mixed number.	$1\dfrac{3}{5} = \dfrac{5(1) + 3}{5} = \dfrac{8}{5}$ $-6\dfrac{1}{3} = -\left(\dfrac{3(6) + 1}{3}\right) = -\dfrac{19}{3}$
Graphing Fractions and Mixed Numbers	When graphing a fraction or mixed number, divide each whole distance on the number line into an equal number of parts determined by the denominator of the fraction to be graphed.	The numbers $\frac{1}{3}$, $\frac{4}{3}$, and $2\frac{2}{3}$ are graphed on the number line.

4.1 **Exercises**

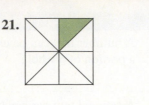

CONCEPTS AND VOCABULARY

1. A(n) _____ is a number that can be used to describe a portion of a whole.

2. The top number in a fraction is called the _____, and the bottom number is the _____.

3. A(n) _____ number can be expressed as a fraction where both the numerator and denominator are integers and the denominator does not equal 0.

4. If $|a| < |b|$ in the fraction $\frac{a}{b}$ with $b \neq 0$, then the fraction is a(n) _____ fraction.

5. If $|a| \geq |b|$ in the fraction $\frac{a}{b}$ with $b \neq 0$, then the fraction is a(n) _____ fraction.

6. When an integer is written with a proper fraction, the resulting number is called a(n) _____.

FRACTIONS

Exercises 7–14: Write the numerator and denominator of the given fraction.

7. $\frac{6}{13}$

8. $\frac{3}{7}$

9. $\frac{12}{5}$

10. $\frac{15}{4}$

11. $\frac{x}{y}$

12. $\frac{m}{n}$

13. $\frac{3p}{14}$

14. $\frac{5w}{7}$

Exercises 15–22: Write the fraction that is represented by the given shading.

15.

16.

17.

18.

19.

20.

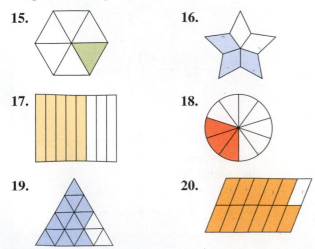

21.

22.

Exercises 23–34: Simplify each fraction, if possible.

23. $\frac{4}{4}$

24. $\frac{-2}{0}$

25. $\frac{0}{-7}$

26. $\frac{10}{10}$

27. $\frac{-13}{1}$

28. $\frac{18}{1}$

29. $\frac{11}{0}$

30. $\frac{0}{-1}$

31. $\frac{-9}{-9}$

32. $\frac{0}{45}$

33. $\frac{53}{1}$

34. $\frac{29}{0}$

IMPROPER FRACTIONS AND MIXED NUMBERS

Exercises 35–40: Write the improper fraction and mixed number represented by the shading.

35.

36.

37.

38.

39.

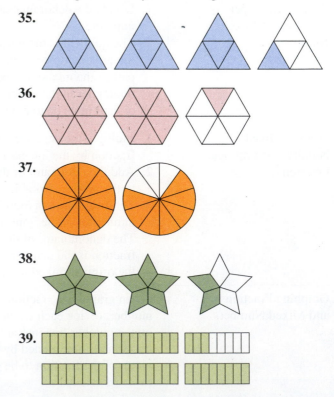

40.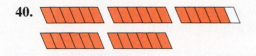

Exercises 41–50: Write the given improper fraction as a mixed number or an integer.

41. $\dfrac{11}{2}$ **42.** $\dfrac{-8}{3}$

43. $\dfrac{17}{6}$ **44.** $\dfrac{50}{5}$

45. $\dfrac{-16}{4}$ **46.** $\dfrac{24}{7}$

47. $\dfrac{91}{8}$ **48.** $\dfrac{73}{9}$

49. $-\dfrac{37}{5}$ **50.** $\dfrac{115}{-11}$

Exercises 51–60: Write the given mixed number as an improper fraction.

51. $5\dfrac{3}{4}$ **52.** $2\dfrac{1}{7}$

53. $-8\dfrac{2}{3}$ **54.** $-1\dfrac{7}{8}$

55. $9\dfrac{5}{8}$ **56.** $10\dfrac{1}{2}$

57. $-35\dfrac{2}{3}$ **58.** $-60\dfrac{4}{5}$

59. $112\dfrac{1}{5}$ **60.** $204\dfrac{5}{6}$

Exercises 61–70: Graph the given fraction or mixed number on a number line.

61. $\dfrac{2}{3}$ **62.** $-\dfrac{1}{4}$

63. $\dfrac{7}{4}$ **64.** $\dfrac{3}{5}$

65. $-\dfrac{4}{5}$ **66.** $-\dfrac{9}{2}$

67. $1\dfrac{3}{5}$ **68.** $-1\dfrac{1}{3}$

69. $-3\dfrac{1}{2}$ **70.** $2\dfrac{1}{4}$

APPLICATIONS

71. *Poverty Level* In 2008, California, New York, and Texas were the only states with more than 2,500,000 families living below the poverty level. What fraction of all states had more than 2,500,000 families living below the poverty level? (*Source:* U.S. Census Bureau.)

72. *Millionaires* In 2009, California, Florida, Illinois, New York, Ohio, Pennsylvania, and Texas were the only states with more than 4,000,000 millionaires. What fraction of all states had more than 4,000,000 millionaires? (*Source:* Phoenix Marketing International.)

73. *Air Quality* There were 9 days of unhealthy air quality in Dallas, Texas in 2009. What fraction of the days in 2009 were *not* unhealthy air quality days in Dallas? (*Source:* Environmental Protection Agency.)

74. *Cell Phone Use* In 2009, it was estimated that 643 of every 1000 Canadians owned a cell phone. Write the fraction of Canadians that did *not* own a cell phone in 2009. (*Source:* International Telecommunications Union.)

75. *Internet Use* It was estimated that 193 of every 1000 people in Asia had Internet access in 2009. What fraction of the Asian population did *not* have Internet access in 2009? (*Source:* Internet World Stats.)

76. *European Union* Of the 27 countries that were members of the European Union in 2010, only Austria, Estonia, Ireland, Italy, and the United Kingdom have names that begin with a vowel. In 2010, what fraction of countries in the European Union had names that started with a vowel?

Exercises 77–80: Truck Sales The following bar graph shows the total number of cement trucks made by a truck manufacturer during selected years.

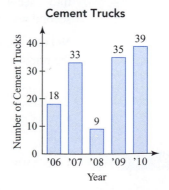

77. For this five-year period, what fraction of the total production occurred in 2010?

78. What fraction of the total production for this five-year period occurred in either 2006 or 2007?

79. Considering only the odd-numbered years, what fraction of the production occurred in 2009?

80. What fraction of production for the first four years occurred in 2006?

WRITING ABOUT MATHEMATICS

81. Give an example of an improper fraction that is not greater than or equal to 1. What must be true about any fraction of this type?

82. Explain how the result of a division problem can be expressed as a mixed number.

83. Describe the steps used to write an improper fraction as a mixed number.

84. Describe the steps used to write a mixed number as an improper fraction.

4.2   Prime Factorization and Lowest Terms

Divisibility Tests • Prime Factorization • Equivalent Fractions • Simplifying Fractions • Using the GCF • Comparing Fractions • Rational Expressions

A LOOK INTO MATH ▶

When the inventory of a particular product sold in a store falls below a certain level, the store manager will order more of the product from the wholesaler. Suppose that it is company policy to reorder a product each time that the inventory reaches one-half of its original amount. In this case, the store manager must understand that there are different ways to express the fraction $\frac{1}{2}$. For example, when 50 tubes of toothpaste are sold from an inventory of 100 tubes, it is time to reorder because $\frac{50}{100} = \frac{1}{2}$. Similarly, when 15 electric blankets are sold from an inventory of 30 blankets, it is time to reorder because $\frac{15}{30} = \frac{1}{2}$. In this section, we will discuss the many ways to express *equivalent* fractions.

NEW VOCABULARY

☐ Factor
☐ Divisible
☐ Prime number
☐ Composite number
☐ Prime factorization
☐ Factor tree
☐ Equivalent fractions
☐ Basic principle of fractions
☐ Greatest common factor
☐ Lowest terms
☐ Cross products
☐ Rational expression

Divisibility Tests

When two whole numbers are multiplied, we call them *factors*. For example,

$$5 \quad \cdot \quad 4 \quad = \quad 20.$$

factor   factor   product

Because division is closely related to multiplication, we know that dividing 20 by 5 results in a quotient of 4 with no remainder. That is, 20 is *divisible* by 5, and 5 is a *factor* of 20. Similarly, dividing 20 by 4 results in the quotient 5 with no remainder. In this case, we say that 20 is *divisible* by 4, and 4 is a *factor* of 20. A **factor** is any whole number that divides into a second whole number exactly with no remainder. When this occurs, we say that the second number is **divisible** by the first.

NOTE: We often say that a factor divides "evenly" into another number, which means that the remainder is 0.

READING CHECK

• What does it mean to say that one number is divisible by another number?

EXAMPLE 1   **Checking for divisibility and factors**

Using the definitions of "divisible" and "factor," answer each question.
(a) Is 52 divisible by 16?   **(b)** Is 13 a factor of 39?

Solution
(a) Dividing 52 by 16 results in 3 r4. The remainder is not 0, so 52 is not divisible by 16.
(b) Because 13 divides into 39 exactly 3 times with no remainder, 13 is a factor of 39.

▪ **Now Try Exercises 15, 17**

It is not always necessary to perform long division to determine if one number is divisible by another. The following rules can be used to test divisibility by 2, 3, or 5.

DIVISIBILITY TESTS

A whole number is divisible by

- 2 if its ones digit is 0, 2, 4, 6, or 8.
- 3 if the sum of its digits is divisible by 3.
- 5 if its ones digit is 0 or 5.

EXAMPLE 2 **Performing divisibility tests**

Use a divisibility test to answer each question.
(a) Is 97 divisible by 2?   (b) Does 3 divide evenly into 144?   (c) Is 5 a factor of 168?

Solution
(a) The number 9**7** is not divisible by 2 because its ones digit is not 0, 2, 4, 6, or 8.
(b) Because the sum of its digits $1 + 4 + 4 = 9$ is divisible by 3, the number 144 is divisible by 3. So, 3 is a factor of 144.
(c) The number 16**8** is not divisible by 5 because its ones digit is neither 0 nor 5.

Now Try Exercises 21, 23, 27

In addition to the tests for 2, 3, and 5, the following divisibility tests may also be helpful when determining divisibility by 4, 6, or 9.

MORE DIVISIBILITY TESTS

A whole number is divisible by

- 4 if the number formed by its last two digits is divisible by 4.
- 6 if it is divisible by 2 and 3.
- 9 if the sum of its digits is divisible by 9.

NOTE: There are divisibility tests for 7 and 8; however, they are often more difficult to apply than simply performing long division.

EXAMPLE 3 **Performing divisibility tests**

Use a divisibility test to answer each question.
(a) Can 124 be divided evenly by 4?   (b) Is 6 a factor of 458?
(c) Does 9 divide evenly into 846?

Solution
(a) The number 1**24** divides evenly by 4 because **24** is divisible by 4.
(b) The number 458 is divisible by 2 because its ones digit is 8. However, it is not divisible by 3 because the sum of its digits $4 + 5 + 8 = 17$ is not divisible by 3. So, 6 is not a factor of 458.
(c) Because the sum of its digits $8 + 4 + 6 = 18$ is divisible by 9, we know that 9 divides evenly into 846.

Now Try Exercises 25, 29, 31

Prime Factorization

When we *factor* a number we write it as a product. Some numbers can be factored in many different ways. For example, 12 can be factored as

$$12 = 1 \cdot 12, \quad 12 = 2 \cdot 6, \quad 12 = 3 \cdot 4, \quad \text{or} \quad 12 = 2 \cdot 2 \cdot 3.$$

The last factorization displayed is important because the factors are all *prime numbers*. A prime number is defined as follows.

PRIME NUMBERS AND COMPOSITE NUMBERS

A **prime number** is a natural number greater than 1 whose *only* factors are 1 and itself.
The first ten prime numbers are 2, 3, 5, 7, 11, 13, 17, 19, 23, and 29.

A **composite number** is a natural number greater than 1 that is not prime.
The first ten composite numbers are 4, 6, 8, 9, 10, 12, 14, 15, 16, and 18.

The natural number 1 is *neither* prime nor composite.

READING CHECK

• How do prime and composite numbers differ?

We can determine if a number is prime by using the following property:

Every composite number is divisible by at least one prime number.

For example, the composite number 68 is divisible by the prime number 2. To see if a number is prime, we divide it by each prime number from 2 up to (but not including) the first prime number whose square is greater than the number. This process is illustrated in the next example.

EXAMPLE 4 | **Determining if a number is prime**

Determine if each number is prime, composite, or neither.
(a) 65 **(b)** 1 **(c)** 47

Solution
(a) Since $7^2 = 49$ and $11^2 = 121$, the first prime number whose square is greater than 65 is 11. We will check to see if 65 is divisible by the prime numbers 2, 3, 5, or 7.

   2: The ones digit is 5, so we know that 65 is not divisible by 2.

   3: The sum $6 + 5 = 11$ is not divisible by 3, so 65 is not divisible by 3.

   5: Since the ones digit is 5, we know that 65 is divisible by 5.

   Since 65 is divisible by the prime number 5, it is a composite number.

(b) As noted in the definition box, the number 1 is neither prime nor composite.
(c) Since $5^2 = 25$ and $7^2 = 49$, the first prime number whose square is greater than 47 is 7. We will check to see if 47 is divisible by the prime numbers 2, 3, or 5.

   2: The ones digit is 7, so 47 is not divisible by 2.

   3: The sum $4 + 7 = 11$ is not divisible by 3, so 47 is not divisible by 3.

   5: Since the ones digit is neither 0 nor 5, we know that 47 is not divisible by 5.

   Since 47 is not divisible by 2, 3, or 5, it is a prime number.

Now Try Exercises 33, 37, 39

Every composite number can be factored into a product of prime numbers. This product, known as the **prime factorization**, is unique. For example, the prime factorization of 12 is $12 = 2 \cdot 2 \cdot 3$. No other product of *prime factors* of 12 can be written that is different from this one. (Note that the factorization $12 = 2 \cdot 3 \cdot 2$ does not represent a different product of primes. Only the *order* of the factors is different.)

One way to write a number as the product of primes is to use a **factor tree**. To demonstrate this method, we will find the prime factorization of 120. Write 120 at the top of the tree and make branches to *any* two factors whose product is 120. Although there are many choices, we will use 10 and 12.

If the number at the end of a branch is prime, circle it and stop working on its branch. If it is composite, continue by making two branches to *any* two factors whose product is the composite number.

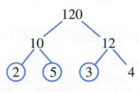

Continue this process until every branch ends in a (circled) prime number.

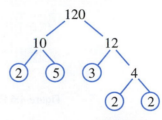

The prime factorization is given by the product of the circled numbers. It is customary to write the prime factors in order from smallest to largest. So, $120 = 2 \cdot 2 \cdot 2 \cdot 3 \cdot 5$. Note that the factorization could also be written using exponents, $120 = 2^3 \cdot 3 \cdot 5$.

READING CHECK

• Why do we use a factor tree?

MAKING CONNECTIONS

Factor Trees

Because multiplication is both commutative and associative, there may be several ways to draw a factor tree for a given number. However, all factor trees for a specified number give the same prime factorization. For example, a different factor tree for 120 is shown here.

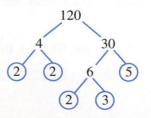

In either case, the prime factorization is the same.

EXAMPLE 5 **Writing prime factorizations of composite numbers**

Find the prime factorization of each composite number.
(a) 75 **(b)** 126

Solution
A factor tree for each number is shown below.

(a)

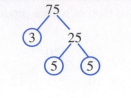

(b)

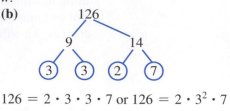

$126 = 2 \cdot 3 \cdot 3 \cdot 7$ or $126 = 2 \cdot 3^2 \cdot 7$

$75 = 3 \cdot 5 \cdot 5$ or $75 = 3 \cdot 5^2$

Now Try Exercises 43, 45

Equivalent Fractions

A Look Into Math for this section (page 212) suggests that the fractions $\frac{50}{100}$, $\frac{15}{30}$, and $\frac{1}{2}$ each represent one-half of a whole amount. To illustrate this concept visually, consider the three shaded regions shown in Figure 4.6.

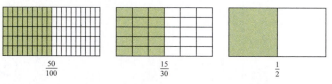

Figure 4.6 Fractions Representing One-Half of a Rectangle

Another way to see that different fractions can name the same number is to construct number lines. Figure 4.7 shows that the fractions $\frac{2}{3}$, $\frac{4}{6}$, and $\frac{6}{9}$ all name the same number.

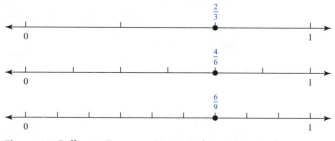

Figure 4.7 Different Fractions Naming the Same Number

READING CHECK

• What principle allows us to write equivalent fractions?

Fractions that name the same number are called **equivalent fractions**. When we write a given fraction as an equivalent fraction, we can use the **basic principle of fractions**. This principle states that if we multiply the numerator and denominator of any fraction by the same nonzero number, the result is an equivalent fraction.

BASIC PRINCIPLE OF FRACTIONS

If the variables a, b, and c represent integers with $b \neq 0$ and $c \neq 0$, then

$$\frac{a \cdot c}{b \cdot c} = \frac{a}{b}.$$

EXAMPLE 6 | **Writing equivalent fractions**

Find a number to replace the question mark so that the fractions are equivalent.

(a) $\dfrac{3}{5} = \dfrac{?}{35}$ (b) $\dfrac{24}{36} = \dfrac{2}{?}$

Solution

(a) For some nonzero number c, the basic principle of fractions gives

$$\frac{3}{5} = \frac{3 \cdot c}{5 \cdot c} = \frac{?}{35}$$

Because $5 \cdot 7 = 35$ is the value of the denominator in the equivalent fraction, we can conclude that c must be **7**. To find the value of the unknown numerator, we also multiply 3 by c to obtain $3 \cdot c = 3 \cdot 7 = 21$. This method can be visualized as follows, where 3 times **7** gives the unknown value of 21.

$$\frac{3}{5} \overset{\times 7}{\underset{\times 7}{=}} \frac{?}{35}$$

(b) The unknown value can be found by reversing the arrows used in the visual method shown in part (a). Because $2 \cdot 12 = 24$, the value of c is **12**.

$$\frac{24}{36} \overset{\times 12}{\underset{\times 12}{=}} \frac{2}{?}$$

Multiplying the unknown value by **12** must result in 36, so the unknown value is 3 because $3 \cdot 12 = 36$.

Now Try Exercises 51, 61

Simplifying Fractions Using the GCF

The largest number that divides evenly into two or more given numbers is known as the **greatest common factor** (GCF). For example, 6 is the GCF of 12 and 18 because 6 divides both 12 and 18 evenly and no *larger* number can be found that also divides both 12 and 18. (Note that 2 is a *common factor* of 12 and 18, but it is not the *greatest* common factor.)

One way to find the GCF of two or more numbers is to list all of the factors for each number and then visually search for the largest factor that is common to each list. For example, the GCF of 36, 48, and 72 is **12**, as shown in the following lists of factors.

Factors of 36: **1, 2, 3, 4, 6, 9, 12, 18, 36**
Factors of 48: **1, 2, 3, 4, 6, 8, 12, 16, 24, 48**
Factors of 72: **1, 2, 3, 4, 6, 8, 9, 12, 18, 24, 36, 72**

EXAMPLE 7 | **Using the listing method to find the GCF**

Use the listing method to find the GCF of 16 and 40.

Solution

First, list the factors of 16 and 40.

Factors of 16: 1, 2, 4, **8**, 16

Factors of 40: 1, 2, 4, 5, **8**, 10, 20, 40

The common factors are 1, 2, 4, and 8 and the GCF is **8**.

Now Try Exercise 65

Often we find that listing all of the factors for a number is quite difficult. If any factors are missed, we may not be able to find the correct GCF. To avoid this problem, it is common to find the GCF of two or more numbers by using the prime factorization of each number. In the next example, we use this method to verify that 12 is the GCF of 36, 48, and 72.

EXAMPLE 8   **Using prime factorization to find the GCF**

Find the GCF of 36, 48, and 72.

Solution
First, find the prime factorization of each number. Use a factor tree, if needed.

$$36 = \mathbf{2} \cdot \mathbf{2} \cdot \mathbf{3} \cdot 3$$
$$48 = \mathbf{2} \cdot \mathbf{2} \cdot 2 \cdot 2 \cdot \mathbf{3}$$
$$72 = \mathbf{2} \cdot \mathbf{2} \cdot 2 \cdot \mathbf{3} \cdot 3$$

The prime factors **2**, **2**, and **3** are common to all three factorizations. The GCF is the product of the common prime factors **2** · **2** · **3** = **12**.

Now Try Exercise 69

READING CHECK

• What does it mean to simplify a fraction to lowest terms?

Because a fraction can be written in many equivalent forms, we often want to *simplify* a fraction to *lowest terms*. A fraction is simplified to **lowest terms** if the GCF of its numerator and denominator is 1. In other words, a fraction is in lowest terms if the only common factor of the numerator and the denominator is 1. We can use the GCF of the numerator and the denominator to simplify a fraction to lowest terms. For example, because the GCF of 30 and 45 is **15**, the basic principle of fractions can be used to simplify the fraction $\frac{30}{45}$:

$$\frac{30}{45} = \frac{2 \cdot \mathbf{15}}{3 \cdot \mathbf{15}} = \frac{2}{3}$$

The procedure for simplifying a fraction to lowest terms can be summarized as follows.

CALCULATOR HELP

To simplify fractions with a calculator, see the Appendix (page AP-4).

SIMPLIFYING FRACTIONS TO LOWEST TERMS

To simplify a fraction to lowest terms, do the following.

STEP 1: Determine a number c that is the GCF of the numerator and denominator.

STEP 2: Write the numerator as a product of two factors, a and c.
Then write the denominator as a product of two factors, b and c.

STEP 3: Apply the basic principle of fractions.

$$\frac{a \cdot c}{b \cdot c} = \frac{a}{b}$$

EXAMPLE 9   **Simplifying fractions to lowest terms**

Simplify each fraction to lowest terms.

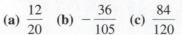

(a) $\dfrac{12}{20}$ (b) $-\dfrac{36}{105}$ (c) $\dfrac{84}{120}$

Solution

(a) STEP 1: Find the GCF of the numerator and denominator.

$$12 = \mathbf{2 \cdot 2} \cdot 3$$
$$20 = \mathbf{2 \cdot 2} \cdot 5$$

The GCF is the product of the common prime factors $\mathbf{2 \cdot 2} = \mathbf{4}$.

STEP 2: Write the numerator as $12 = 3 \cdot \mathbf{4}$ and the denominator as $20 = 5 \cdot \mathbf{4}$.

STEP 3: Apply the basic principle of fractions.

$$\frac{12}{20} = \frac{3 \cdot \mathbf{4}}{5 \cdot \mathbf{4}} = \frac{3}{5}$$

(b) STEP 1: The GCF of the numerator and denominator is $\mathbf{3}$, as shown below.

$$36 = 2 \cdot 2 \cdot \mathbf{3} \cdot 3$$
$$105 = \mathbf{3} \cdot 5 \cdot 7$$

STEP 2: Write the numerator as $36 = 12 \cdot \mathbf{3}$ and the denominator as $105 = 35 \cdot \mathbf{3}$.

STEP 3: Apply the basic principle of fractions.

$$-\frac{36}{105} = -\frac{12 \cdot \mathbf{3}}{35 \cdot \mathbf{3}} = -\frac{12}{35}$$

(c) STEP 1: As shown here, the GCF of the numerator and denominator is $\mathbf{2 \cdot 2 \cdot 3} = \mathbf{12}$.

$$84 = \mathbf{2 \cdot 2 \cdot 3} \cdot 7$$
$$120 = \mathbf{2 \cdot 2} \cdot 2 \cdot \mathbf{3} \cdot 5$$

STEP 2: Write the numerator as $84 = 7 \cdot \mathbf{12}$ and the denominator as $120 = 10 \cdot \mathbf{12}$.

STEP 3: Apply the basic principle of fractions.

$$\frac{84}{120} = \frac{7 \cdot \mathbf{12}}{10 \cdot \mathbf{12}} = \frac{7}{10}$$

Now Try Exercises 73, 75, 77

MAKING CONNECTIONS

Lowest Terms and Canceling

Have you ever heard the word *canceling* used when simplifying a fraction to lowest terms? Canceling is another way to apply the basic principle of fractions. For example, the fraction in Example 9(c) can be simplified by writing the prime factorizations for the numerator and the denominator and then "canceling" the common prime factors from each factorization.

$$\frac{84}{120} = \frac{\overset{1}{\cancel{2}} \cdot \overset{1}{\cancel{2}} \cdot \overset{1}{\cancel{3}} \cdot 7}{\underset{1}{\cancel{2}} \cdot \underset{1}{\cancel{2}} \cdot 2 \cdot \underset{1}{\cancel{3}} \cdot 5} = \frac{7}{10}$$

NOTE: It is important to remember that a $\mathbf{1}$ remains after a factor has been canceled.

Comparing Fractions

When two fractions have denominators that are equal (a common denominator), it is easy to compare the fractions by simply comparing the numerators. For example, $\frac{5}{7} > \frac{2}{7}$ because $5 > 2$. However, when two fractions have different denominators, it can be difficult to compare the sizes of the numbers named by the fractions. To compare fractions with unlike denominators, we write equivalent fractions that have a common denominator and then compare the numerators. This process is demonstrated in the next example.

EXAMPLE 10

Comparing fractions

Place the correct symbol, $<$, $>$, or $=$, in the blank for each pair of fractions.

(a) $\dfrac{7}{9}$ ___ $\dfrac{8}{10}$ **(b)** $\dfrac{13}{16}$ ___ $\dfrac{25}{31}$

Solution

(a) To write a fraction that is equivalent to $\frac{7}{9}$, we use the basic principle of fractions with the value of c equal to the denominator of the second fraction, **10**.

$$\frac{7}{9} = \frac{7 \cdot \mathbf{10}}{9 \cdot \mathbf{10}} = \frac{70}{90}$$

Likewise, to write a fraction that is equivalent to $\frac{8}{10}$, use the basic principle of fractions with the value of c equal to the denominator of the first fraction, **9**.

$$\frac{8}{10} = \frac{8 \cdot \mathbf{9}}{10 \cdot \mathbf{9}} = \frac{72}{90}$$

Now we compare the two fractions $\frac{70}{90}$ and $\frac{72}{90}$. Because $\frac{70}{90} < \frac{72}{90}$, the symbol $<$ can be placed in the blank and we write $\frac{7}{9} < \frac{8}{10}$.

(b) To compare the fractions $\frac{13}{16}$ and $\frac{25}{31}$, we use the basic principle of fractions, as demonstrated in part (a). A more concise process is shown here.

$$\frac{13}{16} = \frac{13 \cdot \mathbf{31}}{16 \cdot \mathbf{31}} = \frac{403}{496} \quad \text{and} \quad \frac{25}{31} = \frac{25 \cdot \mathbf{16}}{31 \cdot \mathbf{16}} = \frac{400}{496}$$

Because $\frac{403}{496} > \frac{400}{496}$, the symbol $>$ can be placed in the blank and we write $\frac{13}{16} > \frac{25}{31}$.

Now Try Exercises 85, 87

READING CHECK

• How is the cross product used to see if two fractions are equivalent?

In the previous example we multiplied the numerator of each fraction by the denominator of the other. These products are called the **cross products** and can be used to see if two fractions are equivalent. For example, the cross product method for comparing $\frac{9}{14}$ and $\frac{12}{19}$ is written as follows.

$$19 \cdot 9 \quad \frac{9}{14} \stackrel{?}{=} \frac{12}{19} \quad 14 \cdot 12$$

Since $19 \cdot 9 = 171$ and $14 \cdot 12 = 168$, the cross products are not equal and the fractions $\frac{9}{14}$ and $\frac{12}{19}$ are not equivalent. This result is based on the following rule for cross products.

> *Two fractions are equivalent if their cross products are equal, and they are not equivalent if their cross products are not equal.*

EXAMPLE 11

Determining if two fractions are equivalent

Use the cross product method to determine if the two fractions are equivalent.

$$\frac{42}{56} \stackrel{?}{=} \frac{27}{36}$$

Solution

Find the cross product as follows.

$$36 \cdot 42 \quad \frac{42}{56} \stackrel{?}{=} \frac{27}{36} \quad 27 \cdot 56$$

Since $36 \cdot 42 = 1512$ and $27 \cdot 56 = 1512$, the cross products are equal and the fractions $\frac{42}{56}$ and $\frac{27}{36}$ are equivalent.

Now Try Exercise 91

Rational Expressions

In Section 1.7, we defined a *term* as a number, a variable, or a product of numbers and variables raised to powers. Any fraction whose numerator and denominator each contain a term or a sum (or difference) of two or more terms is called a **rational expression**. Examples of rational expressions include

$$\frac{5}{9}, \quad \frac{3x}{7}, \quad \frac{2x^2 + 5}{3x - 1}, \quad \frac{10ab}{5c^3}, \quad \text{and} \quad \frac{3x^2 - 3x + 1}{10x^2 + 2x - 6}.$$

Here we will focus on simplifying rational expressions whose numerator and denominator each contain a *single* term. The same basic process that was used earlier to simplify fractions can be used to simplify rational expressions of this type. In the next example, we find the greatest common factor (GCF) for two terms containing variables.

EXAMPLE 12   **Finding the GCF of two terms**

Find the GCF of $12x^3$ and $18x$.

Solution
First, factor each term completely, as follows.

$$12x^3 = \mathbf{2} \cdot 2 \cdot \mathbf{3} \cdot \mathbf{x} \cdot x \cdot x$$
$$18x = \mathbf{2} \cdot \mathbf{3} \cdot 3 \cdot \mathbf{x}$$

The GCF is the product of the common factors $\mathbf{2} \cdot \mathbf{3} \cdot \mathbf{x} = \mathbf{6x}$.

Now Try Exercise 97

READING CHECK

• How is finding the GCF of two terms different from finding the GCF of two numbers?

When simplifying rational expressions whose numerator and denominator each contain a single term, we use the same process as that used in Example 9 to simplify fractions. The next example demonstrates this process.

EXAMPLE 13   **Simplifying rational expressions**

Simplify each rational expression.

(a) $\dfrac{30x^3}{42xy}$ **(b)** $-\dfrac{55a^3b^2}{33a^2b^2}$

Solution
(a) STEP 1: First find the GCF of the numerator and denominator.

$$30x^3 = \mathbf{2} \cdot \mathbf{3} \cdot 5 \cdot \mathbf{x} \cdot x \cdot x$$
$$42xy = \mathbf{2} \cdot \mathbf{3} \cdot 7 \cdot \mathbf{x} \cdot y$$

The GCF is the product of the common factors $\mathbf{2} \cdot \mathbf{3} \cdot \mathbf{x} = \mathbf{6x}$

STEP 2: Use the GCF to write the numerator as $30x^3 = \mathbf{5x^2} \cdot \mathbf{6x}$ and the denominator as $42xy = \mathbf{7y} \cdot \mathbf{6x}$.

STEP 3: Apply the basic principle of fractions.

$$\frac{30x^3}{42xy} = \frac{5x^2 \cdot 6x}{7y \cdot 6x} = \frac{5x^2}{7y}$$

(b) STEP 1: The GCF of the numerator and denominator is $11a^2b^2$, as shown below.

$$55a^3b^2 = 5 \cdot 11 \cdot a \cdot a \cdot a \cdot b \cdot b$$
$$33a^2b^2 = 3 \cdot 11 \cdot a \cdot a \cdot b \cdot b$$

STEP 2: Use the GCF to write the numerator as $55a^3b^2 = 5a \cdot 11a^2b^2$ and the denominator as $33a^2b^2 = 3 \cdot 11a^2b^2$.

STEP 3: Apply the basic principle of fractions.

$$-\frac{55a^3b^2}{33a^2b^2} = -\frac{5a \cdot 11a^2b^2}{3 \cdot 11a^2b^2} = -\frac{5a}{3}$$

Now Try Exercises 111, 113

4.2   Putting It All Together

CONCEPT	COMMENTS	EXAMPLES
Factor	A *factor* is any whole number that divides into a second whole number evenly and results in a quotient with no remainder.	8 is a factor of 32 because $$32 \div 8 = 4\,r0.$$ That is, 32 is *divisible* by 8.
Divisibility Tests	A whole number is divisible by • 2 if its ones digit is 0, 2, 4, 6, or 8. • 3 if the sum of its digits is divisible by 3 • 5 if its ones digit is 0 or 5. • 4 if the number formed by its last two digits is divisible by 4 • 6 if it is divisible by 2 and 3. • 9 if the sum of its digits is divisible by 9.	2784 is divisible by 2 because its ones digit is 4. 432 divides evenly by 3 because the digit sum, 9, is divisible by 3. 5 is a factor of 420 because the ones digit in 420 is 0. 873 divides evenly by 9 because the digit sum, 18, is divisible by 9.
Prime and Composite Numbers	A *prime* number is any natural number greater than 1 whose only factors are 1 and itself. A *composite* number is any natural number greater than 1 that is not prime.	Prime numbers include 2, 5, 19, 83, and 773. Composite numbers include 6, 18, 77, 104, and 875.
Prime Factorization	Every composite number can be factored into a unique product of prime numbers.	$54 = 2 \cdot 3 \cdot 3 \cdot 3$ $735 = 3 \cdot 5 \cdot 7 \cdot 7$
Equivalent Fractions	Fractions that name the same number are called *equivalent fractions*. We can write equivalent fractions using the *basic principle of fractions*. $$\frac{a \cdot c}{b \cdot c} = \frac{a}{b}$$	The fractions $$\frac{3}{4} \text{ and } \frac{21}{28}$$ are equivalent because $$\frac{3}{4} = \frac{3 \cdot 7}{4 \cdot 7} = \frac{21}{28}.$$

CONCEPT	COMMENTS	EXAMPLES
Greatest Common Factor (GCF)	The largest number that divides evenly into two or more numbers is the GCF.	$18 = 2 \cdot 3 \cdot 3$ and $30 = 2 \cdot 3 \cdot 5$ The GCF of 18 and 30 is $2 \cdot 3 = 6$.
Lowest Terms	A fraction is simplified to *lowest terms* if the GCF of its numerator and denominator is 1.	The GCF of 25 and 30 is 5, so $$\frac{25}{30} = \frac{5 \cdot 5}{6 \cdot 5} = \frac{5}{6}.$$
Cross Product Method	We can determine if two fractions are equivalent by comparing their cross products.	$42 \cdot 5 \quad \dfrac{5}{14} \overset{?}{=} \dfrac{15}{42} \quad 14 \cdot 15$ The cross products $42 \cdot 5 = 210$ and $14 \cdot 15 = 210$ are equal, so $\frac{5}{14}$ and $\frac{15}{42}$ are equivalent.
Rational Expression	Any fraction whose numerator and denominator each contain a term or a sum (or difference) of two or more terms is a rational expression.	$\frac{14x^2y}{21xy}$ is a rational expression and $$\frac{14x^2y}{21xy} = \frac{2x \cdot 7xy}{3 \cdot 7xy} = \frac{2x}{3}.$$

4.2 Exercises

MyMathLab · Math XL PRACTICE · WATCH · DOWNLOAD · READ · REVIEW

CONCEPTS AND VOCABULARY

1. A(n) _____ is any whole number that divides into a second whole number evenly and results in a quotient with no remainder.

2. We can say that the whole number 24 is _____ by 8 because 8 is a factor of 24.

3. A whole number that has 0 or 5 as its ones digit is always divisible by _____.

4. If the sum of the digits of a whole number is divisible by 3, then the whole number is divisible by _____.

5. A(n) _____ number is a natural number greater than 1 whose only factors are 1 and itself.

6. A(n) _____ number is a natural number greater than 1 that is not prime.

7. Every composite number can be factored into a product of primes called its _____ factorization.

8. One way to find the prime factorization of a composite number is to make a(n)_____ tree.

9. Two fractions that name the same number are called _____ fractions.

10. The letters GCF are used to represent the _____.

11. The GCF of two numbers is the largest number that divides (one/both) of the numbers.

12. A fraction is written in _____ if the GCF of its numerator and denominator is 1.

13. A method that can be used to see if two fractions are equivalent is the _____ product method.

14. A(n) _____ expression is a fraction whose numerator and denominator each contain a term or a sum (or difference) of two or more terms.

DIVISIBILITY AND FACTORS

Exercises 15–20: Using the definitions of "divisible" and "factor," answer the given question.

15. Is 64 divisible by 12?

16. Is 46 divisible by 14?

17. Is 13 a factor of 65?

18. Is 8 a factor of 54?

19. Is 19 a factor of 57?

20. Is 72 divisible by 18?

Exercises 21–32: Use a divisibility test to answer the given question.

21. Is 913 divisible by 5?

22. Can 4 be divided evenly into 622?

23. Is 186 divisible by 2?

24. Can 3 be divided evenly into 762?

25. Does 387 divide evenly by 9?

26. Is 5 a factor of 275?

27. Does 691 divide evenly by 3?

28. Can 6 be divided evenly into 533?

29. Is 6 a factor of 834?

30. Can 673 be divided evenly by 9?

31. Does 4 divide evenly into 748?

32. Is 191 divisible by 2?

PRIME FACTORIZATION

Exercises 33–40: Determine if the given number is prime, composite, or neither.

33. 43   **34.** 63

35. 57   **36.** 0

37. 1   **38.** 65

39. 91   **40.** 101

Exercises 41–50: Find the prime factorization of the given composite number.

41. 16   **42.** 54

43. 45   **44.** 90

45. 140   **46.** 315

47. 231   **48.** 390

49. 442   **50.** 845

EQUIVALENT FRACTIONS

Exercises 51–62: Find a number to replace the question mark so that the given fractions are equivalent.

51. $\dfrac{7}{9} = \dfrac{?}{27}$ **52.** $\dfrac{5}{11} = \dfrac{20}{?}$

53. $\dfrac{5}{6} = \dfrac{35}{?}$ **54.** $\dfrac{?}{15} = \dfrac{12}{45}$

55. $\dfrac{2}{?} = \dfrac{10}{25}$ **56.** $\dfrac{2}{9} = \dfrac{?}{63}$

57. $\dfrac{?}{4} = \dfrac{36}{48}$ **58.** $\dfrac{1}{?} = \dfrac{7}{42}$

59. $\dfrac{4}{13} = \dfrac{?}{65}$ **60.** $\dfrac{8}{17} = \dfrac{24}{?}$

61. $\dfrac{45}{60} = \dfrac{3}{?}$ **62.** $\dfrac{?}{6} = \dfrac{50}{60}$

SIMPLIFYING FRACTIONS USING THE GCF

Exercises 63–72: Find the GCF of the given numbers.

63. 18 and 40   **64.** 24 and 60

65. 20 and 70   **66.** 36 and 63

67. 24 and 72   **68.** 15 and 75

69. 8, 12, and 20   **70.** 15, 30, and 55

71. 16, 32, and 48   **72.** 12, 36, and 72

Exercises 73–82: Simplify the fraction to lowest terms.

73. $\dfrac{16}{24}$ **74.** $-\dfrac{15}{35}$

75. $\dfrac{26}{65}$ **76.** $\dfrac{64}{72}$

77. $-\dfrac{42}{77}$ **78.** $-\dfrac{48}{56}$

79. $\dfrac{105}{350}$ **80.** $\dfrac{120}{156}$

81. $-\dfrac{200}{450}$ **82.** $\dfrac{320}{600}$

COMPARING FRACTIONS

Exercises 83–90: Place the correct symbol, $<$, $>$, or $=$, in the blank in the given pair of fractions.

83. $\dfrac{9}{13}$ —— $\dfrac{10}{13}$ **84.** $\dfrac{8}{11}$ —— $\dfrac{7}{11}$

85. $\dfrac{12}{17}$ —— $\dfrac{22}{31}$ **86.** $\dfrac{4}{19}$ —— $\dfrac{6}{25}$

87. $\dfrac{27}{34}$ —— $\dfrac{5}{7}$ **88.** $\dfrac{5}{6}$ —— $\dfrac{75}{90}$

89. $\dfrac{72}{84}$ —— $\dfrac{6}{7}$ **90.** $\dfrac{33}{47}$ —— $\dfrac{17}{24}$

Exercises 91–96: Use the cross product method to determine if the two fractions are equivalent.

91. $\dfrac{68}{85} \overset{?}{=} \dfrac{32}{40}$ **92.** $\dfrac{20}{25} \overset{?}{=} \dfrac{28}{35}$

93. $\dfrac{28}{42} \overset{?}{=} \dfrac{30}{36}$ **94.** $\dfrac{21}{70} \overset{?}{=} \dfrac{9}{30}$

95. $\dfrac{33}{77} \overset{?}{=} \dfrac{27}{63}$ **96.** $\dfrac{14}{15} \overset{?}{=} \dfrac{17}{18}$

RATIONAL EXPRESSIONS

Exercises 97–104: Find the GCF of the given terms.

97. $10x^2$ and $16x$ **98.** $15y^2$ and $20y$

99. $14a^2$ and $35a^3$ **100.** $4b$ and $12b^3$

101. $100m^2$ and $50n^2$ **102.** $105x^3$ and $35y^2$

103. $8a^3c^2$ and $24a^2c$ **104.** $10x^2y$ and $25x^4y$

Exercises 105–114: Simplify the rational expression.

105. $\dfrac{6x}{18}$ **106.** $-\dfrac{35a}{55a}$

107. $-\dfrac{12x}{40xy}$ **108.** $\dfrac{18xy}{24y}$

109. $\dfrac{13mn}{26mn}$ **110.** $-\dfrac{5a}{5ab}$

111. $-\dfrac{6x^2y}{3x}$ **112.** $\dfrac{12m^2n^3}{4mn}$

113. $\dfrac{25x^2y}{45xy^3}$ **114.** $-\dfrac{28a^3b^2}{42ab}$

APPLICATIONS

115. *Study Time* A student spends 8 hours studying for a physics exam. What fraction of a 24-hour day is this? Write your answer in lowest terms.

116. *Extra Credit* In a math class with 31 students, 27 students earned extra credit. In a biology class with 21 students, 17 students earned extra credit. Which class had a larger fraction of its students earning extra credit?

117. *Age Puzzle* A man who is between the ages of 35 and 60 says that his age is a prime number and that two years ago it was also a prime number. How old is the man?

118. *Twin Primes* Prime numbers that differ by only 2 are called *twin primes*. For example, 5 and 7 are twin primes. Find all of the twin prime pairs between 50 and 100.

119. *Female Physicians* In 1970, about 8 out of every 100 physicians were female. By 2008, about 28 out of every 100 physicians were female. Write these values as fractions in lowest terms. (*Source:* American Medical Association.)

120. *Female Pharmacists* In 1980, about 18 out of every 100 pharmacists were female. By 2008, there were about 54 female pharmacists for every 100 pharmacists. Write these values as fractions in lowest terms. (*Source:* American Pharmaceutical Association.)

WRITING ABOUT MATHEMATICS

121. Explain how to determine if a natural number is prime or composite.

122. Do some research to find a divisibility test for 11. Show how the test works for the number 719,653.

123. Explain how to simplify a fraction to lowest terms.

124. Explain how the cross product method can be used to determine if two fractions are equivalent.

Checking Basic Concepts

1. Write the numerator and denominator of $\frac{5}{18}$.

2. Write the improper fraction and mixed number represented by the shading.

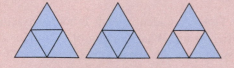

3. Write $\frac{19}{5}$ as a mixed number.

4. Write $6\frac{3}{7}$ as an improper fraction.

5. Is 9762 divisible by 3?

6. Determine if each given number is prime, composite, or neither.
   (a) 91　　　　　**(b)** 103

7. Write the prime factorization of each number.
   (a) 50　　　　　**(b)** 105

8. Find a number to replace the question mark so that the given fractions are equivalent.

$$\frac{5}{12} = \frac{?}{72}$$

9. Simplify each rational expression to lowest terms.
   (a) $\frac{30}{48}$　　　　　**(b)** $-\frac{16xy}{28y}$

10. *United States* The following states have names that begin with vowels: Alabama, Alaska, Arizona, Arkansas, Idaho, Illinois, Indiana, Iowa, Ohio, Oklahoma, Oregon, and Utah. What fraction of the 50 states have names that begin with a vowel? Write your answer in lowest terms.

4.3　Multiplying and Dividing Fractions

Multiplying Fractions ● Multiplying Rational Expressions ● Powers and Square Roots of Fractions ● Dividing Fractions and Rational Expressions ● Applications

A LOOK INTO MATH ▶

Some liquid detergents need to be diluted in water before use. If a particular cleaning product requires that we mix $\frac{1}{3}$ cup of concentrated detergent with 1 gallon of water, we could *multiply* fractions to determine how much concentrate should be mixed with $\frac{1}{2}$ gallon of water. Also, if a bottle of this product contains 8 cups of detergent, we could *divide* by a fraction to determine the total number of 1-gallon mixtures that could be made from the bottle. In this section, we will discuss how to multiply and divide fractions.

NEW VOCABULARY

☐ Reciprocal
☐ Multiplicative inverse

Multiplying Fractions

One way to visualize a product is to draw one vertical and one horizontal number line that each begin at a common location for 0. A shaded rectangle can be used to find a product. For example, a rectangle representing $3 \cdot 4$ is shown in Figure 4.8. The product, 12, is found by counting the number of squares that appear inside the larger rectangle.

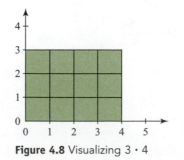

Figure 4.8 Visualizing $3 \cdot 4$

The product of two fractions can be visualized in a similar manner. Figure 4.9 illustrates how to find the product $\frac{2}{3} \cdot \frac{4}{5}$. Because $\mathbf{1} \cdot \mathbf{1} = \mathbf{1}$ whole, the largest rectangle represents $\mathbf{1}$ whole. Since this largest rectangle contains 15 smaller rectangles of equal size and 8 of these rectangles are shaded, the product is $\frac{2}{3} \cdot \frac{4}{5} = \frac{8}{15}$.

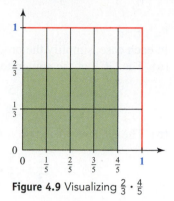

Figure 4.9 Visualizing $\frac{2}{3} \cdot \frac{4}{5}$

READING CHECK

• How are fractions multiplied?

This discussion suggests that the numerator of the product is found by multiplying the numerators of the factors, and the denominator of the product is found by multiplying the denominators of the factors. This procedure is summarized as follows.

STUDY TIP

Much of mathematics builds on previous knowledge. Since we already know how to multiply integers (from Chapter 2), the only new information needed to multiply fractions is to know that we multiply numerators and denominators separately.

MULTIPLYING FRACTIONS

If the variables a, b, c, and d represent numbers with $b \neq 0$ and $d \neq 0$, then the product of the fractions $\frac{a}{b}$ and $\frac{c}{d}$ is given by

$$\frac{a}{b} \cdot \frac{c}{d} = \frac{ac}{bd}.$$

EXAMPLE 1 **Multiplying fractions**

Multiply.

(a) $\dfrac{3}{7} \cdot \dfrac{2}{5}$ **(b)** $-\dfrac{1}{6} \cdot \dfrac{5}{8}$ **(c)** $-\dfrac{2}{3} \cdot \left(-\dfrac{2}{5}\right)$

Solution

(a) Multiply numerators and multiply denominators.

$$\frac{3}{7} \cdot \frac{2}{5} = \frac{3 \cdot 2}{7 \cdot 5} = \frac{6}{35}$$

(b) The product of two numbers with unlike signs is negative.

$$-\frac{1}{6} \cdot \frac{5}{8} = -\frac{1 \cdot 5}{6 \cdot 8} = -\frac{5}{48}$$

CALCULATOR HELP

To multiply fractions with a calculator, see the Appendix (page AP-4).

(c) The product of two numbers with like signs is positive.

$$-\frac{2}{3} \cdot \left(-\frac{2}{5}\right) = \frac{2 \cdot 2}{3 \cdot 5} = \frac{4}{15}$$

Now Try Exercises 7, 11, 15

The three products found in Example 1 are all fractions written in lowest terms. However, this is not necessarily the case with every product of two fractions. Sometimes the resulting fraction needs to be simplified, as illustrated in the next example.

EXAMPLE 2　**Multiplying fractions and simplifying the result**

Multiply and simplify the result.

(a) $\dfrac{5}{6} \cdot \dfrac{4}{5}$　(b) $-\dfrac{2}{3} \cdot \dfrac{15}{22}$

Solution

In each case, simplify the product by applying the basic principle of fractions.

(a) The GCF of the numerator product **20** and the denominator product **30** is **10**.

$$\frac{5}{6} \cdot \frac{4}{5} = \frac{5 \cdot 4}{6 \cdot 5} = \frac{\mathbf{20}}{\mathbf{30}} = \frac{2 \cdot \mathbf{10}}{3 \cdot \mathbf{10}} = \frac{2}{3}$$

(b) The GCF of **30** and **66** is **6**.

$$-\frac{2}{3} \cdot \frac{15}{22} = -\frac{2 \cdot 15}{3 \cdot 22} = -\frac{\mathbf{30}}{\mathbf{66}} = -\frac{5 \cdot \mathbf{6}}{11 \cdot \mathbf{6}} = -\frac{5}{11}$$

▍**Now Try Exercises 21, 23**

MAKING CONNECTIONS

The Basic Principle of Fractions and Multiplying by 1

The basic principle of fractions is a useful application of the identity property for multiplication. In other words, it is a handy way to multiply by 1. For example, the fraction $\frac{20}{30}$ can be simplified as

$$\frac{20}{30} = \frac{2 \cdot \mathbf{10}}{3 \cdot \mathbf{10}} = \frac{2}{3} \cdot \frac{\mathbf{10}}{\mathbf{10}} = \frac{2}{3} \cdot 1 = \frac{2}{3}.$$

When a product of fractions contains relatively large numbers, it may be easier to simplify the product *before* multiplying. This method is summarized in the following box and is demonstrated in the next example.

SIMPLIFYING BEFORE MULTIPLYING FRACTIONS

Let the variables a, b, c, and d represent numbers with $b \neq 0$ and $d \neq 0$. To simplify the product $\frac{a}{b} \cdot \frac{c}{d}$ before multiplying, do the following.

STEP 1: Write $\frac{a}{b} \cdot \frac{c}{d}$ as $\frac{a \cdot c}{b \cdot d}$. Do not multiply.

STEP 2: Replace each composite factor with its prime factorization.

STEP 3: Eliminate all factors that are common to the numerator and denominator by applying the basic principle of fractions.

STEP 4: Multiply the remaining factors in the numerator and multiply the remaining factors in the denominator.

READING CHECK

• Why is it useful to simplify before multiplying fractions?

EXAMPLE 3　**Simplifying before multiplying fractions**

Find each product by simplifying before multiplying.

(a) $\dfrac{12}{25} \cdot \dfrac{35}{36}$　(b) $\dfrac{15}{44} \cdot \dfrac{22}{39}$　(c) $-\dfrac{9}{40} \cdot \left(-\dfrac{8}{27}\right)$

Solution

(a) **STEP 1:** Write the product $\frac{12}{25} \cdot \frac{35}{36}$ as follows.

$$\frac{12 \cdot 35}{25 \cdot 36}$$

STEP 2: Replace every composite factor with its prime factorization.

$$\frac{(2 \cdot 2 \cdot 3) \cdot (5 \cdot 7)}{(5 \cdot 5) \cdot (2 \cdot 2 \cdot 3 \cdot 3)}$$

STEP 3: Regroup the factors and apply the basic principle of fractions.

$$\frac{7 \cdot (2 \cdot 2 \cdot 3 \cdot 5)}{3 \cdot 5 \cdot (2 \cdot 2 \cdot 3 \cdot 5)} = \frac{7}{3 \cdot 5}$$

STEP 4: Multiply the remaining factors in the numerator and in the denominator.

$$\frac{7}{3 \cdot 5} = \frac{7}{15}$$

(b) The entire four-step process can be written in a single line of equivalent expressions. Note that it is not necessary to regroup the common factors (as in Step 3 above) in order to apply the basic principle of fractions. Such grouping can be done mentally.

$$\frac{15}{44} \cdot \frac{22}{39} \overset{\text{Step 1}}{=} \frac{15 \cdot 22}{44 \cdot 39} \overset{\text{Step 2}}{=} \frac{(3 \cdot 5) \cdot (2 \cdot 11)}{(2 \cdot 2 \cdot 11) \cdot (3 \cdot 13)} \overset{\text{Step 3}}{=} \frac{5}{2 \cdot 13} \overset{\text{Step 4}}{=} \frac{5}{26}$$

(c) The product of two negative fractions is positive. Also, note that the basic principle of fractions eliminates *all* of the factors from the numerator. When this happens, it is important to remember that the remaining factor is **1**.

$$-\frac{9}{40} \cdot \left(-\frac{8}{27}\right) \overset{\text{Step 1}}{=} \frac{9 \cdot 8}{40 \cdot 27} \overset{\text{Step 2}}{=} \frac{(3 \cdot 3) \cdot (2 \cdot 2 \cdot 2)}{(2 \cdot 2 \cdot 2 \cdot 5) \cdot (3 \cdot 3 \cdot 3)} \overset{\text{Step 3}}{=} \frac{1}{3 \cdot 5} \overset{\text{Step 4}}{=} \frac{1}{15}$$

Now Try Exercises 25, 31, 35

Multiplying Rational Expressions

In Section 4.2, we simplified rational expressions whose numerator and denominator each contained a single term. A process similar to that used in Example 3 can also be used to multiply such rational expressions. The next example illustrates this process.

EXAMPLE 4 **Multiplying rational expressions**

Multiply the rational expressions. Simplify your result.

(a) $\dfrac{x^2}{3y} \cdot \dfrac{6y^4}{5x}$ (b) $\dfrac{4a^2}{9b^2} \cdot \dfrac{15b}{16a}$ (c) $8m \cdot \left(-\dfrac{3n}{4m^2}\right)$

Solution

(a) **STEP 1:** Write the product $\frac{x^2}{3y} \cdot \frac{6y^4}{5x}$ as follows.

$$\frac{x^2 \cdot 6y^4}{3y \cdot 5x}$$

STEP 2: Replace every factor in $\frac{x^2 \cdot 6y^4}{3y \cdot 5x}$ with its complete factorization.

$$\frac{(x \cdot x) \cdot (2 \cdot 3 \cdot y \cdot y \cdot y \cdot y)}{(3 \cdot y) \cdot (5 \cdot x)}$$

STEP 3: Regroup the factors and apply the basic principle of fractions.

$$\frac{2 \cdot x \cdot y \cdot y \cdot y \cdot (3 \cdot x \cdot y)}{5 \cdot (3 \cdot x \cdot y)} = \frac{2 \cdot x \cdot y \cdot y \cdot y}{5}$$

STEP 4: Multiply the remaining factors in the numerator and in the denominator.

$$\frac{2 \cdot x \cdot y \cdot y \cdot y}{5} = \frac{2xy^3}{5}$$

(b) Once again, the entire four-step process can be written as a single line of equivalent expressions, as shown here.

$$\underset{}{\frac{4a^2}{9b^2} \cdot \frac{15b}{16a}} = \overset{\text{Step 1}}{\frac{4a^2 \cdot 15b}{9b^2 \cdot 16a}} = \overset{\text{Step 2}}{\frac{(2 \cdot 2 \cdot a \cdot a) \cdot (3 \cdot 5 \cdot b)}{(3 \cdot 3 \cdot b \cdot b) \cdot (2 \cdot 2 \cdot 2 \cdot 2 \cdot a)}} = \overset{\text{Step 3}}{\frac{5 \cdot a}{2 \cdot 2 \cdot 3 \cdot b}} = \overset{\text{Step 4}}{\frac{5a}{12b}}$$

(c) Begin by writing $8m$ as the rational expression $\frac{8m}{1}$. Note that the product is negative because we are multiplying rational expressions with unlike signs.

$$\frac{8m}{1} \cdot \left(-\frac{3n}{4m^2}\right) = \overset{\text{Step 1}}{-\frac{8m \cdot 3n}{1 \cdot 4m^2}} = \overset{\text{Step 2}}{-\frac{(2 \cdot 2 \cdot 2 \cdot m) \cdot (3 \cdot n)}{1 \cdot (2 \cdot 2 \cdot m \cdot m)}} = \overset{\text{Step 3}}{-\frac{2 \cdot 3 \cdot n}{1 \cdot m}} = \overset{\text{Step 4}}{-\frac{6n}{m}}$$

Now Try Exercises 43, 45, 49

Powers and Square Roots of Fractions

Recall that a natural number exponent tells us *how many times* to multiply the base of an exponential expression by itself. For example, 5^4 means that we multiply 5 by itself a total of 4 times: $5 \cdot 5 \cdot 5 \cdot 5$. This rule also applies when the base is a fraction; a natural number exponent tells us how many times we multiply the fraction by itself.

EXAMPLE 5   **Raising a fraction to a power**

Evaluate each exponential expression.

(a) $\left(\frac{1}{2}\right)^3$ **(b)** $\left(-\frac{3}{4}\right)^2$ **(c)** $\left(-\frac{2}{3}\right)^3$

Solution

(a) $\left(\frac{1}{2}\right)^3 = \frac{1}{2} \cdot \frac{1}{2} \cdot \frac{1}{2} = \frac{1 \cdot 1 \cdot 1}{2 \cdot 2 \cdot 2} = \frac{1}{8}$

(b) Because there is an even number of negative factors, the product is positive.

$$\left(-\frac{3}{4}\right)^2 = \left(-\frac{3}{4}\right) \cdot \left(-\frac{3}{4}\right) = \frac{3 \cdot 3}{4 \cdot 4} = \frac{9}{16}$$

(c) With an odd number of negative factors, we have a negative product.

$$\left(-\frac{2}{3}\right)^3 = \left(-\frac{2}{3}\right) \cdot \left(-\frac{2}{3}\right) \cdot \left(-\frac{2}{3}\right) = -\frac{2 \cdot 2 \cdot 2}{3 \cdot 3 \cdot 3} = -\frac{8}{27}$$

Now Try Exercises 57, 59, 61

CALCULATOR HELP

To raise a fraction to a power with a calculator, see the Appendix (page AP-4).

When a fraction is squared, the result is a fraction whose numerator and denominator are both perfect squares. For example, squaring the fraction $\frac{5}{9}$ results in

$$\left(\frac{5}{9}\right)^2 = \frac{5}{9} \cdot \frac{5}{9} = \frac{25}{81},$$

where both 25 and 81 are perfect squares. By definition, the positive square root of $\frac{25}{81}$ is the number (fraction) whose square is $\frac{25}{81}$, or $\frac{5}{9}$. That is,

$$\sqrt{\frac{25}{81}} = \frac{5}{9}.$$

The fact that $\sqrt{25} = 5$ and $\sqrt{81} = 9$ suggests the following procedure for finding the positive square root of a fraction.

SQUARE ROOTS OF FRACTIONS

If the variables a and b represent whole numbers with $b \neq 0$ and the fraction $\frac{a}{b}$ is in lowest terms, then

$$\sqrt{\frac{a}{b}} = \frac{\sqrt{a}}{\sqrt{b}}.$$

READING CHECK

• How is the square root of a fraction found?

NOTE: If $\frac{a}{b}$ is not given in lowest terms, simplify it before applying this procedure.

EXAMPLE 6 | **Finding the square root of a fraction**

Find each square root.

(a) $\sqrt{\dfrac{9}{49}}$ **(b)** $\sqrt{\dfrac{45}{80}}$ **(c)** $\sqrt{\dfrac{75}{3}}$

Solution

(a) $\sqrt{\dfrac{9}{49}} = \dfrac{\sqrt{9}}{\sqrt{49}} = \dfrac{3}{7}$

(b) Because the fraction is not given in lowest terms, it should be simplified first.

$$\frac{45}{80} = \frac{9 \cdot 5}{16 \cdot 5} = \frac{9}{16}$$

CALCULATOR HELP

To find the square root of a fraction with a calculator, see the Appendix (page AP-4).

As a result,

$$\sqrt{\frac{45}{80}} = \sqrt{\frac{9}{16}} = \frac{\sqrt{9}}{\sqrt{16}} = \frac{3}{4}.$$

(c) Because $\frac{75}{3} = 25$, the given square root is evaluated as $\sqrt{\frac{75}{3}} = \sqrt{25} = 5$.

Now Try Exercises 65, 69, 71

Dividing Fractions and Rational Expressions

In order to develop a strategy for dividing fractions, we must first understand the concept of a *reciprocal*. Two numbers are **reciprocals** or **multiplicative inverses** of each other if their product is 1. Several numbers and their reciprocals are listed in Table 4.1 on the next page.

TABLE 4.1 Numbers and Their Reciprocals

Number	$\frac{2}{3}$	4	$-\frac{5}{2}$	$\frac{1}{7}$	1
Reciprocal	$\frac{3}{2}$	$\frac{1}{4}$	$-\frac{2}{5}$	7	1

NOTE: If neither a nor b is 0, the reciprocal of the fraction $\frac{a}{b}$ is $\frac{b}{a}$. Zero has no reciprocal.

▶ **REAL-WORLD CONNECTION** Many college degree programs are designed to be completed in 4 years. Since a semester is $\frac{1}{2}$ of a school year, we can use division to find the number of semesters needed to complete such a program.

$$4 \div \frac{1}{2}$$

Because every school year has 2 semesters, we can also find the number of semesters needed to complete the program, using multiplication.

$$4 \cdot 2$$

Since $4 \cdot 2 = 8$, it is reasonable to conclude that $4 \div \frac{1}{2} = 8$. This discussion suggests a strategy for division by a fraction—*multiply by its reciprocal.*

DIVIDING FRACTIONS

If the variables a, b, c, and d represent numbers with $b \neq 0$, $c \neq 0$, and $d \neq 0$, then the quotient found when dividing $\frac{a}{b}$ by $\frac{c}{d}$ is given by

$$\frac{a}{b} \div \frac{c}{d} = \frac{a}{b} \cdot \frac{d}{c}.$$

That is, we multiply the first fraction by the reciprocal of the second fraction.

READING CHECK

• What process is used when dividing by a fraction?

EXAMPLE 7   **Dividing fractions**

Divide.

(a) $\dfrac{4}{3} \div \dfrac{2}{9}$ **(b)** $-\dfrac{5}{6} \div \dfrac{5}{8}$ **(c)** $8 \div \dfrac{4}{3}$ **(d)** $\dfrac{4}{9} \div 2$

Solution

(a) Multiply the first fraction by the reciprocal of the second fraction.

$$\frac{4}{3} \div \frac{2}{9} = \frac{4}{3} \cdot \frac{9}{2} = \frac{4 \cdot 9}{3 \cdot 2} = \frac{(2 \cdot 2) \cdot (3 \cdot 3)}{3 \cdot 2} = \frac{2 \cdot 3}{1} = \frac{6}{1} = 6$$

(b) The quotient of two numbers with unlike signs is negative.

$$-\frac{5}{6} \div \frac{5}{8} = -\frac{5}{6} \cdot \frac{8}{5} = -\frac{5 \cdot 8}{6 \cdot 5} = -\frac{5 \cdot (2 \cdot 2 \cdot 2)}{(3 \cdot 2) \cdot 5} = -\frac{2 \cdot 2}{3} = -\frac{4}{3}$$

CALCULATOR HELP

To divide fractions with a calculator, see the Appendix (page AP-4).

(c) Start by writing 8 in fraction form as $\frac{8}{1}$.

$$\frac{8}{1} \div \frac{4}{3} = \frac{8}{1} \cdot \frac{3}{4} = \frac{8 \cdot 3}{1 \cdot 4} = \frac{(2 \cdot 2 \cdot 2) \cdot 3}{2 \cdot 2} = \frac{2 \cdot 3}{1} = \frac{6}{1} = 6$$

(d) Because $2 = \frac{2}{1}$, the reciprocal of 2 is $\frac{1}{2}$.

$$\frac{4}{9} \div 2 = \frac{4}{9} \cdot \frac{1}{2} = \frac{4 \cdot 1}{9 \cdot 2} = \frac{2 \cdot 2}{(3 \cdot 3) \cdot 2} = \frac{2}{3 \cdot 3} = \frac{2}{9}$$

Now Try Exercises 81, 83, 87, 91

MAKING CONNECTIONS

Simplifying Before Dividing Fractions

Do not simplify a division problem until *after* the problem has been changed to multiplication. For example, we **cannot** eliminate the **2**s from

$$\frac{2}{3} \div \frac{1}{2}.$$

We must first change the division problem to a multiplication problem.

$$\frac{2}{3} \div \frac{1}{2} = \frac{2}{3} \cdot \frac{2}{1} = \frac{2 \cdot 2}{3 \cdot 1} = \frac{4}{3}.$$

Eliminating the 2s before changing to multiplication would give an incorrect answer.

Earlier in this section, we found the product of two rational expressions whose numerator and denominator each contained a single term. We can also find the quotient of two rational expressions of this type. In the next example, we divide rational expressions.

EXAMPLE 8 **Dividing rational expressions**

Divide.

(a) $\dfrac{x^2}{2y} \div \dfrac{8x}{5y}$ **(b)** $-\dfrac{4n}{3m} \div 2n$

Solution

(a) Multiply the first rational expression by the reciprocal of the second expression.

$$\frac{x^2}{2y} \div \frac{8x}{5y} = \frac{x^2}{2y} \cdot \frac{5y}{8x} = \frac{x^2 \cdot 5y}{2y \cdot 8x} = \frac{(x \cdot x) \cdot (5 \cdot y)}{(2 \cdot y) \cdot (2 \cdot 2 \cdot 2 \cdot x)} = \frac{x \cdot 5}{2 \cdot 2 \cdot 2 \cdot 2} = \frac{5x}{16}$$

(b) Because $2n = \frac{2n}{1}$, the reciprocal of $2n$ is $\frac{1}{2n}$.

$$-\frac{4n}{3m} \div 2n = -\frac{4n}{3m} \cdot \frac{1}{2n} = -\frac{4n \cdot 1}{3m \cdot 2n} = -\frac{2 \cdot 2 \cdot n}{(3 \cdot m) \cdot (2 \cdot n)} = -\frac{2}{3 \cdot m} = -\frac{2}{3m}$$

Now Try Exercises 97, 101

Applications

In applications involving fractions, we often encounter phrases such as "one-half **of** the total" or "three-fourths **of** the students." In this context, the word "of" indicates multiplication. For example, we can find two-thirds of three-fifths by multiplying $\frac{2}{3} \cdot \frac{3}{5}$. In the next two examples, we solve application problems involving fractions.

EXAMPLE 9 **Finding a fraction of a class**

In a class of 24 students, two-thirds of the students said that they had studied more than 2 hours for an upcoming exam. Determine the number of students from this class who studied more than 2 hours for the exam.

Solution
Two-thirds **of** 24 can be found by multiplying as follows.

$$\frac{2}{3} \cdot 24 = \frac{2}{3} \cdot \frac{24}{1} = \frac{2 \cdot 24}{3 \cdot 1} = \frac{2 \cdot (2 \cdot 2 \cdot 2 \cdot 3)}{3} = \frac{2 \cdot 2 \cdot 2 \cdot 2}{1} = \frac{16}{1} = 16$$

Therefore, a total of 16 students studied more than 2 hours for the exam.

Now Try Exercise 113

EXAMPLE 10 **Finding a fraction of a fraction**

Suppose that one-half of the fish in a small pond are sunfish. If four-fifths of the sunfish in the pond are under 3 inches in length, what fraction of the fish in the pond are sunfish under 3 inches in length?

Solution
Four-fifths **of** one-half of the fish in the pond are sunfish under 3 inches in length.

$$\frac{4}{5} \cdot \frac{1}{2} = \frac{4 \cdot 1}{5 \cdot 2} = \frac{2 \cdot 2}{5 \cdot 2} = \frac{2}{5}$$

So, $\frac{2}{5}$ of the fish in the pond are sunfish under 3 inches in length.

Now Try Exercise 119

AREA OF A TRIANGLE Sometimes formulas contain fractions. Consider the triangle shown in Figure 4.10. A formula for its area can be discovered by creating a triangle of identical shape and size that can be turned and placed directly above the first triangle, as shown in Figure 4.11.

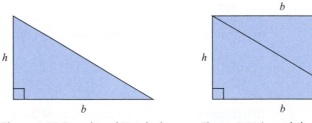

Figure 4.10 Base *b* and Height *h*   **Figure 4.11** Length *b* and Width *h*

Figure 4.11 is a rectangle with length *b* and width *h*. Its area is

$$A = bh.$$

Because the triangle from Figure 4.10 makes up $\frac{1}{2}$ of the area of the rectangle in Figure 4.11, its area is

$$A = \frac{1}{2}bh.$$

In general, the height of a triangle whose base is positioned horizontally is the vertical distance from the base to the highest point in the triangle, as shown in Figure 4.12.

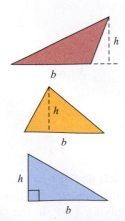

Figure 4.12 Heights of Various Triangles

AREA OF A TRIANGLE

The area A of a triangle with base b and height h is given by the formula

$$A = \frac{1}{2}bh.$$

EXAMPLE 11 **Finding the area of a triangle**

Find the area of each triangle. Assume that the units are inches.

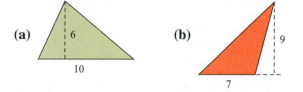

(a)   6   10

(b)   9   7

Solution

(a) Replace b with 10 and replace h with 6 in the formula $A = \frac{1}{2}bh$.

$$A = \frac{1}{2}(10)(6) = \frac{1}{2} \cdot \frac{10}{1} \cdot \frac{6}{1} = \frac{1 \cdot 10 \cdot 6}{2 \cdot 1 \cdot 1} = \frac{(2 \cdot 5) \cdot (2 \cdot 3)}{2} = \frac{5 \cdot 2 \cdot 3}{1} = \frac{30}{1} = 30$$

The area of the triangle is 30 square inches.

(b) Replace b with 7 and replace h with 9 in the formula $A = \frac{1}{2}bh$.

$$A = \frac{1}{2}(7)(9) = \frac{1}{2} \cdot \frac{7}{1} \cdot \frac{9}{1} = \frac{1 \cdot 7 \cdot 9}{2 \cdot 1 \cdot 1} = \frac{7 \cdot (3 \cdot 3)}{2} = \frac{7 \cdot 3 \cdot 3}{2} = \frac{63}{2}$$

The area of the triangle is $\frac{63}{2}$ square inches.

Now Try Exercises 121, 123

4.3 Putting It All Together

CONCEPT	COMMENTS	EXAMPLES
Multiplying Fractions and Rational Expressions	If a, b, c, and d represent numbers (or single terms) with $b \neq 0$ and $d \neq 0$, then $$\frac{a}{b} \cdot \frac{c}{d} = \frac{ac}{bd}.$$	$\dfrac{1}{4} \cdot \dfrac{3}{5} = \dfrac{1 \cdot 3}{4 \cdot 5} = \dfrac{3}{20}$ $\dfrac{2x}{y} \cdot \dfrac{3x}{5y} = \dfrac{2x \cdot 3x}{y \cdot 5y} = \dfrac{6x^2}{5y^2}$
Simplifying Before Multiplying Fractions	1. Write $\frac{a}{b} \cdot \frac{c}{d}$ as $\frac{a \cdot c}{b \cdot d}$. Do not multiply. 2. Replace each factor with its prime factorization. 3. Eliminate all factors that are common to the numerator and denominator by applying the basic principle of fractions. 4. Multiply the remaining factors in the numerator and the remaining factors in the denominator.	$\dfrac{4}{9} \cdot \dfrac{3}{10} = \dfrac{4 \cdot 3}{9 \cdot 10}$ Step 1 $= \dfrac{(2 \cdot 2) \cdot 3}{(3 \cdot 3) \cdot (2 \cdot 5)}$ Step 2 $= \dfrac{2}{3 \cdot 5}$ Step 3 $= \dfrac{2}{15}$ Step 4

continued on next page

continued from previous page

CONCEPT	COMMENTS	EXAMPLES
Powers of Fractions	A natural number exponent on a fraction indicates how many times to multiply the fraction by itself.	$\left(\dfrac{2}{5}\right)^3 = \dfrac{2}{5} \cdot \dfrac{2}{5} \cdot \dfrac{2}{5} = \dfrac{8}{125}$
Square Roots of Fractions	If a and b represent whole numbers with $b \neq 0$ and the fraction $\frac{a}{b}$ is in lowest terms, then $$\sqrt{\dfrac{a}{b}} = \dfrac{\sqrt{a}}{\sqrt{b}}.$$	$\sqrt{\dfrac{16}{25}} = \dfrac{\sqrt{16}}{\sqrt{25}} = \dfrac{4}{5}$
Dividing Fractions and Rational Expressions	If a, b, c, and d represent numbers (or single terms) with $b \neq 0$, $c \neq 0$ and $d \neq 0$, then $$\dfrac{a}{b} \div \dfrac{c}{d} = \dfrac{a}{b} \cdot \dfrac{d}{c}.$$	$\dfrac{2}{3} \div \dfrac{3}{4} = \dfrac{2}{3} \cdot \dfrac{4}{3} = \dfrac{2 \cdot 4}{3 \cdot 3} = \dfrac{8}{9}$ $\dfrac{x}{5} \div \dfrac{3}{2y} = \dfrac{x}{5} \cdot \dfrac{2y}{3} = \dfrac{x \cdot 2y}{5 \cdot 3} = \dfrac{2xy}{15}$
Area of a Triangle	A triangle with base b and height h has area $$A = \dfrac{1}{2}bh.$$	$A = \dfrac{1}{2}(13)(6) = 39$ square units

4.3 Exercises

MyMathLab · Math XL PRACTICE · WATCH · DOWNLOAD · READ · REVIEW

CONCEPTS AND VOCABULARY

1. When multiplying fractions, the numerator of the product is found by multiplying the _____ of the factors, and the denominator of the product is found by multiplying the _____ of the factors.

2. When finding a product of fractions, we can simplify before multiplying by applying the basic principle of _____ to eliminate factors that are common to the numerator and denominator.

3. A natural number exponent on a fraction tells us how many times to _____ the fraction by itself.

4. To find the square root of a fraction, find the square root of the _____ and _____.

5. Fractions $\frac{3}{4}$ and $\frac{4}{3}$ are _____ of each other.

6. In application problems involving fractions, the word "of" often indicates _____.

MULTIPLYING FRACTIONS AND RATIONAL EXPRESSIONS

Exercises 7–38: Multiply. Simplify, if necessary.

7. $\dfrac{1}{4} \cdot \dfrac{3}{5}$

8. $\dfrac{2}{7} \cdot \dfrac{1}{3}$

9. $\dfrac{11}{9} \cdot \dfrac{2}{3}$

10. $\dfrac{7}{4} \cdot \dfrac{1}{8}$

11. $-\dfrac{2}{3} \cdot \dfrac{5}{3}$

12. $-\dfrac{11}{12} \cdot \dfrac{7}{4}$

13. $\dfrac{7}{5} \cdot \left(-\dfrac{9}{4}\right)$

14. $\dfrac{6}{11} \cdot \left(-\dfrac{8}{5}\right)$

15. $-\dfrac{1}{3} \cdot \left(-\dfrac{1}{8}\right)$

16. $-\dfrac{5}{2} \cdot \left(-\dfrac{7}{9}\right)$

17. $9 \cdot \dfrac{2}{3}$

18. $\dfrac{5}{6} \cdot 24$

19. $-6 \cdot \dfrac{3}{4}$

20. $\dfrac{7}{12} \cdot (-16)$

21. $\dfrac{3}{10} \cdot \dfrac{2}{3}$

22. $\dfrac{4}{5} \cdot \dfrac{5}{12}$

23. $-\dfrac{3}{4} \cdot \dfrac{16}{9}$

24. $-\dfrac{8}{9} \cdot \dfrac{3}{10}$

25. $\dfrac{14}{27} \cdot \dfrac{18}{49}$

26. $\dfrac{10}{9} \cdot \dfrac{3}{25}$

27. $\dfrac{5}{24} \cdot \dfrac{12}{25}$

28. $\dfrac{5}{28} \cdot \dfrac{14}{15}$

29. $-\dfrac{12}{5} \cdot \dfrac{5}{4}$

30. $\dfrac{24}{7} \cdot \dfrac{35}{12}$

31. $\dfrac{14}{39} \cdot \dfrac{26}{35}$

32. $-\dfrac{56}{65} \cdot \dfrac{15}{28}$

33. $\dfrac{27}{55} \cdot \left(-\dfrac{25}{18}\right)$

34. $\dfrac{25}{33} \cdot \left(-\dfrac{44}{75}\right)$

35. $-\dfrac{24}{45} \cdot \left(-\dfrac{55}{48}\right)$

36. $-\dfrac{34}{35} \cdot \left(-\dfrac{15}{17}\right)$

37. $\dfrac{5}{7} \cdot \dfrac{14}{15} \cdot \dfrac{3}{8}$

38. $\dfrac{11}{6} \cdot \dfrac{14}{27} \cdot \dfrac{54}{55}$

Exercises 39–56: Multiply the rational expressions. Simplify, if necessary.

39. $\dfrac{x}{2} \cdot \dfrac{5}{y}$

40. $\dfrac{3m}{5n} \cdot \dfrac{2}{7}$

41. $\dfrac{4p}{3q} \cdot \dfrac{2p}{5q}$

42. $\dfrac{x}{3y} \cdot \dfrac{7x}{y}$

43. $\dfrac{3a^2}{4b} \cdot \dfrac{2b^3}{a}$

44. $\dfrac{4z^3}{9y^2} \cdot \dfrac{3y}{2z^2}$

45. $\dfrac{12x^3}{5y^3} \cdot \dfrac{15y^2}{8x}$

46. $\dfrac{14w}{5x^2} \cdot \dfrac{2x}{21w^3}$

47. $-\dfrac{3u^2}{5v} \cdot \dfrac{10u}{9v}$

48. $\dfrac{4x}{5y} \cdot \left(-\dfrac{5y}{4x}\right)$

49. $-5x \cdot \dfrac{7y}{15x^2}$

50. $\dfrac{2a^2}{3b^2} \cdot (-6b)$

51. $\dfrac{3x}{14y} \cdot \dfrac{7y}{9x^2}$

52. $\dfrac{2a^2}{9b^2} \cdot \dfrac{3b}{10a^3}$

53. $-\dfrac{10m^2}{3n} \cdot \left(-\dfrac{6n}{5m^2}\right)$

54. $-\dfrac{2y^2}{5z} \cdot \left(-\dfrac{5y}{8z^2}\right)$

55. $\dfrac{3x}{4y} \cdot \dfrac{2x}{9y} \cdot \dfrac{6y}{5x}$

56. $\dfrac{7y}{5x} \cdot \dfrac{15}{8x} \cdot \dfrac{4x}{21y}$

POWERS AND SQUARE ROOTS OF FRACTIONS

Exercises 57–72: Evaluate the expression.

57. $\left(\dfrac{1}{4}\right)^2$

58. $\left(\dfrac{3}{2}\right)^3$

59. $\left(-\dfrac{3}{5}\right)^2$

60. $\left(-\dfrac{1}{3}\right)^3$

61. $\left(-\dfrac{3}{4}\right)^3$

62. $\left(\dfrac{2}{5}\right)^3$

63. $\left(\dfrac{8}{11}\right)^2$

64. $\left(-\dfrac{7}{9}\right)^2$

65. $\sqrt{\dfrac{9}{25}}$

66. $\sqrt{\dfrac{16}{81}}$

67. $\sqrt{\dfrac{1}{64}}$

68. $\sqrt{\dfrac{36}{121}}$

69. $\sqrt{\dfrac{80}{5}}$

70. $\sqrt{\dfrac{72}{2}}$

71. $\sqrt{\dfrac{20}{45}}$

72. $\sqrt{\dfrac{27}{48}}$

DIVIDING FRACTIONS AND RATIONAL EXPRESSIONS

Exercises 73–80: Write the reciprocal of the number.

73. $\dfrac{3}{5}$

74. $-\dfrac{7}{13}$

75. $-\dfrac{1}{12}$

76. $\dfrac{1}{9}$

77. 15

78. -8

79. 1

80. -1

Exercises 81–96: Divide. Simplify, if necessary.

81. $\dfrac{3}{4} \div \dfrac{9}{20}$

82. $-\dfrac{4}{7} \div \dfrac{16}{21}$

83. $-\dfrac{9}{2} \div \dfrac{27}{40}$

84. $\dfrac{6}{25} \div \dfrac{42}{15}$

85. $\dfrac{33}{8} \div \left(-\dfrac{55}{64}\right)$

86. $\dfrac{20}{27} \div \left(-\dfrac{40}{81}\right)$

87. $9 \div \dfrac{3}{2}$

88. $-10 \div \dfrac{5}{9}$

89. $-7 \div \dfrac{21}{5}$

90. $5 \div \dfrac{45}{4}$

91. $\dfrac{40}{9} \div 8$

92. $-\dfrac{24}{7} \div 6$

93. $-\dfrac{18}{11} \div 4$

94. $\dfrac{30}{7} \div 8$

95. $\dfrac{35}{72} \div \dfrac{65}{36}$

96. $\dfrac{80}{63} \div \dfrac{100}{81}$

Exercises 97–106: Divide. Simplify, if necessary.

97. $\dfrac{3x^2}{14y} \div \dfrac{3x}{7y^2}$

98. $\dfrac{3u}{5v} \div \dfrac{5u^3}{3v^2}$

99. $-\dfrac{6b^2}{5a^2} \div \dfrac{2b}{15a^2}$

100. $\dfrac{3m^2}{16n^2} \div \dfrac{9m^3}{2n^2}$

101. $-\dfrac{15p^2}{6q} \div 3p$

102. $\dfrac{3y}{10x} \div (-6y)$

103. $12x^3 \div \dfrac{4x^2}{3y^2}$

104. $18w^3 \div \dfrac{6w}{5}$

105. $-\dfrac{3m^2}{7n^2} \div \left(-\dfrac{3m^2}{7n^2}\right)$

106. $\dfrac{a^5}{b^3} \div \left(-\dfrac{a^4}{2b^3}\right)$

Exercises 107–112: Find the indicated fractional part.

107. $\dfrac{2}{3}$ of $\dfrac{5}{4}$

108. $\dfrac{1}{8}$ of $\dfrac{6}{7}$

109. $\dfrac{1}{5}$ of -15

110. $\dfrac{2}{9}$ of -36

111. $\dfrac{7}{6}$ of 20

112. $\dfrac{2}{35}$ of 50

APPLICATIONS

113. *Climbing Kilimanjaro* It is estimated that only $\dfrac{9}{20}$ of the people who attempt to climb Mount Kilimanjaro actually reach the summit. If 180 people attempt the climb, how many are expected to reach the summit? (*Source:* Kilitrekker.com)

114. *Male Teachers* About $\dfrac{7}{20}$ of teachers in American public secondary schools are male. If a city has 100 secondary teachers, how many are expected to be male? (*Source:* National Education Association.)

115. *Family Budget* A family spends $\dfrac{3}{20}$ of its monthly budget on utilities. If the family's monthly budget is $3200, how much is spent on the utility bill?

116. *Pension Savings* An employee has $\dfrac{2}{50}$ of his gross weekly pay withheld for his pension. How much is withheld for his pension if the employee's gross weekly pay is $1250?

117. *Baking* If a recipe for a batch of cookies calls for $\dfrac{2}{3}$ cup powdered sugar, how much powdered sugar is needed to make only $\dfrac{1}{2}$ of a batch?

118. *Cleaning Solution* If $\dfrac{3}{4}$ cup of concentrated detergent should be mixed with 1 gallon of water to make a cleaning solution, how much concentrated detergent should be mixed with $\dfrac{4}{5}$ gallon of water?

119. *College Students* In a particular school, two-thirds of all students are taking a biology class. Of the students taking biology, one-eighth are also taking an art class. What fraction of the students are taking both classes?

120. *Horror Movies* At a particular theater, one-twelfth of last year's films were horror movies. If two-thirds of the horror films had an R rating, what fraction of last year's films were R-rated horror movies?

Exercises 121–124: Use the formula $A = \frac{1}{2}bh$ to find the area of the triangle.

121.

122.

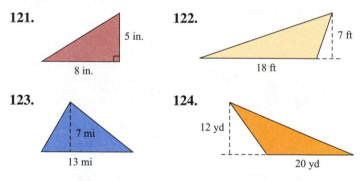

123.

124.

WRITING ABOUT MATHEMATICS

125. Explain the steps necessary to simplify a product of fractions before multiplying.

126. Explain why the number 0 has no reciprocal.

127. A student incorrectly changes a division problem to a multiplication problem by multiplying the reciprocal of the *first* fraction by the second fraction. How does this student's answer compare to the correct answer?

128. Does the commutative property hold when multiplying fractions? Does it hold when dividing fractions? Explain your reasoning and give examples.

4.4 Adding and Subtracting Fractions—Like Denominators

Adding Fractions—Like Denominators • Subtracting Fractions—Like Denominators • Adding and Subtracting Rational Expressions • Applications

A LOOK INTO MATH ▶

In the United States, the coin with a value of 25 cents is called a *quarter* because 25 cents is one-quarter (or one-fourth) of a dollar. A person with 3 quarters has $\frac{3}{4}$ of a dollar. Just as a quarter names a fourth of a dollar, a dime names a tenth of a dollar and a nickel names a twentieth of a dollar. In mathematics, we can name the kind of fraction we have by looking at its denominator. For example, $\frac{2}{7}$ is a fraction given in *sevenths*, and $\frac{4}{9}$ is a fraction given in *ninths*. If two fractions have the same denominator, we say that they are **like fractions** and the fractions have a **common denominator**.

NEW VOCABULARY

☐ Like fractions
☐ Common denominator

Adding Fractions—Like Denominators

▶ **REAL-WORLD CONNECTION** If a person with 5 quarters is given 6 more quarters, then the person has a total of 11 quarters. This statement translates to mathematics as

$$\frac{5}{4} + \frac{6}{4} = \frac{11}{4}.$$

Notice that the numerator of the resulting sum is obtained by adding the numerators of the addends, but the denominator of the resulting sum remains quarters (fourths). When two fractions with a *common denominator* are added, the resulting sum has the same denominator as the fractions being added.

READING CHECK

• How are fractions with like denominators added?

> **ADDING FRACTIONS WITH LIKE DENOMINATORS**
>
> If the variables a, b, and d represent numbers with $d \neq 0$, then
>
> $$\frac{a}{d} + \frac{b}{d} = \frac{a + b}{d}.$$

STUDY TIP

Do you want to know what material will be covered on your next exam? Often, the best place to look is on previously completed assignments and quizzes. If a topic is not discussed in class, is not found on the syllabus, and is not part of your assignments, then your time may be better spent studying other topics.

EXAMPLE 1 **Adding fractions with like denominators**

Add the fractions.

(a) $\dfrac{3}{7} + \dfrac{2}{7}$ (b) $\dfrac{4}{15} + \dfrac{4}{15}$ (c) $-\dfrac{8}{21} + \dfrac{19}{21}$

Solution

(a) $\dfrac{3}{7} + \dfrac{2}{7} = \dfrac{3 + 2}{7} = \dfrac{5}{7}$

(b) $\dfrac{4}{15} + \dfrac{4}{15} = \dfrac{4 + 4}{15} = \dfrac{8}{15}$

(c) Begin by writing $-\frac{8}{21}$ as $\frac{-8}{21}$.

$$-\frac{8}{21} + \frac{19}{21} = \frac{-8}{21} + \frac{19}{21} = \frac{-8 + 19}{21} = \frac{11}{21}$$

Now Try Exercises 7, 11, 13

When finding a sum of fractions, it is often necessary to simplify the resulting fraction to lowest terms. In the next example, we simplify sums of fractions.

EXAMPLE 2 | **Adding fractions with like denominators**

Add the fractions. Simplify the result.

(a) $-\frac{5}{8} + \frac{7}{8}$ **(b)** $\frac{5}{12} + \frac{7}{12}$

Solution

(a) $-\frac{5}{8} + \frac{7}{8} = \frac{-5 + 7}{8} = \frac{2}{8} = \frac{2}{2 \cdot 2 \cdot 2} = \frac{1}{2 \cdot 2} = \frac{1}{4}$

(b) $\frac{5}{12} + \frac{7}{12} = \frac{5 + 7}{12} = \frac{12}{12} = 1$

Now Try Exercises 17, 23

Subtracting Fractions—Like Denominators

▶ **REAL-WORLD CONNECTION** If a person who has 7 dimes gives away 4 dimes, then the person is left with 3 dimes. Because a dime is one-tenth of a dollar, this statement translates to mathematics as

$$\frac{7}{10} - \frac{4}{10} = \frac{3}{10}.$$

The numerator of the resulting difference is found by subtracting the numerators of the two fractions. Notice that the difference is given in dimes (tenths). When fractions with a common denominator are subtracted, the resulting difference has the same denominator.

READING CHECK

• How are fractions with like denominators subtracted?

SUBTRACTING FRACTIONS WITH LIKE DENOMINATORS

If the variables a, b, and d, represent numbers with $d \neq 0$, then

$$\frac{a}{d} - \frac{b}{d} = \frac{a - b}{d}.$$

EXAMPLE 3 | **Subtracting fractions with like denominators**

Subtract.

(a) $\frac{8}{11} - \frac{3}{11}$ **(b)** $\frac{4}{15} - \frac{8}{15}$ **(c)** $-\frac{5}{13} - \frac{4}{13}$

Solution

(a) $\dfrac{8}{11} - \dfrac{3}{11} = \dfrac{8-3}{11} = \dfrac{5}{11}$

(b) $\dfrac{4}{15} - \dfrac{8}{15} = \dfrac{4-8}{15} = \dfrac{-4}{15} = -\dfrac{4}{15}$

(c) Begin by writing $-\frac{5}{13}$ as $\frac{-5}{13}$.

$$-\dfrac{5}{13} - \dfrac{4}{13} = \dfrac{-5}{13} - \dfrac{4}{13} = \dfrac{-5-4}{13} = \dfrac{-9}{13} = -\dfrac{9}{13}$$

▋ **Now Try Exercises 25, 29, 31**

When finding a difference of two fractions, it is often necessary to simplify the resulting fraction to lowest terms. In the next example, we simplify the resulting difference when one fraction is subtracted from another.

EXAMPLE 4   **Subtracting fractions with like denominators**

Subtract. Simplify the result.

(a) $\dfrac{9}{10} - \dfrac{3}{10}$ (b) $\dfrac{5}{18} - \dfrac{13}{18}$ (c) $\dfrac{5}{7} - \left(-\dfrac{2}{7}\right)$

Solution

(a) $\dfrac{9}{10} - \dfrac{3}{10} = \dfrac{9-3}{10} = \dfrac{6}{10} = \dfrac{2 \cdot 3}{2 \cdot 5} = \dfrac{3}{5}$

(b) $\dfrac{5}{18} - \dfrac{13}{18} = \dfrac{5-13}{18} = \dfrac{-8}{18} = -\dfrac{2 \cdot 2 \cdot 2}{2 \cdot 3 \cdot 3} = -\dfrac{2 \cdot 2}{3 \cdot 3} = -\dfrac{4}{9}$

(c) $\dfrac{5}{7} - \left(-\dfrac{2}{7}\right) = \dfrac{5}{7} - \dfrac{-2}{7} = \dfrac{5-(-2)}{7} = \dfrac{5+2}{7} = \dfrac{7}{7} = 1$

▋ **Now Try Exercises 33, 39, 41**

Adding and Subtracting Rational Expressions

The process used to add or subtract rational expressions with like denominators is similar to that used for adding and subtracting fractions with like denominators. We add or subtract the numerators and keep the common denominator. The next two examples illustrate how rational expressions are added and subtracted.

EXAMPLE 5   **Adding and subtracting rational expressions**

Add or subtract as indicated.

(a) $\dfrac{3x}{5y} + \dfrac{x}{5y}$ (b) $\dfrac{4}{7x^2} - \dfrac{6}{7x^2}$ (c) $\dfrac{2x}{3m} + \left(-\dfrac{4y}{3m}\right)$

Solution

(a) $\dfrac{3x}{5y} + \dfrac{x}{5y} = \dfrac{3x+x}{5y} = \dfrac{4x}{5y}$

(b) $\dfrac{4}{7x^2} - \dfrac{6}{7x^2} = \dfrac{4-6}{7x^2} = \dfrac{-2}{7x^2} = -\dfrac{2}{7x^2}$

(c) Begin by writing $-\frac{4y}{3m}$ as $\frac{-4y}{3m}$.

$$\frac{2x}{3m} + \left(-\frac{4y}{3m}\right) = \frac{2x}{3m} + \frac{-4y}{3m} = \frac{2x + (-4y)}{3m} = \frac{2x - 4y}{3m}$$

Note that the terms $2x$ and $4y$ cannot be combined.

Now Try Exercises 55, 57, 59

EXAMPLE 6 Adding and subtracting rational expressions

Add or subtract as indicated. Simplify the result.

(a) $\dfrac{2y}{6x^3} + \dfrac{y}{6x^3}$ (b) $\dfrac{5a^2}{24b} - \dfrac{7a^2}{24b}$

Solution

(a) $\dfrac{2y}{6x^3} + \dfrac{y}{6x^3} = \dfrac{2y + y}{6x^3} = \dfrac{3y}{6x^3} = \dfrac{3 \cdot y}{2 \cdot 3 \cdot x \cdot x \cdot x} = \dfrac{y}{2 \cdot x \cdot x \cdot x} = \dfrac{y}{2x^3}$

(b) $\dfrac{5a^2}{24b} - \dfrac{7a^2}{24b} = \dfrac{5a^2 - 7a^2}{24b} = \dfrac{-2a^2}{24b} = -\dfrac{2 \cdot a \cdot a}{2 \cdot 2 \cdot 2 \cdot 3 \cdot b} = -\dfrac{a \cdot a}{2 \cdot 2 \cdot 3 \cdot b} = -\dfrac{a^2}{12b}$

Now Try Exercises 63, 65

Applications

▶ **REAL-WORLD CONNECTION** The next two examples illustrate how we may need to add or subtract fractions in applications involving real-world data.

EXAMPLE 7 Analyzing population data

In 1960, about $\frac{17}{20}$ of the U.S. population identified itself as non-Hispanic white. By 2060, it is projected that $\frac{9}{20}$ of the population will be non-Hispanic white. Subtract the 2060 value from the 1960 value to find the difference in the fractional portion of the population that identifies itself as non-Hispanic white. (*Source: Pew Research Center—2008.*)

Solution

$$\frac{17}{20} - \frac{9}{20} = \frac{17 - 9}{20} = \frac{8}{20} = \frac{2 \cdot 2 \cdot 2}{2 \cdot 2 \cdot 5} = \frac{2}{5}$$

Over this 100-year period, the fractional portion of the U.S. population that identifies itself as non-Hispanic white will decrease by $\frac{2}{5}$.

Now Try Exercise 71

EXAMPLE 8 Constructing trim for cabinetry

If a cabinet maker glues a piece of trim that is $\frac{5}{16}$ inch thick to another piece that is $\frac{7}{16}$ inch thick to make a single, thicker piece of trim, how thick is the newly formed trim piece?

Solution

To find the total thickness, we add $\frac{5}{16}$ and $\frac{7}{16}$.

$$\frac{5}{16} + \frac{7}{16} = \frac{5 + 7}{16} = \frac{12}{16} = \frac{2 \cdot 2 \cdot 3}{2 \cdot 2 \cdot 2 \cdot 2} = \frac{3}{2 \cdot 2} = \frac{3}{4}$$

The new trim piece is $\frac{3}{4}$ inch thick.

Now Try Exercise 75

4.4   Putting It All Together

CONCEPT	COMMENTS	EXAMPLES
Adding Fractions with Like Denominators	If the variables a, b, and d represent numbers with $d \neq 0$, then $$\frac{a}{d} + \frac{b}{d} = \frac{a+b}{d}.$$	$$\frac{3}{11} + \frac{4}{11} = \frac{3+4}{11} = \frac{7}{11}$$ $$\frac{5}{12} + \frac{3}{12} = \frac{5+3}{12} = \frac{8}{12} = \frac{2}{3}$$
Subtracting Fractions with Like Denominators	If the variables a, b, and d represent numbers with $d \neq 0$, then $$\frac{a}{d} - \frac{b}{d} = \frac{a-b}{d}.$$	$$\frac{7}{9} - \frac{5}{9} = \frac{7-5}{9} = \frac{2}{9}$$ $$\frac{7}{10} - \frac{3}{10} = \frac{7-3}{10} = \frac{4}{10} = \frac{2}{5}$$
Adding and Subtracting Rational Expressions	The process used to add or subtract rational expressions with like denominators is similar to that used for adding and subtracting fractions.	$$\frac{2x}{3y} + \frac{5x}{3y} = \frac{2x+5x}{3y} = \frac{7x}{3y}$$ $$\frac{7a}{12b^2} - \frac{11a}{12b^2} = \frac{7a-11a}{12b^2} = \frac{-4a}{12b^2}$$ $$= -\frac{a}{3b^2}$$

4.4   Exercises

MyMathLab Math XL PRACTICE WATCH DOWNLOAD READ REVIEW

CONCEPTS AND VOCABULARY

1. Two fractions with the same denominator are called _____ fractions and we say that the two fractions have a(n) _____ denominator.

2. When two fractions with a common denominator are added, the numerator of the sum is found by adding the _____ of the two fractions.

3. When adding two fractions with a common denominator, the sum has the same _____ as the two fractions being added.

4. When finding the difference of two fractions with a common denominator, the numerator of the difference is found by subtracting the _____ of the two fractions.

5. When finding the difference of two fractions with a common denominator, the difference has the same _____ as the two fractions.

6. The process used to add or subtract rational expressions with like denominators is similar to that used for adding and subtracting _____.

ADDING AND SUBTRACTING LIKE FRACTIONS

Exercises 7–24: Add. Simplify, if necessary.

7. $\dfrac{1}{5} + \dfrac{2}{5}$

8. $\dfrac{2}{7} + \dfrac{4}{7}$

9. $\dfrac{2}{9} + \left(-\dfrac{7}{9}\right)$

10. $\dfrac{3}{11} + \left(-\dfrac{10}{11}\right)$

11. $\dfrac{12}{25} + \dfrac{12}{25}$

12. $\dfrac{9}{17} + \dfrac{10}{17}$

13. $-\dfrac{5}{13} + \dfrac{15}{13}$

14. $-\dfrac{14}{19} + \dfrac{7}{19}$

15. $\dfrac{1}{10} + \dfrac{7}{10}$

16. $\dfrac{1}{12} + \dfrac{11}{12}$

17. $\dfrac{8}{9} + \left(-\dfrac{2}{9}\right)$

18. $-\dfrac{8}{15} + \dfrac{13}{15}$

19. $-\dfrac{7}{5} + \left(-\dfrac{3}{5}\right)$

20. $\dfrac{11}{6} + \left(-\dfrac{5}{6}\right)$

21. $-\dfrac{17}{20} + \dfrac{17}{20}$ **22.** $-\dfrac{3}{8} + \dfrac{3}{8}$

23. $\dfrac{5}{14} + \dfrac{9}{14}$ **24.** $-\dfrac{5}{12} + \left(-\dfrac{7}{12}\right)$

Exercises 25–42: Subtract. Simplify, if necessary.

25. $\dfrac{6}{7} - \dfrac{4}{7}$ **26.** $\dfrac{4}{5} - \dfrac{1}{5}$

27. $\dfrac{7}{9} - \left(-\dfrac{1}{9}\right)$ **28.** $-\dfrac{3}{15} - \dfrac{8}{15}$

29. $\dfrac{3}{17} - \dfrac{15}{17}$ **30.** $\dfrac{19}{21} - \dfrac{11}{21}$

31. $-\dfrac{2}{11} - \dfrac{3}{11}$ **32.** $-\dfrac{10}{19} - \dfrac{3}{19}$

33. $\dfrac{5}{8} - \dfrac{3}{8}$ **34.** $\dfrac{19}{24} - \dfrac{7}{24}$

35. $\dfrac{4}{9} - \left(-\dfrac{4}{9}\right)$ **36.** $-\dfrac{1}{12} - \dfrac{11}{12}$

37. $\dfrac{1}{10} - \dfrac{3}{10}$ **38.** $\dfrac{11}{3} - \left(-\dfrac{4}{3}\right)$

39. $-\dfrac{1}{18} - \dfrac{5}{18}$ **40.** $\dfrac{5}{27} - \dfrac{5}{27}$

41. $\dfrac{11}{16} - \left(-\dfrac{5}{16}\right)$ **42.** $-\dfrac{5}{14} - \left(-\dfrac{7}{14}\right)$

Exercises 43–54: Add or subtract as indicated. Simplify, if necessary.

43. $\dfrac{22}{27} - \dfrac{13}{27}$ **44.** $-\dfrac{3}{50} + \dfrac{39}{50}$

45. $\dfrac{7}{36} + \left(-\dfrac{25}{36}\right)$ **46.** $-\dfrac{13}{45} - \dfrac{8}{45}$

47. $\dfrac{7}{30} - \left(-\dfrac{13}{30}\right)$ **48.** $\dfrac{28}{75} + \dfrac{8}{75}$

49. $-\dfrac{7}{60} + \dfrac{11}{60}$ **50.** $-\dfrac{25}{48} - \left(-\dfrac{7}{48}\right)$

51. $\dfrac{13}{120} - \dfrac{77}{120}$ **52.** $-\dfrac{21}{80} + \left(-\dfrac{63}{80}\right)$

53. $-\dfrac{38}{63} + \left(-\dfrac{25}{63}\right)$ **54.** $-\dfrac{77}{90} - \left(-\dfrac{77}{90}\right)$

ADDING AND SUBTRACTING RATIONAL EXPRESSIONS

Exercises 55–68: Add or subtract as indicated. Simplify, if necessary.

55. $\dfrac{5x}{9y} + \dfrac{2x}{9y}$ **56.** $-\dfrac{3a}{b} + \dfrac{5a}{b}$

57. $\dfrac{8}{5m^2} + \left(-\dfrac{12}{5m^2}\right)$ **58.** $-\dfrac{3p^2}{5q} - \dfrac{p^2}{5q}$

59. $\dfrac{7x}{y} - \dfrac{3w}{y}$ **60.** $-\dfrac{8a}{7} + \left(-\dfrac{4a}{7}\right)$

61. $\dfrac{2}{3d} - \left(-\dfrac{1}{3d}\right)$ **62.** $\dfrac{3m^2}{10n^3} + \dfrac{2m^2}{10n^3}$

63. $\dfrac{4y}{3x^2} + \dfrac{8y}{3x^2}$ **64.** $-\dfrac{3y}{4w^2} - \dfrac{5y}{4w^2}$

65. $\dfrac{2k^2}{15c} - \dfrac{7k^2}{15c}$ **66.** $-\dfrac{3x}{2} - \left(-\dfrac{7x}{2}\right)$

67. $\dfrac{13x^2}{6} + \dfrac{5x^2}{6}$ **68.** $-\dfrac{4}{x} + \left(-\dfrac{4}{x}\right)$

APPLICATIONS

69. *Carpentry* A board that is $\frac{15}{16}$ inch thick is run through a surfacing machine to reduce its thickness by $\frac{3}{16}$ inch. What is the thickness of the board after this process?

70. *Cutting Hair* A barber cuts hair that is $\frac{7}{2}$ inches long to a length of $\frac{1}{2}$ inch. How many inches of hair are removed?

71. *International Travel* In 2008, about $\frac{3}{250}$ of international travelers entering the United States were from Asia and about $\frac{1}{250}$ were from Africa. Subtract the African value from the Asian value to find the difference. (*Source:* U.S. Department of Commerce.)

72. *Carbon Emissions* The United States accounted for about $\frac{23}{100}$ of worldwide carbon dioxide emissions from the consumption of fossil fuels in 1990. By 2008, this fraction had dropped to $\frac{18}{100}$. Find the difference. (*Source:* U.S. Energy Information Administration.)

73. *Nursing Home Residents* In 2008, about $\frac{19}{5000}$ of all Americans were female nursing home residents and about $\frac{7}{5000}$ of all Americans were male nursing home residents. Find the total fraction of the American population that was residing in nursing homes in 2008. (*Source:* Centers for Disease Control and Prevention.)

74. *Classroom Population* In a large college class, $\frac{57}{300}$ of the students are female and under the age of 24, while $\frac{111}{300}$ of the students are male and under the age of 24. Find the total fraction of students in this class that are under the age of 24.

75. *Largest U.S. Cities* In 2009, Texas was home to $\frac{6}{25}$ of the 25 largest U.S. cities, and California was home to $\frac{4}{25}$ of the 25 largest U.S. cities. What fraction of the 25 largest U.S. cities could be found in these two states together? (*Source:* U.S. Census Bureau.)

76. *Largest World Cities* In 2009, India and China each had $\frac{3}{20}$ of the world's 20 largest cities. What fraction of the world's 20 largest cities could be found in these two countries together? (*Source: The World Atlas.*)

77. *Going Green* A company has a goal of replacing $\frac{47}{50}$ of its incandescent light bulbs with energy-efficient

LED bulbs. If the company has already replaced $\frac{37}{50}$ of its light bulbs, what fraction of the incandescent bulbs still need to be replaced?

78. *Going Organic* A farmer plans to convert $\frac{7}{24}$ of his crop land to organic crops. If the farmer has already converted $\frac{5}{24}$ of his crop land, what fraction of his land has yet to be converted?

79. *Scholarships* A university foundation has set aside $\frac{11}{30}$ of its budget to award scholarships for students with high academic achievement and $\frac{7}{30}$ of its budget to award scholarships for low-income students. What fraction of the foundation's total budget is used for these two groups?

80. *Fuel Consumption* At the beginning of a trip, a car's gas gauge shows that the car has $\frac{3}{4}$ of a tank of gas. At the end of the trip, the gauge indicates that the car has $\frac{1}{4}$ of a tank of gas. What fraction of the tank was used during the trip?

WRITING ABOUT MATHEMATICS

81. Explain how to determine if a sum or difference of fractions needs to be simplified.

82. A student says that two fractions can be added by adding the numerators *and* adding the denominators. Give a real-world example that would disprove this student's method.

| SECTIONS 4.3 and 4.4 | **Checking Basic Concepts** |

1. Multiply or divide as indicated. If necessary, simplify your results.

(a) $\frac{5}{3} \cdot \frac{12}{15}$ (b) $-\frac{7}{2} \cdot \frac{10}{7}$

(c) $\frac{14}{25} \div \frac{7}{10}$ (d) $\frac{2}{7} \div \left(-\frac{8}{21}\right)$

(e) $\frac{2x^2}{9y^2} \cdot \frac{3y}{8x^2}$ (f) $-\frac{15m}{4n} \div \frac{5m}{4n}$

(g) $-12a \div \frac{6}{5a}$ (h) $\frac{3x}{8y^2} \cdot 4y$

2. Evaluate the expression.

(a) $\left(\frac{2}{5}\right)^2$ (b) $\sqrt{\frac{50}{2}}$

3. Add or subtract as indicated. If necessary, simplify your results.

(a) $-\frac{5}{19} + \frac{17}{19}$ (b) $\frac{4}{9} + \left(-\frac{1}{9}\right)$

(c) $\frac{13}{21} - \frac{7}{21}$ (d) $-\frac{1}{14} - \frac{13}{14}$

(e) $\frac{15x^2}{4} + \frac{x^2}{4}$ (f) $\frac{5m^2}{8n^3} - \frac{3m^2}{8n^3}$

4. *Female Teachers* About $\frac{13}{20}$ of teachers in public secondary schools are female. If a city has 60 secondary teachers, how many are expected to be female? (*Source: National Education Association.*)

4.5   Adding and Subtracting Fractions—Unlike Denominators

Least Common Multiple (LCM) • Least Common Denominator (LCD) • Adding and Subtracting Fractions—Unlike Denominators • Applications

A LOOK INTO MATH ▶

In the previous section, we discussed how different coins represent different fractions of a dollar. Our everyday experience tells us that a person with three dimes and four nickels has $\frac{1}{2}$ of a dollar. Because a dime represents $\frac{1}{10}$ of a dollar and a nickel represents $\frac{1}{20}$ of a dollar, the following sum illustrates that three dimes plus four nickels equals $\frac{1}{2}$ of a dollar.

$$\frac{3}{10} + \frac{4}{20} = \frac{1}{2}$$

The addends in this equation do *not* have a common denominator. In this section, we add and subtract fractions that do not have a common denominator.

NEW VOCABULARY

☐ Common multiples
☐ Least common multiple
☐ Listing method
☐ Prime factorization method
☐ Least common denominator

Least Common Multiple (LCM)

▶ **REAL-WORLD CONNECTION**   Suppose that two friends have jobs as sales representatives. One friend calls on customers in the Chicago area every fourth week, and the other friend calls on customers in the Chicago area every sixth week. If they are both in Chicago this week, how many weeks will pass before they are both in Chicago again?

We can answer this question by listing the weeks that each person is in Chicago.

First Person: 4,   8,   **12**,   16,   20,   **24**,   28,   32, ...
Second Person: 6,   **12**,   18,   **24**,   30,   36,   42, ...

After 12 weeks, the two friends will be in Chicago at the same time. The next time is after 24 weeks. The numbers 12 and 24 are two **common multiples** of 4 and 6. However, 12 is the *least common multiple* (LCM) of 4 and 6. The **least common multiple** of two or more numbers is the smallest number that is divisible by each of the given numbers.

STUDY TIP

Even if we know exactly how to do a math problem correctly, a simple computational error will often cause us to get an incorrect answer. In the above example, if we had listed the multiples of 4 and 6 incorrectly, it is unlikely that we would find the correct LCM. Be sure to take your time on simple calculations.

READING CHECK

• How is the listing method used to find the LCM?

The method used above to find the LCM of 4 and 6 is called the **listing method** because multiples of each number are listed and the LCM is chosen from the lists. In the next example, we use the listing method to find the LCM of two numbers.

EXAMPLE 1   **Finding the LCM using the listing method**

Find the least common multiple of 8 and 10.

Solution
List the multiples of each number by multiplying each number by 1, 2, 3, etc.

Multiples of 8:   8,   16,   24,   32,   **40**,   48,   56,   64, ...
Multiples of 10:   10,   20,   30,   **40**,   50,   60,   70, ...

The LCM is the smallest number that is common to both lists. The LCM of 8 and 10 is **40**.

Now Try Exercise 9

The listing method is often convenient for small numbers and could be done mentally, but it is not practical for finding the LCM in every case. For example, to find the LCM of 16 and 27 using the listing method, the two lists would contain at least 43 values. A second method for finding the LCM of two or more numbers is called the **prime factorization method**. The steps for this method are summarized as follows:

FINDING THE LCM USING PRIME FACTORIZATION

To find the least common multiple (LCM) of two or more numbers, do the following.

STEP 1: Find the prime factorization of each number.

STEP 2: List each factor that appears in one or more of the factorizations. If a factor is repeated in any of the factorizations, list this factor the maximum number of times that it is repeated in any one of the factorizations.

STEP 3: Find the product of this list of factors. The result is the LCM.

READING CHECK

- How do we determine how many times a factor should be listed when using prime factorization to find the LCM?

EXAMPLE 2   **Finding the LCM using the prime factorization method**

Find the LCM of the given numbers.
(a) 12 and 18   **(b)** 4, 6, and 10

Solution
(a) STEP 1: Find the prime factorizations of 12 and 18.

$$12 = 2 \cdot 2 \cdot 3$$
$$18 = 2 \cdot 3 \cdot 3$$

STEP 2: List the factors: $2, 2, 3, 3$. Because 2 appears twice in one factorization and one time in the other, we list it twice, which is the maximum number of times that it is repeated in any *one* of the factorizations. For the same reason, 3 is listed twice.

STEP 3: The product of this list is $2 \cdot 2 \cdot 3 \cdot 3 = 36$, so the LCM of 12 and 18 is 36.

(b) STEP 1: Find the prime factorizations of 4, 6, and 10.

$$4 = 2 \cdot 2$$
$$6 = 2 \cdot 3$$
$$10 = 2 \cdot 5$$

STEP 2: List the factors: $2, 2, 3, 5$.
STEP 3: The LCM of 4, 6, and 10 is $2 \cdot 2 \cdot 3 \cdot 5 = 60$.

Now Try Exercises 13, 17

NOTE: Because the LCM of two or more numbers is simply the smallest number that is divisible by each of the given numbers, the LCM of 4, 6, and 10 in Example 2(b) is 60 because 60 is the smallest number that is divisible by 4, 6, and 10.

The steps used to find the LCM of two or more numbers can also be used to find the LCM of two or more variable expressions of a single term. In the next example, we find the LCM of two variable expressions.

EXAMPLE 3 | **Finding the LCM using the factorization method**

Find the LCM for the given variable expressions.
(a) $3x^2$ and $4x^3$ **(b)** $2ab^2$ and $5a^2b$

Solution
(a) STEP 1: Find the complete factorizations of $3x^2$ and $4x^3$.

$$3x^2 = 3 \cdot x \cdot x$$
$$4x^3 = 2 \cdot 2 \cdot x \cdot x \cdot x$$

STEP 2: List the factors: $2, 2, 3, x, x, x$.
STEP 3: The LCM of $3x^2$ and $4x^3$ is $2 \cdot 2 \cdot 3 \cdot x \cdot x \cdot x = 12x^3$.
(b) STEP 1: Find the complete factorizations of $2ab^2$ and $5a^2b$.

$$2ab^2 = 2 \cdot a \cdot b \cdot b$$
$$5a^2b = 5 \cdot a \cdot a \cdot b$$

STEP 2: List the factors: $2, 5, a, a, b, b$.
STEP 3: The LCM of $2ab^2$ and $5a^2b$ is $2 \cdot 5 \cdot a \cdot a \cdot b \cdot b = 10a^2b^2$.

▍ **Now Try Exercises 27, 29**

READING CHECK

• How is the LCD of two fractions found?

Least Common Denominator (LCD)

In this section, we will add and subtract fractions with unlike denominators. To do this, we need to write the fractions as equivalent fractions that have a common denominator. Even though *any* common denominator can be used for this purpose, it is often best to find the *least* common denominator (LCD). The **least common denominator** for a list of fractions is the *least common multiple* of the denominators.

EXAMPLE 4 | **Finding the LCD for a list of fractions**

Find the LCD for the given fractions.
(a) $\dfrac{7}{16}$ and $\dfrac{5}{6}$ **(b)** $\dfrac{2}{5x^2}$ and $\dfrac{3}{4xy}$

Solution
(a) The LCD for the fractions $\frac{7}{16}$ and $\frac{5}{6}$ is the LCM of the denominators 16 and 6. So, we use the three-step process for finding the LCM.
STEP 1: Find the prime factorizations of 16 and 6.

$$16 = 2 \cdot 2 \cdot 2 \cdot 2$$
$$6 = 2 \cdot 3$$

STEP 2: List the factors: $2, 2, 2, 2, 3$.
STEP 3: The LCD is $2 \cdot 2 \cdot 2 \cdot 2 \cdot 3 = 48$.
(b) To find the LCD for $\frac{2}{5x^2}$ and $\frac{3}{4xy}$, we look for the LCM of the denominators $5x^2$ and $4xy$.
STEP 1: Find the complete factorizations of $5x^2$ and $4xy$.

$$5x^2 = 5 \cdot x \cdot x$$
$$4xy = 2 \cdot 2 \cdot x \cdot y$$

STEP 2: List the factors: $2, 2, 5, x, x, y$.
STEP 3: The LCD is $2 \cdot 2 \cdot 5 \cdot x \cdot x \cdot y = 20x^2y$.

▍ **Now Try Exercises 35, 41**

Adding and Subtracting Fractions—Unlike Denominators

When we are faced with a new mathematical problem with no understanding of how to find the solution, it is often helpful to look at similar problems that we *can* solve and then try to relate the new problem to these previously solved problems. For example, we could add and subtract fractions with *unlike* denominators if we could rewrite the fractions as equivalent fractions with *like* denominators. In fact, this is exactly what we do when adding and subtracting fractions with unlike denominators. The entire process is summarized as follows:

> **ADDING OR SUBTRACTING FRACTIONS WITH UNLIKE DENOMINATORS**
>
> To add or subtract fractions with unlike denominators, do the following.
>
> **STEP 1:** Determine the LCD for all fractions involved.
>
> **STEP 2:** Write each fraction as an equivalent fraction with the LCD as its denominator.
>
> **STEP 3:** Add or subtract the newly written like fractions and simplify the result.

READING CHECK

• How is the LCD used when adding fractions with unlike denominators?

In the next example, we will practice the important concept in Step 2.

EXAMPLE 5   **Writing equivalent fractions using the LCD**

Rewrite the fractions as equivalent fractions with the given LCD.

(a) $\dfrac{1}{6}$ and $\dfrac{4}{15}$; LCD: 30   (b) $\dfrac{5}{6}, \dfrac{1}{4},$ and $\dfrac{3}{8}$; LCD: 24

Solution

(a) Because the denominator of $\frac{1}{6}$ is 6, we can obtain a denominator of 30 by multiplying $6 \times \mathbf{5} = 30$. So, we multiply both the numerator and denominator of $\frac{1}{6}$ by $\mathbf{5}$.

$$\overset{\times 5}{\underset{\times 5}{\frac{1}{6} = \frac{?}{30}}}$$

Since $1 \times \mathbf{5} = 5$, the unknown numerator is 5. That is, $\frac{1}{6}$ is equivalent to $\frac{5}{30}$.

Because the denominator of $\frac{4}{15}$ is 15, we obtain a denominator of 30 by multiplying $15 \times \mathbf{2} = 30$. So, we multiply both the numerator and denominator of $\frac{4}{15}$ by $\mathbf{2}$.

$$\overset{\times 2}{\underset{\times 2}{\frac{4}{15} = \frac{?}{30}}}$$

Since $4 \times \mathbf{2} = 8$, the unknown numerator is 8. That is, $\frac{4}{15}$ is equivalent to $\frac{8}{30}$.

(b) A convenient and more concise way to write equivalent fractions is shown below. Note that we obtain a denominator of 24 by multiplying the denominator of $\frac{5}{6}$ by $\mathbf{4}$, multiplying the denominator of $\frac{1}{4}$ by $\mathbf{6}$, and multiplying the denominator of $\frac{3}{8}$ by $\mathbf{3}$.

$$\frac{5}{6} \cdot \frac{\mathbf{4}}{\mathbf{4}} = \frac{20}{24} \qquad \frac{1}{4} \cdot \frac{\mathbf{6}}{\mathbf{6}} = \frac{6}{24} \qquad \frac{3}{8} \cdot \frac{\mathbf{3}}{\mathbf{3}} = \frac{9}{24}$$

Now Try Exercises 49, 51

NOTE: The fractions in Example 5(b) on the previous page were each multiplied by 1, which was written in a form that helped us obtain the desired common denominator.

In Example 6, we use the entire three-step process shown above to add or subtract fractions with unlike denominators. Then, Example 7 illustrates how the same three-step process can be written more concisely.

EXAMPLE 6   **Adding or subtracting fractions with unlike denominators**

Add or subtract as indicated.

(a) $\dfrac{5}{8} - \dfrac{2}{3}$ **(b)** $\dfrac{2}{3} + \dfrac{7}{12} + \dfrac{3}{4}$

Solution

(a) STEP 1: Find the LCD for $\frac{5}{8}$ and $\frac{2}{3}$.

$$8 = 2 \cdot 2 \cdot 2$$
$$3 = 3$$

CALCULATOR HELP

To add or subtract fractions with a calculator, See the Appendix (page AP-5).

The LCD is $2 \cdot 2 \cdot 2 \cdot 3 = 24$.

STEP 2: Multiply each fraction by 1, written in a form that is appropriate for rewriting each fraction with the LCD.

$$\frac{5}{8} \cdot \frac{3}{3} = \frac{15}{24} \qquad \frac{2}{3} \cdot \frac{8}{8} = \frac{16}{24}$$

STEP 3: Subtract the newly written like fractions.

$$\frac{15}{24} - \frac{16}{24} = \frac{15 - 16}{24} = \frac{-1}{24} = -\frac{1}{24}$$

The resulting difference is $-\frac{1}{24}$, which is simplified to lowest terms.

(b) STEP 1: Find the LCD for $\frac{2}{3}$, $\frac{7}{12}$, and $\frac{3}{4}$.

$$3 = 3$$
$$12 = 3 \cdot 2 \cdot 2$$
$$4 = 2 \cdot 2$$

The LCD is $3 \cdot 2 \cdot 2 = 12$.

STEP 2: Multiply each fraction by 1, written in a form that is appropriate for rewriting each fraction with the LCD. Note that the denominator of $\frac{7}{12}$ is already 12.

$$\frac{2}{3} \cdot \frac{4}{4} = \frac{8}{12} \qquad \frac{7}{12} = \frac{7}{12} \qquad \frac{3}{4} \cdot \frac{3}{3} = \frac{9}{12}$$

STEP 3: Add the newly written like fractions and simplify the result.

$$\frac{8}{12} + \frac{7}{12} + \frac{9}{12} = \frac{8 + 7 + 9}{12} = \frac{24}{12} = \frac{2 \cdot 2 \cdot 2 \cdot 3}{2 \cdot 2 \cdot 3} = \frac{2}{1} = 2$$

The sum is 2.

Now Try Exercises 55, 65

EXAMPLE 7   **Adding or subtracting fractions with unlike denominators**

Add or subtract as indicated.

(a) $\dfrac{13}{20} + \dfrac{1}{8}$ (b) $-\dfrac{1}{7} - \dfrac{5}{14}$

Solution

(a) The LCD for $\dfrac{13}{20}$ and $\dfrac{1}{8}$ is 40. Verify this. (Step 1)

$$\dfrac{13}{20} + \dfrac{1}{8} = \dfrac{13}{20} \cdot \dfrac{2}{2} + \dfrac{1}{8} \cdot \dfrac{5}{5} \qquad \text{Multiply each fraction by 1.}$$

$$= \dfrac{26}{40} + \dfrac{5}{40} \qquad \text{Rewrite fractions with the LCD. (Step 2)}$$

$$= \dfrac{26 + 5}{40} \qquad \text{Addition of fractions}$$

$$= \dfrac{31}{40} \qquad \text{Add. (Step 3)}$$

(b) The LCD for $-\dfrac{1}{7}$ and $\dfrac{5}{14}$ is 14. Verify this. (Step 1)

$$-\dfrac{1}{7} - \dfrac{5}{14} = -\dfrac{1}{7} \cdot \dfrac{2}{2} - \dfrac{5}{14} \qquad \text{Multiply by 1.}$$

$$= -\dfrac{2}{14} - \dfrac{5}{14} \qquad \text{Rewrite fractions with the LCD. (Step 2)}$$

$$= \dfrac{-2 - 5}{14} \qquad \text{Subtraction of fractions}$$

$$= -\dfrac{7}{14} \qquad \text{Subtract. (Step 3)}$$

$$= -\dfrac{1}{2} \qquad \text{Simplify. (Step 3)}$$

Now Try Exercises 59, 63

In the next example, we add and subtract rational expressions with unlike denominators.

EXAMPLE 8   **Adding and subtracting rational expressions with unlike denominators**

Add or subtract as indicated.

(a) $\dfrac{5a}{6} + \dfrac{a}{8}$ (b) $\dfrac{3}{4x} - \dfrac{1}{3x^2}$

Solution

(a) The LCD for $\dfrac{5a}{6}$ and $\dfrac{a}{8}$ is 24. Verify this. (Step 1)

$$\dfrac{5a}{6} + \dfrac{a}{8} = \dfrac{5a}{6} \cdot \dfrac{4}{4} + \dfrac{a}{8} \cdot \dfrac{3}{3} \qquad \text{Multiply each fraction by 1.}$$

$$= \dfrac{20a}{24} + \dfrac{3a}{24} \qquad \text{Rewrite fractions with the LCD. (Step 2)}$$

$$= \dfrac{20a + 3a}{24} \qquad \text{Addition of fractions}$$

$$= \dfrac{23a}{24} \qquad \text{Add. (Step 3)}$$

(b) The LCD for $\frac{3}{4x}$ and $\frac{1}{3x^2}$ is $12x^2$. Verify this. (Step 1)

$$\frac{3}{4x} - \frac{1}{3x^2} = \frac{3}{4x} \cdot \frac{\mathbf{3x}}{\mathbf{3x}} - \frac{1}{3x^2} \cdot \frac{\mathbf{4}}{\mathbf{4}} \qquad \text{Multiply each fraction by 1.}$$

$$= \frac{9x}{12x^2} - \frac{4}{12x^2} \qquad \text{Rewrite fractions with the LCD. (Step 2)}$$

$$= \frac{9x - 4}{12x^2} \qquad \text{Subtraction of fractions (Step 3)}$$

Note that $\frac{9x-4}{12x^2}$ cannot be simplified further.

Now Try Exercises 71, 77

Applications

▶ **REAL-WORLD CONNECTION** In the next two examples, we work with unlike fractions in applications involving real-world data.

EXAMPLE 9 **Analyzing ice sheet data**

The Antarctic ice sheet contains about $\frac{9}{10}$ of all the ice in the world, while the Greenland ice sheet contains about $\frac{9}{100}$ of the ice. Find the total fraction of Earth's ice that is contained in these two ice sheets. (*Source:* National Science Foundation, Office of Polar Programs.)

Solution
The LCD required for adding $\frac{9}{10}$ and $\frac{9}{100}$ is 100. (Step 1)

$$\frac{9}{10} + \frac{9}{100} = \frac{9}{10} \cdot \frac{\mathbf{10}}{\mathbf{10}} + \frac{9}{100} \qquad \text{Multiply by 1.}$$

$$= \frac{90}{100} + \frac{9}{100} \qquad \text{Rewrite fractions with the LCD. (Step 2)}$$

$$= \frac{90 + 9}{100} \qquad \text{Addition of fractions}$$

$$= \frac{99}{100} \qquad \text{Add. (Step 3)}$$

So, $\frac{99}{100}$ of Earth's ice is contained in the Antarctic and Greenland ice sheets.

Now Try Exercise 81

EXAMPLE 10 **Finding the perimeter of a 1-acre plot of land**

A 1-acre rectangular plot of land has a length of $\frac{1}{16}$ mile and a width of $\frac{1}{40}$ mile, as shown in the following figure. Find the perimeter of this 1-acre plot.

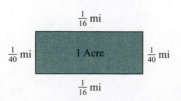

Solution

The perimeter of a rectangle is found by adding the measures of all four sides. The LCD for the fractions $\frac{1}{16}$ and $\frac{1}{40}$ is 80. (Step 1)

$$\frac{1}{16} + \frac{1}{40} + \frac{1}{16} + \frac{1}{40} = \frac{1}{16} \cdot \frac{5}{5} + \frac{1}{40} \cdot \frac{2}{2} + \frac{1}{16} \cdot \frac{5}{5} + \frac{1}{40} \cdot \frac{2}{2} \qquad \text{Multiply by 1.}$$

$$= \frac{5}{80} + \frac{2}{80} + \frac{5}{80} + \frac{2}{80} \qquad \text{Use the LCD. (Step 2)}$$

$$= \frac{5 + 2 + 5 + 2}{80} \qquad \text{Addition of fractions}$$

$$= \frac{14}{80} \qquad \text{Add. (Step 3)}$$

$$= \frac{7}{40} \qquad \text{Simplify. (Step 3)}$$

The perimeter of the 1-acre plot of land is $\frac{7}{40}$ mile.

■ **Now Try Exercise 91**

4.5 Putting It All Together

CONCEPT	COMMENTS	EXAMPLES
Least Common Multiple (LCM)	The least common multiple (LCM) of two or more numbers is the smallest number that is divisible by each of the given numbers.	Common multiples of 3 and 4 include 12, 24, 36, 48, ... The LCM of 3 and 4 is 12.
Prime Factorization Method for LCM	1. Find the prime factorization of each number. 2. List each factor that appears in one or more of the factorizations. If a factor is repeated in any of the factorizations, list this factor the maximum number of times that it is repeated in any one of the factorizations. 3. The product of this list of factors is the LCM.	Find the LCM of 12 and 18. $12 = 2 \cdot 2 \cdot 3$ (Step 1) $18 = 2 \cdot 3 \cdot 3$ (Step 1) $2, 2, 3, 3$ (Step 2) $2 \cdot 2 \cdot 3 \cdot 3 = 36$ (Step 3) The LCM of 12 and 18 is 36.
Least Common Denominator (LCD)	The least common denominator (LCD) of two or more fractions is the least common multiple of the denominators.	Because the LCM of 12 and 18 is 36, the LCD of $\frac{7}{12}$ and $\frac{5}{18}$ is 36.
Adding or Subtracting Fractions with Unlike Denominators	1. Determine the LCD for all fractions involved. 2. Write each fraction as an equivalent fraction with the LCD as its denominator. 3. Add or subtract the newly written like fractions and simplify the result.	Add $\frac{7}{12} + \frac{5}{18}$. The LCD of the fractions $\frac{7}{12}$ and $\frac{5}{18}$ is 36. (Step 1) $\frac{7}{12} + \frac{5}{18} = \frac{7}{12} \cdot \frac{3}{3} + \frac{5}{18} \cdot \frac{2}{2}$ $= \frac{21}{36} + \frac{10}{36}$ (Step 2) $= \frac{31}{36}$ (Step 3)

4.5   Exercises

CONCEPTS AND VOCABULARY

1. Although 12, 24, 36, and 48 are all common multiples of 4 and 6, the _____ common multiple is 12.

2. The LCM of a list of numbers is the smallest number that is _____ by each of the given numbers.

3. When the LCM of two numbers is chosen from lists of the multiples of each number, we are using the _____ method for finding the LCM.

4. When the LCM of two numbers is found by first finding the prime factorization of each number, we are using the _____ method for finding the LCM.

5. If a factor is repeated in any of the factorizations used in the prime factorization method for finding the LCM, we list it the _____ number of times that it is repeated in any factorization.

6. The least common denominator (LCD) for a list of fractions is the _____ of the denominators.

LEAST COMMON MULTIPLE (LCM)

Exercises 7–12: Use the listing method to find the LCM of the given numbers.

7. 4 and 10

8. 6 and 14

9. 12 and 15

10. 9 and 15

11. 6 and 12

12. 8 and 24

Exercises 13–24: Use the prime factorization method to find the LCM of the given numbers.

13. 20 and 16

14. 18 and 24

15. 15 and 90

16. 16 and 80

17. 3, 6, and 15

18. 3, 15, and 21

19. 9, 12, and 45

20. 6, 10, and 27

21. 27 and 45

22. 50 and 75

23. 48 and 81

24. 25 and 42

Exercises 25–32: Use the factorization method to find the LCM of the given variable expressions.

25. $3x$ and $9x$

26. ab^2 and $5ab$

27. $3y^3$ and $8y$

28. $4pq$ and $10p^2q$

29. $6a^3b^2$ and $8ab^2$

30. $4m$ and $12n^2$

31. $4x$, $6y$, and $3z$

32. $8a$, $3ab$, and $4b$

LEAST COMMON DENOMINATOR (LCD)

Exercises 33–46: Find the LCD for the given fractions.

33. $\frac{2}{9}$ and $\frac{5}{12}$

34. $\frac{7}{15}$ and $\frac{1}{6}$

35. $\frac{3}{4}$ and $\frac{9}{14}$

36. $\frac{19}{20}$ and $\frac{5}{8}$

37. $\frac{11}{24}$ and $\frac{17}{30}$

38. $\frac{19}{26}$ and $\frac{13}{24}$

39. $\frac{1}{3}, \frac{4}{5}$, and $\frac{3}{4}$

40. $\frac{1}{2}, \frac{4}{7}$, and $\frac{5}{6}$

41. $\frac{2}{3xy}$ and $\frac{5}{6y^2}$

42. $\frac{7}{10m^2}$ and $\frac{3a}{4mn}$

43. $\frac{5b}{8m}$ and $\frac{3}{16mn^2}$

44. $\frac{x}{13y^2}$ and $\frac{y}{39x}$

45. $\frac{1}{a}, \frac{2}{b}$, and $\frac{3}{c}$

46. $\frac{1}{2x}, \frac{2}{3y}$, and $\frac{3}{4z}$

ADDING AND SUBTRACTING UNLIKE FRACTIONS

Exercises 47–52: Rewrite the fractions as equivalent fractions with the given LCD.

47. $\frac{3}{4}$ and $\frac{1}{6}$ LCD: 12

48. $\frac{5}{8}$ and $\frac{7}{12}$ LCD: 24

49. $\frac{11}{18}$ and $\frac{3}{4}$ LCD: 36

50. $\frac{4}{15}$ and $\frac{7}{9}$ LCD: 45

51. $\frac{1}{2}, \frac{5}{6}$, and $\frac{9}{10}$ LCD: 30

52. $\frac{2}{5}, \frac{11}{12}$, and $\frac{1}{6}$ LCD: 60

Exercises 53–70: Add or subtract as indicated.

53. $\frac{1}{4} + \frac{3}{10}$

54. $\frac{5}{6} + \frac{3}{8}$

55. $\frac{7}{10} - \frac{7}{8}$

56. $\frac{11}{12} - \frac{3}{10}$

57. $-\dfrac{7}{3} + \dfrac{8}{9}$

58. $\dfrac{1}{2} + \left(-\dfrac{5}{8}\right)$

59. $\dfrac{1}{10} + \dfrac{1}{15}$

60. $-\dfrac{7}{18} + \dfrac{5}{6}$

61. $\dfrac{11}{12} + \left(-\dfrac{1}{4}\right)$

62. $\dfrac{19}{24} - \dfrac{5}{8}$

63. $-\dfrac{5}{12} - \dfrac{3}{4}$

64. $\dfrac{17}{30} - \dfrac{5}{18}$

65. $\dfrac{3}{8} + \dfrac{5}{12} + \dfrac{5}{24}$

66. $\dfrac{5}{12} + \dfrac{5}{6} + \dfrac{3}{4}$

67. $\dfrac{1}{5} + \dfrac{3}{10} + \dfrac{1}{4}$

68. $\dfrac{1}{6} + \dfrac{1}{12} + \dfrac{3}{8}$

69. $\dfrac{2}{3} + \dfrac{7}{12} + \dfrac{5}{4}$

70. $\dfrac{4}{5} + \dfrac{7}{10} + \dfrac{3}{2}$

Exercises 71–80: Add or subtract as indicated.

71. $\dfrac{3x}{4} + \dfrac{5x}{6}$

72. $\dfrac{y}{6} - \dfrac{7y}{9}$

73. $\dfrac{7m^2}{14} - \dfrac{m^2}{6}$

74. $-\dfrac{7w}{12} + \dfrac{5w}{8}$

75. $\dfrac{x}{16} + \dfrac{5y}{12}$

76. $\dfrac{a}{9} - \dfrac{b}{12}$

77. $\dfrac{5}{6y} - \dfrac{1}{8y^2}$

78. $\dfrac{3}{a} + \dfrac{4}{b}$

79. $\dfrac{3x}{5y} + \dfrac{5y}{3x}$

80. $\dfrac{m}{2n} - \dfrac{2n}{3m}$

APPLICATIONS

81. *World Population* In 2009, the two most populous countries in the world were China and India. About $\frac{1}{5}$ of the world's population lived in China, while about $\frac{17}{100}$ of the population lived in India. What is the total fraction of the world's population that lived in these two countries in 2009? (*Source: United Nations.*)

82. *World Population* The third and fourth most populous countries in the world in 2009 were the United States and Indonesia. About $\frac{9}{200}$ of the world's population lived in the United States, while about $\frac{17}{500}$ of the population lived in Indonesia. What is the total fraction of the world's population that lived in these two countries in 2009? (*Source: United Nations.*)

83. *United States* The fraction of state names that start with a vowel is $\frac{6}{25}$. If $\frac{9}{50}$ of the state names begin with A, E, I, or U, what fraction of the states have names that begin with O?

84. *European Union* While about $\frac{1}{2}$ of European Union citizens are able to speak English, only about $\frac{3}{20}$ are native English speakers. Determine the fraction of European Union citizens who are non-native English speakers. (*Source: European Commission.*)

85. *Wind Energy* In 2007, about $\frac{13}{50}$ of U.S. wind power was produced in Texas. That same year, California produced about $\frac{3}{20}$ of the country's wind power. What fraction of all U.S wind power was produced in these two states in 2007? (*Source: U.S. Department of Energy.*)

86. *Voter Turnout* In the 2008 presidential election, the nation's highest voter turnout occurred in Minnesota, where $\frac{39}{50}$ of the state's eligible voters went to the polls. In the same presidential election, the national average was $\frac{4}{25}$ lower than the Minnesota fraction. What fraction of the eligible U.S. population voted in 2008? (*Source: Minnesota Secretary of State.*)

87. *Fuel Consumption* At the beginning of a trip, a car's gas gauge shows that the car has $\frac{3}{4}$ of a tank of gas. At the end of the trip, the gauge indicates that the car has $\frac{1}{8}$ of a tank of gas. What fraction of the tank was used during the trip?

88. *Efficient Lighting* In an effort to replace $\frac{23}{30}$ of a home's incandescent light bulbs with energy efficient LED bulbs, a contractor has already replaced $\frac{3}{5}$ of the light bulbs. What fraction of the incandescent bulbs still need to be replaced?

Exercises 89–94: Geometry Find the perimeter of the given figure.

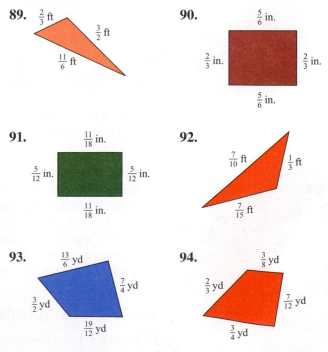

89. $\frac{2}{3}$ ft, $\frac{3}{2}$ ft, $\frac{11}{6}$ ft

90. $\frac{5}{6}$ in., $\frac{2}{3}$ in., $\frac{2}{3}$ in., $\frac{5}{6}$ in.

91. $\frac{11}{18}$ in., $\frac{5}{12}$ in., $\frac{5}{12}$ in., $\frac{11}{18}$ in.

92. $\frac{7}{10}$ ft, $\frac{1}{3}$ ft, $\frac{7}{15}$ ft

93. $\frac{13}{6}$ yd, $\frac{7}{4}$ yd, $\frac{3}{2}$ yd, $\frac{19}{12}$ yd

94. $\frac{3}{8}$ yd, $\frac{2}{3}$ yd, $\frac{7}{12}$ yd, $\frac{3}{4}$ yd

WRITING ABOUT MATHEMATICS

95. Explain how the LCM of two or more numbers could be used to find other multiples of the numbers.

96. Write a detailed paragraph that explains how to add or subtract two fractions with unlike denominators.

97. Use both the listing method and the prime factorization method to find the LCM of 52 and 77. Which method do you prefer?

98. If two fractions are being added and one denominator is a multiple of the other, what can be said about the LCD? Explain your reasoning.

Group Activity   Working with Real Data

Directions: Form a group of 2 to 4 people. Select someone to record the group's responses for this activity. All members of the group should work cooperatively to answer the questions. If your instructor asks for your results, each member of the group should be prepared to respond.

Major U.S. Rivers The mainland United States is made up of all states except Alaska and Hawaii. The two largest rivers within the mainland United States are the Mississippi and the Missouri. The states that each of these rivers passes through (or borders) are listed below.

Mississippi:
Arkansas, Illinois, Iowa, Kentucky, Louisiana, Minnesota, Mississippi, Missouri, Tennessee, and Wisconsin.

Missouri:
Iowa, Kansas, Missouri, Montana, Nebraska, North Dakota, and South Dakota.

(a) Write a simplified fraction that represents the fraction of the mainland states through which the Mississippi river passes.

(b) Write a simplified fraction that represents the fraction of the mainland states through which the Missouri river passes.

(c) Write a simplified fraction that represents the fraction of the mainland states through which both rivers pass.

(d) By adding your answers from parts (a) and (b) and then subtracting your answer from part (c), find the fraction of the mainland states through which either river (or both) passes.

4.6   Operations on Mixed Numbers

Review of Improper Fractions and Rounding • Operations on Mixed Numbers • Adding and Subtracting Mixed Numbers—Method II

A LOOK INTO MATH ▶

Mixed numbers commonly occur in cooking and baking. A recipe may call for $2\frac{2}{3}$ cup of flour or $1\frac{1}{4}$ cup of cooking wine. When we double or triple a recipe, we may need to *multiply* a mixed number by a whole number. Also, some recipes require that we add a particular ingredient at an early stage in the cooking process, and then add more of the ingredient later. In this case, we may need to *add* mixed numbers if we want to find the total amount of the ingredient needed. Other situations may result in the need to *subtract* or *divide* mixed numbers. In this section, we perform arithmetic on mixed numbers.

Review of Improper Fractions and Rounding

In Section 4.1, we discussed how a mixed number can be written as an improper fraction. An abbreviated summary of the process follows.

STUDY TIP

Questions on exams do not always come in the order that they are presented in the text. When studying for an exam, choose review exercises randomly so that the topics are studied in the same random way that they may appear on an exam.

WRITING A MIXED NUMBER AS AN IMPROPER FRACTION

A mixed number of the form

$$\text{Quotient}\ \frac{\text{Remainder}}{\text{Divisor}} \quad \text{can be written as} \quad \frac{\text{Divisor} \cdot \text{Quotient} + \text{Remainder}}{\text{Divisor}}.$$

In the next example we review how to write mixed numbers as improper fractions.

EXAMPLE 1 Writing mixed numbers as improper fractions

Write each mixed number as an improper fraction.

(a) $5\frac{3}{4}$ (b) $-2\frac{3}{5}$

Solution

(a) $5\frac{3}{4} = \frac{4 \cdot 5 + 3}{4} = \frac{20 + 3}{4} = \frac{23}{4}$

(b) Write the improper fraction for the absolute value of $-2\frac{3}{5}$ and then negate the result.

Because $2\frac{3}{5} = \frac{5 \cdot 2 + 3}{5} = \frac{10 + 3}{5} = \frac{13}{5}$, we know that $-2\frac{3}{5} = -\frac{13}{5}$.

Now Try Exercises 7, 13

When performing operations on mixed numbers, it is often helpful to have a rough estimate of the result before completing more complicated computations. To do this, we round the mixed numbers to the nearest integer. For example, the mixed number $3\frac{5}{6}$ can be rounded to 4, and the mixed number $-7\frac{1}{8}$ can be rounded to -7. The following procedure can be used to round a mixed number to the nearest integer.

ROUNDING A MIXED NUMBER TO THE NEAREST INTEGER

To round a mixed number to the nearest integer, do the following.

If the fraction part of the mixed number is

(a) less than $\frac{1}{2}$, drop the fraction part of the mixed number.

(b) $\frac{1}{2}$ or more, drop the fraction part of the mixed number and

  (i) add 1 to the integer part if the mixed number is positive.

  (ii) subtract 1 from the integer part if the mixed number is negative.

NOTE: If twice the numerator of a fraction is greater than or equal to the denominator of the fraction, then the fraction is greater than or equal to $\frac{1}{2}$. For example, the fraction $\frac{7}{13}$ is greater than $\frac{1}{2}$ because $2 \cdot 7 = 14$ and $14 > 13$. Similarly, the fraction $\frac{17}{40}$ is less than $\frac{1}{2}$ because $2 \cdot 17 = 34$ and $34 < 40$.

READING CHECK

• What is the process used to round mixed numbers to the nearest integer?

EXAMPLE 2   **Rounding mixed numbers to the nearest integer**

Round each mixed number to the nearest integer.

(a) $9\dfrac{3}{7}$ (b) $-6\dfrac{8}{15}$

Solution
(a) The fraction part of the given mixed number is $\dfrac{3}{7}$, which is less than $\dfrac{1}{2}$. To round the mixed number $9\dfrac{3}{7}$ to the nearest integer, drop the fraction part to obtain 9.
(b) The fraction part is $\dfrac{8}{15}$, which is greater than $\dfrac{1}{2}$. Because the mixed number $-6\dfrac{8}{15}$ is negative, we round it to the nearest integer by dropping the fraction part and subtracting 1 from the integer part to obtain -7.

Now Try Exercises 15, 17

Operations on Mixed Numbers

Because we have already studied operations on fractions in Sections 4.3–4.5, a new process for operations on mixed numbers is not needed. We just need to rewrite any mixed numbers as improper fractions before performing the operations. The following is a process for performing operations on mixed numbers.

OPERATIONS ON MIXED NUMBERS

To perform operations on mixed numbers, do the following.

STEP 1: Rewrite any mixed numbers or integers as improper fractions.

STEP 2: Perform the required arithmetic as usual.

STEP 3: Rewrite the result as a mixed number, when appropriate.

In the next four examples, we use these steps (in a single line of equivalent expressions) to perform operations on mixed numbers and integers. In each example, we first find an estimation of the result in order to reveal any potential computational errors.

EXAMPLE 3   **Adding mixed numbers**

Add.

(a) $2\dfrac{3}{4} + 6\dfrac{1}{4}$ (b) $-3\dfrac{1}{6} + 5\dfrac{2}{3}$ (c) $\dfrac{4}{9} + 1\dfrac{3}{5}$

Solution
(a) Since $2\dfrac{3}{4}$ rounds to 3 and $6\dfrac{1}{4}$ rounds to 6, an estimate of the sum is $3 + 6 = 9$.

$$2\dfrac{3}{4} + 6\dfrac{1}{4} = \dfrac{11}{4} + \dfrac{25}{4} = \dfrac{11 + 25}{4} = \dfrac{36}{4} = 9$$

The estimation and the computed sum agree.
(b) Here, $-3\dfrac{1}{6}$ rounds to -3 and $5\dfrac{2}{3}$ rounds to 6, so an estimate of the sum is $-3 + 6 = 3$.

$$-3\dfrac{1}{6} + 5\dfrac{2}{3} = -\dfrac{19}{6} + \dfrac{17}{3} = \dfrac{-19}{6} + \dfrac{17}{3}\cdot\dfrac{2}{2} = \dfrac{-19}{6} + \dfrac{34}{6} = \dfrac{-19 + 34}{6} = \dfrac{15}{6}$$

Because the problem involves mixed numbers, we express the result as a mixed number.

$$\frac{15}{6} = 2\frac{3}{6} = 2\frac{1}{2}$$

The computed sum is reasonably close to the estimate.

(c) Because $\frac{4}{9}$ rounds to 0 and $1\frac{3}{5}$ rounds to 2, an estimate of the sum is $0 + 2 = 2$.

$$\frac{4}{9} + 1\frac{3}{5} = \frac{4}{9} + \frac{8}{5} = \frac{4}{9}\cdot\frac{5}{5} + \frac{8}{5}\cdot\frac{9}{9} = \frac{20}{45} + \frac{72}{45} = \frac{20 + 72}{45} = \frac{92}{45}$$

Expressing the result as a mixed number, we get

$$\frac{92}{45} = 2\frac{2}{45}.$$

The computed sum is reasonably close to the estimate.

Now Try Exercises 23, 33, 39

EXAMPLE 4   **Subtracting mixed numbers**

Subtract.

(a) $7\frac{2}{15} - 3\frac{1}{6}$ (b) $-14\frac{1}{4} - \frac{7}{8}$

Solution

(a) Because $7\frac{2}{15}$ rounds to 7 and $3\frac{1}{6}$ rounds to 3, the estimated difference is $7 - 3 = 4$.

$$7\frac{2}{15} - 3\frac{1}{6} = \frac{107}{15} - \frac{19}{6} = \frac{107}{15}\cdot\frac{2}{2} - \frac{19}{6}\cdot\frac{5}{5} = \frac{214}{30} - \frac{95}{30} = \frac{214 - 95}{30} = \frac{119}{30}$$

Expressing the result as a mixed number, we get

$$\frac{119}{30} = 3\frac{29}{30}.$$

The computed sum is reasonably close to the estimate.

(b) Here, $-14\frac{1}{4}$ rounds to -14 and $\frac{7}{8}$ rounds to 1, so the estimate is $-14 - 1 = -15$.

$$-14\frac{1}{4} - \frac{7}{8} = -\frac{57}{4} - \frac{7}{8} = \frac{-57}{4}\cdot\frac{2}{2} - \frac{7}{8} = \frac{-114}{8} - \frac{7}{8} = \frac{-114 - 7}{8} = -\frac{121}{8}$$

As a mixed number, this result is

$$-\frac{121}{8} = -15\frac{1}{8}.$$

The computed sum is reasonably close to the estimate.

Now Try Exercises 27, 35

EXAMPLE 5   **Multiplying mixed numbers**

Multiply.

(a) $9\frac{4}{7}\cdot 3$ (b) $-5\frac{1}{3}\cdot 3\frac{1}{8}$ (c) $\frac{3}{4}\cdot 8\frac{2}{3}$

Solution

(a) Because $9\frac{4}{7}$ rounds to 10, the estimated product is $10 \cdot 3 = 30$.

$$9\frac{4}{7} \cdot 3 = \frac{67}{7} \cdot \frac{3}{1} = \frac{201}{7} = 28\frac{5}{7}$$

The computed product is reasonably close to the estimate.

(b) Because $-5\frac{1}{3}$ rounds to -5 and $3\frac{1}{8}$ rounds to 3, the estimated product is $-5 \cdot 3 = -15$.

$$-5\frac{1}{3} \cdot 3\frac{1}{8} = \frac{-16}{3} \cdot \frac{25}{8} = \frac{-(\mathbf{2} \cdot \mathbf{2} \cdot 2 \cdot 2) \cdot (5 \cdot 5)}{3 \cdot (\mathbf{2} \cdot \mathbf{2} \cdot \mathbf{2})} = \frac{-50}{3} = -16\frac{2}{3}$$

The computed product is reasonably close to the estimate.

(c) Since $\frac{3}{4}$ rounds to 1 and $8\frac{2}{3}$ rounds to 9, the estimated product is $1 \cdot 9 = 9$.

$$\frac{3}{4} \cdot 8\frac{2}{3} = \frac{3}{4} \cdot \frac{26}{3} = \frac{\mathbf{3} \cdot (\mathbf{2} \cdot 13)}{(\mathbf{2} \cdot 2) \cdot \mathbf{3}} = \frac{13}{2} = 6\frac{1}{2}$$

Although the estimated product may appear to be significantly off from the computed product, the computed result is correct. See the Note following this example.

Now Try Exercises 25, 31, 43

NOTE: The use of rounded values in estimating some products or quotients may give us results that do not seem accurate. This is clear when a fraction rounds to 0. In this case, an estimated product or quotient will be either 0 or undefined.

EXAMPLE 6 **Dividing mixed numbers**

Divide.

(a) $9\frac{11}{12} \div 2\frac{1}{8}$ **(b)** $20 \div 3\frac{2}{5}$

Solution

(a) Here $9\frac{11}{12}$ rounds to 10 and $2\frac{1}{8}$ rounds to 2, so the estimated quotient is $10 \div 2 = 5$.

$$9\frac{11}{12} \div 2\frac{1}{8} = \frac{119}{12} \div \frac{17}{8} = \frac{119}{12} \cdot \frac{8}{17} = \frac{(7 \cdot \mathbf{17}) \cdot (\mathbf{2} \cdot \mathbf{2} \cdot 2)}{(\mathbf{2} \cdot \mathbf{2} \cdot 3) \cdot \mathbf{17}} = \frac{14}{3} = 4\frac{2}{3}$$

The computed quotient is reasonably close to the estimate.

(b) Because $3\frac{2}{5}$ rounds to 3, the estimated quotient is $20 \div 3 = \frac{20}{3} = 6\frac{2}{3}$.

$$20 \div 3\frac{2}{5} = \frac{20}{1} \div \frac{17}{5} = \frac{20}{1} \cdot \frac{5}{17} = \frac{(2 \cdot 2 \cdot 5) \cdot 5}{17} = \frac{100}{17} = 5\frac{15}{17}$$

The computed quotient is reasonably close to the estimate.

Now Try Exercises 29, 37

Adding and Subtracting Mixed Numbers—Method II

Although the method just described for adding and subtracting mixed numbers can always be used, there is a second method that may be more efficient. In this alternate method, we add or subtract the integer parts and the fraction parts separately. The next two examples demonstrate this method.

EXAMPLE 7   **Adding mixed numbers—Method II**

Add.

(a) $5\frac{3}{4} + 8\frac{2}{5}$ (b) $4 + 7\frac{2}{3}$

Solution

(a) Begin by aligning the integer and fraction parts vertically. The LCD for 4 and 5 is 20.

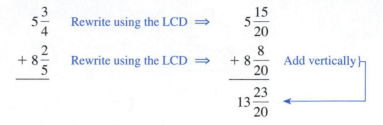

$$5\frac{3}{4} \quad \text{Rewrite using the LCD} \Rightarrow \quad 5\frac{15}{20}$$

$$+8\frac{2}{5} \quad \text{Rewrite using the LCD} \Rightarrow \quad +8\frac{8}{20} \quad \text{Add vertically}$$

$$\overline{\qquad\qquad\qquad\qquad\qquad\qquad 13\frac{23}{20}}$$

Since $\frac{23}{20} = 1\frac{3}{20}$, we simplify the sum as follows.

$$13\frac{23}{20} = 13 + 1\frac{3}{20} = 14\frac{3}{20}$$

(b) Method II is particularly efficient for this sum.

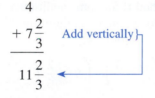

$$4$$
$$+7\frac{2}{3} \quad \text{Add vertically}$$
$$\overline{\qquad\qquad}$$
$$11\frac{2}{3}$$

Now Try Exercises 47, 51

NOTE: To add mixed numbers with different signs, change the addition problem to an equivalent subtraction problem. Subtraction is illustrated in the next example.

EXAMPLE 8   **Subtracting mixed numbers—Method II**

Subtract.

(a) $9\frac{7}{8} - 2\frac{1}{6}$ (b) $13\frac{2}{3} - 5\frac{9}{10}$

Solution

(a) Begin by aligning the integer and fraction parts vertically. The LCD for 8 and 6 is 24.

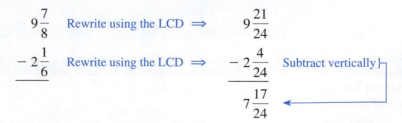

$$9\frac{7}{8} \quad \text{Rewrite using the LCD} \Rightarrow \quad 9\frac{21}{24}$$

$$-2\frac{1}{6} \quad \text{Rewrite using the LCD} \Rightarrow \quad -2\frac{4}{24} \quad \text{Subtract vertically}$$

$$\overline{\qquad\qquad\qquad\qquad\qquad\qquad 7\frac{17}{24}}$$

(b) Align the integer and fraction parts vertically. The LCD for 3 and 10 is 30.

$$13\frac{2}{3} \quad \text{Rewrite using the LCD} \Rightarrow \quad 13\frac{20}{30}$$

$$-5\frac{9}{10} \quad \text{Rewrite using the LCD} \Rightarrow \quad -5\frac{27}{30}$$

$$\overline{\qquad\qquad\qquad\qquad\qquad}$$

Because subtracting $\frac{27}{30}$ from $\frac{20}{30}$ would result in a negative difference, we borrow 1 from the integer part of $13\frac{20}{30}$, to rewrite the mixed number as follows.

Borrow 1　　　Write 1 as $\frac{30}{30}$

$$13\frac{20}{30} = 12 + \mathbf{\color{red}1} + \frac{20}{30} = 12 + \frac{\color{red}\mathbf{30}}{\color{red}\mathbf{30}} + \frac{20}{30} = 12 + \frac{30 + 20}{30} = 12\frac{50}{30}$$

Rewriting $13\frac{20}{30}$ as $12\frac{50}{30}$, we complete the given subtraction as follows.

$$13\frac{20}{30}$$
$$-5\frac{27}{30}$$

Borrow 1 from 13 $\Rightarrow$

$$12\frac{50}{30}$$
$$-5\frac{27}{30}$$ Subtract vertically
$$\overline{7\frac{23}{30}}$$

Now Try Exercises 41, 45

MAKING CONNECTIONS

Multiplying Mixed Numbers—Method II

READING CHECK

• Why is it useful to write mixed numbers as improper fractions before doing any arithmetic?

While we can use Method II for some addition and subtraction, it is not practical for multiplication or division. For example, to multiply $2\frac{3}{4}$ and $5\frac{2}{3}$ using Method II, we must utilize the distributive property as follows.

$$2\frac{3}{4} \cdot 5\frac{2}{3} = \left(2 + \frac{3}{4}\right) \cdot \left(5 + \frac{2}{3}\right) \qquad \text{Rewrite the mixed numbers.}$$

$$= \left(2 + \frac{3}{4}\right) \cdot 5 + \left(2 + \frac{3}{4}\right) \cdot \frac{2}{3} \qquad \text{Distribute $2 + \frac{3}{4}$.}$$

$$= 2 \cdot 5 + \frac{3}{4} \cdot 5 + 2 \cdot \frac{2}{3} + \frac{3}{4} \cdot \frac{2}{3} \qquad \text{Distribute 5 and $\frac{2}{3}$.}$$

$$= 10 + \frac{15}{4} + \frac{4}{3} + \frac{1}{2} \qquad \text{Multiply.}$$

$$= 15\frac{7}{12} \qquad \text{Add.}$$

4.6　Putting It All Together

CONCEPT	COMMENTS	EXAMPLES
Writing Mixed Numbers as Improper Fractions	A mixed number of the form $$\text{Quotient} \frac{\text{Remainder}}{\text{Divisor}}$$ can be written as $$\frac{\text{Divisor} \cdot \text{Quotient} + \text{Remainder}}{\text{Divisor}}.$$	$$9\frac{4}{7} = \frac{7 \cdot 9 + 4}{7} = \frac{67}{7}$$ $$-4\frac{2}{5} = -\frac{5 \cdot 4 + 2}{5} = -\frac{22}{5}$$

CONCEPT	COMMENTS	EXAMPLES

Rounding Mixed Numbers to the Nearest Integer	If the fraction part of the mixed number is
	(a) less than $\frac{1}{2}$, drop the fraction part.
	(b) $\frac{1}{2}$ or more, drop the fraction part and
	  (i) add 1 to the integer part if the mixed number is positive.
	  (ii) subtract 1 from the integer part if the mixed number is negative.
		$3\frac{2}{9}$ rounds to 3.
		
		$-4\frac{7}{8}$ rounds to -5.
		
		$-11\frac{3}{7}$ rounds to -11.

Operations on Mixed Numbers	1. Rewrite any mixed numbers or integers as improper fractions. 2. Perform the required arithmetic as usual. 3. Rewrite the result as a mixed number.	$1\frac{2}{3} \cdot 3\frac{1}{2} = \frac{5}{3} \cdot \frac{7}{2} = \frac{35}{6} = 5\frac{5}{6}$ (Step 1) (Step 2) (Step 3) $5\frac{2}{3} + 2\frac{2}{3} = \frac{17}{3} + \frac{8}{3} = \frac{25}{3} = 8\frac{1}{3}$ (Step 1) (Step 2) (Step 3)
Method II for Adding and Subtracting Mixed Numbers	Add or subtract vertically by adding or subtracting the integer parts and the fraction parts separately.	$\begin{array}{r} 4\frac{1}{5} \\ + 6\frac{2}{5} \\ \hline 10\frac{3}{5} \end{array} \qquad \begin{array}{r} 11\frac{7}{9} \\ - 3\frac{2}{9} \\ \hline 8\frac{5}{9} \end{array}$

4.6 Exercises

MyMathLab | Math XL PRACTICE | WATCH | DOWNLOAD | READ | REVIEW

CONCEPTS AND VOCABULARY

1. To round a mixed number to the nearest integer, we must determine if the fraction part of the mixed number is less than, equal to, or greater than _____.

2. A fraction is less than $\frac{1}{2}$ if twice its numerator is _____ than its denominator.

3. The first step in performing operations on mixed numbers is to rewrite any mixed numbers or integers as _____.

4. When doing arithmetic on mixed numbers, it is often helpful to have a rough _____ of the result before performing more complicated computations.

5. When using Method II for addition or subtraction, we align the mixed numbers _____ before completing the indicated arithmetic.

6. If the fraction in the upper mixed number is less than the fraction in the lower mixed number when subtracting using Method II, we need to _____ 1 from the integer part of the upper mixed number.

IMPROPER FRACTIONS AND ROUNDING

Exercises 7–14: Write the given mixed number as an improper fraction.

7. $4\frac{1}{5}$

8. $11\frac{2}{7}$

9. $-3\frac{7}{15}$

10. $-12\frac{3}{8}$

11. $8\frac{9}{10}$

12. $13\frac{2}{3}$

13. $-15\frac{1}{4}$

14. $-20\frac{4}{5}$

Exercises 15–22: Round the given mixed number to the nearest integer.

15. $7\frac{8}{11}$

16. $12\frac{1}{8}$

17. $-14\frac{6}{13}$

18. $-3\frac{5}{6}$

19. $11\frac{3}{7}$

20. $6\frac{7}{12}$

21. $-9\frac{5}{9}$

22. $-1\frac{9}{25}$

OPERATIONS ON MIXED NUMBERS

Exercises 23–60: Perform the indicated arithmetic.

23. $5\frac{1}{6} + 4\frac{5}{6}$

24. $7\frac{1}{3} - \left(-\frac{5}{8}\right)$

25. $9\frac{7}{8} \cdot 4$

26. $16\frac{3}{28} \div 5\frac{6}{7}$

27. $11\frac{5}{12} - 4\frac{3}{8}$

28. $8\frac{2}{9} + \left(-5\frac{1}{12}\right)$

29. $14\frac{5}{8} \div 2\frac{1}{6}$

30. $11\frac{2}{5} \cdot 3$

31. $-7\frac{1}{6} \cdot 4\frac{2}{3}$

32. $\frac{8}{9} \cdot 5\frac{1}{4}$

33. $-7\frac{3}{4} + 6\frac{5}{8}$

34. $6\frac{3}{7} + 5\frac{4}{7}$

35. $-8\frac{2}{5} - \frac{7}{10}$

36. $5\frac{1}{10} + 9\frac{5}{6}$

37. $32 \div 2\frac{3}{8}$

38. $15\frac{7}{18} - 10\frac{1}{2}$

39. $3\frac{4}{5} + \frac{6}{7}$

40. $17\frac{5}{12} - 9\frac{1}{4}$

41. $14\frac{5}{6} - 3\frac{1}{4}$

42. $3\frac{1}{4} \cdot \left(-5\frac{1}{2}\right)$

43. $10\frac{1}{3} \cdot \frac{5}{7}$

44. $18\frac{1}{5} \div 3\frac{1}{5}$

45. $8\frac{1}{8} - 2\frac{5}{6}$

46. $7 + 6\frac{8}{15}$

47. $7\frac{2}{3} + 6\frac{7}{8}$

48. $\frac{4}{7} + 5\frac{9}{14}$

49. $15\frac{1}{3} \div 3\frac{5}{6}$

50. $19\frac{3}{8} - 6\frac{7}{10}$

51. $2\frac{6}{11} + 9$

52. $7\frac{1}{5} \div 9$

53. $-16\frac{2}{9} \div 1\frac{1}{6}$

54. $-6 \cdot \left(-4\frac{2}{9}\right)$

55. $4\frac{1}{8} - \left(-10\frac{6}{8}\right)$

56. $8\frac{4}{5} + \left(-7\frac{13}{15}\right)$

57. $5\frac{11}{12} \cdot (-4)$

58. $4\frac{2}{3} \div \left(-6\frac{1}{8}\right)$

59. $-5\frac{7}{15} + 8\frac{1}{3}$

60. $-7\frac{1}{12} - 6\frac{2}{3}$

APPLICATIONS

61. *Model Building* An industrial model builder needs several pieces of wire that are $6\frac{3}{4}$ inches long. How many whole pieces of wire of this length can be cut from a spool containing 72 inches of wire? How much wire is left over?

62. *Jogging* An athlete jogs $4\frac{1}{2}$ miles on Tuesday and $7\frac{3}{4}$ miles on Wednesday. What is the total distance for these two days?

63. *Baking* A recipe calls for $2\frac{2}{3}$ cups of flour. How much flour is needed if the recipe is doubled?

64. *Baking* A recipe calls for $1\frac{1}{4}$ cups of sugar. How much sugar is needed if the recipe is halved?

65. *Languages* In 2008, about 185 thousand Americans spoke Hmong as the primary language used in their homes. About $3\frac{2}{5}$ times as many Americans spoke Polish in their homes that year. Find the number of Americans who spoke Polish in their homes in 2008. (*Source:* U.S. Census Bureau.)

66. *Homeschooling* About 240 thousand children of parents with annual incomes less than $25,000 were homeschooled in 2007. If $2\frac{1}{10}$ times as many children of parents with annual incomes above $75,000 were homeschooled that year, find the number of children at this income level that were homeschooled in 2007. (*Source:* U.S. Department of Education.)

67. *Height* If the birth announcement for a baby boy indicates that his length at birth was $17\frac{1}{4}$ inches and another birth announcement for a baby girl states that her length at birth was $19\frac{5}{8}$ inches, how much longer was the girl at birth compared to the boy?

68. *Height* At birth, a baby boy was $18\frac{1}{2}$ inches long. If he grew $5\frac{3}{4}$ inches during his first year, how tall was he on his first birthday?

69. *Painting* A painter needs $1\frac{1}{4}$ gallons of paint for a project. If there is $\frac{2}{3}$ gallon available in a partially used container, how much paint from a new container must be used?

70. *Cleaning Solution* If $1\frac{1}{3}$ cups of liquid concentrate are mixed with 1 gallon of water to make a cleaning solution, how many such mixtures can be made from a 16-cup (1-gallon) jug of concentrate?

Exercises 71–74: Geometry Find the area of the given figure.

71.

$3\frac{1}{5}$ ft

$5\frac{5}{24}$ ft

72.
$4\frac{2}{3}$ in.

$6\frac{5}{6}$ in.

73.
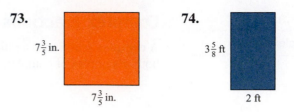
$7\frac{3}{5}$ in.

$7\frac{3}{5}$ in.

74.
$3\frac{5}{8}$ ft

2 ft

WRITING ABOUT MATHEMATICS

75. A student claims that an easy way to multiply two mixed numbers is to simply multiply the integer parts and then multiply the fraction parts. Give a real-world example that would disprove this student's method.

76. Try to devise a "Method II" for dividing one mixed number by another. Give detailed steps. (Refer to the Making Connections on page 262.)

SECTIONS 4.5 and 4.6

Checking Basic Concepts

1. Find the least common multiple.
   (a) 12 and 15   **(b)** $4x$ and $5x^2$

2. Find the least common denominator.
   (a) $\frac{5}{8}$ and $\frac{7}{9}$ **(b)** $\frac{3}{10xy}$ and $\frac{5}{6x^2}$

3. Add or subtract as indicated. If necessary, simplify your results.
   (a) $\frac{1}{8} + \frac{3}{18}$ **(b)** $\frac{8}{15} - \frac{7}{12}$
   (c) $\frac{1}{5} + \frac{1}{4} + \frac{3}{8}$ **(d)** $-\frac{5x}{8} + \frac{7x}{12}$

4. Perform the indicated arithmetic.
   (a) $-5\frac{1}{5} + 4\frac{7}{10}$ **(b)** $\frac{4}{15} - \left(-6\frac{7}{12}\right)$
   (c) $3\frac{5}{12} \cdot 9$ **(d)** $16\frac{5}{6} \div 3\frac{5}{32}$

5. *Fuel Consumption* At the beginning of a trip, a car's gas gauge shows that the car has $\frac{7}{8}$ of a tank of gas. At the end of the trip, the gauge indicates that the car has $\frac{1}{3}$ of a tank of gas. What fraction of a tank was used during the trip?

4.7   Complex Fractions and Order of Operations

Complex Fractions • Simplifying Complex Fractions—Method I • Simplifying Complex Fractions—Method II • Order of Operations

A LOOK INTO MATH ▶

Suppose that $3\frac{1}{2}$ watermelons are shared among 14 people so that each person ends up with $\frac{1}{4}$ watermelon. This situation can be represented mathematically by

$$3\frac{1}{2} \div 14 = \frac{1}{4} \quad \text{or} \quad \frac{3\frac{1}{2}}{14} = \frac{1}{4} \quad \text{or} \quad \frac{3 + \frac{1}{2}}{14} = \frac{1}{4}.$$

In the last two equations, the expression on the left side of the equal sign is an example of a *complex fraction*. In this section, we discuss methods for simplifying complex fractions.

NEW VOCABULARY

☐ Complex fraction

Complex Fractions

A **complex fraction** is a rational expression with fractions in its numerator, denominator, or both. Examples of complex fractions include

$$\frac{1 + \frac{2}{3}}{3 - \frac{5}{6}}, \quad \frac{7}{\frac{x}{3} + \frac{1}{6}}, \quad \text{and} \quad \frac{\frac{3}{x}}{\frac{5x}{7}}.$$

READING CHECK

• What is a complex fraction?

Typically, we want to rewrite a complex fraction as a standard fraction in the form $\frac{a}{b}$, where the variables a and b represent expressions that are not fractions. For some basic complex fractions, this can be accomplished by noting that fractions are divided using the rule $\frac{a}{b} \div \frac{c}{d} = \frac{a}{b} \cdot \frac{d}{c}$ where b, c, and d are not 0. Because the expression $\frac{a}{b} \div \frac{c}{d}$ can be written as the complex fraction

$$\frac{\frac{a}{b}}{\frac{c}{d}},$$

we can summarize this strategy as follows.

SIMPLIFYING BASIC COMPLEX FRACTIONS

If the variables a, b, c, and d represent numbers with $b \neq 0$, $c \neq 0$, and $d \neq 0$, then

$$\frac{\frac{a}{b}}{\frac{c}{d}} = \frac{a}{b} \cdot \frac{d}{c}.$$

EXAMPLE 1   **Simplifying basic complex fractions**

Simplify each complex fraction.

(a) $\dfrac{\frac{4}{5}}{\frac{8}{15}}$ **(b)** $\dfrac{\frac{3}{2x}}{\frac{9}{y}}$

Solution

(a) $\dfrac{\frac{4}{5}}{\frac{8}{15}} = \frac{4}{5} \cdot \frac{15}{8} = \frac{(2 \cdot 2) \cdot (3 \cdot 5)}{5 \cdot (2 \cdot 2 \cdot 2)} = \frac{3}{2}$

(b) $\dfrac{\frac{3}{2x}}{\frac{9}{y}} = \frac{3}{2x} \cdot \frac{y}{9} = \frac{3 \cdot y}{(2 \cdot x) \cdot (3 \cdot 3)} = \frac{y}{2 \cdot x \cdot 3} = \frac{y}{6x}$

Now Try Exercises 5, 13

There are two common methods for simplifying complex fractions. In the first method, we begin by simplifying both the numerator and denominator of the complex fraction and then divide the resulting two fractions. In the second method, we multiply the numerator and denominator of the complex fraction by the LCD of all fractions within the complex fraction.

Simplifying Complex Fractions—Method I

The following steps outline Method I for simplifying complex fractions.

SIMPLIFYING COMPLEX FRACTIONS—METHOD I

To simplify a complex fraction, do the following.

STEP 1: Write the numerator as a single fraction; do the same to the denominator.

STEP 2: Multiply the fraction in the numerator by the reciprocal of the fraction in the denominator. Simplify the result.

EXAMPLE 2 **Simplifying complex fractions using Method I**

Simplify each complex fraction.

(a) $\dfrac{\frac{4}{5} + \frac{2}{3}}{\frac{7}{15} - \frac{2}{5}}$ (b) $\dfrac{\frac{3x}{4}}{1 - \frac{7}{16}}$

Solution

(a) **STEP 1:** Add the fractions $\frac{4}{5} + \frac{2}{3}$ in the numerator. The LCD is 15.

$$\frac{4}{5} + \frac{2}{3} = \frac{4}{5} \cdot \frac{3}{3} + \frac{2}{3} \cdot \frac{5}{5} = \frac{12}{15} + \frac{10}{15} = \frac{12 + 10}{15} = \frac{22}{15}$$

Subtract the fractions $\frac{7}{15} - \frac{2}{5}$ in the denominator. The LCD here is also 15.

$$\frac{7}{15} - \frac{2}{5} = \frac{7}{15} - \frac{2}{5} \cdot \frac{3}{3} = \frac{7}{15} - \frac{6}{15} = \frac{7 - 6}{15} = \frac{1}{15}$$

STEP 2: Multiply $\frac{22}{15}$ by $\frac{15}{1}$, the reciprocal of $\frac{1}{15}$.

$$\frac{\frac{4}{5} + \frac{2}{3}}{\frac{7}{15} - \frac{2}{5}} = \frac{\frac{22}{15}}{\frac{1}{15}} = \frac{22}{15} \cdot \frac{15}{1} = \frac{(2 \cdot 11) \cdot (3 \cdot 5)}{(3 \cdot 5)} = \frac{2 \cdot 11}{1} = \frac{22}{1} = 22$$

(b) **STEP 1:** The numerator of the complex fraction is already written as a single fraction, so we begin by subtracting $1 - \frac{7}{16}$ in the denominator. The LCD is 16.

$$1 - \frac{7}{16} = \frac{1}{1} \cdot \frac{16}{16} - \frac{7}{16} = \frac{16}{16} - \frac{7}{16} = \frac{16 - 7}{16} = \frac{9}{16}$$

STEP 2: Multiply $\frac{3x}{4}$ by $\frac{16}{9}$, the reciprocal of $\frac{9}{16}$.

$$\frac{\frac{3x}{4}}{1 - \frac{7}{16}} = \frac{\frac{3x}{4}}{\frac{9}{16}} = \frac{3x}{4} \cdot \frac{16}{9} = \frac{(3 \cdot x) \cdot (2 \cdot 2 \cdot 2 \cdot 2)}{(2 \cdot 2) \cdot (3 \cdot 3)} = \frac{2 \cdot 2 \cdot x}{3} = \frac{4x}{3}$$

Now Try Exercises 23, 25

Simplifying Complex Fractions—Method II

The steps used in Method II for simplifying complex fractions are summarized as follows.

> ## SIMPLIFYING COMPLEX FRACTIONS—METHOD II
>
> To simplify a complex fraction, do the following.
>
> **STEP 1:** Find the LCD of all fractions within the complex fraction.
>
> **STEP 2:** Multiply both the numerator and the denominator of the complex fraction by the LCD found in Step 1. Simplify the result.

READING CHECK

• Why isn't the value of a complex fraction changed when we multiply its numerator and denominator by the LCD?

NOTE: Multiplying both the numerator and denominator of a complex fraction by the same nonzero number (the LCD) is equivalent to multiplying the complex fraction by 1. Doing so will *not change the value* of the complex fraction.

EXAMPLE 3 **Simplifying complex fractions using Method II**

Simplify each complex fraction.

(a) $\dfrac{\dfrac{5}{8} - \dfrac{1}{2}}{\dfrac{1}{4} + \dfrac{2}{3}}$ (b) $\dfrac{\dfrac{5x}{4}}{\dfrac{1}{2} + 7}$

Solution

(a) **STEP 1:** The LCD for the fractions $\frac{5}{8}, \frac{1}{2}, \frac{1}{4}$, and $\frac{2}{3}$ is 24.

STEP 2: Use the distributive property to multiply the numerator and the denominator of the complex fraction by 24. Note that 24 is written as $\frac{24}{1}$.

$$\frac{\dfrac{5}{8} - \dfrac{1}{2}}{\dfrac{1}{4} + \dfrac{2}{3}} = \frac{24 \cdot \left(\dfrac{5}{8} - \dfrac{1}{2}\right)}{24 \cdot \left(\dfrac{1}{4} + \dfrac{2}{3}\right)} = \frac{\dfrac{24}{1} \cdot \dfrac{5}{8} - \dfrac{24}{1} \cdot \dfrac{1}{2}}{\dfrac{24}{1} \cdot \dfrac{1}{4} + \dfrac{24}{1} \cdot \dfrac{2}{3}}$$

Because 24 is the LCD for all fractions involved, it is divisible by each of the denominators. Dividing each denominator into 24 eliminates the fractions from within the complex fraction. The simplification continues as follows.

$$\frac{\dfrac{24}{1} \cdot \dfrac{5}{8} - \dfrac{24}{1} \cdot \dfrac{1}{2}}{\dfrac{24}{1} \cdot \dfrac{1}{4} + \dfrac{24}{1} \cdot \dfrac{2}{3}} = \frac{3 \cdot 5 - 12 \cdot 1}{6 \cdot 1 + 8 \cdot 2} = \frac{15 - 12}{6 + 16} = \frac{3}{22}$$

(b) **STEP 1:** The LCD for the fractions $\frac{5x}{4}$ and $\frac{1}{2}$ is 4.

STEP 2: Multiply the numerator and the denominator of the complex fraction by 4.

$$\frac{\dfrac{5x}{4}}{\dfrac{1}{2} + 7} = \frac{4 \cdot \dfrac{5x}{4}}{4 \cdot \left(\dfrac{1}{2} + 7\right)} = \frac{\dfrac{4}{1} \cdot \dfrac{5x}{4}}{\dfrac{4}{1} \cdot \dfrac{1}{2} + 4 \cdot 7} = \frac{1 \cdot 5x}{2 + 28} = \frac{5x}{30} = \frac{x}{6}$$

Note that the result $\frac{5x}{30}$ was simplified to $\frac{x}{6}$.

Now Try Exercises 29, 37

Order of Operations

To simplify expressions containing fractions, we use the order of operations agreement.

> ### ORDER OF OPERATIONS
>
> 1. Do all calculations within grouping symbols such as parentheses and radicals, or above and below a fraction bar.
> 2. Evaluate all exponential expressions.
> 3. Do all multiplication and division from *left to right*.
> 4. Do all addition and subtraction from *left to right*.

EXAMPLE 4   **Using the order of operations to simplify expressions with fractions**

Simplify each expression.

(a) $\dfrac{1}{6} + \dfrac{3}{4} \cdot \dfrac{1}{2}$ (b) $\left(\dfrac{5}{6} + \dfrac{1}{4}\right)\left(\dfrac{2}{5} - \dfrac{1}{2}\right)$ (c) $\left(\dfrac{6}{5}\right)^2 \div \left(\dfrac{16}{25} + \dfrac{4}{5}\right)$

Solution

(a) Multiply $\frac{3}{4}$ and $\frac{1}{2}$ before adding.

$$\dfrac{1}{6} + \dfrac{3}{4} \cdot \dfrac{1}{2} = \dfrac{1}{6} + \dfrac{3}{8} \qquad \text{Multiply.}$$

$$= \dfrac{1}{6} \cdot \dfrac{4}{4} + \dfrac{3}{8} \cdot \dfrac{3}{3} \qquad \text{The LCD is 24.}$$

$$= \dfrac{4}{24} + \dfrac{9}{24} \qquad \text{Multiply.}$$

$$= \dfrac{13}{24} \qquad \text{Add.}$$

(b) Complete the arithmetic within parentheses before multiplying.

$$\left(\dfrac{5}{6} + \dfrac{1}{4}\right)\left(\dfrac{2}{5} - \dfrac{1}{2}\right) = \left(\dfrac{5}{6} \cdot \dfrac{2}{2} + \dfrac{1}{4} \cdot \dfrac{3}{3}\right)\left(\dfrac{2}{5} \cdot \dfrac{2}{2} - \dfrac{1}{2} \cdot \dfrac{5}{5}\right) \qquad \text{The LCDs are 12 and 10.}$$

$$= \left(\dfrac{10}{12} + \dfrac{3}{12}\right)\left(\dfrac{4}{10} - \dfrac{5}{10}\right) \qquad \text{Compute four products.}$$

$$= \left(\dfrac{13}{12}\right)\left(-\dfrac{1}{10}\right) \qquad \text{Add; subtract.}$$

$$= -\dfrac{13}{120} \qquad \text{Multiply.}$$

(c) Square $\frac{6}{5}$ and then add within parentheses before performing the division.

$$\left(\dfrac{6}{5}\right)^2 \div \left(\dfrac{16}{25} + \dfrac{4}{5}\right) = \dfrac{36}{25} \div \left(\dfrac{16}{25} + \dfrac{4}{5}\right) \qquad \text{Square } \tfrac{6}{5}.$$

$$= \dfrac{36}{25} \div \left(\dfrac{16}{25} + \dfrac{4}{5} \cdot \dfrac{5}{5}\right) \qquad \text{The LCD is 25.}$$

$$= \dfrac{36}{25} \div \left(\dfrac{16}{25} + \dfrac{20}{25}\right) \qquad \text{Multiply.}$$

$$= \dfrac{36}{25} \div \dfrac{36}{25} \qquad \text{Add.}$$

$$= 1 \qquad a \div a = 1 \text{ when } a \neq 0.$$

Now Try Exercises 41, 45, 51

4.7   Putting It All Together

CONCEPT	COMMENTS	EXAMPLES
Complex Fraction	A complex fraction is a rational expression with fractions in its numerator, denominator, or both.	$\dfrac{3 + \dfrac{5}{6}}{\dfrac{1}{2} - \dfrac{4}{9}}$ and $\dfrac{\dfrac{3x}{5}}{\dfrac{2}{3} - 5}$
Simplifying Basic Complex Fractions	If the variables a, b, c, and d represent numbers with $b \neq 0$, $c \neq 0$, and $d \neq 0$, then $$\dfrac{\dfrac{a}{b}}{\dfrac{c}{d}} = \dfrac{a}{b} \cdot \dfrac{d}{c}.$$	$\dfrac{\dfrac{4}{5}}{\dfrac{5}{6}} = \dfrac{4}{5} \cdot \dfrac{6}{5} = \dfrac{24}{25}$
Simplifying Complex Fractions—Method I	**STEP 1:** Write the numerator as a single fraction; do the same to the denominator. **STEP 2:** Multiply the fraction in the numerator by the reciprocal of the fraction in the denominator. Simplify the result.	$\dfrac{\dfrac{4}{9} + \dfrac{1}{9}}{\dfrac{4}{x} - \dfrac{2}{x}} = \dfrac{\dfrac{5}{9}}{\dfrac{2}{x}} = \dfrac{5}{9} \cdot \dfrac{x}{2} = \dfrac{5x}{18}$
Simplifying Complex Fractions—Method II	**STEP 1:** Find the LCD of all fractions within the complex fraction. **STEP 2:** Multiply both the numerator and the denominator of the complex fraction by the LCD from Step 1. Simplify the result.	$\dfrac{\dfrac{1}{6} + \dfrac{2}{3}}{\dfrac{1}{2} + \dfrac{5}{6}} = \dfrac{6 \cdot \left(\dfrac{1}{6} + \dfrac{2}{3} \right)}{6 \cdot \left(\dfrac{1}{2} + \dfrac{5}{6} \right)} = \dfrac{5}{8}$

4.7   Exercises

MyMathLab   Math XL PRACTICE   WATCH   DOWNLOAD   READ   REVIEW

CONCEPTS AND VOCABULARY

1. A(n) _____ fraction is a rational expression with fractions in its numerator, denominator, or both.

2. One strategy for simplifying a complex fraction with a single fraction in the numerator and a single fraction in the denominator is to multiply the fraction in the numerator by the _____ of the fraction in the denominator.

3. If a complex fraction is being simplified by using Method _____, the first step is to write the numerator as a single fraction and write the denominator as a single fraction.

4. To simplify a complex fraction using Method _____, first find the LCD of all fractions within the complex fraction.

SIMPLIFYING COMPLEX FRACTIONS

Exercises 5–16: Simplify the given complex fraction.

5. $\dfrac{\dfrac{2}{7}}{\dfrac{4}{5}}$

6. $\dfrac{\dfrac{9}{10}}{\dfrac{3}{5}}$

7. $\dfrac{\dfrac{6}{2}}{\dfrac{3}{}}$

8. $\dfrac{\dfrac{8}{4}}{\dfrac{4}{5}}$

9. $\dfrac{\dfrac{12}{5}}{\dfrac{3}{4}}$

10. $\dfrac{\dfrac{18}{25}}{\dfrac{4}{5}}$

11. $\dfrac{\dfrac{3}{4}}{9}$

12. $\dfrac{\dfrac{4}{9}}{8}$

13. $\dfrac{\dfrac{5}{4x}}{\dfrac{y}{10}}$

14. $\dfrac{\dfrac{x}{7}}{\dfrac{x}{14}}$

15. $\dfrac{\dfrac{16}{5x}}{\dfrac{4}{15x}}$

16. $\dfrac{\dfrac{w}{18}}{\dfrac{2}{3w}}$

Exercises 17–28: Simplify the given complex fraction using Method I.

17. $\dfrac{\dfrac{2}{3}+\dfrac{5}{6}}{\dfrac{2}{5}+\dfrac{1}{2}}$

18. $\dfrac{\dfrac{3}{7}-\dfrac{1}{3}}{\dfrac{7}{8}-\dfrac{3}{4}}$

19. $\dfrac{\dfrac{2}{5}+3}{5-\dfrac{3}{4}}$

20. $\dfrac{6-\dfrac{3}{8}}{\dfrac{3}{4}}$

21. $\dfrac{23}{4+\dfrac{3}{5}}$

22. $\dfrac{7-\dfrac{3}{5}}{16}$

23. $\dfrac{\dfrac{5}{3}-\dfrac{1}{4}}{\dfrac{9}{10}+\dfrac{14}{15}}$

24. $\dfrac{\dfrac{9}{5}+\dfrac{1}{10}}{\dfrac{18}{5}-\dfrac{3}{4}}$

25. $\dfrac{\dfrac{10x}{7}}{3-\dfrac{9}{8}}$

26. $\dfrac{\dfrac{x}{5}+\dfrac{2x}{3}}{\dfrac{13}{5}}$

27. $\dfrac{\dfrac{11x}{3}+x}{\dfrac{2y}{9}}$

28. $\dfrac{\dfrac{7b}{12}+\dfrac{b}{6}}{b}$

Exercises 29–40: Simplify the given complex fraction using Method II.

29. $\dfrac{\dfrac{5}{12}-\dfrac{1}{6}}{\dfrac{7}{4}-\dfrac{1}{2}}$

30. $\dfrac{\dfrac{2}{3}+\dfrac{1}{8}}{\dfrac{3}{4}+\dfrac{5}{6}}$

31. $\dfrac{7-\dfrac{3}{2}}{4}$

32. $\dfrac{26}{3+\dfrac{7}{2}}$

33. $\dfrac{\dfrac{7}{10}-\dfrac{1}{6}}{\dfrac{9}{5}+\dfrac{1}{15}}$

34. $\dfrac{\dfrac{2}{9}+\dfrac{5}{6}}{\dfrac{13}{12}-\dfrac{3}{4}}$

35. $\dfrac{2-\dfrac{5}{6}}{\dfrac{21}{8}}$

36. $\dfrac{\dfrac{1}{2}+6}{3-\dfrac{2}{5}}$

37. $\dfrac{\dfrac{13x}{4}}{\dfrac{4}{3}+3}$

38. $\dfrac{\dfrac{11x}{4}-\dfrac{2x}{3}}{\dfrac{25}{6}}$

39. $\dfrac{\dfrac{7a}{5}+a}{\dfrac{3a}{10}}$

40. $\dfrac{\dfrac{5w}{9}+\dfrac{5w}{6}}{\dfrac{10}{3}}$

ORDER OF OPERATIONS

Exercises 41–56: Use the order of operations agreement to simplify the given expression.

41. $\dfrac{13}{6}+\dfrac{2}{3}\cdot\dfrac{5}{4}$

42. $\dfrac{11}{8}-\dfrac{1}{3}\cdot\dfrac{3}{2}$

43. $\dfrac{7}{10}\div\dfrac{3}{5}\cdot\dfrac{9}{14}$

44. $\dfrac{3}{8}\cdot\dfrac{4}{9}\div\dfrac{5}{12}$

45. $\left(\dfrac{3}{4} + \dfrac{2}{3}\right)\left(\dfrac{1}{2} - \dfrac{7}{8}\right)$ **46.** $\left(\dfrac{1}{2} + \dfrac{1}{3}\right)\left(\dfrac{2}{5} - \dfrac{1}{4}\right)$

47. $\dfrac{1}{4^2} \div \left(\dfrac{3}{8} + \dfrac{1}{4}\right)$ **48.** $3^3 \cdot \left(\dfrac{5}{6} - \dfrac{7}{9}\right)$

49. $5 - 10 \div \sqrt{\dfrac{25}{64}}$ **50.** $\left(\dfrac{7}{8} - \dfrac{3}{4}\right)^2 + \sqrt{\dfrac{1}{4}}$

51. $\left(\dfrac{3}{2}\right)^2 \cdot \left(\dfrac{7}{12} - \dfrac{5}{6}\right)$ **52.** $\left(\dfrac{3}{8} - \dfrac{5}{9}\right) \cdot 12$

53. $\dfrac{17}{18} - \dfrac{2}{3} \cdot \left(\dfrac{5}{6} + \dfrac{1}{3}\right)$ **54.** $\dfrac{3}{8} - 3 \cdot \left(\dfrac{1}{6} - \dfrac{2}{9}\right)$

55. $\dfrac{8}{15} \div \dfrac{4}{5} + \dfrac{5}{6} \cdot \dfrac{9}{25}$ **56.** $\dfrac{7}{9} - \dfrac{1}{4} \div \dfrac{3}{8} + \dfrac{1}{3}$

APPLICATIONS

57. *Walking Rate* Use the formula

$$\text{Rate} = \dfrac{\text{Distance}}{\text{Time}}$$

to find the rate for a person who walks $2\frac{1}{2}$ miles in $\frac{4}{5}$ hour. Express your answer as a mixed number.

58. *Flying Distance* A fighter jet is flying at a speed of $\frac{9}{20}$ miles per second. Use the formula

$$\text{Time} = \dfrac{\text{Distance}}{\text{Rate}}$$

to find the time in seconds that it takes for the jet to fly 54 miles.

59. *Average Time* A student completes his math homework in $\frac{5}{6}$ hour and his science homework in $\frac{2}{3}$ hour. Find the average time for completing these two homework assignments.

60. *Serving Size* If a sugar bowl contains $8\frac{1}{2}$ cups of sugar, how many $\frac{1}{4}$-cup scoops of sugar can be taken from the bowl?

61. *Converting Temperature* To convert a temperature F given in degrees Fahrenheit to an equivalent temperature C in degrees Celsius, use the formula

$$C = \dfrac{5}{9}(F - 32).$$

Find the Celsius temperature that is equivalent to a temperature of $82\frac{2}{5}°$F.

62. *Converting Temperature* To convert a temperature C given in degrees Celsius to an equivalent temperature F in degrees Fahrenheit, use the formula

$$F = \dfrac{9}{5}C + 32.$$

Find the Fahrenheit temperature that is equivalent to a temperature of $18\frac{1}{3}°$C.

WRITING ABOUT MATHEMATICS

63. Write a paragraph explaining the steps involved in simplifying a complex fraction using Method I.

64. Write a paragraph explaining the steps involved in simplifying a complex fraction using Method II.

4.8 Solving Equations Involving Fractions

Solving Equations Algebraically • Solving Equations Numerically • Solving Equations Visually • Applications

A LOOK INTO MATH ▶

An adult can often perform physical tasks faster than a child. For example, it may take an adult 3 hours to rake leaves in a yard, while the same task may take a child 6 hours to complete. Working together, the adult and child should be able to rake the lawn in less time than it takes the faster person working alone. So, it should take less than 3 hours. To find the time required for two people to rake the lawn while working together, we use an equation involving fractions. In this section, we focus on solving equations involving fractions.

Solving Equations Algebraically

NEW VOCABULARY

☐ Trapezoid
☐ Base (of a trapezoid)

In Section 3.3, we studied the properties of equality that are used to solve linear equations. These properties are also valid for equations involving fractions. In this subsection, we will review the *addition property of equality* and the *multiplication property of equality* and then apply each property to equations involving fractions.

STUDY TIP

The equations presented in this section are solved algebraically, numerically, and visually. These are the same methods used to solve many equations throughout the course. Remember these methods when you are trying to solve equations.

READING CHECK

• How is the addition property of equality used when solving equations?

THE ADDITION PROPERTY OF EQUALITY Because subtraction can be rewritten as addition of the opposite, there are two ways to state the addition property of equality.

1. $a = b$ is equivalent to $a + c = b + c.$

2. $a = b$ is equivalent to $a - c = b - c.$

The requirements for the variables a, b, and c are simply that they represent numbers. Since fractions are numbers, the addition property of equality can be used to solve equations involving fractions, as demonstrated in the next example.

EXAMPLE 1   **Using the addition property of equality**

Solve each equation and check the solution.

(a) $x + \dfrac{3}{4} = \dfrac{5}{6}$ (b) $2 = y - \dfrac{8}{3}$

Solution

(a) Because $\frac{3}{4}$ is being **added** to x in the given equation, we can isolate x on the left side of the equation by **subtracting** $\frac{3}{4}$ from each side of the equation.

$$x + \frac{3}{4} = \frac{5}{6} \qquad \text{Given equation}$$

$$x + \frac{3}{4} - \frac{3}{4} = \frac{5}{6} - \frac{3}{4} \qquad \text{Subtract } \tfrac{3}{4} \text{ from each side.}$$

$$x = \frac{5}{6} \cdot \frac{2}{2} - \frac{3}{4} \cdot \frac{3}{3} \qquad \text{Simplify; the LCD is 12.}$$

$$x = \frac{10}{12} - \frac{9}{12} \qquad \text{Multiply.}$$

$$x = \frac{1}{12} \qquad \text{Subtract.}$$

To check this solution, replace x with $\frac{1}{12}$ in the *given* equation.

$$x + \frac{3}{4} = \frac{5}{6} \qquad \text{Given equation}$$

$$\frac{1}{12} + \frac{3}{4} \stackrel{?}{=} \frac{5}{6} \qquad \text{Replace } x \text{ with } \tfrac{1}{12}.$$

$$\frac{1}{12} + \frac{3}{4} \cdot \frac{3}{3} \stackrel{?}{=} \frac{5}{6} \qquad \text{The LCD is 12.}$$

$$\frac{1}{12} + \frac{9}{12} \stackrel{?}{=} \frac{5}{6} \qquad \text{Multiply.}$$

$$\frac{10}{12} \stackrel{?}{=} \frac{5}{6} \qquad \text{Add.}$$

$$\frac{5}{6} = \frac{5}{6} \;\checkmark \qquad \text{Simplify; it checks.}$$

(b) Because $\frac{8}{3}$ is being **subtracted** from y in the given equation, we can isolate y on the right side of the equation by **adding** $\frac{8}{3}$ to each side of the equation.

$$2 = y - \frac{8}{3} \qquad \text{Given equation}$$

$$2 + \frac{8}{3} = y - \frac{8}{3} + \frac{8}{3} \qquad \text{Add } \tfrac{8}{3} \text{ to each side.}$$

$$\frac{2}{1} \cdot \frac{3}{3} + \frac{8}{3} = y \qquad \text{The LCD is 3; simplify.}$$

$$\frac{6}{3} + \frac{8}{3} = y \qquad \text{Multiply.}$$

$$\frac{14}{3} = y \qquad \text{Add.}$$

To check this solution, replace y with $\frac{14}{3}$ in the *given* equation.

$$2 = y - \frac{8}{3} \qquad \text{Given equation}$$

$$2 \stackrel{?}{=} \frac{14}{3} - \frac{8}{3} \qquad \text{Replace } y \text{ with } \tfrac{14}{3}.$$

$$2 \stackrel{?}{=} \frac{6}{3} \qquad \text{Subtract.}$$

$$2 = 2 \checkmark \qquad \text{Simplify; it checks.}$$

■ **Now Try Exercises 5, 9**

THE MULTIPLICATION PROPERTY OF EQUALITY There are two ways to state the multiplication property of equality because division (if defined) can be rewritten as multiplication by the reciprocal.

1. $a = b$ is equivalent to $ac = bc$, where $c \neq 0$.

2. $a = b$ is equivalent to $\dfrac{a}{c} = \dfrac{b}{c}$, where $c \neq 0$.

READING CHECK

• How is the multiplication property of equality used when solving equations?

The variables a, b, and c must represent numbers with $c \neq 0$. Since fractions are numbers, the multiplication property of equality can be used to solve equations involving fractions, as illustrated in the next example.

EXAMPLE 2   **Using the multiplication property of equality**

Solve each equation and check the solution.

(a) $\dfrac{2}{3}x = 8$ **(b)** $\dfrac{w}{4} = \dfrac{3}{10}$ **(c)** $-2y = \dfrac{8}{9}$

Solution

(a) Because x is **multiplied** by $\frac{2}{3}$ in the given equation, we can isolate x on the left side of the equation by **dividing** each side of the equation by $\frac{2}{3}$. Recall that dividing by a fraction is the same as multiplying by its reciprocal. That is, to isolate x we multiply each side of the equation by the reciprocal of $\frac{2}{3}$, which is $\frac{3}{2}$.

$$\frac{2}{3}x = 8 \qquad \text{Given equation}$$

$$\frac{3}{2} \cdot \frac{2}{3}x = \frac{3}{2} \cdot \frac{8}{1} \qquad \text{Multiply each side by } \tfrac{3}{2}.$$

$$1x = \frac{3 \cdot 8}{2} \qquad \text{Multiply.}$$

$$x = 12 \qquad \text{Simplify.}$$

To check this solution, replace x with 12 in the *given* equation.

$$\frac{2}{3}x = 8 \qquad \text{Given equation}$$

$$\frac{2}{3} \cdot 12 \overset{?}{=} 8 \qquad \text{Replace } x \text{ with 12.}$$

$$\frac{2}{3} \cdot \frac{12}{1} \overset{?}{=} 8 \qquad \text{Write 12 as } \frac{12}{1}.$$

$$8 = 8 \checkmark \qquad \text{Simplify; it checks.}$$

(b) Because w is **divided** by 4 in the given equation, we can isolate w on the left side of the equation by **multiplying** each side of the equation by 4, written as $\frac{4}{1}$.

$$\frac{w}{4} = \frac{3}{10} \qquad \text{Given equation}$$

$$\frac{4}{1} \cdot \frac{w}{4} = \frac{4}{1} \cdot \frac{3}{10} \qquad \text{Multiply each side by } \frac{4}{1}.$$

$$1w = \frac{4 \cdot 3}{1 \cdot 10} \qquad \text{Multiply.}$$

$$w = \frac{6}{5} \qquad \text{Simplify.}$$

To check this solution, replace w with $\frac{6}{5}$ in the *given* equation.

$$\frac{w}{4} = \frac{3}{10} \qquad \text{Given equation}$$

$$\frac{\frac{6}{5}}{4} \overset{?}{=} \frac{3}{10} \qquad \text{Replace } w \text{ with } \frac{6}{5}.$$

$$\frac{6}{5} \cdot \frac{1}{4} \overset{?}{=} \frac{3}{10} \qquad \text{Invert and multiply.}$$

$$\frac{6 \cdot 1}{5 \cdot 4} \overset{?}{=} \frac{3}{10} \qquad \text{Multiply.}$$

$$\frac{3}{10} = \frac{3}{10} \checkmark \qquad \text{Simplify; it checks.}$$

(c) Because y is **multiplied** by -2 in the given equation, we can isolate y on the left side of the equation by **dividing** each side of the equation by -2. However, rather than dividing the fraction $\frac{8}{9}$ by -2, we will multiply each side by the reciprocal of -2, or $-\frac{1}{2}$.

$$-2y = \frac{8}{9} \qquad \text{Given equation}$$

$$-\frac{1}{2} \cdot (-2y) = -\frac{1}{2} \cdot \frac{8}{9} \qquad \text{Multiply each side by } -\frac{1}{2}.$$

$$1y = -\frac{1 \cdot 8}{2 \cdot 9} \qquad \text{Multiply.}$$

$$y = -\frac{4}{9} \qquad \text{Simplify.}$$

To check this solution, replace y with $-\frac{4}{9}$ in the *given* equation.

$$-2y = \frac{8}{9} \qquad \text{Given equation}$$

$$-2 \cdot \left(-\frac{4}{9}\right) \stackrel{?}{=} \frac{8}{9} \qquad \text{Replace } y \text{ with } -\frac{4}{9}.$$

$$\left(-\frac{2}{1}\right) \cdot \left(-\frac{4}{9}\right) \stackrel{?}{=} \frac{8}{9} \qquad \text{Write } -2 \text{ as } -\frac{2}{1}.$$

$$\frac{2 \cdot 4}{1 \cdot 9} \stackrel{?}{=} \frac{8}{9} \qquad \text{Multiply.}$$

$$\frac{8}{9} = \frac{8}{9} \quad \checkmark \qquad \text{Simplify; it checks.}$$

Now Try Exercises 13, 17, 25

READING CHECK

• How is the LCD used to clear fractions from an equation?

CLEARING FRACTIONS One of the most efficient ways to solve equations involving fractions is to multiply each side of the equation by the LCD of all fractions within the equation. Doing so "clears fractions" from the equation and produces an equivalent equation.

EXAMPLE 3 **Solving equations by clearing fractions**

Solve each equation.

(a) $\dfrac{5x}{8} = \dfrac{15}{4}$ (b) $y - \dfrac{1}{6} = \dfrac{4}{9}$

Solution

(a) Multiply each side of the equation by 8, the LCD of all fractions involved. We write 8 in the form $\frac{8}{1}$ in order to simplify computation.

$$\frac{5x}{8} = \frac{15}{4} \qquad \text{Given equation}$$

$$\frac{8}{1} \cdot \frac{5x}{8} = \frac{15}{4} \cdot \frac{8}{1} \qquad \text{Multiply each side by 8.}$$

$$1 \cdot 5x = 15 \cdot 2 \qquad \text{Simplify each side.}$$

$$5x = 30 \qquad \text{Multiply.}$$

$$\frac{5x}{5} = \frac{30}{5} \qquad \text{Divide each side by 5.}$$

$$x = 6 \qquad \text{Simplify.}$$

(b) Multiply each side of the equation by 18, which is the LCD. Multiplying the left side of the equation by 18 requires the use of the distributive property.

$$y - \frac{1}{6} = \frac{4}{9} \qquad \text{Given equation}$$

$$18 \cdot \left(y - \frac{1}{6}\right) = 18 \cdot \frac{4}{9} \qquad \text{Multiply each side by 18.}$$

$$18y - \frac{18}{1} \cdot \frac{1}{6} = \frac{18}{1} \cdot \frac{4}{9} \qquad \text{Distributive property}$$

$$18y - 3 = 2 \cdot 4 \qquad \text{Simplify each side.}$$

$$18y - 3 = 8 \qquad \text{Multiply.}$$

$$18y - 3 + 3 = 8 + 3 \qquad \text{Add 3 to each side.}$$

$$18y = 11 \qquad \text{Simplify.}$$

$$\frac{18y}{18} = \frac{11}{18} \qquad \text{Divide each side by 18.}$$

$$y = \frac{11}{18} \qquad \text{Simplify.}$$

Now Try Exercises 27, 29

The process of multiplying each side of an equation by a number (the LCD) is equivalent to multiplying *each term* in the equation by that number. With this in mind, the following steps can be used to solve equations involving fractions.

SOLVING EQUATIONS INVOLVING FRACTIONS ALGEBRAICALLY

To solve an equation involving fractions algebraically, do the following.

STEP 1: Use the distributive property to clear parentheses from the equation.

STEP 2: Multiply each term in the equation by the LCD of all fractions involved.

STEP 3: Simplify each term in the equation and combine any like terms.

STEP 4: Solve the resulting equation.

EXAMPLE 4 Solving equations involving fractions algebraically

Solve each equation.

(a) $\dfrac{5}{9}x + \dfrac{1}{3} = \dfrac{5}{6}$ (b) $\dfrac{x}{4} - \dfrac{x}{6} = \dfrac{2}{3}$ (c) $\dfrac{3}{5}\left(2x - \dfrac{1}{2}\right) = \dfrac{7}{10}$

Solution

(a) Because there are no parentheses in the equation, we start with Step 2. Multiply each term in the equation by 18, the LCD of all fractions involved.

$$\frac{5}{9}x + \frac{1}{3} = \frac{5}{6} \qquad \text{Given equation}$$

$$\frac{18}{1} \cdot \frac{5}{9}x + \frac{18}{1} \cdot \frac{1}{3} = \frac{18}{1} \cdot \frac{5}{6} \qquad \text{Multiply each term by 18. (Step 2)}$$

$$10x + 6 = 15 \qquad \text{Simplify each term. (Step 3)}$$

$$10x + 6 - 6 = 15 - 6 \qquad \text{Subtract 6 from each side. (Step 4)}$$

$$10x = 9 \qquad \text{Simplify.}$$

$$\frac{10x}{10} = \frac{9}{10} \qquad \text{Divide each side by 10. (Step 4)}$$

$$x = \frac{9}{10} \qquad \text{Simplify.}$$

(b) Begin with Step 2. Multiply each term in the equation by 12, the LCD.

$$\frac{x}{4} - \frac{x}{6} = \frac{2}{3} \qquad \text{Given equation}$$

$$\frac{12}{1} \cdot \frac{x}{4} - \frac{12}{1} \cdot \frac{x}{6} = \frac{12}{1} \cdot \frac{2}{3} \qquad \text{Multiply each term by 12. (Step 2)}$$

$$3x - 2x = 8 \qquad \text{Simplify each term.}$$

$$x = 8 \qquad \text{Combine like terms. (Steps 3 and 4)}$$

(c) First use the distributive property to clear the parentheses from the equation.

$$\frac{3}{5}\left(2x - \frac{1}{2}\right) = \frac{7}{10}$$ Given equation

$$\frac{3}{5} \cdot \frac{2}{1}x - \frac{3}{5} \cdot \frac{1}{2} = \frac{7}{10}$$ Distributive property (Step 1)

$$\frac{6}{5}x - \frac{3}{10} = \frac{7}{10}$$ Multiply.

$$\frac{10}{1} \cdot \frac{6}{5}x - \frac{10}{1} \cdot \frac{3}{10} = \frac{10}{1} \cdot \frac{7}{10}$$ Multiply each term by 10. (Step 2)

$$12x - 3 = 7$$ Simplify each term. (Step 3)

$$12x - 3 + 3 = 7 + 3$$ Add 3 to each side. (Step 4)

$$12x = 10$$ Simplify.

$$\frac{12x}{12} = \frac{10}{12}$$ Divide each side by 12. (Step 4)

$$x = \frac{5}{6}$$ Simplify.

Now Try Exercises 37, 41, 47

MAKING CONNECTIONS

Clearing Fractions and the Properties of Equality

There are *many correct* ways to solve an equation. For example, the solution to the equation in Example 4(a) can be found without first clearing fractions.

$$\frac{5}{9}x + \frac{1}{3} = \frac{5}{6}$$ Given equation

$$\frac{5}{9}x + \frac{1}{3} - \frac{1}{3} = \frac{5}{6} - \frac{1}{3}$$ Subtract $\frac{1}{3}$ from each side.

$$\frac{5}{9}x = \frac{5}{6} - \frac{1}{3} \cdot \frac{2}{2}$$ The LCD is 6.

$$\frac{5}{9}x = \frac{1}{2}$$ Subtract; simplify.

$$\frac{9}{5} \cdot \frac{5}{9}x = \frac{9}{5} \cdot \frac{1}{2}$$ Multiply each side by $\frac{9}{5}$.

$$x = \frac{9}{10}$$ Simplify.

Solving Equations Numerically

Equations involving fractions can also be solved numerically. In the next example, a table of values is used to solve an equation involving fractions.

EXAMPLE 5 | **Solving an equation numerically**

Complete Table 4.2 for the given values of x and then solve the equation $\frac{3}{4}x - 2 = -\frac{1}{2}$.

TABLE 4.2

x	-1	0	1	2	3	4
$\frac{3}{4}x - 2$						

Solution

Begin by replacing x in $\frac{3}{4}x - 2$ with -1.

$$\frac{3}{4}(-1) - 2 = -\frac{3}{4} - \frac{8}{4} = \frac{-3 - 8}{4} = -\frac{11}{4}$$

We write the result $-\frac{11}{4}$ below -1, as shown in Table 4.3. The remaining values in Table 4.3 are found as follows:

$$\frac{3}{4}(0) - 2 = 0 - 2 = -2 \qquad\qquad \frac{3}{4}(1) - 2 = \frac{3}{4} - \frac{8}{4} = \frac{3 - 8}{4} = -\frac{5}{4}$$

$$\frac{3}{4}(2) - 2 = \frac{3}{2} - \frac{4}{2} = \frac{3 - 4}{2} = -\frac{1}{2} \qquad \frac{3}{4}(3) - 2 = \frac{9}{4} - \frac{8}{4} = \frac{9 - 8}{4} = \frac{1}{4}$$

$$\frac{3}{4}(4) - 2 = 3 - 2 = 1$$

TABLE 4.3

x	-1	0	1	2	3	4
$\frac{3}{4}x - 2$	$-\frac{11}{4}$	-2	$-\frac{5}{4}$	$-\frac{1}{2}$	$\frac{1}{4}$	1

Table 4.3 reveals that the left side of the given equation equals $-\frac{1}{2}$ when $x = 2$. Therefore, the solution is **2**.

Now Try Exercise 51

READING CHECK

• How does a solution that is found visually compare to a solution found algebraically or numerically?

Solving Equations Visually

When a graph is provided for an equation involving fractions, an estimate of the solution can be found visually. As we have seen earlier, solutions found visually must be checked. In the next example, we estimate and check a solution to an equation involving fractions.

EXAMPLE 6 | **Solving an equation visually**

A graph of the expression $\frac{3}{2}x + 1$ is shown in Figure 4.13. Use the graph to solve the equation $\frac{3}{2}x + 1 = 4$ visually.

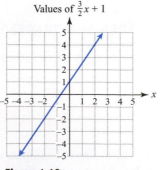

Values of $\frac{3}{2}x + 1$

Figure 4.13

Solution

Locate 4 on the vertical axis and move **horizontally** to the right to the graphed line. From this position, move **vertically** downward to the horizontal axis, as shown in Figure 4.14. The solution appears to be 2.

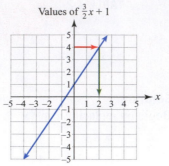

Values of $\frac{3}{2}x + 1$

Figure 4.14

To see if 2 is the correct solution, check it in the given equation as $\frac{2}{1}$.

$$\frac{3}{2}x + 1 = 4 \qquad \text{Given equation}$$

$$\frac{3}{2}\left(\frac{2}{1}\right) + 1 \overset{?}{=} 4 \qquad \text{Replace } x \text{ with } \frac{2}{1}.$$

$$3 + 1 \overset{?}{=} 4 \qquad \text{Simplify.}$$

$$4 = 4 \checkmark \quad \text{The solution checks.}$$

The solution to the equation $\frac{3}{2}x + 1 = 4$ is 2.

Now Try Exercise 55

Applications

A **trapezoid** is a four-sided geometric shape with one pair of parallel sides. Each parallel side is called a **base** of the trapezoid. An example of a trapezoid is shown in Figure 4.15.

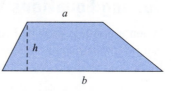

Figure 4.15 Trapezoid with Height *h* and Bases *a* and *b*

To compute the area of a trapezoid, we use the following formula.

AREA OF A TRAPEZOID

The area *A* of a trapezoid with height *h* and bases *a* and *b* is given by

$$A = \frac{1}{2}(a + b)h.$$

In the next example, we use the trapezoid area formula to set up an equation whose solution is an unknown measure for a given trapezoid.

EXAMPLE 7   **Finding an unknown measure for a trapezoid**

Find the length of base a for the trapezoid in the following figure if its area is 4 square feet.

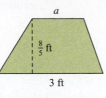

a

$\frac{8}{5}$ ft

3 ft

Solution
Let $A = 4$, $b = 3$, and $h = \frac{8}{5}$ in the trapezoid area formula: $A = \frac{1}{2}(a + b)h$.

$$4 = \frac{1}{2}(a + 3)\frac{8}{5} \qquad \text{Equation to solve}$$

$$4 = \frac{1}{2} \cdot \frac{8}{5}(a + 3) \qquad \text{Commutative property}$$

$$4 = \frac{4}{5}(a + 3) \qquad \text{Multiply.}$$

$$4 = \frac{4}{5}a + \frac{4}{5} \cdot \frac{3}{1} \qquad \text{Distributive property}$$

$$4 = \frac{4}{5}a + \frac{12}{5} \qquad \text{Multiply.}$$

$$5 \cdot 4 = \frac{5}{1} \cdot \frac{4}{5}a + \frac{5}{1} \cdot \frac{12}{5} \qquad \text{Multiply each term by 5.}$$

$$20 = 4a + 12 \qquad \text{Simplify each term.}$$

$$20 - 12 = 4a + 12 - 12 \qquad \text{Subtract 12 from each side.}$$

$$8 = 4a \qquad \text{Simplify.}$$

$$\frac{8}{4} = \frac{4a}{4} \qquad \text{Divide each side by 4.}$$

$$2 = a \qquad \text{Simplify.}$$

The length of base a is 2 feet.

Now Try Exercise 73

▶ **REAL-WORLD CONNECTION** While U.S. temperatures are usually reported using the Fahrenheit temperature scale, many other countries throughout the world use the Celsius temperature scale. In the next example, an equation involving fractions is solved in order to convert a Fahrenheit temperature to its equivalent Celsius temperature.

EXAMPLE 8   **Finding a Celsius temperature**

The formula

$$F = \frac{9}{5}C + 32$$

gives the relationship between F degrees Fahrenheit and C degrees Celsius. Use the formula to find the Celsius temperature that is equivalent to $57\frac{1}{5}°F$.

Solution

Write $57\frac{1}{5}$ as the improper fraction $\frac{286}{5}$ and substitute it in the formula for F.

$$\frac{286}{5} = \frac{9}{5}C + 32 \qquad \text{Equation to solve}$$

$$\frac{5}{1} \cdot \frac{286}{5} = \frac{5}{1} \cdot \frac{9}{5}C + 5 \cdot 32 \qquad \text{Multiply each term by 5.}$$

$$286 = 9C + 160 \qquad \text{Simplify each term.}$$

$$286 - 160 = 9C + 160 - 160 \qquad \text{Subtract 160 from each side.}$$

$$126 = 9C \qquad \text{Simplify.}$$

$$\frac{126}{9} = \frac{9C}{9} \qquad \text{Divide each side by 9.}$$

$$14 = C \qquad \text{Simplify.}$$

Therefore $57\frac{1}{5}°F$ is equivalent to $14°C$.

▌ **Now Try Exercise 77**

EXAMPLE 9   **Finding time to complete a task when working together**

If an adult can rake a lawn in 3 hours and the same task takes a child 6 hours to complete, how long will it take for these two people working together to rake the lawn?

Solution

The adult can rake the entire lawn in 3 hours, which means that the adult rakes $\frac{1}{3}$ of the lawn per hour. By the same reasoning, the child rakes $\frac{1}{6}$ of the lawn per hour. If we let x represent the number of hours they work together, then the adult will rake $\frac{1}{3}x$ of the lawn in x hours, and the child will rake $\frac{1}{6}x$ of the lawn in x hours. Together, they will rake

$$\frac{1}{3}x + \frac{1}{6}x$$

of the lawn in x hours. The job is complete when the fraction of the lawn that is raked reaches 1 (one entire lawn is raked). To find out how long this will take the two people working together, solve the equation $\frac{1}{3}x + \frac{1}{6}x = 1$.

$$\frac{1}{3}x + \frac{1}{6}x = 1 \qquad \text{Equation to solve}$$

$$\frac{6}{1} \cdot \frac{1}{3}x + \frac{6}{1} \cdot \frac{1}{6}x = 6 \cdot 1 \qquad \text{Multiply each term by 6.}$$

$$2x + x = 6 \qquad \text{Simplify each term.}$$

$$3x = 6 \qquad \text{Combine like terms.}$$

$$\frac{3x}{3} = \frac{6}{3} \qquad \text{Divide each side by 3.}$$

$$x = 2 \qquad \text{Simplify.}$$

Together, they can rake the lawn in 2 hours.

▌ **Now Try Exercise 79**

4.8   Putting It All Together

CONCEPT	COMMENTS	EXAMPLES
Properties of Equality	$a = b$ is equivalent to $a + c = b + c.$ $a = b$ is equivalent to $a - c = b - c.$ $a = b$ is equivalent to $ac = bc$ for $c \neq 0.$ $a = b$ is equivalent to $\dfrac{a}{c} = \dfrac{b}{c}$ for $c \neq 0.$	$$x - \frac{1}{5} = \frac{3}{5}$$ $$x - \frac{1}{5} + \frac{1}{5} = \frac{3}{5} + \frac{1}{5}$$ $$x = \frac{4}{5}$$
Solving Equations Algebraically	To solve equations algebraically, 1. Use the distributive property to clear parentheses from the equation. 2. Multiply each term in the equation by the LCD of all fractions involved. 3. Simplify each term in the equation and combine any like terms. 4. Solve the resulting equation.	$$4\left(x - \frac{5}{8}\right) = \frac{2}{3}$$ $$4x - \frac{5}{2} = \frac{2}{3} \quad \text{(Step 1)}$$ $$6 \cdot 4x - \frac{6}{1} \cdot \frac{5}{2} = \frac{6}{1} \cdot \frac{2}{3} \quad \text{(Step 2)}$$ $$24x - 15 = 4 \quad \text{(Step 3)}$$ $$24x = 19$$ $$x = \frac{19}{24} \quad \text{(Step 4)}$$
Solving Equations Numerically	To solve equations numerically, complete a table of values for various values of the variable and then select the solution from the table, if possible.	The solution to $\frac{1}{2}x - \frac{3}{2} = -1$ is **1**. <table><tr><td>x</td><td>−1</td><td>0</td><td>1</td></tr><tr><td>$\frac{1}{2}x - \frac{3}{2}$</td><td>−2</td><td>$-\frac{3}{2}$</td><td>**−1**</td></tr></table>
Solving Equations Visually	To solve equations visually, use a graph of the equation to estimate a solution and check the estimated solution in the given equation to be sure that it is correct.	The solution to $-\frac{2}{5}x - \frac{4}{5} = -2$ is 3. Values of $-\frac{2}{5}x - \frac{4}{5}$

4.8 Exercises

MyMathLab   Math XL PRACTICE   WATCH   DOWNLOAD   READ   REVIEW

CONCEPTS AND VOCABULARY

1. The _____ property of equality states that adding (subtracting) the same number to (from) each side of an equation gives an equivalent equation.

2. The _____ property of equality allows us to multiply or divide each side of an equation by the same nonzero number to obtain an equivalent equation.

3. The _____ property can be used to clear parentheses from an equation.

4. To clear fractions from an equation, multiply each term by the _____ of all fractions involved.

USING PROPERTIES OF EQUALITY

Exercises 5–26: Use the properties of equality to solve the given equation.

5. $x + \dfrac{1}{4} = \dfrac{9}{10}$

6. $m + \dfrac{5}{3} = \dfrac{7}{8}$

7. $w - \dfrac{5}{3} = -\dfrac{9}{5}$

8. $y - \dfrac{1}{6} = \dfrac{3}{4}$

9. $-1 = p - \dfrac{5}{12}$

10. $x + \dfrac{9}{7} = 2$

11. $x + 3 = \dfrac{17}{5}$

12. $-\dfrac{14}{3} = m - 4$

13. $\dfrac{3}{7}y = 6$

14. $\dfrac{9}{2}n = 8$

15. $5 = -\dfrac{2}{3}x$

16. $10 = -\dfrac{1}{7}x$

17. $\dfrac{d}{7} = \dfrac{9}{28}$

18. $\dfrac{r}{6} = -\dfrac{5}{16}$

19. $-\dfrac{k}{6} = \dfrac{1}{20}$

20. $-\dfrac{x}{8} = -\dfrac{1}{10}$

21. $-\dfrac{11}{6} = -\dfrac{c}{6}$

22. $\dfrac{3}{8} = -\dfrac{p}{4}$

23. $5x = \dfrac{10}{9}$

24. $6w = -\dfrac{12}{5}$

25. $-\dfrac{10}{3} = 8m$

26. $\dfrac{16}{5} = -10x$

CLEARING FRACTIONS

Exercises 27–48: Use the LCD to clear fractions and solve the given equation.

27. $\dfrac{5n}{6} = \dfrac{3}{8}$

28. $\dfrac{4q}{15} = -\dfrac{12}{5}$

29. $y + \dfrac{2}{9} = \dfrac{1}{3}$

30. $x + 3 = \dfrac{2}{5}$

31. $-\dfrac{7}{2}q = 3$

32. $-\dfrac{2b}{3} = \dfrac{8}{15}$

33. $-\dfrac{14a}{3} = -\dfrac{7}{9}$

34. $y - \dfrac{1}{18} = \dfrac{7}{9}$

35. $\dfrac{5}{8}x - \dfrac{1}{2} = \dfrac{3}{4}$

36. $\dfrac{7}{12}k + \dfrac{2}{3} = \dfrac{1}{4}$

37. $\dfrac{5}{6} + \dfrac{2}{3}w = \dfrac{3}{8}$

38. $\dfrac{3}{10}d + \dfrac{5}{6} = \dfrac{14}{15}$

39. $\dfrac{13}{18} = \dfrac{1}{12}q + \dfrac{2}{9}$

40. $\dfrac{7}{2} = \dfrac{1}{6} - \dfrac{10}{9}p$

41. $\dfrac{x}{4} + \dfrac{x}{10} = \dfrac{3}{5}$

42. $\dfrac{y}{12} - \dfrac{y}{4} = \dfrac{7}{9}$

43. $\dfrac{13b}{8} - \dfrac{7b}{12} = \dfrac{5}{6}$

44. $\dfrac{9a}{4} - \dfrac{11a}{16} = \dfrac{5}{8}$

45. $4\left(\dfrac{5}{2}n - \dfrac{2}{3}\right) = \dfrac{5}{6}$

46. $5\left(\dfrac{1}{2}r + \dfrac{3}{5}\right) = \dfrac{3}{4}$

47. $\dfrac{4}{5}\left(3x - \dfrac{1}{3}\right) = \dfrac{8}{15}$

48. $\dfrac{1}{3}(6m + 7) = \dfrac{5}{6}$

SOLVING EQUATIONS NUMERICALLY

Exercises 49–54: Solve the equation numerically by completing the given table of values.

49. $\dfrac{1}{3}x - 3 = 0$

x	1	3	5	7	9
$\frac{1}{3}x - 3$					

50. $-\dfrac{2}{3}x + 4 = 2$

x	0	1	2	3	4
$-\frac{2}{3}x + 4$					

51. $-\frac{4}{5}x + 3 = \frac{19}{5}$

x	-2	-1	0	1	2
$-\frac{4}{5}x + 3$					

52. $\frac{3}{4}x - 2 = -\frac{7}{2}$

x	-4	-3	-2	-1	0
$\frac{3}{4}x - 2$					

53. $\frac{1}{2}\left(x - \frac{2}{3}\right) = \frac{5}{3}$

x	-4	-2	0	2	4
$\frac{1}{2}\left(x - \frac{2}{3}\right)$					

54. $-\frac{1}{4}\left(x + \frac{1}{2}\right) = \frac{7}{8}$

x	-10	-8	-6	-4	-2
$-\frac{1}{4}\left(x + \frac{1}{2}\right)$					

SOLVING EQUATIONS VISUALLY

Exercises 55–60: Solve the equation visually.

55. $-\frac{5}{2}x - 2 = 3$ **56.** $\frac{3}{2}x - 1 = -1$

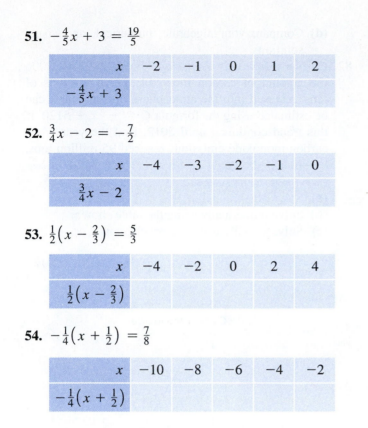

57. $\frac{2}{3}x - \frac{5}{3} = 1$ **58.** $-\frac{3}{4}x - \frac{1}{4} = 2$

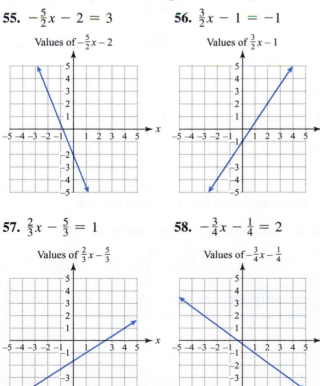

59. $-2x + \frac{5}{2} = \frac{1}{2}$ **60.** $3x + \frac{5}{2} = -\frac{1}{2}$

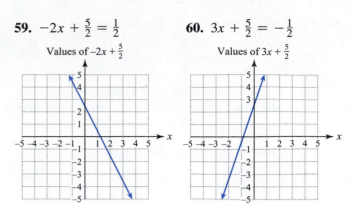

YOU DECIDE THE METHOD

Exercises 61–72: Solve the given equation.

61. $w + \frac{15}{4} = 1$ **62.** $\frac{x}{8} + \frac{x}{12} = \frac{25}{6}$

63. $-\frac{p}{9} = -\frac{7}{6}$ **64.** $8y = \frac{12}{5}$

65. $\frac{1}{4}d + \frac{1}{6} = -\frac{8}{3}$ **66.** $2\left(\frac{5}{4}n - \frac{1}{2}\right) = \frac{5}{4}$

67. $\frac{9a}{20} - \frac{5a}{4} = \frac{7}{10}$ **68.** $\frac{5r}{8} = -\frac{7}{12}$

69. $\frac{3}{2}\left(3x + \frac{2}{3}\right) = \frac{9}{10}$ **70.** $-\frac{9}{7}n = 18$

71. $4\left(3 - \frac{5}{6}m\right) = \frac{32}{3}$ **72.** $7 - \frac{7a}{2} = \frac{7}{6}$

APPLICATIONS

*Exercises 73–76: **Geometry** Use the trapezoid area formula, $A = \frac{1}{2}(a + b)h$, to find the unknown measure.*

73. $A = \frac{29}{16}$ square feet   **74.** $A = \frac{13}{15}$ square inch

75. $A = \frac{63}{4}$ square inches   **76.** $A = \frac{77}{5}$ square feet

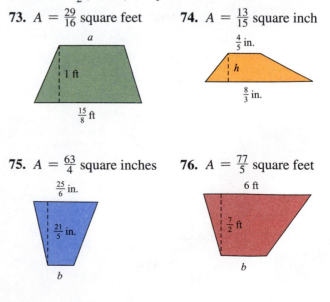

77. *Converting Temperature* The formula

$$F = \frac{9}{5}C + 32$$

gives the relationship between F degrees Fahrenheit and C degrees Celsius. Find the Celsius temperature that is equivalent to $46\frac{2}{5}°F$.

78. *Converting Temperature* The formula

$$C = \frac{5}{9}(F - 32)$$

gives the relationship between C degrees Celsius and F degrees Fahrenheit. Find the Fahrenheit temperature that is equivalent to $33\frac{1}{3}°C$.

79. *Working Together* If one pump can inflate an air mattress in 4 minutes and a second pump can inflate the same mattress in 12 minutes, find the amount of time that would be needed for the two pumps working together to inflate the mattress.

80. *Working Together* An experienced painter can paint a house in 6 days, while a less experienced painter can paint the same house in 10 days. How many days would it take the two painters working together to paint the house?

81. *Gaming Revenue* The revenue R, in billions of dollars, for all gaming can be estimated by the formula $R = \frac{9}{2}x - 8939$, where x is a year from 2000 to 2008. Find the year when gaming revenue was 88 billion by solving the equation $88 = \frac{9}{2}x - 8939$. (*Source:* American Gaming Association.)

(a) Solve algebraically.
(b) Solve numerically using the table shown.
(c) Solve visually using the graph shown.

x	2000	2002	2004	2006	2008
$\frac{9}{2}x - 8939$					

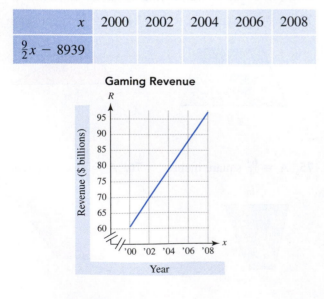

Gaming Revenue

(d) Compare your algebraic, numerical, and visual solutions.

82. *Carbon Monoxide Emissions* From 2002 to 2006 the amount of carbon monoxide C, in millions of tons, released into the atmosphere during year x can be estimated using the formula $C = -\frac{5}{2}x + 5120$. If this trend continued until 2012, find the year when carbon monoxide emissions reached 95 million tons, by solving the equation $95 = -\frac{5}{2}x + 5120$. (*Source:* Environmental Protection Agency.)

(a) Solve algebraically.
(b) Solve numerically using the table shown.
(c) Solve visually using the graph shown.

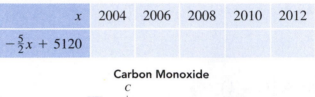

x	2004	2006	2008	2010	2012
$-\frac{5}{2}x + 5120$					

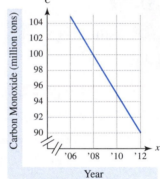

Carbon Monoxide

(d) Compare your algebraic, numerical, and visual solutions.

WRITING ABOUT MATHEMATICS

83. Would you rather solve the equation

$$\frac{7}{15}x + \frac{1}{6} = \frac{3}{10}$$

by clearing fractions first, or would using properties of equality work better for you? Explain.

84. Write and explain the steps needed to solve

$$\frac{4}{9}\left(5x - \frac{1}{3}\right) = \frac{7}{6}$$

by clearing fractions first *before* applying the distributive property to clear the parentheses.

Checking Basic Concepts

1. Simplify each complex fraction.

(a) $\dfrac{\dfrac{4}{5}}{\dfrac{8}{15}}$

(b) $\dfrac{\dfrac{2w}{5} - \dfrac{w}{3}}{\dfrac{9}{5}}$

(c) $\dfrac{\dfrac{4}{9} + \dfrac{1}{6}}{\dfrac{7}{12} - \dfrac{1}{4}}$

(d) $\dfrac{\dfrac{11x}{6}}{\dfrac{3}{4} + 2}$

2. Simplify each expression.

(a) $\dfrac{5}{8} - \dfrac{2}{3} \cdot \dfrac{1}{2}$

(b) $\dfrac{3}{2} \cdot \left(1 - \dfrac{5}{6}\right)$

3. Solve each equation algebraically.

(a) $y - \dfrac{3}{8} = \dfrac{1}{4}$

(b) $6 = -\dfrac{1}{5}x$

(c) $\dfrac{4}{9}d + \dfrac{5}{6} = \dfrac{7}{2}$

(d) $\dfrac{9a}{4} - \dfrac{7a}{12} = \dfrac{5}{3}$

4. Solve $-\dfrac{1}{2}(x - 1) = 4$ numerically.

x	-11	-9	-7	-5	-3
$-\frac{1}{2}(x - 1)$					

5. Solve $-\dfrac{3}{4}x + \dfrac{1}{2} = 2$ visually.

Values of $-\frac{3}{4}x + \frac{1}{2}$

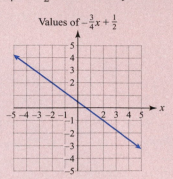

6. *Converting Temperature* The formula

$$C = \frac{5}{9}(F - 32)$$

gives the relationship between C degrees Celsius and F degrees Fahrenheit. Find the Fahrenheit temperature that is equivalent to $23\frac{1}{3}$°C.

CHAPTER 4   Summary

SECTION 4.1 ■ INTRODUCTION TO FRACTIONS AND MIXED NUMBERS

Fraction

A fraction is a number that describes a portion of a whole.

Example: $\dfrac{5}{8}$

Rational Number

A rational number can be expressed as $\dfrac{p}{q}$, where p and q are integers with $q \neq 0$.

Examples: $\dfrac{2}{3}, -\dfrac{4}{9}, 15 = \dfrac{15}{1}$

Properties of Fractions

The identity and zero properties for fractions are as follows.

1. When $n \neq 0$, $\frac{n}{n}$ simplifies to 1.
2. $\frac{n}{1}$ simplifies to n.
3. When $n \neq 0$, $\frac{0}{n}$ simplifies to 0.
4. $\frac{n}{0}$ is undefined.

Examples: **1.** $\dfrac{3}{3} = 1,$ **2.** $\dfrac{7}{1} = 7,$ **3.** $\dfrac{0}{5} = 0,$ **4.** $\dfrac{8}{0}$ is undefined.

Proper and Improper Fractions	The fraction $\frac{a}{b}$ with $b \neq 0$ is a proper fraction whenever $|a| < |b|$, and it is an improper fraction whenever $|a| \geq |b|$.

Examples:

Proper	Improper
$\dfrac{6}{11}, -\dfrac{3}{8}, \dfrac{20}{29}$	$-\dfrac{7}{2}, \dfrac{16}{5}, \dfrac{45}{45}$

Mixed Number

A mixed number is an integer written with a proper fraction.

Examples: $7\dfrac{3}{8}, -12\dfrac{1}{2}, 8\dfrac{23}{24}$

Writing an Improper Fraction as a Mixed Number

STEP 1: Divide the denominator into the numerator.

STEP 2: The quotient is the integer part of the mixed number. The fraction part of the mixed number has the remainder in its numerator and the divisor in its denominator.

$$\text{Mixed Number} = \text{Quotient} \frac{\text{Remainder}}{\text{Divisor}}$$

Examples: $\dfrac{5}{2} = 2\dfrac{1}{2}, \dfrac{29}{3} = 9\dfrac{2}{3}$

Writing a Mixed Number as an Improper Fraction

STEP 1: Multiply the denominator of the fraction by the integer.

STEP 2: Add the numerator of the fraction to the product from Step 1.

STEP 3: The sum found in Step 2 is the numerator of the improper fraction. The denominator of the improper fraction is the denominator from the original mixed number. So, a mixed number of the form

$$\text{Quotient} \frac{\text{Remainder}}{\text{Divisor}} \text{ is written as } \frac{\text{Divisor} \cdot \text{Quotient} + \text{Remainder}}{\text{Divisor}}.$$

Examples: $3\dfrac{7}{9} = \dfrac{9 \cdot 3 + 7}{9} = \dfrac{34}{9}, -6\dfrac{1}{5} = -\left(\dfrac{5 \cdot 6 + 1}{5}\right) = -\dfrac{31}{5}$

Graphing Fractions and Mixed Numbers

When graphing a fraction, divide each whole distance on the number line into an equal number of parts determined by the denominator of the fraction.

Examples:

SECTION 4.2 ■ PRIME FACTORIZATION AND LOWEST TERMS

Factor

A factor is any whole number that divides into a second whole number and results in a quotient with no remainder.

Example: 7 is a factor of 21 because $21 \div 7 = 3 \, \text{r} \, 0$. We can also say that 21 is divisible by 7.

Divisibility Tests

A whole number is divisible by

- 2 if its ones digit is 0, 2, 4, 6, or 8.
- 3 if the sum of its digits is divisible by 3.
- 4 if the number formed by its last two digits is divisible by 4.
- 5 if its ones digit is 0 or 5.
- 6 if it is divisible by 2 and 3.
- 9 if the sum of its digits is divisible by 9.

Examples: 5476 is divisible by 2 because its ones digit is 6.
9 is a factor of 621 because the digit sum 9 is divisible by 9.
34,865 divides evenly by 5 because its ones digit is 5.

Prime and Composite Numbers

A prime number is any natural number greater than 1 whose only factors are 1 and itself. A composite number is a number greater than 1 that is not prime.

Examples:	**Prime**	**Composite**
	2, 7, 89, and 503	8, 14, 39, and 117

Prime Factorization

Every composite number can be factored into a unique product of primes.

Examples: $108 = 2 \cdot 2 \cdot 3 \cdot 3 \cdot 3$, $910 = 2 \cdot 5 \cdot 7 \cdot 13$

Equivalent Fractions

Fractions that name the same number are called equivalent fractions. To write equivalent fractions, use the basic principle of fractions.

Example: The fractions $\frac{2}{3}$ and $\frac{10}{15}$ are equivalent because $\frac{2}{3} = \frac{2 \cdot 5}{3 \cdot 5} = \frac{10}{15}$.

Greatest Common Factor (GCF)

The largest number that divides evenly into two or more numbers is the GCF.

Example: The GCF of 25 and 40 is 5.

Lowest Terms

A fraction is in lowest terms if the GCF of its numerator and denominator is 1.

Example: The fraction $\frac{15}{60}$ written in lowest terms is $\frac{1}{4}$.

Simplifying to Lowest Terms

To simplify a fraction to lowest terms,

1. Determine a number c that is the GCF of the numerator and denominator.
2. Write the numerator as a product of two factors, a and c. Then write the denominator as a product of two factors, b and c.
3. Apply the basic principle of fractions.

Example: For $\frac{14}{35}$, the GCF of 14 and 35 is 7, so $c = 7$. In simplest form,

$$\frac{14}{35} = \frac{2 \cdot 7}{5 \cdot 7} = \frac{2}{5}.$$

Cross Product Method

We can determine if two fractions are equivalent by comparing their cross products.

Example: Because $12 \cdot 21 = 28 \cdot 9$ we know that $\frac{21}{28} = \frac{9}{12}$.

Rational Expressions

Any fraction whose numerator and denominator each consist of an expression that can be written as the sum or difference of two or more terms

Examples: $\dfrac{12x^2}{5}$, $\dfrac{3x + 7}{2x - 1}$,   and $\dfrac{5ab}{3c^2}$

SECTION 4.3 ■ MULTIPLYING AND DIVIDING FRACTIONS

Multiplying Fractions and Rational Expressions

If a, b, c, and d represent numbers (or single terms) with $b \neq 0$ and $d \neq 0$, then

$$\frac{a}{b} \cdot \frac{c}{d} = \frac{ac}{bd}.$$

Examples: $\dfrac{1}{5} \cdot \dfrac{3}{4} = \dfrac{1 \cdot 3}{5 \cdot 4} = \dfrac{3}{20}$, $\dfrac{3x}{2y} \cdot \dfrac{x}{4} = \dfrac{3x \cdot x}{2y \cdot 4} = \dfrac{3x^2}{8y}$

Simplifying Before Multiplying	**STEP 1:** Write $\frac{a}{b} \cdot \frac{c}{d}$ as $\frac{a \cdot c}{b \cdot d}$. Do not multiply.
	STEP 2: Replace each composite factor with its prime factorization.
	STEP 3: Eliminate all factors that are common to the numerator and denominator by applying the basic principle of fractions.
	STEP 4: Multiply the remaining factors in the numerator and multiply the remaining factors in the denominator.
	Example: $\dfrac{1}{3} \cdot \dfrac{6}{7} = \dfrac{1 \cdot 6}{3 \cdot 7} = \dfrac{1 \cdot (2 \cdot 3)}{3 \cdot 7} = \dfrac{1 \cdot 2}{7} = \dfrac{2}{7}$
Powers of Fractions	For natural number exponent n, multiply the fraction by itself n times.
	Example: $\left(\dfrac{2}{3}\right)^3 = \dfrac{2}{3} \cdot \dfrac{2}{3} \cdot \dfrac{2}{3} = \dfrac{8}{27}$
Square Roots of Fractions	If a and b represent whole numbers with $b \neq 0$ and the fraction $\frac{a}{b}$ is in lowest terms, then
	$$\sqrt{\dfrac{a}{b}} = \dfrac{\sqrt{a}}{\sqrt{b}}.$$
	Example: $\sqrt{\dfrac{81}{100}} = \dfrac{\sqrt{81}}{\sqrt{100}} = \dfrac{9}{10}$
Reciprocal	Two numbers are reciprocals (or multiplicative inverses) of each other if their product is 1.
	Example: $\dfrac{5}{7}$ and $\dfrac{7}{5}$ are reciprocals because $\dfrac{5}{7} \cdot \dfrac{7}{5} = 1$.
Dividing Fractions and Rational Expressions	If a, b, c, and d represent numbers (or single terms) with $b \neq 0$, $c \neq 0$, and $d \neq 0$, then the quotient found when dividing $\frac{a}{b}$ by $\frac{c}{d}$ is given by
	$$\dfrac{a}{b} \div \dfrac{c}{d} = \dfrac{a}{b} \cdot \dfrac{d}{c}.$$
	Examples: $\dfrac{2}{7} \div \dfrac{3}{5} = \dfrac{2}{7} \cdot \dfrac{5}{3} = \dfrac{10}{21}$, $\dfrac{3x}{2} \div \dfrac{5}{x} = \dfrac{3x}{2} \cdot \dfrac{x}{5} = \dfrac{3x^2}{10}$
Area of a Triangle	The area A of a triangle with base b and height h is given by $A = \frac{1}{2}bh$.
	Example: For the triangle shown, $$A = \dfrac{1}{2}(12)(5) = 30 \text{ square units.}$$

SECTION 4.4 ■ ADDING AND SUBTRACTING FRACTIONS—LIKE DENOMINATORS

Adding or Subtracting Fractions with Like Denominators	If a, b, and d represent numbers with $d \neq 0$, then
	$$\dfrac{a}{d} + \dfrac{b}{d} = \dfrac{a+b}{d} \quad \text{and} \quad \dfrac{a}{d} - \dfrac{b}{d} = \dfrac{a-b}{d}.$$
	Examples: $\dfrac{3}{7} + \dfrac{2}{7} = \dfrac{3+2}{7} = \dfrac{5}{7}$, $\dfrac{7}{12} - \dfrac{5}{12} = \dfrac{7-5}{12} = \dfrac{2}{12} = \dfrac{1}{6}$
Adding and Subtracting Rational Expressions	The process used to add or subtract rational expressions with like denominators is similar to that used for adding and subtracting fractions with like denominators.
	Examples: $\dfrac{2w}{5x} + \dfrac{w}{5x} = \dfrac{2w+w}{5x} = \dfrac{3w}{5x}$, $\dfrac{3x}{4} - \dfrac{x}{4} = \dfrac{3x-x}{4} = \dfrac{2x}{4} = \dfrac{x}{2}$

SECTION 4.5 ■ ADDING AND SUBTRACTING FRACTIONS—UNLIKE DENOMINATORS

Least Common Multiple (LCM)

The least common multiple (LCM) of two or more numbers is the smallest number that is divisible by each of the given numbers.

Example: Common multiples of 5 and 6 include 30, 60, 90, 120,…. The least common multiple is 30.

The Listing Method for Finding the LCM

The LCM of two or more numbers can be found by listing multiples of each number and then finding the smallest common multiple found in all lists.

Example: The LCM of 10 and 15 is 30, which is found as follows:
Multiples of 10: 10, 20, **30**, 40, 50, 60, …
Multiples of 15: 15, **30**, 45, 60, 75, 90, …

Prime Factorization Method for Finding the LCM

STEP 1: Find the prime factorization of each number.

STEP 2: List each factor that appears in one or more factorizations. If a factor is repeated in any of the factorizations, list this factor the maximum number of times that it is repeated in any one of the factorizations.

STEP 3: Find the product of this list of factors. The result is the LCM.

Example: The LCM of 18 and 24 is found as follows:
$18 = 2 \cdot 3 \cdot 3$
$24 = 2 \cdot 2 \cdot 2 \cdot 3$
The LCM is $2 \cdot 2 \cdot 2 \cdot 3 \cdot 3 = 72$.

Least Common Denominator (LCD)

The least common denominator (LCD) of two or more fractions is the LCM of the denominators.

Example: The LCM of 18 and 24 is 72, so the LCD of $\frac{5}{18}$ and $\frac{11}{24}$ is 72.

Adding or Subtracting Fractions with Unlike Denominators

STEP 1: Determine the LCD for all fractions involved.

STEP 2: Write each fraction as an equivalent fraction with the LCD as its denominator.

STEP 3: Add or subtract the newly written like fractions and simplify.

Example: $\dfrac{3}{8} + \dfrac{1}{6} = \dfrac{3}{8} \cdot \dfrac{3}{3} + \dfrac{1}{6} \cdot \dfrac{4}{4} = \dfrac{9}{24} + \dfrac{4}{24} = \dfrac{9 + 4}{24} = \dfrac{13}{24}$

SECTION 4.6 ■ OPERATIONS ON MIXED NUMBERS

Writing a Mixed Number as an Improper Fraction

A mixed number of the form

$\text{Quotient} \dfrac{\text{Remainder}}{\text{Divisor}}$ can be written as $\dfrac{\text{Divisor} \cdot \text{Quotient} + \text{Remainder}}{\text{Divisor}}$.

Examples: $4\dfrac{5}{7} = \dfrac{7 \cdot 4 + 5}{7} = \dfrac{33}{7}$, $\quad -2\dfrac{3}{8} = -\dfrac{8 \cdot 2 + 3}{8} = -\dfrac{19}{8}$

Rounding a Mixed Number to the Nearest Integer

If the fraction part of the mixed number is

(a) less than $\frac{1}{2}$, drop the fraction part of the mixed number.

(b) $\frac{1}{2}$ or more, drop the fraction part of the mixed number and
  (i) add 1 to the integer part if the mixed number is positive.
  (ii) subtract 1 from the integer part if the mixed number is negative.

Examples: $4\dfrac{1}{6}$ rounds to 4; $\quad -8\dfrac{4}{7}$ rounds to -9.

Operations on Mixed Numbers

STEP 1: Rewrite any mixed numbers or integers as improper fractions.

STEP 2: Perform the required arithmetic as usual.

STEP 3: Rewrite the result as a mixed number, when appropriate.

Examples: $2\dfrac{5}{8} \cdot 3\dfrac{5}{9} = \dfrac{21}{8} \cdot \dfrac{32}{9} = \dfrac{7 \cdot 4}{1 \cdot 3} = \dfrac{28}{3} = 9\dfrac{1}{3}$

$4\dfrac{2}{5} - 2\dfrac{3}{5} = \dfrac{22}{5} - \dfrac{13}{5} = \dfrac{22 - 13}{5} = \dfrac{9}{5} = 1\dfrac{4}{5}$

Method II for Adding or Subtracting Mixed Numbers

Add or subtract the integer and fraction parts vertically.

Examples:

$$\begin{array}{r} 3\dfrac{1}{7} \\ + 4\dfrac{3}{7} \\ \hline 7\dfrac{4}{7} \end{array} \qquad \begin{array}{r} 13\dfrac{8}{9} \\ - 5\dfrac{7}{9} \\ \hline 8\dfrac{1}{9} \end{array}$$

SECTION 4.7 ■ COMPLEX FRACTIONS AND ORDER OF OPERATIONS

Complex Fraction

A complex fraction is a rational expression with fractions in its numerator, denominator, or both.

Examples: $\dfrac{4 + \dfrac{5}{6}}{\dfrac{8}{9} - 7}, \qquad \dfrac{\dfrac{3x}{7} - \dfrac{1}{2}}{\dfrac{2}{5} + \dfrac{3x}{5}}$

Simplifying Basic Complex Fractions

If a, b, c, and d represent numbers with $b \neq 0$, $c \neq 0$, and $d \neq 0$, then

$$\dfrac{\dfrac{a}{b}}{\dfrac{c}{d}} = \dfrac{a}{b} \cdot \dfrac{d}{c}.$$

Example: $\dfrac{\dfrac{3}{7}}{\dfrac{4}{9}} = \dfrac{3}{7} \cdot \dfrac{9}{4} = \dfrac{27}{28}$

Method I: Simplifying Complex Fractions

To simplify a complex fraction,

1. Write the numerator as a single fraction; do the same to the denominator.
2. Multiply the fraction in the numerator by the reciprocal of the fraction in the denominator. This is called *invert and multiply*. Simplify the result.

Example: $\dfrac{\dfrac{3}{7} + \dfrac{1}{7}}{\dfrac{5}{y} - \dfrac{2}{y}} = \dfrac{\dfrac{4}{7}}{\dfrac{3}{y}} = \dfrac{4}{7} \cdot \dfrac{y}{3} = \dfrac{4y}{21}$

Method II: Simplifying Complex Fractions

To simplify a complex fraction,

1. Find the LCD of all fractions within the complex fraction.
2. Multiply both the numerator and the denominator of the complex fraction by the LCD found in Step 1. Simplify the result.

Example: $\dfrac{\dfrac{5}{6} - \dfrac{1}{2}}{\dfrac{2}{3} + \dfrac{1}{6}} = \dfrac{6 \cdot \left(\dfrac{5}{6} - \dfrac{1}{2}\right)}{6 \cdot \left(\dfrac{2}{3} + \dfrac{1}{6}\right)} = \dfrac{5 - 3}{4 + 1} = \dfrac{2}{5}$

SECTION 4.8 ■ SOLVING EQUATIONS INVOLVING FRACTIONS

Properties of Equality

$a = b$ is equivalent to $a + c = b + c$.

$a = b$ is equivalent to $a - c = b - c$.

$a = b$ is equivalent to $ac = bc$ for $c \neq 0$.

$a = b$ is equivalent to $\dfrac{a}{c} = \dfrac{b}{c}$ for $c \neq 0$.

Example: $x + \dfrac{2}{5} = \dfrac{4}{5}$ is equivalent to $x + \dfrac{2}{5} - \dfrac{2}{5} = \dfrac{4}{5} - \dfrac{2}{5}$, so $x = \dfrac{2}{5}$.

Solving Equations Algebraically

STEP 1: Use the distributive property to clear parentheses from the equation.

STEP 2: Multiply each term by the LCD of all fractions involved.

STEP 3: Simplify each term and combine any like terms.

STEP 4: Solve the resulting equation.

Example:

$$3\left(x + \dfrac{5}{6}\right) = \dfrac{7}{8}$$

$$3x + \dfrac{5}{2} = \dfrac{7}{8} \qquad \text{(Step 1)}$$

$$8 \cdot 3x + \dfrac{8}{1} \cdot \dfrac{5}{2} = \dfrac{8}{1} \cdot \dfrac{7}{8} \qquad \text{(Step 2)}$$

$$24x + 20 = 7 \qquad \text{(Step 3)}$$

$$24x = -13$$

$$x = -\dfrac{13}{24} \qquad \text{(Step 4)}$$

Solving Equations Numerically

To solve some equations involving fractions, complete a table of values for various values of the variable and then select the solution from the table.

Example: The solution to the equation $\frac{2}{3}x + 4 = \mathbf{8}$ is **6**.

x	-2	0	2	4	**6**	8
$\frac{2}{3}x + 4$	$\frac{8}{3}$	4	$\frac{16}{3}$	$\frac{20}{3}$	**8**	$\frac{28}{3}$

Solving Equations Visually

To solve equations visually, use a graph of the equation to estimate a solution and then check the estimated solution to be sure that it is correct.

Example: The solution to the equation $-\frac{2}{5}x + \frac{4}{5} = \mathbf{2}$ is $\mathbf{-3}$.

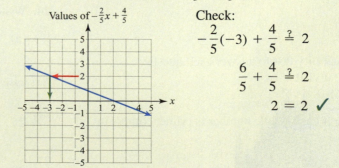

Values of $-\frac{2}{5}x + \frac{4}{5}$

Check:

$$-\dfrac{2}{5}(-3) + \dfrac{4}{5} \overset{?}{=} 2$$

$$\dfrac{6}{5} + \dfrac{4}{5} \overset{?}{=} 2$$

$$2 = 2 \ \checkmark$$

CHAPTER 4   Review Exercises

SECTION 4.1

1. Write the numerator and denominator of each fraction.

   (a) $\dfrac{7}{18}$ (b) $\dfrac{x}{5}$

2. Write the fraction represented by each shading.

   (a)   (b)

   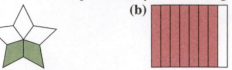

3. Simplify each fraction, if possible.

   (a) $\dfrac{0}{6}$ (b) $\dfrac{-2}{1}$

   (c) $\dfrac{-5}{-5}$ (d) $\dfrac{8}{0}$

4. Write the improper fraction and mixed number represented by each shading.

   (a)

   (b)

   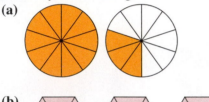

5. Write each improper fraction as a mixed number.

   (a) $\dfrac{19}{3}$ (b) $\dfrac{-14}{5}$

6. Write each mixed number as an improper fraction.

   (a) $-6\dfrac{2}{5}$ (b) $7\dfrac{7}{9}$

7. Graph each fraction or mixed number on a number line.

   (a) $-2\dfrac{1}{2}$ (b) $\dfrac{4}{5}$

SECTION 4.2

8. Use a divisibility test to answer each question.
   (a) Is 742 divisible by 3?
   (b) Is 4565 divisible by 5?

9. Determine if each whole number is prime, composite, or neither.
   (a) 39   (b) 59
   (c) 1   (d) 81

10. Find the prime factorization of each number.
    (a) 40   (b) 110

11. Find a number to replace the question mark so that the given fractions are equivalent.

    (a) $\dfrac{3}{?} = \dfrac{15}{25}$ (b) $\dfrac{3}{7} = \dfrac{?}{63}$

12. Find the GCF of the given numbers.
    (a) 12 and 42   (b) 16, 24, and 48

13. Simplify each fraction to lowest terms.

    (a) $-\dfrac{20}{65}$ (b) $\dfrac{64}{88}$

14. Place the correct symbol, $<$, $>$, or $=$, in the blank for each given pair of fractions.

    (a) $\dfrac{5}{8}$ —— $\dfrac{9}{13}$ (b) $\dfrac{11}{16}$ —— $\dfrac{21}{32}$

15. Find the GCF of the given terms.
    (a) $3x^2$ and $12x$ (b) $4x^2y^3$ and $6xy^2$

16. Simplify each rational expression.

    (a) $\dfrac{16xy}{20y}$ (b) $-\dfrac{9x^2y}{3y}$

SECTION 4.3

Exercises 17–22: Multiply. Simplify, if necessary.

17. $\dfrac{5}{6} \cdot \dfrac{5}{8}$ 18. $-\dfrac{2}{3} \cdot \left(-\dfrac{5}{8}\right)$

19. $-\dfrac{18}{7} \cdot \dfrac{7}{6}$ 20. $\dfrac{20}{33} \cdot \dfrac{22}{35}$

21. $\dfrac{15w^2}{8x} \cdot \dfrac{4x^2}{25w^3}$ 22. $-4x \cdot \dfrac{9y}{16x^2}$

23. Evaluate each expression.

    (a) $\left(-\dfrac{1}{4}\right)^3$ (b) $\sqrt{\dfrac{49}{100}}$

24. Write the reciprocal of each number.

    (a) $-\dfrac{4}{9}$ (b) $\dfrac{1}{10}$

Exercises 25–28: Divide. Simplify, if necessary.

25. $\dfrac{7}{25} \div \dfrac{21}{10}$ 26. $-\dfrac{7}{2} \div \dfrac{35}{36}$

27. $\dfrac{3x^2}{28y} \div \dfrac{6x}{7y^2}$ 28. $\dfrac{21p^3}{8q} \div 7p$

Exercises 29 and 30: Find the fractional part.

29. $\frac{4}{9}$ of -45 **30.** $\frac{3}{4}$ of $\frac{5}{18}$

SECTION 4.4

Exercises 31–36: Add or subtract as indicated. Simplify your result if necessary.

31. $\frac{3}{8} + \frac{5}{8}$ **32.** $-\frac{1}{10} + \frac{9}{10}$

33. $-\frac{2}{3} - \left(-\frac{8}{3}\right)$ **34.** $-\frac{7}{16} - \frac{3}{16}$

35. $\frac{7y^2}{8x} - \frac{5y^2}{8x}$ **36.** $\frac{5}{6m} + \left(-\frac{5}{6m}\right)$

SECTION 4.5

37. Find the LCM of the given numbers or expressions.
(a) 12 and 20   (b) 6, 8, and 16
(c) $4x^2$ and $3x$ (d) $2m$, $4mn$, and $6n$

38. Find the least common denominator for the fractions.
(a) $\frac{3}{16}$ and $\frac{7}{12}$ (b) $\frac{x}{3y^2}$ and $\frac{9}{8y}$

Exercises 39 and 40: Rewrite the fractions as equivalent fractions with the given LCD.

39. $\frac{1}{6}$ and $\frac{8}{9}$ LCD: 18

40. $\frac{5}{6}, \frac{2}{7}$, and $\frac{3}{14}$ LCD: 42

Exercises 41–46: Add or subtract as indicated. Simplify your result if necessary.

41. $\frac{3}{4} + \frac{1}{6}$ **42.** $-\frac{3}{10} - \frac{1}{8}$

43. $\frac{2}{5} - \left(-\frac{4}{15}\right)$ **44.** $-\frac{7}{18} + \frac{5}{6}$

45. $\frac{5y^2}{2} - \frac{y^2}{6}$ **46.** $\frac{3}{4m} + \frac{5}{8m}$

SECTION 4.6

Exercises 47 and 48: Round the given mixed number to the nearest integer.

47. $-3\frac{7}{9}$ **48.** $5\frac{3}{10}$

Exercises 49–56: Perform the indicated arithmetic.

49. $5\frac{1}{4} + 7\frac{5}{6}$ **50.** $-4\frac{3}{10} \cdot 5$

51. $8\frac{2}{5} - 5\frac{7}{15}$ **52.** $-3\frac{5}{18} + 8\frac{5}{6}$

53. $-3\frac{1}{8} \cdot 4\frac{2}{5}$ **54.** $3\frac{3}{8} \div \frac{3}{4}$

55. $6\frac{2}{9} \div 2\frac{1}{3}$ **56.** $-4\frac{1}{3} - 3\frac{8}{9}$

SECTION 4.7

Exercises 57–60: Simplify the given complex fraction.

57. $\dfrac{\frac{5}{4}}{\frac{15}{8}}$ **58.** $\dfrac{\frac{4}{3} + \frac{1}{8}}{\frac{5}{6} - \frac{1}{4}}$

59. $\dfrac{\frac{9x}{2} - \frac{3x}{4}}{\frac{15}{8}}$ **60.** $\dfrac{\frac{12y}{5}}{\frac{8}{3} + 4}$

Exercises 61 and 62: Simplify the given expression.

61. $\frac{7}{9} - \frac{2}{3} \cdot \frac{1}{6}$ **62.** $1 + \frac{3}{8} \cdot \left(\frac{1}{6} + \frac{1}{2}\right)$

SECTION 4.8

Exercises 63–70: Solve the equation algebraically.

63. $y - \frac{1}{8} = \frac{5}{6}$ **64.** $8 = -\frac{1}{9}w$

65. $\frac{7}{12}k + \frac{5}{8} = \frac{3}{4}$ **66.** $\frac{x}{3} + \frac{x}{15} = \frac{4}{5}$

67. $5\left(n - \frac{1}{4}\right) = \frac{5}{3}$ **68.** $-\frac{16}{3} = m - 2$

69. $\frac{7a}{18} - \frac{5a}{6} = \frac{8}{9}$ **70.** $\frac{7}{4} = \frac{1}{12} - \frac{5}{9}p$

Exercises 71 and 72: Solve the equation numerically by completing the given table of values.

71. $\frac{1}{5}x - 4 = -3$

x	-3	-1	1	3	5
$\frac{1}{5}x - 4$					

72. $\frac{3}{2}x - 1 = \frac{7}{2}$

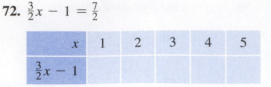

x	1	2	3	4	5
$\frac{3}{2}x - 1$					

Exercises 73 and 74: Solve the equation visually.

73. $-\frac{5}{3}x - \frac{1}{3} = 3$ **74.** $-\frac{3}{5}x + \frac{2}{5} = -2$

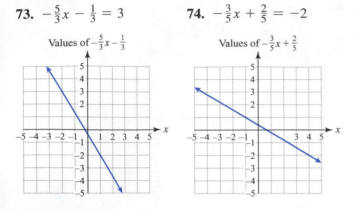

APPLICATIONS

75. *Cable TV* In 2010, it was estimated that 7 of every 100 households in India had cable TV. Find the fraction of India's households that did *not* have cable TV in 2010. (*Source:* International Telecommunications Union.)

76. *Phone Numbers* Out of 25 students in the red classroom at a preschool, 14 students could recite their parents' phone number. In the blue room, 9 students out of 17 could recite their parents' phone number. Which class had a larger fraction of students who knew their parents' phone number?

77. *Cleaning Solution* If $\frac{2}{3}$ cup of concentrated detergent should be mixed with 1 gallon of water to make a cleaning solution, how much concentrated detergent should be mixed with $\frac{3}{4}$ gallon of water?

78. *Going Organic* A farmer plans to convert $\frac{11}{15}$ of his crop land to organic crops. If the farmer has already converted $\frac{7}{15}$ of his crop land, what fraction of his land has yet to be converted to organic crops?

79. *Voter Turnout* In the 2008 presidential election, the fraction of Iowa's eligible voters that went to the polls was $\frac{2}{25}$ higher than the national average. If $\frac{31}{50}$ of the nation's voters went to the polls, what fraction of Iowa's eligible voters turned out to vote? (*Source:* Iowa Secretary of State.)

80. *Height* At birth, a baby boy was $17\frac{3}{4}$ inches long. If he grew $7\frac{1}{2}$ inches during his first year, how tall was he on his first birthday?

81. *Converting Currency* At one time, the formula for converting E euros (€) to D dollars was

$$D = \frac{33}{25}E.$$

Use the formula to find the number of dollars that could be purchased for €100.

82. *Working Together* Experienced roofers can put shingles on a house in 8 hours, while less experienced roofers can shingle the same house in 12 hours. How many hours would it take the two roofing crews working together to shingle the house?

CHAPTER 4 Test

1. Write the improper fraction and mixed number represented by the shading.

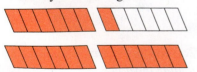

2. Write $4\frac{5}{7}$ as an improper fraction.

3. Write $\frac{22}{3}$ as a mixed number.

4. Determine if the given whole number is prime, composite, or neither.
  (a) 16 **(b)** 0
  (c) 29 **(d)** 63

5. Find the prime factorization of 240.

6. Find a number to replace the question mark so that the given fractions are equivalent.
  (a) $\frac{3}{?} = \frac{15}{25}$ **(b)** $\frac{3}{7} = \frac{?}{63}$

7. Find the GCF of each pair.
  (a) 16 and 72 **(b)** $14x^2$ and $21x$

8. Simplify to lowest terms.
  (a) $-\frac{45}{60}$ **(b)** $\frac{28xy^2}{7y^2}$

9. Multiply or divide as indicated. Simplify your result.

(a) $\dfrac{4}{3} \cdot \left(-\dfrac{9}{8}\right)$ (b) $\dfrac{6}{25} \div \dfrac{24}{35}$

(c) $\dfrac{30x^2}{7y} \div 6x$ (d) $\dfrac{8y}{15x} \cdot (-5x^2)$

10. Evaluate each expression.

(a) $\sqrt{\dfrac{81}{121}}$ (b) $\left(-\dfrac{5}{6}\right)^2$

11. Add or subtract as indicated. Simplify your result.

(a) $\dfrac{5}{24} + \dfrac{3}{8}$ (b) $-\dfrac{2}{15} - \dfrac{1}{10}$

(c) $-\dfrac{7}{10} + \left(-\dfrac{5}{6}\right)$ (d) $\dfrac{9w}{4x} + \dfrac{7w}{4x}$

12. Perform the indicated arithmetic.

(a) $7\dfrac{1}{4} + 4\dfrac{1}{6}$ (b) $7 \cdot \left(-3\dfrac{1}{5}\right)$

13. Simplify the expression.

(a) $\dfrac{\dfrac{1}{6} + \dfrac{7}{8}}{\dfrac{7}{4} - \dfrac{2}{3}}$ (b) $\dfrac{6}{5} \cdot \left(\dfrac{5}{6} - \dfrac{2}{3}\right) + \dfrac{4}{5}$

14. Solve the equation algebraically.

(a) $-\dfrac{3}{4}y = 12$ (b) $\dfrac{3}{4} = 3\left(m - \dfrac{1}{2}\right)$

(c) $\dfrac{x}{6} + \dfrac{x}{12} = \dfrac{5}{4}$ (d) $\dfrac{7w}{2} - \dfrac{5w}{6} = \dfrac{4}{3}$

15. Solve the equation $\frac{4}{3}x + \frac{1}{2} = \frac{19}{6}$ numerically by completing the given table of values.

x	-2	-1	0	1	2
$\frac{4}{3}x + \frac{1}{2}$					

16. Solve the equation $\frac{3}{4}x - \frac{5}{4} = 1$ visually.

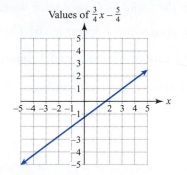

Values of $\frac{3}{4}x - \frac{5}{4}$

17. *Family Budget* A family spends $\frac{4}{25}$ of its monthly budget on food. If the family has a total of $4200 to spend each month, how much is spent on food?

18. *Gaming Revenue* The revenue R, in billions of dollars, for all gaming can be estimated by the formula $R = \frac{9}{2}x - 8939$, where x is a year from 2000 to 2008. Find the year when gaming revenue was 61 billion by solving the equation $61 = \frac{9}{2}x - 8939$.

(*Source:* American Gaming Association.)

CHAPTERS 1–4   Cumulative Review Exercises

1. Write *thirty-six thousand, two hundred eighty-five* in standard form.

2. Translate the given word phrase into a mathematical expression and then compute the result.

The product of 15 and 4

3. Divide $3567 \div 16$.

4. Evaluate the algebraic expression $y \div x$ when $x = 7$ and $y = 35$.

5. Estimate the sum $993 + 2002 + 696$ by rounding each value to the nearest hundred.

6. Evaluate $11 - (7 + 3) \div 2$.

7. Simplify the expression $5x - (2x + 1)$.

8. Simplify $-(-(-(-4)))$.

9. State the addition property illustrated by the equation $3 + (4 + 6) = (3 + 4) + 6$.

10. Simplify the expression $-7 - (-5) + 8$.

11. Multiply $2(-3)(-1)(5)(-2)$.

12. Evaluate $\sqrt{-3^2 + 45} \div (2 + 4)$.

13. Complete the table and solve $3x - 5 = -5$.

x	-2	-1	0	1	2
$3x - 5$					

14. Simplify the expression $-5(2y - 1) + 12y$.

Exercises 15 and 16: Translate each sentence into an equation using x as the variable. Do not solve the equation.

15. Seven equals a number decreased by 13.

16. Double the sum of a number and 3 equals -14.

17. Is -2 a solution to the equation $-x + 3 = -1$?

18. Is the equation $-x^3 + 5 = 2$ linear?

19. Solve the equation $-6y + (-16) = 2y$.

20. Solve the linear equation $-2x - 3 = 3$ visually.

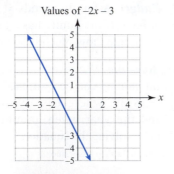

21. Write the improper fraction and mixed number represented by the shading.

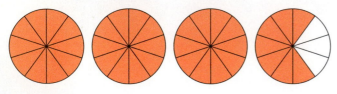

22. Find the prime factorization of 198.

Exercises 23–28: Perform the indicated arithmetic.

23. $-\dfrac{8}{7} \div \dfrac{32}{21}$

24. $\dfrac{3}{14} \cdot \dfrac{7}{9}$

25. $\dfrac{4}{7} + \left(-\dfrac{4}{7}\right)$

26. $\dfrac{13}{10} - \dfrac{5}{6}$

27. $2\dfrac{1}{3} + 6\dfrac{5}{8}$

28. $-\dfrac{1}{6} + \dfrac{2}{3} \cdot \dfrac{9}{10}$

Exercises 29 and 30: Solve the equation algebraically.

29. $-\dfrac{14}{3} = x - 5$

30. $\dfrac{5}{10}y + \dfrac{3}{8} = \dfrac{7}{4}$

31. *Street Vendor* A person bought 9 items from a street vendor and received $13 in change. If each item was priced at $3, how much did the person give to the vendor when paying?

32. *Medicare Costs* Total Medicare costs C, in billions of dollars, can be computed using the formula

$$C = 18x - 35{,}750,$$

where x is a year from 1996 to 2008. Replace C in the formula with 286 and solve the resulting equation to find the year when Medicare costs were $286 billion. (*Source:* Office of Management and Budget.)

33. *Rectangle Dimensions* The length of a rectangle is 5 inches longer than its width. If the perimeter of the rectangle is 26 inches, find its length and width.

34. *Efficient Lighting* In an effort to replace $\dfrac{17}{30}$ of a home's incandescent light bulbs with energy-efficient LED bulbs, a contractor has already replaced $\dfrac{1}{6}$ of the light bulbs. What fraction of the incandescent bulbs still need to be replaced?

5

Decimals

If we did the things we are capable of, we would astound ourselves.

—THOMAS EDISON

One of the world's largest shopping malls is the Mall of America in Bloomington, Minnesota. Each year, more than 40 million people visit the indoor shopping complex. With the average visitor spending 1.52 times the national average for retail shoppers, the Mall of America's contribution to Minnesota's economy is more than $1.8 billion annually.

The walking distance around one of the mall's four levels is about 0.57 mile, with more than 520 stores and approximately 4.3 miles of store front footage. The 4.2-million-square-foot complex also offers more than 50 restaurants, 14 movie screens, a 6.7-acre indoor amusement park, and a 1.2-million-gallon aquarium.

Even though Minnesota's winters are cold, the Mall of America has no central heating system. The mall is heated by the body heat of visitors, the heat from lighting, and solar heat from more than 1.2 miles of skylights.

In the previous paragraphs, numerical information was given about the Mall of America. Did you notice how much of this information was written in *decimal* form? Whether you are spending money at the Mall of America or at your local retail store, decimal numbers are sure to be involved. In this chapter, we discuss decimal numbers and their role in mathematics.

Source: Mall of America.

5.1   Introduction to Decimals

Decimal Notation • Writing Decimals in Words • Writing Decimals as Fractions • Graphing Decimals on a Number Line • Comparing Decimals • Rounding Decimals

A LOOK INTO MATH ▶

It is difficult to imagine just how complicated our modern money system would be if stores and banks used mixed numbers rather than decimals. A bag of chips could cost 3\frac{9}{50}$, a large coffee drink might cost 4\frac{3}{20}$, and a checking account could have a balance of 246\frac{17}{25}$. While computations on mixed numbers are often quite tedious, the same computations on decimal numbers are less time-consuming. In this section, we learn how to write fractions and mixed numbers in decimal notation.

NEW VOCABULARY

☐ Decimal notation
☐ Decimal point
☐ Decimal number (decimal)

Decimal Notation

Numbers expressed in **decimal notation** have an integer part and a fractional part separated by a **decimal point**. Numbers written in this way are called **decimal numbers**, or simply **decimals**. Examples of decimals include

$$12.36, \quad -0.198, \quad -545.90132, \quad \text{and} \quad 0.003.$$

In a decimal number, the integer part is written to the left of the decimal point, and the fractional part is written to the right of the decimal point.

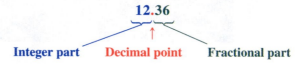

Integer part   Decimal point   Fractional part

STUDY TIP

By this time in the semester, it is likely that you know some of your classmates. Have you started or joined a study group? Be sure not to miss the opportunity to study math with your classmates.

READING CHECK

• What two parts of a decimal number are separated by the decimal point?

Each digit in a decimal has a place value. The place value names for the integer part are identical to those for whole numbers discussed in Section 1.1, and the place values for the fractional part have names based on fractions whose denominators are powers of 10. For example, Figure 5.1 shows the decimal 1783.954 written in a place value chart.

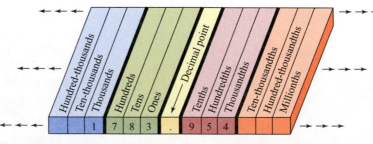

Figure 5.1 Place Value Chart for the Decimal Number 1783.954

NOTE: There are many ways to write a whole number in decimal form. For example, the following numbers all represent the whole number 72.

$$72 \qquad 72.0 \qquad 72.00 \qquad 72.00000$$

MAKING CONNECTIONS

Decimal Place Values and Periods

Except for the ones place value, the names for the place values to the right of the decimal point are similar to the names to the left of the decimal point. This pattern continues indefinitely in both directions. However, we do *not* use commas on the right side of a decimal point.

Writing Decimals in Words

▶ **REAL-WORLD CONNECTION** Try reading the following sentence aloud.

I bought a new iPod game for $2.99.

Did you read the price as "two point nine nine dollars?" Probably not. It is more natural to read the price as "two dollars and ninety-nine cents." We separate the whole number 2 from the fractional part 99 (cents) by reading the decimal point as "and." Because 99 cents is the same as 99 hundredths of a dollar, we write the decimal 2.99 in word form as

two and ninety-nine hundredths.

By first writing a decimal in expanded form, we can determine a general procedure for writing a decimal in words. Consider the expanded form of the decimal 57.125.

Tens	Ones	Tenths	Hundredths	Thousandths

$$5 \cdot 10 \ + \ 7 \cdot 1 \ + \ 1 \cdot \frac{1}{10} \ + \ 2 \cdot \frac{1}{100} \ + \ 5 \cdot \frac{1}{1000}$$

We can substitute the word "and" for the plus sign in the position of the decimal point and then simplify the remaining expressions by multiplying and adding.

$$5 \cdot 10 + 7 \cdot 1 \quad \text{and} \quad 1 \cdot \frac{1}{10} + 2 \cdot \frac{1}{100} + 5 \cdot \frac{1}{1000} \qquad \text{Insert the word "and."}$$

$$50 + 7 \quad \text{and} \quad \frac{1}{10} + \frac{2}{100} + \frac{5}{1000} \qquad \text{Multiply.}$$

$$50 + 7 \quad \text{and} \quad \frac{1}{10} \cdot \frac{100}{100} + \frac{2}{100} \cdot \frac{10}{10} + \frac{5}{1000} \qquad \text{The LCD is 1000.}$$

$$50 + 7 \quad \text{and} \quad \frac{100}{1000} + \frac{20}{1000} + \frac{5}{1000} \qquad \text{Multiply.}$$

$$57 \quad \text{and} \quad \frac{125}{1000} \qquad \text{Add.}$$

The decimal **57.125** is written in words as

*fifty-seven **and** one hundred twenty-five thousandths*.

This discussion suggests the following procedure for writing a decimal in words.

READING CHECK

- What word is used for the decimal point when writing a decimal in words?

WRITING A DECIMAL IN WORDS

To write a decimal in words, do the following.

1. Write the integer part in words.
2. Include the word "and" for the decimal point.
3. Write the word form of the whole number formed by the fractional part, followed by the place value of the rightmost digit.

NOTE: If the integer part of a decimal is 0, we do not write the words "zero and" before writing the fractional part in words. For example, the decimal 0.922 is written in words as *nine hundred twenty-two thousandths*.

EXAMPLE 1   **Writing decimals in words**

Write each decimal in words.
(a) 3451.92 **(b)** −3.687 **(c)** 0.05

Solution

(a) The decimal 3451.92 is written in words as

three thousand four hundred fifty-one and ninety-two hundredths.

(b) We write −3.687 as

negative three and six hundred eighty-seven thousandths.

(c) The integer part and the word "and" are not written. We write 0.05 in words as

five hundredths.

Now Try Exercises 7, 9, 13

▶ **REAL-WORLD CONNECTION** When writing a check, the whole number part of the amount is written in word form, followed by the word "and" for the decimal point. However, the fractional part of the amount is written as a fraction with a denominator of 100. This is illustrated in the next example.

EXAMPLE 2   **Writing a check**

Fill in the blank line on the check in the following figure.

Electronics Company	1001
301 Technology Drive, Anytown, USA 12345	
	Date: _2/2/2012_
Pay to the order of _The Power Company_	$ 103.47
	Dollars
Electric Bill	_Gavin Groehler_
Memo	Authorized Signature
⑈01001⑈ ⑆111222333⑆ 444555⑈	

Solution

The decimal 103.47 is written on the check in words as shown in the following figure.

Electronics Company	1001
301 Technology Drive, Anytown, USA 12345	
	Date: _2/2/2012_
Pay to the order of _The Power Company_	$ 103.47
One hundred three and $\frac{47}{100}$	Dollars
Electric Bill	_Gavin Groehler_
Memo	Authorized Signature
⑈01001⑈ ⑆111222333⑆ 444555⑈	

Now Try Exercise 17

Writing Decimals as Fractions

In the previous subsection, we saw that the decimal 57.125 could be written as 57 and $\frac{125}{1000}$. By removing the word "and" from this expression and writing the fractional part in lowest terms, we obtain the mixed number $57\frac{1}{8}$. To write a decimal as a mixed number or fraction, use the following procedure.

WRITING A DECIMAL AS A MIXED NUMBER OR FRACTION

To write a decimal as a mixed number or fraction, do the following.

1. Write the integer part of the decimal as the integer part of the mixed number.
2. Take the whole number formed by the fractional part of the decimal and place it in the numerator of a fraction whose denominator is a power of 10 that corresponds to the place value of the rightmost digit.
3. Simplify the fraction to lowest terms.

NOTE: Since we do not write 0 as the integer part of a mixed number, any decimal with 0 as its integer part is written as a proper fraction rather than a mixed number.

EXAMPLE 3   **Writing decimals as mixed numbers or fractions**

Write each decimal as a mixed number or fraction in lowest terms.
(a) 3.025   **(b)** −19.48   **(c)** 0.78

Solution

(a) $3.0\underset{\text{Thousandths}}{25} = 3\frac{25}{1000} = 3\frac{1 \cdot 25}{40 \cdot 25} = 3\frac{1}{40}$

(b) $-19.\underset{\text{Hundredths}}{48} = -19\frac{48}{100} = -19\frac{12 \cdot 4}{25 \cdot 4} = -19\frac{12}{25}$

(c) $0.\underset{\text{Hundredths}}{78} = \frac{78}{100} = \frac{39 \cdot 2}{50 \cdot 2} = \frac{39}{50}$

Now Try Exercises 23, 25, 27

Graphing Decimals on a Number Line

Graphing decimals on a number line is similar to graphing fractions and mixed numbers. To get the desired accuracy, we use equally spaced tick marks to break the distance between consecutive values into 10 equal parts. For example, the decimal 3.68 is located between 3 and 4 on the number line. An estimate of its position is shown in Figure 5.2.

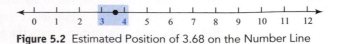

Figure 5.2 Estimated Position of 3.68 on the Number Line

For a more accurate graph of 3.68, we can enlarge the shaded portion of the number line in Figure 5.2 and insert equally spaced tick marks to break the distance from **3** to **4** into 10 equal parts, as shown in Figure 5.3 on the next page.

Figure 5.3 More Accurate Position of 3.68 on the Number Line

We can graph the *exact* position of 3.68 by enlarging the shaded portion of the number line in Figure 5.3 and inserting equally spaced tick marks to break the distance from **3.6** to **3.7** into 10 equal parts. The resulting number line is shown in Figure 5.4.

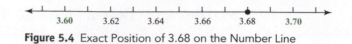

Figure 5.4 Exact Position of 3.68 on the Number Line

When graphing decimals on a number line, it is *not* necessary to progress through a series of enlarged number lines as demonstrated above. In the next example, decimals are graphed exactly on the first try.

<table>
<tr><td>**EXAMPLE 4**</td><td>**Graphing decimals on a number line**</td></tr>
</table>

Graph each decimal on a number line.
(a) 6.43   **(b)** −2.357

Solution
(a) We look at the next-to-last digit in the given decimal to determine that 6.**4**3 is between 6.**4** and 6.**5**. Use equally spaced tick marks to break the distance between **6.4** and **6.5** into 10 equal parts and graph the exact position of 6.43, as shown in the figure.

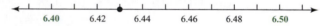

READING CHECK

• When graphing a decimal, how is its position on the number line determined?

(b) The next-to-last digit in −2.3**5**7 indicates that it is between −2.3**5** and −2.3**6**. Break the distance between **−2.35** and **−2.36** into 10 equal parts and graph the exact position of −2.357. Remember that −2.36 is to the left of −2.35 on the number line.

$$\longleftarrow \underset{-2.360}{|} \underset{}{|} \underset{-2.358}{|} \underset{}{|} \underset{-2.356}{|} \underset{}{\bullet} \underset{-2.354}{|} \underset{}{|} \underset{-2.352}{|} \underset{}{|} \underset{-2.350}{|} \longrightarrow$$

Now Try Exercises 39, 43

Comparing Decimals

When two decimals are graphed on the same number line, the decimal to the left is *less than* the decimal to the right and the decimal to the right is *greater than* the decimal to the left. However, using a number line to compare decimals such as 23.67843 and 23.68 is not very practical because these numbers are difficult to graph on the same number line. A method for comparing decimals by comparing their digits is summarized as follows.

COMPARING TWO POSITIVE DECIMALS

To compare two positive decimals with the same number of digits to the left of the decimal point, compare digits in corresponding places from left to right until unequal digits are found. The decimal with the greater of these digits is the larger number.

NOTE: This process can also be used to compare two negative decimals as long as you remember that the decimal with the larger of the unequal digits is *less than* the decimal with the smaller of the unequal digits.

EXAMPLE 5

Comparing two decimals

Place the correct symbol, $<$ or $>$, in the blank between the given decimals.
(a) 31.673 ___ 31.689   **(b)** 7.4569 ___ 7.45   **(c)** -52.7899 ___ -52.7893

Solution

(a) As we move from left to right, the corresponding digits 3, 1, and 6 are equal. The first unequal digits are in the hundredths place.

Because $7 < 8$, we conclude that $31.673 < 31.689$.

(b) Begin by writing 7.45 as 7.4500 so that it has the same number of digits to the right of the decimal point as 7.4569. The first unequal digits for 7.45**6**9 and 7.45**0**0 are in the thousandths place. Since $6 > 0$, we know that $7.4569 > 7.45$.

(c) The first unequal digits for -52.789**9** and -52.789**3** are in the ten-thousandths place. Because $9 > 3$ and the decimals are negative, we have $-52.7899 < -52.7893$.

Now Try Exercises 47, 49, 59

Rounding Decimals

Number lines are helpful when rounding decimals to a given place value. For example, we round the decimal 3.568 to the nearest **hundredth** by graphing it on a number line with steps of **0.01** (one-hundredths), as shown in Figure 5.5. Because 3.568 is **closer to 3.57** than to 3.56, we round **up to 3.57**.

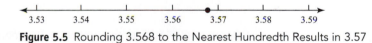

Figure 5.5 Rounding 3.568 to the Nearest Hundredth Results in 3.57

Since graphing decimals can be tedious, we usually round decimals in a way that is similar to rounding whole numbers.

ROUNDING DECIMALS

To round a decimal to the nearest whole number or to a place value to the right of the decimal point, do the following.

STEP 1: Identify the first digit to the *right* of the given place value.

STEP 2: If this digit is

      (a) less than 5, do not change the digit in the given place value.

      (b) 5 or more, add 1 to the digit in the given place value.

STEP 3: Drop all digits to the right of the given place value.

READING CHECK

• How are decimals rounded to a given place value?

EXAMPLE 6 **Rounding decimals**

Round 345.61489 to the given place value.
(a) tenth (b) thousandth

Solution

(a) **STEP 1:** The digit in the tenths place is 6. The first digit to the right of the **6** is **1**.

$$345.\textbf{61}489$$

STEP 2: Because 1 is less than 5, we do not change the digit in the tenths place.

$$345.\textbf{61}489$$

STEP 3: Drop all digits to the right of the tenths place.

$$345.6$$

Rounding 345.61489 to the nearest tenth results in 345.6.

(b) The thousandths digit in the decimal 345.61489 is **4**, and the first digit to the right of the thousandths place is **8** (Step 1). Because 8 is more than 5, we add 1 to the digit in the thousandths place (Step 2), and we drop all digits to its right (Step 3). The result of rounding 345.61489 to the nearest thousandth is 345.615.

Now Try Exercises 61, 67

In the next example, part (a) shows how to round a decimal to the nearest whole number, and part (b) demonstrates how to round up when a 9 is located in the given place value.

EXAMPLE 7 **Rounding decimals in money**

Round each amount to the given value.
(a) $45.77, dollar (b) $0.498, cent

Solution

(a) Rounding $45.**77** to the nearest dollar means that we are rounding to the nearest whole number. The specified place value is the ones place. Because **7** is more than 5, we add 1 to the digit in the ones place. After dropping all digits to the right of the ones place and the decimal point, the result is $46.

(b) Rounding $0.4**98** to the nearest cent means rounding to the nearest hundredth (of a dollar). Because **8** is more than 5, add 1 to the digit in the hundredths place. However, adding 1 to **9** gives **10** so we must carry **1** to the next digit to the left. The result is $0.50.

Now Try Exercises 73, 75

5.1 Putting It All Together

CONCEPT	COMMENTS	EXAMPLES
Decimal Notation	Numbers represented in decimal notation have an integer part and a fractional part separated by a decimal point. Numbers written in this way are called decimals.	**41.78** Integer part Decimal point Fractional part

CONCEPT	COMMENTS	EXAMPLES
Writing a Decimal in Words	1. Write the integer part in words. 2. Include the word "and" for the decimal point. 3. Write the word form of the whole number formed by the fractional part, followed by the place value of the rightmost digit.	The decimal 3.79 is written as three and seventy-nine hundredths. The decimal 0.063 is written as sixty-three thousandths.
Writing a Decimal as a Mixed Number or Fraction	1. Write the integer part of the decimal as the integer part of the mixed number. 2. Take the whole number formed by the fractional part of the decimal and place it in the numerator of a fraction whose denominator is a power of 10 that corresponds to the place value of the rightmost digit. 3. Simplify the fraction to lowest terms.	$0.58 = \dfrac{58}{100} = \dfrac{29}{50}$ $76.375 = 76\dfrac{375}{1000} = 76\dfrac{3}{8}$
Graphing Decimals on a Number Line	A decimal can be graphed on a number line by using equally spaced tick marks to break the distance between values into 10 equal parts.	The graph of 3.41 is shown below. 3.40 3.42 3.44 3.46 3.48 3.50
Comparing Decimals	To compare two positive decimals with the same number of digits to the left of the decimal point, compare digits in corresponding places from left to right until unequal digits are found. The decimal with the greater of these digits is the larger number.	$382.792 < 382.795$ $0.29016 > 0.29009$
Rounding Decimals	To round a decimal to the nearest whole number or to a place value to the right of the decimal point, 1. Identify the first digit to the *right* of the given place value. 2. If this digit is (a) less than 5, do not change the digit in the given place value. (b) 5 or more, add 1 to the digit in the given place value. 3. Drop all digits to the right of the given place value.	Rounding 8.719 to the nearest hundredth results in 8.72. Rounding 0.5189 to the nearest tenth results in 0.5.

5.1 Exercises

CONCEPTS AND VOCABULARY

1. Numbers in _____ notation have an integer part and a fractional part separated by a decimal point.

2. Numbers written in decimal notation are called _____ numbers, or simply _____.

3. When writing decimals in words, the decimal point is written as the word "_____."

4. When writing decimals as fractions, the denominator of the fractional part is a power of _____ that corresponds to the place value of the rightmost digit.

5. To graph a decimal on a number line, use equally spaced tick marks to break the distance between consecutive values into _____ equal parts.

6. When comparing decimals, compare digits in corresponding places from left to right until _____ digits are found.

WRITING DECIMALS IN WORDS

Exercises 7–16: Write the given decimal in words.

7. 0.56

8. −0.1

9. 7.116

10. 6.0009

11. −58.7

12. 0.000135

13. −2.001003

14. −6.39

15. 501.0012

16. 6002.009

Exercises 17 and 18: Fill in the blank line on the check.

17.

Electronics Company   1001
301 Technology Drive, Anytown, USA 12345
Date: _2/2/2012_
Pay to the order of _The Phone Company_ $ |129.68|
_____ Dollars
Phone Bill   _Aaron Groehler_
Memo   Authorized Signature
⑈01001⑈ ⑆111222333⑆ 444555⑈

18.

Electronics Company   1001
301 Technology Drive, Anytown, USA 12345
Date: _2/2/2012_
Pay to the order of _Electronics Superstore_ $ |2387.19|
_____ Dollars
Laptop   _Sydney Groehler_
Memo   Authorized Signature
⑈01001⑈ ⑆111222333⑆ 444555⑈

WRITING DECIMALS AS FRACTIONS

Exercises 19–34: Write the given decimal as a proper fraction or mixed number in lowest terms.

19. 0.3

20. 0.25

21. −0.04

22. −0.625

23. 0.85

24. 0.54

25. −8.2

26. 7.075

27. 12.75

28. −14.19

29. 23.205

30. −22.0875

31. −1.028

32. 3.448

33. 6.5125

34. 18.4375

GRAPHING DECIMALS

Exercises 35–44: Graph the decimal on a number line.

35. 8.3

36. 4.7

37. 26.76

38. 54.22

39. 0.315

40. 2.593

41. −2.1

42. −9.9

43. −5.74

44. −7.08

COMPARING DECIMALS

Exercises 45–60: Place the correct symbol, $<$ or $>$, in the blank between the given decimals.

45. 0.7 ___ 0.9

46. 4.42 ___ 4.402

47. 3.4998 ___ 3.5

48. 0.89 ___ 0.8903

49. 23.654 ___ 23.645

50. 6.0003 ___ 6.003

51. 0.19546 ___ 0.19548

52. 14.0101 ___ 14.0111

53. −3.9 ___ −4.0

54. −5.69 ___ −5.70

55. −1.593 ___ 1.593

56. 23.87 ___ −23.87

57. −7.999 ___ 8.000

58. 5.99 ___ −6.00

59. −14.5903 ___ −14.5913

60. −560.9 ___ −560.1

ROUNDING DECIMALS

Exercises 61–72: Round the given decimal to the indicated place value.

61. 0.3821   tenth

62. 3.7241   thousandth

63. 52.00764   hundredth

64. 265.802   tenth

65. −7.009367   ten-thousandth

66. −0.1111115   millionth

67. 9.00304   thousandth

68. 2.020406   hundred-thousandth

69. −1.1060213   millionth

70. 693.003   hundredth

71. 5.738291   hundred-thousandth

72. −3.016489   ten-thousandth

Exercises 73–78: Round to the given value.

73. $3.78   dollar

74. $23.9087   cent

75. $143.298   cent

76. $8.30   dollar

77. $19.89   dollar

78. $0.9987   cent

APPLICATIONS

79. *Tall Building* The Empire State Building in New York City is among the tallest buildings in the world, with an approximate height of 1453.71 feet. Write this decimal in words. (*Source:* World Almanac, 2008.)

80. *Student Age* A college student calculates that he is about 19.3087 years old. Write this decimal in words.

81. *High Jump* As of 2009, the world record holder in the men's high jump was Javier Sotomayor of Cuba, with a jump of just over 8 feet. In international competition, heights are measured in meters. Write his record jump of 2.45 meters as a mixed number. (*Source:* World Almanac, 2008.)

82. *Baseball Stats* Baseball great Harmon Killebrew had a lifetime batting average of 0.256. Write this decimal as a fraction. (*Source:* Major League Baseball.)

83. *Sprinters* In the 1996 Olympic games, American Michael Johnson ran the 200-meter dash in 19.32 seconds. Twelve years later, Jamaican Usain Bolt ran the 200-meter dash in 19.30 seconds. Which sprinter was faster? (*Source:* Olympics.)

84. *Size of Bacteria* A bacterium in dish A measures 0.0000476 inch across, and a bacterium in dish B measures 0.0000467 inch across. Which of these bacteria is smaller?

85. *Gasoline Prices* In February 2009, the national average price of a gallon of regular unleaded gasoline was $1.909. Round this number to the nearest cent. (*Source:* Energy Information Administration.)

86. *Driving Distance* According to MapQuest, the driving distance from a restaurant in East Sandwich, Massachusetts to a hotel in Boston is about 40.78 miles. Round this value to the nearest mile.

87. *Vegetarian Diets* The fraction of U.S. adults that follow a vegetarian diet can be written in decimal form as 0.032. Write this decimal as a fraction in lowest terms. (*Source:* Vegetarian Times, 2010)

88. *Women College Students* In 2010, the fraction of college students who were women attending part-time could be expressed in decimal form as 0.377. Write this decimal in word form. (*Source:* National Center for Education Statistics.)

WRITING ABOUT MATHEMATICS

89. Explain how to round the decimal 12.9995 to the nearest thousandth.

90. Name two numbers that are located between 3.4729 and 3.4730 on a number line.

91. Explain how to compare two decimals that have the same number of digits to the left of the decimal point.

92. Is the process for rounding negative decimals to a given place value different than the process used for rounding positive decimals? Explain.

5.2   Adding and Subtracting Decimals

Estimating Decimal Sums and Differences • Adding and Subtracting Decimals •
Evaluating and Simplifying Expressions • Applications

A LOOK INTO MATH ▶   Have you ever been shopping and suddenly realized that you have only a few dollars with you? When this happens, we often try to keep a running estimate of the total cost of the items we want. If you have only $6, can you purchase a soda for $1.59, two bags of chips for $0.99 each, and a newspaper for $1.50? By estimating 1.59 + 0.99 + 0.99 + 1.50, we should see that $6 is enough for this purchase. In this section, we find estimates of decimal sums and differences before finding the exact results.

STUDY TIP

Do you have enough time to study your notes and complete your assignments? One way to manage your time is to make a list of your time commitments and determine the amount of time that each activity requires. Remember to include time for eating, sleeping, and relaxing.

Estimating Decimal Sums and Differences

Before we find exact values of decimal sums and differences, it is often helpful to first find an estimate. We can do this in several ways. For example, an estimation of the sum

$$148.19$$
$$+ \;\; 21.03$$

can be done by first rounding 148.19 to 148 and 21.03 to 21. With these rounded values, our estimate is found by adding as follows.

$$148$$
$$+ \;\; 21$$

However, this sum is not as convenient to find mentally as an estimate using the rounded values 150 and 20.

$$150$$
$$+ \;\; 20$$
$$\overline{170}$$

The actual value of 148.19 + 21.03 is 169.22. (A procedure for finding an exact sum will be discussed in the next subsection.) In this case, an estimate of 170 is reasonably close to the actual value of 169.22.

    In the next example, estimates of a decimal sum and difference are found. In each case, estimates that can be computed mentally are provided.

READING CHECK

• Why is it useful to estimate a sum or difference?

EXAMPLE 1   **Estimating a decimal sum and difference**

Estimate the sum or difference.
(a) 38.79 + 408.25   (b) 307.8 − 89.73

Solution

(a) If **38.79** is rounded to **40** and **408.25** to **410**, the estimated sum is **40** + **410** = 450. A different estimate can be found by rounding **38.79** to **40** and **408.25** to **400**. In this case, the estimated sum is **40** + **400** = 440. (The actual sum is 447.04.)

(b) If **307.8** is rounded to **310** and **89.73** to **90**, the estimated difference is **310** − **90** = 220. A different estimate can be found by rounding **307.8** to **300** and **89.73** to **100**. In this case, the estimated sum is **300** − **100** = 200. (The actual difference is 218.07.)

Now Try Exercises 5, 9

NOTE: In Example 1(b), the estimates 310 − 100 = 210 and 300 − 90 = 210 are also reasonable. All four of these estimates provide a reasonable approximation.

Adding and Subtracting Decimals

In Section 5.1, decimals were written as mixed numbers, and in Section 4.6, mixed numbers were added and subtracted. One way to add or subtract decimals is to first change them to mixed numbers. For example, the sum 11.17 + 53.42 can be found as follows.

$$11.17 + 53.42 = 11\frac{17}{100} + 53\frac{42}{100} = 64\frac{59}{100} = 64.59$$

Although this method works, there is an easier way to find the sum. Consider the same sum when it is stacked vertically.

$$\begin{array}{r} 11.17 \\ + 53.42 \\ \hline 64.59 \end{array}$$

The correct sum can be found by adding the digits in corresponding place values. Adding or subtracting decimals is similar to adding or subtracting whole numbers and integers.

CALCULATOR HELP

To add or subtract decimals with a calculator, see the Appendix (page AP-5).

ADDING AND SUBTRACTING DECIMALS

To add or subtract decimals, do the following.

1. Stack the decimals vertically with the decimal points aligned. Extra zeros may be written to the right of the last digit on any decimal number so that corresponding digits align neatly.
2. Add or subtract digits in corresponding place values. Carry or borrow as needed.
3. Place a decimal point in the result so that it is aligned vertically with the decimal points in the numbers stacked above it.

EXAMPLE 2 **Adding positive decimals**

Estimate the sum and then add the decimals.
(a) 123.45 + 56.394 **(b)** 17.2 + 134.971 + 1.84 **(c)** 68 + 12.75

Solution

(a) One possible estimate of the sum is 120 + 60 = 180. If the actual sum differs significantly from 180, a computational error may have occurred.

$$\begin{array}{r} 1 \\ 123.45\textbf{0} \longleftarrow \text{Insert a 0.} \\ + 56.394 \\ \hline 179.844 \longleftarrow \text{The result is reasonably close to the estimate.} \end{array}$$

(b) An estimate of the sum $17.2 + 134.971 + 1.84$ is $20 + 130 + 0 = 150$.

$$
\begin{array}{r}
\overset{1}{1}\overset{2}{\ }1 \\
17.2\mathbf{00} \longleftarrow \text{Insert two 0s.} \\
134.971 \\
+\ \ 1.84\mathbf{0} \longleftarrow \text{Insert a 0.} \\
\hline
154.011 \longleftarrow \text{The result is reasonably close to the estimate.}
\end{array}
$$

(c) One estimate of the sum $68 + 12.75$ is $70 + 10 = 80$. Note that in the whole number 68, a decimal point can be inserted to the right of the ones digit.

$$
\begin{array}{r}
1 \\
68.\mathbf{00} \longleftarrow \text{Insert a decimal point and two 0s.} \\
+\ 12.75 \\
\hline
80.75 \longleftarrow \text{The result is reasonably close to the estimate.}
\end{array}
$$

Now Try Exercises 19, 21, 27

EXAMPLE 3   **Subtracting positive decimals**

Estimate the difference and then subtract the decimals.
(a) $687.248 - 35.09$ **(b)** $9274.63 - 510$

Solution

(a) One possible estimate of the difference is $690 - 40 = 650$.

$$
\begin{array}{r}
\overset{1}{\ }\overset{14}{\ } \\
687.2\cancel{4}8 \\
-\ \ 35.09\mathbf{0} \longleftarrow \text{Insert a 0.} \\
\hline
652.158 \longleftarrow \text{The result is reasonably close to the estimate.}
\end{array}
$$

(b) An estimate of the difference is $9300 - 500 = 8800$.

$$
\begin{array}{r}
\overset{8}{\ }\overset{12}{\ } \\
\cancel{9}\cancel{2}74.63 \\
-\ \ 510.\mathbf{00} \longleftarrow \text{Insert a decimal point and two 0s.} \\
\hline
8764.63 \longleftarrow \text{The result is reasonably close to the estimate.}
\end{array}
$$

Now Try Exercises 23, 25

The rules used when adding and subtracting integers are also used when adding and subtracting signed decimals. In the two next examples, we add and subtract signed decimals.

EXAMPLE 4   **Adding signed decimals**

Estimate the sum and then add the decimals.
(a) $-453.62 + 27.119$ **(b)** $-99.3 + (-402.597)$

Solution

(a) To find the sum of two numbers with unlike signs, subtract the smaller absolute value from the larger absolute value and give the sum the sign of the number with the larger absolute value. One possible estimate of the sum is $-450 + 30 = -420$.

$$
\begin{array}{r}
\overset{4}{\ }\overset{13}{\ }\overset{1}{\ }\overset{10}{\ } \\
4\cancel{5}\cancel{3}.6\cancel{2}\cancel{0} \longleftarrow \text{Insert a 0 in the absolute value of } -453.62. \\
-\ \ 27.119 \longleftarrow \text{Subtract the absolute value of } 27.119. \\
\hline
426.501 \longleftarrow \text{Difference of the absolute values}
\end{array}
$$

Since $|-453.62| > |27.119|$, the result is negative: $-453.62 + 27.119 = -426.501$.

(b) To find the sum of two numbers with like signs, add absolute values and give the sum the common sign. One possible estimate of the sum is $-100 + (-400) = -500$.

$$
\begin{array}{r}
\overset{11}{} \\
99.3\mathbf{00} \\
+\ 402.597 \\
\hline
501.897
\end{array}
$$

    99.3**00** ⟵——— Insert two 0s in the absolute value of -99.3.
+ 402.597 ⟵——— Add the absolute value of -402.597.
    501.897 ⟵——— Sum of the absolute values

The addends are negative, so the sum is negative: $-99.3 + (-402.597) = -501.897$.

▌ **Now Try Exercises 31, 33**

EXAMPLE 5   ## Subtracting signed decimals

Estimate the difference and then subtract the decimals.
(a) $-2.7 - 163.902$ **(b)** $-6044.2 - (-39)$

Solution
(a) Begin by changing the difference to the sum $-2.7 + (-163.902)$. One possible estimate of this sum (and the given difference) is $-3 + (-164) = -167$.

        $\overset{1}{}$
    2.**700** ⟵——— Insert two 0s in the absolute value of -2.7.
+ 163.902 ⟵——— Add the absolute value of -163.902.
    166.602 ⟵——— Sum of the absolute values

Since the addends are both negative, this sum is -166.602. The given difference is $-2.7 - 163.902 = -166.602$.

(b) First change the difference to the sum $-6044.2 + 39$. An estimate of this sum (and the given difference) is $-6040 + 40 = -6000$.

      $\overset{3\ 14}{}$
    6 0 $\cancel{4}\cancel{4}$. 2 ⟵——— Write the absolute value of -6044.2.
−   3 9 . **0** ⟵——— Insert a decimal point and a 0 in the absolute value of 39.
    6 0 0 5 . 2 ⟵——— Difference of the absolute values

Since $|-6044.2| > |39|$, this difference is negative, and the given difference is $-6044.2 - (-39) = -6005.2$.

▌ **Now Try Exercises 35, 39**

READING CHECK

• How does addition or subtraction of decimals differ from addition or subtraction of integers?

Evaluating and Simplifying Expressions

▶ **REAL-WORLD CONNECTION** If the price of a ham sandwich is h dollars and the price of a soft drink is d dollars, then we can find the total cost of the two items by using the algebraic expression given by

$$h + d.$$

To evaluate an algebraic expression for decimal values of the variables, we use the same process as that used to evaluate algebraic expressions for integer or whole number values of the variables. That is, we replace each variable in the expression with its given decimal value. In the next example, we evaluate the expression $h + d$ for given decimal values of h and d.

| EXAMPLE 6 | **Evaluating an expression** |

If a ham sandwich costs \$3.39 and a soft drink costs \$1.79, evaluate the expression $h + d$ for $h = 3.39$ and $d = 1.79$ to find the total cost (excluding tax) of the two items.

Solution
When $h = \mathbf{3.39}$ and $d = \mathbf{1.79}$, the expression $h + d$ becomes $\mathbf{3.39} + \mathbf{1.79}$. The total cost for the ham sandwich and soft drink is \$5.18, as shown here.

$$
\begin{array}{r}
\mathbf{1\ 1} \\
3.39 \\
+\ \ 1.79 \\
\hline
\$5.18
\end{array}
$$

> **Now Try Exercise 59**

Recall that like terms in algebraic expressions can be combined. Two or more terms are like terms if they contain the same variables raised to the same powers. In the next example, like terms with decimal coefficients are combined.

| EXAMPLE 7 | **Simplifying algebraic expressions** |

Simplify the expression by combining like terms.
(a) $3.7x + 4 - 2.1x$ **(b)** $3.68y^2 - 7.2x - 1.4y^2$

Solution
(a) Begin by changing the subtraction to addition of the opposite.

$$
\begin{aligned}
3.7x + 4 \,\mathbf{-}\, 2.1x &= 3.7x + 4 \,\mathbf{+}\, (\mathbf{-}2.1x) && \text{Add the opposite.} \\
&= 3.7\mathbf{x} + (-2.1\mathbf{x}) + 4 && \text{Commutative property} \\
&= (3.7 + (-2.1))\mathbf{x} + 4 && \text{Distributive property} \\
&= 1.6x + 4 && \text{Simplify.}
\end{aligned}
$$

(b) First change each subtraction to addition of the opposite.

$$
\begin{aligned}
3.68y^2 - 7.2x - 1.4y^2 &= 3.68y^2 \,\mathbf{+}\, (\mathbf{-}7.2x) \,\mathbf{+}\, (\mathbf{-}1.4y^2) && \text{Add the opposite.} \\
&= 3.68\mathbf{y^2} + (-1.4\mathbf{y^2}) + (-7.2x) && \text{Commutative property} \\
&= (3.68 + (-1.4))\mathbf{y^2} + (-7.2x) && \text{Distributive property} \\
&= 2.28y^2 - 7.2x && \text{Simplify.}
\end{aligned}
$$

> **Now Try Exercises 63, 67**

Applications

▶ **REAL-WORLD CONNECTION**   The next three examples demonstrate everyday situations that require addition and subtraction of decimals. The first of these examples uses decimal addition to analyze changes in the stock market. The second example uses decimal subtraction to calculate the distance traveled by a car. In the last example both addition and subtraction of decimals are used to find the balance in a checkbook register.

| EXAMPLE 8 | **Analyzing stock market changes** |

In March 2009, the Dow Jones Industrial Average began a particular trade day at 7170.06 and finished the day 53.92 points higher. Determine its value at the end of the day.

Solution

To determine the ending value, we need to add $7170.06 + 53.92$.

$$
\begin{array}{r}
\overset{1}{} \\
7170.06 \\
+\ \ \ 53.92 \\
\hline
7223.98.
\end{array}
$$

The Dow Jones Industrial Average ended the day at 7223.98.

Now Try Exercise 73

EXAMPLE 9   Computing miles traveled

At the beginning of a trip, a car's odometer displayed 78,904.7 and at the end of the trip, it displayed 79,351.9. What was the total number of miles traveled on this trip?

Solution

To determine the total number of miles traveled, subtract $79,351.9 - 78,904.7$.

$$
\begin{array}{r}
8\ \ 13\ \ 4\ \ 11 \\
7\,9,3\,5\,1.9 \\
-\ 7\,8,9\,0\,4.7 \\
\hline
4\,4\,7.2
\end{array}
$$

A total of 447.2 miles was traveled on the trip.

Now Try Exercise 81

EXAMPLE 10   Finding a checking account balance

Find the final balance in the following checkbook register.

Date	Number	Description	Debits		Credits		Balance
10/7		Deposit			897	61	2347.83
10/9	5671	Electric Company	73	92			
10/10	5672	Credit Card Bill	267	14			
10/11		Deposit			207	03	
10/15	5673	Pizza Restaurant	23	76			

Solution

To find the final balance, we must subtract from the initial balance of $2347.83 all debits (payments) and add all credits (deposits). This is done from left to right, as follows.

$$
\begin{aligned}
\text{Final Balance} &= 2347.83 - 73.92 - 267.14 + 207.03 - 23.76 \\
&= 2273.91 - 267.14 + 207.03 - 23.76 \\
&= 2006.77 + 207.03 - 23.76 \\
&= 2213.80 - 23.76 \\
&= 2190.04
\end{aligned}
$$

The final balance in the checking account is $2190.04.

Now Try Exercise 89

5.2 Putting It All Together

CONCEPT	COMMENTS	EXAMPLES
Estimating Decimal Sums and Differences	By rounding decimal numbers to convenient values, a sum or difference can be estimated before the actual computation is performed.	The sum $31.2 + 194.6$ can be estimated as $30 + 190 = 220$ or $30 + 200 = 230$.
Adding or Subtracting Decimals	1. Stack the decimals vertically with the decimal points aligned. Extra zeros may be written to the right of the last digit of any decimal number so that corresponding digits align neatly. 2. Add or subtract digits in corresponding place values. Carry or borrow as needed. 3. Place a decimal point in the result so that it is aligned vertically with the decimal points in the numbers stacked above it.	The difference $458.657 - 25.47$ is found as follows. $$\begin{array}{r} 5\ 15 \\ 458.6\cancel{5}7 \\ -\ \ 25.470 \\ \hline 433.187 \end{array}$$

5.2 Exercises

MyMathLab · Math XL PRACTICE · WATCH · DOWNLOAD · READ · REVIEW

CONCEPTS AND VOCABULARY

1. Before we find an exact value of a decimal sum or difference, it is often helpful to first _____ the result.

2. When finding an estimate of a decimal sum or difference, the decimals should be _____ so that the estimate can be found mentally when possible.

3. When adding or subtracting, stack decimals vertically with the _____ aligned.

4. A decimal point is placed in the result of a sum or difference so that it is aligned _____ with the decimal points in the numbers stacked above it.

ESTIMATING SUMS AND DIFFERENCES

Exercises 5–16: Estimate the given sum or difference by rounding the decimals. Answers may vary.

5. $22.13 + 397.79$

6. $168.3 + 42.19$

7. $1479.67 - 293.74$

8. $262.78 - 57.9$

9. $689.236 - 90.793$

10. $14.3 + 2008.91$

11. $1652.917 + 349.6$

12. $8003.6 - 261.77$

13. $-302.56 + 73.9$

14. $2012.73 + (-478.22)$

15. $-41.83 - 129.6$

16. $-61.332 - (-39.74)$

Exercises 17–40: Estimate the sum or difference and then add or subtract the decimals. Estimates may vary.

17. $0.458 + 9.93$

18. $16.8 - 0.12$

19. $614.28 + 58.619$

20. $77.94 + 222.71$

21. $52 + 781.62$

22. $1198.2 + 406.39$

23. $786.429 - 28.17$

24. $153.4 - 78.95$

25. $6949 - 401.82$

26. $570.65 - 310.55$

27. $14.38 + 158.9 + 0.2$

28. $207.8 + 1.3 + 41.35$

29. $2.457 + 670.1 + 28.3$ **30.** $8.8 + 21.37 + 99.6$

31. $-48.97 + 211.67$ **32.** $309.723 + (-2.67)$

33. $-178.97 + (-19.4)$ **34.** $-8.83 + (-397.2)$

35. $-683.01 - 12.6$ **36.** $-679 - 296.7$

37. $19.04 - (-1279.9)$ **38.** $709.3 - (-7.97)$

39. $-51.61 - (-107.93)$ **40.** $-293.4 - (-11)$

Exercises 41–52: Find the actual sum or difference.

41. $-9.623 + 143.83$ **42.** $35.97 + (-192.78)$

43. $19.2 - (-781.76)$ **44.** $-13 - (-273.42)$

45. $-360.89 - 2793.4$ **46.** $998.99 + 676.99$

47. $-3.02 + (-58.672)$ **48.** $-16.9 + (-16.9)$

49. $27.901 + 1207.998$ **50.** $0.987 - (-3.457)$

51. $6004.003 - 997.448$ **52.** $100.003 - 2997.33$

EVALUATING AND SIMPLIFYING EXPRESSIONS

Exercises 53–56: Evaluate the expression $x - y$ for the given values of the variables.

53. $x = 4.702, y = 2.104$ **54.** $x = 506.4, y = 27.33$

55. $x = 0.693, y = -0.78$ **56.** $x = -12.8, y = 56$

Exercises 57–60: If a hotdog costs h dollars and a bag of chips costs c dollars, evaluate the expression $h + c$ to find the total cost of the two items.

57. Hotdog: $1.79, chips: $0.99

58. Hotdog: $3.49, chips: $1.39

59. Hotdog: $2.89, chips: $1.25

60. Hotdog: $4.79, chips: $2.55

Exercises 61–70: Simplify the algebraic expression.

61. $4.3y + 6.73y$ **62.** $12.99x - 5.2x$

63. $6.9w - 5 + 2.7w$ **64.** $16.5m + 2b - 8.32m$

65. $8.9n^2 - 7.1n^2$ **66.** $45.3x^3 + 10.7x^3$

67. $0.5p^2 + 13.6 - 3.1p^2$ **68.** $-0.45y^3 - 0.37y^3$

69. $4.3x + 8.1y - 2.7x - 5.4y$

70. $8.6a + 0.3 - 5.7b - 7.9a + 9.2b$

APPLICATIONS

71. *Steel Imports* The U.S. imported 31.8 million tons of steel in 2008 and 13.5 million tons in 2009. Find the total amount of steel imported in these two years. (*Source:* American Iron and Steel Institute.)

72. *Steel Exports* The U.S. exported 13.5 million tons of steel in 2008 and 9.3 million tons in 2009. Find the total amount of steel exported in these two years. (*Source:* American Iron and Steel Institute.)

73. *Stock Prices* If the price of a particular stock begins the day at $18.37 and rises in value by $1.15 for the day, what is the ending price of the stock?

74. *Stock Prices* If the price of a stock begins the day at $189.32 and decreases by $12.67 for the day, what is the ending price of the stock?

75. *Patents* There were 80.9 thousand patents issued to U.S. corporations in 2008 and 3.3 thousand more similar patents issued a year later. How many patents were issued to U.S. corporations in 2009? (*Source:* U.S. Patent and Trademark Office.)

76. *Lunch Cost* Find the total cost of a lunch consisting of a sandwich for $2.59, a drink for $0.99, and an apple for $0.59.

Exercises 77–80: Water Consumption The following bar graph shows the daily U.S. per capita (per person) water consumption, in thousands of gallons, during selected years. (*Source:* U.S. Geological Survey.)

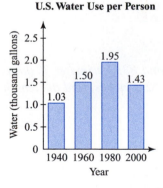

U.S. Water Use per Person

77. Which of the years shown had the highest per capita water consumption?

78. How much lower was the 2000 per capita water consumption compared to 1980?

79. How much higher was the 1980 per capita water consumption compared to 1940?

80. In which 20-year period was the increase in per capita water consumption the largest, from 1940 to 1960 or from 1960 to 1980?

81. *Driving Distance* At the beginning of a trip, a car's odometer displayed 38,410.2 and at the end of the trip, it displayed 38,707.1. What was the total number of miles traveled on this trip?

82. *Walking Distance* A pedometer is a device that measures walking distance. If a person's pedometer reads 2.54 miles at a small bridge and later reads 4.07 miles at a park bench, how far did the person walk between these two locations?

Exercises 83–86: Perimeter Find the perimeter of the given figure.

83.

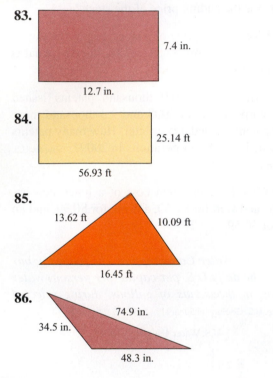

7.4 in.

12.7 in.

84.

25.14 ft

56.93 ft

85.

13.62 ft   10.09 ft

16.45 ft

86.

74.9 in.

34.5 in.

48.3 in.

Exercises 87 and 88: Perimeter If the perimeter of the given figure is 52.5 inches, find the length represented by the variable x.

87.

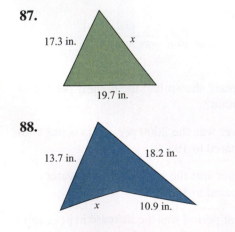

17.3 in.   x

19.7 in.

88.

13.7 in.   18.2 in.

x   10.9 in.

Exercises 89 and 90: Checkbook Register Find the final balance in the given checkbook register.

89.

Date	Num.	Description	Debits		Credits		Balance
1/2		Deposit			153	16	704.93
1/3	8322	Convenience Store	35	61			
1/3	8323	Retail Clothing Store	159	94			
1/7		Deposit			399	83	
1/9	8324	Seafood Restaurant	41	36			

90.

Date	Num.	Description	Debits		Credits		Balance
1/2		Deposit			488	51	2783.06
1/3	5310	Bait & Tackle Shop	79	23			
1/3		Deposit			953	86	
1/7	5311	Health Club	97	22			
1/9	5312	Grocery Store	178	45			

91. *Internet Use* In 2009, the average Internet user spent 1.883 hours each month on Google and 5.400 hours each month on Facebook. How much more time was spent on Facebook? (*Source:* Nielsen/NetRatings.)

92. *Kentucky Derby* The winner of the 2002 Kentucky Derby was War Emblem with a time of 121.13 seconds. In 2007, Street Sense won the Kentucky Derby with a time of 122.17 seconds. How much faster was War Emblem? (*Source:* Daily Racing Form.)

WRITING ABOUT MATHEMATICS

93. Discuss the difficulties of adding decimals as shown.

$$\begin{array}{r} 1503.98 \\ 1.3964 \\ 0.998 \\ + \quad 2.7 \\ \hline \end{array}$$

94. Explain the purpose of estimating a sum or difference before finding its actual value.

95. Give three examples from everyday life when you may need to add or subtract decimals.

96. Is it reasonable to apply the commutative, associative, and distributive properties to decimals? If so, give an example of each property.

**SECTIONS
5.1 and 5.2**

Checking Basic Concepts

1. Write the decimal 23.097 in words.

2. Write each decimal as a fraction or mixed number in simplest form.
   (a) -5.6 (b) 0.52

3. Graph the decimal 34.27 on a number line.

4. Place the correct symbol, $<$ or $>$, in the blank between the given decimals.
   (a) 15.47 ___ 15.56   (b) 0.8901 ___ 0.08901

5. Round 0.27839 to the nearest thousandth.

6. Round $8.83 to the nearest dollar.

7. Estimate each sum or difference by rounding the decimals. Answers may vary.
   (a) $149.87 + 21.32$ (b) $6993.7 - 201.6$

8. Find each sum or difference.
   (a) $28.64 + 7.38$ (b) $837.52 - 402.93$
   (c) $-159.2 + 87.54$ (d) $-29 - (-311.62)$

9. Simplify each algebraic expression.
   (a) $0.34x + 4.7x$ (b) $3.9y^2 + 2 - 0.4y^2$

10. *Lunch Cost* Find the total cost of a lunch consisting of a salad for $4.79, a drink for $1.89, and a slice of pie for $1.39.

5.3   Multiplying and Dividing Decimals

Estimating Decimal Products and Quotients • Multiplying Decimals • Dividing Decimals • Writing Fractions as Decimals • Evaluating and Simplifying Expressions • Applications

A LOOK INTO MATH ▶

Some restaurants offer buffet-style meals for a set price. A group of friends can determine the total price for the group by simply multiplying the number of people by the buffet price. For example, if the price per person is $8.69, the total cost for sixteen friends is found by multiplying (8.69)(16). In a different situation, four friends with a total bill of $58.72 can split the bill evenly by dividing 58.72 ÷ 4. These two situations demonstrate the need for multiplying and dividing decimals. In this section, we learn how to multiply and divide decimals.

NEW VOCABULARY

☐ Repeating decimal
☐ Repeat bar

Estimating Decimal Products and Quotients

It is often helpful to first find an estimate of a product or quotient of decimals before computing the actual result. Remember that there may be more than one reasonable way to round decimals when finding an estimate.

EXAMPLE 1   **Estimating a decimal product and quotient**

Estimate the product or quotient.
(a) $(6.93)(11.12)$
(b) $47.4 \div 4.8$

Solution

(a) One estimate is obtained if **6.93** is rounded to **7** and **11.12** is rounded to **11**. In this case, the estimated product is $7 \cdot 11 = 77$. A second estimate can be found by rounding **6.93** to **7** and **11.12** to **10**. In this case, the estimated product is $7 \cdot 10 = 70$. (The actual product is 77.0616.)

(b) If **47.4** is rounded to **50** and **4.8** to **5**, the estimated quotient is **50** ÷ **5** = 10. A second estimate can be found by rounding **47.4** to **45** and **4.8** to **5**. In this case, the estimated quotient is **45** ÷ **5** = 9. (The actual quotient is 9.875.)

Now Try Exercises 9, 11

NOTE: To avoid confusing the multiplication dot with a decimal point, a product such as $38.67 \cdot 13.2$ is written as 38.67(13.2) or (38.67)(13.2).

MAKING CONNECTIONS

Estimating Products and Quotients

Recall that a decimal such as 0.3 represents a fraction of a whole. If we round 0.3 to the nearest whole number in the product (2079.6)(0.3), then an estimate such as $2100 \cdot 0 = 0$ could result. Obviously, this estimate is not helpful. However, if we write 0.3 as the fraction $\frac{3}{10}$, then a better estimate of $2100 \cdot \frac{3}{10} = 630$ results. The actual product is 623.88.

Multiplying Decimals

Decimals are multiplied in ways that are similar to how whole numbers are multiplied. One difference is that we must decide where to write the decimal point in the product.

READING CHECK

- How does multiplication of decimals differ from multiplication of integers?

MULTIPLYING A DECIMAL BY A DECIMAL Decimals can be written as fractions, and we already know how to multiply fractions. So, to find a product such as (0.4)(0.7) we can start by writing each factor as a fraction.

1 decimal place
1 decimal place

$$(0.4)(0.7) = \frac{4}{10} \cdot \frac{7}{10} \qquad \text{Write the decimals as fractions.}$$

$$= \frac{28}{100} \qquad \text{Multiply.}$$

$$= 0.28 \qquad \text{Write the fraction as a decimal.}$$

2 decimal places

Note that the factor 0.4 has **1** decimal place, the factor 0.7 has **1** decimal place, and the product 0.28 has **1** + **1** = **2** decimal places. Next, we multiply (0.04)(0.007).

2 decimal places
3 decimal places

$$(0.04)(0.007) = \frac{4}{100} \cdot \frac{7}{1000} \qquad \text{Write the decimals as fractions.}$$

$$= \frac{28}{100,000} \qquad \text{Multiply.}$$

$$= 0.00028 \qquad \text{Write the fraction as a decimal.}$$

5 decimal places

In this case, the factor 0.04 has **2** decimal places, the factor 0.007 has **3** decimal places, and the product 0.00028 has **2** + **3** = **5** decimal places.

This discussion suggests that the number of decimal places in the product is equal to the *sum* of the number of decimal places in the factors. The process for multiplying decimals can be summarized as follows.

STUDY TIP

This procedure is another example of how mathematics builds on concepts that have already been studied. Try to get in the regular habit of reviewing topics from earlier parts of the text.

MULTIPLYING DECIMALS

To multiply decimals, do the following.

1. Multiply the decimals as though they were whole numbers. All decimal points may be ignored during computation.
2. Place a decimal point in the product so that the number of decimal places in the product is equal to the sum of the number of decimal places in the given factors.

EXAMPLE 2 | **Multiplying decimals**

Multiply.
(a) $(9.3)(4)$ **(b)** $0.34(7.3)$ **(c)** $28.59(1.65)$

Solution

(a)
$$\begin{array}{r} 9.3 \leftarrow \textbf{1 decimal place} \\ \underline{\times\ 4} \leftarrow \textbf{0 decimal places} \\ 37.2 \leftarrow \textbf{1 decimal place} \end{array}$$

(b)
$$\begin{array}{r} 0.34 \leftarrow \textbf{2 decimal places} \\ \underline{\times\ 7.3} \leftarrow \textbf{1 decimal place} \\ 102 \\ 2\ 380 \\ \hline 2.482 \leftarrow \textbf{3 decimal places} \end{array}$$

(c)
$$\begin{array}{r} 28.59 \leftarrow \textbf{2 decimal places} \\ \underline{\times\ 1.65} \leftarrow \textbf{2 decimal places} \\ 1\ 4295 \\ 17\ 1540 \\ 28\ 5900 \\ \hline 47.1735 \leftarrow \textbf{4 decimal places} \end{array}$$

Now Try Exercises 17, 21, 25

EXAMPLE 3 | **Multiplying signed decimals**

Multiply.
(a) $(-3.8)(6.7)$ **(b)** $(-8.2)(-0.51)$

Solution

The product of two decimals with unlike signs is negative, and the product of two decimals with like signs is positive. To avoid confusion, negative signs are omitted during computation and inserted in the final product as necessary.

CALCULATOR HELP

To multiply decimals with a calculator, see the Appendix (page AP-5).

(a)
$$\begin{array}{r} 3.8 \leftarrow \textbf{1 decimal place} \\ \underline{\times\ 6.7} \leftarrow \textbf{1 decimal place} \\ 2\ 66 \\ 22\ 80 \\ \hline 25.46 \leftarrow \textbf{2 decimal places} \end{array}$$
The product is -25.46.

(b)
$$\begin{array}{r} 8.2 \leftarrow \textbf{1 decimal place} \\ \underline{\times\ 0.51} \leftarrow \textbf{2 decimal places} \\ 82 \\ 4\ 100 \\ \hline 4.182 \leftarrow \textbf{3 decimal places} \end{array}$$
The product is 4.182.

Now Try Exercises 29, 31

MULTIPLYING A DECIMAL BY A POWER OF 10 A simple pattern arises when a decimal is multiplied by a power of 10. (To review powers of 10, see Section 1.4.) Consider the resulting products when 3.92 is multiplied by 10, 100, and 1000.

$$\begin{array}{r} 3.92 \\ \underline{\times\ \ \ 10} \\ 0\ 00 \\ 39\ 20 \\ \hline 39.20 \end{array} \qquad \begin{array}{r} 3.92 \\ \underline{\times\ 100} \\ 0\ 00 \\ 00\ 00 \\ 392\ 00 \\ \hline 392.00 \end{array} \qquad \begin{array}{r} 3.92 \\ \underline{\times\ 1000} \\ 0\ 00 \\ 00\ 00 \\ 000\ 00 \\ 3920\ 00 \\ \hline 3920.00 \end{array}$$

Note that the number **10** has **1** zero, and the product 39.2 is found by moving the decimal point in 3.92 to the right **1** place. Similarly, the number **100** has **2** zeros, and the product 392 is found by moving the decimal point in 3.92 to the right **2** places. Finally, the number **1000** has **3** zeros, and the product 3920 is found by first inserting a zero as a placeholder and then moving the decimal point in 3.920 to the right **3** places. This discussion suggests the following procedure for multiplying a decimal by a power of 10.

READING CHECK

• How do we know how many places to move the decimal point when multiplying by a power of 10?

MULTIPLYING A DECIMAL BY A POWER OF 10

The product of a decimal and a (natural number) power of 10 is found by moving the decimal point to the right the same number of places as the number of zeros in the power of 10. If needed, insert zeros at the end of the decimal as placeholders.

NOTE: Powers of 10 that are less than 1 will be discussed later in the text. The procedure presented here works only for powers of 10 such as 10, 100, 1000, etc.

EXAMPLE 4   **Multiplying decimals by powers of 10**

Multiply.
(a) $10(65.98)$ **(b)** $0.83(1000)$ **(c)** $10,000(-2.678452)$

Solution
(a) Because 1**0** has **1** zero, move the decimal point in 65.98 to the right **1** place.

$$10(65.98) = 659.8$$

(b) Insert a zero at the end of 0.83 and move the decimal point to the right 3 places.

$$0.830(1000) = 830$$

(c) $10,000(-2.678452) = -26,784.52$

Now Try Exercises 33, 39, 41

Dividing Decimals

First we will consider a process for dividing a decimal by a whole number.

DIVIDING A DECIMAL BY A WHOLE NUMBER As with multiplying decimals, we can divide decimals by first writing them as fractions and then performing the division. For example, to find the quotient $54.3 \div 3$, we begin by writing the divisor and dividend as improper fractions.

$$54.3 \div 3 = 54\frac{3}{10} \div 3 \qquad \text{Write 54.3 as a mixed number.}$$

$$= \frac{543}{10} \div \frac{3}{1} \qquad \text{Write as improper fractions.}$$

$$= \frac{543}{10} \cdot \frac{1}{3} \qquad \text{Multiply by the reciprocal.}$$

$$= \frac{(181 \cdot \mathbf{3}) \cdot 1}{10 \cdot \mathbf{3}} \qquad \text{Factor before multiplying.}$$

$$= \frac{181}{10} \qquad \text{Simplify.}$$

$$= 18\frac{1}{10} \qquad \text{Write as a mixed number.}$$

$$= 18.1 \qquad \text{Write as a decimal.}$$

After some effort, we have shown that $54.3 \div 3 = 18.1$. Now consider the results using long division to find this quotient. When the divisor is a whole number, the decimal point in the quotient is written directly above the decimal point in the dividend.

$$
\begin{array}{r}
18.1 \\
3\overline{)54.3} \\
-3 \\
\hline
24 \\
-24 \\
\hline
03 \\
-3 \\
\hline
0
\end{array}
$$

When using long division, the correct result is found directly without the need to convert the decimal to a fraction.

DIVIDING A DECIMAL BY A WHOLE NUMBER

To divide a decimal by a whole number, do the following.

1. Using long division, divide as though the dividend is a whole number. The decimal point in the dividend may be ignored during computation.
2. Place a decimal point in the quotient so that it is directly above the decimal point in the original dividend.
3. When needed, insert extra 0s to the right of the last digit in the dividend.

EXAMPLE 5   **Dividing a decimal by a whole number**

Divide. Check your result using multiplication.
(a) $57 \div 6$ **(b)** $3.72 \div 5$

Solution
The dividend in part (a) is a whole number, so a decimal point is written after its rightmost digit. In part (b), a leading 0 is inserted to the left of the decimal point in the quotient.

Leading 0 ────────┐

(a)
$$
\begin{array}{r}
9.5 \\
6\overline{)57.0} \leftarrow \text{Insert a 0.} \\
-54 \\
\hline
30 \\
-30 \\
\hline
0
\end{array}
$$

(b)
$$
\begin{array}{r}
0.744 \\
5\overline{)3.720} \leftarrow \text{Insert a 0.} \\
-35 \\
\hline
22 \\
-20 \\
\hline
20 \\
-20 \\
\hline
0
\end{array}
$$

CALCULATOR HELP

To divide decimals with a calculator, see the Appendix (page AP-5).

Check:
$$
\begin{array}{r}
9.5 \\
\times 6 \\
\hline
57.0
\end{array}
$$

Check:
$$
\begin{array}{r}
0.744 \\
\times 5 \\
\hline
3.720
\end{array}
$$

Now Try Exercises 43, 45

The digits to the right of the decimal point in some decimals continue without end in a repeating pattern. Such decimals are called **repeating decimals**. Examples include

$$0.33333\ldots, \quad 235.659659659\ldots, \quad \text{and} \quad 14.0323232\ldots.$$

A **repeat bar** is used to write repeating decimals more concisely. Using repeat bars, the repeating decimals shown at the bottom of page 323 are written as

$$0.\overline{3}, \quad 235.\overline{659}, \quad \text{and} \quad 14.0\overline{32}.$$

The quotient in the next example is a repeating decimal.

EXAMPLE 6 **Finding a repeating decimal when dividing**

Divide. $64.5 \div 9$

Solution

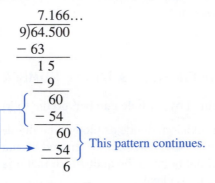

The quotient is $7.1\overline{6}$.

Now Try Exercise 51

DIVIDING A DECIMAL BY A DECIMAL When dividing a decimal by a decimal, a small adjustment can be made so that the process for dividing a decimal by a whole number can be used to find the quotient. Consider the following series of equivalent expressions.

READING CHECK

• How does division of decimals differ from division of integers?

$$4.65 \div 2.5 = \frac{4.65}{2.5} = \frac{4.65(10)}{2.5(10)} = \frac{46.5}{25} = 46.5 \div 25$$

By multiplying both the divisor and the dividend by a power of 10, the quotient $4.65 \div 2.5$ becomes the quotient $46.5 \div 25$. Since multiplying a decimal by a power of 10 simply moves the decimal point to the right, the following procedure can be used to divide a decimal by a decimal.

DIVIDING A DECIMAL BY A DECIMAL

To divide a decimal by a decimal, do the following.

1. Make the divisor a whole number by moving its decimal point to the right.
2. Move the decimal point in the dividend to the right the same number of places.
3. Find the quotient using the procedure for dividing a decimal by a whole number.

EXAMPLE 7 **Dividing a decimal by a decimal**

Divide.
(a) $40.39 \div 3.5$ **(b)** $2.9 \div 0.04$

Solution

(a) $40.39 \div 3.5 = 403.9 \div 35$

Move decimal points one place to the right.

$$
\begin{array}{r}
11.54 \\
35\overline{)403.90} \\
-35 \\
\hline
53 \\
-35 \\
\hline
18\,9 \\
-17\,5 \\
\hline
1\,40 \\
-1\,40 \\
\hline
0
\end{array}
$$

(b) $2.90 \div 0.04 = 290 \div 4$

Move decimal points two places to the right.

$$
\begin{array}{r}
72.5 \\
4\overline{)290.0} \\
-28 \\
\hline
10 \\
-8 \\
\hline
2\,0 \\
-2\,0 \\
\hline
0
\end{array}
$$

▌ **Now Try Exercises 55, 57**

DIVIDING A DECIMAL BY A POWER OF 10 When multiplying a decimal by a power of 10, the decimal point moves to the *right*. When dividing a decimal by a power of 10, the decimal point moves to the *left*. To see this, consider the resulting quotients when 54.8 is divided by 10, 100, and 1000.

$$
\begin{array}{r}
5.48 \\
10\overline{)54.80} \\
-50 \\
\hline
4\,8 \\
-4\,0 \\
\hline
80 \\
-80 \\
\hline
0
\end{array}
\qquad
\begin{array}{r}
0.548 \\
100\overline{)54.800} \\
-50\,0 \\
\hline
4\,80 \\
-4\,00 \\
\hline
800 \\
-800 \\
\hline
0
\end{array}
\qquad
\begin{array}{r}
0.0548 \\
1000\overline{)54.8000} \\
-50\,00 \\
\hline
4\,800 \\
-4\,000 \\
\hline
8000 \\
-8000 \\
\hline
0
\end{array}
$$

The number of places that the decimal point moves to the left is equal to the number of zeros in the power of 10. This process is summarized as follows.

READING CHECK

• How do we know how many places to move the decimal point when dividing by a power of 10?

> ### DIVIDING A DECIMAL BY A POWER OF 10
>
> The quotient of a decimal and a (natural number) power of 10 is found by moving the decimal point to the left the same number of places as the number of zeros in the power of 10. If needed, insert zeros at the beginning of the decimal as placeholders.

EXAMPLE 8   **Dividing decimals by powers of 10**

Divide.

(a) $245.92 \div 10$ (b) $3.9 \div 1000$ (c) $-617 \div 100$

Solution

(a) Because **10** has **1** zero, move the decimal point in 245.92 to the left **1** place.

$$245.92 \div 10 = 24.592$$

(b) Insert two zeros at the beginning of 3.9 and move the decimal point to the left 3 places.

$$003.9 \div 1000 = 0.0039$$

(c) $-617 \div 100 = -6.17$

▌ **Now Try Exercises 63, 67, 69**

Writing Fractions as Decimals

There are two methods for writing fractions as decimals. The first method involves finding an equivalent fraction whose denominator is a power of 10. The second method uses long division to find a decimal representation.

METHOD I: WRITING THE DENOMINATOR AS A POWER OF 10 Some fractions can be written as decimals by multiplying the numerator and denominator by the same number to result in an equivalent fraction whose denominator is a power of 10. This method is illustrated in the next example.

EXAMPLE 9   **Writing a fraction as a decimal—Method I**

Write each fraction as a decimal.

(a) $\dfrac{3}{4}$ (b) $-\dfrac{7}{125}$

Solution

(a) Because $4 \cdot 25 = 100$, we multiply the numerator and denominator of $\frac{3}{4}$ by **25** to get an equivalent fraction whose denominator is a power of 10.

$$\frac{3}{4} = \frac{3 \cdot 25}{4 \cdot 25} = \frac{75}{100} = 0.75$$

(b) Because $125 \cdot 8 = 1000$, we multiply the numerator and denominator of $-\frac{7}{125}$ by **8** to get an equivalent fraction whose denominator is a power of 10.

$$-\frac{7}{125} = -\frac{7 \cdot 8}{125 \cdot 8} = -\frac{56}{1000} = -0.056$$

Now Try Exercises 73, 77

METHOD II: USING LONG DIVISION Method I works well if the denominator of the given fraction is a convenient factor of a power of 10, but it is not practical for writing fractions such as $\frac{14}{37}$ in decimal form. In this case, long division is used to write the decimal form.

EXAMPLE 10   **Writing a fraction as a decimal—Method II**

Write each fraction as a decimal.

(a) $\dfrac{2}{3}$ (b) $\dfrac{7}{12}$ (c) $-\dfrac{5}{8}$

Solution

(a) Use long division to divide 3 into 2.

$$
\begin{array}{r}
0.666\ldots \\
3\overline{)2.000} \\
-\ 18 \\
\hline
20 \\
-\ 18 \\
\hline
20 \\
-\ 18 \\
\hline
2
\end{array}
$$

The decimal form is $0.\overline{6}$.

(b) Use long division to divide 12 into 7.

$$
\begin{array}{r}
0.5833\ldots \\
12\overline{)7.0000} \\
-\ 60 \\
\hline
1\ 00 \\
-\ 96 \\
\hline
40 \\
-\ 36 \\
\hline
40 \\
-\ 36 \\
\hline
4
\end{array}
$$

The decimal form is $0.58\overline{3}$.

(c) Ignoring the negative sign, we use long division to divide 8 into 5.

$$
\begin{array}{r}
0.625 \\
8\overline{)5.000} \\
-\,4\,8 \\
\hline
20 \\
-\,16 \\
\hline
40 \\
-\,40 \\
\hline
0
\end{array}
$$

Inserting the negative sign results in -0.625.

Now Try Exercises 79, 81, 83

NOTE: We could have used Method I to write the fraction in Example 10(c) as a decimal.

While Method I works for *some* fractions, Method II works for *all* fractions. The two methods can be summarized as follows.

READING CHECK

• Which method for writing fractions as decimals works for all fractions?

WRITING FRACTIONS AS DECIMALS

Method I: Multiply the numerator and denominator of the fraction by the same (nonzero) number to get an equivalent fraction whose denominator is a power of 10. Then write the new fraction as a decimal.

Method II: Use long division to divide the fraction's denominator into its numerator.

WRITING MIXED NUMBERS AS DECIMALS A mixed number can be written as a decimal by first writing the integer part of the mixed number to the left of a decimal point. Then Method I or Method II can be used to write the fractional part of the mixed number as the decimal digits to the right of the decimal point. The next example illustrates this process.

EXAMPLE 11 **Writing a mixed number as a decimal**

Write the mixed number $-11\frac{7}{12}$ as a decimal.

Solution

Because -11 is the integer part of the mixed number, we write -11 to the left of the decimal point and then use Method II to write the fraction $\frac{7}{12}$ as the decimal digits to the right of the decimal point. In Example 10(b), we used Method II to find that $\frac{7}{12} = 0.58\overline{3}$. The decimal form of the mixed number $-11\frac{7}{12}$ is $-11.58\overline{3}$.

Now Try Exercise 87

Evaluating and Simplifying Expressions

Variable expressions that involve multiplication or division can be evaluated by replacing the variable(s) with decimal numbers. The next example illustrates this process.

EXAMPLE 12 **Evaluating an expression**

Evaluate each expression for $x = 2.5$ and $y = 50.3$.

(a) xy **(b)** $\dfrac{y}{x}$

Solution

(a) Replacing x with **2.5** and y with **50.3** in the expression xy results in $(2.5)(50.3)$.

$$
\begin{array}{r}
50.3 \\
\times\ 2.5 \\
\hline
25\ 15 \\
100\ 60 \\
\hline
125.75
\end{array}
$$

Evaluating the expression xy for $x = 2.5$ and $y = 50.3$ results in 125.75.

(b) Replacing x with **2.5** and y with **50.3** in the expression $\frac{y}{x}$ results in $\frac{50.3}{2.5}$.

$$50.3 \div 2.5 = 503 \div 25$$

Move decimal points one place right.

$$
\begin{array}{r}
20.12 \\
25\overline{)503.00} \\
-\,50 \\
\hline
3 \\
-\,0 \\
\hline
3\,0 \\
-\,2\,5 \\
\hline
50 \\
-\,50 \\
\hline
0
\end{array}
$$

Evaluating the expression $\frac{y}{x}$ for $x = 2.5$ and $y = 50.3$ results in 20.12.

Now Try Exercises 91, 95

We can also multiply or divide an algebraic expression by a decimal, as illustrated in the next example.

EXAMPLE 13 **Simplifying algebraic expressions**

Simplify each expression.

(a) $3.4(1.8x + 4)$ **(b)** $\dfrac{4.8y}{3}$

Solution

(a) Use the distributive property first.

$$3.4(1.8x + 4) = 3.4(1.8x) + 3.4(4) = 6.12x + 13.6$$

(b) Divide 3 into 4.8.

$$\frac{4.8y}{3} = \frac{4.8}{3} \cdot \frac{y}{1} = 1.6y$$

Now Try Exercises 99, 103

Applications

▶ **REAL-WORLD CONNECTION** In an effort to offer cars with improved gas mileage, automobile manufacturers have created gas/electric hybrid cars. Gas mileage M is measured in miles per gallon and is computed using the formula

$$M = \frac{m}{g},$$

where m is the number of miles driven and g is the number of gallons of gasoline used.

EXAMPLE 14 **Computing gas mileage**

Find the mileage for a hybrid vehicle that travels 308.2 miles on 6.7 gallons of gas.

Solution

Divide the number of gallons into the number of miles. Move each decimal point one place to the right before using long division.

$$
\begin{array}{r}
46 \\
67\overline{)3082} \\
-\,268 \\
\hline
402 \\
-\,402 \\
\hline
0
\end{array}
$$

The hybrid vehicle's mileage is 46 miles per gallon.

Now Try Exercise 109

EXAMPLE 15 **Finding the cost of catering**

A catering company charges $8.89 for each person attending a company dinner. What is the total cost if 76 people attend the dinner?

Solution

Multiply the number of people by the per-person cost.

$$
\begin{array}{r}
8.89 \\
\times\ 76 \\
\hline
53\ 34 \\
622\ 30 \\
\hline
675.64
\end{array}
$$

The total cost for the dinner is $675.64.

Now Try Exercise 117

5.3 Putting It All Together

CONCEPT	COMMENTS	EXAMPLES
Estimating Decimal Products and Quotients	By rounding decimal numbers to convenient values, a product or quotient can be estimated before the actual computation is performed.	The product $(7.04)(10.9)$ can be estimated as $7 \cdot 11 = 77$.
Multiplying Decimals	1. Multiply the decimals as though they were whole numbers. All decimal points may be ignored during computation. 2. Place a decimal point in the product so that the number of decimal places in the product is equal to the sum of the number of decimal places in the given factors.	$\begin{array}{r} 13.7 \\ \times\ 8.1 \\ \hline 1\ 37 \\ 109\ 60 \\ \hline 110.97 \end{array}$

continued on next page

continued from previous page

CONCEPT	COMMENTS	EXAMPLES
Multiplying a Decimal by a Power of 10	Move the decimal point to the right the same number of places as the number of zeros in the power of 10. If needed, insert zeros at the end of the decimal as placeholders.	$100(5.673) = 567.3$ $10(0.42) = 4.2$
Dividing a Decimal by a Whole Number	1. Divide as though the dividend is a whole number. The decimal point in the dividend may be ignored during computation. 2. Place a decimal point in the quotient so that it is directly above the decimal point in the original dividend. 3. When needed, place extra 0s to the right of the last digit in the dividend.	The quotient $151.9 \div 7$ is found as follows. $$\begin{array}{r} 21.7 \\ 7\overline{)151.9} \\ -14 \\ \hline 11 \\ -7 \\ \hline 4\,9 \\ -4\,9 \\ \hline 0 \end{array}$$
Dividing a Decimal by a Decimal	1. Make the divisor a whole number by moving its decimal point to the right. 2. Move the decimal point in the dividend to the right the same number of places. 3. Find the quotient using the procedure for dividing a decimal by a whole number.	$0.68 \div 0.2 = 6.8 \div 2$ $$\begin{array}{r} 3.4 \\ 2\overline{)6.8} \\ -6 \\ \hline 8 \\ -8 \\ \hline 0 \end{array}$$
Dividing a Decimal by a Power of 10	Move the decimal point to the left the same number of places as the number of zeros in the power of 10. If needed, insert zeros at the beginning of the decimal as placeholders.	$114.2 \div 10 = 11.42$ $27.3 \div 100 = 0.273$
Writing Fractions as Decimals	**Method I:** Multiply the numerator and denominator of the fraction by the same (nonzero) number to get an equivalent fraction whose denominator is a power of 10. Then write the new fraction as a decimal. **Method II:** Use long division to divide the fraction's denominator into its numerator.	**Method I:** Write $\frac{3}{5}$ as a decimal. $$\frac{3}{5} = \frac{3 \cdot 2}{5 \cdot 2} = \frac{6}{10} = 0.6$$ **Method II:** Write $\frac{1}{2}$ as a decimal. $$\begin{array}{r} 0.5 \\ 2\overline{)1.0} \\ -1\,0 \\ \hline 0 \end{array}$$

5.3   Exercises

MyMathLab   Math XL PRACTICE   WATCH   DOWNLOAD   READ   REVIEW

CONCEPTS AND VOCABULARY

1. Before we find an exact value of a decimal product or quotient, it is often helpful to _____ the result.

2. The number of decimal places in a product for decimal numbers is equal to the _____ of the number of decimal places in the given factors.

3. When multiplying a decimal by a power of 10, move the decimal point to the _____ as many places as there are zeros in the power of 10.

4. When dividing a decimal by a whole number, place the decimal point in the quotient so that it is directly _____ the decimal point in the dividend.

5. When dividing a decimal by a decimal, start by making the divisor a whole number by moving its decimal point to the _____.

6. If the decimal point in the divisor is moved 2 places to make the divisor a whole number, then how many places should the decimal point in the dividend be moved?

7. When dividing a decimal by a power of 10, move the decimal point to the _____ as many places as there are zeros in the power of 10.

8. The method that can *always* be used to write a fraction as a decimal involves using long division to divide the fraction's _____ into its _____.

ESTIMATING PRODUCTS AND QUOTIENTS

Exercises 9–16: Estimate the given product or quotient by rounding the decimals. Answers may vary.

9. $(2.1)(26.97)$
10. $0.92(543.11)$
11. $11.984 \div 4.031$
12. $101.5 \div 19.93$
13. $489.67(5.23)$
14. $(6.99)(10.01)$
15. $87.035 \div 0.996$
16. $52.099 \div 12.9001$

MULTIPLYING DECIMALS

Exercises 17–32: Multiply.

17. $9(13.7)$
18. $6(41.3)$
19. $(0.7)(4.2)$
20. $(0.3)(86.1)$
21. $(5.9)(0.67)$
22. $(6.32)(9.5)$
23. $3.99(4)$
24. $(2.59)(18)$
25. $10.01(3.44)$
26. $(15.02)(3.04)$
27. $-3(14.6)$
28. $11(-7.9)$
29. $(-3.9)(1.8)$
30. $(8.3)(-4.3)$
31. $(-12.3)(-0.17)$
32. $(-7.91)(-9.4)$

Exercises 33–42: Multiply.

33. $10(12.489)$
34. $10(-0.399)$
35. $(4.679)(-100)$
36. $(100)(0.035)$
37. $(100)(4.1)$
38. $(-100)(0.007)$
39. $1000(-0.0098)$
40. $(1000)(1.4)$
41. $10,000(3.4498)$
42. $10,000(-0.006)$

DIVIDING DECIMALS

Exercises 43–62: Divide.

43. $47 \div 5$
44. $98 \div 4$
45. $8.91 \div 3$
46. $56.42 \div 7$
47. $-103.2 \div 8$
48. $-4.96 \div 5$
49. $14 \div 3$
50. $16 \div 9$
51. $22.4 \div 6$
52. $102.7 \div 9$
53. $35.88 \div 26$
54. $377.06 \div 34$
55. $38.35 \div 2.6$
56. $77.184 \div 3.6$
57. $6.9 \div 0.08$
58. $5.1 \div 0.04$
59. $-0.94 \div 2.5$
60. $-6.72 \div 1.44$
61. $720 \div (-1.2)$
62. $350 \div (-2.5)$

Exercises 63–72: Divide.

63. $17.79 \div 10$
64. $-452.6 \div 10$
65. $63.4 \div (-100)$
66. $0.3 \div 100$
67. $7894 \div 100$
68. $-5904 \div 100$
69. $7.6 \div (-1000)$
70. $8.6 \div 1000$
71. $1 \div 10,000$
72. $-7 \div 10,000$

WRITING FRACTIONS AS DECIMALS

Exercises 73–84: Write the fraction as a decimal.

73. $\dfrac{1}{4}$ **74.** $\dfrac{4}{5}$

75. $\dfrac{3}{8}$ **76.** $\dfrac{11}{20}$

77. $-\dfrac{6}{25}$ **78.** $-\dfrac{9}{125}$

79. $\dfrac{1}{3}$ **80.** $\dfrac{4}{9}$

81. $\dfrac{4}{15}$ **82.** $\dfrac{19}{30}$

83. $-\dfrac{73}{200}$ **84.** $-\dfrac{7}{250}$

Exercises 85–90: Write the mixed number as a decimal.

85. $3\dfrac{1}{5}$ **86.** $-10\dfrac{7}{8}$

87. $-9\dfrac{8}{15}$ **88.** $92\dfrac{3}{20}$

89. $17\dfrac{2}{3}$ **90.** $-2\dfrac{2}{9}$

EVALUATING AND SIMPLIFYING EXPRESSIONS

Exercises 91–94: Evaluate the expression xy for the given values of the variables.

91. $x = 4.6, y = 10.8$ **92.** $x = 103.6, y = 0.7$

93. $x = 0.44, y = -5.31$ **94.** $x = -19.9, y = 8$

Exercises 95–98: Evaluate the expression $\frac{y}{x}$ for the given values of the variables.

95. $x = 0.3, y = 12.66$ **96.** $x = 13.2, y = 117.48$

97. $x = 1.6, y = -0.08$ **98.** $x = -1.3, y = 39$

Exercises 99–106: Simplify the algebraic expression.

99. $2.7(1.3x + 8)$ **100.** $-0.5(6.4y - 7)$

101. $-5.1(4y + 3.6)$ **102.** $0.8(0.2y + 0.1)$

103. $\dfrac{6.4x}{8}$ **104.** $\dfrac{33.46y}{1.4}$

105. $\dfrac{-88w}{1.1}$ **106.** $\dfrac{-36m}{0.9}$

APPLICATIONS

107. *Burning Calories* The average 168-pound person burns 8.9 calories per minute while backpacking. How many calories would a person of this weight burn while backpacking for 65 minutes? (*Source:* The Calorie Control Council.)

108. *Exchange Rates* In early 2009, a person could exchange 1 U.S. dollar for 1.237 Canadian dollars. How many Canadian dollars could a person get for 180 U.S. dollars?

109. *Gas Mileage* Find the mileage for a Smart car that travels 198.9 miles on 3.9 gallons of gas.

110. *Gas Mileage* Find the mileage for a gas-powered scooter that travels 58.4 miles on 0.8 gallon.

111. *Buying Groceries* If the unit price for a box of cereal is $0.24 per ounce, how much does a 19.5-ounce box of the cereal cost?

112. *Running a Relay* Each runner in a relay took 14.2 seconds for her leg of the race. If there were four runners, what was the total time?

113. *Estimated Taxes* A small business owner expects to owe $1056.48 next year in estimated taxes. If estimated taxes are paid in four equal payments, how much is each payment?

114. *Car Payments* An automobile manufacturer offers a zero-interest loan on new car purchases. If a buyer finances $27,891 for 5 years (60 equal payments), how much is the monthly payment?

115. *Home Loans* Most people take out a mortgage when they purchase a home. A mortgage is a loan to be paid back in equal payments that include interest. If the monthly payments for a $235,000 mortgage are $1275.93, how much is paid to the bank over 360 months (30 years)?

116. *Home Loans* If a bank offered a zero-interest loan for the $235,000 mortgage in Exercise 115, what would the equal monthly payments be for a 360-month (30-year) loan?

117. *Feeding the Team* A head coach takes his football team to a buffet restaurant for lunch. What is the total bill for 54 people if the restaurant charges $9.49 per person?

118. *Feeding the Team* A coach pays $515.78 to take 41 people to dinner at a buffet restaurant. How much does the restaurant charge per person for a buffet dinner?

WRITING ABOUT MATHEMATICS

119. Find two methods for estimating the product

$$89.7(0.41).$$

Which of your methods is most accurate?

120. Find a fast way to multiply the decimal 186.34 by each of the following fractions.

$$\frac{1}{10}, \frac{1}{100}, \text{ and } \frac{1}{1000}$$

How does your method compare to that used to multiply by powers of 10 such as 10, 100, and 1000?

5.4 Real Numbers, Square Roots, and Order of Operations

Real Numbers • Revisiting Square Roots • Order of Operations • Applications

A LOOK INTO MATH ▶

There are infinitely many numbers between any two numbers on a number line. So far in this text, we have used number lines to graph natural numbers, whole numbers, integers, and rational numbers. All of these numbers have decimal representations whose digits either terminate (end) or repeat indefinitely. In this section, we will discuss the *real* numbers, which include all of the sets of numbers listed above and also include a new set called the *irrational* numbers.

STUDY TIP

Some computations in this section are quite involved. Students with a firm grasp of multiplication facts are likely to complete these computations with more ease and accuracy than students who lack such knowledge. For a review of basic multiplication facts, see Table 1.6 in Section 1.3.

NEW VOCABULARY

☐ Irrational number
☐ Real number

Real Numbers

As defined in Section 4.1, a rational number is any number that can be written as a fraction whose numerator and denominator are both integers (and the denominator is not 0). Note that *all* natural numbers, whole numbers, and integers can be written as rational numbers by using 1 as the denominator. Are there numbers that are not rational numbers? The answer is yes. Such numbers are called irrational numbers. Examples include

$$\sqrt{7}, \quad \pi, \quad \text{and} \quad 3.1010010001\ldots.$$

The symbol π is the Greek letter pi. It is used to denote a specific irrational number that is frequently encountered when working with circles. The digits in the decimal representation of π neither terminate nor repeat. In fact, every irrational number has a *nonterminating, nonrepeating* decimal representation. Decimal approximations for the irrational numbers $\sqrt{7}$ and π are

$$\sqrt{7} \approx 2.64575131 \quad \text{and} \quad \pi \approx 3.14159265.$$

An **irrational number** cannot be expressed as a fraction whose numerator and denominator are both integers. In other words, it is *not* rational. Furthermore, its decimal representation is nonterminating and nonrepeating.

The **real numbers** are numbers that can be written as decimals. They include all natural numbers, whole numbers, integers, rational numbers, and irrational numbers. The set of real numbers is represented visually in Figure 5.6 on the next page.

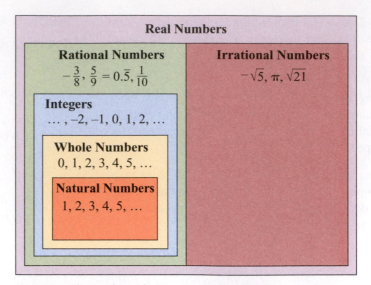

Figure 5.6 The Real Numbers

- What are the two sets of numbers that include all real numbers?

Note that the numbers $-\sqrt{5}$ and $\sqrt{21}$ are listed as irrational numbers. In fact, if the square root of a whole number is not a whole number, then it is an irrational number. In the next example, real numbers are identified as rational numbers, irrational numbers, integers, whole numbers, and natural numbers.

EXAMPLE 1 **Identifying real numbers**

Identify the numbers in the given list that belong to each of the following sets of numbers.
(a) natural numbers   **(b)** whole numbers   **(c)** integers
(d) rational numbers   **(e)** irrational numbers

$$\sqrt{11}, \quad -\frac{2}{3}, \quad 0, \quad 4.2, \quad \sqrt{9}, \quad \text{and} \quad -13$$

Solution
(a) Because 9 is a perfect square, $\sqrt{9} = 3$. The only natural number in the list is $\sqrt{9}$.
(b) Whole numbers include 0 and any natural numbers. The whole numbers are 0 and $\sqrt{9}$.
(c) Integers include whole numbers and any numbers that are opposites of whole numbers. The integers are 0, $\sqrt{9}$, and −13.
(d) Rational numbers include the integers and any fractions, terminating decimals, or repeating decimals. The rational numbers are $-\frac{2}{3}$, 0, 4.2, $\sqrt{9}$, and −13.
(e) Any number that is not rational is irrational. The only irrational number is $\sqrt{11}$.

Now Try Exercise 9

Revisiting Square Roots

Up to this point in the text, square roots such as $\sqrt{10}$ and $\sqrt{23}$ have been approximated to the nearest whole number. There are, however, several methods for computing more accurate approximations for square roots of whole numbers that are not perfect squares. Two such methods are discussed in this subsection.

THE GUESS-AND-CHECK METHOD The process for finding a square root using the guess-and-check method can be summarized as follows.

CALCULATOR HELP

To approximate a square root with a calculator, see the Appendix (page AP-5).

APPROXIMATING SQUARE ROOTS BY GUESS-AND-CHECK

To approximate a square root, do the following.

1. Use a number line to make a reasonable guess for the square root.
2. Make a table of values based on your guess. Compute the square of each value in the table until the given radicand is between consecutive values in the table.
3. If the desired accuracy for the approximation has been found, stop. Otherwise, make a more accurate guess of the square root and go back to Step 2.

EXAMPLE 2   **Approximating a square root using guess-and-check**

Approximate $\sqrt{11}$ to the nearest hundredth.

Solution
The radicand **11** is located on the number line between the perfect squares **9** and **16**, as shown in Figure 5.7.

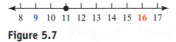

Figure 5.7

READING CHECK

• How is the guess-and-check method used to estimate a square root?

Because 11 is closer to 9 than to 16, it is reasonable to think that $\sqrt{11}$ is closer to $\sqrt{9} = 3$ than to $\sqrt{16} = 4$. A reasonable initial guess for $\sqrt{11}$ might be 3.3. To check this guess, complete a table such as Table 5.1. For example, the square of 3.2 is found by multiplying $(3.2)(3.2) = 10.24$. The other squares are found in a similar manner.

TABLE 5.1

Guess	3.2	3.3	3.4	3.5
Guess Squared	10.24	**10.89**	**11.56**	12.25

The radicand 11 is between **10.89** and **11.56** but is closer to 10.89. Our next guess for $\sqrt{11}$ might be 3.32. To check this guess, complete a table such as Table 5.2.

TABLE 5.2

Guess	3.31	**3.32**	3.33	3.34
Guess Squared	10.9561	**11.0224**	11.0889	11.1556

Table 5.2 reveals that the square of **3.32** is closest to 11. Based on this observation, we can conclude that $\sqrt{11} \approx 3.32$ when rounded to the nearest hundredth.

Now Try Exercise 17

READING CHECK

• What is the primary use of the Babylonian algorithm?

THE BABYLONIAN ALGORITHM Nearly 4000 years ago, the Babylonians discovered a technique for finding accurate approximations for square roots. The process involves only basic arithmetic and is outlined as follows.

THE BABYLONIAN ALGORITHM FOR SQUARE ROOTS

To approximate $\sqrt{Q}$ where Q represents a whole number, do the following.

1. Make a reasonable guess for the square root. Assign this value to the variable A.
2. Compute the quotient $Q \div A$.*
3. Add A to the result from Step 2.
4. Divide the result of Step 3 by 2. The result is an approximation for $\sqrt{Q}$.*

For a more accurate approximation, assign the result from Step 4 to the variable A and repeat the process beginning at Step 2.

* To avoid complicated arithmetic, these results may be rounded to 4 decimal places.

NOTE: When the initial guess is reasonably accurate, the number of correct digits in the approximation for $\sqrt{Q}$ will double each time through the algorithm.

EXAMPLE 3 **Approximating a square root using the Babylonian algorithm**

Go through the Babylonian algorithm two times to approximate the value of $\sqrt{19}$. Give four decimal places in your answer.

Solution

STEP 1: Since the radicand **19** is closer to **16** than to 25, a reasonable initial guess for $\sqrt{19}$ is a value that is slightly more than $\sqrt{16} = 4$. Let A be 4.4.

STEP 2: The quotient $Q \div A$ results in $19 \div 4.4 = 4.3\overline{18}$. For ease of computation, round this value to 4.3182.

STEP 3: Adding A to the result from Step 2 gives $4.3182 + 4.4 = 8.7182$.

STEP 4: Dividing the result from Step 3 by 2 gives $8.7182 \div 2 = 4.3591$.

Let $A = 4.3591$ and repeat the process beginning at Step 2.

STEP 2: The quotient $Q \div A$ results in $19 \div 4.3591 \approx 4.3587$.

STEP 3: Adding A to the result from Step 2 gives $4.3587 + 4.3591 = 8.7178$.

STEP 4: Dividing the result from Step 3 by 2 gives $8.7178 \div 2 = 4.3589$.

By the Babylonian algorithm, $\sqrt{19} \approx 4.3589$ when rounded to four decimal places.

Now Try Exercise 25

NOTE: To nine decimal places, $\sqrt{19} \approx 4.358898944$. Two or three passes through the Babylonian algorithm can give amazingly accurate results. However, to obtain better approximations, results from Steps 2 and 4 should *not* be rounded along the way.

Order of Operations

The order of operations agreement that applies to decimals is the same as the agreement used for whole numbers, integers, and fractions.

ORDER OF OPERATIONS

1. Do all calculations within grouping symbols such as parentheses and radicals, or above and below a fraction bar.
2. Evaluate all exponential expressions.
3. Do all multiplication and division from *left to right*.
4. Do all addition and subtraction from *left to right*.

EVALUATING NUMERICAL EXPRESSIONS The next two examples demonstrate how the order of operations agreement applies to numerical expressions involving decimals.

EXAMPLE 4 **Evaluating numerical expressions**

Evaluate each expression.
(a) $6.8 + 4.1(3) - 2$ **(b)** $9.83 + 17.8 \div 10$

Solution
(a) Perform multiplication before addition or subtraction.

$$6.8 + \mathbf{4.1(3)} - 2 = 6.8 + \mathbf{12.3} - 2 \quad \text{Multiply.}$$
$$= 19.1 - 2 \quad \text{Add.}$$
$$= 17.1 \quad \text{Subtract.}$$

(b) Perform division before addition.

$$9.83 + \mathbf{17.8 \div 10} = 9.83 + 1.78 \quad \text{Divide.}$$
$$= 11.61 \quad \text{Add.}$$

Now Try Exercises 35, 37

EXAMPLE 5 **Evaluating numerical expressions involving grouping symbols**

Evaluate each expression.

(a) $-3.2 + 5(4.1 - 0.6)$ **(b)** $-14.8 \div \sqrt{21.2 - 5.2}$ **(c)** $\dfrac{0.6 + 2.7 \div 3}{0.03(1000)}$

Solution
(a) The expression within parentheses is evaluated first.

$$-3.2 + 5(\mathbf{4.1 - 0.6}) = -3.2 + 5(\mathbf{3.5}) \quad \text{Subtract within parentheses.}$$
$$= -3.2 + 17.5 \quad \text{Multiply.}$$
$$= 14.3 \quad \text{Add.}$$

(b) The expression under the radical is evaluated first.

$$-14.8 \div \sqrt{\mathbf{21.2 - 5.2}} = -14.8 \div \sqrt{\mathbf{16}} \quad \text{Subtract under radical.}$$
$$= -14.8 \div \mathbf{4} \quad \text{Evaluate } \sqrt{16}.$$
$$= -3.7 \quad \text{Divide.}$$

(c) Use the order of operations both above and below the fraction bar.

$$\frac{0.6 + \mathbf{2.7 \div 3}}{\mathbf{0.03(1000)}} = \frac{0.6 + \mathbf{0.9}}{\mathbf{30}} \quad \text{Divide above; multiply below.}$$
$$= \frac{1.5}{30} \quad \text{Add.}$$
$$= 0.05 \quad \text{Divide.}$$

Now Try Exercises 39, 41, 45

EVALUATING ALGEBRAIC EXPRESSIONS The order of operations agreement often is needed when algebraic expressions are evaluated for decimal values of the variable(s). The next example demonstrates this situation.

EXAMPLE 6

Evaluating algebraic expressions

Evaluate each expression for $x = 4.2$ and $y = 2.1$.
(a) $-4x + y$ **(b)** $5(y - 0.3) - x$

Solution

(a) Start by replacing x with **4.2** and y with **2.1** in the given expression.

$$\begin{array}{lll} -4x + y &= -4(\mathbf{4.2}) + \mathbf{2.1} & x = 4.2 \text{ and } y = 2.1. \\ &= -16.8 + 2.1 & \text{Multiply } -4(4.2). \\ &= -14.7 & \text{Add.} \end{array}$$

(b) Replace x with **4.2** and y with **2.1** in the given expression.

$$\begin{array}{lll} 5(y - 0.3) - x &= 5(\mathbf{2.1} - 0.3) - \mathbf{4.2} & x = 4.2 \text{ and } y = 2.1. \\ &= 5(1.8) - 4.2 & \text{Evaluate } (2.1 - 0.3). \\ &= 9 - 4.2 & \text{Multiply.} \\ &= 4.8 & \text{Subtract.} \end{array}$$

Now Try Exercises 51, 53

Applications

▶ **REAL-WORLD CONNECTION** Each year, U.S. airlines transport billions of bags. Table 5.3 lists the number of bags per thousand that were mishandled by selected U.S. airlines from January to October 2009.

TABLE 5.3 Mishandled Bags per Thousand

Airline	American	Continental	Delta	US Airways
Bags	4.35	2.72	4.90	2.97

Source: U.S. Department of Transportation.

In the next example, we compute the average number of mishandled bags per thousand for these four airlines.

EXAMPLE 7

Finding the average number of mishandled bags

Find the average number of bags per thousand that were mishandled by the four airlines in Table 5.3.

Solution

To find the average of **4.35**, **2.72**, **4.90**, and **2.97**, add the numbers together and then divide the result by 4 because there are **4** numbers in the list.

$$\text{Average} = \frac{\mathbf{4.35} + \mathbf{2.72} + \mathbf{4.90} + \mathbf{2.97}}{\mathbf{4}} = \frac{14.94}{4} = 3.735$$

From January to October 2009, these four airlines mishandled an average of 3.735 bags per thousand.

Now Try Exercise 59

In the next example, the order of operations agreement is applied when finding the area of a trapezoid whose sides have decimal lengths.

EXAMPLE 8   **Finding the area of a trapezoid**

Find the area of the trapezoid shown in the following figure.

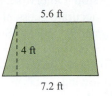

5.6 ft

4 ft

7.2 ft

Solution

The formula for the area of a trapezoid is

$$A = \frac{1}{2}(a + b)h,$$

where a and b are the bases and h is the height.

$$A = \frac{1}{2}(5.6 + 7.2)(4) \quad a = 5.6, b = 7.2, \text{ and } h = 4$$

$$= \frac{1}{2}(12.8)(4) \qquad \text{Add } (5.6 + 7.2).$$

$$= 0.5(12.8)(4) \qquad \text{Write } \frac{1}{2} \text{ as } 0.5.$$

$$= 25.6 \qquad \text{Multiply.}$$

The area of the trapezoid is 25.6 square feet.

Now Try Exercise 63

5.4   Putting It All Together

CONCEPT	COMMENTS	EXAMPLES
Irrational Numbers	Irrational numbers cannot be expressed as fractions whose numerator and denominator are both integers. If the square root of a whole number is not a whole number, then it is an irrational number.	$\sqrt{13} \approx 3.605551275$ $\pi \approx 3.141592654$
Real Numbers	Real numbers can be written as decimals. They include natural numbers, whole numbers, integers, rational numbers, and irrational numbers.	$-\sqrt{5},\ -2,\ 0,\ \dfrac{5}{7},\ \text{and } 4.\overline{9}$

continued on next page

continued from previous page

CONCEPT	COMMENTS	EXAMPLES
Approximating Square Roots by Guess-and-Check	1. Use a number line to make a reasonable guess for the square root. 2. Make a table of values based on your guess. Compute the square of each value in the table until the given radicand is between consecutive values in the table. 3. If the desired accuracy for the approximation has been found, stop. Otherwise, make a more accurate guess of the square root and go back to Step 2.	Approximate $\sqrt{8}$ as follows. $\sqrt{4}=2 \qquad \sqrt{9}=3$ 0 1 2 3 **4** 5 6 7 **8** **9** 10 A reasonable guess is 2.8. Guess · 2.7 · 2.8 · 2.9 Guess Squared · 7.29 · 7.84 · 8.41 To the nearest tenth, $\sqrt{8} \approx 2.8$.
Approximating Square Roots with the Babylonian Algorithm	To approximate $\sqrt{Q}$, do the following. 1. Make a reasonable guess for the square root. Assign this value to the variable A. 2. Compute the quotient $Q \div A$. 3. Add A to the result from Step 2. 4. Divide the result of Step 3 by 2. The result is an approximation for $\sqrt{Q}$. For a more accurate approximation, assign the result from Step 4 to the variable A and repeat the process beginning at Step 2.	Approximate $\sqrt{10}$ as follows. 1. A reasonable guess is 3.2. 2. $10 \div 3.2 = 3.125$ 3. $3.125 + 3.2 = 6.325$ 4. $6.325 \div 2 = 3.1625$ Rounded to two decimal places, $\sqrt{10} \approx 3.16$.

5.4 Exercises

MyMathLab · Math XL PRACTICE · WATCH · DOWNLOAD · READ · REVIEW

CONCEPTS AND VOCABULARY

1. A(n) _____ number cannot be expressed as a fraction whose numerator and denominator are both integers.

2. Numbers that can be written as decimals are called _____ numbers.

3. The _____ numbers include natural numbers, whole numbers, integers, rational numbers, and irrational numbers.

4. Which of the numbers, $\sqrt{35}$ or $\sqrt{36}$, is *not* an irrational number?

5. Which of the methods for approximating square roots involves making tables of values?

6. The _____ algorithm can be used to approximate the square root of a number.

REAL NUMBERS

Exercises 7–10: Identify the numbers from the given list that belong to each of the following sets of numbers.

  (a) *natural numbers*   (b) *whole numbers*
  (c) *integers*   (d) *rational numbers*
  (e) *irrational numbers*

7. $-\dfrac{5}{8}, 3, 0, \sqrt{4}, \sqrt{5}$

8. $8, -\dfrac{2}{3}, 3.\overline{8}, -1.1, \sqrt{6}$

9. $\sqrt{10}, 0, \frac{9}{3}, -1.\overline{2}, 6.4$

10. $-\sqrt{4}, 3, \frac{8}{2}, 7.\overline{57}, \sqrt{20}$

SQUARE ROOTS

Exercises 11–16: Approximate the given square root to the nearest tenth using the guess-and-check method.

11. $\sqrt{5}$ **12.** $\sqrt{3}$

13. $\sqrt{15}$ **14.** $\sqrt{22}$

15. $\sqrt{83}$ **16.** $\sqrt{62}$

Exercises 17–22: Approximate the given square root to the nearest hundredth using the guess-and-check method.

17. $\sqrt{14}$ **18.** $\sqrt{17}$

19. $\sqrt{42}$ **20.** $\sqrt{57}$

21. $\sqrt{99}$ **22.** $\sqrt{88}$

Exercises 23–28: Go through the Babylonian algorithm one time to approximate the given square root. Give two decimal places in your answer.

23. $\sqrt{6}$ **24.** $\sqrt{8}$

25. $\sqrt{18}$ **26.** $\sqrt{28}$

27. $\sqrt{78}$ **28.** $\sqrt{92}$

Exercises 29–34: Go through the Babylonian algorithm two times to approximate the given square root. Give four decimal places in your answer.

29. $\sqrt{2}$ **30.** $\sqrt{12}$

31. $\sqrt{30}$ **32.** $\sqrt{44}$

33. $\sqrt{55}$ **34.** $\sqrt{85}$

ORDER OF OPERATIONS

Exercises 35–50: Evaluate the expression.

35. $3.9 - 6(0.3) + 7.7$ **36.** $-5.9 + 9.4(0.8)$

37. $7 - 10(0.37) \div 2$ **38.** $8.7 + 99 \div 10$

39. $4.6 - 3(2 - 5.1)$ **40.** $\sqrt{4.1 + 4.9} \div 5$

41. $5 \div \sqrt{18.6 - 2.6}$ **42.** $|4.2 - 8.7| + 1$

43. $\dfrac{3.2 + 6.9}{5.4 - 4.8}$ **44.** $\dfrac{100(0.034)}{6}$

45. $\dfrac{1.3 - 2.8 \div 7}{300 \div 1000}$ **46.** $\dfrac{12.4 - 5(1.6)}{-2.8 - 5.2}$

47. $2 - 0.4^2 + 5.34$ **48.** $4(5.3 - 4.8)^2$

49. $13.9 - |4.2 - 8.7| + (0.7 + 0.4)^2$

50. $(\sqrt{11.5 + 24.5} - |2.1 - 9.3|) \div 0.4$

Exercises 51–56: Evaluate the expression for $x = -1.5$ and $y = 4.6$.

51. $3x - y$ **52.** $2y + 5x$

53. $3.4(5.5 + x) - y$ **54.** $y - (3x + 1.8)$

55. $\dfrac{10y - 30}{x}$ **56.** $\dfrac{10x + 32.48}{y}$

APPLICATIONS

57. *Flight of a Baseball* If a baseball is hit upward with a velocity of 66 feet per second, its height h in feet above the ground after t seconds can be approximated using the formula

$$h = -16t^2 + 66t + 3.$$

Find the height of the ball after 0.5 second.

58. *Flight of a Projectile* If a projectile is fired upward with a velocity of 88 feet per second, its height h in feet above the ground after t seconds can be approximated using the formula

$$h = -16t^2 + 88t.$$

Find the projectile's height after 2.5 seconds.

59. *Endowment for the Arts* The following table shows the funding, in millions of dollars, from the National Endowment for the Arts during selected years. Find the average funding during this time.

Year	2007	2008	2009	2010
Funding	124.6	144.7	155.0	167.5

Source: National Endowment for the Arts.

60. *Philanthropy* The following table shows grant funding in billions of dollars from the Bill and Melinda Gates Foundation during selected years. Find the average funding during this time. Round to one decimal place.

Year	2006	2007	2008	2009
Funding	1.6	2.0	2.8	3.5

Source: Bill and Melinda Gates Foundation.

Exercises 61–64: Geometry Use a formula from the list provided to find the area of the figure.

Rectangle: $A = lw$,   Trapezoid: $A = 0.5(a + b)h$,

Triangle: $A = 0.5bh$,   Square: $A = s^2$

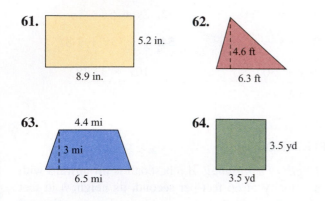

61.   5.2 in.   8.9 in.

62.   4.6 ft   6.3 ft

63.   4.4 mi   3 mi   6.5 mi

64.   3.5 yd   3.5 yd

65. *Gas Mileage* The formula for computing a car's gas mileage M is

$$M = \frac{E - B}{G},$$

where E is the ending odometer reading, B is the beginning odometer reading, and G is the number of gallons of gasoline used for the trip. Find the mileage for a trip where $E = 38{,}989.5$, $B = 38{,}423.1$, and $G = 19.2$.

66. *Gas Mileage* Refer to Exercise 65. Find the mileage for a trip where $E = 87{,}261.8$, $B = 86{,}658.7$, and $G = 18.5$.

67. *SUV Sales* Sales of sport utility vehicles increased dramatically during the 1990s. The number N in millions of SUVs sold during year x is approximated by

$$N = 0.275x - 546.5.$$

Determine the number of SUVs sold in 1996. (*Source:* Autodata Corporation.)

68. *Target Heart Rate* For general health and weight loss, a person who is x years old should maintain a minimum target heart rate of T beats per minute during extended exercise, where

$$T = -0.5x + 120.$$

Determine the minimum target heart rate for a person who is 48 years old.

69. *Building a Play Set* An engineer who is designing a rope swing for a backyard play set determines that the rope must be $\sqrt{48}$ feet in length. Approximate this length to the nearest hundredth of a foot.

70. *Quality Control* For safety reasons, the company that makes the play set in Exercise 69 requires that all calculations are checked by a second engineer. The second engineer determines that the rope length should be $4 \cdot \sqrt{3}$ feet. Approximate this length to the nearest hundredth of a foot. Do the engineers agree?

WRITING ABOUT MATHEMATICS

71. A student claims that the number

$$\frac{1.3}{4}$$

is irrational because it is not written as a fraction whose numerator and denominator are both integers. Do you agree with this observation? Explain.

72. Would you rather use the guess-and-check method or the Babylonian algorithm to approximate $\sqrt{12}$ to six decimal places? Explain.

Checking Basic Concepts

1. Multiply.
(a) $(3.6)(2.1)$ (b) $-7(34.9)$
(c) $10(2.63)$ (d) $(3.41)(-5.6)$

2. Divide.
(a) $28.48 \div 8$ (b) $82.4 \div 9$
(c) $63.5 \div 100$ (d) $-0.64 \div 1.6$

3. Write each fraction or mixed number as a decimal.
(a) $\dfrac{9}{20}$ (b) $\dfrac{7}{15}$

(c) $-3\dfrac{1}{8}$ (d) $9\dfrac{5}{6}$

4. Simplify each expression.
(a) $2.5(7.6x + 3)$ (b) $\dfrac{-28y}{0.4}$

5. Use the guess-and-check method to approximate the value of $\sqrt{24}$ to the nearest tenth.

6. Go through the Babylonian algorithm one time to approximate $\sqrt{20}$. Give two decimal places in your answer.

7. Simplify each expression.
(a) $6.3 + 4 \div 10$ (b) $\dfrac{5.2 - 1.8}{0.2 + 1.6}$

8. Evaluate the expression for $x = 4.5$ and $y = 0.3$.
$$100y - (3x + 22.8)$$

9. *Estimated Taxes* A small business owner must pay $4678.48 next year in estimated taxes. If the estimated taxes are paid in four equal payments, how much is each payment?

10. *Flight of a Golf Ball* If a golf ball is hit upward with a velocity of 66 feet per second, its height h in feet above the ground after t seconds can be approximated using the formula
$$h = -16t^2 + 66t.$$
Determine the height of the golf ball 3.5 seconds after it is hit.

5.5 Solving Equations Involving Decimals

Solving Equations Algebraically • Solving Equations Numerically •
Solving Equations Visually • Applications

A LOOK INTO MATH ▶

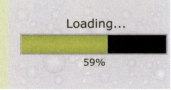
Loading...
59%

When downloading files from the Internet, computers often display a progress bar. To provide this information, the computer must continuously solve equations involving the size of the file, the current connection speed, and the fraction of the file that remains to be downloaded. Equations of this type involve decimals. To get computers to solve such equations, software engineers use the same properties of equality as those found in this text. In this section, we use algebraic, numerical, and visual methods to solve equations involving decimals.

STUDY TIP

Once again, the equations presented in this section are solved algebraically, numerically, and visually. These are the same methods used in previous chapters to solve equations.

Solving Equations Algebraically

There are two common algebraic methods used to solve equations involving decimals. In the first method, the properties of equality are used directly with all decimals involved. In the second method, decimals are "cleared" before using the properties of equality.

WORKING DIRECTLY WITH DECIMALS In the next example, we use properties of equality (from Section 3.3) to solve equations involving decimals.

| EXAMPLE 1 | **Solving equations by working directly with decimals** |

Solve each equation and check the solution.
(a) $x + 7.9 = 18.6$ **(b)** $-4.8y = 31.2$

Solution
(a) Because 7.9 is being **added** to x in the given equation, we can isolate x on the left side of the equation by **subtracting 7.9** from each side of the equation.

$$x + 7.9 = 18.6 \qquad \text{Given equation}$$
$$x + 7.9 - \mathbf{7.9} = 18.6 - \mathbf{7.9} \qquad \text{Subtract 7.9 from each side.}$$
$$x = 10.7 \qquad \text{Simplify.}$$

To check this solution, replace x with 10.7 in the given equation.

$$x + 7.9 = 18.6 \qquad \text{Given equation}$$
$$\mathbf{10.7} + 7.9 \stackrel{?}{=} 18.6 \qquad \text{Replace } x \text{ with 10.7.}$$
$$18.6 = 18.6 \checkmark \qquad \text{Add; the solution checks.}$$

(b) Since –4.8 is being **multiplied** with y in the given equation, we isolate y on the left side of the equation by **dividing** each side of the equation by $-\mathbf{4.8}$.

$$-4.8y = 31.2 \qquad \text{Given equation}$$
$$\frac{-4.8y}{-\mathbf{4.8}} = \frac{31.2}{-\mathbf{4.8}} \qquad \text{Divide each side by } -4.8.$$
$$y = -6.5 \qquad \text{Simplify.}$$

To check this solution, replace y with –6.5 in the given equation.

$$-4.8y = 31.2 \qquad \text{Given equation}$$
$$-4.8(\mathbf{-6.5}) \stackrel{?}{=} 31.2 \qquad \text{Replace } y \text{ with } -6.5.$$
$$31.2 = 31.2 \checkmark \qquad \text{Multiply; the solution checks.}$$

Now Try Exercises 7, 11

In the next example, equations are solved by applying both the addition and multiplication properties of equality.

| EXAMPLE 2 | **Solving equations by working directly with decimals** |

Solve each equation and check the solution.
(a) $2.6w + 22.9 = 34.6$ **(b)** $4.2(m - 7) = 1.7m + 2.1$

Solution
(a) Start by **subtracting 22.9** from each side of the equation.

$$2.6w + 22.9 = 34.6 \qquad \text{Given equation}$$
$$2.6w + 22.9 - \mathbf{22.9} = 34.6 - \mathbf{22.9} \qquad \text{Subtract 22.9 from each side.}$$
$$2.6w = 11.7 \qquad \text{Simplify.}$$
$$\frac{2.6w}{\mathbf{2.6}} = \frac{11.7}{\mathbf{2.6}} \qquad \text{Divide each side by 2.6.}$$
$$w = 4.5 \qquad \text{Simplify.}$$

To check this solution, replace w with 4.5 in the given equation.

$2.6w + 22.9 = 34.6$	Given equation
$2.6(\mathbf{4.5}) + 22.9 \stackrel{?}{=} 34.6$	Replace w with 4.5.
$11.7 + 22.9 \stackrel{?}{=} 34.6$	Multiply.
$34.6 = 34.6$ ✓	Add; the solution checks.

(b) Start by applying the distributive property on the left side of the equation.

$4.2(m - 7) = 1.7m + 2.1$	Given equation
$4.2m - 29.4 = 1.7m + 2.1$	Distributive property
$4.2m - 29.4 + \mathbf{29.4} = 1.7m + 2.1 + \mathbf{29.4}$	Add 29.4 to each side.
$4.2m = 1.7m + 31.5$	Simplify.
$4.2m - \mathbf{1.7m} = 1.7m - \mathbf{1.7m} + 31.5$	Subtract $1.7m$ from each side.
$2.5m = 31.5$	Simplify.
$\dfrac{2.5m}{\mathbf{2.5}} = \dfrac{31.5}{\mathbf{2.5}}$	Divide each side by 2.5.
$m = 12.6$	Simplify.

Check this solution by replacing m with 12.6 in the given equation.

$4.2(\mathbf{m} - 7) = 1.7\mathbf{m} + 2.1$	Given equation
$4.2(\mathbf{12.6} - 7) \stackrel{?}{=} 1.7(\mathbf{12.6}) + 2.1$	Replace m with 12.6.
$4.2(5.6) \stackrel{?}{=} 1.7(12.6) + 2.1$	Subtract.
$23.52 \stackrel{?}{=} 21.42 + 2.1$	Multiply.
$23.52 = 23.52$ ✓	Add; the solution checks.

Now Try Exercises 13, 25

CLEARING DECIMALS In Section 4.8, the LCD was used to clear fractions from equations. In a similar way, we can multiply each side of an equation by a power of 10 to clear decimals from an equation. Recall that multiplying each *side* of an equation by a (nonzero) number is equivalent to multiplying each *term* in the equation by that number.

SOLVING EQUATIONS BY CLEARING DECIMALS

To solve an equation by clearing decimals, do the following.

STEP 1: Use the distributive property to clear parentheses from the equation.

STEP 2: Note the maximum number of decimal places in any number in the equation. Multiply every term in the equation by a power of 10 with that many zeros.

STEP 3: Simplify each term in the equation and combine any like terms.

STEP 4: Solve the resulting equation.

EXAMPLE 3   **Solving equations by clearing decimals**

Solve each equation.
(a) $5.2(x - 3) = 14.3$ **(b)** $4.2x - 2.84 = 6.2x + 2.5$

Solution

(a) The maximum number of decimal places in any number is **one**. For Step 2, multiply each term by the power of 10 with **one** zero, or multiply by **10**.

$$5.2(x - 3) = 14.3 \qquad \text{Given equation}$$
$$5.2x - 15.6 = 14.3 \qquad \text{Distributive property (Step 1)}$$
$$\mathbf{10}(5.2x) - \mathbf{10}(15.6) = \mathbf{10}(14.3) \qquad \text{Multiply by 10. (Step 2)}$$
$$52x - 156 = 143 \qquad \text{Simplify. (Step 3)}$$
$$52x - 156 + \mathbf{156} = 143 + \mathbf{156} \qquad \text{Add 156 to each side. (Step 4)}$$
$$52x = 299 \qquad \text{Simplify.}$$
$$\frac{52x}{\mathbf{52}} = \frac{299}{\mathbf{52}} \qquad \text{Divide each side by 52. (Step 4)}$$
$$x = 5.75 \qquad \text{Simplify.}$$

(b) The maximum number of decimal places in any number is **two**. For Step 2, multiply each term by **100**. Since there are no parentheses, Step 1 can be skipped.

$$4.2x - 2.84 = 6.2x + 2.5 \qquad \text{Given equation}$$
$$\mathbf{100}(4.2x) - \mathbf{100}(2.84) = \mathbf{100}(6.2x) + \mathbf{100}(2.5) \qquad \text{Multiply by 100. (Step 2)}$$
$$420x - 284 = 620x + 250 \qquad \text{Simplify. (Step 3)}$$
$$420x - 284 + \mathbf{284} = 620x + 250 + \mathbf{284} \qquad \text{Add 284 to each side. (Step 4)}$$
$$420x = 620x + 534 \qquad \text{Simplify.}$$
$$420x - \mathbf{620x} = 620x - \mathbf{620x} + 534 \qquad \text{Subtract 620x from each side. (Step 4)}$$
$$-200x = 534 \qquad \text{Simplify.}$$
$$\frac{-200x}{\mathbf{-200}} = \frac{534}{\mathbf{-200}} \qquad \text{Divide each side by } -200. \text{ (Step 4)}$$
$$x = -2.67 \qquad \text{Simplify.}$$

Now Try Exercises 27, 29

READING CHECK

• How is clearing decimals similar to clearing fractions?

MAKING CONNECTIONS

Clearing Decimals with a Power of 10

To understand why we use powers of 10 to clear decimals, consider that rewriting

$$3.74x + 19.6 = 8.1$$

as an equivalent equation involving fractions results in

$$\frac{374}{100}x + \frac{196}{10} = \frac{81}{10}.$$

We clear fractions from this equation by multiplying each side of the equation by the LCD of all fractions involved. Note that the LCD is 100 and has the same number of zeros as there are decimal places in the number 3.74 in the given equation.

Solving Equations Numerically

Equations involving decimals can also be solved numerically. In the next example, a table of values is used to solve an equation involving decimals.

EXAMPLE 4   **Solving an equation numerically**

Complete Table 5.4 for the given values of x and then solve the equation $2.6x - 3.1 = 4.7$.

TABLE 5.4

x	−1	0	1	2	3	4
$2.6x - 3.1$						

Solution

Begin by replacing x in $2.6x - 3.1$ with **−1**. Since $2.6(-1) - 3.1$ evaluates to -5.7, the left side of the equation is equal to -5.7 when $x = -1$. We write the result -5.7 below **−1**, as shown in Table 5.5. Likewise, we write -3.1 below **0** because the expression $2.6(0) - 3.1$ evaluates to -3.1. The remaining values are found similarly and are shown in Table 5.5.

TABLE 5.5

x	**−1**	**0**	1	2	**3**	4
$2.6x - 3.1$	−5.7	−3.1	−0.5	2.1	4.7	7.3

Table 5.5 reveals that the left side of the given equation equals 4.7 when $x = $ **3**. Therefore, the solution is **3**.

Now Try Exercise 31

When an equation has a decimal solution, the top row of the table of values will need to include decimal values. This is demonstrated in the next example.

EXAMPLE 5   **Solving an equation numerically**

Complete Table 5.6 for the given values of x and then solve the equation $4x + 5 = 12.2$.

TABLE 5.6

x	1.5	1.6	1.7	1.8	1.9	2
$4x + 5$						

Solution
The completed table is shown in Table 5.7.

TABLE 5.7

x	1.5	1.6	1.7	**1.8**	1.9	2
$4x + 5$	11	11.4	11.8	12.2	12.6	13

The left side of the given equation equals 12.2 when $x = $ **1.8**. The solution is **1.8**.

Now Try Exercise 35

READING CHECK

• Why do we need to check a solution found visually?

Solving Equations Visually

When a graph is provided for an equation involving decimals, an estimate of the solution can be found visually. In the next example, we estimate and then check a solution to an equation involving decimals.

EXAMPLE 6   **Solving an equation visually**

A graph of the expression $1.7x - 2.3$ is shown in the following figure. Use the graph to solve the equation $1.7x - 2.3 = -5.7$ visually.

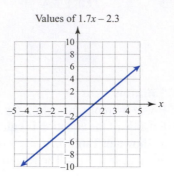

Values of $1.7x - 2.3$

Solution

Approximate the position of -5.7 on the vertical axis and move **horizontally** to the left to the graphed line. From this position, move **vertically** upward to the horizontal axis. The solution appears to be -2, as shown in the following figure.

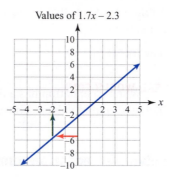

Values of $1.7x - 2.3$

To see if -2 is the correct solution, check it in the given equation.

$$1.7x - 2.3 = -5.7 \qquad \text{Given equation}$$
$$1.7(-2) - 2.3 \overset{?}{=} -5.7 \qquad \text{Replace } x \text{ with } -2.$$
$$-3.4 - 2.3 \overset{?}{=} -5.7 \qquad \text{Multiply.}$$
$$-5.7 = -5.7 \ \checkmark \quad \text{Subtract; the solution checks.}$$

The solution to the equation $1.7x - 2.3 = -5.7$ is -2.

▌ **Now Try Exercise 39**

Applications

▶ **REAL-WORLD CONNECTION** Some cellular phone companies offer text messaging plans that charge a flat fee for a set number of text messages. However, subscribers who go over the designated number are usually charged a fee for *each* additional text message. For example, a plan may charge $15 for up to 400 text messages, and 5 cents for each additional text message. A subscriber who sends (or receives) more than 400 messages can compute the total cost C of x messages using the formula

$$C = 0.05(x - 400) + 15.$$

In the next example, we use this formula to analyze the cost of text messaging.

EXAMPLE 7

Analyzing the cost of text messaging

Find the number of text messages that correspond to total charges of $18.95 by replacing C in the formula $C = 0.05(x - 400) + 15$ with 18.95 and then solving the resulting equation.

Solution

$$
\begin{aligned}
0.05(x - 400) + 15 &= 18.95 && \text{Given equation (rewritten)} \\
0.05x - 20 + 15 &= 18.95 && \text{Distributive property (Step 1)} \\
0.05x - 5 &= 18.95 && \text{Combine like terms.} \\
\mathbf{100}(0.05x) - \mathbf{100}(5) &= \mathbf{100}(18.95) && \text{Multiply by 100. (Step 2)} \\
5x - 500 &= 1895 && \text{Simplify. (Step 3)} \\
5x - 500 + \mathbf{500} &= 1895 + \mathbf{500} && \text{Add 500 to each side. (Step 4)} \\
5x &= 2395 && \text{Simplify.} \\
\frac{5x}{\mathbf{5}} &= \frac{2395}{\mathbf{5}} && \text{Divide each side by 5. (Step 4)} \\
x &= 479 && \text{Simplify.}
\end{aligned}
$$

The charges are $18.95 for 479 text messages.

Now Try Exercise 51

▶ **REAL-WORLD CONNECTION** Unsolicited bulk email messages known as spam arrive in email inboxes every day. The average *daily* number of spam messages N in billions during year x (after 2004) can be computed using the formula

$$N = 22.4x - 44{,}878.6.$$

In the next example, we use this formula to analyze email spam. (*Source:* SpamUnit.)

EXAMPLE 8

Analyzing email spam

Use the formula $N = 22.4x - 44{,}878.6$ to find the year when the average daily number of email spam messages reached 145.4 billion.

Solution

Replace N in the formula with 145.4 and solve the resulting equation.

$$
\begin{aligned}
22.4x - 44{,}878.6 &= 145.4 && \text{Given equation (rewritten)} \\
\mathbf{10}(22.4x) - \mathbf{10}(44{,}878.6) &= \mathbf{10}(145.4) && \text{Multiply by 10. (Step 2)} \\
224x - 448{,}786 &= 1454 && \text{Simplify. (Step 3)} \\
224x - 448{,}786 + \mathbf{448{,}786} &= 1454 + \mathbf{448{,}786} && \text{Add 448,786 to each side. (Step 4)} \\
224x &= 450{,}240 && \text{Simplify.} \\
\frac{224x}{\mathbf{224}} &= \frac{450{,}420}{\mathbf{224}} && \text{Divide each side by 224. (Step 4)} \\
x &= 2010 && \text{Simplify.}
\end{aligned}
$$

The average daily number of email spam messages reached 145.4 billion in 2010.

Now Try Exercise 57

EXAMPLE 9 **Finding a Celsius temperature**

The formula

$$F = 1.8C + 32$$

gives the relationship between F degrees Fahrenheit and C degrees Celsius. Use the formula to find the Celsius temperature that is equivalent to 98.6°F.

Solution

Substitute 98.6 in the formula for F and solve the resulting equation.

$$1.8C + 32 = 98.6 \qquad \text{Given equation (rewritten)}$$
$$\mathbf{10}(1.8C) + \mathbf{10}(32) = \mathbf{10}(98.6) \qquad \text{Multiply by 10. (Step 2)}$$
$$18C + 320 = 986 \qquad \text{Simplify. (Step 3)}$$
$$18C + 320 - \mathbf{320} = 986 - \mathbf{320} \qquad \text{Subtract 320 from each side. (Step 4)}$$
$$18C = 666 \qquad \text{Simplify.}$$
$$\frac{18C}{\mathbf{18}} = \frac{666}{\mathbf{18}} \qquad \text{Divide each side by 18. (Step 4)}$$
$$C = 37 \qquad \text{Simplify.}$$

Therefore, 98.6°F is equivalent to 37°C.

Now Try Exercise 59

5.5 Putting It All Together

CONCEPT	COMMENTS	EXAMPLES
Solving Equations Algebraically by Clearing Decimals	1. Use the distributive property to clear parentheses from the equation. 2. Note the maximum number of decimal places for any number in the equation. Multiply every term in the equation by a power of 10 with that many zeros. 3. Simplify each term in the equation and combine any like terms. 4. Solve the resulting equation.	$2.4(x + 5) = 18.6$ $2.4x + 12 = 18.6$ $24x + 120 = 186$ $24x + 120 - 120 = 186 - 120$ $24x = 66$ $\dfrac{24x}{24} = \dfrac{66}{24}$ $x = 2.75$
Solving Equations Numerically	A table of values can be used to solve some equations involving decimals. Complete the table for various values of the variable and then select the solution from the table.	The solution to the equation $5.2x + 3 = -2.2$ is -1. | x	-3	-2	-1
---	---	---	---
$5.2x + 3$	-12.6	-7.4	-2.2	

CONCEPT	COMMENTS	EXAMPLES
Solving Equations Visually	Use a graph of the equation to estimate a solution. Then check the estimated solution in the given equation to be sure that it is correct.	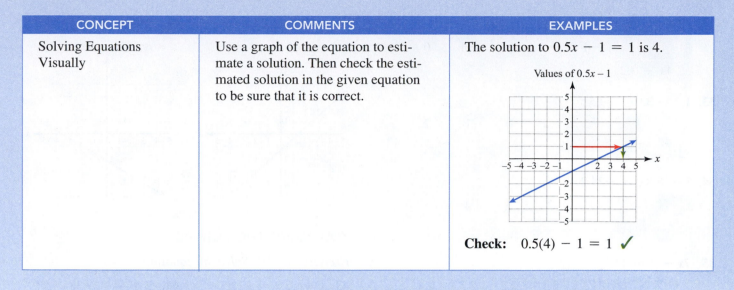

The solution to $0.5x - 1 = 1$ is 4.

Check: $0.5(4) - 1 = 1$ ✓

5.5 Exercises

CONCEPTS AND VOCABULARY

1. Name the three methods discussed in this section for solving equations.

2. When we multiply each side of an equation by a power of 10, we are clearing _____.

3. The power of 10 used to clear decimals has the same number of zeros as the maximum number of _____ in any number in the equation.

4. When a table of values is used to solve an equation, the equation is being solved _____.

SOLVING EQUATIONS ALGEBRAICALLY

Exercises 5–30: Solve the equation algebraically.

5. $x - 4.8 = 11.7$

6. $m + 12.6 = 3.2$

7. $y + 18.1 = 5.7$

8. $w - 0.9 = 13.7$

9. $2.8n = 9.8$

10. $4.25q = -51$

11. $-8.6x = -42.14$

12. $-3.3b = -23.43$

13. $4.5m + 1.1 = 15.5$

14. $8y + 0.1 = 21.7$

15. $16.12 = 3.4w - 7$

16. $3.1k + 7 = 0.8$

17. $-5.7x + 2.1 = 4$

18. $9x - 4.6 = 8$

19. $3.6 - 4b = 15.8$

20. $-6.7 = 5 - 3w$

21. $3(x + 2.1) = 17.4$

22. $4.1(y - 2) = 8.2$

23. $0.6(n - 3.7) = 9$

24. $6(q + 1.6) = 29.4$

25. $2.5(p - 4) = 2.9p + 6.1$

26. $4.8(m + 7) = 7.9m + 2.6$

27. $6.4(y + 12) = 73.6$

28. $10.7(6w - 25) = 2.14$

29. $3.2k - 5.64 = 2.9k + 7.8$

30. $7.8p - 3 = 4.2p + 3.84$

SOLVING EQUATIONS NUMERICALLY

Exercises 31–36: Solve the equation numerically by completing the given table of values.

31. $4.8x - 2.5 = 7.1$

x	0	1	2	3	4
$4.8x - 2.5$					

32. $2.4x + 3.2 = 0.8$

x	-4	-3	-2	-1	0
$2.4x + 3.2$					

33. $1.7 - 3.9x = 9.5$

x	-2	-1	0	1	2
$1.7 - 3.9x$					

34. $3 - 6.8x = -24.2$

x	1	2	3	4	5
$3 - 6.8x$					

35. $7x - 3 = 11.7$

x	2	2.1	2.2	2.3	2.4
$7x - 3$					

36. $5x - 12 = -7.5$

x	0.6	0.7	0.8	0.9	1
$5x - 12$					

SOLVING EQUATIONS VISUALLY

Exercises 37–42: Solve the equation visually.

37. $-0.8x - 2 = 2$

Values of $-0.8x - 2$

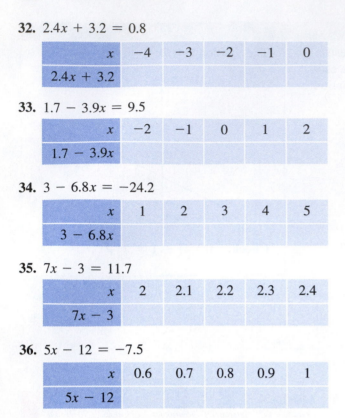

38. $0.6x + 2 = 5$

Values of $0.6x + 2$

39. $0.7x + 0.4 = 2.5$

Values of $0.7x + 0.4$

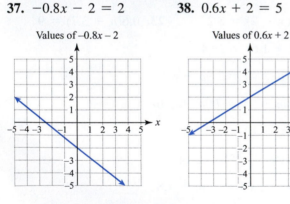

40. $-0.4x + 1.1 = 1.5$

Values of $-0.4x + 1.1$

41. $-0.9x - 1.3 = -4$

Values of $-0.9x - 1.3$

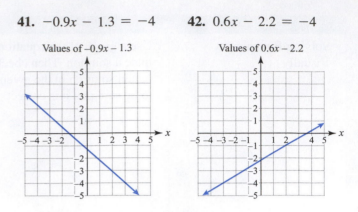

42. $0.6x - 2.2 = -4$

Values of $0.6x - 2.2$

YOU DECIDE THE METHOD

Exercises 43–50: Solve the equation.

43. $x + 13.7 = 22.85$

44. $5.7b = -19.38$

45. $8.1y - 3.8 = 20.5$

46. $4.9 - 4n = -1.9$

47. $2.6(k - 5) = 9.1$

48. $6(w - 1.3) = 24$

49. $8.7(4w + 7) = 19.14$

50. $6.2m + 11.94 = 1.8m - 9.4$

APPLICATIONS

51. *Text Messaging* Suppose that the cost C of sending or receiving x text messages is given by the formula

$$C = 0.1(x - 200) + 8, \text{ where } x \geq 200.$$

Find the number of text messages that correspond to total charges of \$29.70 by replacing C in the formula with 29.7 and solving the resulting equation.

52. *Cell Phone Plan* Suppose that the cost C of talking for x minutes on a cell phone is given by the formula

$$C = 0.25(x - 500) + 30, \text{ where } x \geq 500.$$

Find the number of minutes that correspond to total charges of \$55.75 by replacing C in the formula with 55.75 and solving the resulting equation.

Exercises 53–56: Geometry Use a formula from the list provided and the given area to find the unknown measure.

Rectangle: $A = lw$, Trapezoid: $A = 0.5(a + b)h$,
Triangle: $A = 0.5bh$

53. $A = 40.95$ square feet

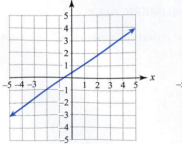

12.6 ft

54. $A = 11$ square inch

4.4 in.

55. $A = 41.7$ square inches   **56.** $A = 51.06$ square feet

Capital Investment

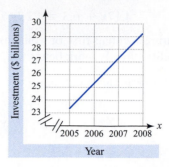

57. *Movie Tickets* The national average price P (in current dollars) for admission to a movie during year x can be estimated using the formula

$$P = 0.2x - 394.6,$$

where x is a year after 1995. Find the year when the the average price of a movie ticket was \$7.20. (*Source:* Motion Picture Association of America.)

58. *Drive-in Theaters* The number N of drive-in movie theaters in the United States during year x can be estimated using the formula

$$N = -18.5x + 37,740,$$

where x is a year after 1990. Find the year when there were 629 drive-in theaters. (*Source:* National Association of Theatre Owners.)

59. *Converting Temperature* The formula

$$F = 1.8C + 32$$

gives the relationship between F degrees Fahrenheit and C degrees Celsius. Find the Celsius temperature that is equivalent to $57.2°F$.

60. *Converting Temperature* The formula

$$C = \frac{(F - 32)}{1.8}$$

gives the relationship between C degrees Celsius and F degrees Fahrenheit. Find the Fahrenheit temperature that is equivalent to $46.5°C$.

61. *Venture Capital* The amount of money M, in billions of dollars, invested by venture capital firms during year x can be estimated by $M = 1.95x - 3886.45$, where x is a year from 2005 to 2008. Find the year when \$27.2 billion were invested by solving the equation $27.2 = 1.95x - 3886.45$. (*Source:* Pricewaterhouse Coopers.)
 (a) Solve algebraically.
 (b) Solve numerically using the table shown.
 (c) Solve visually using the graph shown.

x	2005	2006	2007	2008
$1.95x - 3886.45$				

(d) Compare the algebraic, numerical, and visual solutions.

62. *Poverty Threshold* The poverty threshold P, in thousands of dollars, for a family of four during year x can be estimated using the formula $P = 0.5x - 982.5$, where x is a year from 2003 to 2009. Find the year when the poverty threshold for a family of four was \$22 thousand by solving $22 = 0.5x - 982.5$. (*Source:* U.S Census Bureau.)
 (a) Solve algebraically.
 (b) Solve numerically using the table shown.
 (c) Solve visually using the graph shown.

	x	2003	2005	2007	2009
$0.5x - 982.5$					

Poverty Threshold

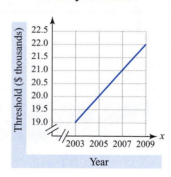

(d) Compare the algebraic, numerical, and visual solutions.

WRITING ABOUT MATHEMATICS

63. Compare the powers of 10 used to clear decimals and the LCD used to clear fractions. What is the relationship between these two concepts?

64. Would you rather solve the equation

$$4.17x + 6.3 = 22.146$$

by clearing decimals first, or would using the properties of equality directly work better for you? Explain.

65. The solution to the equation

$$9x + 7 = 11$$

is $0.\overline{4}$. Discuss the difficulties that occur when solving this equation numerically with a table of values.

66. The solution to the equation

$$3.2x + 6 = 9.5$$

is 1.09375. Discuss the difficulties that occur when solving this equation visually with a graph.

Group Activity   Working with Real Data

Directions: Form a group of 2 to 4 people. Select someone to record the group's responses for this activity. All members of the group should work cooperatively to answer the questions. If your instructor asks for your results, each member of the group should be prepared to respond.

iPod Capacity The amount of storage space available on an iPod touch is measured in units called gigabytes (GB). Most songs, photos, and applications stored on an iPod have file sizes measured in megabytes (MB). There are 1000 megabytes in 1 gigabyte.

(a) A typical 3-minute music file (one song) requires about 3.46 MB of storage space. How many megabytes are required to store 1794 songs that average 3 minutes in length?

(b) A typical digital photo requires about 0.6 MB of storage space on an iPod. How many megabytes are required to store 673 such photos?

(c) A typical 2-hour movie requires about 1.35 GB of storage. How many gigabytes are required to store 7 such movies?

(d) A student wants to buy an iPod with enough memory to store all of the files listed in parts (a), (b), and (c), as well as 3.74 GB in applications. How many *giga*bytes of storage space are needed to hold all of the files?

(e) If the iPod is available in either a 16-GB or 32-GB model, which model should the student buy?

5.6   Applications from Geometry and Statistics

Circles: Circumference and Area • Area of Composite Regions • The Pythagorean Theorem • Mean, Median, and Mode • Weighted Mean and GPA

A LOOK INTO MATH ▶

Sometimes, a rainbow appears after a storm. This colorful arc is actually a portion of a *circle*. Although it is not possible for an actual rainbow to appear as a complete circle, it is possible to create a circular rainbow by spraying a garden hose into the air on a sunny day. In this section, we discuss circles and other applications involving decimals.

Circles: Circumference and Area

A **circle** is a collection of points that are all the same distance from a central point. The central point is called the **center** of the circle, and the distance between the center and any point on the circle is called the **radius**. The distance across the circle on a straight line through its center is called the **diameter** of the circle. See Figure 5.8.

NEW VOCABULARY

☐ Circle
☐ Center
☐ Radius
☐ Diameter
☐ Circumference
☐ Composite region
☐ Semicircle
☐ Right triangle
☐ Right angle
☐ Legs (of a right triangle)
☐ Hypotenuse
☐ Measures of central
   tendency
☐ Mean (arithmetic mean)
☐ Median
☐ Mode
☐ Weighted mean
☐ Grade point average (GPA)

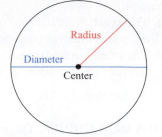

Figure 5.8 Parts of a Circle

NOTE: The diameter of any circle is always exactly twice the radius. If r represents the radius and d represents the diameter, then the following relationships hold.

$$d = 2r \quad \text{and} \quad r = \frac{1}{2}d$$

CIRCUMFERENCE OF A CIRCLE The distance around an enclosed region is usually called the perimeter. However, the distance around a circle is called the **circumference**. For *any* circle, dividing the circumference C by the diameter d always results in the irrational number π, where $\pi \approx 3.14159265$. We can write this relationship as the equation

$$\frac{C}{d} = \pi.$$

STUDY TIP

This section contains many computations involving decimals. If calculator use is not allowed during exams, do not use one while completing your homework.

READING CHECK

• How are the radius and diameter related?

The multiplication property of equality can be used to rewrite this equation so that it gives the circumference of a circle. Because the variable d represents the diameter of the circle, which is a nonzero number, we can multiply each side of the equation by d, as follows.

$$\frac{C}{d} = \pi \qquad \text{Given equation}$$

$$\frac{C}{d} \cdot \frac{d}{1} = \pi \cdot d \qquad \text{Multiply each side by } d.$$

$$C = \pi d \qquad \text{Simplify.}$$

Because $d = 2r$, we have the following formulas for the circumference of a circle.

CIRCUMFERENCE OF A CIRCLE

The circumference C of a circle with radius r and diameter d is

$$C = \pi d \quad \text{or} \quad C = 2\pi r.$$

EXAMPLE 1   **Finding the circumference of a circle**

Find the exact circumference of the given circle and then approximate the circumference using 3.14 as an approximation for π.

(a)   **(b)**

Solution
(a) The exact circumference is $C = \pi d = \pi(7) = 7\pi$ inches. Using 3.14 for π results in an approximate circumference of $C \approx 7(3.14) = 21.98$ inches.
(b) The exact circumference is $C = 2\pi r = 2\pi(5) = 10\pi$ feet. The approximate circumference is $C \approx 10(3.14) = 31.4$ feet.

Now Try Exercises 15, 21

AREA OF A CIRCLE The formula for finding the area of a circle also involves the irrational number π. While a more complete discussion of the area of a circle can be found in Chapter 10, we state the formula here, as follows.

AREA OF A CIRCLE

The area A of a circle with radius r is
$$A = \pi r^2.$$

EXAMPLE 2   **Finding the area of a circle**

Find the exact area of the given circle and then approximate the area using 3.14 as an approximation for π.

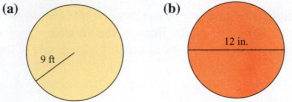
(a)   9 ft   (b)   12 in.

Solution
(a) The exact area is $A = \pi r^2 = \pi(9^2) = 81\pi$ square feet. An approximation for the area of the circle is $A \approx 81(3.14) = 254.34$ square feet.
(b) The radius is half of the diameter, or 6 inches. If $r = 6$, the exact area of the circle is $A = \pi r^2 = \pi(6^2) = 36\pi$ square inches. Using 3.14 for π, we calculate the approximate area as $A \approx 36(3.14) = 113.04$ square inches.

Now Try Exercises 23, 29

Area of Composite Regions

READING CHECK

• What is a composite region?

A region that consists of more than one geometric shape is called a **composite region**. For example, the composite region shown in Figure 5.9 includes a rectangle and a half circle. (The geometric name for a half circle is a **semicircle**.)

Figure 5.9 A Composite Region

To find the area of a composite region, more than one area formula is usually needed, as shown in the next example.

EXAMPLE 3   **Finding areas of composite regions**

Find the area of the shaded region. Use 3.14 to approximate π, if needed.

(a) 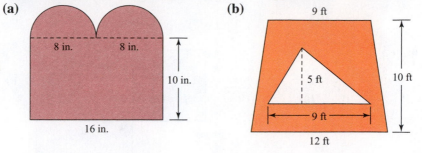   **(b)**

Solution

(a) The region comprises a rectangle and two semicircles. Together, the semicircles make one complete circle with a radius of 4 inches. The total shaded area is the sum of the areas of the rectangle and the two semicircles.

Rectangle: $A = lw = 16(10) = 160$ square inches

Semicircles: $A = \pi r^2 = \pi(4^2) = 16\pi \approx 16(3.14) = 50.24$ square inches

The total area is approximately $160 + 50.24 = 210.24$ square inches.

(b) The area of the shaded region can be found by subtracting the area of the white triangle from the area of the trapezoid.

Trapezoid: $A = \dfrac{1}{2}(a + b)h = \dfrac{1}{2}(9 + 12) \cdot 10 = 105$ square feet

Triangle: $A = \dfrac{1}{2}bh = \dfrac{1}{2}(9)(5) = 22.5$ square feet

The shaded area is exactly $105 - 22.5 = 82.5$ square feet.

Now Try Exercises 33, 35

The Pythagorean Theorem

Over 2500 years ago, the Greek mathematician Pythagoras showed that a special relationship exists among the lengths of the sides of a *right triangle*. In a **right triangle**, one of the angles is a **right angle** with a measure of 90° (degrees). To indicate that an angle is a right angle, it is marked with a small square. (Note that each corner of a square forms a right angle.)

Right Triangle

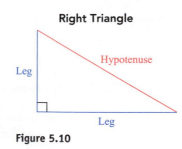

Figure 5.10

The two shorter sides that form the right angle are called the **legs** of the right triangle, and the longest side, which is opposite the right angle, is called the **hypotenuse**. See Figure 5.10. The Pythagorean theorem states that for any right triangle, the sum of the squares of the lengths of the two legs is always equal to the square of the length of the hypotenuse. A visual illustration of the Pythagorean theorem is shown in Figure 5.11 on the next page.

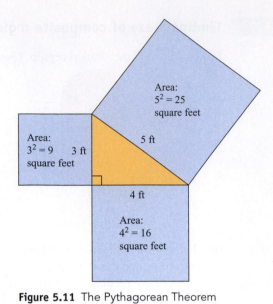

Figure 5.11 The Pythagorean Theorem

The squares formed off the legs of the right triangle in Figure 5.11 have a total area of $9 + 16 = 25$ square feet, which equals the area of the square formed off the hypotenuse. The Pythagorean theorem is summarized as follows.

READING CHECK

• What kind of triangle must we have in order to apply the Pythagorean theorem?

THE PYTHAGOREAN THEOREM

If a and b represent the lengths of the legs of a right triangle and c represents the length of the hypotenuse, then

$$a^2 + b^2 = c^2.$$

The next example shows how the Pythagorean theorem can be used to find an unknown length of one side of a right triangle when the lengths of the other two sides are known.

EXAMPLE 4   **Using the Pythagorean theorem**

Find the length of the unknown side of the right triangle. If the answer is not a whole number, approximate it to one decimal place.

(a)

(b)

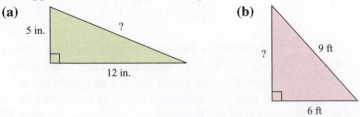

Solution

(a) Let $a = 5$ and $b = 12$ in the equation $a^2 + b^2 = c^2$ and then solve for c.

$$a^2 + b^2 = c^2 \qquad \text{Pythagorean theorem}$$
$$5^2 + 12^2 = c^2 \qquad \text{Replace } a \text{ with 5 and } b \text{ with 12.}$$
$$25 + 144 = c^2 \qquad \text{Square 5 and 12.}$$
$$169 = c^2 \qquad \text{Add.}$$

Even though the whole number 169 has both a positive and a negative square root, the value of c must be positive because it represents a length. For this reason, we are interested only in the positive (or principal) square root.

$$c = \sqrt{169} \qquad \text{Positive square root}$$
$$c = 13 \qquad \text{Simplify.}$$

The length of the hypotenuse is 13 inches.

(b) Let $b = 6$ and $c = 9$ in the equation $a^2 + b^2 = c^2$ and then solve for a.

$$a^2 + b^2 = c^2 \qquad \text{Pythagorean theorem}$$
$$a^2 + 6^2 = 9^2 \qquad \text{Replace } b \text{ with 6 and } c \text{ with 9.}$$
$$a^2 + 36 = 81 \qquad \text{Square 6 and 9.}$$
$$a^2 = 45 \qquad \text{Subtract 36 from each side.}$$
$$a = \sqrt{45} \qquad \text{Positive square root}$$
$$a \approx 6.7 \qquad \text{Approximate.}$$

The unknown length is approximately 6.7 feet.

Now Try Exercises 39, 43

▶ **REAL-WORLD CONNECTION** Quilters often use pinwheel patterns such as the one shown in the margin to create decorative quilts. To make this pattern, right triangles must be cut from selected fabrics and stitched onto the quilt. The two legs of the right triangles used in a pinwheel pattern often have equal measure. The next example illustrates how the length of the legs can be found using the Pythagorean theorem.

EXAMPLE 5   **Making a pinwheel pattern**

The fabric used in a pinwheel pattern is cut into right triangles with a hypotenuse length of 4 inches. If the two legs must be equal in measure, find the length of one leg of the right triangle to the nearest tenth of an inch.

Solution

Since the legs must have the same length, we will call the unknown length x, as shown in the following triangle.

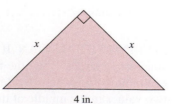

4 in.

Use the Pythagorean theorem to find x.

$$x^2 + x^2 = c^2 \qquad \text{Pythagorean theorem}$$
$$x^2 + x^2 = 4^2 \qquad \text{Replace } c \text{ with 4.}$$
$$2x^2 = 16 \qquad \text{Combine like terms; simplify.}$$
$$x^2 = 8 \qquad \text{Divide each side by 2.}$$
$$x = \sqrt{8} \qquad \text{Positive square root}$$
$$x \approx 2.8 \qquad \text{Approximate.}$$

Each leg of the triangle is about 2.8 inches long.

Now Try Exercise 73

Mean, Median, and Mode

In statistics, the *mean*, *median*, and *mode* are used to measure the "middle" of a list of numbers. For this reason, they are known as **measures of central tendency**. Each of these measures can be used in a specific way to describe a data set (list of numbers).

FINDING THE MEAN The **mean** or **arithmetic mean** of a list of numbers is commonly called the *average* of the numbers.

THE MEAN

The mean of a list of numbers is defined as follows.

$$\text{Mean} = \frac{\text{the sum of a list of values}}{\text{the number of values in the list}}$$

EXAMPLE 6 | **Finding the mean of a data set**

Table 5.8 lists the hourly wages paid to a school bus driver for selected years. Find the mean hourly wage for the driver over this 6-year period.

TABLE 5.8 Hourly Wages for a Bus Driver

Year	2007	2008	2009	2010	2011	2012
Wages	$13.49	$13.85	$14.39	$14.81	$15.06	$15.58

Solution
To find the mean hourly wage, add the numbers together and then divide the result by 6 because there are **6** numbers in the list.

$$\text{Mean} = \frac{13.49 + 13.85 + 14.39 + 14.81 + 15.06 + 15.58}{6} = \frac{87.18}{6} = 14.53$$

The mean hourly wage over this period of time is $14.53.

Now Try Exercise 75

READING CHECK

• Why is it important to order a list of data before finding its median?

FINDING THE MEDIAN If an *ordered* list contains an odd number of values written from smallest to largest, then the **median** is the middle value in the list. If the list contains an even number of values written from smallest to largest, then the median is the mean of the two values in the middle of the list.

THE MEDIAN

The median of an *ordered* list of values is defined as follows.

- For an odd number of values, the median is the middle value in the list.
- For an even number of values, the median is the mean of the two middle values in the list.

EXAMPLE 7 | **Finding the median of a data set**

Find the median of each data set.
(a) 4.7, 3.9, 2.4, 6.8, 5.2, 7.1, 4.4   **(b)** 57, 62, 48, 55, 71, 63, 39, 65, 70, 48

Solution
(a) Write the values in order from smallest to largest. Since the list contains an odd number of values, the median is the middle value in the *ordered* list.

$$2.4, 3.9, 4.4, \mathbf{4.7}, 5.2, 6.8, 7.1$$

Three values   Three values

The median is **4.7**.
(b) Write the values in order from smallest to largest. The list contains an even number of values, so the median is the mean of the middle two values in the *ordered* list.

$$39, 48, 48, 55, \mathbf{57}, \mathbf{62}, 63, 65, 70, 71$$

Four values   Four values

The median is the mean of **57** and **62**, or

$$\frac{57 + 62}{2} = \frac{119}{2} = 59.5.$$

Now Try Exercises 51, 53

FINDING THE MODE The **mode** in a list of numbers is the value that occurs most often. A data set may have more than one mode or no mode at all. For larger data sets, the mode is often easier to find if the numbers are first written in order from smallest to largest.

THE MODE

The mode of a list of values is the value that occurs most often in the list. It is possible for a data set to have more than one mode. When none of the values in a data set occurs more than once, the data set has no mode.

EXAMPLE 8 | **Finding the mode of a data set**

If possible, find the mode(s) of each data set.
(a) 3.8, 2.1, 4.0, 3.8, 2.7, 2.8   **(b)** 7, 4, 6, 6, 3, 4, 9, 4, 5, 8, 6

Solution
(a) Written in order, the list is 2.1, 2.7, 2.8, **3.8**, **3.8**, 4.0. The mode is 3.8.
(b) In order, the list is 3, **4**, **4**, **4**, 5, **6**, **6**, **6**, 7, 8, 9. There are two modes, 4 and 6.

Now Try Exercises 59, 63

Weighted Mean and GPA

▶ **REAL-WORLD CONNECTION** Suppose that a teacher wants to find the mean of the following list of test scores.

74, 74, 74, 74, 76, 76, 76, 80, 80, 85, 85, 85, 85, 88, 88, 88, 90, 90

Finding the sum of 18 scores could take a long time. However, the total can be found more quickly if we note that the score 74 is repeated 4 times, the score 76 is repeated 3 times, and so on. In this way, the mean can be found as follows.

$$\text{Mean} = \frac{4(74) + 3(76) + 2(80) + 4(85) + 3(88) + 2(90)}{4 + 3 + 2 + 4 + 3 + 2}$$

This computation is called a **weighted mean** because each score is *weighted* by the number of times that it occurs. For example, even though 74 is a lower score than 80, it has more *weight* in computing the mean because it occurs more often. This computation gives the correct mean because the numerator is the sum of the 18 test scores and the denominator is the number of test scores.

For college students, one of the most useful applications of the weighted mean is in the computation of **grade point average (GPA)**. Many colleges and universities use the following grade point values when computing GPA.

A: 4 grade points,   B: 3 grade points,   C: 2 grade points,

D: 1 grade point,   and   F: 0 grade points

Based on these grade point values, a student's GPA can be computed as follows.

COMPUTING GRADE POINT AVERAGE (GPA)

If the variables *a, b, c, d,* and *f* represent the *number of credits* earned with a grade of A, B, C, D, and F, respectively, then the grade point average (GPA) is given by

$$\text{GPA} = \frac{4a + 3b + 2c + 1d + 0f}{a + b + c + d + f}.$$

READING CHECK

• Why is GPA an example of weighted mean?

EXAMPLE 9   ### Finding a student's grade point average (GPA)

A student's grade report is shown in Table 5.9. Find the student's GPA.

TABLE 5.9 Grade Report

Course	Credits	Grade
Math	4	B
English	3	A
Astronomy	4	D
Psychology	3	A
Soccer	2	B

Solution

The student has **6** credits with a grade of A, **6** credits with a grade of B, and **4** credits with a grade of D. There are **0** credits for both grades C and F.

$$\text{GPA} = \frac{4(6) + 3(6) + 2(0) + 1(4) + 0(0)}{6 + 6 + 0 + 4 + 0} = \frac{24 + 18 + 4}{16} = \frac{46}{16} = 2.875$$

Now Try Exercise 65

5.6 Putting It All Together

CONCEPT	COMMENTS	EXAMPLES
Circle	A *circle* is a collection of points that are all the same distance from a central point called the *center*. The distance between the center and any point on the circle is called the *radius*, and the distance across the circle on a straight line through the center is called the *diameter*.	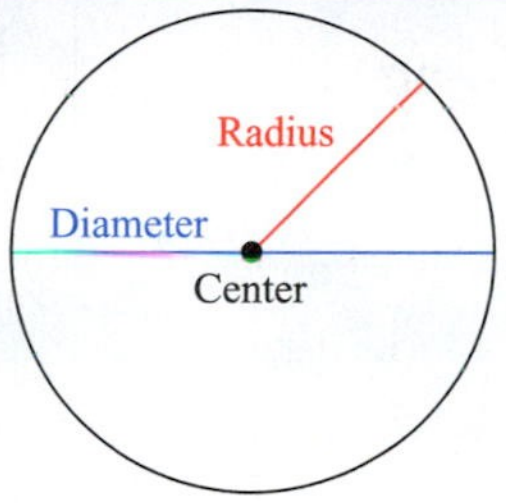
Circumference of a Circle	The distance around a circle is called its *circumference*. For a circle with radius r and diameter d, the circumference is given by $$C = \pi d \quad \text{or} \quad C = 2\pi r.$$	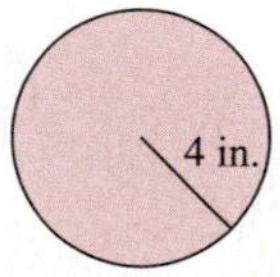 $C = 2\pi(4) = 8\pi \approx 25.12$ inches
Area of a Circle	For a circle with radius r, the area is given by $$A = \pi r^2.$$	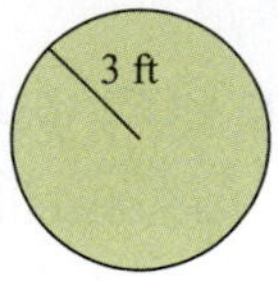 $A = \pi(3^2) = 9\pi \approx 28.26$ square feet
Pythagorean Theorem	If a and b are the lengths of the legs of a right triangle and c is the length of the hypotenuse, $$a^2 + b^2 = c^2.$$	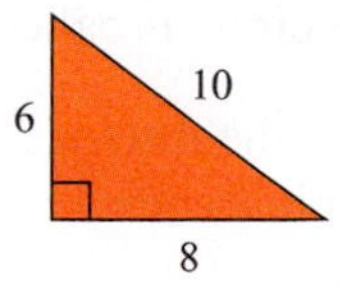 $6^2 + 8^2 = 10^2$
The Mean	The mean of a list of numbers is commonly called the average of the numbers. $$\text{Mean} = \frac{\text{the sum of a list of values}}{\text{number of values in the list}}$$	Find the arithmetic mean of the list $$3, 6, 2, 7, 6.$$ $$\text{Mean} = \frac{3 + 6 + 2 + 7 + 6}{5} = 4.8$$

continued on next page

continued from previous page

CONCEPT	COMMENTS	EXAMPLES
The Median	The median of an *ordered* list of values is defined as follows. • For an odd number of values, the median is the middle value in the list. • For an even number of values, the median is the mean of the two middle values in the list.	The median of the list $$2, 5, 6, 8, 9$$ is the middle value, 6. The median of the list $$2, 5, 6, 8, 9, 11$$ is the mean of 6 and 8, $\frac{6+8}{2} = 7$.
The Mode	The mode of a list of values is the value that occurs most often in the list. It is possible for a data set to have more than one mode or no mode.	The mode of the list $$3, 6, 4, 7, 3, 5$$ is 3 because it occurs most often.
Grade Point Average (GPA)	If the variables *a, b, c, d,* and *f* represent the *number of credits* earned with a grade of A, B, C, D, and F, respectively, then the grade point average is given by $$GPA = \frac{4a + 3b + 2c + 1d + 0f}{a + b + c + d + f}.$$	A student who earns 9 credits of As, 5 credits of Bs, and 2 credits of Cs has a GPA of $$\frac{4(9) + 3(5) + 2(2) + 1(0) + 0(0)}{9 + 5 + 2 + 0 + 0}$$ $$= \frac{55}{16} = 3.4375.$$

5.6 Exercises

MyMathLab | Math XL PRACTICE | WATCH | DOWNLOAD | READ | REVIEW

CONCEPTS AND VOCABULARY

1. A(n) _____ is a collection of points that are all the same distance from a central point.

2. The distance between the center of a circle and any point on the circle is called the _____.

3. The _____ is the distance across a circle on a straight line through its center.

4. The diameter is always twice the _____.

5. The perimeter of a circle is usually called the _____ of the circle.

6. An approximation of the number _____ is 3.14.

7. A half circle is called a(n) _____.

8. The Pythagorean theorem describes the relationship among the legs and hypotenuse of a(n) _____ triangle.

9. The sides of a right triangle that form the right angle are the _____ of the triangle.

10. The _____ is the side that is opposite the right angle in a right triangle.

11. The _____ of a list of numbers is commonly called the average of the numbers.

12. The _____ of an ordered list of numbers is either the middle value in the list or the mean of the two middle values in the list.

13. The _____ of a list of numbers is the value that occurs most often in the list.

14. Grade point average is an example of a(n) _____ mean.

CIRCLES

Exercises 15–22: Find the exact circumference of the circle and then approximate the circumference using 3.14 as an approximation for π.

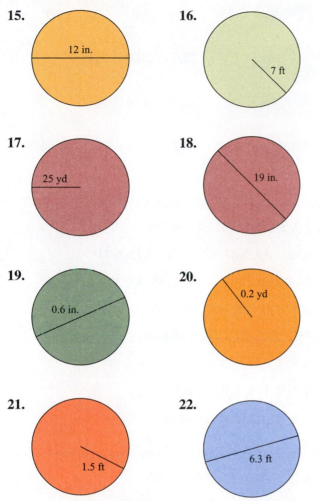

15. 12 in.

16. 7 ft

17. 25 yd

18. 19 in.

19. 0.6 in.

20. 0.2 yd

21. 1.5 ft

22. 6.3 ft

Exercises 23–30: Find the exact area of the given circle and then approximate the area using 3.14 as an approximation for π.

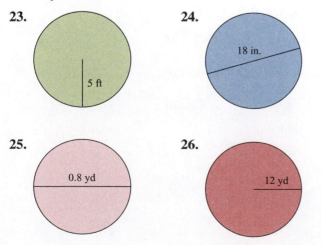

23. 5 ft

24. 18 in.

25. 0.8 yd

26. 12 yd

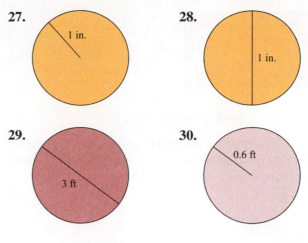

27. 1 in.

28. 1 in.

29. 3 ft

30. 0.6 ft

COMPOSITE REGIONS

Exercises 31–38: Find the area of the shaded region. Use 3.14 to approximate π, if needed.

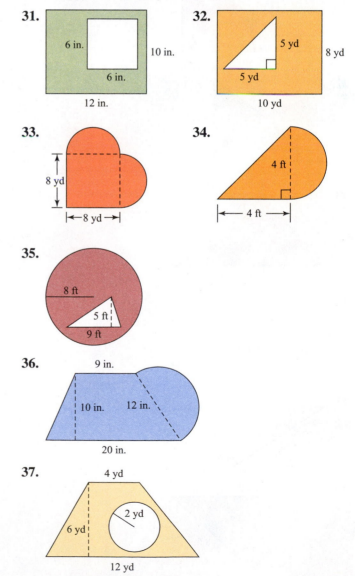

31. 6 in.   10 in.   6 in.   12 in.

32. 5 yd   8 yd   5 yd   10 yd

33. 8 yd   8 yd

34. 4 ft   4 ft

35. 8 ft   5 ft   9 ft

36. 9 in.   10 in.   12 in.   20 in.

37. 4 yd   2 yd   6 yd   12 yd

38.

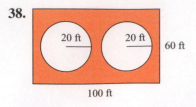

THE PYTHAGOREAN THEOREM

Exercises 39–46: Find the length of the unknown side of the right triangle. If the answer is not a whole number, approximate it to one decimal place.

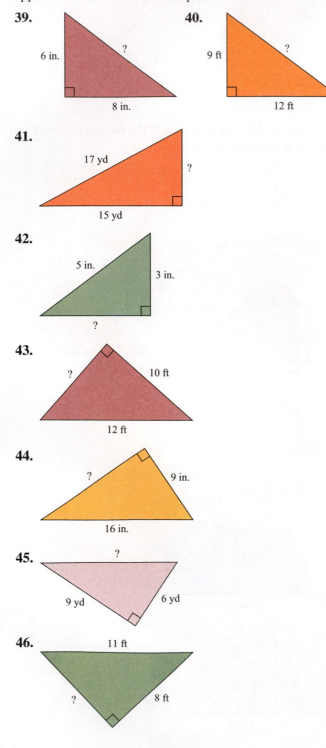

39. 6 in. ? 8 in.

40. 9 ft ? 12 ft

41. 17 yd ? 15 yd

42. 5 in. 3 in. ?

43. ? 10 ft 12 ft

44. ? 9 in. 16 in.

45. ? 9 yd 6 yd

46. 11 ft ? 8 ft

MEAN, MEDIAN, AND MODE

Exercises 47–50: Find the mean of the data presented in the given table of values.

47.

Age	72	68	70	70	69

48.

Mileage	19.7	22.6	20.8	21.5

49.

Score	27.4	32.2	30.9	26.3

50.

Price	0.89	0.97	0.58	0.85	0.66

Exercises 51–58: Find the median of the data set.

51. 2.0, 3.4, 5.4, 1.8, 3.2, 4.4, 6.1

52. 6.98, 7.13, 5.87, 6.53, 7.02, 7.23, 6.74

53. 18, 42, 31, 56, 11, 45, 17, 55, 43, 39, 22, 89

54. 100, 300, 200, 400, 500, 300

55. 25.2, 29.8, 27.3, 24.1, 28.0, 46.9

56. 9, 8, 7, 6, 5, 4, 3, 2, 1

57. 0, 1, 0, 1, 0, 1, 0, 1, 3429

58. 1, 8945, 7356, 6559, 8882

Exercises 59–64: If possible, find the mode(s) of the data set.

59. 3.2, 4.7, 3.3, 3.9, 4.4, 4.7, 3.5

60. 6.99, 5.99, 4.78, 5.99, 6.23, 5.87

61. 17, 14, 23, 11, 15, 19, 22, 28

62. 150, 170, 120, 160, 180, 140, 100, 130

63. 5, 7, 5, 4, 9, 9, 3, 1, 6, 8

64. 11.6, 12.4, 13.2, 13.2, 11.6

Exercises 65–68: Find the GPA for the grade report.

65.

Course	Credits	Grade
Nursing I	3	A
Sociology	3	A
French	3	B
Chemistry	4	A
History	3	D

66.

Course	Credits	Grade
Speech	3	B
Math	4	B
English	4	C
Psychology	3	A
PE	2	C

67.

Course	Credits	Grade
Chemistry	4	A
Calculus II	4	A
Physics	4	A
Biology	3	B

68.

Course	Credits	Grade
Psychology	3	F
English	3	C
Study Skills	3	C
Biology	3	D

APPLICATIONS

69. *Wind Turbines* The blades on the E-126 wind turbine turn through a large circular region. Approximate the circumference of the circular region if one blade is 206 feet long. Use 3.14 for π. Note that blade length equals the radius. (*Source:* WindBlatt, Enercon Magazine for Wind Energy, 2007.)

70. *Wind Turbines* Refer to the previous exercise. When turning, the blades of a wind turbine sweep through a circular region. Using 3.14 for π, find the approximate area of the circular region for the blades of the E-126 wind turbine.

71. *Running Track* A running track is constructed with semicircle curves, as shown in the figure. Approximate the area of the region enclosed by the track. Use 3.14 for π.

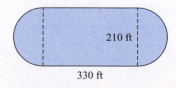

210 ft

330 ft

72. *Running Track* Using 3.14 for π, approximate the distance around the track in the previous exercise.

73. *Ladder Length* A ladder is leaning against a wall so that the foot of the ladder is 8 feet from the wall and the top of the ladder is 15 feet above the ground, as shown in the figure. How long is the ladder?

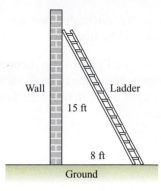

Wall Ladder

15 ft

8 ft

Ground

74. *Laptop Screen* A rectangular laptop screen measures 11 inches by 8 inches. Find the measure of the diagonal. Use 1 decimal place in your answer.

Exercises 75–80: Television The table lists the longest-running TV series of all time (as of 2008). Use the information in the table to find the following.

TV Series	Seasons
60 Minutes	39
Walt Disney Shows	33
Ed Sullivan Show	24
Gunsmoke	20
Red Skelton Show	20
Meet the Press	18
What's My Line?	18
I've Got a Secret	17
Lassie	17
Lawrence Welk Show	17

(*Source:* Nielsen Media research)

75. The mean number of seasons for the 10 TV series

76. The mean number of seasons for the top 5 TV series

77. The median number of seasons for the 10 TV series

78. The median number of seasons for the top 5 TV series

79. The mode for all 10 TV series

80. The mode for the top 5 TV series

WRITING ABOUT MATHEMATICS

81. Would the mean, median, or mode work best to describe the annual income in a community? Explain your reasoning.

82. Explain why the circumference and diameter of a circle cannot *both* be integers.

83. A student uses the Pythagorean theorem to determine that the longest side of the triangle shown in the right-hand column has a length of 5 feet. Is the student correct? Explain.

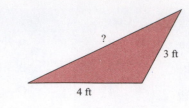

84. A student claims that the median of the following data set is 9.

$$3, 12, 9, 4, 7$$

Explain why this student is not correct.

SECTIONS 5.5 and 5.6

Checking Basic Concepts

1. Solve the equation algebraically.
 (a) $6.5w + 7.1 = 27.9$
 (b) $2(x - 4.8) = 26.6$
 (c) $4.2(m + 4) = 6.4m - 1.6$

2. Solve the equation $3.4x + 7.5 = 0.7$ numerically.

x	-4	-3	-2	-1	0
$3.4x + 7.5$					

3. Solve the equation $-0.75x + 1 = -2$ visually.

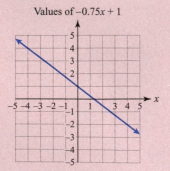

Values of $-0.75x + 1$

4. Approximate the circumference and the area of the circle shown. Use 3.14 for π.

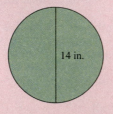

14 in.

5. Find the length of the unknown side of the right triangle. If the answer is not a whole number, then approximate it to 1 decimal place.

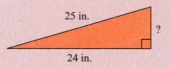

25 in.

24 in.

?

6. Find the mean, median, and mode for the data set.

$$14, 6, 3, 11, 6, 4, 10, 9$$

7. Find the GPA for the given grade report.

Course	Credits	Grade
Physics	4	C
Calculus I	5	C
Ecology	3	A
Biology	4	B

8. *Drive-in Theaters* The number N of drive-in movie theaters in the United States during year x can be estimated using the formula

$$N = -18.5x + 37,740,$$

where x is a year after 1990. Find the year when there were 555 drive-in movie theaters. (*Source: National Association of Theatre Owners.*)

CHAPTER 5   Summary

SECTION 5.1 ■ INTRODUCTION TO DECIMALS

Decimal Notation	Numbers written in decimal notation have an integer part and a fractional part separated by a decimal point. They are called decimals.

Examples: 4.983, 0.64, −3.2

Writing Decimals in Words

1. Write the integer part in words.
2. Include the word "and" for the decimal point.
3. Write the word form of the whole number formed by the fractional part, followed by the place value of the rightmost digit.

Example: The decimal 6.29 is written as "six and twenty-nine hundredths."

Writing Decimals as Mixed Numbers or Fractions

1. Write the integer part of the decimal as the integer part of the mixed number.
2. Take the whole number formed by the fractional part of the decimal and place it in the numerator of a fraction whose denominator is a power of 10 that corresponds to the place value of the rightmost digit.
3. Simplify the fraction to lowest terms.

Examples: $0.65 = \dfrac{65}{100} = \dfrac{13}{20}$, $14.125 = 14\dfrac{125}{1000} = 14\dfrac{1}{8}$

Graphing Decimals

A decimal can be graphed on a number line by using equally spaced tick marks to break the distance between values into 10 equal parts.

Example: Graph the decimal 5.28.

5.20 5.22 5.24 5.26 5.28 5.30

Comparing Decimals

To compare two positive decimals with the same number of digits to the left of the decimal point, compare digits in corresponding places from left to right until unequal digits are found. The decimal with the greater of these digits is the larger number.

Examples: $652.734 < 652.743$, $0.57401 > 0.57104$

Rounding Decimals

To round a decimal to the nearest whole number or to a place value to the right of the decimal point,

STEP 1: Identify the first digit to the *right* of the given place value.
STEP 2: If this digit is:
  (a) less than 5, do not change the digit in the given place value.
  (b) 5 or more, add 1 to the digit in the given place value.

STEP 3: Drop all digits to the right of the given place value.

Examples: Rounding 16.5682 to the nearest hundredth results in 16.57.
Rounding 0.7142 to the nearest tenth results in 0.7.

SECTION 5.2 ■ ADDING AND SUBTRACTING DECIMALS

Estimating Decimal Sums and Differences

By rounding decimal numbers to convenient values, a sum or difference can be estimated before the actual computation is performed.

Example: The sum $31.2 + 194.6$ can be estimated as $30 + 190 = 220$ or as $30 + 200 = 230$.

Adding and Subtracting Decimals

1. Stack the decimals vertically with the decimal points aligned. Extra zeros may be written to the right of the last digit of any decimal number so that corresponding digits align neatly.

steps continued on next page

2. Add or subtract digits in corresponding place values. Carry or borrow as needed.

3. Place a decimal point in the result so that it is aligned vertically with the decimal points in the numbers stacked above it.

Example: The sum $251.873 + 55.48$ is shown.

$$
\begin{array}{r}
1\ 1\ 1 \\
251.873 \\
+\ 55.480 \\
\hline
307.353
\end{array}
$$

SECTION 5.3 ■ MULTIPLYING AND DIVIDING DECIMALS

Estimating Decimal Products and Quotients

By rounding decimal numbers to convenient values, a product or quotient can be estimated before the actual computation is performed.

Example: The quotient $72.15 \div 7.94$ can be estimated as $72 \div 8 = 9$.

Multiplying Decimals

1. Multiply the decimals as though they were whole numbers. All decimal points may be ignored during computation.

2. Place a decimal point in the product so that the number of decimal places in the product equals the sum of the number of decimal places in the factors.

Example: The product $14.2(7.9)$ is shown.

$$
\begin{array}{r}
14.2 \\
\times\ 7.9 \\
\hline
12\ 78 \\
99\ 40 \\
\hline
112.18
\end{array}
$$

Multiplying Decimals by Powers of 10

Move the decimal point to the right the same number of places as the number of zeros in the power of 10. If needed, insert zeros at the end of the decimal as placeholders.

Example: $1000(2.6935) = 2693.5$

Dividing Decimals by Whole Numbers

1. Using long division, divide as though the dividend is a whole number. The decimal point in the dividend may be ignored during computation.

2. Place a decimal point in the quotient so that it is directly above the decimal point in the original dividend.

3. When needed, place extra 0s to the right of the last digit in the dividend.

Example: The quotient $259.2 \div 8$ is shown.

$$
\begin{array}{r}
32.4 \\
8)\overline{259.2} \\
-\ 24 \\
\hline
19 \\
-\ 16 \\
\hline
3\ 2 \\
-\ 3\ 2 \\
\hline
0
\end{array}
$$

Dividing Decimals by Decimals

1. Make the divisor a whole number by moving its decimal point to the right.

2. Move the decimal point in the dividend to the right the same number of places.

3. Find the quotient using the procedure for dividing a decimal by a whole number.

Example: To find the quotient $0.48 \div 0.6$, move the decimal points one place to the right to result in the quotient $4.8 \div 6$.

$$
\begin{array}{r}
0.8 \\
6)\overline{4.8} \\
-\ 0 \\
\hline
4\ 8 \\
-\ 4\ 8 \\
\hline
0
\end{array}
$$

Dividing Decimals by Powers of 10	Move the decimal point to the left the same number of places as the number of zeros in the power of 10. If needed, insert zeros at the beginning of the decimal as placeholders.

Example: $4257.9 \div 100 = 42.579$

Writing Fractions as Decimals	**Method I:** Multiply the numerator and denominator of the fraction by the same (nonzero) number to get an equivalent fraction whose denominator is a power of 10. Then write the new fraction as a decimal.
	Method II: Use long division to divide the denominator into the numerator.

Examples: Method I: Write $\frac{11}{20}$ as a decimal.

$$\frac{11}{20} = \frac{11 \cdot 5}{20 \cdot 5} = \frac{55}{100} = 0.55$$

Method II: Write $\frac{1}{5}$ as a decimal.

$$\begin{array}{r} 0.2 \\ 5\overline{)1.0} \\ -1\,0 \\ \hline 0 \end{array}$$

SECTION 5.4 ■ REAL NUMBERS, SQUARE ROOTS, AND ORDER OF OPERATIONS

Irrational Numbers	Irrational numbers cannot be expressed as fractions whose numerator and denominator are both integers.

Examples: $\pi \approx 3.141592654$, $\sqrt{19} \approx 4.358898944$

Real Numbers	Real numbers can be written as decimals. They include all natural numbers, whole numbers, integers, rational numbers, and irrational numbers.

Examples: $-\sqrt{7}$, -5, 0, $\frac{3}{4}$,   and $6.\overline{3}$

Approximating Square Roots by Guess-and-Check	1. Use a number line to make a reasonable guess for the square root.
	2. Make a table of values based on your guess. Compute the square of each value in the table until the given radicand is between consecutive values in the table.
	3. If the desired accuracy for the approximation has been found, stop. Otherwise, make a more accurate guess of the square root and go back to Step 2.

Example: The radicand **6** is shown between the perfect squares **4** and **9**.

$$\sqrt{4} = 2 \qquad \sqrt{9} = 3$$

```
◄─┼─┼─┼─┼─┼─┼─┼─┼─┼─┼─►
  0  1  2  3  4  5  6  7  8  9  10
```

A reasonable guess for $\sqrt{6}$ is 2.5.

Guess	2.4	2.5	2.6
Guess Squared	5.76	6.25	6.76

To the nearest tenth, $\sqrt{6} \approx 2.4$.

Approximating Square Roots with the Babylonian Algorithm	To approximate $\sqrt{Q}$, do the following.

1. Make a reasonable guess for the square root. Assign this value to the variable A.
2. Compute the quotient $Q \div A$.
3. Add A to the result from Step 2.
4. Divide the result of Step 3 by 2. The result is an approximation for $\sqrt{Q}$.

For a more accurate approximation, assign the result from Step 4 to the variable A and repeat the process beginning at Step 2.

Example: 1. A reasonable guess for $\sqrt{40}$ is 6.4.

2. $40 \div 6.4 = 6.25$
3. $6.25 + 6.4 = 12.65$
4. $12.65 \div 2 = 6.325$
Rounded to three decimal places, $\sqrt{40} \approx 6.325$.

SECTION 5.5 ■ SOLVING EQUATIONS INVOLVING DECIMALS

Solving Equations Algebraically By Clearing Decimals

1. Use the distributive property to clear parentheses from the equation.
2. Note the maximum number of decimal places in any number in the equation. Multiply every term by a power of 10 with that many zeros.
3. Simplify each term in the equation and combine any like terms.
4. Solve the resulting equation.

Example:

$$3.5x + 9 = 18.1$$
$$35x + 90 = 181$$
$$35x + 90 - 90 = 181 - 90$$
$$35x = 91$$
$$\frac{35x}{35} = \frac{91}{35}$$
$$x = 2.6$$

Solving Equations Numerically

A table of values can be used to solve some equations involving decimals. Complete the table for various values of the variable and then select the solution from the table.

Example: The solution to $6.1x + 4 = -8.2$ is -2.

x	-3	-2	-1
$6.1x + 4$	-14.3	-8.2	-2.1

Solving Equations Visually

Use a graph of the equation to estimate a solution. Check the estimated solution in the given equation to be sure that it is correct.

Example: The solution to $0.5x - 2 = -4$ is -4.

Values of $0.5x - 2$

Check: $0.5(-4) - 2 = -4$ ✓

SECTION 5.6 ■ APPLICATIONS FROM GEOMETRY AND STATISTICS

Circumference of a Circle

For a circle with radius r and diameter d, the circumference is given by $C = \pi d$ or $C = 2\pi r$.

Example: $C = \pi(9)$

$$= 9\pi$$
$$\approx 28.26 \text{ inches}$$

9 in.

Area of a Circle

For a circle with radius r, the area is given by $A = \pi r^2$.

Example: $A = \pi(6^2)$

$$= 36\pi$$
$$\approx 113.04 \text{ square feet}$$

6 ft

Pythagorean Theorem

If a and b are the lengths of the legs of a right triangle and c is the length of the hypotenuse, then $a^2 + b^2 = c^2$.

Example: $3^2 + 4^2 = 5^2$

$$9 + 16 = 25$$
$$25 = 25$$

3

5

4

The Mean

The mean of a list of numbers is commonly called the average.

$$\text{Mean} = \frac{\text{the sum of a list of values}}{\text{the number of values in the list}}$$

Example: The mean of the numbers 7, 3, 6, 2, 5, and 4 is

$$\text{Mean} = \frac{7 + 3 + 6 + 2 + 5 + 4}{6} = 4.5.$$

The Median

The median of an *ordered* list of values is defined as follows.
- For an odd number of values, the median is the middle value in the list.
- For an even number of values, the median is the mean of the two middle values in the list.

Examples: For the list 2, 4, 5, 7, 9, the median is 5.
For the list 3, 5, 8, 9, the median is $\frac{5 + 8}{2} = 6.5$.

The Mode

The mode of a list of values is the value that occurs most often in the list. It is possible for a data set to have more than one mode or no mode.

Example: For the list 3, 3, 5, 5, 5, 6, 6, 7, the mode is 5.

Grade Point Average (GPA)

If the variables $a, b, c, d,$ and f represent the *number of credits* earned with a grade of A, B, C, D, and F, respectively, then the grade point average is given by

$$\text{GPA} = \frac{4a + 3b + 2c + 1d + 0f}{a + b + c + d + f}.$$

Example: A student who earns 8 credits of As, 4 credits of Bs, 1 credit of Cs, and 2 credits of Ds has a GPA of

$$\frac{4(8) + 3(4) + 2(1) + 1(2) + 0(0)}{8 + 4 + 1 + 2 + 0} = \frac{48}{15} = 3.2.$$

CHAPTER 5   Review Exercises

SECTION 5.1

Exercises 1 and 2: Write the given decimal in words.

1. 0.76   **2.** −5.206

Exercises 3 and 4: Write the given decimal as a proper fraction or mixed number in lowest terms.

3. −0.08   **4.** 37.25

Exercises 5 and 6: Graph the decimal on a number line.

5. 7.8   **6.** −5.13

Exercises 7 and 8: Place the correct symbol, < or >, in the blank between the given decimals.

7. 41.684 _____ 41.648   **8.** −9.90 _____ −9.89

Exercises 9 and 10: Round to the given place value.

9. −4.008287 to the nearest ten-thousandth

10. 3591.014 to the nearest hundredth

Exercises 11 and 12: Round to the given value.

11. $12.08   dollar

12. $41.807   cent

SECTION 5.2

Exercises 13–18: Estimate the sum or difference and then add or subtract the decimals. Estimates may vary.

13. 290.115 − 39.97   **14.** 870.38 − 210.4

15. 14.97 + 655.009   **16.** 57.94 + 782.13

17. 302.41 + (−12.07)   **18.** −21.3 − (−199.1)

Exercises 19–24: Simplify the algebraic expression.

19. $6.9y + 3.87y$ **20.** $105.6x^3 + 40.1x^3$

21. $3.8q − 1.1 + 3.2q$ **22.** $1.3n + 4b + 6.11n$

23. $7.3x + 6.1y − 3.7x − 1.4y$

24. $8.4a + 0.9 − 6.7b − 8.1a + 3.2b$

SECTION 5.3

Exercises 25 and 26: Estimate the product or quotient by rounding the decimals. Answers may vary.

25. (4.1)(29.97)   **26.** 201.3 ÷ 10.03

Exercises 27–38: Find the given product or quotient.

27. (−3.9)(0.4)   **28.** −7.56 ÷ 4

29. 583.8 ÷ 3   **30.** (100)(0.163)

31. 84.072 ÷ 6.2   **32.** (0.92)(6.8)

33. 6.99(2.5)   **34.** 9246 ÷ 100

35. 8.4 ÷ 0.06   **36.** 10.9 ÷ 1.6

37. 1000(−0.0138)   **38.** (5.05)(2.08)

Exercises 39–42: Write the given fraction or mixed number as a decimal.

39. $-\dfrac{11}{25}$ **40.** $\dfrac{13}{30}$

41. $8\dfrac{2}{15}$ **42.** $64\dfrac{17}{20}$

Exercises 43–46: Simplify the algebraic expression.

43. $2.2(3.5x + 8)$ **44.** $−0.2(7.5y − 3)$

45. $\dfrac{7.2x}{12}$ **46.** $\dfrac{−48n}{0.8}$

SECTION 5.4

Exercises 47 and 48: Identify the numbers from the given list that belong to each of the following sets of numbers.
(a) natural numbers   (b) whole numbers
(c) integers   (d) rational numbers
(e) irrational numbers

47. $\dfrac{5}{8}, 2, 0, \sqrt{9}, -\sqrt{7}$

48. $7, -\dfrac{2}{5}, 3.\overline{6}, −7.4, \sqrt{5}$

Exercises 49 and 50: Approximate the square root to the nearest hundredth using the guess-and-check method.

49. $\sqrt{13}$ **50.** $\sqrt{21}$

Exercises 51 and 52: Go through the Babylonian algorithm one time to approximate the given square root. Give two decimal places in your answer.

51. $\sqrt{7}$ **52.** $\sqrt{28}$

Exercises 53–58: Evaluate the expression.

53. $8 - 10(0.49) \div 4$

54. $\sqrt{8.1 + 16.9} \div 5$

55. $\dfrac{4.6 + 8.6}{9.4 - 7.8}$

56. $\dfrac{12.3 - 1.8 \div 6}{0.004(1000)}$

57. $10.4 - |1.2 - 7.7| + (0.6 + 0.4)^2$

58. $(\sqrt{24.5 + 24.5} - |6.1 - 9.3|) \div 0.4$

Exercises 59 and 60: Evaluate the given expression for
$x = -2.5$ *and* $y = 4.9$.

59. $9x - 2y$

60. $y - (2x + 0.7)$

SECTION 5.5

Exercises 61–70: Solve the equation algebraically.

61. $x + 6.9 = 23.8$

62. $-8.1b = -45.36$

63. $10.12 = 5.2w - 6$

64. $6.4w = 3 + 4w$

65. $7.6x = -75.24$

66. $2.05(y - 1) = 8.2$

67. $5(x + 2.7) = 37.5$

68. $8y - 0.1 = 0.9$

69. $2.2(p - 5) = 2.7p + 4.1$

70. $19.6p - 4 = 14.3p + 7.66$

*Exercises 71 and 72: Solve the equation numerically by
completing the given table of values.*

71. $5.9x - 7.5 = 10.2$

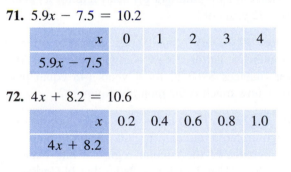

x	0	1	2	3	4
$5.9x - 7.5$					

72. $4x + 8.2 = 10.6$

x	0.2	0.4	0.6	0.8	1.0
$4x + 8.2$					

Exercises 73 and 74: Solve the equation visually.

73. $-0.8x - 1 = 3$

74. $0.7x + 1.4 = 3.5$

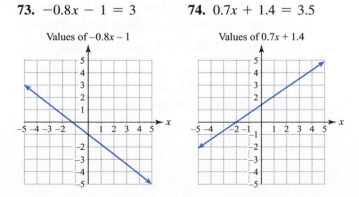

Values of $-0.8x - 1$

Values of $0.7x + 1.4$

SECTION 5.6

*Exercises 75 and 76: Find the exact circumference of the
circle and then approximate the circumference using 3.14
as an approximation for π.*

75.

76.

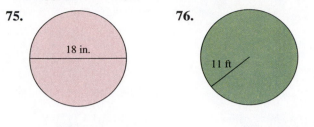

18 in.

11 ft

*Exercises 77 and 78: Find the exact area of the circle and
then approximate the area using 3.14 as an approximation for π.*

77.

78.

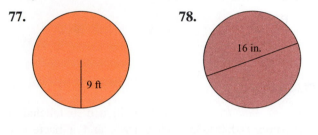

9 ft

16 in.

*Exercises 79 and 80: Find the area of the shaded region.
Use 3.14 to approximate π, if needed.*

79.

80.

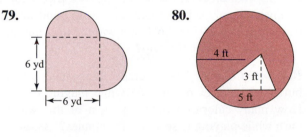

6 yd

6 yd

4 ft

3 ft

5 ft

*Exercises 81 and 82: Find the length of the unknown side
of the right triangle. If the answer is not a whole number,
approximate it to one decimal place.*

81.

82.

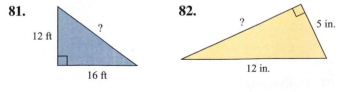

12 ft

?

16 ft

?

5 in.

12 in.

*Exercises 83 and 84: Find the mean, median, and mode
of the given data set.*

83. 4.2, 4.7, 4.3, 4.7, 5.4, 5.7, 4.6

84. 9, 13, 12, 13, 15, 10, 15, 12, 10, 15

Exercises 85 and 86: Find the GPA for the grade report.

85.

Course	Credits	Grade
Literature	4	B
Math	3	B
Biology	3	B
Physics	3	C
PE	2	B

86.

Course	Credits	Grade
Chemistry	3	F
English	3	B
Study Skills	3	D
Biology	3	D

APPLICATIONS

87. *Size of Bacteria* A bacterium in a dish labeled A measures 0.0000635 inch across, and a bacterium in a dish labeled B measures 0.0000653 inch across. Which of these bacteria is smaller?

88. *Driving Distance* At the beginning of a trip, a car's odometer displayed 96,533.2, and at the end of the trip, it displayed 97,003.6. What was the total number of miles traveled on this trip?

89. *Burning Calories* The average 175-pound person burns 4.2 calories per minute while playing frisbee. How many calories would a person of this weight burn while playing frisbee for 45 minutes? (*Source:* The Calorie Control Council.)

90. *Flight of a Projectile* If a projectile is fired upward with a velocity of 72 feet per second, its height h in feet above the ground after t seconds can be approximated using the formula

$$h = -16t^2 + 72t.$$

Find the projectile's height after 1.5 seconds.

91. *Converting Temperature* The formula

$$C = \frac{(F - 32)}{1.8}$$

gives the relationship between C degrees Celsius and F degrees Fahrenheit. Find the Fahrenheit temperature that is equivalent to 34.5°C.

92. *Wire Length* A wire is stretched from the top of a vertical pole to a point on level ground that is 7 feet from the base of the pole. If the wire is 25 feet long, how tall is the pole?

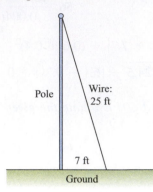

93. *Dinner Cost* A dinner consists of an entree for $16.59, a drink for $4.99, and dessert for $4.79. Find the total cost of the dinner.

94. *Baseball Stats* Baseball great Tony Oliva had a lifetime batting average of 0.304. Write this decimal as a fraction. (*Source:* Major League Baseball.)

95. *Target Heart Rate* For general health and weight loss, a person who is x years old should maintain a minimum target heart rate of T beats per minute during extended exercise, where

$$T = -0.5x + 120.$$

Determine the minimum target heart rate for a person who is 32 years old.

96. *Monthly Payments* A mail order company offers a zero-interest loan on a new computer purchase. If a buyer finances $967.20 for 4 years (48 equal payments), how much is the monthly payment?

97. *Wind Turbines* The blades on a wind turbine turn through a large circular region. Approximate the circumference of the circular region if one blade is 86 feet long. Use 3.14 for π. Note that blade length equals the radius.

98. *Text Messaging* Suppose that the cost C of sending or receiving x text messages is given by the formula

$$C = 0.1(x - 300) + 7.5, \text{ where } x \geq 300.$$

Find the number of text messages that corresponds to total charges of $26.80 by replacing C in the formula with 26.8 and solving the resulting equation.

CHAPTER 5   Test

Exercises 1 and 2: Write the given decimal as a fraction or mixed number in lowest terms.

1. -0.85

2. 13.625

3. Place the correct symbol, $<$ or $>$, in the blank.

$$153.674 \text{\_\_\_\_} 153.746$$

4. Round 91.58068 to the nearest thousandth.

Exercises 5 and 6: Add or subtract the decimals.

5. $194.127 - 63.78$

6. $156.14 + 3552.09$

Exercises 7–10: Multiply or divide the decimals.

7. $3.49(6.5)$

8. $(100)(4.138)$

9. $692.4 \div 10$

10. $2.4 \div 0.05$

Exercises 11 and 12: Simplify the algebraic expression.

11. $2.5(6x + 0.4)$

12. $-\dfrac{75x}{1.5}$

Exercises 13 and 14: Write the given fraction or mixed number as a decimal.

13. $17\dfrac{1}{6}$

14. $-\dfrac{11}{40}$

15. Identify the numbers from the given list that belong to each of the following sets of numbers.
   (a) natural numbers   **(b)** whole numbers
   (c) integers   **(d)** rational numbers
   (e) irrational numbers

$$9, -\frac{2}{3}, \sqrt{4}, 0, \pi$$

16. Approximate $\sqrt{21}$ to two decimal places.

Exercises 17 and 18: Evaluate the expression.

17. $6.3 - 10(0.9) \div 5$

18. $\dfrac{9.7 - 2.4 \div 6}{0.03(100)}$

Exercises 19 and 20: Solve the equation algebraically.

19. $3.25(y - 2.1) = 6.5$

20. $1.1w = 6 + 0.3w$

21. Solve the equation $7x - 1.8 = -5.3$ numerically.

x	-0.8	-0.7	-0.6	-0.5	-0.4
$7x - 1.8$					

22. Solve the equation $0.9x + 1 = 2.8$ visually.

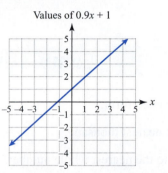
Values of $0.9x + 1$

23. Using 3.14 for π, approximate the circumference and the area of the circle shown.

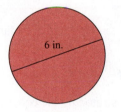
6 in.

24. Approximate the length of the unknown side of the right triangle to 1 decimal place.

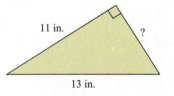
11 in.   ?
13 in.

25. Find the mean, median, and mode of the data set.

$$12.6, \ 10.4, \ 13.1, \ 11.8, \ 12.6, \ 10.9$$

26. Find the GPA for the given grade report.

Course	Credits	Grade
Earth Sci.	4	C
Calculus	5	B
English	4	B
PE	3	B

27. *Flight of a Baseball* If a baseball is hit upward with a velocity of 66 feet per second, its height h in feet above the ground after t seconds can be approximated using the formula

$$h = -16t^2 + 66t + 2.5.$$

Find the height of the ball after 1.5 seconds.

28. *Target Heart Rate* For general health and weight loss, a person who is x years old should maintain a minimum target heart rate of T beats per minute during extended exercise, where

$$T = -0.5x + 120.$$

Determine the age of a person whose minimum target heart rate is $T = 89$ beats per minute.

CHAPTERS 1–5 Cumulative Review Exercises

1. Write 61,005 in expanded form.

2. Round 48,113 to the nearest thousand.

3. Approximate $\sqrt{65}$ to the nearest whole number.

4. Estimate the sum $904 + 92 + 497$ by rounding each number to the nearest hundred.

5. Evaluate $13 - (8 + 6) \div 7$.

6. Place the correct symbol, $<$, $>$, or $=$, in the blank between the numbers: $-|-9|$ _____ -9.

7. Simplify the expression $-(-(-7))$.

8. State the addition property illustrated by the equation $3 + 6 = 6 + 3$.

9. Simplify the expression $7x - (3x - 8)$.

10. Evaluate $\sqrt{-7^2 + 53} \div (-3 + 2)$.

11. Complete the table and solve $5x - 4 = -9$.

x	-2	-1	0	1	2
$5x - 4$					

12. Is 5 a solution to $4 + w \div 3 = 3$?

Exercises 13 and 14: Translate each sentence into an equation using x as the variable. Do not solve the equation.

13. A number decreased by 12 equals -5.

14. Double the sum of a number and 4 equals -10.

Exercises 15 and 16: Solve the equation.

15. $2 + 3x - 21 = 2x - 3 - 7x$

16. $2 + (5x - 19) = 4x - (1 + 3x)$

17. The product of -3 and the sum of a number and 4 is the same as the number decreased by 4. Find the unknown number.

18. Is the equation $-3x + 11 = 2$ linear?

19. Solve the equation $-x - 2 = -4$ visually.

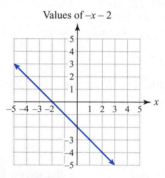

Values of $-x - 2$

20. Write the improper fraction and mixed number represented by the shading.

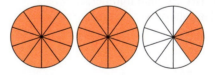

Exercises 21 and 22: Perform the indicated arithmetic.

21. $-\dfrac{4}{7} \div \dfrac{8}{21}$

22. $\dfrac{9}{10} - \dfrac{5}{8}$

23. Write $\frac{4}{15}$ as a decimal.

24. Evaluate the expression $5.3 - 100(0.8) \div 5$.

25. Approximate $\sqrt{45}$ to two decimal places.

26. Place the correct symbol, $<$ or $>$, in the blank between the numbers.

$$87.0254 \text{ \_\_\_\_\_ } 87.0524$$

27. Multiply 2.4(6.15).

28. Solve the equation $5.1(y - 1) = 11.22$.

29. *Geometry* Find the perimeter of the figure.

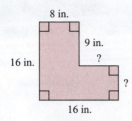

8 in.

9 in.

?

16 in.

?

16 in.

30. *Finances* A checking account starts the month with a balance of $3987. Each of the following positive entries (deposits) and negative entries (withdrawals) are made in the checking register.

$$-467, -26, 900, \text{ and } -1532$$

What is the ending balance?

31. *Hospitals* The number of hospitals H with more than 100 beds during year x can be estimated by the formula $H = -33x + 69,105$, where x is a year from 2000 to 2008. This formula is represented visually in the following figure. In which year were there 2874 hospitals of this type? Find your answer algebraically, numerically, and visually. (*Source:* AHA Hospital Statistics.)

Hospitals

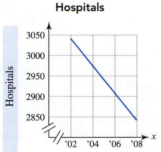

Year

32. *Classroom Population* In a large college class, $\frac{113}{200}$ of the students are female and under the age of 22, while $\frac{59}{200}$ of the students are male and under the age of 22. Find the total fraction of students in this class that are under the age of 22.

33. *Movie Tickets* The national average price P (in current dollars) for admission to a movie during year x can be estimated using the formula

$$P = 0.2x - 394.6,$$

where x is a year after 1995. Find the year when the the average price of a movie ticket was $6.40. (*Source:* Motion Picture Association of America.)

34. *Education and Pay* A 2003 survey of employed college graduates found that those whose highest degree is a bachelor's earned, on average, $7000 per year less than graduates whose highest degree is a master's. If those with a bachelor's degree earned $47,000 per year, on average, how much did those with a masters's degree earn?   (*Source:* National Science Foundation.)

6 Ratios, Proportions, and Measurement

The dictionary is the only place where success comes before work.

— MARK TWAIN

The Global Positioning System, or GPS, is a collection of 24 satellites originally developed by the U.S. military for navigational purposes. The orbits of the GPS satellites are arranged so that at least four satellites are "visible" at all times from any location on Earth. Data transmitted by these satellites are used by a GPS receiver to determine the receiver's location on the planet. To compute the receiver's approximate location, it is necessary to know the *rate* at which radio signals travel. Such information is expressed as a *ratio*. Since some inaccuracy is involved, proportions and other calculations are used to find the exact location of the receiver.

Ratios and proportions have many uses, from computing GPS locations to simple calculations for shopping and cooking. In this chapter, we discuss the mathematics of ratios and proportions.

Source: Marshall, B., and T. Harris, *How GPS Receivers Work.*

6.1 Ratios and Rates

Writing Ratios as Fractions • Unit Ratios • Writing Rates as Fractions • Unit Rates • Unit Pricing

In the multistate lottery game Powerball, there are 100 winning tickets (any prize) for every 3511 losing tickets. We say that the *ratio* of winning tickets to losing tickets is 100 to 3511. This ratio is commonly called the *odds* of winning a prize. For the $10,000 prize, there are 25 winning tickets for every 18,078,616 tickets that are not the $10,000 winner. This ratio is expressed as 25 to 18,078,616. In mathematics, ratios are often expressed as fractions in simplest form. In this section, we write ratios and *rates* as fractions and discuss their meaning in various real-world situations. (*Source:* Powerball.)

NEW VOCABULARY

☐ Ratio
☐ Unit ratio
☐ Rate
☐ Unit rate
☐ Unit pricing

READING CHECK

• What is a ratio?

Writing Ratios as Fractions

A **ratio** is a comparison of two quantities. There are three common ways to write a ratio: (1) in English using the word "to," (2) using a colon (:), and (3) as a fraction. For example, there are 3 apples and 6 bananas shown in Figure 6.1.

Figure 6.1 Apples and Bananas

The ratio of apples to bananas can be written as

$$3 \text{ to } 6, \quad 3:6, \quad \text{or} \quad \frac{3}{6}.$$

When a ratio is written as a fraction, we often express the fraction in simplest form. The ratio $\frac{3}{6}$ can be written in simplest form as $\frac{1}{2}$ because

$$\frac{3}{6} = \frac{1 \cdot 3}{2 \cdot 3} = \frac{1}{2}.$$

WRITING A RATIO AS A FRACTION

Write the quantity found before the word "to" (or before the colon) in the numerator of the fraction, and write the quantity found after the word "to" (or after the colon) in the denominator of the fraction. Then simplify the fraction.

EXAMPLE 1 | **Writing ratios as fractions in simplest form**

Write each ratio as a fraction in simplest form.
(a) 10 to 24 **(b)** $6\frac{1}{2} : 1\frac{2}{3}$ **(c)** 2.5 to 6.25

Solution

(a) A ratio of **10** to **24** is written as

$$\frac{\mathbf{10}}{\mathbf{24}} = \frac{5 \cdot \mathbf{2}}{12 \cdot \mathbf{2}} = \frac{5}{12}.$$

(b) For the ratio $6\frac{1}{2} : 1\frac{2}{3}$, first write $6\frac{1}{2}$ as $\frac{13}{2}$ and write $1\frac{2}{3}$ as $\frac{5}{3}$.

$$\frac{\frac{13}{2}}{\frac{5}{3}} = \frac{13}{2} \div \frac{5}{3} = \frac{13}{2} \cdot \frac{3}{5} = \frac{39}{10}.$$

(c) For the ratio 2.5 to 6.25, we will need to clear decimals from the fraction.

$$\frac{2.5}{6.25} = \frac{2.5 \cdot \mathbf{100}}{6.25 \cdot \mathbf{100}} = \frac{250}{625} = \frac{2 \cdot \mathbf{125}}{5 \cdot \mathbf{125}} = \frac{2}{5}$$

Now Try Exercises 9, 15, 21

MAKING CONNECTIONS

Ratios and Mixed Numbers

In Example 1(b), the ratio was written as the improper fraction $\frac{39}{10}$. Even though $\frac{39}{10}$ is equal to the mixed number $3\frac{9}{10}$, we do *not* write a ratio as a mixed number. Remember that a ratio is a comparison of *two* quantities.

EXAMPLE 2   **Flying a kite**

When a person flying a kite reels in $8\frac{1}{2}$ feet of string, the kite gets $6\frac{4}{5}$ feet closer to the ground, as illustrated in the figure. Write the ratio of the amount of string reeled in to the decrease in the kite's height as a fraction in simplest form.

Solution

First write $8\frac{1}{2}$ as $\frac{17}{2}$ and write $6\frac{4}{5}$ as $\frac{34}{5}$. The ratio is written as

$$\frac{\frac{17}{2}}{\frac{34}{5}} = \frac{17}{2} \cdot \frac{5}{34} = \frac{5}{4}.$$

This means that when 5 feet of string is reeled in, the kite gets 4 feet closer to the ground.

Now Try Exercise 61

EXAMPLE 3   **Analyzing the names of the states**

Of the 50 U.S. states, 12 have names starting with a vowel. Write the ratio of the number of state names starting with a vowel to the number of state names starting with a consonant as a fraction in simplest form.

Solution

There are 12 states having names that begin with a vowel. The rest, $50 - 12 = 38$, have names that begin with a consonant. The ratio 12 to 38 is written as

$$\frac{12}{38} = \frac{6 \cdot 2}{19 \cdot 2} = \frac{6}{19}.$$

So, there are 6 states having names starting with a vowel for every 19 states having names starting with a consonant.

▌ **Now Try Exercise 67**

READING CHECK

• What is one reason for computing unit ratios?

Unit Ratios

▶ **REAL-WORLD CONNECTION**   Students who would like a more personal educational experience may be interested in knowing the student-to-instructor ratio at a college. In order to make comparisons, it is often helpful to compute such ratios as *unit ratios*. A **unit ratio** is a ratio expressed as a fraction with a denominator of 1. In the next example, we compare the student-to-instructor ratios at two colleges.

EXAMPLE 4   **Using unit ratios to compare two colleges**

While researching local schools, a student found that the student-to-instructor ratio at the state college is 945 to 18 and the student-to-instructor ratio at the community college is 2145 to 75. Find the unit ratio at each school and interpret the results.

Solution

One way to find a unit ratio is to write the ratio as a simplified fraction and then divide the denominator into the numerator. We do this for each college, as shown.
For the state college, the unit ratio is

$$\frac{945}{18} = \frac{105 \cdot 9}{2 \cdot 9} = \frac{105}{2} = \frac{52.5}{1} = 52.5.$$

For the community college, the unit ratio is

$$\frac{2145}{75} = \frac{143 \cdot 15}{5 \cdot 15} = \frac{143}{5} = \frac{28.6}{1} = 28.6.$$

This means that there are 52.5 students for each instructor at the state college and there are 28.6 students for each instructor at the community college.

▌ **Now Try Exercise 71**

NOTE: Although part of a whole student is not possible, the unit ratios in Example 4 are very helpful in comparing the two colleges.

FINDING A UNIT RATIO

To find a unit ratio, divide the denominator into the numerator.

Writing Rates as Fractions

A **rate** is a ratio that is used to compare different kinds of quantities. Because the quantities in a rate have different units, the units are expressed as part of the ratio. For example, a car that travels 15 miles in 12 minutes has a rate of

$$\frac{15 \text{ miles}}{12 \text{ minutes}} = \frac{5 \text{ miles}}{4 \text{ minutes}}.$$

EXAMPLE 5 **Writing rates as fractions in simplest form**

Write each rate as a fraction in simplest form.
(a) A person eats 8 hot dogs in 60 seconds.
(b) A person earns $72 in 9 hours.

Solution
(a) A rate of **8** hot dogs in **60** seconds written as a fraction is

$$\frac{8 \text{ hot dogs}}{60 \text{ seconds}} = \frac{2 \text{ hot dogs}}{15 \text{ seconds}}.$$

The person eats at a rate of 2 hot dogs every 15 seconds.
(b) A rate of **72** dollars in **9** hours written as a fraction is

$$\frac{72 \text{ dollars}}{9 \text{ hours}} = \frac{8 \text{ dollars}}{1 \text{ hour}}.$$

The person's pay rate is $8 per hour.

Now Try Exercises 31, 39

Unit Rates

The rate in Example 5(b) is called a *unit rate*. It is common to use the word "per" to indicate division in a unit rate.

▶ **REAL-WORLD CONNECTION** As with unit ratios, a **unit rate** is a rate expressed as a fraction with a denominator of 1. Note the use of the word "per" in the following everyday unit rates.

- **The speed (rate) of a car is measured in miles per hour.**
- **Heart rate is recorded in beats per minute.**
- **Salary (pay rate) is given in dollars per hour.**
- **Mileage (gas consumption rate) is given in miles per gallon.**

Unit rates and unit ratios are computed in the same way.

FINDING A UNIT RATE

To find a unit rate, divide the denominator into the numerator.

NOTE: To express a unit rate more concisely, the symbol "/" can be used in place of the word "per" and the units can be abbreviated. For example, we can write the unit rate 17 miles per gallon as 17 mi/gal.

EXAMPLE 6 **Writing unit rates**

Write each rate as a unit rate.
(a) A person earns $580 in 40 hours.
(b) A person runs a 110-meter hurdle race in 16 seconds.

Solution

(a) $\dfrac{580 \text{ dollars}}{40 \text{ hours}} = \dfrac{29 \text{ dollars}}{2 \text{ hours}} = \$14.50/\text{hr}$

The person earns $14.50 per hour.

(b) $\dfrac{110 \text{ meters}}{16 \text{ seconds}} = \dfrac{55 \text{ meters}}{8 \text{ seconds}} = 6.875 \text{ m/sec}$

The person runs at an *average* rate of 6.875 meters per second.

■ **Now Try Exercises 43, 47**

STUDY TIP

Concepts are often easier to remember when they are associated with something that you already do. Unit rates are used in comparison shopping to find unit pricing.

Unit Pricing

▶ **REAL-WORLD CONNECTION** One practical use of unit rates occurs in consumer mathematics. Comparison shopping is much easier when unit pricing is involved. For example, it is difficult to tell whether a 4-ounce bag of snack mix for $1.79 is a better buy than a 10.5-ounce bag of the same snack mix for $3.49. To see which is a better buy, we use the following formula for **unit pricing**.

READING CHECK

• How is unit pricing used in comparison shopping?

FINDING UNIT PRICING

If the price of q units of a product is p, the unit price U is given by

$$U = \frac{p}{q}.$$

In the next example, we compare two pricing options using unit pricing.

EXAMPLE 7 **Determining the better buy**

Use unit pricing to determine the better buy.
    3.5-ounce bag of popcorn for $2.10   or   8.5 ounce bag of popcorn for $4.93

Solution
Compare the unit prices.

3.5-ounce bag   **8.5-ounce bag**

$U = \dfrac{p}{q} = \dfrac{\$2.10}{3.5 \text{ ounces}} = \$0.60/\text{oz}$ $U = \dfrac{p}{q} = \dfrac{\$4.93}{8.5 \text{ ounces}} = \$0.58/\text{oz}$

Because the unit price for the 8.5-ounce bag is lower, it is the better buy.

■ **Now Try Exercise 75**

In the next example, we analyze a situation in which unit pricing cannot be used to determine a better buy.

EXAMPLE 8   **Analyzing unit pricing**

For a 16-ounce jar of organic peanut butter that sells for $5.28 and a 7.5-ounce can of pinto beans that sells for $0.90, do the following.
(a) Find the unit pricing for each product.
(b) Explain why the better buy cannot be determined in this case.

Solution

(a) **Peanut butter**   **Pinto beans**

$$U = \frac{p}{q} = \frac{\$5.28}{16 \text{ ounces}} = \$0.33/\text{oz} \qquad U = \frac{p}{q} = \frac{\$0.90}{7.5 \text{ ounces}} = \$0.12/\text{oz}$$

(b) A better buy cannot be determined because we are not comparing size options for the *same* product.

Now Try Exercise 79

6.1   Putting It All Together

CONCEPT	COMMENTS	EXAMPLES
Ratio	A ratio is a comparison of two quantities. When writing a ratio as a fraction, the first number in the ratio is written as the numerator, and the second number is written as the denominator.	If a bowl contains two pears and four limes, the ratio of pears to limes can be written as $$2 \text{ to } 4, \quad 2:4, \quad \text{or} \quad \tfrac{2}{4} = \tfrac{1}{2}.$$
Unit Ratio	A unit ratio is a ratio expressed as a fraction with a denominator of 1. Unit ratios are helpful in making comparisons. To find a unit ratio, divide the denominator into the numerator.	If the student-to-instructor ratio at a college is 496 to 20, the unit ratio is $$\frac{496}{20} = 24.8,$$ or 24.8 students for each instructor.
Rate	A rate is a ratio that is used to compare different kinds of quantities. The units for each quantity are expressed as part of the rate.	A person who earns $19 in 3 hours has an earning rate of $$\frac{19 \text{ dollars}}{3 \text{ hours}}.$$
Unit Rate	A unit rate is a rate expressed as a fraction with a denominator of 1. Unit rates are helpful in making comparisons. To find a unit rate, divide the denominator into the numerator.	If a person earns $67.50 in 9 hours, his hourly pay rate is $$\frac{67.50 \text{ dollars}}{9 \text{ hours}} = \$7.50/\text{hr}.$$
Unit Pricing	If the price of q units of a product is p, then the unit price U is given by $$U = \frac{p}{q}.$$	A 3-ounce bottle of perfume at $17.79 has a unit price of $$U = \frac{p}{q} = \frac{\$17.79}{3 \text{ ounces}} = \$5.93/\text{oz}.$$

6.1 Exercises

MyMathLab Math XL PRACTICE WATCH DOWNLOAD READ REVIEW

CONCEPTS AND VOCABULARY

1. A(n) _____ is a comparison of two quantities.

2. A unit ratio is a ratio expressed as a fraction with a denominator of _____.

3. To find a unit ratio, divide the _____ into the _____.

4. A(n) _____ is a ratio that is used to compare different kinds of quantities.

5. A(n) _____ rate is a rate expressed as a fraction with a denominator of 1.

6. Unit _____ is a special type of unit rate that is helpful in comparison shopping.

RATIOS

Exercises 7–22: Write the given ratio as a fraction in simplest form.

7. 6 to 18	8. 7 to 42
9. 12 : 32	10. 25 : 60
11. $\frac{1}{2}$ to $\frac{5}{8}$	12. $\frac{4}{3}$ to $\frac{6}{7}$
13. $\frac{12}{25} : \frac{8}{15}$	14. $\frac{16}{9} : \frac{10}{27}$
15. $1\frac{2}{3}$ to $4\frac{1}{6}$	16. $3\frac{3}{4}$ to $1\frac{1}{5}$
17. $6\frac{1}{4} : 1\frac{7}{8}$	18. $2\frac{3}{4} : 1\frac{5}{6}$
19. 8.5 to 6.8	20. 3.2 to 10.4
21. 5.25 : 4.5	22. 8.4 : 2.45

Exercises 23–30: Find the unit ratio.

23. 21 to 7	24. 20 to 80
25. 15 : 24	26. 72 : 16
27. 165 to 30	28. 48 to 160
29. 115 : 115	30. 250 : 150

RATES

Exercises 31–40: Write the given rate as a fraction in simplest form.

31. It rains 6 inches in 9 hours.

32. It snows 16 inches in 12 hours.

33. A person earns $39 in 6 hours.

34. A bicycle travels 18 miles in 48 minutes.

35. A theater has 60 seats in 5 rows.

36. A lab has 10 microscopes for 20 students.

37. An office has 3 copiers for 51 employees.

38. There are 25 calculators for 100 students.

39. There are 8 slices of pizza for 4 people.

40. A car travels 350 miles in 6 hours.

Exercises 41–50: Write the given rate as a unit rate.

41. A person earns $105 in 12 hours.

42. A jet travels 210 miles in 30 minutes.

43. A car goes 128 miles on 5 gallons of gas.

44. It snows 32 inches in 20 hours.

45. It rains 4 inches in 5 hours.

46. There are 12 corn plants for every 12 feet.

47. A vendor makes $40.50 selling 18 drinks.

48. There are 63 players on 7 teams.

49. A heart beats 186 times in 3 minutes.

50. A person pays $100 for 8 gallons of paint.

UNIT PRICING

Exercises 51–56: Find the unit price.

51. A 3.5-ounce bag of crackers for $1.47

52. A 6-ounce bottle of perfume for $259.50

53. A 2-pound bag of pistachios for $7.90

54. A 3-liter bottle of store-brand soda for $1.89

55. A 2000-pound load of dirt for $8.50

56. A 16-ounce jar of jam for $3.52

APPLICATIONS

57. *Online Newspapers* In 2007, online newspaper sites had 60 million unique visitors. By 2008, there were 72 million such visitors. Write the ratio of online newspaper visitors in 2007 to the number in 2008 as a fraction in simplest form. (*Source:* Nielsen Online.)

58. *School Districts* There were 133 school districts in Alabama in 2008. By comparison, Arkansas had 252 school districts that year. Write the ratio of the number of school districts in Alabama to the number of school districts in Arkansas as a fraction in simplest form. (*Source:* U.S. Census Bureau.)

59. *School Children* An elementary school class has 12 boys and 18 girls. Write the ratio of boys to girls as a fraction in simplest form.

60. *Animal Shelters* An animal shelter has 36 dogs and 16 cats. Write the ratio of dogs to cats as a fraction in simplest form.

61. *Bicycle Wheels* The wheels on a bicycle rotate $4\frac{1}{2}$ times with $1\frac{1}{4}$ rotations of the pedals. Write the ratio of wheel rotations to pedal rotations as a fraction in simplest form.

62. *Baking* A recipe calls for $3\frac{2}{3}$ cups of flour and $2\frac{3}{4}$ cups of sugar. Write the ratio of flour to sugar as a fraction in simplest form.

63. *American Flag* An official American flag has a length of $4\frac{3}{4}$ feet when the width is $2\frac{1}{2}$ feet. Write the ratio of the length to the width as a fraction in simplest form.

64. *American Flag* The blue rectangle containing the stars on an American flag is called the *union*. The union on an official American flag will have a width of $17\frac{1}{2}$ inches when the flag has a width of $32\frac{1}{2}$ inches. Write the ratio of the width of the union to the width of the flag as a fraction in simplest form.

65. *Farming* The following table lists the number of farms in thousands for selected states in 2009. Use the data in the table to write each requested ratio.

State	Minnesota	South Carolina	Texas	Wyoming
Farms	81	27	250	11

Source: U.S. Department of Agriculture.

  (a) The number of farms in South Carolina to the number of farms in Minnesota

  (b) The number of farms in Wyoming to the number of farms in Texas

  (c) The number of farms in South Carolina to the total number of farms in these four states

66. *Union Activity* The following table lists the number of union-related work stoppages for selected years in companies with 1000 or more workers. Use the data in the table to write each requested ratio.

Year	2006	2007	2008	2009
Stoppages	20	21	15	5

Source: U.S. Bureau of Labor Statistics.

  (a) The number of stoppages in 2009 to the number in 2008

  (b) The number of stoppages in 2008 to the number in 2006

  (c) The number of stoppages in 2007 to the total number of stoppages in these four years

67. *Raffle Tickets* A math club sold 1000 raffle tickets. If 15 of the tickets are winning tickets, write the ratio of the number of winning tickets to the number of losing tickets as a fraction in simplest form.

68. *Vegetarian Meals* A banquet is prepared for 500 guests. If 28 of the guests have vegetarian meals, write the ratio of vegetarian meals to non-vegetarian meals as a fraction in simplest form.

69. *Equity Loans* Some banks will provide an equity loan if the loan-to-value ratio is less than 0.8. Compute the loan-to-value ratio as a unit ratio for a person who owns a $250,000 home and wants an equity loan for $180,000. Does this person qualify for the loan?

70. *Equity Loans* A bank will provide an equity loan if the loan-to-value ratio is less than 0.75. Compute the loan-to-value ratio as a unit ratio for a person who owns a $160,000 home and wants an equity loan for $125,000. Does this person qualify for the loan?

71. *Comparing Colleges* A public university has a student-to-instructor ratio of 692 to 16, while the local community college has a student-to-instructor ratio of 455 to 14. Find the unit ratio at each school. Interpret the results for the community college.

72. *Comparing Schools* A public high school has a student-to-teacher ratio of 396 to 15, while a private high school has a student-to-teacher ratio of 465 to 25. Find the unit ratio at each school. Interpret the results for the public high school.

73. *Keyboarding* A receptionist can key a 165-word email in 2.5 minutes, while an office manager can key a 254-word email in 4 minutes. Find the unit rate for each person. Who is faster?

74. *Downloading* A laptop computer can download a 1.8-MB song in 9 seconds, while a desktop computer can download a 3.4-MB song in 16 seconds. Find the unit rate for each computer. Which is faster?

75. *Shopping* Find the unit price for each size option.

> Large jar of jam: 14.5 ounces for $3.19
> Small jar of jam: 8 ounces for $1.64

Which is the better buy?

76. *Coffee* Find the unit price for each size option.

> Large coffee drink: 26 ounces for $4.81
> Small coffee drink: 14 ounces for $2.66

Which is the better buy?

77. *Pharmacy* Which option is the better buy?

> Generic allergy pills: 30 pills for $7.68
> Brand name allergy pills: 16 pills for $5.84

78. *Doughnuts* Which option is the better buy?

> Iced doughnuts: 36 doughnuts for $17.64
> Filled doughnuts: 60 doughnuts for $28.20

79. *Unit Pricing* A grocery coupon can be used to buy a 12.5-ounce jar of beets for $1.75 or a 7.8-ounce box of mints for $2.73.

(a) Find the unit price for each product.
(b) Explain why we cannot determine the better buy in this situation.

80. *Unit Pricing* A grocery coupon can be used to buy a 6.5-ounce can of peas for $0.52 or a 2.4-ounce bag of chips for $1.08.

(a) Find the unit price for each product.
(b) Explain why we cannot determine the better buy in this situation.

WRITING ABOUT MATHEMATICS

81. A friend finds two unit prices as $0.14/oz and $1.89/oz. Is this enough information to determine the better buy? Explain.

82. A manufacturer of shampoo changes the size of the bottle from 33 ounces to 29 ounces but does not change the price. Will the unit price increase or decrease? Explain.

83. Give an example other than unit pricing where unit rates can be used for comparison purposes.

84. Give three examples of using rates in everyday life.

6.2 Proportions and Similar Figures

**Identifying Proportions • Solving a Proportion for an Unknown Value •
Similar Figures • Applications**

A LOOK INTO MATH ▶

Some wilderness maps have a *scale* of $1\frac{1}{2}$ inches to 1 mile. This means that objects $1\frac{1}{2}$ inches apart on the map are 1 mile apart on the ground. A canoeist can use the map's scale as a ratio for finding actual distance by setting up an equation called a *proportion*. For example, a *portage* (a trail over which a canoe and gear are carried) may have a length of $\frac{3}{4}$ inch on the map. Because this measurement is half the first number in the map's scale, the actual distance must be half the second number in the map's scale. So, the portage is $\frac{1}{2}$ of 1 mile or $\frac{1}{2}$ mile in length. In this section, we discuss proportions and how they are used to solve many types of problems.

NEW VOCABULARY

☐ Proportion
☐ Similar figures

Identifying Proportions

A **proportion** is a statement indicating that two ratios are equal. An equation for a proportion can be written and read as follows.

PROPORTION

If $\frac{a}{b}$ and $\frac{c}{d}$ are ratios that are equal in value, then the equation

$$\frac{a}{b} = \frac{c}{d}$$

is a proportion, where $b \neq 0$ and $d \neq 0$. This proportion can be read as

"*a* is to *b* as *c* is to *d*."

When two ratios are set equal to each other, the equation may be true or false. If the equation is true, then it is a proportion. Since ratios are written as fractions, we can use *cross products* to determine if two ratios are equal.

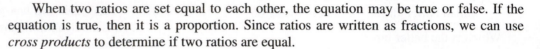

$3 \cdot 20 \qquad \frac{3}{4} \overset{?}{=} \frac{15}{20} \qquad 4 \cdot 15$

Because $3 \cdot 20 = 60$ and $4 \cdot 15 = 60$, the cross products are equal, and the equation is a proportion.

THE CROSS PRODUCT RULE

For $b \neq 0$ and $d \neq 0$, if the cross products ad and bc are equal, then

$$\frac{a}{b} = \frac{c}{d}$$

is a proportion.

EXAMPLE 1   **Determining whether an equation represents a proportion**

Determine if the equation is a proportion.

(a) $\frac{8}{18} \overset{?}{=} \frac{4}{9}$ **(b)** $\frac{6}{5.1} \overset{?}{=} \frac{4}{3.4}$ **(c)** $\frac{3\frac{1}{2}}{4\frac{3}{4}} \overset{?}{=} \frac{6\frac{2}{3}}{9\frac{1}{4}}$

Solution
(a) Start by finding the cross products.

$8 \cdot 9 \qquad \frac{8}{18} \overset{?}{=} \frac{4}{9} \qquad 18 \cdot 4$

Because $8 \cdot 9 = 72$ and $18 \cdot 4 = 72$, the cross products are equal, and the equation is a proportion.

(b) Find the cross products.

$6 \cdot 3.4 \qquad \frac{6}{5.1} \overset{?}{=} \frac{4}{3.4} \qquad 5.1 \cdot 4$

Because $6 \cdot 3.4 = 20.4$ and $5.1 \cdot 4 = 20.4$, the cross products are equal, and the equation is a proportion.

(c) Write the mixed numbers as improper fractions and then find the cross products.

Because $\frac{7}{2} \cdot \frac{37}{4} = \frac{259}{8} = 32.375$ and $\frac{19}{4} \cdot \frac{20}{3} = \frac{95}{3} = 31.\overline{6}$, the cross products are not equal, and the equation is not a proportion.

Now Try Exercises 7, 11, 19

Solving a Proportion for an Unknown Value

A proportion is an equation with four values: two numerators and two denominators. When one of these values is unknown, we can find its value using cross products. In the next two examples, we solve proportions for unknown values.

EXAMPLE 2   **Solving a proportion for an unknown value**

Solve each proportion for the unknown value. Check your solutions.

(a) $\dfrac{3}{5} = \dfrac{x}{20}$ **(b)** $\dfrac{2.4}{3} = \dfrac{3.6}{w}$

Solution

(a) Use cross products to write an equation.

$$\frac{3}{5} = \frac{x}{20} \qquad \text{Given proportion}$$

$$3 \cdot 20 = 5 \cdot x \qquad \text{Cross products are equal.}$$

$$60 = 5x \qquad \text{Simplify.}$$

$$\frac{60}{5} = \frac{5x}{5} \qquad \text{Divide each side by 5.}$$

$$12 = x \qquad \text{Simplify.}$$

To check this solution, replace x with 12 in the given proportion and use cross products to verify that the resulting equation is a proportion.

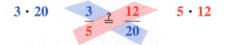

Because $3 \cdot 20 = 60$ and $5 \cdot 12 = 60$, the cross products are equal, and the equation is a proportion. The solution checks.

(b) Use cross products to write an equation.

$$\frac{2.4}{3} = \frac{3.6}{w} \qquad \text{Given proportion}$$

$$2.4 \cdot w = 3 \cdot 3.6 \qquad \text{Cross products are equal.}$$

$$2.4w = 10.8 \qquad \text{Simplify.}$$

$$\frac{2.4w}{2.4} = \frac{10.8}{2.4} \qquad \text{Divide each side by 2.4.}$$

$$w = 4.5 \qquad \text{Simplify.}$$

To check this solution, replace w with 4.5 in the given proportion and use cross products to verify that the resulting equation is a proportion.

$$2.4 \cdot 4.5 \qquad \frac{2.4}{3} \overset{?}{=} \frac{3.6}{4.5} \qquad 3 \cdot 3.6$$

Because $2.4 \cdot 4.5 = 10.8$ and $3 \cdot 3.6 = 10.8$, the cross products are equal, and the equation is a proportion. The solution checks.

■ **Now Try Exercises 23, 25**

EXAMPLE 3 **Solving a proportion for an unknown value**

Solve each proportion.

(a) $\dfrac{-2}{9} = \dfrac{m}{36}$ **(b)** $\dfrac{\frac{1}{2}}{n} = \dfrac{12}{\frac{3}{4}}$

Solution

(a)

$$\frac{-2}{9} = \frac{m}{36} \qquad \text{Given proportion}$$

$$-2 \cdot 36 = 9 \cdot m \qquad \text{Cross products are equal.}$$

$$-72 = 9m \qquad \text{Simplify.}$$

$$\frac{-72}{9} = \frac{9m}{9} \qquad \text{Divide each side by 9.}$$

$$-8 = m \qquad \text{Simplify.}$$

The value of m is -8.

(b)

$$\frac{\frac{1}{2}}{n} = \frac{12}{\frac{3}{4}} \qquad \text{Given proportion}$$

$$\frac{1}{2} \cdot \frac{3}{4} = 12 \cdot n \qquad \text{Cross products are equal.}$$

$$\frac{3}{8} = 12n \qquad \text{Simplify.}$$

$$\frac{1}{12} \cdot \frac{3}{8} = \frac{1}{12} \cdot 12n \qquad \text{Multiply each side by } \frac{1}{12}.$$

$$\frac{1}{32} = n \qquad \text{Simplify.}$$

The value of n is $\frac{1}{32}$.

■ **Now Try Exercises 31, 39**

Similar Figures

When comparing everyday objects, the word "similar" means that the objects resemble each other but are not exactly the same. For example, we might say that two houses are similar. However, in geometry, the word "similar" has a more specific meaning.

▶ **REAL-WORLD CONNECTION** Some computer programs can enlarge or reduce an image while keeping the new image *proportional* to the original. For example, Figure 6.2 shows two images that are proportional and two that are not.

Proportional   **Not Proportional**

Figure 6.2 Computer Images

Two geometric figures are **similar figures** if the measures of the corresponding angles are equal and the measures of the corresponding sides are proportional. In other words, the figures are similar if they have *exactly* the same shape but not necessarily the same size.

SIMILAR FIGURES

When two geometric figures are similar,

the measures of the corresponding sides are proportional.

Consider the similar triangles shown in Figure 6.3.

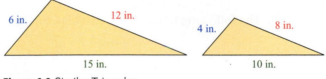

Figure 6.3 Similar Triangles

Because the triangles are similar, the corresponding sides are proportional.

$$\frac{\text{large triangle measure}}{\text{small triangle measure}} \qquad \frac{6}{4} = \frac{12}{8} = \frac{15}{10}$$

In the next two examples, we use the proportional measures of similar figures to find the length of an unknown side.

EXAMPLE 4   **Finding the length of an unknown side**

The following triangles are similar. Find the measure of x.

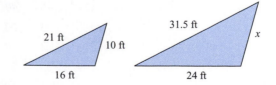

Solution

If the measures from the smaller triangle are written in the numerators of three ratios and the corresponding measures from the larger triangle are written in the corresponding denominators, the following proportions result.

$$\frac{16}{24} = \frac{21}{31.5} = \frac{10}{x}$$

To solve for x, set *either* of the known ratios equal to the ratio with the unknown value. We will set $\frac{16}{24}$ equal to $\frac{10}{x}$. For ease of computation, we simplify $\frac{16}{24}$ to $\frac{2}{3}$ in the second step.

$$\frac{16}{24} = \frac{10}{x} \qquad \text{Set up a proportion.}$$

$$\frac{2}{3} = \frac{10}{x} \qquad \text{In simplest terms, } \frac{16}{24} = \frac{2}{3}.$$

$$2 \cdot x = 3 \cdot 10 \qquad \text{Cross products are equal.}$$

$$2x = 30 \qquad \text{Simplify.}$$

$$\frac{2x}{2} = \frac{30}{2} \qquad \text{Divide each side by 2.}$$

$$x = 15 \qquad \text{Simplify.}$$

The measure of the unknown side is 15 feet.

∎ **Now Try Exercise 45**

NOTE: It makes no difference which *known* ratio we use when solving for an unknown length. In Example 4, the same result is found by setting up the proportion as $\frac{21}{31.5} = \frac{10}{x}$.

EXAMPLE 5 **Finding the length of an unknown side**

The following trapezoids are similar. Find the measure of m.

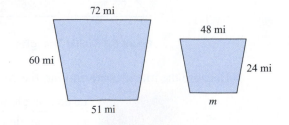

72 mi · 48 mi · 60 mi · 24 mi · 51 mi · m

Solution
The only corresponding measures that provide a *known* ratio are the measures of the top lengths of each trapezoid. If the measures from the larger trapezoid are written in the numerators of two ratios and the corresponding measures from the smaller trapezoid are written in the corresponding denominators, then the following proportion results.

$$\frac{72}{48} = \frac{51}{m} \qquad \text{Set up a proportion.}$$

$$\frac{3}{2} = \frac{51}{m} \qquad \text{In simplest terms, } \frac{72}{48} = \frac{3}{2}.$$

$$3 \cdot m = 2 \cdot 51 \qquad \text{Cross products are equal.}$$

$$3m = 102 \qquad \text{Simplify.}$$

$$\frac{3m}{3} = \frac{102}{3} \qquad \text{Divide each side by 3.}$$

$$m = 34 \qquad \text{Simplify.}$$

The measure of the unknown side is 34 miles.

∎ **Now Try Exercise 51**

READING CHECK

• How many known values must a proportion have in order to find an unknown value?

Applications

Proportions can be used in many applications to find an unknown value. When two ratios are set equal to each other, there are two numerators and two denominators. Whenever we know three of these values, we can find the fourth value by computing the cross products. The next three examples illustrate how proportions are used to find unknown values in application problems.

EXAMPLE 6   **Reading a map**

Every 2 inches on a map represents a distance of 75 miles on the ground. Find the actual distance between two towns if the corresponding distance on the map is 5 inches.

Solution

If we let x represent the unknown ground distance, then the given information can be written as a proportion in word form:

(Map)	(Ground)		(Map)	(Ground)
2 inches is to **75 miles**		**as**	**5 inches** is to **x miles**.	

This proportion can be written as an equation by setting the two ratios equal to each other.

$$\frac{2}{75} = \frac{5}{x}$$

To find the actual distance, solve this proportion for x.

$$\frac{2}{75} = \frac{5}{x} \qquad \text{Proportion to solve}$$

$$2 \cdot x = 5 \cdot 75 \qquad \text{Cross products are equal.}$$

$$2x = 375 \qquad \text{Simplify.}$$

$$\frac{2x}{2} = \frac{375}{2} \qquad \text{Divide each side by 2.}$$

$$x = 187.5 \qquad \text{Simplify.}$$

The actual distance between the two towns is 187.5 miles.

Now Try Exercise 55

EXAMPLE 7   **Providing drinking water for a recreational event**

The organizers of a disc golf tournament found that 40 participants will drink 65 bottles of water. If they are expecting 56 participants at the next tournament, how many bottles of water should be provided?

Solution

If we let x represent the unknown number of water bottles, the given information can be written as a proportion in words:

(Participants)	(Water)		(Participants)	(Water)
40 people is to **65 bottles**		**as**	**56 people** is to **x bottles**.	

This proportion can be written as an equation by setting the two ratios equal to each other.

$$\frac{40}{65} = \frac{56}{x}$$

To find the required number of bottles of water, solve this proportion for x.

$$\frac{40}{65} = \frac{56}{x}$$ Proportion to solve

$$\frac{8}{13} = \frac{56}{x}$$ In simplest terms, $\frac{40}{65} = \frac{8}{13}$.

$$8 \cdot x = 13 \cdot 56$$ Cross products are equal.

$$8x = 728$$ Simplify.

$$\frac{8x}{8} = \frac{728}{8}$$ Divide each side by 8.

$$x = 91$$ Simplify.

The organizers should provide 91 bottles of water.

Now Try Exercise 61

EXAMPLE 8   **Finding the height of a clock tower**

Similar triangles can be used to find the height of a tall object. A clock tower has a shadow measuring 70 feet, while a nearby sign post has a shadow measuring 3.5 feet. If the sign post is 8 feet tall, how tall is the clock tower?

Solution
To see how similar triangles are used to solve this problem, consider the following picture.

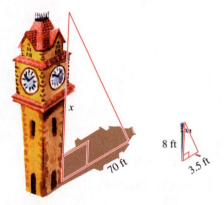

This proportion can be written as

(Post shadow)   **(Tower shadow)**   **(Post height)**   **(Tower height)**
3.5 feet   is to   **70 feet**   as   **8 feet**   is to   **x feet**.

Write this proportion as an equation and solve for x.

$$\frac{3.5}{70} = \frac{8}{x}$$ Proportion to solve

$$3.5 \cdot x = 70 \cdot 8$$ Cross products are equal.

$$3.5x = 560$$ Simplify.

$$\frac{3.5x}{3.5} = \frac{560}{3.5}$$ Divide each side by 3.5.

$$x = 160$$ Simplify.

The clock tower is 160 feet tall.

Now Try Exercise 75

6.2   Putting It All Together

CONCEPT	COMMENTS	EXAMPLES
Proportions	If $\frac{a}{b}$ and $\frac{c}{d}$ are ratios that are equal in value, then $$\frac{a}{b} = \frac{c}{d}$$ is a proportion, where $b \neq 0$ and $d \neq 0$. This proportion is read as "a is to b as c is to d."	The proportion $$\frac{2}{3} = \frac{30}{45}$$ is read as "2 is to 3 as 30 is to 45."
Cross Product Rule	For $b \neq 0$ and $d \neq 0$, if the cross products ad and bc are equal, then $$\frac{a}{b} = \frac{c}{d}$$ represents a proportion.	The equation $$\frac{4}{3} = \frac{20}{15}$$ represents a proportion because $4 \cdot 15 = 3 \cdot 20.$
Similar Figures	When two figures are similar, the measures of the corresponding sides are proportional.	4 in. 3 in. 6 in. 12 in. 9 in. 18 in. $$\frac{4}{12} = \frac{6}{18} = \frac{3}{9}$$

6.2   Exercises

MyMathLab   Math XL PRACTICE   WATCH   DOWNLOAD   READ   REVIEW

CONCEPTS AND VOCABULARY

1. A(n) _____ is a statement indicating that two ratios are equal.

2. The proportion "a is to b as c is to d" is written as the equation _____ involving two ratios.

3. In the proportion $\frac{a}{b} = \frac{c}{d}$, the products ad and bc are called the _____.

4. The equation $\frac{a}{b} = \frac{c}{d}$ represents a proportion if the cross products are _____.

5. Geometric figures are _____ if they have exactly the same shape but not necessarily the same size.

6. When two geometric figures are similar, the measures of the corresponding sides are _____.

PROPORTIONS

Exercises 7–22: Determine if the given equation is a proportion.

7. $\frac{5}{11} \overset{?}{=} \frac{15}{33}$

8. $\frac{28}{9} \overset{?}{=} \frac{6}{2}$

9. $\frac{18}{10} \overset{?}{=} \frac{24}{15}$

10. $\frac{9}{21} \overset{?}{=} \frac{6}{14}$

11. $\frac{5.2}{2} \overset{?}{=} \frac{15.6}{6}$

12. $\frac{6}{2.4} \overset{?}{=} \frac{5}{1.8}$

13. $\frac{7.5}{2.5} \overset{?}{=} \frac{12.9}{4.3}$

14. $\frac{1}{6.4} \overset{?}{=} \frac{11}{68.2}$

15. $\dfrac{\frac{1}{3}}{15} \overset{?}{=} \dfrac{\frac{2}{5}}{21}$

16. $\dfrac{10}{\frac{5}{6}} \overset{?}{=} \dfrac{42}{\frac{7}{2}}$

17. $\dfrac{\frac{5}{9}}{\frac{3}{5}} \overset{?}{=} \dfrac{\frac{5}{6}}{\frac{9}{10}}$

18. $\dfrac{\frac{1}{2}}{\frac{3}{4}} \overset{?}{=} \dfrac{\frac{4}{5}}{\frac{5}{6}}$

19. $\dfrac{4\frac{1}{2}}{2\frac{1}{4}} \overset{?}{=} \dfrac{3\frac{5}{6}}{1\frac{7}{10}}$

20. $\dfrac{4\frac{2}{3}}{1\frac{1}{6}} \overset{?}{=} \dfrac{5\frac{1}{5}}{1\frac{3}{10}}$

21. $\dfrac{2\frac{2}{3}}{4\frac{1}{5}} \overset{?}{=} \dfrac{3\frac{1}{3}}{5\frac{2}{5}}$

22. $\dfrac{3\frac{5}{6}}{1\frac{11}{12}} \overset{?}{=} \dfrac{6\frac{1}{5}}{3\frac{1}{10}}$

Exercises 23–28: Solve the proportion for the unknown value. Check your solution.

23. $\dfrac{2}{7} = \dfrac{10}{x}$

24. $\dfrac{y}{6} = \dfrac{7}{21}$

25. $\dfrac{4}{2.8} = \dfrac{9}{m}$

26. $\dfrac{4.5}{w} = \dfrac{15}{2.5}$

27. $\dfrac{\frac{1}{2}}{2} = \dfrac{k}{\frac{4}{3}}$

28. $\dfrac{6}{\frac{2}{3}} = \dfrac{9}{g}$

Exercises 29–44: Solve the proportion.

29. $\dfrac{9}{2} = \dfrac{4.5}{m}$

30. $\dfrac{3.5}{g} = \dfrac{14}{8}$

31. $\dfrac{10}{y} = \dfrac{-5}{6}$

32. $\dfrac{-16}{8} = \dfrac{a}{-4}$

33. $\dfrac{-12}{y} = \dfrac{-5}{8}$

34. $\dfrac{-15}{8} = \dfrac{d}{4}$

35. $\dfrac{5}{-2.5} = \dfrac{n}{-1.8}$

36. $\dfrac{4.8}{b} = \dfrac{-1.6}{3}$

37. $\dfrac{-3.6}{-2.4} = \dfrac{x}{4.6}$

38. $\dfrac{3.5}{k} = \dfrac{2.1}{-1.2}$

39. $\dfrac{x}{\frac{4}{3}} = \dfrac{\frac{9}{2}}{9}$

40. $\dfrac{\frac{1}{10}}{-\frac{2}{5}} = \dfrac{y}{-4}$

41. $\dfrac{\frac{5}{9}}{-\frac{5}{6}} = \dfrac{n}{\frac{9}{10}}$

42. $\dfrac{\frac{1}{6}}{d} = \dfrac{\frac{4}{5}}{-\frac{6}{7}}$

43. $\dfrac{1\frac{1}{5}}{-6\frac{2}{3}} = \dfrac{a}{4\frac{1}{6}}$

44. $\dfrac{3\frac{4}{5}}{1\frac{3}{16}} = \dfrac{\frac{4}{5}}{w}$

SIMILAR FIGURES

Exercises 45–52: Find the measure of x for the similar figures shown.

45.

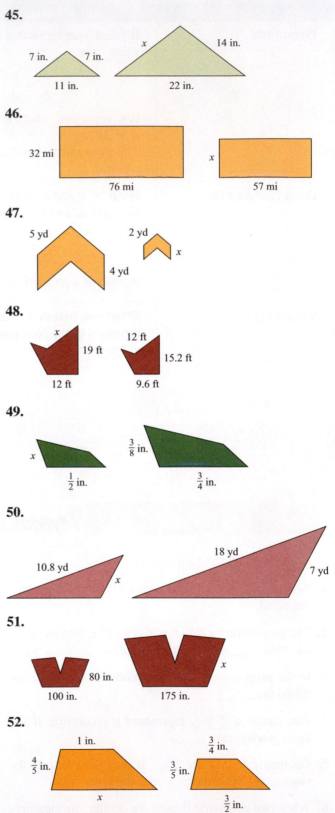

46.

47.

48.

49.

50.

51.

52.

APPLICATIONS

53. *Paid Vacation* An employee earns 6 days of paid vacation after working for 21 weeks. How many days of paid vacation does this employee earn after working for 49 weeks?

54. *Reading Rate* If a student reads 81 pages of a book in 90 minutes, how long will it take this student to read 117 pages?

55. *Wilderness Maps* Every $1\frac{1}{2}$ inches on a wilderness map represents 1 mile. If the actual distance between two lakes is 8 miles, how far apart are the lakes on the map?

56. *Construction Plans* Every $\frac{1}{4}$ inch on a floor plan represents 1 foot. If a wall on the floor plan measures $3\frac{1}{2}$ inches, how long is the actual wall?

57. *Grocery Shopping* If 4 limes sell for $0.76, how much will 6 limes cost?

58. *Wardrobe* A person has 8 shirts for every 5 pairs of pants. If this person has a total of 40 shirts in the closet, how many pairs of pants are there?

59. *Going Green* A survey determined that 80 of 480 students walk to school each day. If this ratio holds in a math class with 24 students, how many of the math students walk to school?

60. *Voting* A survey determined that 6 out of 10 voters intend to vote "no" on a controversial issue. If there are 2400 voters in the district, how many would we expect to vote "no"?

61. *Feeding Teenagers* If 30 teenagers can eat 72 slices of pizza, how many slices are needed to feed 120 teenagers?

62. *Mixing Fuel* The mixing ratio of gas to oil for a lawn mower is 50 to 1. How many gallons of gas should be mixed with 0.05 gallon of oil?

63. *Baking* A recipe that serves 6 people requires $1\frac{1}{2}$ cups of sugar. Using this recipe, how many cups of sugar are needed to serve 10 people?

64. *Basketball* A basketball player makes 16 out of 20 free throws. How many free throws is this player expected to make in 65 attempts?

65. *iPod Capacity* A 32-gigabyte iPod Touch can hold about 7000 songs. How many songs can be stored on an 8-gigabyte iPod Touch?

66. *Model Car* A model car is scaled so that 2 inches on the model are equal to $3\frac{3}{4}$ feet on the actual car. How tall is the model if the actual car is 5 feet tall?

67. *Living in Poverty* About 21 of every 200 Delaware citizens were living in poverty in 2008. How many people would you expect to have found living in poverty in a Delaware community with a population of 100,000 in 2008? (*Source:* U.S. Census Bureau.)

68. *Prison Time* About 9 out of every 200 U.S. citizens who have not been to prison before age 20 will end up serving time during their lifetime. A community has 1000 residents who are 20 years old and have never been to prison. How many of these residents are expected to serve time during their lifetime? (*Source:* Bureau of Justice Statistics.)

69. *Business Majors* From 2000 to 2008, about 8 of every 50 college freshmen chose to major in business. How many business majors would you expect to find in a group of 1500 college freshmen during this time? (*Source:* Higher Education Research Institute.)

70. *Education* In 2009, about 6 out of every 20 people in the U.S. who were over the age of 25 had completed four or more years of college. If a community has 1860 residents over the age of 25 with four or more years of college, how many residents over the age of 25 live there? (*Source:* U.S. Census Bureau.)

71. *Population Growth* Between 2005 and 2010, the country with the highest birth rate was Niger, where 36 babies were born for every 5 women. The country with the lowest birth rate was China (Macao SAR), where 9 babies were born for every 10 women. How many babies were born for every 1000 women in each country? (*Source:* United Nations.)

72. *Internet Use* In 2008, about 32 of every 50 Internet users visited sites owned or run by Microsoft. That same year, about 19 of every 25 Internet users visited sites owned or run by Google. For every 1000 Internet users, how many visited sites owned or run by each of these companies? (*Source:* ComScore.)

73. *Olympic Pool* A swimming pool that is 16 meters long and 8 meters wide is geometrically similar to an Olympic swimming pool. If an Olympic pool is 50 meters long, how wide is it?

74. *Board Games* The travel version of a popular board game is 8 inches long and 6 inches wide. If the larger version of the game is geometrically similar and has a length of 28 inches, what is its width?

75. *Tree Height* A tree casts a 65-foot shadow, while a nearby tree that is 11 feet tall casts a 5-foot shadow. How tall is the larger tree?

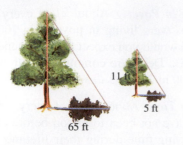

11 ft
5 ft
65 ft

76. *Antenna Height* Refer to the previous exercise. If a 6-foot person casts a 10-foot shadow, how tall is an antenna that casts a 75-foot shadow?

77. *Digital Images* The focal length of a camera is the distance from the lens to the image sensor. A camera with a focal length of 2 inches is used to take a picture of a large letter F that is 24 inches tall. If the letter F is 80 inches from the lens of the camera, as shown in the figure at the top of the next column, how tall is the F on the image sensor?

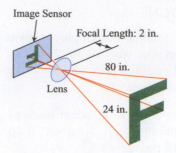

Image Sensor
Focal Length: 2 in.
80 in.
Lens
24 in.

78. *Photography* Refer to the previous exercise. A camera with a 3-inch focal length is used to take a picture of a child standing 16 feet in front of the camera. If the child's image is 0.75 inch tall on the image sensor, how tall is the child?

WRITING ABOUT MATHEMATICS

79. In order to determine how many pages he can read in 1 hour, a student who reads 50 pages in 75 minutes sets up the following proportion.

$$\frac{50}{75} = \frac{x}{1}$$

Is this proportion correct? Explain.

80. Are all squares geometrically similar? Explain.

SECTIONS 6.1 and 6.2

Checking Basic Concepts

1. Write each ratio as a fraction in simplest form.
(a) $16:36$ (b) $\frac{8}{9}:\frac{3}{18}$

2. Write each rate as a fraction in simplest form.
(a) It rains 6 inches in 15 hours.
(b) A car travels 582 miles in 8 hours.

3. Write each rate as a unit rate.
(a) A person earns $148 in 16 hours.
(b) It snows 30 inches in 20 hours.

4. Compute each unit price.
(a) A 2-pound bag of chips for $5.90
(b) A 16-ounce jar of pickles for $2.88

5. Determine if each ratio is a proportion.
(a) $\frac{4}{26} \stackrel{?}{=} \frac{6}{39}$ (b) $\frac{5}{2.5} \stackrel{?}{=} \frac{3}{1.6}$

6. Solve each proportion.
(a) $\frac{18}{y} = \frac{-6}{10}$ (b) $\frac{\frac{3}{10}}{-\frac{6}{5}} = \frac{m}{-4}$

7. The following figures are similar. Find x.

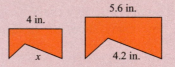

4 in.
5.6 in.
x
4.2 in.

8. *Shopping* Find the unit price for each option.

Large drink: 28 ounces for $1.96
Small drink: 12 ounces for $1.14

Which is the better buy?

9. *Baking* A recipe that serves 8 people requires $1\frac{1}{3}$ cups of milk. How many cups of milk would be needed to serve 10 people?

10. *Tree Height* A tree casts a 35-foot shadow, while a nearby lamp post that is 24 feet tall casts a 10-foot shadow. How tall is the tree?

6.3   The American System of Measurement

Units of Length • Units of Area • Units of Capacity and Volume • Units of Weight

A LOOK INTO MATH ▶

The Omnibus Trade and Competitiveness Act of 1988 named the metric system as the "preferred system of weights and measures" for U.S. trade and commerce. But today, decades after this law was passed, the United States remains one of only three countries that have not officially adopted the metric system. The other two countries are Liberia and Myanmar. Even though many Americans recognize a 2-liter bottle of soda and a 10-kilometer road race, we are more familiar with measurements such as gallons and miles. In this section, we discuss many of the units used in the American system of measurement.

NEW VOCABULARY

☐ Inch, foot, yard, mile
☐ Unit fraction
☐ Capacity
☐ Volume
☐ Ounce, cup, pint, quart, gallon
☐ Ounce, pound, ton

Units of Length

The **inch**, **foot**, **yard**, and **mile** are the four most commonly used units of *length* in the American system of measurement. Inches (in.), feet (ft), yards (yd), and miles (mi) are related as follows.

AMERICAN UNITS OF LENGTH

$$1 \text{ ft} = 12 \text{ in.}$$
$$1 \text{ yd} = 36 \text{ in.} \quad \text{and} \quad 1 \text{ yd} = 3 \text{ ft}$$
$$1 \text{ mi} = 1760 \text{ yd} \quad \text{and} \quad 1 \text{ mi} = 5280 \text{ ft}$$

STUDY TIP

Dimensional analysis is used throughout the remainder of this chapter and can be valuable in other classes, especially the sciences. Take some extra time to fully learn this technique. Get extra help and practice if needed.

To convert from one unit of length to another, we use a method called *dimensional analysis*. This method is based on the fact that multiplying a measurement by 1 does not change the value of the measurement. Since 1 can be written in many forms, it is important to write it in the form of a *unit fraction*. A **unit fraction** is a fraction that is equivalent to 1. Examples include

$$\frac{12 \text{ in.}}{1 \text{ ft}}, \quad \frac{1 \text{ yd}}{36 \text{ in.}}, \quad \frac{3 \text{ ft}}{1 \text{ yd}}, \quad \frac{5280 \text{ ft}}{1 \text{ mi}}, \quad \text{and} \quad \frac{1 \text{ mi}}{1760 \text{ yd}}.$$

Note that the numerator value and the denominator value represent the same measurement, just with different units. Thus a unit fraction is always equal to one.

When a length is multiplied by a unit fraction, the units in the denominator of the unit fraction are *divided out* and the units in the numerator of the unit fraction are *multiplied in*.

READING CHECK

• How is a unit fraction used when converting from one unit of measure to another?

CONVERTING FROM ONE UNIT OF MEASURE TO ANOTHER

To convert from one unit of measure to another, do the following.

STEP 1: Write the given measure (including its units) over 1.

STEP 2: Multiply by a unit fraction whose denominator contains the given (unwanted) units and whose numerator contains the new (desired) units.

It may be necessary to multiply by more than one unit fraction to get the desired units in the final result.

EXAMPLE 1   **Converting from one unit of length to another**

Convert as indicated.
(a) 75 inches to feet   (b) $\frac{3}{4}$ mile to yards   (c) 6.5 yards to inches

Solution
(a) The given (unwanted) units are inches, and the new (desired) units are feet. To make this conversion, write 75 inches over 1 and then multiply by a unit fraction with inches in the denominator and feet in the numerator.

$$75 \text{ in.} = \frac{75 \text{ in.}}{1} \cdot \frac{1 \text{ ft}}{12 \text{ in.}} = \frac{75}{12} \text{ ft} = \frac{25 \cdot 3}{4 \cdot 3} \text{ ft} = \frac{25}{4} \text{ ft} = 6\frac{1}{4} \text{ ft, or 6.25 ft}$$

(b) We are given miles and want to convert to yards. To do this, write $\frac{3}{4}$ mile over 1 and then multiply by a unit fraction with miles in the denominator and yards in the numerator.

$$\frac{3}{4} \text{ mi} = \frac{\frac{3}{4} \text{ mi}}{1} \cdot \frac{1760 \text{ yd}}{1 \text{ mi}} = \frac{3}{4} \cdot \frac{1760}{1} \text{ yd} = \frac{3 \cdot (440 \cdot 4)}{1 \cdot 4} \text{ yd} = \frac{1320}{1} \text{ yd} = 1320 \text{ yd}$$

(c) The given units are yards, and we want to convert to inches.

$$6.5 \text{ yd} = \frac{6.5 \text{ yd}}{1} \cdot \frac{36 \text{ in.}}{1 \text{ yd}} = 6.5 \cdot 36 \text{ in.} = 234 \text{ in.}$$

▌ **Now Try Exercises 11, 15, 19**

If you do not know the direct relationship between two units of measure, then it may be necessary to use more than one unit fraction during a conversion. The next example illustrates this process.

EXAMPLE 2   **Converting by using more than one unit fraction**

Convert 1.4 miles to inches by using more than one unit fraction.

Solution
Since most people do not know how many inches are in a mile, we will first convert miles to feet and then convert feet to inches. To convert from miles to feet, we use a unit fraction with feet in the numerator and miles in the denominator. Then, to convert from feet to inches, we use a unit fraction with inches in the numerator and feet in the denominator.

$$1.4 \text{ mi} = \frac{1.4 \text{ mi}}{1} \cdot \frac{5280 \text{ ft}}{1 \text{ mi}} \cdot \frac{12 \text{ in.}}{1 \text{ ft}} = \frac{1.4 \cdot 5280 \cdot 12}{1} \text{ in.} = 88{,}704 \text{ in.}$$

Miles "out," feet "in"   Feet "out," inches "in"

▌ **Now Try Exercise 21**

Units of Area

While there are 3 feet in 1 yard, this does *not* mean that there are 3 square feet in 1 square yard. Figure 6.4 can be used to visualize the correct relationship between square feet and square yards.

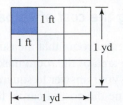

Figure 6.4 Square Feet in a Square Yard

As Figure 6.4 shows, there are 9 square feet in one square yard. When converting from one square unit to another, we can use the relationships for length that we already know. For example, we can write square yards as either

$$\mathbf{yd^2} \quad \text{or} \quad \mathbf{yd \cdot yd}.$$

READING CHECK

• When converting from one square unit of measure to another, how many times is a unit fraction used?

To see that there are 9 square feet in 1 square yard, we find a known unit fraction that involves feet and yards and then use it twice.

$$\mathbf{1\ yd^2} = \frac{1\ \cancel{yd} \cdot \cancel{yd}}{1} \cdot \frac{3\ ft}{1\ \cancel{yd}} \cdot \frac{3\ ft}{1\ \cancel{yd}} = \mathbf{3\ ft \cdot 3\ ft = 9\ ft^2}$$

Unit fraction used twice

EXAMPLE 3   **Converting from one unit of area to another**

Convert as indicated.
(a) 6 square feet to square inches   **(b)** 108 square feet to square yards

Solution
(a) Use a unit fraction relating feet and inches, twice.

$$6\ ft^2 = \frac{6\ \cancel{ft} \cdot \cancel{ft}}{1} \cdot \frac{12\ in.}{1\ \cancel{ft}} \cdot \frac{12\ in.}{1\ \cancel{ft}} = 6 \cdot 12\ in. \cdot 12\ in. = 864\ in^2$$

(b) Use a unit fraction relating feet and yards, twice.

$$108\ ft^2 = \frac{108\ \cancel{ft} \cdot \cancel{ft}}{1} \cdot \frac{1\ yd}{3\ \cancel{ft}} \cdot \frac{1\ yd}{3\ \cancel{ft}} = \frac{108}{3 \cdot 3}\ yd \cdot yd = 12\ yd^2$$

Now Try Exercises 25, 29

MAKING CONNECTIONS

Converting Units of Area Directly

We can also derive the appropriate unit fraction for converting units of area directly. For example, a unit fraction for converting square feet to square inches is found by squaring the unit fraction used to convert feet to inches.

$$\left(\frac{12\ in.}{1\ ft}\right)^2 = \frac{12\ in.}{1\ ft} \cdot \frac{12\ in.}{1\ ft} = \frac{12\ in. \cdot 12\ in.}{1\ ft \cdot 1\ ft} = \frac{144\ in^2}{1\ ft^2}$$

So, to convert square feet to square inches directly, use the unit fraction

$$\frac{144\ in^2}{1\ ft^2}.$$

▶ **REAL-WORLD CONNECTION**   While it is common to measure the floor of a room in square feet, some brands of carpet are sold by the square yard. To find the cost of carpet for a room, we can convert square feet to square yards, as demonstrated in the next example.

| EXAMPLE 4 | **Computing the cost of carpet for a room** |

For a rectangular room that measures 12 feet by 16.5 feet, do the following.
(a) Find the area of the room in square feet.
(b) Convert your answer to square yards.
(c) Determine the cost of carpet for the room if the carpet sells for $38 per square yard.

Solution
(a) The area of the room is

$$A = lw = 16.5 \text{ ft} \cdot 12 \text{ ft} = 198 \text{ ft}^2.$$

(b) Use a unit fraction relating feet and yards, twice.

$$198 \text{ ft}^2 = \frac{198 \,\cancel{ft} \cdot \cancel{ft}}{1} \cdot \frac{1 \text{ yd}}{3 \,\cancel{ft}} \cdot \frac{1 \text{ yd}}{3 \,\cancel{ft}} = \frac{198}{3 \cdot 3} \text{ yd} \cdot \text{yd} = 22 \text{ yd}^2$$

(c) The total cost of the carpet is

$$\text{Cost} = (\text{yardage})(\text{price}) = \frac{22 \,\cancel{yd^2}}{1} \cdot \frac{\$38}{1 \,\cancel{yd^2}} = 22 \cdot \$38 = \$836.$$

■ **Now Try Exercise 71**

Units of Capacity and Volume

READING CHECK

• Explain how capacity and volume differ.

Even though the words *capacity* and *volume* are often used interchangeably, they do not have exactly the same meaning. **Capacity** is a measure of the amount of substance that a container can hold, while **volume** is a measure of the actual amount of the substance present. For example, if a 16-gallon gas tank is half full, the capacity of the tank is 16 gallons, and the volume of gas in the tank is 8 gallons.

In the American system of measurement, the units used to measure *both* capacity and volume of liquids are (fluid) **ounce**, **cup**, **pint**, **quart**, and **gallon**. Ounces (oz), cups (c), pints (pt), quarts (qt), and gallons (gal) are related as follows.

AMERICAN UNITS OF CAPACITY AND VOLUME

1 c = 8 oz	1 pt = 2 c
1 qt = 2 pt	1 gal = 4 qt

Figure 6.5 shows some common containers that hold a cup, pint, quart, and gallon.

1 c   1 pt   1 qt   1 gal
Figure 6.5 Common Containers

EXAMPLE 5   **Converting from one unit of capacity to another**

Convert as indicated.
(a) 152 ounces to pints   (b) 2.5 gallons to cups

Solution
(a) First, convert from ounces to cups and then from cups to pints.

$$152 \text{ oz} = \frac{152 \text{ oz}}{1} \cdot \frac{1 \text{ c}}{8 \text{ oz}} \cdot \frac{1 \text{ pt}}{2 \text{ c}} = \frac{152}{8 \cdot 2} \text{ pt} = \frac{19}{2} \text{ pt} = 9\frac{1}{2} \text{ pt or } 9.5 \text{ pt}$$

(b) Convert from gallons to quarts, then quarts to pints, and finally from pints to cups.

$$2.5 \text{ gal} = \frac{2.5 \text{ gal}}{1} \cdot \frac{4 \text{ qt}}{1 \text{ gal}} \cdot \frac{2 \text{ pt}}{1 \text{ qt}} \cdot \frac{2 \text{ c}}{1 \text{ pt}} = 2.5 \cdot 4 \cdot 2 \cdot 2 \text{ c} = 40 \text{ c}$$

Now Try Exercises 41, 49

Units of Weight

The units used to measure weight in the American system of measurement are the **ounce**, **pound**, and **ton**. The weight of a box of breakfast cereal is measured in ounces, while the weight of a person is measured in pounds. Tons are used for the weight of very heavy objects such as trucks. Ounces (oz), pounds (lb), and tons (T) are related as follows.

AMERICAN UNITS OF WEIGHT

$$1 \text{ lb} = 16 \text{ oz} \qquad 1 \text{ T} = 2000 \text{ lb}$$

EXAMPLE 6   **Converting from one unit of weight to another**

Convert as indicated.
(a) 200 ounces to pounds   (b) 0.04 ton to ounces

Solution
(a) Use a unit fraction relating ounces and pounds.

$$200 \text{ oz} = \frac{200 \text{ oz}}{1} \cdot \frac{1 \text{ lb}}{16 \text{ oz}} = \frac{200}{16} \text{ lb} = 12\frac{1}{2} \text{ lb or } 12.5 \text{ lb}$$

(b) Convert from tons to pounds and then from pounds to ounces.

$$0.04 \text{ T} = \frac{0.04 \text{ T}}{1} \cdot \frac{2000 \text{ lb}}{1 \text{ T}} \cdot \frac{16 \text{ oz}}{1 \text{ lb}} = 0.04 \cdot 2000 \cdot 16 \text{ oz} = 1280 \text{ oz}$$

Now Try Exercises 53, 57

MAKING CONNECTIONS

Ounces of Capacity and Ounces of Weight

The term *ounce* is used to name two different units of measure: a unit of capacity and a unit of weight. For this reason, we must clearly understand whether a problem involves capacity or weight. To avoid confusion, some people use the term *fluid ounces* (fl oz) for capacity and the term *ounces* (oz) for weight.

6.3   Putting It All Together

CONCEPT	COMMENTS	EXAMPLES
Unit Fraction	A unit fraction is a fraction that is equivalent to 1. Unit fractions can be used to convert from one unit of measure to another.	$\dfrac{3\text{ ft}}{1\text{ yd}}$, $\dfrac{1\text{ gal}}{4\text{ qt}}$, and $\dfrac{2000\text{ lb}}{1\text{ T}}$
Units of Length	Inches, feet, yards, and miles **Relationships:** 1 ft = 12 in. 1 yd = 36 in. and 1 yd = 3 ft 1 mi = 1760 yd and 1 mi = 5280 ft	Convert 24 inches to feet. $\dfrac{24\text{ in.}}{1} \cdot \dfrac{1\text{ ft}}{12\text{ in.}} = \dfrac{24}{12}\text{ ft} = 2\text{ ft}$
Units of Area	Units of area are squares of the units of length. To convert from one unit of area to another, use the corresponding unit fraction relating units of length, twice.	Convert 2 square yards to square feet. $\dfrac{2\text{ yd} \cdot \text{yd}}{1} \cdot \dfrac{3\text{ ft}}{1\text{ yd}} \cdot \dfrac{3\text{ ft}}{1\text{ yd}} = 18\text{ ft}^2$
Units of Capacity and Volume	Ounce, cup, pint, quart, and gallon **Relationships:** 1 c = 8 oz 1 pt = 2 c 1 qt = 2 pt 1 gal = 4 qt	Convert 4 gallons to quarts. $\dfrac{4\text{ gal}}{1} \cdot \dfrac{4\text{ qt}}{1\text{ gal}} = 16\text{ qt}$
Units of Weight	Ounce, pound, and ton **Relationships:** 1 lb = 16 oz 1 T = 2000 lb	Convert 64 ounces to pounds. $\dfrac{64\text{ oz}}{1} \cdot \dfrac{1\text{ lb}}{16\text{ oz}} = \dfrac{64}{16}\text{ lb} = 4\text{ lb}$

6.3   Exercises

MyMathLab   MathXL PRACTICE   WATCH   DOWNLOAD   READ   REVIEW

CONCEPTS AND VOCABULARY

1. The foot, yard, and mile are units of _____.

2. A(n) _____ fraction is a fraction that is equivalent to 1.

3. When using a unit fraction to convert from one unit of measure to another, write the given (unwanted) unit in the _____ of the unit fraction.

4. When using a unit fraction to convert from one unit of measure to another, write the new (desired) unit in the _____ of the unit fraction.

5. When converting from one unit of _____ to another, use the corresponding unit fraction relating units of length, twice.

6. The amount of a substance that a container can hold is called the container's _____.

7. The actual amount of a substance that is present in a container is called the _____ of the substance.

8. The units of _____ and _____ are the ounce, cup, pint, quart, and gallon.

9. The ounce, pound, and ton are units of _____.

10. The term _____ can be used to name two different units of measure: a unit of capacity and a unit of weight.

UNITS OF LENGTH AND AREA

Exercises 11–24: Convert the length as indicated.

11. 48 inches to feet

12. 6.5 feet to inches

13. 880 yards to miles

14. 4.5 miles to yards

15. $\frac{2}{3}$ yard to inches

16. 156 inches to yards

17. 87 feet to yards

18. $17\frac{1}{3}$ yards to feet

19. 1.2 miles to feet

20. 2112 feet to miles

21. 0.025 mile to inches

22. 47,520 inches to miles

23. 1 inch to feet

24. 1 inch to yards

Exercises 25–32: Convert the area as indicated.

25. 10 square feet to square inches

26. 216 square inches to square feet

27. 0.01 square mile to square yards

28. 92,928 square yards to square miles

29. 324 square feet to square yards

30. $7\frac{1}{3}$ square yards to square feet

31. 278,784 square feet to square miles

32. 0.02 square mile to square feet

UNITS OF CAPACITY AND VOLUME

Exercises 33–52: Convert the capacity as indicated.

33. 48 ounces to cups

34. $12\frac{1}{2}$ cups to ounces

35. $\frac{1}{2}$ quart to pints

36. 17 pints to quarts

37. 10.25 gallons to quarts

38. 18 quarts to gallons

39. 26 cups to pints

40. $11\frac{1}{2}$ pints to cups

41. 100 ounces to pints

42. 3.25 quarts to ounces

43. $\frac{3}{4}$ gallon to pints

44. 0.5 gallon to ounces

45. 60 cups to gallons

46. 96 ounces to quarts

47. $\frac{1}{4}$ quart to cups

48. 24 cups to quarts

49. 5 gallons to cups

50. 1.25 pints to ounces

51. 800 ounces to quarts

52. 1 pint to gallons

UNITS OF WEIGHT

Exercises 53–64: Convert the weight as indicated.

53. 440 ounces to pounds

54. 0.008 ton to pounds

55. $7\frac{5}{8}$ pounds to ounces

56. 200,000 ounces to tons

57. 0.00025 ton to ounces

58. 15,000 pounds to tons

59. $\frac{1}{100}$ ton to pounds

60. 496 ounces to pounds

61. 1,760,000 ounces to tons

62. 9 pounds to ounces

63. 22,500 pounds to tons

64. 0.4 ton to ounces

APPLICATIONS

65. *Soda Bottle* A person buys a 20-ounce bottle of soda from a vending machine. How many pints of soda did the person buy?

66. *Jumbo Bottle* A jumbo bottle of soda contains about 101 ounces of soda. Does this bottle hold more or less than a gallon of soda?

67. *Canadian Football* A standard Canadian football field is 330 feet long. How many yards is this?

68. *Checked Luggage* Many airlines restrict the size of a checked luggage item so that the sum of its length, width, and height can be no more than $5\frac{1}{6}$ feet. What is this restriction in inches?

69. *Weight in Stones* The *stone* is a unit of weight in the Imperial system of measurement. One stone is equivalent to 14 pounds. If a person weighs 154 pounds, find his weight in stones.

70. *Length in Rods* The *rod* is a unit of length often used on wilderness maps to describe the length of a portage (trail between bodies of water). A rod is equivalent to 16.5 feet. Find the length in rods of a portage that is 1 mile long.

71. *Cost of Carpet* For a square room that measures 15 feet on a side, do the following.
(a) Find the area of the room in square feet.
(b) Convert your answer to square yards.
(c) Determine the cost of carpet for the room if the carpet sells for $32 per square yard.

72. *Cost of Carpet* For a rectangular room that measures 15 feet by 18 feet, do the following.
(a) Find the area of the room in square feet.
(b) Convert your answer to square yards.
(c) Determine the cost of carpet for the room if the carpet sells for $46 per square yard.

73. *Serving Size* The label on a 1-gallon jug of fruit juice states that the serving size is 8 oz. How many servings are in the jug?

74. *Serving Size* The label on a 2-pound bag of chips states that the serving size is 4 oz. How many servings are in the bag?

75. *Cutting Rope* A spool of rope contains 75 yards of rope. How many 4-foot pieces of rope can be cut from the spool? How much rope is left over?

76. *Friendship Bracelets* A child uses four 18-inch pieces of string to make a friendship bracelet. If the string is available on a spool with 105 feet of string, how many bracelets can the child make? How much string is left over?

77. *Freight* Freight cars on some trains can each haul 100 tons of cargo. How many 400-pound containers can be hauled by this type of freight car, assuming that all containers fit on the car?

78. *Filling an Aquarium* A person is filling a 30-gallon aquarium by pouring 1 cup of water into the aquarium at a time. How many times will this person need to fill the cup and pour it into the aquarium?

WRITING ABOUT MATHEMATICS

79. A student converts 3 pounds to cups as follows.

$$3\text{ lb} = \frac{3\text{ lb}}{1} \cdot \frac{16\text{ oz}}{1\text{ lb}} \cdot \frac{1\text{ c}}{8\text{ oz}} = 6\text{ c}$$

Explain the error in this computation.

80. A classmate tells you that there are 36 square inches in 1 square yard because there are 36 inches in 1 yard. How do you convince your classmate that this is incorrect?

81. Explain how to derive a unit fraction used to convert inches to miles directly.

82. Explain how to derive a unit fraction used to convert ounces to gallons directly.

6.4   The Metric System of Measurement

Metric Prefixes • Units of Length • Units of Area • Units of Capacity and Volume • Units of Mass

A LOOK INTO MATH ▶ The metric system of measurement is used by most countries. It is the system of measurement used for international trade, science, medicine, and sports such as those of the Olympics. Since the metric system is based on the number 10, computations and conversions within the metric system are simpler than similar computations and conversions within the American system of measurement. In this section, we discuss metric units of measure.

Metric Prefixes

Within the metric system, each type of measure has a **base unit**. For example, the base metric unit for length is the *meter*. All metric measures of length have names that are based on

NEW VOCABULARY

☐ Base unit
☐ Meter, liter, gram
☐ Mass
☐ Weight

STUDY TIP

Rather than comparing metric units to American units, it is more useful to develop a working knowledge of metric units. Here are some things to try:

Find the distance from your home to your college campus in kilometers. Learn what your weight is in kilograms. Take note of the size of a 1-liter bottle of water.

the meter. Similar statements can be made for the *liter* and the ... for capacity and mass, respectively. The name of a metric measure base metric units ing the prefixes in Table 6.1 to its base unit. derived by applying-

READING CHECK

• What are the three base units in the metric system?

TABLE 6.1 **Metric Prefixes**

Prefix	kilo	hecto	deka	Base	deci	centi	milli
Size Relative to Base Unit	× 1000	× 100	× 10		× $\frac{1}{10}$	× $\frac{1}{100}$	× $\frac{1}{1000}$

Different prefixes in Table 6.1 are used depending on a measurement's size *relative* to the base unit. For example, a measurement that is 100 times the length of a meter is called a hectometer; a measurement that is $\frac{1}{1000}$ of a gram is called a milligram; and a measurement that is 1000 times the volume of a liter is called a kiloliter.

In the next example, the prefixes from Table 6.1 are used to name metric measurements.

EXAMPLE 1   **Naming metric measurements**

Write the name of the metric unit of measurement described.

(a) 10 times a meter   (b) $\frac{1}{10}$ of a liter   (c) $\frac{1}{100}$ of a gram

Solution

(a) The prefix that means to multiply by 10 is "deka." Since the base unit is the meter, the correct name of this unit of length is the dekameter.

(b) The prefix "deci" means multiply by $\frac{1}{10}$. The base unit is the liter, so the correct name of this unit of capacity is the deciliter.

(c) "Centi" is the prefix that means multiply by $\frac{1}{100}$. Since the base unit is the gram, the correct name of this unit of mass is the centigram.

Now Try Exercises 9, 11, 15

Units of Length

The base unit of length in the metric system is the **meter**. Rather than comparing the meter to a unit of measure in the American system of measurement, it is more helpful to practice holding out your hands a distance of about one meter, as illustrated in Figure 6.6.

Figure 6.6 A Meter

Meters can be used to measure things such as the height of a diving platform, the length of a playground, distances on a running track, and room dimensions.

Although any of the metric prefixes can be used with the meter, the three most common measures associated with the meter are kilometers, centimeters, and millimeters. We use kilometers to measure long distances, such as the distance between two towns. Centimeters are used for smaller measurements, such as the height of a dog and the width of a piece of notebook paper. Finally, millimeters are used to measure very small items, such as the length of a wood screw or a small amount of rainfall. Figure 6.7 shows the relative sizes of the kilometer, centimeter, and millimeter.

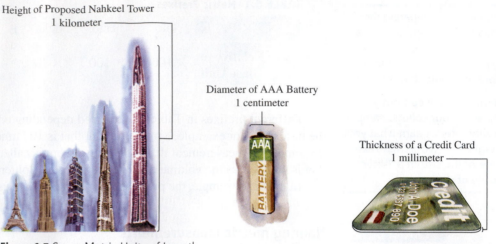

Figure 6.7 Some Metric Units of Length

To convert from one metric unit of length to another, we can first see how each unit relates to the base unit. Kilometers (km), hectometers (hm), dekameters (dam), decimeters (dm), centimeters (cm), and millimeters (mm) relate to the meter (m) as follows.

METRIC UNITS OF LENGTH

1 km = 1000 m	1 hm = 100 m	1 dam = 10 m
1 m = 10 dm	1 m = 100 cm	1 m = 1000 mm

The next example illustrates how unit fractions and dimensional analysis can be used to convert one metric unit of length to another by first converting to the base unit, meters.

EXAMPLE 2   **Converting one unit of length to another**

Convert 0.61 hectometer to decimeters.

Solution

First, convert from hectometers to meters and then from meters to decimeters.

$$0.61 \text{ hm} = \frac{0.61 \text{ hm}}{1} \cdot \frac{100 \text{ m}}{1 \text{ hm}} \cdot \frac{10 \text{ dm}}{1 \text{ m}} = 0.61(100)(10) \text{ dm} = 610 \text{ dm}$$

Now Try Exercise 17

Notice that the answer in Example 2 can be found by simply moving the decimal point in the given measurement. Every unit fraction used in converting metric units has the effect of multiplying or dividing by a power of 10. To determine which direction and the number of places to move the decimal point, consider the following *ordered* list of metric units of length.

km   hm   dam   m   dm   cm   mm

When converting from hectometers to decimeters, we move the decimal point in the given measurement **3** places to the **right**.

$$0.61 \text{ hm} = 0.610 \text{ hm} = 610 \text{ dm}$$

EXAMPLE 3   **Converting from one unit of length to another**

Convert as indicated.
(a) 102,300 millimeters to dekameters   **(b)** 0.045 kilometer to meters

Solution

(a)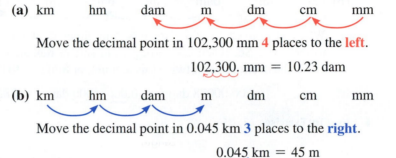

Move the decimal point in 102,300 mm **4** places to the **left**.

$$102,300. \text{ mm} = 10.23 \text{ dam}$$

(b) km   hm   dam   m   dm   cm   mm

Move the decimal point in 0.045 km **3** places to the **right**.

$$0.045 \text{ km} = 45 \text{ m}$$

Now Try Exercises 21, 23

Units of Area

When we converted units of area within the American system of measurement, we used unit fractions relating corresponding units of length, *twice*. This concept can be extended to the metric system—when converting metric units of area, the decimal point is moved *twice* the number of places needed to convert the corresponding units of length.

To illustrate this procedure, we will convert 1 square centimeter to square millimeters by first noting the direction and number of places that the decimal point is moved when converting the corresponding units of length, centimeters to millimeters.

km   hm   dam   m   dm   cm   mm

Since the decimal point is moved **1** place to the **right** when centimeters are converted to millimeters, it must be moved twice this distance, or **2** places to the **right**, when square centimeters are converted to square millimeters.

$$1 \text{ cm}^2 = 1.00 \text{ cm}^2 = 100 \text{ mm}^2$$

Figure 6.8 shows that 1 square centimeter is equivalent to 100 square millimeters.

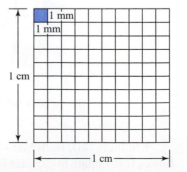

Figure 6.8 100 mm² in 1 cm² (not to scale)

EXAMPLE 4 **Converting from one unit of area to another**

Convert as indicated.
(a) 1,200,000 square centimeters to square meters
(b) 0.00078 square dekameters to square millimeters

Solution

(a) km   hm   dam   m   dm   cm   mm

When converting from centimeters to meters, we move the decimal point 2 places to the left. So, when converting from square centimeters to square meters, we move the decimal point twice this amount, or **4** places to the **left**.

$$1,200,000.\ cm^2 = 120\ m^2$$

(b) km   hm   dam   m   dm   cm   mm

When we convert from dekameters to millimeters, we move the decimal point 4 places to the right. So, when converting from square dekameters to square millimeters, we move the decimal point twice this amount, or **8** places to the **right**.

$$0.00078\ dam^2 = 0.00078000\ dam^2 = 78,000\ mm^2$$

Now Try Exercises 33, 35

Units of Capacity and Volume

READING CHECK

• Name two things from everyday life that can be measured in liters.

The base unit of capacity and volume in the metric system is the **liter**, which is defined as the amount of liquid that would fill a cube measuring 10 centimeters along each edge. As with the meter, any of the metric prefixes can be used with the liter. However, it is most common to measure liquid capacity and volume in either liters or milliliters. Liters are often used to measure amounts of liquid such as gasoline in a tank, soda in a large bottle, or water in an aquarium. Milliliters can be used to measure very small volumes such as doses of liquid medicine. The relative sizes of the liter and milliliter are shown in Figure 6.9.

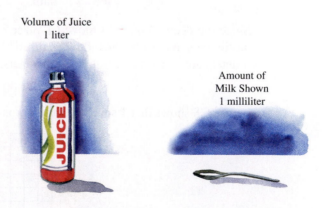

Volume of Juice
1 liter

Amount of
Milk Shown
1 milliliter

Figure 6.9 Some Metric Units of Capacity and Volume

Kiloliters (kl), hectoliters (hl), dekaliters (dal), deciliters (dl), centiliters (cl), and milliliters (mL) relate to the liter (L) as follows.

METRIC UNITS OF CAPACITY AND VOLUME

1 kl = 1000 L	1 hl = 100 L	1 dal = 10 L
1 L = 10 dl	1 L = 100 cl	1 L = 1000 mL

Because the liter is defined as the volume of a cube measuring 10 centimeters along each edge, there are $10 \cdot 10 \cdot 10 = 1000$ cubic centimeters (cc) in one liter. Based on this observation, we can conclude that 1 cubic centimeter is equivalent to 1 milliliter, because there are also 1000 milliliters in one liter. Since

$$1 \text{ L} = 1000 \text{ mL} = 1000 \text{ cc},$$

the units mL and cc may be used interchangeably. The units of capacity and volume are listed as follows.

kl   hl   dal   L   dl   cl   mL (cc)

EXAMPLE 5 **Converting from one unit of capacity to another**

Convert as indicated.
(a) 12 deciliters to milliliters   **(b)** 2500 centiliters to hectoliters

Solution
(a) kl   hl   dal   L   dl   cl   mL (cc)

Move the decimal point in 12 dl **2** places to the **right**.

$$12 \text{ dl} = 12.00 \text{ dl} = 1200 \text{ mL}$$

(b) kl   hl   dal   L   dl   cl   mL (cc)

Move the decimal point in 2500 cl **4** places to the **left**.

$$2500. \text{ cl} = 0.25 \text{ hl}$$

Now Try Exercises 39, 47

Units of Mass

READING CHECK

- Name two things from everyday life that can be measured in kilograms.
- Name two things from everyday life that can be measured in grams.

Although the terms *weight* and *mass* are sometimes used interchangeably, they do not have exactly the same meaning. While **mass** is a measure of the amount of matter in an object, **weight** is a measure of the force on an object due to gravity. The mass of an object does not change, but its weight may change depending on its distance from the center of Earth.

The base unit of mass in the metric system is the **gram**, which is defined as the mass of one milliliter of water. While any of the metric prefixes can be used with the gram, it is most common to measure mass in kilograms, grams, or milligrams.

Kilograms are used to measure the mass of heavy objects, such as building materials, people, and pets. Grams are used to measure the mass of relatively light objects, such as snack foods and dry ingredients in recipes. Finally, milligrams are often used to measure the mass of medicine found in small pills or tablets. The relative sizes of the kilogram, gram, and milligram are shown in Figure 6.10 on the next page.

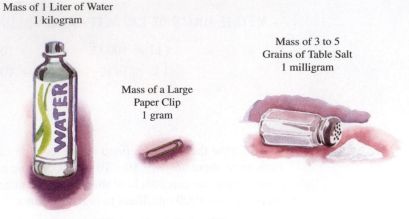

Figure 6.10 Some Metric Units of Mass

Kilogram (kg), hectogram (hg), dekagram (dag), decigram (dg), centigram (cg), and milligram (mg) relate to the gram (g) as follows.

METRIC UNITS OF MASS

1 kg = 1000 g	1 hg = 100 g	1 dag = 10 g
1 g = 10 dg	1 g = 100 cg	1 g = 1000 mg

Units of mass are listed as follows.

$$\text{kg} \quad \text{hg} \quad \text{dag} \quad \text{g} \quad \text{dg} \quad \text{cg} \quad \text{mg}$$

EXAMPLE 6 | **Converting from one unit of mass to another**

Convert as indicated.
(a) 0.00034 kilogram to decigrams **(b)** 10,000 milligrams to grams

Solution
(a) kg   hg   dag   g   dg   cg   mg

Move the decimal point in 0.00034 kg **4** places to the **right**.

$$0.00034 \text{ kg} = 3.4 \text{ dg}$$

(b) kg   hg   dag   g   dg   cg   mg

Move the decimal point in 10,000 mg **3** places to the **left**.

$$10,000. \text{ mg} = 10 \text{ g}$$

Now Try Exercises 51, 57

8. Temperature is measured in degrees _____ in the American system and in degrees _____ in the metric system.

UNITS OF LENGTH

Exercises 9–14: Convert the length as indicated. Round answers to 2 decimal places.

9. 17 inches to centimeters

10. 98 centimeters to inches

11. 240 yards to meters

12. 35 meters to yards

13. 84 miles to kilometers

14. 316 kilometers to miles

Exercises 15–18: Convert the given length as indicated. Round answers to 2 decimal places.

15. 5800 centimeters to yards
    (a) Centimeters → inches → yards
    (b) Centimeters → meters → yards

16. 8000 yards to kilometers
    (a) Yards → meters → kilometers
    (b) Yards → miles → kilometers

17. 4500 meters to miles
    (a) Meters → kilometers → miles
    (b) Meters → yards → miles

18. 0.5 yard to centimeters
    (a) Yards → inches → centimeters
    (b) Yards → meters → centimeters

Exercises 19–28: Convert the length as indicated. Round answers to 2 decimal places. Answers may vary slightly.

19. 27 feet to meters

20. 180 inches to meters

21. 3400 centimeters to feet

22. 0.335 kilometer to yards

23. 0.25 meter to inches

24. 0.0005 kilometer to inches

25. 15,840 feet to kilometers

26. 5.8 meters to feet

27. 0.003 kilometer to feet

28. 11 feet to centimeters

UNITS OF CAPACITY AND VOLUME

Exercises 29–32: Convert the capacity or volume as indicated. Round answers to 2 decimal places.

29. 16 ounces to milliliters

30. 7.5 liters to quarts

31. 56 quarts to liters

32. 1500 milliliters to ounces

Exercises 33–40: Convert the capacity or volume as indicated. Round answers to 2 decimal places. Answers may vary slightly.

33. 75 liters to gallons

34. 1 cup to milliliters

35. 16 pints to liters

36. 0.4 pint to milliliters

37. 956 milliliters to cups

38. 3.6 gallons to liters

39. 55,000.2 milliliters to gallons

40. 4.5 liters to pints

UNITS OF MASS (WEIGHT)

Exercises 41–44: Convert the mass or weight as indicted. Round answers to 2 decimal places.

41. 420 grams to ounces

42. 98 pounds to kilograms

43. 61.2 kilograms to pounds

44. 3 ounces to grams

Exercises 45–52: Convert the mass or weight as indicated. Round answers to 2 decimal places. Answers may vary slightly.

45. 0.135 ounce to milligrams

46. 3515.4 grams to pounds

47. 2.6 kilograms to ounces

48. 100,000 milligrams to pounds

49. 1.45 pounds to grams

50. 7560 milligrams to ounces

51. 0.0064 pound to milligrams

52. 480 ounces to kilograms

UNITS OF TEMPERATURE

Exercises 53–60: Convert the temperature algebraically.
Give exact answers in decimal form.

53. 77°F to Celsius

54. 55°C to Fahrenheit

55. 3°C to Fahrenheit

56. 98°F to Celsius

57. −15°F to Celsius

58. −9°C to Fahrenheit

59. 100°C to Fahrenheit

60. 98.6°F to Celsius

Exercises 61–66: Use the following table to convert the
temperature numerically. If there are two choices, use the
higher value.

°F	98.0	98.1	98.2	98.3	98.4	98.5	98.6	98.7	98.8	98.9
°C	36.7	36.7	36.8	36.8	36.9	36.9	37.0	37.1	37.1	37.2
°F	99.0	99.1	99.2	99.3	99.4	99.5	99.6	99.7	99.8	99.9
°C	37.2	37.3	37.3	37.4	37.4	37.5	37.6	37.6	37.7	37.7
°F	100.0	100.1	100.2	100.3	100.4	100.5	100.6	100.7	100.8	100.9
°C	37.8	37.8	37.9	37.9	38.0	38.1	38.1	38.2	38.2	38.3
°F	101.0	101.1	101.2	101.3	101.4	101.5	101.6	101.7	101.8	101.9
°C	38.3	38.4	38.4	38.5	38.6	38.6	38.7	38.7	38.8	38.8

61. 98.5°F to Celsius

62. 38.0°C to Fahrenheit

63. 38.8°C to Fahrenheit

64. 101.1°F to Celsius

65. 98.6°F to Celsius

66. 36.8°C to Fahrenheit

Exercises 67–72: Use the following graph to convert
the temperatures at the top of the next column visually.
Answers may vary slightly.

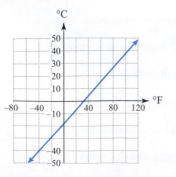

67. 20°C to Fahrenheit

68. 83°F to Celsius

69. −10°F to Celsius

70. −20°C to Fahrenheit

71. 50°F to Celsius

72. 46°C to Fahrenheit

APPLICATIONS

73. *Soda Bottle* To the nearest milliliter, how many milliliters are in a 20-ounce soda bottle?

74. *Printer Paper* A standard sheet of printer paper is 8.5 inches by 11 inches. Give these dimensions to the nearest tenth of a centimeter.

75. *Checked Luggage* Some airlines charge a fee for each checked luggage item that weighs more than 22.7 kilograms. To the nearest whole number, how many pounds is this?

76. *Long Jump* The world record in the long jump was set in 1991 by Mike Powell, with a jump of 8.95 meters. Is this jump over 30 feet? (*Source:* International Association of Athletics Federations.)

77. *Body Temperature* The highest body temperature ever recorded for a person surviving occurred in 1980 when Willie Jones had a body temperature of 46.5°C. Convert this temperature to Fahrenheit. (*Source: Guinness Book of World Records.*)

78. *Human Body* The average adult human has about 6 quarts of blood circulating throughout the body. Find the amount of blood in liters, to the nearest hundredth. (*Source:* NOVA, *Cut to the Heart.*)

79. *Triple Jump* The world record in the triple jump was set in 1995 by Jonathan Edwards, with a jump of 18.29 meters. Using the sequence

Meters → centimeters → inches → feet,

is this jump over 60 feet? (*Source:* International Association of Athletics Federations.)

80. *Blue Whale* An adult blue whale can weigh more than 172,700 kilograms. Find this weight to the nearest ton. (*Source:* Smithsonian National Zoological Park.)

81. *Super Soaker* An Aquashock Hydroblitz Super Soaker has a 101-ounce water reservoir. To the nearest whole number, how many liters of water does this water gun hold? (*Source:* Hasbro.)

82. *Kalahari Desert* Summertime temperatures in the Kalahari Desert can reach 45°C. Find the equivalent Fahrenheit temperature.

83. *Long Hair* The person with the world's longest hair is Xie Qiuping of China. Her hair is 5.627 meters long. Is her hair over 20 feet long? (*Source: Guinness Book of World Records.*)

84. *Obesity* An adult male who is 5 feet 10 inches tall is considered obese if his weight is above 95.5 kilograms. What is this weight to the nearest pound? (*Source: Centers for Disease Control.*)

85. *iPod Nano* The fourth-generation iPod Nano is 90.7 millimeters long and 38.7 millimeters wide. What are these two measurements to the nearest tenth of an inch? (*Source: Apple.*)

86. *Gas Tank* A car's gas tank holds 18.6 gallons of gasoline. How many liters of gasoline does the tank hold? Round to the nearest tenth.

87. *Mars Temperatures* The temperature at the poles on Mars can dip to −220°F. What is the equivalent Celsius temperature? (*Source: NASA.*)

88. *The Moon* The average diameter of the moon is about 3474 kilometers. What is this diameter to the nearest mile? (*Source: NASA.*)

89. *Weight of Milk* A gallon of milk weighs about 8.6 pounds. How much does a gallon of milk weigh to the nearest tenth of a kilogram?

90. *Liquid Air* Depending on its chemical makeup, air will liquefy when cooled to about −195°C. What is the equivalent Fahrenheit temperature?

WRITING ABOUT MATHEMATICS

91. When a conversion is completed in two different ways, the two results may be slightly different. What can be done to be sure results agree more closely?

92. Give two examples of when it might be practical to convert temperatures visually using a graph.

Group Activity   Working with Real Data

Directions: Form a group of 2 to 4 people. Select someone to record the group's responses for this activity. All members of the group should work cooperatively to answer the questions. If your instructor asks for your results, each member of the group should be prepared to respond.

Track and Field Before international standards for track and field were adopted in the United States, races were measured in yards rather than meters. After standards were put in place, many race distances changed.

(a) When international standards were adopted, the 100-yard sprint was replaced by the 100-meter sprint. Which is the shorter race?

(b) The 440-yard "quarter mile" was replaced by a race that is 400 meters long. Did this race get longer or shorter?

(c) The race that corresponds to the mile is now 1500 meters long. Which is longer, the mile race or the 1500-meter race?

(d) The high hurdles race was 120 yards long before standards were adopted. Now the corresponding race is 110 meters long. Using either the sequence

$$\text{meters} \rightarrow \text{centimeters} \rightarrow \text{inches} \rightarrow \text{yards}$$

or

$$\text{yards} \rightarrow \text{inches} \rightarrow \text{centimeters} \rightarrow \text{meters},$$

determine which distance is longer. Note that the relationship 1 m ≈ 1.09 yd is too inaccurate to determine which race is longer.

6.6   Time and Speed

Units of Time • Units of Speed • Applications

A LOOK INTO MATH ▶

When measuring time, we can measure very short intervals of time using metric-sounding units such as *nanoseconds* and *milliseconds*, as well as longer periods of time using units such as *minutes*, *hours*, *days*, *months*, and *years*. All of these units are used in both the American and metric systems of measurement. In this section, we will discuss the various units for time and extend the use of these units to the measurement of speed.

Units of Time

NEW VOCABULARY

☐ Second, minute, hour, day, week, month, year
☐ Speed

The **second** (sec), **minute** (min), **hour** (hr), **day** (d), **week** (wk), **month** (mo), and **year** (yr) are the units most commonly used for measuring time, and these units are related to each other as follows.

STUDY TIP

Look back at your progress so far this semester. Are there parts of your study process that need some adjustment? Are your notes and assignments organized? Are you spending enough time on homework and practice problems?

UNITS OF TIME

$$1 \text{ min} = 60 \text{ sec} \qquad 1 \text{ hr} = 60 \text{ min} \qquad 1 \text{ d} = 24 \text{ hr}$$

$$1 \text{ wk} = 7 \text{ d} \qquad 1 \text{ yr} = 365\frac{1}{4} \text{ d (or 365.25 d)}$$

NOTE: While we often hear that there are 52 weeks in 1 year, the relationship is not exactly accurate because a 52-week period is actually $7 \cdot 52 = 364$ days long, which is not equal to 1 year, or 365.25 days. Also, months should not be used in relationships because the number of days in a month can vary.

EXAMPLE 1   **Converting from one unit of time to another**

Convert as indicated. Express answers in decimal form when needed.
(a) 3.5 minutes to seconds   **(b)** 102 hours to days   **(c)** 5 days to minutes

Solution

(a) $3.5 \text{ min} = \dfrac{3.5 \text{ min}}{1} \cdot \dfrac{60 \text{ sec}}{1 \text{ min}} = 3.5(60) \text{ sec} = 210 \text{ sec}$

(b) $102 \text{ hr} = \dfrac{102 \text{ hr}}{1} \cdot \dfrac{1 \text{ d}}{24 \text{ hr}} = \dfrac{102}{24} \text{ d} = 4.25 \text{ d}$

(c) $5 \text{ d} = \dfrac{5 \text{ d}}{1} \cdot \dfrac{24 \text{ hr}}{1 \text{ d}} \cdot \dfrac{60 \text{ min}}{1 \text{ hr}} = 5 \cdot 24 \cdot 60 \text{ min} = 7200 \text{ min}$

Now Try Exercises 5, 7, 13

Units of Speed

READING CHECK

• Speed is a rate that involves what two quantities?

In Section 6.1, a *rate* was defined as a ratio used to compare two different kinds of quantities. If one quantity is distance and the other is time, then the rate represents *speed*. **Speed** is a rate that gives a distance traveled in an amount of time. A unit of speed is expressed as a ratio with a unit of distance in the numerator and a unit of time in the denominator.

▶ **REAL-WORLD CONNECTION** One recognizable example of speed in everyday life can be found posted along roads and highways on speed limit signs. Every driver knows the possible consequences of driving at speeds above the posted speed limit. In the United States, we measure the speed of a vehicle in miles per hour, or mi/hr (also abbreviated as mph). In countries using the metric system, the speed of a vehicle is measured in kilometers per hour, or km/hr. Other examples of speed include

ft/sec,   **m/min**,   **mi/sec**,   and   **cm/sec**.

It is possible to convert from one unit of speed to another by converting the distance unit, the time unit, or both units. The next example illustrates how to convert from one unit of speed to another using dimensional analysis.

EXAMPLE 2 **Converting units of speed**

Convert as indicated.
(a) 120 miles per hour to miles per minute   **(b)** 66 feet per second to miles per hour

Solution

(a) The distance unit does not change, so we only need to convert the time unit from hours to minutes. Since the hours unit is in the denominator of the given speed, we need a unit fraction that relates hours and minutes and has hours in the numerator.

$$120 \text{ mi/hr} = \frac{120 \text{ mi}}{1 \text{ hr}} \cdot \frac{1 \text{ hr}}{60 \text{ min}} = \frac{120 \text{ mi}}{60 \text{ min}} = \frac{2 \text{ mi}}{1 \text{ min}} = 2 \text{ mi/min}$$

(b) The distance unit must be converted from **feet to miles**, and the time unit must be converted from **seconds to hours**.

$$66 \text{ ft/sec} = \frac{66 \text{ ft}}{1 \text{ sec}} \cdot \frac{1 \text{ mi}}{5280 \text{ ft}} \cdot \frac{60 \text{ sec}}{1 \text{ min}} \cdot \frac{60 \text{ min}}{1 \text{ hr}} = \frac{66 \cdot 60 \cdot 60}{5280} \text{ mi/hr} = 45 \text{ mi/hr}$$

▌ **Now Try Exercises 17, 25**

Applications

Sometimes, speed is more easily interpreted when the units change. The next example illustrates how to convert from one unit of speed to another.

EXAMPLE 3 **Analyzing speed limit**

After crossing the border into Canada, an American driver uses cruise control set at 60 miles per hour. If the posted speed limit on the Canadian road is 90 kilometers per hour, is the American driving over the speed limit?

Solution

To determine if the driver is speeding, convert 60 miles per hour to kilometers per hour.

$$60 \text{ mi/hr} \approx \frac{60 \text{ mi}}{1 \text{ hr}} \cdot \frac{1 \text{ km}}{0.62 \text{ mi}} = \frac{60}{0.62} \text{ km/hr} \approx 96.8 \text{ km/hr}$$

The American is driving nearly 97 kilometers per hour, which is over the speed limit.

▌ **Now Try Exercise 41**

6.6   Putting It All Together

CONCEPT	COMMENTS	EXAMPLES
Units of Time	Second (sec), minute (min), hour (hr), day (d), week (wk), month (mo), and year (yr) ***Relationships:*** $1 \text{ min} = 60 \text{ sec}$, $1 \text{ hr} = 60 \text{ min}$, $1 \text{ d} = 24 \text{ hr}$, $1 \text{ wk} = 7 \text{ d}$, $1 \text{ yr} = 365\frac{1}{4} \text{ d}$ (or 365.25 d)	Convert 3.5 days to hours. $$\frac{3.5 \text{ d}}{1} \cdot \frac{24 \text{ hr}}{1 \text{ d}} = 84 \text{ hr}$$
Units of Speed	A unit of speed is expressed as a ratio with a unit of distance in the numerator and a unit of time in the denominator.	Convert 10,560 feet per hour to miles per hour. $$\frac{10,560 \text{ ft}}{1 \text{ hr}} \cdot \frac{1 \text{ mi}}{5280 \text{ ft}} = 2 \text{ mi/hr}$$

6.6   Exercises

MyMathLab   Math XL PRACTICE   WATCH   DOWNLOAD   READ   REVIEW

CONCEPTS AND VOCABULARY

1. Units of time include seconds, minutes, hours, _____, weeks, months, and years.

2. There are _____ seconds in one minute, and _____ minutes in one hour.

3. A rate that gives distance traveled in an amount of time is called _____.

4. A unit of speed has a unit of _____ in its numerator and a unit of _____ in its denominator.

UNITS OF TIME

Exercises 5–16: Convert as indicated. Express answers in decimal form when needed.

5. 5.25 hours to minutes

6. 3 days to hours

7. 90 minutes to hours

8. 63 days to weeks

9. 4.1 hours to seconds

10. 2 years to days

11. 2.5 weeks to hours

12. 0.25 day to minutes

13. 99,360 seconds to days

14. 19,800 seconds to hours

15. 4383 hours to years

16. 252 hours to weeks

UNITS OF SPEED

Exercises 17–32: Convert as indicated. Express answers in decimal form when needed.

17. 34 yards per second to feet per second

18. 4 kilometers per hour to meters per hour

19. 1200 centimeters per day to meters per day

20. 18 inches per week to feet per week

21. 12 meters per min to meters per second

22. 144 miles per hour to miles per min

23. 3.2 inches per day to inches per week

24. 4 millimeters per hour to millimeters per day

25. 30 miles per hour to feet per second

26. 50 feet per day to inches per hour

27. 7 kilometers per minute to meters per hour

28. 42 meters per hour to centimeters per minute

29. 200 feet per minute to yards per hour

30. 110 feet per second to miles per hour

31. 3 meters per second to kilometers per hour

32. 95 millimeters per minute to meters per hour

Exercises 33–36: Convert as indicated. Round answers to the nearest whole number.

33. 11 meters per second to feet per second

34. 70 miles per hour to kilometers per hour

35. 92 kilometers per hour to miles per hour

36. 190 feet per second to meters per second

APPLICATIONS

37. *NASCAR Speeds* Some NASCAR drivers reach average speeds of 3 miles per minute. Convert this speed to miles per hour.

38. *NASCAR Speeds* On slower speedways, NASCAR drivers reach average speeds of 120 miles per hour. Convert this speed to miles per minute.

39. *Animal Speeds* With top speeds of about 103 feet per second, the cheetah is the fastest land animal. To the nearest whole number, what is the top speed of a cheetah in miles per hour?

40. *Human Speeds* Usain Bolt set a world record in the 100-meter sprint during the 2008 Olympics by running an average speed of 34 feet per second. To the nearest whole number, what was Bolt's average speed in miles per hour?

41. *Speed Limits* An American drives 30 miles per hour after crossing the border into a Canadian town. If the posted speed limit is 50 kilometers per hour, is the driver exceeding the speed limit?

42. *Speed Limits* A Canadian drives 75 kilometers per hour after crossing the border into an American town. If the posted speed limit is 45 miles per hour, is the driver exceeding the speed limit?

WRITING ABOUT MATHEMATICS

43. A student writes the following relationship:

$$1 \text{ mo} \stackrel{?}{=} 4 \text{ wk.}$$

Explain why conversions made using this relationship will probably be inaccurate.

44. Without doing a conversion, explain how you would know that a vehicle traveling at 30 miles per hour is moving faster than a vehicle traveling at 30 kilometers per hour.

SECTIONS 6.5 and 6.6

Checking Basic Concepts

1. Convert as indicated. Answers may vary.
   (a) 250 inches to meters
   (b) 0.5 kilometer to feet
   (c) 5 cups to milliliters
   (d) 1.5 liters to pints
   (e) 1250 grams to pounds
   (f) 704 ounces to kilograms

2. Convert each temperature as indicated.
   (a) −5°C to Fahrenheit
   (b) 158°F to Celsius

3. Convert 54 meters per minute to centimeters per second.

4. Convert 3.5 yards per week to inches per day.

5. *Human Body* The average adult human brain weighs about 1300 grams. Give the weight of the average adult human brain to the nearest tenth of a pound. (*Source: Brain Facts and Figures.*)

6. *Speed Limits* A Canadian drives 80 kilometers per hour in an American town. If the posted speed limit is 50 miles per hour, is the driver going over the speed limit?

CHAPTER 6 Summary

SECTION 6.1 ■ RATIOS AND RATES

Ratio

To write a ratio as a fraction, write the quantity found before the word "to" (or before the colon) in the numerator of the fraction, and write the quantity found after the word "to" (or after the colon) in the denominator of the fraction. Then simplify the fraction.

Examples: 3 to 9, 3 : 9, or $\frac{3}{9} = \frac{1}{3}$

Unit Ratio

A unit ratio is a ratio expressed as a fraction with a denominator of 1. To find a unit ratio, divide the denominator into the numerator.

Example: If the student-to-instructor ratio at a college is 420 to 15, the unit ratio is $\frac{420}{15} = 28$, or 28 students for each instructor.

Rate

A rate is a ratio used to compare different kinds of quantities. The units for each quantity are expressed as part of a ratio.

Example: A person spending $23 in 2 hours spends at a rate of $\frac{23 \text{ dollars}}{2 \text{ hours}}$.

Unit Rate

A unit rate is a rate expressed as a fraction with denominator 1. To find a unit rate, divide the denominator into the numerator.

Example: A person earning $90.75 in 11 hours has an hourly pay rate of $\frac{90.75 \text{ dollars}}{11 \text{ hours}} = \$8.25/\text{hr}$.

Unit Pricing

If the price of q units of a product is p, the unit price U is given by $U = \frac{p}{q}$.

Example: A 20-ounce bottle of soda that sells for $1.60 has a unit price of $U = \frac{p}{q} = \frac{\$1.60}{20 \text{ oz}} = \$0.08/\text{oz}$.

SECTION 6.2 ■ PROPORTIONS AND SIMILAR FIGURES

Proportions

If $\frac{a}{b}$ and $\frac{c}{d}$ are ratios that are equal in value, then $\frac{a}{b} = \frac{c}{d}$ is a proportion, where $b \neq 0$ and $d \neq 0$. This proportion is read "a is to b as c is to d."

Example: The proportion $\frac{4}{5} = \frac{16}{20}$ is read "4 is to 5 as 16 is to 20."

Cross Product Rule

For $b \neq 0$ and $d \neq 0$, if the cross products ad and bc are equal, then $\frac{a}{b} = \frac{c}{d}$ is a proportion.

Example: $\frac{7}{8} = \frac{14}{16}$ is a proportion because $7 \cdot 16 = 8 \cdot 14$.

Similar Figures

When two geometric figures are similar, the measures of the corresponding sides are proportional.

Example:

SECTION 6.3 ■ THE AMERICAN SYSTEM OF MEASUREMENT

American Units of Length

Inches, feet, yards, and miles

Relationships: 1 ft = 12 in., 1 yd = 36 in., 1 yd = 3 ft,

1 mi = 1760 yd, and 1 mi = 5280 ft

Example: Convert 48 inches to feet. 48 in. $= \frac{48 \text{ in.}}{1} \cdot \frac{1 \text{ ft}}{12 \text{ in.}} = \frac{48}{12} \text{ ft} = 4 \text{ ft}$

Unit Fraction	A unit fraction is a fraction that is equivalent to 1.

Examples: $\frac{12\ \text{in.}}{1\ \text{ft}}$, $\frac{1\ \text{lb}}{16\ \text{oz}}$, and $\frac{2\ \text{c}}{1\ \text{pt}}$

American Units of Area	Units of area are the squares of units of length. To convert one unit of area to another, use the corresponding unit fraction relating units of length, twice.

Example: Convert 3 square feet to square inches.

$$3\ \text{ft}^2 = \frac{3\ \cancel{\text{ft}} \cdot \cancel{\text{ft}}}{1} \cdot \frac{12\ \text{in.}}{1\ \cancel{\text{ft}}} \cdot \frac{12\ \text{in.}}{1\ \cancel{\text{ft}}} = 432\ \text{in}^2$$

American Units of Capacity and Volume	Ounce, cup, pint, quart, and gallon

Relationships: 1 c = 8 oz, 1 pt = 2 c, 1 qt = 2 pt, and 1 gal = 4 qt

Example: Convert 6 cups to pints. $6\ \text{c} = \frac{6\ \cancel{\text{c}}}{1} \cdot \frac{1\ \text{pt}}{2\ \cancel{\text{c}}} = \frac{6}{2}\ \text{pt} = 3\ \text{pt}$

American Units of Weight	Ounce, pound, and ton

Relationships: 1 lb = 16 oz, and 1 T = 2000 lb

Example: Convert 3.5 tons to pounds. $3.5\ \text{T} = \frac{3.5\ \cancel{\text{T}}}{1} \cdot \frac{2000\ \text{lb}}{1\ \cancel{\text{T}}} = 7000\ \text{lb}$

SECTION 6.4 ■ THE METRIC SYSTEM OF MEASUREMENT

Metric Prefixes	Kilo, hecto, deka, deci, centi, and milli Each prefix designates the size of a measurement relative to the base unit.

Examples: *Deka* means "10 times the base" and *centi* means "$\frac{1}{100}$ of the base."

Metric Units of Length	The base metric unit of length is the *meter*.

Relationships: 1 km = 1000 m, 1 hm = 100 m, 1 dam = 10 m,

1 m = 10 dm, 1 m = 100 cm, and 1 m = 1000 mm

Example: Convert 2.2 meters to millimeters.

km hm dam m dm cm mm

2.2 m = 2.200 m = 2200 mm

Metric Units of Area	Units of area are the squares of units of length. To convert one unit of area to another, move the decimal point *twice* the number of places needed to convert the corresponding units of length.

Example: Convert 1400 square meters to square hectometers.

km hm dam m dm cm mm

1400. m² = 0.14 hm²

Metric Units of Capacity and Volume	The base metric unit of capacity and volume is the *liter*.

Relationships: 1 kl = 1000 L, 1 hl = 100 L, 1 dal = 10 L,

1 L = 10 dl, 1 L = 100 cl, and 1 L = 1000 mL

Example: Convert 4.65 deciliters to milliliters.

kl hl dal L dl cl mL

4.65 dl = 465 mL

Metric Units of Mass

The base metric unit of mass is the *gram*.

Relationships: 1 kg = 1000 g,   1 hg = 100 g,   1 dag = 10 g,

1 g = 10 dg,   1 g = 100 cg,   and   1 g = 1000 mg

Example: Convert 0.5 centigram to grams.

kg   hg   dag   g   dg   cg   mg

0.5 cg = 000.5 cg = 0.005 g

SECTION 6.5 ■ AMERICAN–METRIC CONVERSIONS; TEMPERATURE

Units of Length

For converting units of length, use the relationships

1 in. = 2.54 cm,   1 m ≈ 1.09 yd,   and   1 km ≈ 0.62 mi.

Example: Convert 7 meters to yards. $7 \text{ m} \approx \frac{7 \text{ m}}{1} \cdot \frac{1.09 \text{ yd}}{1 \text{ m}} = 7.63 \text{ yd}$

Units of Capacity and Volume

For converting units of capacity and volume, use the relationships

1 L ≈ 1.06 qt   and   1 oz ≈ 29.57 mL.

Example: Convert 5 liters to quarts. $5 \text{ L} \approx \frac{5 \text{ L}}{1} \cdot \frac{1.06 \text{ qt}}{1 \text{ L}} = 5.3 \text{ qt}$

Units of Mass (Weight)

For converting units of mass (weight), use the relationships

1 kg ≈ 2.2 lb   and   1 oz ≈ 28.35 g.

Example: Convert 8.8 pounds to kilograms. $8.8 \text{ lb} \approx \frac{8.8 \text{ lb}}{1} \cdot \frac{1 \text{ kg}}{2.2 \text{ lb}} = 4 \text{ kg}$

Units of Temperature

In the American system, temperature is measured in degrees Fahrenheit, and in the metric system, it is measured in degrees Celsius. To convert from one system to the other, use the formulas $F = \frac{9}{5}C + 32$ and $C = \frac{5}{9}(F - 32)$. Note that temperature conversions can also be completed numerically, using a table of values, and visually, using a graph.

Example: Convert 86°F to Celsius.

Algebraic:

$C = \frac{5}{9}(86 - 32)$

$= \frac{5}{9}(54)$

$= 30°C$

Visual:

Numerical:

°F	85.9	**86.0**	86.1	86.2
°C	29.9	**30.0**	30.1	30.1

SECTION 6.6 ■ TIME AND SPEED

Units of Time

Second (sec), minute (min), hour (hr), day (d), week (wk), month (mo), and year (yr)

Relationships: 1 min = 60 sec,   1 hr = 60 min,   1 d = 24 hr,

1 wk = 7 d, 1 yr = $365\frac{1}{4}$ d (or 365.25 d)

Example: Convert 105 minutes to hours.

$$105 \text{ min} = \frac{105 \text{ min}}{1} \cdot \frac{1 \text{ hr}}{60 \text{ min}} = 1.75 \text{ hr}$$

Units of Speed

A unit of speed is expressed as a ratio with a distance in the numerator and time in the denominator.

Examples: Convert 450 feet per minute to yards per minute.

$$450 \text{ ft/min} = \frac{450 \text{ ft}}{1 \text{ min}} \cdot \frac{1 \text{ yd}}{3 \text{ ft}} = 150 \text{ yd/min}$$

CHAPTER 6   Review Exercises

SECTION 6.1

Exercises 1–4: Write the given ratio as a fraction in simplest form.

1. 6 to 24

2. 1.2 to 3.3

3. $\frac{15}{8} : \frac{45}{4}$

4. $1\frac{1}{5} : 3\frac{3}{4}$

Exercises 5 and 6: Find the unit ratio.

5. 12 to 15

6. 140 : 30

Exercises 7 and 8: Write the given rate as a fraction in simplest form.

7. An athlete runs 12 miles in 64 minutes.

8. There are 30 laptops for 120 students.

Exercises 9 and 10: Write the given rate as a unit rate.

9. A person earns $58.80 in 8 hours.

10. A heart beats 296 times in 4 minutes.

Exercises 11 and 12: Find the unit price.

11. A 20-pound block of ice for $2.86

12. A 2-liter bottle of store-brand soda for $0.98

SECTION 6.2

Exercises 13–16: Determine if the given equation is a proportion.

13. $\frac{10}{80} \stackrel{?}{=} \frac{2}{16}$

14. $\frac{\frac{1}{4}}{18} \stackrel{?}{=} \frac{\frac{1}{6}}{24}$

15. $\frac{2}{8.5} \stackrel{?}{=} \frac{11}{48.5}$

16. $\frac{8\frac{1}{2}}{2\frac{5}{6}} \stackrel{?}{=} \frac{5\frac{1}{10}}{1\frac{7}{10}}$

Exercises 17–20: Solve the proportion.

17. $\frac{24}{x} = \frac{-3}{2}$

18. $\frac{1.7}{b} = \frac{-5.1}{3}$

19. $\frac{\frac{3}{10}}{-\frac{2}{5}} = \frac{y}{-4}$

20. $\frac{1\frac{4}{5}}{-9} = \frac{a}{5\frac{5}{6}}$

Exercises 21 and 22: For the similar figures, find the measure of x.

21.

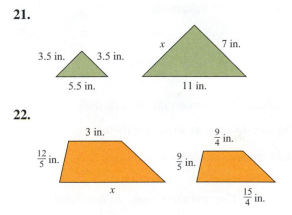

22.

SECTION 6.3

Exercises 23–26: Convert the length as indicated.

23. 12.5 feet to inches

24. 1.8 miles to feet

25. 38,016 inches to miles

26. 144 inches to yards

Exercises 27 and 28: Convert the area as indicated.

27. 270 square feet to square yards

28. 0.04 square mile to square feet

Exercises 29–32: Convert the volume as indicated.

29. 14 gallons to quarts

30. 112 ounces to quarts

31. $8\frac{1}{2}$ pints to cups

32. 2.75 gallons to cups

Exercises 33–36: Convert the weight as indicated.

33. 7.5 pounds to ounces

34. 392 ounces to pounds

35. $\frac{1}{250}$ ton to pounds

36. 1,312,000 ounces to tons

SECTION 6.4

Exercises 37 and 38: Write the name of the metric unit of measurement described.

37. $\frac{1}{1000}$ of a gram

38. 10 times a liter

Exercises 39–42: Convert the length as indicated.

39. 0.75 meter to centimeters

40. 11.4 decimeters to dekameters

41. 12.6 kilometers to hectometers

42. 0.0019 hectometer to decimeters

Exercises 43 and 44: Convert the area as indicated.

43. 780 square decimeters to square meters

44. 0.00085 square kilometer to square meters

Exercises 45–48: Convert the volume as indicated.

45. 52,530 dekaliters to kiloliters

46. 0.075 deciliter to milliliters

47. 7690 centiliters to dekaliters

48. 0.1 milliliter to deciliters

Exercises 49–52: Convert the mass as indicated.

49. 0.087 dekagram to decigrams

50. 28,000 grams to kilograms

51. 0.00077 kilogram to centigrams

52. 0.021 hectogram to grams

SECTION 6.5

Exercises 53 and 54: Convert the specified length using the given sequences. Round answers to 2 decimal places.

53. 7850 yards to kilometers
   (a) Yards → meters → kilometers
   (b) Yards → miles → kilometers

54. 2500 meters to miles
   (a) Meters → kilometers → miles
   (b) Meters → yards → miles

Exercises 55–58: Convert the length as indicated. Round answers to 2 decimal places. Answers may vary slightly.

55. 220 inches to meters

56. 0.25 kilometer to yards

57. 36,960 feet to kilometers

58. 9 feet to centimeters

Exercises 59–62: Convert the capacity or volume as indicated. Round answers to 2 decimal places. Answers may vary slightly.

59. 100 liters to gallons

60. 0.2 pint to milliliters

61. 10.25 gallons to liters

62. 39,250 milliliters to gallons

Exercises 63–66: Convert the given mass or weight as indicated. Round answers to 2 decimal places. Answers may vary slightly.

63. 6514 grams to pounds

64. 0.44 pound to milligrams

65. 3780 milligrams to ounces

66. 1.05 pounds to grams

Exercises 67 and 68: Convert the temperature algebraically. Give exact answers in decimal form.

67. 9°C to Fahrenheit

68. −13°F to Celsius

Exercises 69 and 70: Use the following table to convert the given temperature numerically. If there are two choices, use the higher value.

°F	98.2	98.3	98.4	98.5	98.6	98.7
°C	36.8	36.8	36.9	36.9	37.0	37.1

69. 98.3°F to Celsius

70. 36.9°C to Fahrenheit

Exercises 71 and 72: Use the following graph to convert the given temperature visually. Answers may vary slightly.

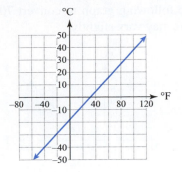

71. −30°F to Celsius

72. 25°C to Fahrenheit

SECTION 6.6

Exercises 73–76: Convert as indicated. Express answers in decimal form when needed.

73. 3 days to hours

74. 0.4 day to minutes

75. 16,920 seconds to hours

76. 420 hours to weeks

Exercises 77–80: Convert as indicated. Express answers in decimal form when appropriate.

77. 7.5 kilometers per hour to meters per hour

78. 30 inches per week to feet per week

79. 80 feet per day to inches per hour

80. 8 kilometers per minute to meters per hour

Exercises 81 and 82: Convert as indicated. Round your answers to the nearest whole number.

81. 75 miles per hour to kilometers per hour

82. 120 meters per second to feet per second

APPLICATIONS

83. *Comparing Colleges* A big university has a student-to-instructor ratio of 1548 to 30, while a small college has a student-to-instructor ratio of 764 to 20. Find the unit ratio at each school. Interpret the results for the small college.

84. *Going Green* A survey determined that 60 of 360 students take public transportation to school each day. How many of every 48 students take public transportation to school each day?

85. *Baling Twine* A farmer uses two 80-inch pieces of twine to tie up a bale of hay. If the twine is available on a spool containing 4500 feet of twine, how many bales can be tied up with one spool? How many inches of twine are left over?

86. *Serving Size* If the label on a 4-liter jug of sports drink states that the serving size is 200 milliliters, how many servings are in the jug?

87. *Obesity* An adult male who is 6 feet 2 inches tall is considered obese if his weight is above 106 kilograms. What is this weight to the nearest pound? (*Source:* Centers for Disease Control.)

88. *Speed Limits* An American drives 35 miles per hour after crossing the border into a Mexican town. If the posted speed limit is 55 kilometers per hour, is the driver going over the speed limit?

89. *Tree Height* A tree casts a 54-foot shadow, while a nearby tree that is 13 feet tall casts a 6-foot shadow. How tall is the larger tree?

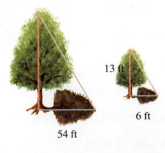

13 ft

6 ft

54 ft

90. *Unit Price* Find the unit price for each size option.

Large coffee drink: 24 ounces for $4.68

Small coffee drink: 16 ounces for $3.28

Which is the better buy?

91. *Checked Luggage* Some airlines charge a fee for each checked luggage item that weighs more than 22.7 kilograms. How many grams is this?

92. *Fruit Drink* A large jug contains 145 ounces of fruit drink. Is this more or less than a gallon?

93. *Warm Day* Temperature in Death Valley can sometimes reach 122°F. Find an equivalent temperature on the Celsius scale.

94. *Liquid Oxygen* Oxygen will liquefy when cooled to about −183°C. Find an equivalent temperature on the Fahrenheit scale.

CHAPTER 6   Test

1. Write the ratio 6 : 18 as a fraction in simplest form.

2. An athlete runs 8 miles in 64 minutes. Write this rate as a unit rate.

3. A 16-ounce box of crackers sells for $3.44. Find the unit price of the crackers.

4. Is $\frac{12}{15} \stackrel{?}{=} \frac{20}{24}$ a proportion?

Exercises 5 and 6: Solve the proportion.

5. $\dfrac{-18}{x} = \dfrac{-2}{3}$

6. $\dfrac{\frac{3}{10}}{7} = \dfrac{w}{-\frac{5}{9}}$

7. The triangles are similar. Find the value of *x*.

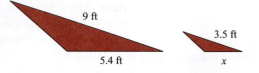

9 ft

3.5 ft

5.4 ft   x

Exercises 8–15: Convert the measurement as indicated.

8. 6.5 yards to inches

9. 540 square inches to square feet

10. $8\frac{1}{2}$ gallons to pints

11. 35,000 pounds to tons

12. 1.54 centimeters to meters

13. 540,000 square meters to square kilometers

14. 0.075 liter to milliliters

15. 2400 decigrams to hectograms

Exercises 16–18: Convert as indicated. Round answers to 2 decimal places. Answers may vary slightly.

16. 350 yards to kilometers

17. 430 milliliters to cups

18. 2.5 pounds to grams

19. Use a formula to convert 15°C to Fahrenheit.

20. Use the following graph to convert 70°F to Celsius. Answers may vary slightly.

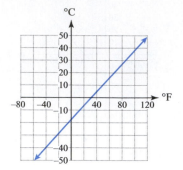

Exercises 21 and 22: Convert as indicated. Express answers in decimal form when appropriate.

21. 41,040 minutes to days

22. 3 days to seconds

Exercises 23 and 24: Convert as indicated. Round your answers to the nearest whole number.

23. 1 inch per second to feet per hour

24. 2 miles per day to meters per day

25. *Unit Price* Find the unit price for each size option.

Large fruit smoothie: 24 ounces for $3.12

Small fruit smoothie: 10 ounces for $1.48

Which is the better buy?

26. *Serving Size* If the label on a 2.2-quart bottle of milk states that the serving size is 250 milliliters, how many full servings are in the bottle?

27. *Temperature* If it was 78°F on Monday and 27°C on Tuesday, which day was warmer?

CHAPTERS 1–6   Cumulative Review Exercises

1. Write the whole number 34,206 in word form.

2. Are the terms $5ab$, $3xy$ like or unlike?

3. Graph the whole numbers 2 and 5 on a number line.

4. Round 730,187 to the nearest ten-thousand.

5. Place the correct symbol, $<$ or $>$, in the blank between the integers: -17 _____ -9.

6. Evaluate the expression $x - y$ for the given values of the variables: $x = -5$, $y = 11$.

7. Find the product: $2(-1)(3)(-2)(-5)$.

8. Solve the equation $-3x + 28 = 25$ visually.

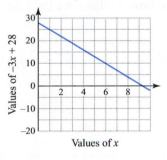

9. Translate the phrase to an algebraic expression using the variable x: *the product of 4 and her age*.

10. Solve algebraically: $-(2q + 3) + 7q = 17$.

11. Solve the linear equation numerically by completing the given table of values: $2x - 5 = -1$.

x	-2	-1	0	1	2
$2x - 5$					

12. When the sum of 6 and a number is divided by 3 the result is 5. Find the number.

13. Write the improper fraction and mixed number represented by the given shading.

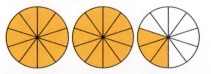

14. Use a divisibility test to answer each question.
    (a) Is 942 divisible by 3?
    (b) Is 4766 divisible by 4?

15. Add and simplify your result, if possible.
$$-\frac{11}{18} + \frac{5}{6}$$

16. Solve the equation. Simplify your result.
$$5\left(n - \frac{1}{6}\right) = \frac{8}{3}$$

17. Round 594.032 to the nearest hundredth

18. Solve algebraically: $3(p - 5) = 2.7p + 4.2$.

19. Find the exact area of the circle and then approximate the area using 3.14 as an approximation for π.

10 ft

20. Calculate the GPA for the grade report.

Course	Credits	Grade
Chemistry	4	C
English	3	B
Study Skills	1	B
Biology	4	A

21. If a car travels 174 miles in 3 hours, write its speed as a unit rate.

22. Solve the proportion.
$$\frac{15}{x} = \frac{-3}{4}$$

23. *Heart Rate* The average heart rate R, in beats per minute (bpm), of a person weighing W kilograms can be approximated by
$$R = \frac{400\sqrt{W}}{W}.$$
Find the heart rate for a 25-kilogram child.

24. *Music Sales* The total music sales S, in thousands of dollars, for a small band during year x can be found using the formula

$$S = 7(x - 2005) + 8.$$

Find the year when music sales reached $43,000 by solving the equation $43 = 7(x - 2005) + 8$ visually using the graph provided.

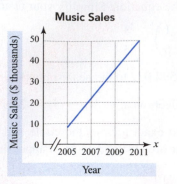

Music Sales

25. *Triangle Dimensions* The longest side of a triangle measures 1 inch less than triple the measure of the shortest side. The measure of the third side is 2 inches more than double the measure of the shortest side. If the perimeter of the triangle is 25 inches, find the measure of the shortest side of the triangle.

26. *Height* At birth, a baby boy was $21\frac{1}{4}$ inches long. If he grew $8\frac{1}{2}$ inches during his first year, how tall was he on his first birthday?

27. *Text Messaging* Suppose that the cost C of sending or receiving x text messages is given by the formula

$$C = 0.05(x - 500) + 10.5, \text{ where } x \geq 500.$$

Find the number of text messages that correspond to total charges of $22.45 by replacing C in the formula with 22.45 and solving the resulting equation.

28. *Wilderness Maps* Every $1\frac{1}{2}$ inches on a wilderness map represents 1 mile. If the actual distance between two lakes is 18 miles, how far apart are the lakes on the map?

7

Percents

Obstacles are those frightful things you see when you take your eyes off the goal.

—HENRY FORD

Health professionals know the importance of proper pacing during exercise. Since fitness levels vary from person to person, no single exercise pace is right for everyone. While a physically fit person's heart rate may remain near resting levels during mild exercise, the same activity may cause an inactive person's heart rate to increase significantly. Regardless of fitness level, maintaining a *target heart rate* during exercise will help maximize the benefits of physical activity.

Target heart rate is a range between 50 and 85 *percent* of a person's maximum heart rate. The proper pace for exercise occurs within this range. Because maximum heart rate depends on age, people can determine their target heart rates before beginning an exercise program. You can find your own target heart rate if you understand percents.

In this chapter, we learn how to solve problems involving percents, which occur in real-world situations such as exercise science, sales tax, interest rates, and exam scores.

Source: The American Heart Association.

7.1   Introduction to Percents; Circle Graphs

**Percent Notation • Writing Percents as Fractions or Decimals •
Writing Fractions or Decimals as Percents • Circle Graphs • Applications**

When a realtor helps to sell a home, the sellers pay a *commission* for the realtor's services. A commission is a fee paid to the realtor that is a percentage of the home's purchase price. For example, if a realtor charges 6% commission on a home that sells for $380,000, then the seller gets $357,200 and the realtor gets $22,800. In this section, we learn about percent notation and the relationships among percents, fractions, and decimals.

Percent Notation

We can remember the meaning of the word *percent* by breaking it into the words *per* and *cent*. Recall that *per* is associated with division and means "divide by." The word *cent* comes from the Latin word *centum*, or 100. (There are 100 *cents* in $1.) So, the word **percent** means "divide by 100" or "out of 100." The symbol for percent is %. The shading in Figure 7.1 represents 67% because 67 out of 100 squares are shaded.

Figure 7.1 Visualizing 67%

Another way to represent the shading in Figure 7.1 is to write the fraction $\frac{67}{100}$. Since the fraction $\frac{67}{100}$ is equal to the decimal 0.67, we can write 67% as a fraction or a decimal.

Writing Percents as Fractions or Decimals

Recall that dividing by a number is the same as multiplying by its reciprocal. So, dividing by 100 is the same as multiplying by $\frac{1}{100}$ or 0.01. Mathematically, we can write

the expression $x\%$ represents the fraction $\frac{x}{100}$ or the decimal $0.01x$.

We use the following procedure to write a percent as a fraction or decimal.

> **WRITING PERCENTS AS FRACTIONS OR DECIMALS**
>
> To write $x\%$ as a fraction, write $\frac{x}{100}$. Simplify the fraction, if needed.
>
> To write $x\%$ as a decimal, remove the % symbol and then multiply $0.01x$.

NOTE: We can write $x\%$ as a decimal by moving the decimal point in the number x two places to the *left* and removing the % symbol.

EXAMPLE 1 **Writing percents as fractions**

Write each percent as a fraction or mixed number.

(a) 24%   (b) 130%

Solution

(a) $24\% = \dfrac{24}{100} = \dfrac{6 \cdot 4}{25 \cdot 4} = \dfrac{6}{25}$

(b) $130\% = \dfrac{130}{100} = \dfrac{13 \cdot 10}{10 \cdot 10} = \dfrac{13}{10} \text{ or } 1\dfrac{3}{10}$

Now Try Exercises 13, 19

EXAMPLE 2 **Writing percents as fractions**

Write each percent as a fraction in simplest form.

(a) 8.5%   (b) $66\frac{2}{3}\%$

Solution

(a) The fraction $\frac{8.5}{100}$ is not in simplest form because its numerator is a decimal number. To clear the decimal, multiply by 1 in the form $\frac{2}{2}$.

$$8.5\% = \frac{8.5}{100} = \frac{8.5}{100} \cdot \frac{2}{2} = \frac{17}{200}$$

(b) Begin by writing the mixed number $66\frac{2}{3}$ as the improper fraction $\frac{200}{3}$.

$$66\frac{2}{3}\% = \frac{200}{3}\% = \frac{\frac{200}{3}}{100} = \frac{200}{3} \cdot \frac{1}{100} = \frac{2 \cdot 100}{3 \cdot 100} = \frac{2}{3}$$

Now Try Exercises 17, 23

EXAMPLE 3 **Writing percents as decimals**

Write each percent as a decimal.

(a) 38%   (b) 119%

Solution

(a) $38\% = 0.01(38) = 0.38$

(b) $119\% = 0.01(119) = 1.19$

Now Try Exercises 25, 29

EXAMPLE 4 **Writing percents as decimals**

Write each percent as a decimal.

(a) 2.9%   (b) $45\frac{1}{4}\%$

Solution

(a) $2.9\% = 0.01(2.9) = 0.029$

(b) Begin by writing the mixed number $45\frac{1}{4}$ as the decimal 45.25.

$$45\frac{1}{4}\% = 45.25\% = 0.01(45.25) = 0.4525$$

Now Try Exercises 27, 39

Fractions or Decimals as cents

n we write a percent as a fraction or decimal *divide* by 100 and remove the % symbol. To reverse the process and write a fraction cimal as a percent, we *multiply* by 100 and attach a % symbol. *In other words, we mult by 100%. Since*

$$100\% = \frac{1}{00} = 1,$$

multiplying by 100% is the same as mult ying by 1.

WRITING FRACTIONS OR DECIMALS AS PERCENTS

To write a fraction or decimal as a percent, multiply by 100%.

NOTE: We can write the decimal number x as a percent by moving the decimal point in x two places to the *right* and attaching a % symbol.

EXAMPLE 5 **Writing fractions as percents**

Write each fraction or mixed number as a percent.

(a) $\dfrac{2}{5}$ (b) $2\dfrac{3}{4}$ (c) $\dfrac{5}{6}$

Solution

(a) $\dfrac{2}{5} = \dfrac{2}{5} \cdot 100\% = \dfrac{2}{5} \cdot \dfrac{100}{1}\% = \dfrac{200}{5}\% = 40\%$

(b) Begin by writing the mixed number $2\frac{3}{4}$ as the improper fraction $\frac{11}{4}$.

$$2\frac{3}{4} = \frac{11}{4} = \frac{11}{4} \cdot 100\% = \frac{11}{4} \cdot \frac{100}{1}\% = \frac{1100}{4}\% = 275\%$$

(c) $\dfrac{5}{6} = \dfrac{5}{6} \cdot 100\% = \dfrac{5}{6} \cdot \dfrac{100}{1}\% = \dfrac{500}{6}\% = 83\dfrac{1}{3}\% = 83.\overline{3}\%$

Now Try Exercises 41, 45, 49

EXAMPLE 6 **Writing decimals as percents**

Write each decimal as a percent.
(a) 0.37   (b) 0.9   (c) 0.085   (d) 1.77

Solution
(a) $0.37 = 0.37(100\%) = 37\%$
(b) $0.9 = 0.9(100\%) = 90\%$
(c) $0.085 = 0.085(100\%) = 8.5\%$
(d) $1.77 = 1.77(100\%) = 177\%$

Now Try Exercises 55, 57, 61, 63

Circle Graphs

Another way to visualize percents is to shade parts of a circle that has been divided into 100 equal parts. For example, 17 out of 100 parts are shaded in Figure 7.2. So, the shaded portion represents 17% of the entire circle. This type of graph is called a **circle graph**.

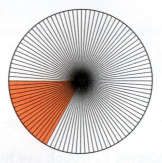

Figure 7.2 Shading 17% on a Circle Graph

Circle graphs used to compare percent data show only the shaded parts of the graph. They do not show all 100 pieces. For example, the circle graph in Figure 7.3 shows the percentage of students using various modes of transportation to get to a particular college.

READING CHECK

• What is the total of the percents in a circle graph?

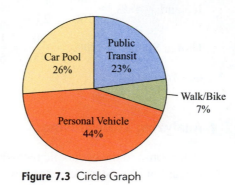

Figure 7.3 Circle Graph

EXAMPLE 7   **Reading a circle graph**

Figure 7.4 shows the market share for leading Internet search engines in February 2009. (*Source:* Compete.com)

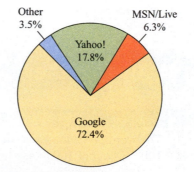

Figure 7.4 Internet Search Engine Market Share

(a) Which Internet search engine had the largest market share?
(b) Write the market share held by Yahoo! as a fraction in simplest form.
(c) Write the market share held by Other as a decimal.

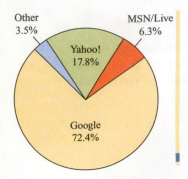

Other
3.5%

MSN/Live
6.3%

Yahoo!
17.8%

Google
72.4%

Figure 7.4 Internet Search Engine Market Share

Solution

(a) The largest market share was held by Google because its shaded region is the largest.

(b) Since $0.8 \cdot 5$ equals a whole number, we can write the market share held by Yahoo! as a fraction by multiplying $\frac{17.8}{100}$ by $\frac{5}{5}$ to clear the decimal.

$$17.8\% = \frac{17.8}{100} = \frac{17.8}{100} \cdot \frac{5}{5} = \frac{89}{500}$$

(c) Written as a decimal, the market share held by Other is $3.5\% = 0.01(3.5) = 0.035$.

Now Try Exercise 71

Applications

The next two examples illustrate how percents are used regularly in everyday life.

EXAMPLE 8 **Comparing Facebook use in two countries**

In 2009, 48% of Iceland's population used Facebook, while 20% of U.S. residents used this social networking Web site. Write each percent as a fraction in simplest form. (*Source:* Facebook.)

Solution

Iceland: $48\% = \frac{48}{100} = \frac{12 \cdot 4}{25 \cdot 4} = \frac{12}{25}$

United States: $20\% = \frac{20}{100} = \frac{1 \cdot 20}{5 \cdot 20} = \frac{1}{5}$

Now Try Exercise 77

EXAMPLE 9 **Analyzing store closings**

In response to the economic recession, Starbucks closed $\frac{1}{20}$ of its U.S. stores in 2008. What percent of U.S. Starbucks stores were closed that year? (*Source:* Starbucks.)

Solution

$$\frac{1}{20} = \frac{1}{20} \cdot 100\% = \frac{1}{20} \cdot \frac{100}{1}\% = \frac{100}{20}\% = 5\%$$

Starbucks closed 5% of its U.S. stores in 2008.

Now Try Exercise 75

7.1 Putting It All Together

CONCEPT	COMMENTS	EXAMPLES
Percents	The word *percent* means "divide by 100" or "out of 100."	39% means "39 divided by 100" and is written as 39%, $\frac{39}{100}$, or 0.39.
Writing Percents as Fractions or Decimals	To write $x\%$ as a fraction, write $\frac{x}{100}$ and then simplify the fraction, if needed.	$14\% = \frac{14}{100} = \frac{7}{50}$
	To write $x\%$ as a decimal, remove the % symbol and then multiply $0.01x$.	$14\% = 0.01(14) = 0.14$

CONCEPT	COMMENTS	EXAMPLES
Writing Fractions or Decimals as Percents	To write a fraction or decimal as a percent, multiply by 100%.	$\dfrac{4}{5} = \dfrac{4}{5} \cdot 100\% = \dfrac{400}{5}\% = 80\%$ $0.64 = 0.64(100\%) = 64\%$
Circle Graph	A circle graph can be used to display percent data visually.	The following circle graph shows the percentage of employees who work late on a given day.

7.1 Exercises

MyMathLab Math XL PRACTICE WATCH DOWNLOAD READ REVIEW

CONCEPTS AND VOCABULARY

1. The word _____ means "divide by 100."

2. The symbol for percent is _____.

3. To write $x\%$ as a fraction, write _____ and simplify the result, if needed.

4. To write $x\%$ as a decimal, remove the % symbol and then multiply _____.

5. We can write $x\%$ as a decimal by moving the decimal point in the number x two places to the _____ and removing the % symbol.

6. To write a fraction or decimal as a percent, multiply by _____.

7. We can write the decimal number x as a percent by moving the decimal point in the number x two places to the _____ and attaching a % symbol.

8. By shading parts of a circle, percent data can be displayed visually in a(n) _____.

PERCENTS AS FRACTIONS AND DECIMALS

Exercises 9–24: Write the percent as a fraction or mixed number in simplest form.

9. 28%   10. 8%

11. $3\frac{3}{4}\%$ 12. $8\frac{1}{3}\%$

13. 55%   14. 75%

15. 4%   16. 2%

17. 7.5%   18. 16.5%

19. 116%   20. 250%

21. 8.25%   22. 10.4%

23. $33\frac{1}{3}\%$ 24. $9\frac{3}{5}\%$

Exercises 25–40: Write the percent as a decimal.

25. 58%   26. 14%

27. $9\frac{1}{2}\%$ 28. $12\frac{1}{4}\%$

29. 173%   **30.** 206%

31. 6%   **32.** 8%

33. $\frac{1}{4}$%   **34.** $\frac{4}{5}$%

35. 0.3%   **36.** 0.2%

37. 116%   **38.** 250%

39. 8.4%   **40.** 1.3%

FRACTIONS AND DECIMALS AS PERCENTS

Exercises 41–54: Write the fraction or mixed number as a percent.

41. $\frac{7}{10}$ **42.** $\frac{5}{8}$

43. $\frac{9}{40}$ **44.** $\frac{6}{25}$

45. $5\frac{1}{2}$ **46.** $4\frac{2}{5}$

47. $\frac{9}{50}$ **48.** $\frac{3}{16}$

49. $\frac{2}{3}$ **50.** $\frac{4}{15}$

51. $\frac{7}{12}$ **52.** $\frac{8}{9}$

53. $1\frac{1}{4}$ **54.** $2\frac{1}{8}$

Exercises 55–68: Write the decimal as a percent.

55. 0.81   **56.** 0.57

57. 0.01   **58.** 0.03

59. 1.6   **60.** 2.1

61. 0.072   **62.** 0.049

63. 2.99   **64.** 3.43

65. 0.005   **66.** 0.004

67. 0.0401   **68.** 0.0802

Exercises 69 and 70: Complete the given table by finding the two missing values in each row.

69.

Percent	Decimal	Fraction
80%		
		$\frac{3}{10}$
	0.65	
22.5%		
	2.6	
		$\frac{2}{5}$

70.

Percent	Decimal	Fraction
		$\frac{23}{20}$
108%		
	0.625	
		$\frac{5}{16}$
	1.8	
2.5%		

CIRCLE GRAPHS

71. The following circle graph shows how Maine disposed of its municipal solid waste in 2008. (*Source: Maine State Planning Office.*)

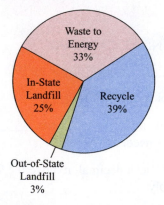

(a) Which disposal method received the largest share of the solid waste?

(b) What fraction of the solid waste went to in-state landfills? Write this fraction in simplest form.

(c) Write the share of solid waste that was converted to energy as a decimal.

72. The following circle graph shows revenue sources for the Oregon Department of Fish and Wildlife from 2007 to 2009. (*Source:* Oregon Department of Fish and Wildlife.)

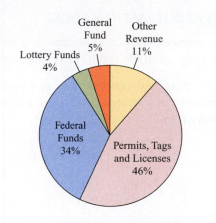

(a) Which source provided the least revenue?

(b) What fraction of the revenue came from the general fund? Write this fraction in simplest form.

(c) Write the share of revenue that came from federal funds as a decimal.

73. The following circle graph shows electricity use for a typical Florida home in 2009. (*Source:* U.S. Department of Energy.)

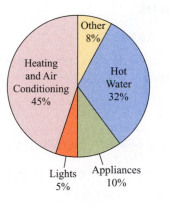

(a) What percent of the electricity was used to power appliances?

(b) What fraction of the electricity was used to heat water? Write this fraction in simplest form.

(c) Write the share of electricity that was used for lighting as a decimal.

74. The following circle graph shows the methods for producing New England's electricity in 2009. (*Source:* New England Wind Fund.)

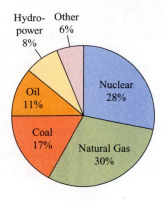

(a) What source accounted for $\frac{3}{10}$ of the electricity in 2009?

(b) What fraction of the electricity came from nuclear power? Write this fraction in simplest form.

(c) Write the share of electricity that was produced from hydropower as a decimal.

APPLICATIONS

75. *Bone Marrow* In 2009, about $\frac{7}{300}$ of the American population was registered with the National Marrow Donor Program. Write this fraction as a percent. (*Source:* National Marrow Donor Program.)

76. *Obesity* A 2009 study found that $\frac{17}{50}$ of the American population was obese. Write this fraction as a percent. (*Source:* National Center for Health Statistics.)

77. *Shopping* A bargain table at an electronics store advertises "60% off the regular price." Write this percent as a fraction in simplest form.

78. *Online Books* An online book seller claims that it charges "45% less" for its books. Write this percent as a fraction in simplest form.

79. *Cell Phones* In 2009, the portion of the Swedish population that had cell phone subscriptions could be written in decimal form as 1.1. Write this decimal as a percent. Interpret your answer. (*Source:* Ny Teknik.)

80. *Web Browsers* In 2009, the portion of the online population that used Firefox to browse the Internet could be written in decimal form as 0.2251. Write this decimal as a percent. (*Source:* Nielsen Online.)

81. *Hybrid Car* A student claims that her hybrid car gets 23.6% better mileage than her old car. Write this percent as a decimal.

82. *Computer Quality* A computer maker claims that only 0.35% of its hard drives fail within the first year. Write this percent as a decimal.

83. *Historic Election* In the 2008 presidential election, Barack Obama received 52% of the popular vote. What fraction of the voting public is this? Write this fraction in simplest form. (*Source:* CNN.)

84. *Body Fat* Some male athletes have body fat levels as low as 6%. Write this percent as a fraction in simplest form. (*Source:* American Council on Exercise.)

85. *Gender* A class with 30 students has 18 women. What percent of the class are women?

86. *Video Games* A child has completed 36 levels of a video game that has 72 levels. What percent of the levels have been completed?

WRITING ABOUT MATHEMATICS

87. If an improper fraction is written as a percent, what can we say about the result? Give an example.

88. In converting between decimals and percents, when do we move the decimal point to the left and when do we move it to the right?

7.2   Using Equations to Solve Percent Problems

Basic Percent Statements • Translating Percent Statements to Equations • Solving Percent Problems

A LOOK INTO MATH ▶

As noted in the chapter opener, percents are used to find a person's target heart rate. Computations involving percents are also found in many other real-world situations.

- A state may have a 6.5% sales tax rate.
- A person may leave a 15% tip at a restaurant.
- The federal income tax rate may jump from 28% to 33%.

To solve percent applications, we must first translate percent statements into equations or proportions. In this section, we focus on translating percent statements to equations.

NEW VOCABULARY

☐ Basic percent statement form
☐ Basic percent equation
☐ Percent problem

STUDY TIP

If you are having difficulty with your studies, you may be able to find help at the student support services office on your campus.

Basic Percent Statements

A percent needs to be placed in context in order to have meaning in real-world situations. For example, 25% does not have meaning without the word "**of**." Phrases such as

$$\text{25\% \textbf{of the students}} \quad \text{or} \quad \text{25\% \textbf{of the price}}$$

make more sense. To solve percent problems, we first write the given information in **basic percent statement form**.

BASIC PERCENT STATEMENT FORM

A percent statement is in *basic percent statement form* when it is written as follows.

A percent of the whole is a part.

For example,   20% of 400 is 80.
                    Percent Whole Part

The following shows how percent statements can be written in basic percent statement form.

Percent Statement	Basic Percent Statement Form
5 is 10% of 50	10% of 50 is 5
30 out of 200 is 15%	15% of 200 is 30

In each statement, the *whole* always follows the word "of," the *percent* is always followed by the % symbol or the word "percent," and the other number in the statement is always the *part*. In the next example, we write percent statements in basic percent statement form.

EXAMPLE 1   **Writing percent statements in basic percent statement form**

Write each percent statement in basic percent statement form.
(a) 8 is 25% of 32.   **(b)** 9 out of 18 is 50%.

Solution

(a) First, write the statement in the following form: A **percent** of a **whole** is a **part**. Note that the number **32** is the *whole* because it follows the word "of," the *percent* is **25**, and the *part* is the remaining number, **8**. The basic percent statement form is

$$25\% \text{ of } 32 \text{ is } 8.$$

(b) The number **18** is the *whole* because it follows the word "of," the *percent* is **50**, and the *part* is **9**. The basic percent statement form is

$$50\% \text{ of } 18 \text{ is } 9.$$

Now Try Exercises 9, 13

Translating Percent Statements to Equations

Recall from Section 4.3 that the word "of" means *multiply* when it follows a fraction. The phrase

$$\frac{1}{4} \text{ of the price} \quad \text{means} \quad \frac{1}{4} \cdot \text{the price}.$$

Since the fraction $\frac{1}{4}$ is equal to 25%, we can also write

$$25\% \text{ of the price} \quad \text{means} \quad 25\% \cdot \text{the price}.$$

We also know that the word "is" means *equals*. Therefore, any statement in basic percent statement form can be translated to a **basic percent equation**.

BASIC PERCENT EQUATION

The statement *a percent of the whole is a part* translates to the equation

$$\text{percent} \cdot \text{whole} = \text{part}.$$

Be sure to write the percent as either a decimal or a fraction.

NOTE: Before a percent equation can be solved for an unknown value, the percent must be written as either a decimal or a fraction.

When translating a percent statement to an equation, we should first write the statement in basic percent statement form, as illustrated in the next example.

EXAMPLE 2 | **Translating percent statements to equations**

Translate each percent statement to a basic percent equation.
(a) 24 is 40% of 60.   (b) 11 out of 44 is 25%.

Solution
(a) First, write the statement in basic percent statement form.

$$\textbf{40\% of 60 is 24.}$$

Using 0.4 for 40%, this statement translates to the equation

$$\mathbf{0.4 \cdot 60 = 24}.$$

(b) In basic percent statement form, we have **25**% of **44** is **11**. Using 0.25 for 25%, this statement translates to the equation $\mathbf{0.25 \cdot 44 = 11}$.

Now Try Exercises 19, 23

Solving Percent Problems

When the percent, whole, or part of a percent statement is replaced by the words "what percent" or "what number," the result is a *percent problem*. A **percent problem** is a question asking us to find the percent, whole, or part in a percent statement. Each of the following questions is an example of a percent problem.

READING CHECK

• What is a percent problem?

- 25% of what number is 50?
- What number is 32% of 90?
- 18 out of 72 is what percent?

Percent problems can be translated to equations by using a variable for the unknown value, as shown in the next example.

EXAMPLE 3 | **Translating percent problems to equations**

Translate each percent problem to a basic percent equation. Do not solve the equation.
(a) What percent of 80 is 6?   (b) What number is 20% of 300?

Solution
(a) The question is already written in basic percent statement (question) form.

$$\textbf{What percent of 80 is 6?}$$

Using x to represent the unknown percent, this question translates to the equation

$$\mathbf{x \cdot 80 = 6},$$

which can also be written as $80x = 6$.

(b) First, we write the question in basic percent statement (question) form.

$$\textbf{20\% of 300 is what number?}$$

Using x for the unknown value and 0.2 for 20%, this question translates to the equation

$$\mathbf{0.2 \cdot 300 = x}.$$

Now Try Exercises 29, 31

The unknown value in a percent problem can be the *percent*, the *whole*, or the *part*. We find the unknown value by solving the corresponding equation. In the next example, we find the value of an unknown *part*.

EXAMPLE 4 **Solving percent problems for an unknown part**

Find each unknown value.
(a) 5% of 80 is what number? **(b)** What number is 16% of 87.5?

Solution
(a) The question is in the following form: **5%** of **80** is **what number**?

$$5\% \cdot 80 = x \quad \text{Translate to an equation.}$$
$$0.05 \cdot 80 = x \quad \text{Write 5\% as 0.05.}$$
$$4 = x \quad \text{Multiply.}$$

(b) Write the question in the following form: **16%** of **87.5** is **what number**?

$$16\% \cdot 87.5 = x \quad \text{Translate to an equation.}$$
$$0.16 \cdot 87.5 = x \quad \text{Write 16\% as 0.16.}$$
$$14 = x \quad \text{Multiply.}$$

Now Try Exercises 37, 41

In the next example, we find the value of an unknown *whole*.

EXAMPLE 5 **Solving percent problems for an unknown whole**

Find each unknown value.

(a) 72% of what number is 90? **(b)** $9\frac{1}{5}$ is 8% of what number?

Solution
(a) The question is in the following form: **72%** of **what number** is **90**?

$$72\% \cdot x = 90 \quad \text{Translate to an equation.}$$
$$0.72x = 90 \quad \text{Write 72\% as 0.72.}$$
$$\frac{0.72x}{0.72} = \frac{90}{0.72} \quad \text{Divide each side by 0.72.}$$
$$x = 125 \quad \text{Simplify.}$$

(b) Write the question in the following form: **8%** of **what number** is $9\frac{1}{5}$?

$$8\% \cdot x = 9\frac{1}{5} \quad \text{Translate to an equation.}$$
$$0.08x = 9.2 \quad \text{Write 8\% as 0.08 and } 9\frac{1}{5} \text{ as 9.2.}$$
$$\frac{0.08x}{0.08} = \frac{9.2}{0.08} \quad \text{Divide each side by 0.08.}$$
$$x = 115 \quad \text{Simplify.}$$

Now Try Exercises 45, 47

When finding an unknown *percent*, we must convert the solution to the corresponding equation from a decimal to a percent, as shown in the next example.

EXAMPLE 6 **Solving percent problems for an unknown percent**

Find each unknown value.
(a) What percent of 84 is 28? **(b)** 7.8 is what percent of 65?
(c) What percent is 52 of 13?

Solution

(a) The question is in the following form: **What percent** of **84** is **28**?

$$x \cdot 84 = 28 \qquad \text{Translate to an equation.}$$

$$\frac{x \cdot 84}{84} = \frac{28}{84} \qquad \text{Divide each side by 84.}$$

$$x = 0.\overline{3} \qquad \text{Simplify.}$$

$$x = 33.\overline{3}\% \qquad \text{Convert } 0.\overline{3} \text{ to } 33.\overline{3}\%.$$

Note that the last step is completed by multiplying $0.\overline{3}$ by 100%.

(b) Write the question in the following form: **What percent** of **65** is **7.8**?

$$x \cdot 65 = 7.8 \qquad \text{Translate to an equation.}$$

$$\frac{x \cdot 65}{65} = \frac{7.8}{65} \qquad \text{Divide each side by 65.}$$

$$x = 0.12 \qquad \text{Simplify.}$$

$$x = 12\% \qquad \text{Convert 0.12 to 12\%.}$$

(c) Write the question in the following form: **What percent** of **13** is **52**?

$$x \cdot 13 = 52 \qquad \text{Translate to an equation.}$$

$$\frac{x \cdot 13}{13} = \frac{52}{13} \qquad \text{Divide each side by 13.}$$

$$x = 4 \qquad \text{Simplify.}$$

$$x = 400\% \qquad \text{Convert 4 to 400\%.}$$

Now Try Exercises 49, 53, 57

7.2   Putting It All Together

CONCEPT	COMMENTS	EXAMPLES
Basic Percent Statement Form	A percent statement is in *basic percent statement form* when it is written as follows. A percent of the whole is a part.	The statement 7 is 10% of 70 can be written as **10%** of **70** is **7**. **Percent Whole Part**
Basic Percent Equation	The statement *a percent of a whole is a part* translates to the equation percent · whole = part.	The statement 25% of 60 is 15 translates to the equation $25\% \cdot 60 = 15.$
Percent Problem	A percent problem is a question asking us to find the percent, whole, or part in a percent statement.	What number is 40% of 90? 30% of what number is 12?
Solving a Percent Problem	A percent problem can be solved by solving the corresponding basic percent equation.	20% of 75 is what number? $20\% \cdot 75 = x$ Write an equation. $0.2 \cdot 75 = x$ Write 20% as 0.2. $15 = x$ Multiply.

7.2 Exercises

MyMathLab | Math XL PRACTICE | WATCH | DOWNLOAD | READ | REVIEW

CONCEPTS AND VOCABULARY

1. A percent statement is in basic percent statement form when it is written as: A percent of a(n) _____ is a(n) _____.

2. When writing a statement in basic percent statement form, the *whole* always follows the word _____.

3. In basic percent statement form, the word "of" means _____, and the word "is" means _____.

4. The statement *a percent of a whole is a part* translates to the equation _____.

5. A percent _____ is a question asking us to find the percent, whole, or part in a percent statement.

6. When finding an unknown *percent*, we must convert the solution to the corresponding equation from a(n) _____ to a percent.

BASIC PERCENT STATEMENTS

Exercises 7–16: Write the given percent statement in basic percent statement form.

7. 15% of 120 is 18.

8. 40% of 50 is 20.

9. 63 is 150% of 42.

10. 55 is 250% of 22.

11. 2.5% is 19 of 760.

12. 12% is 6 of 50.

13. 4 out of 5 is 80%.

14. 7 out of 10 is 70%.

15. 17 of 20 is 85%.

16. 26 of 104 is 25%.

BASIC PERCENT EQUATIONS

Exercises 17–26: Translate the given percent statement to a basic percent equation.

17. 18% of 40 is 7.2.

18. 62% of 130 is 80.6.

19. 49 is 140% of 35.

20. 126 is 210% of 60.

21. 64% is 48 of 75.

22. 67.5% is 54 of 80.

23. 3 out of 24 is 12.5%.

24. 18 out of 20 is 90%.

25. 15 of 24 is 62.5%.

26. 32 of 80 is 40%.

PERCENT PROBLEMS

Exercises 27–36: Translate the given percent problem to a basic percent equation. Do not solve the equation.

27. 68% of what number is 17?

28. $3\frac{1}{2}$% of 8 is what number?

29. What percent of 95 is 39.9?

30. 44 is what percent of 40?

31. What number is 104% of 70?

32. 198 is 99% of what number?

33. 48 out of 60 is what percent?

34. 76 out of what number is 47.5%?

35. What percent is 3 of 9?

36. 27.5 of 75 is what percent?

Exercises 37–60: Find the unknown value.

37. 6% of 50 is what number?

38. $3\frac{2}{5}$ out of what number is 40%?

39. 3.9 is what percent of 11.7?

40. What percent of 65.4 is 327?

41. What number is 48% of 12.5?

42. 7 of 9 is what percent?

43. 7.6% of 9850 is what number?

44. 3 out of 4 is what percent?

45. 105% of what number is 63?

46. What number is 115% of 60?

47. $4\frac{1}{2}$ is 3% of what number?

48. 96% of what number is 72?

49. What percent of 42 is 28?

50. 66 is 30% of what number?

51. 71 is 2% of what number?

52. 600 out of 960 is what percent?

53. 10.5 is what percent of 60?

54. 45 of 50 is what percent?

55. 66 out of what number is 240%?

56. 18 is what percent of 48?

57. What percent is 13 of 65?

58. 18% of 250 is what number?

59. 19 out of what number is 25%?

60. What number is $8\frac{1}{3}\%$ of 84?

WRITING ABOUT MATHEMATICS

61. If we find 140% of a number, will the result be larger or smaller than the given number? Explain.

62. An advertisement for computer memory claims to cut prices up to 119%. Is this possible? Explain.

SECTIONS 7.1 and 7.2 — Checking Basic Concepts

1. Write each percent as a fraction in simplest form.
 (a) 76% **(b)** $6\frac{2}{3}\%$ **(c)** 9.6%

2. Write each percent as a decimal.
 (a) $6\frac{1}{2}\%$ **(b)** 214% **(c)** 2.8%

3. Write each fraction or mixed number as a percent.
 (a) $\dfrac{3}{25}$ **(b)** $3\frac{1}{2}$ **(c)** $\dfrac{7}{8}$

4. Write each decimal as a percent.
 (a) 0.29 **(b)** 0.073 **(c)** 2.97

5. Write each percent statement in basic percent statement form.
 (a) 49 is 350% of 14. **(b)** 11 of 25 is 44%.

6. Translate each percent statement to a basic percent equation.
 (a) 54 is 180% of 30. **(b)** 9 of 36 is 25%.

7. Translate each percent problem to a basic percent equation. Do not solve the equation.
 (a) 64 is 75% of what number?
 (b) What percent is 5 of 25?

8. Find each unknown value.
 (a) 12 out of 60 is what percent?
 (b) $2\frac{1}{4}$ is 5% of what number?
 (c) What number is 109% of 120?

9. *Historic Election* In losing the 2008 presidential election, John McCain received 46% of the popular vote. What fraction of the voting public voted for John McCain? Write this fraction in simplest form. (*Source:* CNN.)

10. *Economic Status* The following circle graph shows how Americans were classified in a study. (*Source: Society in Focus.*)

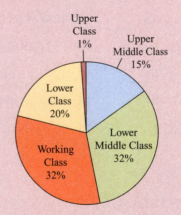

 (a) Which class is smallest?
 (b) What fraction of Americans were classified as lower middle class? Write this fraction in simplest form.
 (c) Write the percent of lower class Americans as a decimal.

7.3   Using Proportions to Solve Percent Problems

Review of Proportions • **Translating Percent Statements to Proportions** •
Solving Percent Problems

A LOOK INTO MATH ▶

When a sale item is marked "50% off," the retail price has been cut by $\frac{1}{2}$. Although many shoppers make the mental connection between 50% and $\frac{1}{2}$ without setting up a mathematical proportion, the fractions associated with "15% off" or "30% off" may be more difficult to find. In this section, we use proportions to solve percent problems.

Review of Proportions

In Section 6.2, we learned that a proportion is a statement that two ratios are equal. For example, since the ratios $\frac{3}{5}$ and $\frac{18}{30}$ are equal, we can write the proportion

$$\frac{3}{5} = \frac{18}{30}.$$

In a proportion, the *cross products* are always equal.

$$3 \cdot 30 \quad \frac{3}{5} = \frac{18}{30} \quad 5 \cdot 18$$

In this case, the cross products $3 \cdot 30$ and $5 \cdot 18$ are both equal to 90.

When a proportion contains a variable for an unknown value, we use the cross products to solve the proportion. For example,

$$\frac{x}{5} = \frac{18}{30} \qquad \text{Given proportion}$$

$$x \cdot 30 = 5 \cdot 18 \qquad \text{Cross products are equal.}$$

$$30x = 90 \qquad \text{Simplify.}$$

$$\frac{30x}{30} = \frac{90}{30} \qquad \text{Divide each side by 30.}$$

$$x = 3 \qquad \text{Simplify.}$$

READING CHECK

• How are cross products used to solve a proportion?

Percent problems can be solved using proportions. To do so, we must first learn how to translate percent statements to proportions.

Translating Percent Statements to Proportions

Consider the basic percent equation

$$\textbf{percent} \cdot \textbf{whole} = \textbf{part},$$

where the words "percent," "whole," and "part" represent numbers. Dividing each side of this equation by the (nonzero) number "whole" gives

$$\frac{\textbf{percent} \cdot \textbf{whole}}{\textbf{whole}} = \frac{\textbf{part}}{\textbf{whole}},$$

which simplifies to

$$\textbf{percent} = \frac{\textbf{part}}{\textbf{whole}}.$$

Since a percent can be written as a ratio with denominator **100**, we get the proportion

$$\frac{\text{percent}}{100} = \frac{\text{part}}{\text{whole}}.$$

NOTE: In the equations on page 457 and above, the **percent** (in blue) is a decimal or fraction, and the **percent** (in purple) is a ratio with a denominator of 100.

TRANSLATING PERCENT STATEMENTS TO PROPORTIONS

The percent statement *a percent of the whole is a part* translates to the proportion

$$\frac{\text{percent}}{100} = \frac{\text{part}}{\text{whole}}.$$

Be sure that the percent is written without the % symbol.

NOTE: The % symbol is not needed because we are dividing the percent by 100.

STUDY TIP

Remember to look back at material that you have already learned. The placements of the *whole*, *percent*, and *part* in a percent statement are the same here as they were in Section 7.2.

Recall that for any percent statement, the *whole* always follows the word "of," the *percent* is always followed by the % symbol or the word "percent," and the other number in the statement is always the *part*. The next example shows how percent statements are translated to proportions.

EXAMPLE 1   **Translating percent statements to proportions**

Translate each percent statement to a proportion.
(a) 27 is 30% of 90.   **(b)** 14 out of 56 is 25%.

Solution
(a) We translate the percent statement to the proportion

$$\frac{\text{percent}}{100} = \frac{\text{part}}{\text{whole}}.$$

The number **90** is the *whole* because it follows the word "of," the *percent* is **30**, and the *part* is **27**. The proportion is

$$\frac{30}{100} = \frac{27}{90}.$$

(b) The number **56** is the *whole*, the *percent* is **25**, and the *part* is **14**. The proportion is

$$\frac{25}{100} = \frac{14}{56}.$$

Now Try Exercises 7, 11

Percent problems can be translated to proportions by using a variable for the unknown value, as shown in the next example.

EXAMPLE 2   **Translating percent problems to proportions**

Translate each percent problem to a proportion. Do not solve the proportion.
(a) 12 out of 30 is what percent?   (b) 60 is 35% of what number?

Solution
(a) The number **30** is the *whole* because it follows the word "of," the unknown *percent* can be represented by **x**, and the *part* is the remaining number, **12**. The proportion is

$$\frac{x}{100} = \frac{12}{30}.$$

(b) The *whole* is unknown. We represent it with **x**. The *percent* is **35**, and the *part* is **60**. The proportion is

$$\frac{35}{100} = \frac{60}{x}.$$

Now Try Exercises 19, 23

READING CHECK

• How do we use proportions to find an unknown value in a percent problem?

Solving Percent Problems

The unknown value in a percent problem can be the *percent*, the *whole*, or the *part*. Its value can be found by solving the corresponding proportion. In the next example, we find the value of an unknown *part*.

EXAMPLE 3   **Solving percent problems for an unknown part**

Find each unknown value.
(a) What number is 32% of 62.5?   (b) 140% of 75 is what number?

Solution
(a) The *whole* is **62.5**, the *percent* is **32**, and the *part* is **unknown**.

$$\frac{32}{100} = \frac{x}{62.5} \qquad \text{Translate to a proportion.}$$

$$32 \cdot 62.5 = 100 \cdot x \qquad \text{Cross products are equal.}$$

$$2000 = 100x \qquad \text{Simplify.}$$

$$\frac{2000}{100} = \frac{100x}{100} \qquad \text{Divide each side by 100.}$$

$$20 = x \qquad \text{Simplify.}$$

(b) The *whole* is **75**, the *percent* is **140**, and the *part* is **unknown**.

$$\frac{140}{100} = \frac{x}{75} \qquad \text{Translate to a proportion.}$$

$$140 \cdot 75 = 100 \cdot x \qquad \text{Cross products are equal.}$$

$$10{,}500 = 100x \qquad \text{Simplify.}$$

$$\frac{10{,}500}{100} = \frac{100x}{100} \qquad \text{Divide each side by 100.}$$

$$105 = x \qquad \text{Simplify.}$$

Now Try Exercises 25, 31

In the next example, we find the value of an unknown *whole*.

EXAMPLE 4 | **Solving percent problems for an unknown whole**

Find each unknown value.

(a) $7\frac{1}{2}$ is 6% of what number?   (b) 85% of what number is 98.6?

Solution

(a) The *whole* is **unknown**, the *percent* is **6**, and the *part* is $7\frac{1}{2}$, which we write as **7.5** because multiplying 7.5 by 100 can be done mentally.

$$\frac{6}{100} = \frac{7.5}{x} \qquad \text{Translate to a proportion.}$$

$$6 \cdot x = 100 \cdot 7.5 \qquad \text{Cross products are equal.}$$

$$6x = 750 \qquad \text{Simplify.}$$

$$\frac{6x}{6} = \frac{750}{6} \qquad \text{Divide each side by 6.}$$

$$x = 125 \qquad \text{Simplify.}$$

(b) The *whole* is **unknown**, the *percent* is **85**, and the *part* is **98.6**.

$$\frac{85}{100} = \frac{98.6}{x} \qquad \text{Translate to a proportion.}$$

$$85 \cdot x = 100 \cdot 98.6 \qquad \text{Cross products are equal.}$$

$$85x = 9860 \qquad \text{Simplify.}$$

$$\frac{85x}{85} = \frac{9860}{85} \qquad \text{Divide each side by 85.}$$

$$x = 116 \qquad \text{Simplify.}$$

▌ **Now Try Exercises 35, 39**

The next example shows how to find an unknown *percent*.

NOTE: When finding an unknown *percent*, the solution to the corresponding proportion is already in percent form. Remember to include the % symbol.

EXAMPLE 5 | **Solving percent problems for an unknown percent**

Find each unknown value.
(a) What percent of 93 is 62?   (b) 12.6 is what percent of 90?

Solution

(a) The *whole* is **93**, the *percent* is **unknown**, and the *part* is **62**.

$$\frac{x}{100} = \frac{62}{93} \qquad \text{Translate to a proportion.}$$

$$x \cdot 93 = 100 \cdot 62 \qquad \text{Cross products are equal.}$$

$$93x = 6200 \qquad \text{Simplify.}$$

$$\frac{93x}{93} = \frac{6200}{93} \qquad \text{Divide each side by 93.}$$

$$x = 66.\overline{6}\% \qquad \text{Simplify and include the % symbol.}$$

(b) The *whole* is **90**, the *percent* is **unknown**, and the *part* is **12.6**.

$$\frac{x}{100} = \frac{12.6}{90} \qquad \text{Translate to a proportion.}$$

$$x \cdot 90 = 100 \cdot 12.6 \qquad \text{Cross products are equal.}$$

$$90x = 1260 \qquad \text{Simplify.}$$

$$\frac{90x}{90} = \frac{1260}{90} \qquad \text{Divide each side by 90.}$$

$$x = 14\% \qquad \text{Simplify and include the \% symbol.}$$

■ **Now Try Exercises 41, 47**

7.3 Putting It All Together

CONCEPT	COMMENTS	EXAMPLES
Proportion	A proportion is a statement that two ratios are equal.	$\dfrac{45}{100} = \dfrac{9}{20}$
Cross Products	In a proportion, the cross products are equal.	$1 \cdot 15 \quad \dfrac{1}{5} = \dfrac{3}{15} \quad 5 \cdot 3$ $1 \cdot 15$ and $3 \cdot 5$ both equal 15.
Translating Percent Statements to Proportions	The percent statement *a percent of the whole is a part* translates to the proportion $$\frac{percent}{100} = \frac{part}{whole}.$$	The percent statement $$50\% \text{ of } 90 \text{ is } 45$$ translates to the proportion $$\frac{50}{100} = \frac{45}{90}.$$
Solving a Percent Problem	A percent problem can be solved by solving the corresponding proportion.	10% of 40 is what number? $\dfrac{10}{100} = \dfrac{x}{40}$ Write a proportion. $10 \cdot 40 = 100 \cdot x$ Equal cross products $400 = 100x$ Simplify. $\dfrac{400}{100} = \dfrac{100x}{100}$ Divide by 100. $4 = x$ Simplify.

7.3   Exercises

MyMathLab

Math XL
PRACTICE   WATCH   DOWNLOAD   READ   REVIEW

CONCEPTS AND VOCABULARY

1. A(n) _____ is a statement that two ratios are equal.

2. In a proportion, the cross products are _____.

3. In a percent statement, the _____ is always followed by the % symbol or the word "percent."

4. The percent statement *a percent of a whole is a part* translates to the proportion _____.

PROPORTIONS

Exercises 5–14: Translate the given percent statement to a proportion.

5. 24% of 70 is 16.8.

6. 4% of 755 is 30.2.

7. 45 is 180% of 25.

8. 138 is 230% of 60.

9. 40% is 34 of 85.

10. 13.5% is 27 of 200.

11. 8 out of 64 is 12.5%.

12. 19 out of 20 is 95%.

13. 25 of 40 is 62.5%.

14. 21 of 30 is 70%.

PERCENT PROBLEMS

Exercises 15–24: Translate the given percent problem to a proportion. Do not solve the proportion.

15. 48% of what number is 19?

16. $5\frac{1}{2}$% of 22 is what number?

17. What percent of 114 is 38.4?

18. 55 is what percent of 50?

19. 24 out of 80 is what percent?

20. What number is 165% of 20?

21. 39 out of what number is 42.5%?

22. What percent is 2 of 8?

23. 297 is 99% of what number?

24. 13.5 of 92 is what percent?

Exercises 25–48: Find the unknown value.

25. What number is 48% of 37.5?

26. $6\frac{4}{5}$ out of what number is 40%?

27. 14.6 is what percent of 21.9?

28. 16% of 25 is what number?

29. What percent of 90.5 is 362?

30. 5 of 6 is what percent?

31. 108% of 25 is what number?

32. 6 out of 10 is what percent?

33. 130% of what number is 52?

34. What number is 192% of 50?

35. $2\frac{1}{4}$ is 9% of what number?

36. 36 out of what number is 45%?

37. 52 is 40% of what number?

38. 63 is 5% of what number?

39. 88% of what number is 30.8?

40. 696 out of 960 is what percent?

41. What percent of 48 is 16?

42. 32 of 50 is what percent?

43. 52 out of what number is 160%?

44. 49 is what percent of 56?

45. What percent is 19 of 95?

46. 26% of 250 is what number?

47. 14.1 is what percent of 60?

48. What number is $5\frac{1}{3}$% of 225?

WRITING ABOUT MATHEMATICS

49. Write your own percent problem and solve it using an equation. Then solve it using a proportion. Which method do you prefer? Explain.

50. In your own words, describe how to find the *percent*, the *whole*, and the *part* in a percent statement.

7.4 Applications: Sales Tax, Discounts, and Net Pay

Basic Percent Applications • **Sales Tax and Total Price** • **Discounts and Sale Price** • **Commission and Net Pay** • **Percent Change**

A LOOK INTO MATH ▶

In the early days of Internet shopping, many online stores were not required to charge sales tax. Although tax laws for Internet sales have changed, one thing remains the same—sales tax rates vary by state and are calculated using percents. In this section, we solve many types of percent applications.

NEW VOCABULARY

☐ Sales tax
☐ Total price
☐ Discount
☐ Sale price
☐ Commission
☐ Net pay
☐ Withholdings
☐ Gross pay
☐ Percent change
☐ Percent increase
☐ Percent decrease

Basic Percent Applications

When solving basic percent applications, we may need to find the *part*, the *whole*, or the *percent*. Each of these types of percent applications is discussed in this section, beginning with finding the *part*.

FINDING A MISSING PART As we learned in Section 7.2, the *whole* in a percent problem follows the word "of," and the *percent* is followed by the % symbol or the word "percent." In the next example, we identify the *whole* and the *percent* so that an equation can be used to find the missing *part*.

EXAMPLE 1

Finding the number of downloaded country music songs

If country music accounts for 32% of all downloads from an online music store, how many country songs are sold on a day with a total of 1425 songs downloaded?

Solution

To use an equation to solve this percent problem, we write the question in basic percent statement form.

$$\underset{\textsf{Percent}}{32\%} \text{ of } \underset{\textsf{Whole}}{1425} \text{ is } \underset{\textsf{Part}}{\textbf{what number}}?$$

$$32\% \cdot 1425 = x \quad \text{Translate to an equation.}$$
$$0.32 \cdot 1425 = x \quad \text{Write 32\% as 0.32.}$$
$$456 = x \quad \text{Multiply.}$$

On a day when 1425 songs are downloaded, we can predict that 456 are country songs.

Now Try Exercise 13

FINDING A MISSING WHOLE In the previous example, we used an equation to find a missing *part*. In the next example, we set up and solve a proportion to find the missing *whole* in a percent application.

Finding a selected portion of the Hawaiian population

A 2008 survey found that 25% of Hawaii's population over the age of 4 spoke a language other than English at home. If 297 thousand Hawaiians fit this description, how many Hawaiians over the age of 4 were there in 2008? (*Source:* U.S. Census Bureau.)

Solution
Begin by writing the given information in the following form:

25% of **what number** is **297** thousand?

The *whole* is **unknown**, the *percent* is **25**, and the *part* is **297** thousand.

$$\frac{25}{100} = \frac{297}{x} \qquad \text{Translate to a proportion.}$$

$$25 \cdot x = 100 \cdot 297 \qquad \text{Cross products are equal.}$$

$$25x = 29{,}700 \qquad \text{Simplify.}$$

$$\frac{25x}{25} = \frac{29{,}700}{25} \qquad \text{Divide each side by 25.}$$

$$x = 1188 \qquad \text{Simplify.}$$

Since the information is given in thousands, the number of Hawaiians over age 4 in 2008 was $1188 \cdot 1000 = 1{,}188{,}000$.

▌**Now Try Exercise 19**

FINDING A MISSING PERCENT In the next example, we will solve a proportion to find the missing *percent* in a percent application.

Finding the percent of teens who recycle

In a survey of 420 teenagers, 35 said that they *never* recycle. What percent of teenagers surveyed did some kind of recycling?

Solution
Since 35 teens stated that they never recycle, the remaining $420 - 35 = 385$ do some kind of recycling. Write this information in the following form:

What percent of **420** is **385**?

The *whole* is **420**, the *percent* is **unknown**, and the *part* is **385**.

$$\frac{x}{100} = \frac{385}{420} \qquad \text{Translate to a proportion.}$$

$$x \cdot 420 = 100 \cdot 385 \qquad \text{Cross products are equal.}$$

$$420x = 38{,}500 \qquad \text{Simplify.}$$

$$\frac{420x}{420} = \frac{38{,}500}{420} \qquad \text{Divide each side by 420.}$$

$$x = 91.\overline{6}\% \qquad \text{Simplify.}$$

Of the teens surveyed, $91.\overline{6}\%$ did some kind of recycling.

▌**Now Try Exercise 25**

Sales Tax and Total Price

READING CHECK

- How does sales tax affect the price paid for an item?

Most states require retailers and service providers to collect sales tax when certain kinds of items or services are sold. **Sales tax** is a percent of the purchase price, which is added to the purchase price. The **total price** paid is the sum of the purchase price and the sales tax.

▶ **REAL-WORLD CONNECTION** Many states have a sales tax rate of **6%**. If the purchase price of an item is $**24.50**, then the sales tax is

$$0.06(24.50) = \$1.47,$$

and the total price of the item is

$$24.50 + 1.47 = \$25.97.$$

Sales tax and total price can be found using the following equations.

SALES TAX AND TOTAL PRICE

If P represents the purchase price of an item and r (written as a decimal) represents the sales tax rate, then the amount of sales tax S is given by

$$S = rP.$$

The total price T that is paid for the item is

$$T = P + S.$$

STUDY TIP

The box above is the first of four formula boxes in this section. Try to avoid simply memorizing the formulas. Instead, study and practice the formulas until you have a firm grasp of the situations that require their use.

NOTE: Taxes and dollar amounts are rounded to the nearest cent.

EXAMPLE 4 **Finding sales tax and total price**

If a cell phone costs $149 and the sales tax rate is 4%, find the sales tax and the total price.

Solution
Written as a decimal, the sales tax rate is $r = $ **0.04**, and the purchase price is $P = $ **149**. The sales tax is $S = $ **0.04**(**149**) $= $ **$5.96** and the total price is **149** $+$ **5.96** $=$ $154.96.

▮ **Now Try Exercise 29**

MAKING CONNECTIONS

Calculating Total Price Directly

In Example 4, the buyer pays the purchase price, which is 100% of $149, and then pays an additional 4% of $149 as sales tax. In other words, the buyer pays 100% + 4%, or 104% of the purchase price. To calculate the total price directly, add the sales tax rate to 100% to get 1.04 and then multiply by the purchase price, 149, to get the total price of $154.96.

$$1.04(149) = \$154.96$$

EXAMPLE 5 **Finding the sales tax rate**

When a student bought a new Nintendo Wii game with a purchase price of $50, the sales tax was $3.75. Find the sales tax rate.

Solution

The purchase price is $P = 50$, and the sales tax is $S = 3.75$. Let r represent the unknown sales tax rate.

$$S = rP \qquad \text{Sales tax equation}$$
$$3.75 = r \cdot 50 \qquad S = 3.75 \text{ and } P = 50.$$
$$\frac{3.75}{50} = \frac{r \cdot 50}{50} \qquad \text{Divide each side by 50.}$$
$$0.075 = r \qquad \text{Simplify.}$$

The sales tax rate is 0.075 or 7.5%.

Now Try Exercise 33

READING CHECK

• How does a discount affect the price paid for an item?

Discounts and Sale Price

To attract customers, retailers offer discounted prices. A **discount** is a percent of the original price that is subtracted from the original price to give the **sale price**.

A discount amount and the sale price can be found using the following equations.

DISCOUNTS AND SALE PRICE

If O represents the *original* price of an item and r (written as a decimal) represents the discount rate, then the discount D is given by

$$D = rO.$$

The sale price is the new purchase price P, where

$$P = O - D.$$

EXAMPLE 6 **Finding the discount and sale price**

If new shoes priced at $120 are on sale for 30% off, find the discount and the sale price.

Solution

Written as a decimal, the discount rate is $r = 0.3$, and the original price is $O = 120$. The discount is $D = 0.3(120) = \$36$, and the sale price is $120 - 36 = \$84$.

Now Try Exercise 37

MAKING CONNECTIONS

Calculating Sale Price Directly

In Example 6, a discount of 30% of $120 is subtracted from the original price, which is 100% of $120. In other words, the buyer pays $100\% - 30\%$, or 70% of the original price. To calculate the sale price directly, subtract the discount rate from 100% to get 0.70 and then multiply by the original price, 120, to get the sale price of $84.

$$0.70(120) = \$84$$

Commission and Net Pay

Workers can be paid in several ways. Some earn hourly wages, while others receive pay based on an annual salary. Some people who work in sales earn a **commission**, which is a percent of total sales. A commission amount can be found as follows.

READING CHECK

• How is a salesperson's commission calculated?

COMMISSION

If T represents the total sales and r (written as a decimal) represents the commission rate, then the commission C is given by

$$C = rT.$$

EXAMPLE 7   **Finding commission**

A real estate broker earns a commission of 6% on total sales. If the broker sells a home for $212,000, find her commission.

Solution
Written as a decimal, the commission rate is $r = \textbf{0.06}$, and total sales are $T = \textbf{212,000}$. The commission is $C = \textbf{0.06}(\textbf{212,000}) = \$12,720$.

Now Try Exercise 45

EXAMPLE 8   **Finding total sales**

A salesperson receives a commission of $876. If his commission rate is 3%, find the total sales.

Solution
Written as a decimal, the commission rate is $r = \textbf{0.03}$, and the commission is $C = \textbf{876}$. Let T represent the unknown sales tax rate.

$$C = rT \qquad \text{Commission equation}$$
$$876 = 0.03T \qquad C = 876 \text{ and } r = 0.03.$$
$$\frac{876}{0.03} = \frac{0.03T}{0.03} \qquad \text{Divide each side by 0.03.}$$
$$29{,}200 = T \qquad \text{Simplify.}$$

Total sales are $29,200.

Now Try Exercise 49

READING CHECK

• Which amount is larger, net pay or gross pay?

Most paycheck amounts are less than hourly wages, salary, or commission would suggest. The final dollar amount on a paycheck is called the *net pay* or *take-home pay*. **Net pay** is the amount remaining after **withholdings** such as taxes, insurance, and other deductions have been subtracted from the **gross pay**. Withholdings are calculated as a percent of gross pay.

NET PAY

If G represents gross pay and r (written as a decimal) represents the total withholding rate, then the total withholdings W are given by

$$W = rG.$$

The net pay N is

$$N = G - W.$$

EXAMPLE 9 **Finding withholdings and net pay**

A factory worker has 26% of her weekly pay withheld for taxes, insurance, and a pension plan. If her weekly gross pay is $860, find her total withholdings and net pay.

Solution

Written as a decimal, the withholding rate is $r = 0.26$, and the gross pay is $G = 860$. The withholdings are $W = 0.26(860) = \$223.60$, and the net pay is $860 - 223.60 = \$636.40$.

Now Try Exercise 53

Percent Change

▶ **REAL-WORLD CONNECTION** When prices increase or decrease, the amount of the increase or decrease is often not as important as the *percent change* in price. For example, the price of a new car may increase from $24,000 to $24,600, while the cost of tuition at a two-year college may increase from $4800 to $5400. Although each amount went up $600, the price of the car increased by 2.5%, and the cost of tuition increased by 12.5%. The increase in tuition is much more dramatic than the increase in the price of the new car.

If an amount changes from an **old amount** to a **new amount**, then the **percent change** is

$$\frac{\text{new amount} - \text{old amount}}{\text{old amount}} \cdot 100.$$

READING CHECK

• How do we know if a percent change represents a percent increase or a percent decrease?

NOTE: We multiply by 100 to change the decimal representation to a percent.

If the new amount is more than the old amount, the percent change is *positive* and is called a **percent increase** because the amount increases. If the new amount is less than the old amount, the percent change is *negative* and is called a **percent decrease** because the amount decreases.

EXAMPLE 10 **Finding percent increase**

A worker's annual salary increases from $40,000 to $41,200. Find the percent increase.

Solution

The **old amount** is **40,000**, and the **new amount** is **41,200**. The percent change is

$$\frac{41{,}200 - 40{,}000}{40{,}000} \cdot 100 = \frac{1200}{40{,}000} \cdot 100 = 0.03 \cdot 100 = 3\%.$$

Since the percent change is positive, we have a 3% increase.

Now Try Exercise 59

EXAMPLE 11 **Finding percent decrease**

A CEO reduces her company's workforce from 250 to 220. Find the percent decrease.

Solution

The **old amount** is **250**, and the **new amount** is **220**. The percent change is

$$\frac{220 - 250}{250} \cdot 100 = \frac{-30}{250} \cdot 100 = -0.12 \cdot 100 = -12\%.$$

Since the percent change is negative, we have a 12% decrease.

Now Try Exercise 63

NOTE: In Example 11, we write that the percent *change* is -12%, or the percent *decrease* is 12%. It is *not* correct to say that the percent decrease is -12%.

7.4   Putting It All Together

STUDY TIP

Study and practice the formulas in this section until you have a firm grasp of the situations that require their use. It is not necessary to memorize every formula.

CONCEPT	COMMENTS	EXAMPLES
Basic Percent Applications	When solving a basic percent application, we may need to find a missing *part*, *whole*, or *percent*.	If 8% of 150 students surveyed say that they like tofu, then $$0.08(150) = 12$$ students like tofu.
Sales Tax and Total Price	If S is sales tax, P is purchase price, and r (written as a decimal) is the sales tax rate, then $$S = rP.$$ If T is total price, then $$T = P + S.$$	If there is a 5% sales tax rate, then an $800 canoe has $$S = 0.05(800) = \$40$$ in sales tax, and the total price is $$T = 800 + 40 = \$840.$$
Discounts and Sale Price	If D is a discount, O is the original price, and r (written as a decimal) is the discount rate, then $$D = rO.$$ If P is purchase price, then $$P = O - D.$$	If a $200 camera is marked 25% off, then the discount is $$D = 0.25(200) = \$50,$$ and the purchase price is $$P = 200 - 50 = \$150.$$
Commission	If C is commission, T is total sales, and r (written as a decimal) is the commission rate, then $$C = rT.$$	If a person has a 4% commission rate on $7000 in total sales, then the commission is $$C = 0.04(7000) = \$280.$$
Net Pay	If W is withholdings, G is gross pay, and r (written as a decimal) is the withholding rate, then $$W = rG.$$ If N is net pay, then $$N = G - W.$$	If 18% of a worker's $1200 gross pay is withheld, then withholdings are $$W = 0.18(1200) = \$216,$$ and the net pay is $$N = 1200 - 216 = \$984.$$
Percent Change	If an amount changes from an **old amount** to a **new amount**, then the percent change is $$\frac{\textbf{new amount} - \textbf{old amount}}{\textbf{old amount}} \cdot 100.$$	When a price increases from $40 to $50, the percent change is $$\frac{50 - 40}{40} \cdot 100 = 25\%.$$ When a price decreases from $50 to $40, the percent change is $$\frac{40 - 50}{50} \cdot 100 = -20\%.$$

7.4 Exercises

CONCEPTS AND VOCABULARY

1. When solving basic percent applications, we may be asked to find the *part*, the *whole*, or the _____.

2. A percent of the purchase price that is added to the purchase price is called _____ tax.

3. The _____ is the sum of the purchase price and the sales tax.

4. A percent of the purchase price that is subtracted from the purchase price is called a(n) _____.

5. The _____ is the result of subtracting the discount from the purchase price.

6. When a person's pay is a percent of the total sales, then the person is earning a(n) _____.

7. The amount of pay before taxes, insurance, and other deductions are subtracted is called _____.

8. Taxes, insurance, and other deductions that are subtracted from gross pay are called _____.

9. A worker's _____ is the amount remaining after withholdings are subtracted from gross pay.

10. A positive percent change means that we have a percent _____, whereas a negative percent change means that we have a percent _____.

PERCENT APPLICATIONS

11. *Cash Back* A credit card company offers a 1.5% annual cash back rebate on all purchases. If a person uses this credit card for $18,600 in purchases, how much is the rebate?

12. *Nutrition* If a daily diet should include 35 grams of dietary fiber, how many grams of dietary fiber are in a bowl of whole grain cereal with 6% of the recommended amount?

13. *Internet Use* The Internet was used by 30% of Colombia's population in 2008. If 2400 Colombians were surveyed, how many in the survey were using the Internet in 2008? (*Source:* Business News Americas.)

14. *iPhone Apps* In 2009, the game Tap Tap Revenge was installed on 32% of all iPhones. If 400 iPhone users were surveyed, how many of these users had Tap Tap Revenge installed? (*Source:* comScore.)

15. *Smoking Verdict* In 2009, a jury awarded $5.3 million to a man whose wife died from lung cancer after smoking two packs of cigarettes every day for more than 50 years. However, the tobacco company Phillip Morris USA was found to be only 36.5% responsible. How much of the total award was the tobacco company required to pay? (*Source:* CNN.)

16. *Cheating* According to a recent survey of 12,000 high school students, 74% admitted to cheating on an exam at some point during the past year. How many of the students surveyed admitted to cheating? (*Source:* Josephson Institute of Ethics.)

17. *The Human Body* An adult male's body is about 60% water. If a man's body has 111 pounds of water, what is his total body weight?

18. *Cat Owners* A recent survey found that 34% of U.S. households (or 38.42 million) own at least one cat. Find the total number of U.S. households at the time of the survey. (*Source:* American Pet Products Manufacturers Association.)

19. *Steroid Use* In a 2009 survey, 1.5% of high school seniors admitted to using steroids within the last year. If 300 high school seniors admitted to steroid use, how many seniors were surveyed? (*Source:* National Institute on Drug Abuse.)

20. *Smoking Doctors* About 23% of Chinese doctors recently surveyed were smokers. If 817 doctors were smokers, how many doctors were surveyed? Round to the nearest whole number. (*Source:* University of California–Los Angeles.)

21. *Salary* A worker's salary increases by 3.5%. If the raise amounts to $2450, how much was the worker being paid before the raise?

22. *Enrollment* The spring semester's enrollment at a small college was 138 students more than the fall enrollment. If this represents an increase of 12%, how many students were there in the fall?

23. *New Year's Resolutions* On New Year's Eve, 200 people made New Year's resolutions. By the end of January, 14 people had kept their resolutions. What percent of the people kept their resolutions?

24. *Appalachian Trail* In a survey of 8000 hikers, 224 had hiked on the Appalachian Trail. What percent of the hikers surveyed had hiked on the Appalachian Trail?

25. *Saving for College* In 2008, the parents of 800 college freshmen were asked how much they had saved for their child's college education. The parents of 520 freshmen had saved less than $5000. What percent of parents saved less than $5000? (*Source:* College Savings Foundation.)

26. *Fuel Economy* Of 1540 cars and trucks in a parking lot, only 77 had mileage ratings of more than 30 miles per gallon. What percent of the vehicles had mileage ratings of more than 30 miles per gallon?

27. *Graduate School* At a small private college, 124 of 465 seniors intend to go to graduate school. What percent is this?

28. *State Names* The states with names that begin with a vowel are Alabama, Alaska, Arizona, Arkansas, Idaho, Illinois, Indiana, Iowa, Ohio, Oklahoma, Oregon, and Utah. What percent of the 50 U.S. states have names that begin with a vowel?

29. *Digital Music* If the purchase price of a portable MP3 player is $64 and the sales tax rate is 6.5%, find the sales tax and the total price.

30. *Online Shopping* A hammock company offers free shipping on orders over $150. If a hanging hammock chair sells for $199 and the sales tax rate is 7%, find the sales tax and the total price.

31. *Dining Out* The bill for two diners at an upscale restaurant comes to $148.80. If the restaurant is in a state with a 7.5% sales tax rate, find the sales tax and the total cost (without tip).

32. *iPod Application* A business application for an iPod Touch has a purchase price of $9.75. If the sales tax rate is 4%, find the sales tax and the total cost of the application.

33. *HD Video Recorder* A shopper paid $12.75 sales tax on an HD video recorder with a purchase price of $318.75. Find the sales tax rate.

34. *School Laptop* A student bought a new laptop computer with a purchase price of $980.50. If the sales tax was $58.83, find the sales tax rate.

35. *Wireless Router* A homeowner buys a wireless router for a home computer network and pays $10.40 in sales tax. If the sales tax rate is 6.5%, find the purchase price of the router.

36. *Hotel Rooms* In some states, higher tax rates are charged for hotel rooms. If a traveler pays $35.70 in taxes for a one-night stay and the tax rate is 17%, find the purchase price for one night at the hotel.

37. *Golf Clubs* A set of golf clubs is marked at 25% off. If the regular price is $560, find the discount and the sale price.

38. *Online Shopping* A toy company offers free shipping on orders over $100. If an electric train that sells for $189 is marked at 15% off, find the discount and the sale price.

39. *Clearance Sale* A lamp on a clearance table is marked at 70% off. If the regular price is $76.50, find the discount and the sale price.

40. *Swing Set* A backyard swing set is marked at 75% off. If the regular price is $858.80, find the discount and the sale price.

41. *Prepaid Minutes* A cellular customer receives a $33.75 discount when she buys 1500 prepaid minutes. If the regular price for 1500 minutes is $187.50, what is the discount rate?

42. *Airline Tickets* An airline offers a $50 discount on tickets regularly priced at over $375. If a customer buys a ticket with a $400 regular price, what is the discount rate?

43. *School Supplies* A student got a $10.50 discount on a new backpack. If the discount rate is 30%, find the original price of the backpack.

44. *Private School* A parent gets a $225 discount for prepaying his child's entire annual tuition at a private elementary school. If this represents a 5% discount, find the original cost of annual tuition.

45. *Real Estate* A realtor earns a commission of 3% on total sales. If the realtor sells a home for $380,000, find the commission.

46. *Interior Design* An interior designer receives 8% commission on all sales related to a design project. If a corporate design project has total sales of $198,500, find the commission.

47. *Electronics* If a home electronics salesperson earns 5% commission on total sales, find the commission for a month with $87,400 in total sales.

48. *Home Delivery* A frozen-food company pays its sales representatives 11% commission on all home delivery sales. Find the commission for $3200 in home delivery sales.

49. *Telemarketing* A telemarketing company pays its employees 9% commission. Find total sales for an employee who gets $783 in commission.

50. *Home Furnishings* A person selling home furnishings makes $2370 in commission. If the commission rate is 7.5%, find the total sales.

51. *Livestock Feed* If a livestock feed salesperson earns a $990 commission for $16,500 in total sales, find the commission rate.

52. *Building Supplies* If a building supplies salesperson earns a $1340 commission for $26,800 in total sales, find the commission rate.

53. *Net Pay* A worker has 24% of her weekly pay withheld for taxes, insurance, and a pension plan. If her weekly gross pay is $740, find her total withholdings and net pay.

54. *Net Pay* Find the total withholdings and net pay for a worker who has 18% of his $520 weekly pay withheld for taxes and insurance.

55. *Withholdings* What percent of gross pay is being withheld for a worker who has $320 withheld from $2560 in gross pay?

56. *Withholdings* A worker has $472 withheld from $2950 in gross pay. What percent of gross pay is being withheld?

57. *Gross Pay* A worker has $858 withheld from her pay every two weeks. If this represents 22% of her gross pay, what is her gross pay?

58. *Gross Pay* A worker has 28% of his salary withheld each month. Find his gross pay if he has $1820 withheld each month.

59. *Soda Prices* The cost of a 16-ounce soda from a vending machine increases from $1.25 to $1.50. Find the percent increase.

60. *Home Prices* If the average price of a 3-bedroom home decreases from $220,000 to $187,000, find the percent decrease.

61. *Stock Prices* From February to May 2008, the price of one share of Apple Corporation stock rose from $119.46 to $188.75. To the nearest percent, find the percent increase. (*Source:* NASDAQ.)

62. *Salary* If a worker's annual salary increases from $64,200 to $67,410, what is the percent increase in the worker's salary?

63. *Workforce* A company reduces its global workforce from 18,500 employees to 16,280 employees. Find the percent decrease.

64. *Gas Prices* From July to December 2008, the average price of regular unleaded gasoline in the U.S. dropped from $4.10 per gallon to $1.64 per gallon. Find the percent decrease. (*Source:* Energy Information Administration.)

WRITING ABOUT MATHEMATICS

65. What happens to a price that increases by 100%? Is it possible for a price to decrease by 100%? If so, explain what happens to the price.

66. If the price of an item is discounted by x%, explain how to find the sale price without first finding the discount amount.

Checking Basic Concepts

1. Translate each percent problem to a proportion. Do not solve the proportion.
   (a) 64% of what number is 144?
   (b) 80 is what percent of 60?
   (c) What number is 99% of 70?

2. Find each unknown value.
   (a) What percent of 18.5 is 37?
   (b) 12 out of what number is 80%?
   (c) $2\frac{3}{4}$ is 11% of what number?
   (d) 1 of 6 is what percent?

3. *Nutrition* If a woman's daily diet should include 75 milligrams of vitamin C, how many grams of vitamin C are in a cup of yogurt with 20% of the recommended amount for women?

4. *Enrollment* The spring semester's enrollment at a college was 722 students more than the fall enrollment. If this represents an increase of 19%, how many students were there in the fall?

5. *Online Shopping* If accounting software sells for $130 and the sales tax rate is 5.5%, find the sales tax and the total price.

6. *Clearance Sale* A book on a clearance table is marked at 80% off. If the regular price is $15.20, find the discount and the sale price.

7. *Home Theater Sales* If a home theater salesperson earns 8% commission on total sales, what is her commission for a month with $63,900 in total sales?

8. *Hybrid Car Price* If the price of a hybrid car decreases from $24,500 to $24,010, find the percent decrease.

7.5   Applications: Simple and Compound Interest

Key Definitions • Simple Interest • Compound Interest

A LOOK INTO MATH ▶

Most people cannot pay cash for a house, car, or other expensive item. When people borrow money, they *pay interest* for the use of the money over the loan period. Similarly, people can *earn interest* when they invest money. Interest rates are usually a percent of the loan or investment amount. In this section, we will discuss simple and compound interest.

NEW VOCABULARY

☐ Principal
☐ Interest
☐ Interest rate
☐ Simple interest
☐ Annual interest rate
☐ Total value
☐ Compound interest
☐ Annual percentage rate (APR)

Key Definitions

An amount of money that is borrowed or invested is called the **principal**. Lenders or investors are paid a fee for use of the principal. The fee, which is a percent of the principal, is called **interest**. The percent used to calculate the fee is called the **interest rate**. These three terms are summarized as follows.

PRINCIPAL, INTEREST, AND INTEREST RATE

Principal: The initial amount of an investment or loan

Interest: A fee paid to the lender or investor for use of the principal

Interest Rate: The percent used to calculate interest

STUDY TIP

If you are unfamiliar with the new vocabulary in this section, spend extra time learning the meaning of new words. Try writing the definitions in your own words or discussing their meanings with a classmate.

Simple Interest

Interest that is based only on the original principal is called **simple interest**. Such interest is paid at the end of a loan period, which is written in *years*. Because the interest rate is *per year*, we say that it is an **annual interest rate**.

▶ **REAL-WORLD CONNECTION** If a person borrows $100 from a friend and agrees to pay 4% simple interest, then the borrower pays $100 \cdot 0.04 = \$4$ in interest after one year. If the loan period is changed to six months, then the borrower pays only one-half of the $4 interest, or $100 \cdot 0.04 \cdot \frac{1}{2} = \$2$. Similarly, if the loan period is changed to two years, the borrower pays twice the $4 interest, or $100 \cdot 0.04 \cdot 2 = \$8$. This discussion suggests that the amount of interest is found by multiplying the principal, interest rate, and time.

READING CHECK

- Describe in words how simple interest is found.

SIMPLE INTEREST

The total amount of simple interest I is given by

$$I = Prt,$$

where

$$P = \text{Principal},$$
$$r = \text{annual interest rate (written as a decimal), and}$$
$$t = \text{time (in years)}.$$

EXAMPLE 1 **Finding simple interest**

Find the simple interest for the given values of P, r, and t.
(a) $P = \$120$, $r = 6\%$, and $t = 1$ year   (b) $P = \$1600$, $r = 4.5\%$, and $t = 9$ months
(c) $P = \$9520$, $r = 3\%$, and $t = 5$ years

Solution
(a) Substitute the values $P = \mathbf{120}$, $r = \mathbf{0.06}$, and $t = \mathbf{1}$ into the simple interest formula.

$$I = \mathbf{Prt} = \mathbf{120} \cdot \mathbf{0.06} \cdot \mathbf{1} = \$7.20$$

(b) Since 9 months is $\frac{9}{12} = \frac{3}{4}$ year, we have $P = \mathbf{1600}$, $r = \mathbf{0.045}$, and $t = \frac{3}{4}$.

$$I = \mathbf{Prt} = \mathbf{1600} \cdot \mathbf{0.045} \cdot \frac{3}{4} = \$54$$

(c) $P = \mathbf{9520}$, $r = \mathbf{0.03}$, and $t = \mathbf{5}$, so

$$I = \mathbf{Prt} = \mathbf{9520} \cdot \mathbf{0.03} \cdot \mathbf{5} = \$1428.$$

CALCULATOR HELP

To use a calculator to find simple interest, see the Appendix (page AP-6).

Now Try Exercises 11, 13, 15

EXAMPLE 2 **Finding simple interest**

If a student borrows $3400 at 9% simple interest for 6 months to pay for tuition, how much must the student pay in interest?

Solution

The amount of the loan is $P = 3400$, the interest rate is $r = 0.09$, and the time is 6 months, or $t = \frac{6}{12} = \frac{1}{2}$ year.

$$I = Prt = 3400 \cdot 0.09 \cdot \tfrac{1}{2} = \$153$$

▮ Now Try Exercise 19

At the end of a loan period, a borrower repays the original principal and also pays the interest. Similarly, an investor receives both the original principal and the interest earned by the investment. In either case, the **total value** of the loan or investment is the sum of the principal and the interest.

$$\text{Total Value} = P + I$$

EXAMPLE 3 ▶ **Finding the total value of an investment**

If a person invests \$1250 in an account that pays 5% simple interest, find the total value of the investment after 18 months.

Solution

First, find the amount of interest. The amount of the investment is $P = 1250$, the interest rate is $r = 0.05$, and the time is 18 months, or $t = \frac{18}{12} = \frac{3}{2}$ year.

$$I = Prt = 1250 \cdot 0.05 \cdot \tfrac{3}{2} = \$93.75$$

The total value of the investment is the sum of the principal and the interest, or

$$1250 + 93.75 = \$1343.75.$$

▮ Now Try Exercise 23

If the amount of simple interest is known, then we can find an unknown principal, interest rate, or time, as shown in the next example.

EXAMPLE 4 ▶ **Finding an unknown amount of time**

A person pays \$216 in interest when borrowing \$2400 at 4.5% simple interest. Find the length of time for the loan (the loan period).

Solution

The interest is $I = 216$, the principal is $P = 2400$, and the interest rate is $r = 0.045$. Replace these variables in the equation $I = Prt$ and solve for t.

$I = Prt$	Simple interest formula
$216 = 2400 \cdot 0.045 \cdot t$	$I = 216, P = 2400, r = 0.045$
$216 = 108t$	Simplify.
$\dfrac{216}{108} = \dfrac{108t}{108}$	Divide each side by 108.
$2 = t$	Simplify.

The loan period or time of the loan is 2 years.

▮ Now Try Exercise 29

Compound Interest

While simple interest is paid or earned at the end of the *loan period*, **compound interest** is paid or earned at the end of each *compounding period* when it is added to the principal. The resulting sum becomes the principal for the next compounding period. The interest rate for compound interest is usually given as an **annual percentage rate** or **APR**.

Table 7.1 lists the number of times that interest is compounded (added to the principal) each year for common compounding periods.

TABLE 7.1 Compounding Periods

Compounding	Number of Times Compounded per Year
Annually	1
Semiannually	2
Quarterly	4
Monthly	12
Daily	365

Suppose that $1000 is invested at 5% APR compounded *annually* for 4 years. At the end of each compounding period (each year), interest will be added to the principal to result in a new principal amount. Since the interest is compounded annually, the simple interest formula can be used to calculate the interest each compounding period, as shown in Table 7.2.

TABLE 7.2 $1000 Invested at 5% Compounded Annually for 4 Years

Year	Beginning Principal	Interest $I = Prt$	Ending Amount Becomes New Principal for Next Year
1	$1000.00	$1000.00 \cdot 0.05 \cdot 1 = 50.00$	$1000.00 + 50.00 = $**$1050.00$**
2	**$1050.00**	$1050.00 \cdot 0.05 \cdot 1 = 52.50$	$1050.00 + 52.50 = $**$1102.50$**
3	**$1102.50**	$1102.50 \cdot 0.05 \cdot 1 \approx 55.13$	$1102.50 + 55.13 = $**$1157.63$**
4	**$1157.63**	$1157.63 \cdot 0.05 \cdot 1 \approx 57.88$	$1157.63 + 57.88 = $**$1215.51$**

NOTE: In Table 7.2, the total interest after 4 years is **1215.51** − 1000 = $215.51. However, simple interest for the same situation, $1000 invested at 5% for 4 years, is only $I = 1000 \cdot 0.05 \cdot 4 = 200. In general, compound interest earns more than simple interest because it *earns interest on previously earned interest*.

READING CHECK

• How does compound interest differ from simple interest?

The process used to find the final amount in Table 7.2 is not practical for most compound interest problems. For example, compounding interest monthly for 6 years would require a table with 72 rows of computation! For such problems, the following compound interest formula is used to find the final amount directly.

COMPOUND INTEREST

The final amount A in an account paying compound interest is given by

$$A = P\left(1 + \frac{r}{n}\right)^{nt},$$

where

$$P = \text{Principal,}$$
$$r = \text{annual interest rate (written as a decimal),}$$
$$t = \text{time (in years), and}$$
$$n = \text{number of compounding periods per year.}$$

NOTE: For compound interest, the final amount A is the *total value* of the account. There is no need to add the principal to this amount as we did for simple interest.

MAKING CONNECTIONS

Simple Interest and Compound Interest

Whether we are working with simple interest or compound interest, we may want to find either the *total value* (final amount) of the loan or investment, or just the *interest* that is earned or paid. To do this, we use the following formulas.

Simple Interest	**Compound Interest**
$I = Prt$	$A = P\left(1 + \dfrac{r}{n}\right)^{nt}$
Total Value $= P + I$	Interest $= A - P$

In the next example, we use the compound interest formula to find the final amount that was computed earlier in Table 7.2.

EXAMPLE 5 **Finding a total amount after compounding interest**

If \$1000 is invested in an account that pays 5% APR compounded annually, find the amount in the account after 4 years.

Solution

The principal is $P = \mathbf{1000}$, the interest rate is $r = \mathbf{0.05}$, and the time is $t = \mathbf{4}$. Because the interest is compounded annually (one time per year), $n = \mathbf{1}$.

CALCULATOR HELP

To use a calculator to find compound interest, see the Appendix (page AP-6).

$$A = P\left(1 + \frac{r}{n}\right)^{nt} \qquad \text{Compound interest formula}$$

$$A = 1000\left(1 + \frac{0.05}{1}\right)^{1 \cdot 4} \qquad P = 1000, r = 0.05, t = 4, n = 1$$

$$A = 1000(1 + 0.05)^4 \qquad \text{Divide 0.05 by 1, and multiply 1 and 4.}$$

$$A = 1000(1.05)^4 \qquad \text{Add 1 and 0.05.}$$

$$A \approx 1215.51 \qquad \text{Evaluate the exponent and then multiply.}$$

The total amount in the account after 4 years is \$1215.51.

Now Try Exercise 35

EXAMPLE 6

Finding a total amount after compounding interest

If $2500 is invested in an account that pays 6% APR compounded quarterly, find the amount in the account after 2 years.

Solution

The principal is $P = 2500$, the interest rate is $r = 0.06$, and the time is $t = 2$. Because the interest is compounded quarterly (four times per year), $n = 4$.

$$A = P\left(1 + \frac{r}{n}\right)^{nt} \qquad \text{Compound interest formula}$$

$$A = 2500\left(1 + \frac{0.06}{4}\right)^{4 \cdot 2} \qquad P = 2500, r = 0.06, t = 2, n = 4$$

$$A = 2500(1 + 0.015)^8 \qquad \text{Divide 0.06 by 4 and multiply 4 and 2.}$$

$$A = 2500(1.015)^8 \qquad \text{Add 1 and 0.015.}$$

$$A \approx 2816.23 \qquad \text{Evaluate the exponent and then multiply.}$$

The total amount in the account after 2 years is $2816.23.

Now Try Exercise 37

NOTE: Before using a calculator to evaluate the compound interest formula, be sure that you first find the value of the exponent nt by multiplying n and t.

7.5 Putting It All Together

CONCEPT	COMMENTS	EXAMPLES
Principal, Interest, and Interest Rate	The *principal* is the initial amount of an investment or loan, *interest* is a fee paid to the lender or investor for use of the principal, and the *interest rate* is the percent used to calculate interest.	If $800 is invested in an account that pays 5% simple interest for 1 year, then the principal is $800, the interest is $800 \cdot 0.05 \cdot 1 = \$40$, and the interest rate is 5%.
Simple Interest	The amount of simple interest I is given by $$I = Prt,$$ where P = principal, r = annual interest rate (written as a decimal), and t = time (in years).	If $200 is borrowed at 4% simple interest for 3 years, then the interest is $$I = 200 \cdot 0.04 \cdot 3 = \$24.$$
Compound Interest	The amount A in an account paying compound interest is given by $$A = P\left(1 + \frac{r}{n}\right)^{nt},$$ where P = principal, r = annual interest rate (written as a decimal), t = time (in years), and n = the number of compounding periods per year.	If $300 is invested in an account that pays 2% interest compounded monthly, then the amount in the account after 5 years is $$A = 300\left(1 + \frac{0.02}{12}\right)^{12 \cdot 5}$$ $$\approx \$331.52.$$

7.5   Exercises

MyMathLab Math XL PRACTICE WATCH DOWNLOAD READ REVIEW

CONCEPTS AND VOCABULARY

1. An amount of money that is borrowed or invested is called _____.

2. The fee that lenders or investors are paid for use of the principal is called _____.

3. The percent used to calculate the interest on a loan or investment is called the _____.

4. Interest that is based only on the original principal is called _____ interest.

5. An interest rate that is *per year* is called a(n) _____ interest rate.

6. When using the formula $I = Prt$, the interest rate must be written as a(n) _____, and the time must be given in _____.

7. The total value of a loan or investment is the sum of the _____ and _____.

8. Interest that is paid or earned at the end of each compounding period is called _____ interest.

9. The interest rate for compound interest is usually given as a(n) _____, or APR.

10. When using the formula $A = P(1 + \frac{r}{n})^{nt}$, the interest rate must be written as a(n) _____, and the time must be given in _____.

SIMPLE INTEREST

Exercises 11–16: Find the simple interest for the given values of P, r, and t.

11. $P = \$400$, $r = 8\%$, and $t = 1$ year

12. $P = \$600$, $r = 5\%$, and $t = 1$ year

13. $P = \$1200$, $r = 3.5\%$, and $t = 6$ months

14. $P = \$840$, $r = 6.5\%$, and $t = 3$ months

15. $P = \$3250$, $r = 2\%$, and $t = 8$ years

16. $P = \$4750$, $r = 7\%$, and $t = 9$ years

Exercises 17–22: Find the simple interest.

17. A mechanic borrows $2500 at 7% simple interest for 9 months to pay for tools.

18. A homeowner borrows $15,000 at 6.5% simple interest for 2 years to pay for a home theater.

19. A student lends $300 to a friend for 1 month at 4% simple interest.

20. A truck driver borrows $18,000 at 4.5% simple interest for 15 months to pay for new equipment.

21. A student borrows $40,000 from a wealthy aunt for 10 years at 3.5% simple interest.

22. A businessman lends $100,000 to a start-up company for 4 years at 7.5% simple interest.

Exercises 23–28: For the given initial investment, simple interest rate, and time of investment, find the total value of the investment.

23. $640 at 8% for 21 months

24. $12,000 at 3% for 4 months

25. $1600 at 5% for 9 years

26. $80,000 at 2.5% for 8 years

27. $900 at 1.5% for 20 months

28. $4000 at 6.5% for 11 years

Exercises 29–34: Find the unknown value.

29. $504 interest is paid when $3600 is borrowed at 3.5% simple interest. Find the time for the loan.

30. When money is invested at 8% simple interest for 3 years, the interest is $528. Find the principal.

31. $150 simple interest is paid when $5000 is borrowed for 6 months. Find the simple interest rate.

32. $90 interest is paid when $4800 is borrowed at 2.5% simple interest. Find the time for the loan.

33. When money is invested at 7.5% simple interest for 4 months, the interest is $175. Find the principal.

34. $162 simple interest is paid when $1800 is borrowed for 6 years. Find the simple interest rate.

COMPOUND INTEREST

Exercises 35–44: Find the final amount.

35. $1650 invested at 4% APR compounded annually for 3 years

36. $22,000 invested at 5% APR compounded annually for 4 years

37. $1000 invested at 3% APR compounded quarterly for 10 years

38. $8000 invested at 2.5% APR compounded quarterly for 2 years

39. $200 invested at 6.5% APR compounded semiannually for 5 years

40. $12,000 invested at 6% APR compounded semiannually for 15 years

41. $4000 invested at 7% APR compounded daily for 6 years

42. $2500 invested at 2% APR compounded daily for 13 years

43. $1500 invested at 8% APR compounded monthly for 54 years

44. $10,000 invested at 4.5% APR compounded monthly for 9 years

WRITING ABOUT MATHEMATICS

45. Would you rather invest money in an account that pays 5% APR compounded daily or one that pays 5% APR compounded quarterly? Explain.

46. When $100 is invested in an account that pays 6% APR compounded quarterly for 5 years, a student determined that the final amount in the account is

$$A = 100\left(1 + \frac{6}{4}\right)^{4 \cdot 5} \approx \$9{,}094{,}947{,}018.$$

How do you know that this is not correct? Find the error in the calculation.

Group Activity   Working with Real Data

Directions: Form a group of 2 to 4 people. Select someone to record the group's responses for this activity. All members of the group should work cooperatively to answer the questions. If your instructor asks for your results, each member of the group should be prepared to respond.

Federal Debt In 2009, the U.S. public held $7.4 trillion in federal debt. At the same time, 30-year treasury bonds were paying 4.5% interest.
(a) The U.S. population was about 304 million in 2009. How much debt was held by each U.S. citizen in 2009?
(b) If interest was compounded daily and the government intended to pay back the entire amount, including interest, at the end of 30 years, how much would be paid back?

(c) If the U.S. population were to grow to 400 million by 2039, when the debt was to be paid, how much would the government owe each U.S. citizen in 2039?
(d) Suppose that the interest rate was 5.5%, rather than 4.5%, in 2009. How much would the federal government owe after 30 years?

(e) Rework part (c) if the interest rate was 5.5% as computed in part (d).

7.6   Probability and Percent Chance

Key Definitions • Probability and Percent Chance • The Complement of an Event

A LOOK INTO MATH ▶ Weather forecasters often report the *probability* of rain or snow as a percent chance. For example, on a day with a 90% chance of rain, it is quite *likely* that it will rain. Other probabilities, such as those used to describe the likelihood of winning games of chance, can also be expressed as percents. In this section, we will discuss probability and percent chance.

NEW VOCABULARY

☐ Experiment
☐ Outcome
☐ Event
☐ Probability
☐ Percent chance
☐ Impossible event
☐ Certain event
☐ Complement

Key Definitions

We often think of the word *experiment* as it relates to science. In mathematics, however, an **experiment** is an activity with an observable result. Examples of experiments include the following.

- Tossing a coin and observing whether heads or tails results
- Observing the weather and noting any precipitation
- Rolling a six-sided die and observing whether the result is odd or even

A result of an experiment is called an **outcome** of the experiment. Any outcome or group of outcomes for an experiment is called an **event**. Examples of events that are related to the experiments listed above include the following.

- A result of tails when a coin is tossed
- No precipitation on a given day
- A result of 1, 3, or 5 when a six-sided die is rolled

READING CHECK

- What is the difference between an outcome and an event?

STUDY TIP

If the vocabulary for this section is new to you, try to come up with your own examples for each new word. By listing several examples of your own, the words will have more meaning for you.

EXAMPLE 1   **Listing all possible outcomes for an experiment**

When a four-sided die such as the one shown in Figure 7.5 is rolled, the result is the number that shows at the bottom of each of the upright sides. (The result in Figure 7.5 is 3.) List all of the possible outcomes for an experiment in which a four-sided die is rolled.

Figure 7.5 Four-Sided Die

Solution
A four-sided die has four possible outcomes: 1, 2, 3, and 4.

Now Try Exercise 11

Probability and Percent Chance

The **probability** of an event is a measure of its likelihood of occurring. The probability of an event must be a number from 0 to 1, which can be written as a decimal, fraction, or percent. A probability written as a percent is called a **percent chance**. Examples of numbers that could represent a probability include

$$0, \quad \frac{1}{2}, \quad 0.65, \quad 30\%, \quad \text{and} \quad 1.$$

Because a probability must be a number from 0 to 1, numbers that could *not* represent a probability include

$$10, \quad 6\frac{1}{3}, \quad -0.33, \quad 160\%, \quad \text{and} \quad -1.$$

READING CHECK

- What is the probability of an impossible event?
- What is the probability of a certain event?

An event with probability 0 is called an **impossible event** because it cannot occur. For example, the probability of rolling a 13 on a six-sided die is 0 (or 0%) because there is no way to get a 13 on a six-sided die. An event with probability 1 is called a **certain event** because it is certain to occur. For example, the probability of getting heads *or* tails when tossing a coin is 1 (or 100%) because a coin is certain to land as either heads or tails.

EXAMPLE 2 **Determining whether an event is impossible or certain**

State whether each event is impossible or certain and then give its probability.
(a) Choosing a white marble from a bag that contains only black marbles
(b) Getting a result that is less than 9 when rolling a six-sided die

Solution
(a) Since it is impossible to choose a white marble from a bag that contains only black marbles, the event is impossible, and its probability is 0.
(b) Since every number on a six-sided die is less than 9, the event is certain to occur, and its probability is 1.

Now Try Exercises 17, 19

Many events have probabilities that are neither 0 nor 1. To compute the probability of such events, use the following formula.

THE PROBABILITY OF AN EVENT

$$\text{probability of an event} = \frac{\text{number of ways the event can occur}}{\text{number of possible outcomes for the experiment}}$$

NOTE: Probabilities computed using this formula are called *theoretical* probabilities, based on the idea that all outcomes for the experiment are equally likely to occur.

EXAMPLE 3 **Computing probability when tossing a coin**

When a coin is tossed, the result is either heads or tails, as shown in Figure 7.6. Find the probability of getting heads when a coin is tossed. Express your answer as a percent chance.

Figure 7.6 Heads and Tails

Solution
When a coin is tossed, only **1** of the possible outcomes is heads. Since a coin can land as either heads or tails, the total number of possible outcomes is **2**.

$$\text{probability of heads} = \frac{\text{number of ways the event can occur}}{\text{number of possible outcomes}} = \frac{\textbf{1}}{\textbf{2}}$$

Since $\frac{1}{2} = 50\%$, there is a 50% chance of getting heads when a coin is tossed.

Now Try Exercise 21

EXAMPLE 4 **Computing probability when rolling a die**

A six-sided die is shown in Figure 7.7. Find the probability that a number greater than 4 results when a six-sided die is rolled. Express your answer as a fraction in simplest form.

Figure 7.7 Six-Sided Die

Solution

Only **2** of the possible outcomes on a six-sided die are greater than 4, namely 5 and 6. The total number of possible outcomes for a six-sided die is **6**.

$$\text{probability the result is greater than } 4 = \frac{\text{number of ways the event can occur}}{\text{number of possible outcomes}} = \frac{2}{6} = \frac{1}{3}$$

The probability that the result is greater than 4 when rolling a six-sided die is $\frac{1}{3}$.

Now Try Exercise 27

A standard deck of 52 playing cards contains 4 suits with 13 cards each. The two black suits are Clubs and Spades, and the two red suits are Hearts and Diamonds. Each suit includes an Ace, 2, 3, 4, 5, 6, 7, 8, 9, 10, Jack, Queen, and King. The Jack, Queen, and King of each suit are known as *face cards*. See Figure 7.8.

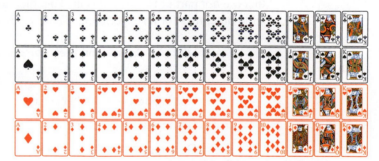

Figure 7.8 A Standard Deck of 52 Playing Cards

EXAMPLE 5 **Computing probabilities involving playing cards**

A single card is randomly drawn from a standard deck of playing cards. Expressing your answer as a fraction in simplest form, what is the probability of getting a
(a) King of Clubs?   **(b)** red 5?   **(c)** face card?

Solution

(a) Only **1** card is the King of Clubs. The total number of possible outcomes is **52** because there are 52 cards in a standard deck.

$$\text{probability of King of Clubs} = \frac{\text{number of ways the event can occur}}{\text{number of possible outcomes}} = \frac{1}{52}$$

The probability of drawing the King of Clubs is $\frac{1}{52}$.

(b) There are **2** red 5s (5 of Hearts and 5 of Diamonds) out of a total of **52** cards.

$$\text{probability of red 5} = \frac{\text{number of ways the event can occur}}{\text{number of possible outcomes}} = \frac{\mathbf{2}}{\mathbf{52}} = \frac{1}{26}$$

The probability of drawing a red 5 is $\frac{1}{26}$.

(c) There are **12** face cards (Jack, Queen, and King in 4 suits) out of **52** cards.

$$\text{probability of face card} = \frac{\text{number of ways the event can occur}}{\text{number of possible outcomes}} = \frac{\mathbf{12}}{\mathbf{52}} = \frac{3}{13}$$

The probability of drawing a face card is $\frac{3}{13}$.

Now Try Exercises 33, 35, 37

MAKING CONNECTIONS

A Career Involving Probability

An *actuary* is a person who uses probability and statistics to assess risk in the insurance and finance industries. Many universities have undergraduate and graduate degree programs in *actuarial science*. In 2010, a study published by the website CareerCast ranked actuary as the #1 job in the United States.

The Complement of an Event

▶ **REAL-WORLD CONNECTION** If there are 36 chairs in a classroom and we know that 3 of the chairs are not occupied, then without counting students in the classroom, we know that there are 33 students in class. In other words, the number of students in class is found by counting the ones that are *not* in class. This example illustrates the idea of the *complement*.

In mathematics, the **complement** of an event includes all outcomes of an experiment that are *not* part of the given event. Together, an event and its complement make up all possible outcomes for an experiment. Table 7.3 lists some events and their complements.

TABLE 7.3 Events and Their Complements

Event	Complement
Rolling an even number on a die	Rolling an odd number on a die
Getting tails on a coin toss	Getting heads on a coin toss
Drawing a Club, Spade, or Heart	Drawing a Diamond

Because an event and its complement make up all possible outcomes for an experiment, the sum of the probability of an event and the probability of its complement is 100% or 1. This idea is used in the following formulas for the probability of the complement.

THE PROBABILITY OF THE COMPLEMENT

probability of an event's complement = 1 − probability of the event

or

probability of an event's complement = 100% − percent chance of the event

---

READING CHECK

• How is the complement of an event computed?

| EXAMPLE 6 | **Finding the probability of the complement** |

Use the given information to find the requested probability or percent chance.
(a) If there is a 30% chance of rain for a given day, what is the chance of no rain?
(b) The probability that a student will pass a test is $\frac{43}{50}$. What is the probability that the student will not pass the test?

Solution

(a) The event "no rain" is the complement of the event "rain."

probability of no rain $= 100\% -$ probability of rain $= 100\% - 30\% = 70\%$

There is a 70% chance that it will not rain.

(b) The event "not pass" is the complement of the event "pass."

$$\text{probability of not pass} = 1 - \text{probability of pass} = 1 - \frac{43}{50} = \frac{7}{50}$$

The probability that the student will not pass the test is $\frac{7}{50}$.

Now Try Exercises 43, 45

7.6   Putting It All Together

CONCEPT	COMMENTS	EXAMPLES
Experiment, Outcome, and Event	An *experiment* is an activity with an observable result. A result of an experiment is called an *outcome* of the experiment. Any outcome or group of outcomes for an experiment is called an *event*.	Rolling a six-sided die is an example of an experiment with possible outcomes of 1, 2, 3, 4, 5, and 6. Getting an odd number when rolling a die is an example of an event.
Probability and Percent Chance	The *probability* of an event is a measure of its likelihood of occurring. A probability written as a percent is called a *percent chance*. $$\text{probability} = \frac{\text{number of ways event can occur}}{\text{number of possible outcomes}}$$	Because 3 numbers are even on a six-sided die, the probability of rolling an even number is $$\text{probability of even} = \frac{3}{6} = \frac{1}{2}.$$
Impossible and Certain Events	An *impossible* event has probability 0, and a *certain* event has probability 1.	Getting a 9 on a six-sided die is an impossible event, while getting a result less than 8 is a certain event.
Complement of an Event	The complement of an event includes all outcomes of an experiment that are *not* part of the given event. The probability of an event's complement is equal to $$1 - \text{probability of the event}$$ or $$100\% - \text{percent chance of the event}.$$	If there is a 40% chance of rain on a particular day, then there is a $100\% - 40\% = 60\%$ chance that it will not rain.

7.6 Exercises

CONCEPTS AND VOCABULARY

1. A(n) _____ is an activity with an observable result.

2. The result of an experiment is called a(n) _____ of the experiment.

3. A(n) _____ is any outcome or group of outcomes for an experiment.

4. The _____ of an event is a measure of its likelihood of occurring.

5. When a probability is written as a percent, it is called a(n) _____.

6. An event with probability 0 is called a(n) _____.

7. An event with probability 1 is called a(n) _____.

8. The _____ of an event includes all outcomes that are *not* part of the given event.

OUTCOMES FOR AN EXPERIMENT

Exercises 9–14: List all possible outcomes for the given experiment.

9. An answer is picked randomly on a true/false test.

10. A woman guesses the gender of her unborn child.

11. An eight-sided die is rolled.

12. A marble is chosen from a bag containing only red, yellow, and green marbles.

13. A teacher randomly chooses a weekday for the next pop quiz.

14. A prime number less than 10 is randomly chosen.

PROBABILITY AND PERCENT CHANCE

Exercises 15–20: State whether the event is impossible or certain and then give its probability.

15. Randomly choosing a month with more than 20 days

16. Naming a day of the week that starts with the letter *b*

17. Rolling a 14 on a six-sided die

18. Getting either heads or tails when tossing a coin

19. Choosing a red marble from a bag that contains only red marbles

20. Drawing the 23 of Clubs from a standard deck

Exercises 21–26: Find the probability of the given event. Express your answer as a percent chance.

21. Guessing correctly on a true/false test question

22. Getting a number that is 3 or less when rolling a four-sided die

23. Choosing a heart from a standard deck of cards

24. Guessing correctly on a multiple choice test question with five possible answers

25. Choosing a white sock from a drawer that contains 8 white socks and 12 black socks

26. Choosing a blue marble from a bag that contains 3 blue marbles and 7 red marbles

Exercises 27–32: Find the probability of the given event. Express your answer as a fraction in simplest form.

27. Rolling a 1 or 2 on a six-sided die

28. Getting a number that is 7 or more when rolling an eight-sided die

29. Guessing *incorrectly* on a true/false test question

30. Guessing *incorrectly* on a multiple choice test question with four possible answers

31. Choosing a green marble from a bag that contains 4 blue marbles and 6 green marbles

32. Rolling a prime number on a twelve-sided die

Exercises 33–38: A card is drawn from a standard deck of cards. Expressing your answer as a fraction in simplest form, what is the probability of getting the specified card?

33. The 9 of Spades

34. A 9

35. A black face card

36. An even-numbered card

37. A black card

38. A red Ace

Exercises 39–42: The following spinner is used to play a board game. Expressing your answer as a fraction in simplest form, what is the probability that the spinner will land on the specified space?

39. Lose Turn

40. A number

41. A number greater than 200

42. A number less than 200

COMPLEMENTS

Exercises 43–48: Use the given information to find the requested probability or percent chance.

43. There is an 80% chance of snow for a given day. What is the chance that it will not snow?

44. There is a 70% chance that a flight will be on time. What is the chance that it will not be on time?

45. The probability that a ski resort will have enough snow to open on or before November 1st is $\frac{7}{10}$. What is the probability that the ski resort will open after November 1st?

46. The probability of drawing a black King from a standard deck is $\frac{1}{26}$. What is the probability of drawing a card that is not a black King?

47. If there is a 60% chance that a baseball pitcher will strike out a batter, what is the chance that the pitcher will not strike out the batter?

48. The probability that a bus is early or on time is $\frac{7}{12}$. What is the probability that the bus is late?

WRITING ABOUT MATHEMATICS

49. What is the complement of the following event?

At least one window is open

Explain your reasoning.

50. Give a real-world example of one impossible event and one certain event.

SECTIONS 7.5 and 7.6	**Checking Basic Concepts**

1. Find the simple interest when lending $200 to a friend for 3 months at 5% simple interest.

2. If $1200 is invested in an account that pays 4.5% simple interest, what is the total amount in the account after 6 months?

3. If $90 interest is paid when $8000 is borrowed at 1.5% simple interest, find the time for the loan.

4. If $18,000 is invested in an account that pays 6% APR compounded monthly, find the amount in the account after 6 years.

5. Find the probability of getting an 8 when rolling a six-sided die.

6. Find the probability of choosing a face card that is a Spade when selecting a card from a standard deck. Write your answer as a fraction in simplest form.

7. Find the probability of guessing *incorrectly* on a multiple choice test question with five possible answers. Give your answer as a percent.

8. The probability that a student will pass a biology exam is $\frac{9}{11}$. What is the probability that the student will fail the exam?

CHAPTER 7   Summary

SECTION 7.1 ■ INTRODUCTION TO PERCENTS; CIRCLE GRAPHS

Percent Notation

The word *percent* means "divide by 100" or "out of 100."

Example: 7% means "7 divided by 100," and is written as 7%, $\frac{7}{100}$, or 0.07.

Percents as Fractions

To write $x\%$ as a fraction, write $\frac{x}{100}$ and then simplify the fraction, if needed.

Example: $26\% = \dfrac{26}{100} = \dfrac{13}{50}$

Percents as Decimals

To write $x\%$ as a decimal, remove the % symbol and then multiply $0.01x$.

Example: $89\% = 0.01(89) = 0.89$

Fractions or Decimals as Percents

To write a fraction or decimal as a percent, multiply by 100%.

Examples: $\dfrac{2}{5} = \dfrac{2}{5} \cdot 100\% = \dfrac{200}{5}\% = 40\%$

$0.99 = 0.99(100\%) = 99\%$

Circle Graph

A circle graph can be used to visually display percent data.

Examples: The following circle graph shows the percentage of sales for each pizza variety sold at a small pizza vendor.

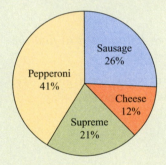

SECTION 7.2 ■ USING EQUATIONS TO SOLVE PERCENT PROBLEMS

Basic Percent Statement Form

The following percent statement is written in basic percent statement form.

A percent of the whole is a part.

Example: The percent statement 12 is 15% of 80 is written in basic percent statement form as 15% of 80 is 12.

Basic Percent Equation

A percent of a whole is a part translates to the equation

percent · whole = part.

Example: 25% of 60 is 15 translates to the equation $25\% \cdot 60 = 15$.

Percent Problem

A percent problem is a question asking us to find the percent, whole, or part in a percent statement.

Example: What number is 40% of 90?

Solving a Percent Problem	Solve a percent problem by solving the corresponding basic percent equation.

Example: 30% of 90 is what number?

$$30\% \cdot 90 = x \quad \text{Write an equation.}$$
$$0.30 \cdot 90 = x \quad \text{Write 30\% as 0.30.}$$
$$27 = x \quad \text{Multiply.}$$

SECTION 7.3 ■ USING PROPORTIONS TO SOLVE PERCENT PROBLEMS

Proportion

A proportion is a statement that two ratios are equal.

Example: $\dfrac{35}{100} = \dfrac{7}{20}$

Cross Products

In a proportion, the cross products are equal.

Example: $1 \cdot 18 \quad \dfrac{1}{2} = \dfrac{9}{18} \quad 2 \cdot 9$

$1 \cdot 18$ and $2 \cdot 9$ both equal 18.

Translating Percent Statements to Proportions

A percent of the whole is a part translates to the proportion

$$\frac{\text{percent}}{100} = \frac{\text{part}}{\text{whole}}.$$

Example: 40% of 50 is 20 translates to the proportion

$$\frac{40}{100} = \frac{20}{50}.$$

Solving a Percent Problem

A percent problem can be solved by solving the corresponding proportion.

Example: 20% of 60 is what number?

$$\frac{20}{100} = \frac{x}{60} \quad \text{Write a proportion.}$$
$$20 \cdot 60 = 100 \cdot x \quad \text{Cross products are equal.}$$
$$1200 = 100x \quad \text{Simplify.}$$
$$\frac{1200}{100} = \frac{100x}{100} \quad \text{Divide by 100.}$$
$$12 = x \quad \text{Simplify.}$$

SECTION 7.4 ■ APPLICATIONS: SALES TAX, DISCOUNTS, AND NET PAY

Basic Percent Applications

When solving a basic percent application, we may need to find a missing *part*, *whole*, or *percent*.

Example: If 12% of 150 students surveyed say that they like carrots, then $0.12(150) = 18$ students like carrots.

Sales Tax and Total Price

If S is sales tax, P is purchase price, and r (written as a decimal) is sales tax rate, then $S = rP$. If T is total price, then $T = P + S$.

Example: If there is a 6% sales tax rate, then a $400 digital camera requires $S = 0.06(400) = \$24$ in sales tax, and the total price of the digital camera is $T = 400 + 24 = \$424$.

Discounts and Sale Price	If D is a discount, O is original price, and r (written as a decimal) is discount rate, then $D = rO$. If P is purchase price, then $P = O - D$.

Example: If a $900 HDTV is marked at 25% off, then the discount is $D = 0.25(900) = \$225$, and the purchase price is $P = 900 - 225 = \$675$.

Commission

If C is commission, T is total sales, and r (written as a decimal) is commission rate, then $C = rT$.

Example: If a person has a 3% commission rate on $18,000 in total sales, then the commission is $C = 0.03(18,000) = \$540$.

Net Pay

If W is withholdings, G is gross pay, and r (written as a decimal) is the withholding rate, then $W = rG$. If N is net pay, then $N = G - W$.

Example: If 22% of a factory worker's $1400 gross pay is withheld, then the withholdings are $W = 0.22(1400) = \$308$, and the net pay is $N = 1400 - 308 = \$1092$.

Percent Change

If an amount changes from an **old amount** to a **new amount**, then the percent change is

$$\frac{\textbf{new amount} - \textbf{old amount}}{\textbf{old amount}} \cdot 100.$$

Examples: When a price increases from $60 to $72, the percent change is

$$\frac{72 - 60}{60} \cdot 100 = 20\%.$$

When a price decreases from $60 to $45, the percent change is

$$\frac{45 - 60}{60} \cdot 100 = -25\%.$$

SECTION 7.5 ■ APPLICATIONS: SIMPLE AND COMPOUND INTEREST

Principal, Interest, and Interest Rate

The *principal* is the initial amount of an investment or loan, *interest* is a fee paid to the lender or investor for use of the principal, and the *interest rate* is the percent used to calculate interest.

Example: If $100 is invested in an account that pays 4% simple interest for 1 year, then the principal is $100, the interest is $100 \cdot 0.04 \cdot 1 = \$4$, and the interest rate is 4%.

Simple Interest

The amount of simple interest I is given by $I = Prt$, where P = principal, r = annual interest rate (written as a decimal), and t = time (in years).

Example: If $300 is borrowed at 5% simple interest for 6 years, then the interest is $I = 300 \cdot 0.05 \cdot 6 = \$90$.

Compound Interest

The amount A in an account paying compound interest is given by

$$A = P\left(1 + \frac{r}{n}\right)^{nt},$$

where P = principal, r = annual interest rate (written as a decimal), t = time (in years), and n = the number of compounding periods per year.

Example: If $1600 is invested in an account that pays 8% interest compounded monthly, then the amount in the account after 4 years is

$$A = 1600\left(1 + \frac{0.08}{12}\right)^{12\cdot4} \approx \$2201.07.$$

SECTION 7.6 ■ PROBABILITY AND PERCENT CHANCE

Experiment, Outcome, and Event

An *experiment* is an activity with an observable result. A result of an experiment is called an *outcome* of the experiment. Any outcome or group of outcomes for an experiment is called an *event*.

Example: Rolling a 12-sided die is an example of an experiment that has possible outcomes of 1, 2, 3, 4, 5, 6, 7, 8, 9, 10, 11, and 12. Getting a number less than 5 when rolling this die is an example of an event.

Probability and Percent Chance

The *probability* of an event is a measure of its likelihood of occurring. A probability written as a percent is called a *percent chance*.

$$\text{probability} = \frac{\text{number of ways event can occur}}{\text{number of possible outcomes}}$$

Example: Because 3 numbers are odd on a six-sided die, the probability of rolling an odd number is

$$\text{probability of odd} = \frac{3}{6} = \frac{1}{2}. \text{ (The percent chance is 50%.)}$$

Impossible and Certain Events

An *impossible* event has probability 0, and a *certain* event has probability 1.

Example: Getting a 5 on a four-sided die is an impossible event, while getting a result less than 5 is a certain event.

Complement of an Event

The complement of an event includes all outcomes of an experiment that are *not* part of the given event. The probability of an event's complement equals

$$1 - \text{probability of the event} \quad \text{or} \quad 100\% - \text{percent chance of the event.}$$

Example: If there is a 15% chance of fog on a particular day, then there is a $100\% - 15\% = 85\%$ chance that there will be no fog on that day.

CHAPTER 7   Review Exercises

SECTION 7.1

Exercises 1–4: Write the percent as a fraction or mixed number in simplest form.

1. 68%

2. 180%

3. $6\frac{1}{4}\%$

4. 7.2%

Exercises 5–8: Write the percent as a decimal.

5. 29%

6. 0.4%

7. 625%

8. $35\frac{3}{4}\%$

Exercises 9–12: Write the fraction or mixed number as a percent.

9. $\frac{17}{25}$

10. $3\frac{1}{5}$

11. $\frac{7}{9}$

12. $\frac{5}{16}$

Exercises 13–16: Write the decimal as a percent.

13. 0.16

14. 0.049

15. 7.02

16. 0.0305

Exercises 17 and 18: The following circle graph shows expenses for a student at a private college.

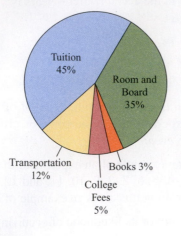

17. Which expense was greatest?

18. What fraction of the student's expenses went toward transportation?

SECTION 7.2

Exercises 19 and 20: Write the given percent statement in basic percent statement form.

19. 52 is 130% of 40.   **20.** 9 out of 12 is 75%.

Exercises 21 and 22: Translate the given percent statement to a basic percent equation.

21. 77.5% is 62 of 80.   **22.** 50% of 184 is 92.

Exercises 23 and 24: Translate the given percent problem to a basic percent equation. Do not solve the equation.

23. 75% of what number is 46.5?

24. 48 out of 72 is what percent?

Exercises 25–28: Use a basic percent equation to find the unknown value.

25. What number is 16% of 85?

26. $10\frac{1}{2}$ out of what number is 30%?

27. 4.1 is what percent of 12.3?

28. 3 out of 5 is what percent?

SECTION 7.3

Exercises 29 and 30: Translate the given percent statement to a proportion.

29. 96 is 240% of 40.   **30.** 28 of 35 is 80%.

Exercises 31 and 32: Translate the given percent problem to a proportion. Do not solve the proportion.

31. What number is 165% of 20?

32. What percent is 3 of 9?

Exercises 33–36: Use a proportion to find the unknown value.

33. What number is 13% of 60?

34. 560 out of 840 is what percent?

35. 34.6 is what percent of 69.2?

36. 4.5% of what number is 198?

SECTION 7.4

37. *Nutrition* If a daily diet should include 35 grams of dietary fiber, how many grams of dietary fiber are in a serving of whole grain bread with 8% of the recommended amount?

38. *Cheating* Of 12,000 high school students surveyed, 26% did *not* cheat on an exam during the past year. How many of the students surveyed did not cheat? (*Source:* Josephson Institute of Ethics.)

39. *Going Green* Only 82 of the 2050 students on a college campus used a bicycle to get to campus. What percent of the students rode a bike to school?

40. *Saving for College* In 2008, the parents of 800 college freshmen were asked how much they had saved for their child's college education. The parents of 280 freshmen had saved more than $5000. What percent of parents saved more than $5000? (*Source:* College Savings Foundation.)

41. *Smartphone* If the purchase price of a smartphone is $198 and the sales tax rate is 5.5%, find the sales tax and the total price.

42. *Tax Rate* A student bought a car stereo with a purchase price of $646.50. If the sales tax was $25.86, find the sales tax rate.

43. *Microwave Oven* A microwave oven is marked at 20% off. If the regular price is $380, find the discount and the sale price.

44. *Prepaid Minutes* A cellular customer receives a $42.50 discount when he buys 2000 prepaid minutes. If the regular price for 2000 minutes is $250, what is the discount rate?

45. *Conference Fees* A professor gets a $15 discount for prepaying conference fees at least 60 days in advance. If this represents an 8% discount, find the original cost of the conference.

46. *Clothing* A salesperson at a women's clothing store earns 7.5% commission on total sales. Find the commission for a month with $7,400 in total sales.

47. *Office Supplies* If an office supplies salesperson earns a $126 commission for $1050 in total sales, find the commission rate.

48. *Net Pay* Find the total withholdings and net pay for a worker who has 24% of his $820 weekly pay withheld for taxes and insurance.

49. *Withholdings* What percent of gross pay is being withheld for a worker who has $420 withheld from $1890 in gross pay?

50. *Gross Pay* A worker has $456 withheld from her pay every two weeks. If this represents a 20% withholding rate, find her gross pay.

51. *Snack Prices* The cost of a candy bar from a vending machine increases from $0.75 to $1.00. Find the percent increase.

52. *Depreciation* If driving a new car off the lot decreases its value from $22,600 to $18,080, find the percent decrease.

SECTION 7.5

Exercises 53 and 54: Find the simple interest for the given values of P, r, and t.

53. $P = \$480$, $r = 4.5\%$, and $t = 4$ months

54. $P = \$38,650$, $r = 4\%$, and $t = 6$ years

Exercises 55 and 56: Find the simple interest.

55. A homeowner borrows $28,000 at 4.5% simple interest for 2 years to pay for a swimming pool.

56. A student borrows $64,000 from a wealthy uncle for 12 years at 3.5% simple interest.

Exercises 57 and 58: For the given initial investment, simple interest rate, and time of investment, find the total value of the investment.

57. $4500 at 6% for 8 years.

58. $1200 at 2.5% for 18 months.

Exercises 59 and 60: Find the unknown value.

59. When money is invested at 6% simple interest for 5 years, the interest is $1170. Find the principal.

60. $45 interest is paid when $2400 is borrowed at 2.5% simple interest. Find the time for the loan.

Exercises 61–64: Find the final amount.

61. $2500 invested at 8% APR compounded monthly for 48 years

62. $7000 invested at 3.5% APR compounded quarterly for 7 years

63. $34,000 invested at 5% APR compounded semiannually for 11 years

64. $400 invested at 7% APR compounded daily for 19 years

SECTION 7.6

Exercises 65 and 66: List all possible outcomes for the given experiment.

65. A marble is chosen from a bag containing only red, black, and blue marbles.

66. A prime number less than 20 is randomly chosen.

Exercises 67 and 68: State whether the event is impossible or certain and then give its probability.

67. Rolling a number less than 14 on a six-sided die.

68. Naming a U.S. state that starts with the letter *z*.

Exercises 69 and 70: Find the probability of the given event. Express your answer as a percent chance.

69. Guessing correctly on a multiple choice test question with four possible answers

70. Choosing a white marble from a jar that contains 9 white marbles and 15 black marbles

Exercises 71 and 72: Find the probability of the given event. Express your answer as a fraction in simplest form.

71. Rolling a 5 or 6 on a six-sided die

72. Guessing *incorrectly* on a multiple choice test question with five possible answers

Exercises 73 and 74: A card is drawn from a standard deck of cards. Expressing your answer as a fraction in simplest form, what is the probability of getting the specified card?

73. a Jack

74. a red, even-numbered card

Exercises 75 and 76: Use the given information to find the requested probability or percent chance.

75. There is a 75% chance that farmers will be able to harvest crops in the next week. What is the chance that they will not be able to harvest crops in the next week?

76. The probability that it will be warm enough for a water park to open on or before June 1st is $\frac{8}{11}$. What is the probability that the park will open after June 1st?

APPLICATIONS

77. *Obesity* A 2009 study found that $\frac{33}{50}$ of the American population was *not* obese. Write this fraction as a percent. (*Source:* National Center for Health Statistics.)

78. *Shopping* A bargain table at a sporting goods store advertises "40% off the regular price." Write this percent as a fraction.

79. *Hybrid Car* A car dealer claims that a hybrid car gets 42.8% better mileage than a large SUV. Write this percent as a decimal.

80. *Gender* A class with 32 students has 20 women. What percent of the class do women represent?

CHAPTER 7 Test

Exercises 1 and 2: Write the given percent as a fraction or mixed number in simplest form.

1. 55%

2. 260%

Exercises 3 and 4: Write the given fraction or decimal as a percent.

3. $\frac{19}{20}$

4. 0.078

5. Translate the following percent problem to a basic percent equation. Do not solve the equation.

    What percent of 160 is 32?

6. Translate the following percent problem to a proportion. Do not solve the proportion.

    What number is 12 percent of 90?

Exercises 7 and 8: Use a basic percent equation to find the unknown value.

7. 30% of what number is 42?

8. What percent of 225 is 75?

Exercises 9 and 10: Use a proportion to find the unknown value.

9. 38 out of 40 is what percent?

10. 10.5% of what number is 84?

11. Find the simple interest when $1400 is borrowed for 6 months at 4.5% interest.

12. If $20,000 is invested in an account that pays 6% simple interest, find the total value of the investment after 5 years.

13. If an investment of $1500 earns $90 simple interest after 2 years, find the interest rate.

14. If $2000 is invested in an account that pays 6% APR compounded quarterly, find the total amount in the account after 3 years.

15. List all possible outcomes when a positive even number less than 11 is randomly selected.

16. State whether the following event is an impossible or certain event and give its probability.

    Naming a month that starts with the letter *q*.

17. Find the probability of getting a 3, 4, or 5 when rolling a six-sided die. Express your answer as a fraction in simplest form.

18. Find the probability of guessing correctly on a multiple choice test question with six possible answers. Express your answer as a percent chance.

19. A card is randomly selected from a standard deck. Find the probability that the card is a black card with the number 4, 5, 6, or 7. Express your answer as a fraction in simplest form.

20. There is a 38% chance that a student has blonde hair. What is the chance that a student is not blonde?

21. *The Human Body* An adult female's body is about 55% water. If a woman's body has 77 pounds of water, what is her total body weight?

22. *Salary* A worker's salary increases by 4.5%. If the amount of the raise amounts to $2880, how much was the worker being paid before the raise?

23. *Car Stereo* If speakers for a car stereo sell for $349 and the sales tax rate is 5%, find the sales tax and the total price.

24. *Clearance Sale* A video game on a clearance table is marked at 60% off. If the regular price is $19.25, find the discount and the sale price.

25. *Home Design* An interior designer receives 12% commission on all sales related to a design project. If a home design project has total sales of $38,500, find the commission.

26. *Stock Price* If a stock price increases from $58.40 per share to $67.16 per share, find the percent increase.

CHAPTERS 1–7   Cumulative Review Exercises

1. Write *forty-six thousand, three hundred ninety-one* in standard form.

2. Round 658,255 to the nearest thousand.

3. Evaluate $18 - (3 \cdot 4 - (24 \div 2^2) + 1)$.

4. Simplify $7w + 2(w - 5)$.

5. Simplify $-7 - 13 + (-1)$.

6. Find the product $3(-1)(3)(-2)(-4)$.

7. Evaluate the expression $w + ((3 - v)^2 \div 2w)$, when $v = 3$ and $w = -2$.

8. Use the following graph to solve $-3x + 28 = 25$ visually.

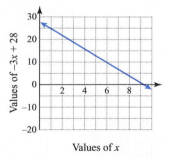

Values of x

9. Simplify $-2(5t + 1) + 3(t - 2)$.

10. Translate the following sentence into an equation using x as the variable. Do not solve the equation.

   Twice the sum of a number and 5 gives -14.

11. Solve $10 + 5x = 4x - 6$ algebraically.

12. Complete the following table of values to solve the equation $-3x + 6 = 3$ numerically.

x	-2	-1	0	1	2
$-3x + 6$					

13. Find the prime factorization of 105.

14. Add $-\frac{5}{7} + \frac{9}{14}$.

15. Subtract $7\frac{3}{5} - 1\frac{4}{15}$.

16. Solve $4(x - \frac{2}{3}) = \frac{5}{4}$ algebraically.

17. Round 468.049 to the nearest hundredth.

18. For the given list of numbers, identify any of the following types of numbers.
   (a) natural numbers   **(b)** whole numbers
   (c) integers   **(d)** rational numbers
   (e) irrational numbers

$$9, -\frac{1}{5}, 4.\overline{6}, -2.1, \sqrt{7}$$

19. Solve $5(x + 4.7) = 53.5$ algebraically.

20. Find the exact area of the circle and then approximate the area using 3.14 as an approximation for π.

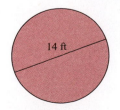

14 ft

21. Convert 2.6 miles to feet.

22. Convert 0.089 deciliter to milliliters.

23. Convert $-31°F$ to Celsius.

24. Convert 5 kilometers per minute to meters per hour.

25. Write 137% as a decimal.

26. What percent of 144 is 48?

27. What number is 15% of 840?

28. A card is randomly selected from a standard deck. Find the probability that the card is a red card with the number 2, 7, or 9. Express your answer as a fraction in simplest form.

29. *Heart Rate* The average heart rate R in beats per minute (bpm) of an animal weighing W pounds can be approximated by

$$R = \frac{885\sqrt{W}}{W}.$$

Find the heart rate for a 25-pound coyote.

30. *Music Sales* The total music sales S in thousands of dollars for a small band during year x can be computed using the formula

$$S = 7(x - 2005) + 8.$$

Find the year when the music sales were $29,000 by solving the equation $29 = 7(x - 2005) + 8$ visually, using the following graph.

Music Sales

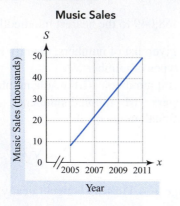

Year

31. *Altitude and Temperature* If the temperature on the ground is 75°F, then the air temperature T at an altitude of x miles is given by $T = 75 - 19x$. Find the altitude where the air temperature is $T = 18°F$.
(*Source:* Miller, A., and R. Anthes, *Meteorology.* 5th ed.)

32. *Height* At birth, a baby boy was $19\frac{3}{4}$ inches long. If he grew $6\frac{1}{2}$ inches during his first year, how tall was he on his first birthday?

33. *Text Messaging* Suppose that the cost C of sending or receiving x text messages is given by the formula

$$C = 0.1(x - 400) + 7.5, \text{ where } x \geq 400.$$

Find the number of text messages that correspond to total charges of $19.60 by replacing C in the formula with 19.6 and solving the resulting equation.

34. *Unit Price* Find the unit price for each size option.

Large fruit drink: 24 ounces for $4.20
Small fruit drink: 16 ounces for $2.96

Which is the better buy?

35. *Coffee Maker* A coffee maker is marked at 25% off. If the regular price is $38, find the discount and the sale price.

36. *Going Green* A company reduces solid waste from 4600 tons per year to 4048 tons per year. Find the percent decrease.

Exponents and Polynomials

First say to yourself what you would be; and then do what you have to do.

— EPICTETUS

xtreme thrill-seekers do not have to leave Earth's atmosphere to experience weightlessness. A specially designed jet following a curved path can provide passengers with about 30 seconds in zero gravity. The weightlessness experienced by passengers begins when the aircraft moves over the peak of the curve and starts to descend as shown in the following figure.

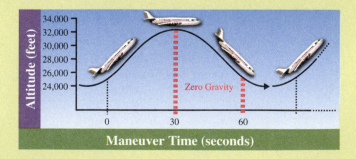

Mathematical equations that describe this type of curve involve expressions called *polynomials*. In this chapter, we will discuss polynomials in more detail.

Source: Alan Boyle, msnbc.com,
"Zero-gravity flights go mainstream."

8.1

8.1   Rules for Exponents

Review of Exponents • The Product Rule • The Power Rule • The Power of a Product Rule

A LOOK INTO MATH ▶

When walking across campus, students often try to find a *shortcut* that will take them from one place to another in the fewest steps. The same idea can be applied to mathematical tasks. Sometimes a property or rule can be used to find an answer without having to "do it the long way." In this section, we discuss rules that can be used as shortcuts for simplifying exponential expressions.

Review of Exponents

STUDY TIP

For a more in-depth review of exponents, review your notes for Sections 1.4 and 2.4.

Natural number exponents describe a fast way to multiply. The exponent tells us how many times the base is used as a repeated factor.

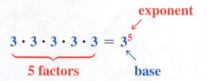

The repeated multiplication on the left side of this equation is written on the right side of the equation as an *exponential expression* with *base* **3** and *exponent* **5**. The base can be an integer, variable, or other algebraic expression. For example,

$$8 \cdot 8 \cdot 8 = 8^3, \quad y \cdot y \cdot y \cdot y = y^4, \quad \text{and} \quad (-5y) \cdot (-5y) = (-5y)^2.$$

EXAMPLE 1 **Writing an exponential expression as repeated multiplication**

Write each exponential expression as repeated multiplication.
(a) $(-6)^2$ **(b)** x^4 **(c)** $(-7y)^3$ **(d)** -8^2

Solution
(a) $(-6)^2 = (-6) \cdot (-6)$
(b) $x^4 = x \cdot x \cdot x \cdot x$
(c) $(-7y)^3 = (-7y) \cdot (-7y) \cdot (-7y)$
(d) The negative sign is not part of the base in -8^2. Parentheses are used to show that we square 8 before finding the opposite.

$$-8^2 = -(8 \cdot 8)$$

Now Try Exercises 9, 11, 13, 15

The Product Rule

Using repeated multiplication, we can find a shortcut for simplifying the product of two exponential expressions *with the same base*. For example, we write the product $3^4 \cdot 3^2$ as

$$3 \cdot 3 \cdot 3 \cdot 3 \cdot 3 \cdot 3.$$

Notice that the base 3 is being multiplied a total of **6** times, which is found by adding the exponents $4 + 2 = 6$.

$$3^4 \cdot 3^2 = 3^{4+2} = 3^6$$

READING CHECK

- What must be true about the bases before the product rule can be applied?

THE PRODUCT RULE FOR EXPONENTS

For any real number base a and natural number exponents m and n,

$$a^m \cdot a^n = a^{m+n}.$$

EXAMPLE 2   Using the product rule for exponents

Multiply.
(a) $x^2 \cdot x^7$ **(b)** $y^4 \cdot y$ **(c)** $b^3 \cdot b^2 \cdot b^2$

Solution

(a) $x^2 \cdot x^7 = x^{2+7} = x^9$

(b) Remember, when no exponent is written, it is assumed to be a **1**.

$$y^4 \cdot y = y^4 \cdot y^1 = y^{4+1} = y^5$$

(c) $b^3 \cdot b^2 \cdot b^2 = b^{3+2+2} = b^7$

Now Try Exercises 17, 21, 25

When exponential expressions containing coefficients other than 1 are multiplied, the commutative property of multiplication is helpful. For example, the product $5x^3 \cdot 7x^4$ can be written as $5 \cdot x^3 \cdot 7 \cdot x^4$. Then the commutative property allows us to write

$$5 \cdot x^3 \cdot 7 \cdot x^4 = (5 \cdot 7) \cdot (x^3 \cdot x^4) = 35 \cdot x^{3+4} = 35x^7.$$

First we multiply the coefficients and then apply the product rule for exponents.

EXAMPLE 3   Multiplying coefficients and using the product rule for exponents

Multiply.
(a) $4z^3 \cdot 6z^5$ **(b)** $7y^2 \cdot (-9y^3)$ **(c)** $-2k^3 \cdot k^2 \cdot (-4k^5)$

Solution

(a) $4z^3 \cdot 6z^5 = (4 \cdot 6) \cdot (z^3 \cdot z^5) = 24 \cdot z^{3+5} = 24z^8$

(b) $7y^2 \cdot (-9y^3) = [7 \cdot (-9)] \cdot (y^2 \cdot y^3) = -63 \cdot y^{2+3} = -63y^5$

(c) Remember, since no coefficient is written for k^2, it is assumed to be a **1**.

$$-2k^3 \cdot 1k^2 \cdot (-4k^5) = [-2 \cdot 1 \cdot (-4)] \cdot (k^3 \cdot k^2 \cdot k^5) = 8 \cdot k^{3+2+5} = 8k^{10}$$

Now Try Exercises 27, 31, 33

The Power Rule

When simplifying an exponential expression raised to a power, we can again use repeated multiplication to find a shortcut. For example, we can write the expression $(y^4)^3$ as

$$\underbrace{(y^4) \cdot (y^4) \cdot (y^4)}_{\textbf{3 factors}}.$$

Now, using the product rule for exponents, we write $(y^4) \cdot (y^4) \cdot (y^4)$ as

$$y^{4+4+4} = y^{12}.$$

Because we are adding the exponent **4** a total of **3** times, the exponent 12 in the resulting expression is found by multiplying the exponents **4** and **3**.

$$(y^4)^3 = y^{4 \cdot 3} = y^{12}$$

READING CHECK

• How does the power rule differ from the product rule?

THE POWER RULE FOR EXPONENTS

For any real number base a and natural number exponents m and n,

$$(a^m)^n = a^{m \cdot n}.$$

EXAMPLE 4 **Using the power rule for exponents**

Simplify.
(a) $(q^4)^2$ **(b)** $(y^2)^5 \cdot (y^3)^2$

Solution
(a) $(q^4)^2 = q^{4 \cdot 2} = q^8$
(b) Use the power rule twice and then multiply the expressions using the product rule.

$$(y^2)^5 \cdot (y^3)^2 = y^{2 \cdot 5} \cdot y^{3 \cdot 2} = y^{10} \cdot y^6 = y^{10+6} = y^{16}$$

Now Try Exercises 35, 39

MAKING CONNECTIONS

The Product and Power Rules

In Chapter 1, we learned that a **product** such as $5 \cdot 3$ is a fast way to **add** $5 + 5 + 5$. We also learned that a **power** such as 2^3 is a fast way to **multiply** $2 \cdot 2 \cdot 2$. These two ideas may help you remember that we **add** exponents when using the **product** rule, and we **multiply** exponents when using the **power** rule.

The Power of a Product Rule

How do we simplify the expression $(7x)^3$, which is a power of a product? Once again, we use repeated multiplication to find a shortcut.

$$(7x)^3 = \underbrace{(7x) \cdot (7x) \cdot (7x)}_{\textbf{3 factors}}$$

Using the commutative property for multiplication, we can regroup the factors.

$$(7 \cdot 7 \cdot 7) \cdot (x \cdot x \cdot x)$$

Because there are **3** factors of 7 and **3** factors of x, we can write

$$7^3 \cdot x^3.$$

So, we can raise a product to a power by raising each factor to the power.

THE POWER OF A PRODUCT RULE FOR EXPONENTS

For any real numbers a and b and natural number exponent n,

$$(ab)^n = a^n b^n.$$

EXAMPLE 5 Using the power of a product rule for exponents

Simplify.
(a) $(2w)^4$ **(b)** $(xy^2)^3$

Solution
(a) $(2w)^4 = 2^4 \cdot w^4 = 16w^4$
(b) Use the power of a product rule and then the power rule.

$$(xy^2)^3 = x^3 \cdot (y^2)^3 = x^3 \cdot y^{2\cdot3} = x^3 y^6$$

Now Try Exercises 43, 47

Example 5(b) suggests that it may be necessary to use more than one rule for exponents when simplifying an expression. In the next example, more than one rule is applied.

EXAMPLE 6 Using more than one rule for exponents

Simplify.
(a) $(5x^4 y^3)^2$ **(b)** $(-2a^2 b^3)^4$ **(c)** $(2m^2 n)^3 (4mn^3)^2$

Solution

(a)
$$(5x^4 y^3)^2 = 5^2 \cdot (x^4)^2 \cdot (y^3)^2 \qquad \text{Power of a product rule}$$
$$= 5^2 \cdot x^{4\cdot2} \cdot y^{3\cdot2} \qquad \text{Power rule}$$
$$= 25x^8 y^6 \qquad \text{Simplify.}$$

(b) In the first step, be sure to raise -2 to the fourth power.
$$(-2a^2 b^3)^4 = (-2)^4 \cdot (a^2)^4 \cdot (b^3)^4 \qquad \text{Power of a product rule}$$
$$= (-2)^4 \cdot a^{2\cdot4} \cdot b^{3\cdot4} \qquad \text{Power rule}$$
$$= 16a^8 b^{12} \qquad \text{Simplify.}$$

(c)
$$(2m^2 n)^3 (4mn^3)^2 = [2^3 \cdot (m^2)^3 \cdot n^3] \cdot [4^2 \cdot m^2 \cdot (n^3)^2] \qquad \text{Power of a product rule}$$
$$= 2^3 \cdot m^{2\cdot3} \cdot n^3 \cdot 4^2 \cdot m^2 \cdot n^{3\cdot2} \qquad \text{Power rule}$$
$$= (2^3 \cdot 4^2) \cdot (m^{2\cdot3} \cdot m^2) \cdot (n^3 \cdot n^{3\cdot2}) \qquad \text{Group like bases.}$$
$$= 8 \cdot 16 \cdot (m^6 \cdot m^2) \cdot (n^3 \cdot n^6) \qquad \text{Simplify.}$$
$$= 8 \cdot 16 \cdot m^{6+2} \cdot n^{3+6} \qquad \text{Product rule}$$
$$= 128m^8 n^9 \qquad \text{Multiply and simplify.}$$

Now Try Exercises 51, 53, 57

8.1 Putting It All Together

CONCEPT	COMMENTS	EXAMPLES
Exponent	A natural number exponent indicates the number of times that the base is used as a repeated factor.	Exponent ↘ $5^4 = 5 \cdot 5 \cdot 5 \cdot 5$ ↗ Base
The Product Rule	For any real number a and natural number exponents m and n, $$a^m \cdot a^n = a^{m+n}.$$	$2^3 \cdot 2^6 = 2^{3+6} = 2^9$ $x^4 \cdot x^3 = x^{4+3} = x^7$
The Power Rule	For any real number a and natural number exponents m and n, $$(a^m)^n = a^{m \cdot n}.$$	$(4^2)^5 = 4^{2 \cdot 5} = 4^{10}$ $(y^3)^4 = y^{3 \cdot 4} = y^{12}$
The Power of a Product Rule	For any real numbers a and b and natural number exponent n, $$(ab)^n = a^n b^n.$$	$(xy)^3 = x^3 y^3$ $(3y^3)^2 = 3^2(y^3)^2 = 3^2(y^{3 \cdot 2}) = 9y^6$

8.1 Exercises

MyMathLab | Math XL PRACTICE | WATCH | DOWNLOAD | READ | REVIEW

CONCEPTS AND VOCABULARY

1. In the exponential expression x^6, the base is _____, and the exponent is _____.

2. If a is a real number and m and n are natural number exponents, then $a^m \cdot a^n =$ _____.

3. If a is a real number and m and n are natural number exponents, then $(a^m)^n =$ _____.

4. If a and b are real numbers and n is a natural number exponent, then $(ab)^n =$ _____.

5. To simplify $x^3 \cdot x^4$, we (add/multiply) the exponents.

6. To simplify $(x^3)^4$, we (add/multiply) the exponents.

EXPONENTS

Exercises 7–16: Write the given exponential expression as repeated multiplication.

7. 8^3

8. 11^4

9. y^6

10. w^5

11. $(-7)^3$

12. $(-19)^2$

13. -3^2

14. -7^4

15. $(-2x)^4$

16. $(-8y)^3$

RULES FOR EXPONENTS

Exercises 17–34: Multiply.

17. $x^2 \cdot x^3$

18. $y^4 \cdot y^5$

19. $m^8 \cdot m^5$

20. $n^6 \cdot n^9$

21. $w \cdot w^7$

22. $b^4 \cdot b$

23. $x^3 \cdot x \cdot x^7$

24. $k^3 \cdot k^6 \cdot k$

25. $a^5 \cdot a^7 \cdot a^2$

26. $z^6 \cdot z^3 \cdot z^8$

27. $4x^3 \cdot 7x^2$

28. $12b^4 \cdot 2b^5$

29. $6y \cdot 9y^5$

30. $9w^3 \cdot 7w$

31. $-3m^6 \cdot 6m^5$

32. $8n^3 \cdot (-4n^2)$

33. $-3x \cdot x^2 \cdot (-2x^3)$

34. $y^2 \cdot (-3y) \cdot 3y^4$

Exercises 35–42: Simplify.

35. $(x^5)^2$

36. $(w^4)^4$

37. $(n^3)^9$

38. $(y^2)^7$

39. $(m^4)^2 \cdot (m^2)^3$

40. $(a^2)^5 \cdot (a^6)^3$

41. $(q^3)^3 \cdot q^2$

42. $p \cdot (p^4)^2$

Exercises 43–50: Simplify.

43. $(3x)^2$

44. $(2r)^3$

45. $(-4y^4)^3$

46. $(-2n^7)^4$

47. $(x^3y)^2$

48. $(mn^3)^4$

49. $(v^6w^4)^2$

50. $(p^4q^2)^5$

MORE THAN ONE RULE

Exercises 51–60: Simplify.

51. $(2x^2y^3)^4$

52. $(7m^5n^3)^2$

53. $(-3a^5b^3)^3$

54. $(-2c^2d^4)^5$

55. $(-9x^8y^3)^2$

56. $(-2v^3w^4)^4$

57. $(5pq^3)^2(2p^2q)^3$

58. $(2xy^4)^2(x^3y)^2$

59. $(-2a^2b^2)^3(3a^4b^3)^2$

60. $(-2x^3y^2)^2(-2x^2y^3)^3$

GEOMETRY

Exercises 61–64: **Area** *Write a simplified expression for the area of the given figure.*

61. Rectangle: $A = lw$

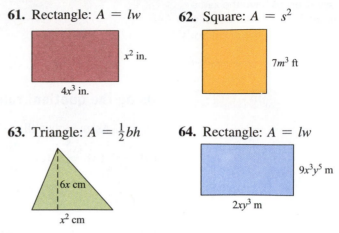

x^2 in.

$4x^3$ in.

62. Square: $A = s^2$

$7m^3$ ft

63. Triangle: $A = \frac{1}{2}bh$

$6x$ cm

x^2 cm

64. Rectangle: $A = lw$

$9x^3y^5$ m

$2xy^3$ m

WRITING ABOUT MATHEMATICS

65. When do we add exponents and when do we multiply exponents? Explain in your own words.

66. When a negative integer is raised to a power, how do we know if the result is positive or negative? Give an example for each situation.

8.2	**Negative Exponents and Scientific Notation**

The Quotient Rule • The Zero Power Rule • Negative Exponents • Scientific Notation • Multiplying and Dividing with Scientific Notation

A LOOK INTO MATH ▶

In 2009, astronomers discovered a large planet that orbits a distant star. The planet, named WASP-17b, is about 5,880,000,000,000,000 miles from Earth. Also in 2009, the H1N1 virus was identified in a worldwide influenza pandemic. A typical flu virus measures about 0.00000468 inch across. In this section, we will discuss how such very large and very small numbers can be written in *scientific notation*. (*Source: Scientific American.*)

The Quotient Rule

Consider the division problem

$$\frac{5^4}{5^2} = \frac{5 \cdot 5 \cdot 5 \cdot 5}{5 \cdot 5} = \frac{5}{5} \cdot \frac{5}{5} \cdot 5 \cdot 5 = 1 \cdot 1 \cdot 5^2 = 5^2.$$

Because there are two more 5s in the numerator than in the denominator, we can find the exponent for the result by subtracting the exponent in the denominator from the exponent in the numerator. So, we write

$$\frac{5^4}{5^2} = 5^{4-2} = 5^2.$$

NEW VOCABULARY

☐ Scientific notation

STUDY TIP

Remember to ask questions in class. Other students are likely to have the same questions and can benefit from hearing the instructor's response.

THE QUOTIENT RULE FOR EXPONENTS

For any nonzero real number base a and integer exponents m and n,

$$\frac{a^m}{a^n} = a^{m-n}.$$

EXAMPLE 1 **Using the quotient rule for exponents**

Divide. Assume that all variables represent nonzero numbers.

(a) $\dfrac{9^7}{9^5}$ (b) $\dfrac{x^8}{x^3}$ (c) $\dfrac{12y^4}{4y}$

Solution

(a) $\dfrac{9^7}{9^5} = 9^{7-5} = 9^2 = 81$

(b) $\dfrac{x^8}{x^3} = x^{8-3} = x^5$

(c) When a variable has no exponent, the exponent is assumed to be **1**.

$$\frac{12y^4}{4y} = \frac{12}{4} \cdot \frac{y^4}{y^1} = 3 \cdot y^{4-1} = 3y^3$$

Now Try Exercises 7, 9, 17

The Zero Power Rule

Consider the pattern in Table 8.1, where values for decreasing powers of 2 are shown.

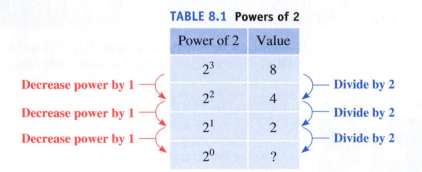

TABLE 8.1 Powers of 2

Power of 2	Value
2^3	8
2^2	4
2^1	2
2^0	?

Each time the power decreases by 1, the resulting value is divided by 2. For this pattern to continue, we need to define 2^0 to be 1 because dividing 2 by 2 results in a value of 1. This discussion suggests that $2^0 = 1$, which is generalized in the following rule.

THE ZERO POWER RULE

For any nonzero real number base a,

$$a^0 = 1.$$

The expression 0^0 is undefined.

EXAMPLE 2   **Using the zero power rule**

Simplify each expression. Assume that all variables represent nonzero numbers.
(a) 8^0 **(b)** $(-2)^0$ **(c)** -3^0 **(d)** $10x^0$

Solution
(a) $8^0 = 1$
(b) The base is -2, so the expression simplifies to $(-2)^0 = 1$.
(c) The base is 3. By the order of operations agreement, we raise 3 to the 0 power before negating the result. The expression simplifies to $\mathbf{-3^0} = -(\mathbf{1}) = -1$.
(d) $10x^0 = 10 \cdot \mathbf{x^0} = 10 \cdot \mathbf{1} = 10$

Now Try Exercises 21, 23, 25, 27

MAKING CONNECTIONS

The Quotient and Zero Power Rules

If a is a nonzero number, then a raised to a power is also a nonzero number. Since the result is 1 whenever a nonzero number is divided by itself, we know that

$$\frac{a^7}{a^7} = 1.$$

Also, by the quotient rule, we also know that

$$\frac{a^7}{a^7} = a^{7-7} = a^0.$$

Because the expression $\frac{a^7}{a^7}$ is equal to 1 and is also equal to a^0, this discussion offers another example suggesting that $a^0 = 1$ when a is a nonzero number.

Negative Exponents

We can continue the pattern shown in Table 8.1 to obtain a new table, which includes zero and negative exponents. In Table 8.2, each new value is obtained by dividing the previous value by 2.

TABLE 8.2 Powers of 2

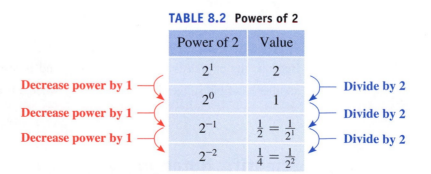

Power of 2	Value
2^1	2
2^0	1
2^{-1}	$\frac{1}{2} = \frac{1}{2^1}$
2^{-2}	$\frac{1}{4} = \frac{1}{2^2}$

Table 8.2 shows that $2^{-1} = \frac{1}{2^1}$ and $2^{-2} = \frac{1}{2^2}$. In other words, a negative power of 2 is the same as the reciprocal of the corresponding positive power of 2.

READING CHECK

• How is the reciprocal of the base used in computing a negative power?

NEGATIVE INTEGER EXPONENTS

For any nonzero real number base a and positive integer n,

$$a^{-n} = \frac{1}{a^n}.$$

Another way to state this rule is to say that a^{-n} is the reciprocal of a^n.

EXAMPLE 3 **Evaluating negative exponents**

Simplify each expression using the definition of negative integer exponents. Assume that all variables represent nonzero numbers.

(a) 2^{-3} **(b)** y^{-1} **(c)** $3x^{-5}$

Solution

(a) $2^{-3} = \dfrac{1}{2^3} = \dfrac{1}{8}$

(b) $y^{-1} = \dfrac{1}{y^1} = \dfrac{1}{y}$

(c) Only x (not $3x$) is the base for the exponent -5.

$$3x^{-5} = 3 \cdot x^{-5} = \frac{3}{1} \cdot \frac{1}{x^5} = \frac{3}{x^5}$$

Now Try Exercises 31, 33, 37

MAKING CONNECTIONS

The Quotient Rule and Negative Exponents

Using repeated multiplication, the expression $\dfrac{a^3}{a^5}$ can be simplified as

$$\frac{a^3}{a^5} = \frac{a \cdot a \cdot a}{a \cdot a \cdot a \cdot a \cdot a} = \frac{1}{a \cdot a} = \frac{1}{a^2}.$$

Also, the quotient rule suggests that

$$\frac{a^3}{a^5} = a^{3-5} = a^{-2}.$$

Because the expression $\dfrac{a^3}{a^5}$ is equal to $\dfrac{1}{a^2}$ and is also equal to a^{-2}, this discussion offers another example suggesting that $a^{-n} = \dfrac{1}{a^n}$ when a is a nonzero number.

Scientific Notation

In this section's A Look Into Math, we learned that the distance to the planet WASP-17b is 5,880,000,000,000,000 miles and the size of a typical virus is 0.00000468 inch. These numbers can be written in *scientific notation* as 5.88×10^{15} and 4.68×10^{-6}, respectively.

SCIENTIFIC NOTATION

A positive real number a is in **scientific notation** when it is written as

$$b \times 10^n,$$

where b is greater than or equal to 1 but less than 10.

EXAMPLE 4   **Recognizing scientific notation**

State whether each number is written in scientific notation.
(a) 14.8×10^{-3} **(b)** 5.2×2^4 **(c)** 8.3×10^{-5}

Solution
(a) The number is not written in scientific notation because 14.8, or b in the definition, is not between 1 and 10.
(b) The number is not written in scientific notation because the base for the exponent 4, or n in the definition, is not 10.
(c) The number is written in scientific notation.

Now Try Exercises 41, 45, 47

In Section 5.3 we learned that multiplying a number by a positive power of 10 moves the number's decimal point to the *right*. For example, we move the decimal point 2 places to the right when multiplying $6 \cdot 10^2$ because $10^2 = 100$ has 2 zeros.

$$6 \cdot 10^2 = 6.00(100) = 600$$

Similarly, since $10^5 = 100,000$ has 5 zeros, the *standard form* of 6.9×10^5 is found by moving the decimal point in 6.9 to the **right** 5 places.

$$6.90000$$

So, $6.9 \times 10^5 = 690,000$.

Likewise, dividing a number by a positive power of 10 moves that number's decimal point to the *left*. However, multiplying a number by a *negative* power of 10 such as 10^{-5} is the same as dividing it by 10^5.

$$10^{-5} = \frac{1}{10^5} = \frac{1}{100,000}$$

We can find the standard form of the number 6.9×10^{-5} by moving the decimal point in 6.9 to the **left** 5 places.

$$00006.9$$

So, $6.9 \times 10^{-5} = 0.000069$. This discussion suggests the following procedure for converting scientific notation to standard form.

CONVERTING SCIENTIFIC NOTATION TO STANDARD FORM

To write $b \times 10^n$ in standard form, move the decimal point in b

- $|n|$ places to the right if n is positive, or
- $|n|$ places to the left if n is negative.

EXAMPLE 5 **Converting scientific notation to standard form**

Write each number in standard form.
(a) 1.89×10^{-3} (b) 3.229×10^7

Solution
(a) The exponent is -3, so we move the decimal point in 1.89 to the **left** 3 places.

$$001.89$$

So, $1.89 \times 10^{-3} = 0.00189$.
(b) The exponent is **7**. Move the decimal point to the **right** 7 places.

$$3.2290000$$

So, $3.229 \times 10^7 = 32{,}290{,}000$.

Now Try Exercises 49, 51

To convert a number in standard form to scientific notation, use the following procedure.

CONVERTING STANDARD FORM TO SCIENTIFIC NOTATION

If a is a positive number expressed as a decimal,

1. Move the decimal point in a until it becomes a number b that is greater than or equal to 1 but less than 10.
2. Let the whole number n be the number of places the decimal point was moved.
3. Write a in scientific notation as follows.

  • If $a \geq 10$, then $a = b \times 10^n$.
  • If $a < 1$, then $a = b \times 10^{-n}$.
  • If $a \geq 1$ and $a < 10$, then $a = a \times 10^0$.

EXAMPLE 6 **Converting standard form to scientific notation**

Write each number in scientific notation.
(a) 308,000,000 (U.S. population in early 2010)
(b) 0.004 in. (average thickness of human hair)

Solution
(a) Move the (assumed) decimal point $n = 8$ places to get $b = \mathbf{3.08}$, which is between 1 and 10.

$$3.08000000.$$

Since $308{,}000{,}000 \geq 10$, it is written in scientific notation as $\mathbf{3.08} \times 10^8$.
(b) Move the decimal point $n = 3$ places to get $b = \mathbf{4}$.

$$0.004.$$

Since $0.004 < 1$, it is written in scientific notation as $\mathbf{4} \times 10^{-3}$.

Now Try Exercises 57, 59

Multiplying and Dividing with Scientific Notation

We can use the product and quotient rules for exponents, as well as the properties of real numbers, to find the products and quotients of numbers written in scientific notation.

EXAMPLE 7   **Multiplying and dividing numbers in scientific notation**

Multiply or divide as indicated. Write answers in standard form and in scientific notation.

(a) $(3 \times 10^5)(8 \times 10^4)$

(b) $\dfrac{8.4 \times 10^{-5}}{2.1 \times 10^2}$

Solution

(a)
$$
\begin{aligned}
(3 \times 10^5)(8 \times 10^4) &= (3 \cdot 8) \times (10^5 \cdot 10^4) && \text{Properties of real numbers} \\
&= 24 \times 10^{5+4} && \text{Multiply; product rule} \\
&= 24 \times 10^9 && \text{Simplify.} \\
&= 24{,}000{,}000{,}000 && \text{Standard form}
\end{aligned}
$$

The result is 24,000,000,000 and is written in scientific notation as 2.4×10^{10}.

(b)
$$
\begin{aligned}
\frac{8.4 \times 10^{-5}}{2.1 \times 10^2} &= \frac{8.4}{2.1} \times \frac{10^{-5}}{10^2} && \text{Properties of real numbers} \\
&= 4 \times 10^{-5-2} && \text{Divide; quotient rule} \\
&= 4 \times 10^{-7} && \text{Simplify.} \\
&= 0.0000004 && \text{Standard form}
\end{aligned}
$$

The result is 0.0000004 and is written in scientific notation as 4×10^{-7}.

CALCULATOR HELP

To work with scientific notation on a calculator, see the Appendix (page AP-6).

■ **Now Try Exercises 63, 69**

8.2   Putting It All Together

CONCEPT	COMMENTS	EXAMPLES
The Quotient Rule	For any nonzero real number base a and integer exponents m and n, $$\frac{a^m}{a^n} = a^{m-n}.$$	$\dfrac{5^6}{5^4} = 5^{6-4} = 5^2 = 25$ $\dfrac{x^{11}}{x^5} = x^{11-5} = x^6$
The Zero Power Rule	For any nonzero real number base a, $$a^0 = 1$$	$7^0 = 1$ $(-14)^0 = 1$
Negative Integer Exponents	For any nonzero real number base a and positive integer n, $$a^{-n} = \frac{1}{a^n}.$$	$9^{-2} = \dfrac{1}{9^2} = \dfrac{1}{81}$ $y^{-7} = \dfrac{1}{y^7}$
Scientific Notation	A positive real number a is in scientific notation when it is written as $$b \times 10^n,$$ where b is greater than or equal to 1 but less than 10.	The numbers $$3.9 \times 10^{13} \text{ and } 4.32 \times 10^{-5}$$ are in scientific notation. The numbers $$207 \times 10^{-3} \text{ and } 4.8 \times 3^6$$ are *not* in scientific notation.

continued on next page

continued from previous page

CONCEPT	COMMENTS	EXAMPLES
Converting Scientific Notation to Standard Form	To write $b \times 10^n$ in standard form, move the decimal point in b • $|n|$ places to the right if n is positive, or • $|n|$ places to the left if n is negative.	Write 6.1×10^{-3} in standard form. 006.1 So, $6.1 \times 10^{-3} = 0.0061$.
Converting Standard Form to Scientific Notation	If a is a positive number expressed as a decimal, 1. Move the decimal point in a until it becomes a number b that is greater than or equal to 1 but less than 10. 2. Let the whole number n be the number of places the decimal point was moved. 3. Write a in scientific notation as follows. • If $a \geq 10$, then $a = b \times 10^n$. • If $a < 1$, then $a = b \times 10^{-n}$. • If $a \geq 1$ and $a < 10$, then $a = a \times 10^0$.	Write 34,700 in scientific notation. 3.4700. Since $34,700 \geq 10$, it is written in scientific notation as 3.47×10^4.

8.2 Exercises

MyMathLab

Math XL
PRACTICE   WATCH   DOWNLOAD   READ   REVIEW

CONCEPTS AND VOCABULARY

1. If a is a nonzero real number and m and n are integer exponents, then $\frac{a^m}{a^n} =$ _____.

2. If a is a nonzero real number, then $a^0 =$ _____.

3. If a is a nonzero real number and n is a positive integer, then $a^{-n} =$ _____.

4. The positive real number a is written in _____ when it is written as $b \times 10^n$, where $b \leq 1$ and $b < 10$, and n is an integer.

5. When converting $b \times 10^n$ to standard form, move the decimal point to the right $|n|$ places if n is _____.

6. If the decimal number a is less than 1, then the exponent in the scientific notation for a is _____.

RULES FOR EXPONENTS

Exercises 7–40: Simplify the given expression using the definition for negative exponents. Assume that all variables represent nonzero numbers.

7. $\dfrac{3^8}{3^5}$

8. $\dfrac{6^9}{6^7}$

9. $\dfrac{w^{11}}{w^4}$

10. $\dfrac{y^{17}}{y^3}$

11. $-\dfrac{a^{12}}{a^9}$

12. $-\dfrac{x^4}{x^2}$

13. $\dfrac{x^5}{x}$

14. $\dfrac{m^4}{m}$

15. $\dfrac{10x^3}{2x^2}$

16. $\dfrac{24y^6}{8y^5}$

17. $\dfrac{18w^5}{9w^2}$

18. $\dfrac{20n^9}{5n^3}$

19. $\dfrac{-36b^{13}}{12b^7}$

20. $\dfrac{48k^{12}}{-8k^5}$

21. 7^0

22. 1^0

23. $(-12)^0$

24. $(-6)^0$

25. -9^0

26. -42^0

27. $32x^0$

28. $16y^0$

29. $-14m^0$

30. $-25n^0$

31. 6^{-2}

32. 4^{-3}

33. g^{-1}

34. k^{-1}

35. x^{-5}

36. y^{-9}

37. $7a^{-3}$

38. $15b^{-4}$

39. $-24n^{-6}$

40. $-13m^{-5}$

SCIENTIFIC NOTATION

Exercises 41–48: State whether the given number is written in scientific notation.

41. 167.4×10^{13}

42. 38.99×10^{-2}

43. 9.2×10^7

44. 2.75×10^{-3}

45. 7.88×6^3

46. 1.3×4^{-2}

47. 3.1×10^{-12}

48. 8.92×10^6

Exercises 49–56: Write the number in standard form.

49. 6.43×10^9

50. 3.7×10^{-5}

51. 5.9×10^{-4}

52. 2.85×10^2

53. 4.11×10^4

54. 8×10^{-2}

55. 3×10^0

56. 4.5×10^0

Exercises 57–62: Write the number in scientific notation.

57. 86,400 (number of seconds in one day)

58. \$40,000,000,000 (net worth of Bill Gates in 2009)

59. 0.0000000068 (probability of Powerball jackpot)

60. 0.0043 in. (thickness of a dollar bill)

61. 5,947,200,000,000,000,000,000,000 kg (Earth's mass)

62. 0.000125 lb (weight of a small feather)

Exercises 63–70: Multiply or divide as indicated. Write your answers in standard form and in scientific notation.

63. $(4 \times 10^3)(9 \times 10^2)$

64. $(8 \times 10^{-6})(2 \times 10^3)$

65. $\dfrac{2.2 \times 10^8}{1.1 \times 10^3}$

66. $\dfrac{9 \times 10^2}{3 \times 10^{-4}}$

67. $(6.1 \times 10^{-7})(2 \times 10^4)$

68. $(3 \times 10^6)(5 \times 10^{-1})$

69. $\dfrac{7.5 \times 10^{-3}}{3 \times 10^2}$

70. $\dfrac{4.2 \times 10^2}{1.4 \times 10^8}$

APPLICATIONS

71. *Age of Mars* In 2008, NASA landed the Phoenix Lander on Mars, a planet that is 8,600,000,000 years old. Write the age of Mars in scientific notation. (*Source:* NASA.)

72. *Computer Processor* The Intel Atom processor is just 0.000000045 meter in length. Write this number in scientific notation. (*Source:* Intel Corp.)

73. *Recycling* In 2008, about 86,000,000,000 pounds of paper were recycled in the United States. Write this number in scientific notation. (*Source:* Environmental Protection Agency.)

74. *iPod Nano* The iPod Nano is 0.24 inch thick. Write this number in scientific notation. (*Source:* Apple Corp.)

75. *National Debt* The population of the United States reached 310,000,000 at the end of 2010. That year, the national debt was \$13,020,000,000,000. (*Source:* National Public Radio.)
(**a**) Write the 2010 population and the national debt in scientific notation.
(**b**) Using scientific notation, divide the national debt by the population to determine how much debt there was for each U.S. citizen in 2010.

76. *National Debt* The population of Canada reached 33,000,000 in 2010. That year, the Canadian national debt reached \$528,000,000,000. (*Source:* Canadian Department of Finance.)
(**a**) Write the 2010 Canadian population and national debt in scientific notation.
(**b**) Using scientific notation, divide the national debt by the population to determine the amount of debt for each Canadian citizen in 2010.

WRITING ABOUT MATHEMATICS

77. When converting from standard form to scientific notation, how do you determine whether the exponent on 10 is positive or negative?

78. Give a real-world example of a very large quantity and write the number in scientific notation. Do the same for a very small quantity.

1. Multiply.
   (a) $y^7 \cdot y^3$ (b) $-2m^4 \cdot 9m^2$

2. Simplify.
   (a) $(x^2)^6$ (b) $b^5 \cdot (b^3)^2$
   (c) $(2x^5)^3$ (d) $(-n^3)^4$
   (e) $(-7x^2y^3)^2$ (f) $(-2ab^2)^3(-3a^4b)^2$

3. Simplify each expression using the definition for negative exponents. Assume that all variables represent nonzero numbers.
   (a) $\dfrac{w^{14}}{w^6}$ (b) $\dfrac{-32n^7}{4n^3}$
   (c) $(-19)^0$ (d) $16m^{-3}$

4. Write the number in standard form.
   (a) 2.8×10^{-6} (b) 5.18×10^5

5. Write the number in scientific notation.
   (a) 34,000,000   (b) 0.0000095

6. Multiply or divide as indicated. Write your answers in standard form and in scientific notation.
   (a) $\dfrac{4.5 \times 10^2}{9 \times 10^{-3}}$ (b) $(2 \times 10^{-4})(7 \times 10^{-1})$

8.3 Adding and Subtracting Polynomials

Monomials and Polynomials • Evaluating Polynomials • Adding Polynomials in One Variable • Subtracting Polynomials in One Variable

A LOOK INTO MATH ▶ Many video games award higher scores to players who can complete a level of the game in the shortest time. To perform a given task such as steering a car or shooting a target, a player must first *react* to the visual image on the screen and then *perform* the appropriate task. *Reaction time* and *performance time* can be represented by *polynomial* expressions. We can then add those expressions to find the total time. In this section, we learn about polynomial expressions and how to add and subtract them.

Monomials and Polynomials

NEW VOCABULARY

☐ Monomial
☐ Degree of a monomial
☐ Coefficient of a monomial
☐ Polynomial
☐ Degree of a polynomial
☐ Descending order

A **monomial** is a number, variable, or product of numbers and variables raised to natural number powers. In this section, we will focus on monomials that have, at most, one variable. Examples of monomials include

$$12, \quad x^4, \quad 3y^5, \quad \text{and} \quad -\frac{2}{3}a^2.$$

Monomials do not contain division by a variable. For example, the expression $\frac{6}{w}$ is not a monomial. Also, expressions that contain addition or subtraction signs are *not* monomials.

The **degree of a monomial** with one variable is the exponent on that variable. If a monomial is a nonzero number (with no variable), it has degree 0. The **coefficient of a monomial** is the number in the monomial. Table 8.3 shows properties of several monomials.

READING CHECK

• How is the degree of a monomial determined?

TABLE 8.3 Properties of Monomials

Monomial	-7	$3x$	$-y^3$	b^5
Degree	0	1	3	5
Coefficient	-7	3	-1	1

NOTE: When there is no number written in a monomial, its coefficient is either 1 or -1, as shown in the last two columns of Table 8.3.

A **polynomial** is the sum of one or more monomials. Each monomial is called a *term* of the polynomial. Addition and subtraction signs separate terms, where subtraction implies that a term is negative. Examples of polynomials with one variable include

$$7x^3, \quad 2y - 8, \quad 3w^2 + 2w - 6, \quad \text{and} \quad -5a^5 - 2a^2 + a - 1.$$

These polynomials have 1, 2, 3, and 4 terms, respectively. The **degree of a polynomial** is the degree of the term (or monomial) with the highest degree.

READING CHECK

• How is the degree of a polynomial determined?

EXAMPLE 1   **Identifying properties of polynomials**

Determine if the expression is a polynomial. If so, state the number of terms in the polynomial and give its degree.

(a) $8x^5 + 3x^3 + 2x + 1$ **(b)** $-3y^2 + 7$ **(c)** $2a^3 - \dfrac{1}{a}$

Solution
(a) The expression $8x^5 + 3x^3 + 2x + 1$ is a polynomial with 4 terms. The first term, $8x^5$, has degree 5 because the exponent on the variable is 5. Similarly, the second term, $3x^3$, has degree 3, and the third term, $2x$, has degree 1. The last term, 1, has degree 0 because it is a nonzero number. The term with the highest degree is $8x^5$, so the polynomial has degree **5**.
(b) The polynomial $-3y^2 + 7$ has 2 terms. The first term, $-3y^2$, has degree 2 because the exponent on the variable is 2. The second term, 7, has degree 0 because it is a nonzero number. The term with the highest degree is $-3y^2$, so the polynomial has degree **2**.
(c) The expression $2a^3 - \frac{1}{a}$ is not a polynomial because it contains division by a variable.

Now Try Exercises 11, 13, 15

STUDY TIP

When evaluating expressions, don't forget to use the order of operations agreement. Also, remember to use parentheses when replacing the variable with the given value.

Evaluating Polynomials

Polynomials are evaluated in the same way as other algebraic expressions. In the next example, we evaluate polynomials for a given value of the variable.

EXAMPLE 2   **Evaluating polynomials**

Evaluate each polynomial for $x = -2$.
(a) $7x^3 - 3x^2 + 5x$ **(b)** $-4x^2 - x$

Solution
(a) Replace x with -2 in $7x^3 - 3x^2 + 5x$.

$$
\begin{aligned}
7(-2)^3 - 3(-2)^2 + 5(-2) &= 7(-8) - 3(4) + 5(-2) \qquad \text{Evaluate the exponents.} \\
&= -56 - 12 + (-10) \qquad \text{Multiply.} \\
&= -78 \qquad \text{Simplify.}
\end{aligned}
$$

(b) Replace x with -2 in $-4x^2 - x$.

$$-4(-2)^2 - (-2) = -4(4) - (-2) \quad \text{Evaluate the exponent.}$$
$$= -16 - (-2) \quad \text{Multiply.}$$
$$= -14 \quad \text{Simplify.}$$

Now Try Exercises 19, 21

▶ **REAL-WORLD CONNECTION** For the first few seconds after a baseball player hits a "fly ball," the height, in feet, of the ball can be found by evaluating a polynomial such as

$$-16x^2 + 96x + 3,$$

where x represents the number of seconds after the ball is hit. In the next example, we find the height of a baseball.

EXAMPLE 3 **Finding the height of a baseball**

Use the polynomial expression $-16x^2 + 96x + 3$ to find the height, in feet, of a baseball after 2 seconds and after 3 seconds.

Solution
To find the height of the baseball after 2 seconds, replace x with **2** in $-16x^2 + 96x + 3$.

$$-16(2)^2 + 96(2) + 3 = -16(4) + 96(2) + 3 \quad \text{Evaluate the exponent.}$$
$$= -64 + 192 + 3 \quad \text{Multiply.}$$
$$= 131 \quad \text{Simplify.}$$

After 2 seconds, the height of the baseball is 131 feet.

To find the height of the baseball after 3 seconds, replace x with **3** in $-16x^2 + 96x + 3$.

$$-16(3)^2 + 96(3) + 3 = -16(9) + 96(3) + 3 \quad \text{Evaluate the exponent.}$$
$$= -144 + 288 + 3 \quad \text{Multiply.}$$
$$= 147 \quad \text{Simplify.}$$

After 3 seconds, the height of the baseball is 147 feet.

Now Try Exercise 81

Adding Polynomials in One Variable

Recall that the commutative and associative properties of addition allow us to rearrange a sum in any order. For example, if we write each subtraction in the polynomial

$$2x - 5 - 4x + 10$$

as an addition, by "adding the opposite," we get the expression

$$2x + (-5) + (-4x) + 10.$$

This *sum* can be rearranged and written as

$$2x + (-4x) + (-5) + 10.$$

Finally, if we change back to the original two subtractions, we get

$$2x - 4x - 5 + 10.$$

Comparing the first and last polynomials in this example, we see that

$$2x - 5 - 4x + 10 \quad \text{is equal to} \quad 2x - 4x - 5 + 10.$$

EXAMPLE 8 | **Writing the opposite of a polynomial**

Write the opposite of each polynomial.
(a) $-7x^2 - 8x + 17$ **(b)** $y^4 + 3y^2 - 4y + 9$

Solution
(a) Change the sign of each term to its opposite. The opposite of $-7x^2 - 8x + 17$ is $7x^2 + 8x - 17$.
(b) The opposite of $y^4 + 3y^2 - 4y + 9$ is $-y^4 - 3y^2 + 4y - 9$.

Now Try Exercises 57, 59

Now that we know how to write the opposite of a polynomial, we use the following rule for subtracting one polynomial from another.

SUBTRACTING POLYNOMIALS

To subtract one polynomial from another, add the opposite of the second polynomial to the first polynomial.

EXAMPLE 9 | **Subtracting polynomials**

Subtract.
(a) $(8x + 15) - (2x - 9)$ **(b)** $(7y^2 - y) - (-y^2 - y + 1)$

Solution
(a) Add the opposite of $2x - 9$ to $8x + 15$.

$$
\begin{aligned}
(8x + 15) - (2x - 9) &= (8x + 15) + (-2x + 9) &&\text{Add the opposite.} \\
&= 8x + 15 + (-2x) + 9 &&\text{Remove parentheses.} \\
&= [8x + (-2x)] + (15 + 9) &&\text{Group like terms.} \\
&= 6x + 24 &&\text{Combine like terms.}
\end{aligned}
$$

(b)
$$
\begin{aligned}
(7y^2 - y) - (-y^2 - y + 1) &= (7y^2 - y) + (y^2 + y - 1) &&\text{Add the opposite.} \\
&= 7y^2 - y + y^2 + y - 1 &&\text{Remove parentheses.} \\
&= (7y^2 + y^2) + (-y + y) - 1 &&\text{Group like terms.} \\
&= 8y^2 - 1 &&\text{Combine like terms.}
\end{aligned}
$$

Now Try Exercises 63, 67

Polynomials can also be subtracted without writing the regrouped terms, as illustrated in the next example.

EXAMPLE 10 | **Subtracting polynomials**

Subtract $(18x^2 - 9x + 3) - (5x^2 + 9)$.

Solution
Add the opposite of $5x^2 + 9$ to $18x^2 - 9x + 3$.

$$(18x^2 - 9x + 3) + (-5x^2 - 9)$$

Remove parentheses and combine like terms as shown.

$$18x^2 \quad - \quad 9x \quad + \quad 3 \quad - \quad 5x^2 \quad - \quad 9$$

The resulting sum is the polynomial

$$13x^2 - 9x - 6.$$

Now Try Exercises 73

Because polynomial subtraction can be written as polynomial addition, we can also use a vertical format to subtract polynomials, as shown in the next example.

EXAMPLE 11   **Subtracting polynomials vertically**

Subtract $-5x^2 - 3x + 4$ from $9x^2 - 2x - 1$ using a vertical format.

Solution
Since $-5x^2 - 3x + 4$ is being subtracted *from* $9x^2 - 2x - 1$, we write the subtraction in a vertical format as shown and then change the subtraction to addition of the opposite.

$$
\begin{array}{r}
9x^2 - 2x - 1 \\
-(-5x^2 - 3x + 4)
\end{array}
\qquad
\xrightarrow[\text{opposite}]{\text{Add the}}
\qquad
\begin{array}{r}
9x^2 - 2x - 1 \\
+\ 5x^2 + 3x - 4 \\
\hline
14x^2 +\ \ x - 5
\end{array}
$$

Now Try Exercise 79

8.3   Putting It All Together

CONCEPT	COMMENTS	EXAMPLES
Monomials	A monomial is a number, variable, or product of numbers and variables raised to natural number powers.	9, $\ y^2$, $\ -2x^5$, and $\ -\frac{4}{5}x^{11}$
Degree and Coefficient of a Monomial	The degree of a monomial with one variable is the exponent on that variable. If the variable in a monomial has no exponent, its degree is 1. If a monomial has no variable, its degree is 0. The number in a monomial is called its coefficient.	$-5x^3$ has degree 3 and coefficient -5. $2y$ has degree 1 and coefficient 2. x^9 has degree 9 and coefficient 1.
Polynomials	A polynomial is the sum of one or more monomials.	$-5x^3 + 2x^2 - 4x + 7$ $7x^2 - 9$
Degree of a Polynomial	The degree of a polynomial is the degree of the term (or monomial) with the highest degree.	The degree of $4x^3 - 2x + 5$ is 3, and the degree of $4y - 7y^2 + 6$ is 2.

CONCEPT	COMMENTS	EXAMPLES
Rearranging Terms of a Polynomial	Using the commutative and associative properties of addition, we can rearrange the terms of a polynomial in any order. Make sure that the appropriate sign stays with its corresponding term.	The polynomial $-3x^2 - 4x + 8$ may be written as $-4x - 3x^2 + 8$ or as $8 - 4x - 3x^2$.
Adding Polynomials	To add polynomials, combine like terms.	$(6y^2 + 8y - 3) + (y^2 - 3y)$ $= 6y^2 + 8y - 3 + y^2 - 3y$ $= (6y^2 + y^2) + (8y - 3y) - 3$ $= 7y^2 + 5y - 3$
The Opposite of a Polynomial	To write the opposite of a polynomial, change the sign of each term to its opposite.	The opposite of $-2x^2 - 7x + 1$ is $2x^2 + 7x - 1$, and the opposite of $4 - 8x^2$ is $-4 + 8x^2$.
Subtracting Polynomials	To subtract one polynomial from another, add the opposite of the second polynomial to the first polynomial.	$(5y^2 + 1) - (8y^2 - 2y)$ $= (5y^2 + 1) + (-8y^2 + 2y)$ $= 5y^2 + 1 - 8y^2 + 2y$ $= (5y^2 - 8y^2) + 2y + 1$ $= -3y^2 + 2y + 1$

8.3 Exercises

MyMathLab   Math XL PRACTICE   WATCH   DOWNLOAD   READ   REVIEW

CONCEPTS AND VOCABULARY

1. A(n) _____ is a number, variable, or product of numbers and variables raised to powers.

2. For a monomial with one variable, the _____ of the monomial is the exponent on the variable.

3. The number in a monomial is called the _____ of the monomial.

4. When there is no number written in a monomial, its coefficient is either _____ or _____.

5. A(n) _____ is the sum of one or more monomials.

6. Each monomial in a polynomial is called a(n) _____ of the polynomial.

7. The terms of a polynomial are written in _____ order if the exponents on the variables in the terms decrease as the terms are written from left to right.

8. To write the opposite of a polynomial, change the _____ of each term to its opposite.

9. To subtract one polynomial from another, add the _____ of the second polynomial to the first polynomial.

10. Polynomials can be added and subtracted vertically by aligning _____ terms.

POLYNOMIALS

Exercises 11–18: Determine if the expression is a polynomial. If so, state the number of terms in the polynomial and give its degree.

11. $4x^5 - 2x^3$

12. $x^3 - 7x^2 + 6x - 5$

13. $4x^3 - \dfrac{1}{x}$

14. $\dfrac{2}{x + 1}$

15. $1 + x - x^2 - x^3 - x^4$

16. -12

17. $132x$

18. $8 - x$

Exercises 19–26: Evaluate the polynomial for the given value of x.

19. $5x^2 - 2x$, for $x = -1$

20. $2x^3 + x - 5$, for $x = 2$

21. $-2x^3 - 2x + 19$, for $x = 1$

22. $-3x - x^2$, for $x = -5$

23. $-x^3 + x$, for $x = -3$

24. $7x - x^2 + 10$, for $x = -2$

25. $\frac{1}{2}x^2 + 17$, for $x = 8$

26. $\frac{1}{3}x^2 - 2x$, for $x = -6$

ADDING POLYNOMIALS

Exercises 27–34: Write the polynomial so that its terms are in descending order.

27. $3 + 2x$

28. $7 - y^2$

29. $4 - 3n^2 + 2n$

30. $-5w + 2 + w^2$

31. $3x^3 - 5x + x^2$

32. $-y^2 - 3y^3 + 2y$

33. $n - 3n^2 - 2 + 4n^3$

34. $3a^2 - 1 + 2a - 6a^3$

Exercises 35–48: Add.

35. $(3y - 2) + (7y + 8)$

36. $(2x + 4) + (9x - 5)$

37. $(-w + 5) + (w - 4)$

38. $(2n - 8) + (2n + 8)$

39. $(-3a^2 + 2a - 1) + (4a^2 + 2a)$

40. $(2k^2 - 3k) + (8k^2 + 4k + 5)$

41. $(-3x^2 + 5x) + (2x^2 + 7)$

42. $(10y^2 - 5) + (y + 8)$

43. $(16w^3 + 4w - 3) + (-9w^3 + 3w^2 - 5)$

44. $(11m^3 - 3m^2 + 2) + (-5m^3 + m - 8)$

45. $(3x^3 + 7) + (-2x^2 - 5x)$

46. $(8 + 8y^2) + (9y - 2y^3)$

47. $(3x^2 + 2x - 5) + (-3x^2 - 2x + 5)$

48. $(-5y^2 - 3y + 1) + (5y^2 + 3y - 1)$

Exercises 49–54: Add using a vertical format.

49. $(-a + 1) + (2a - 9)$

50. $(y - 4) + (6y + 5)$

51. $(-x^2 + 5x - 3) + (x^2 + 10)$

52. $(10m^2 - 2m) + (-8m^2 - 6m - 9)$

53. $(12w^2 - w + 11) + (-4w^2 - 3w - 2)$

54. $(-3k^2 + 2k - 2) + (-5k^2 - 2k + 1)$

SUBTRACTING POLYNOMIALS

Exercises 55–62: Write the opposite of the polynomial.

55. $-4x - 13$

56. $2y + 5$

57. $2m^2 + 3m - 19$

58. $14h^2 - 7h + 2$

59. $-y^3 + 3y^2 + 2y - 6$

60. $b^3 - b^2 + 2b - 5$

61. $165n^2 - 12n - 17$

62. $-89w^2 - 210w + 5$

Exercises 63–74: Subtract.

63. $(8a - 2) - (3a + 8)$

64. $(2b + 6) - (3b - 2)$

65. $(9w + 1) - (9w - 7)$

66. $(m - 4) - (-m - 4)$

67. $(7n^2 - 3) - (n^2 - 4n + 11)$

68. $(5x^3 + 1) - (-9x^2 - 8x)$

69. $(-5x^2 + 3x) - (4x^2 + 19)$

70. $(11y^2 - 7) - (10y + 9)$

71. $(-12m^3 - 4m - 3) - (-9m^3 - 3m^2 + 5)$

72. $(q^3 + q^2 + 4) - (-5q^3 + q - 3)$

73. $(-5k^2 + 2k - 6) - (4k^2 + 7)$

74. $(x^2 + 2x - 4) - (x^2 + 2x - 5)$

Exercises 75–80: Subtract using a vertical format.

75. $(-6g - 1) - (g + 9)$

76. $(y + 11) - (y + 3)$

77. $(x^2 - 5x + 17) - (x^2 + 13)$

78. $(4n^2 - 7n) - (-2n^2 - 6n + 1)$

79. $(10w^2 - 3w + 16) - (-4w^2 + w - 2)$

80. $(-6y^2 + 2y - 2) - (-5y^2 + 2y - 1)$

APPLICATIONS

81. *Height of a Baseball* For the first few seconds after a baseball is hit, the polynomial

$$-16x^2 + 88x + 2$$

gives the height of the baseball in feet after x seconds. Find the height of the baseball after 1 second and after 3 seconds.

82. *Height of a Golf Ball* For the first few seconds after a golf ball is hit, the polynomial

$$-16x^2 + 105x$$

gives the height of the golf ball in feet after x seconds. Find the height of the golf ball after 2 seconds and after 4 seconds.

83. *Heart Rate* For the first few minutes after an athlete stops exercising, the polynomial

$$2x^2 - 30x + 160$$

gives the athlete's heart rate in beats per minute after x minutes. What is the heart rate 5 minutes after exercise stops?

84. *Insect Population* The polynomial

$$x^3 - 60x^2 + 860x + 4200, \text{ where } x \le 50$$

gives the insect population x days after the beginning of a scientific study. Determine the insect population after 2 days.

85. *Falling Object* The polynomial

$$-16x^2 + 144, \text{ where } x \le 3$$

gives the height in feet x seconds after an object is dropped from a height of 144 feet. Find the height of the object after 3 seconds. Interpret your answer.

86. *Falling Object* The polynomial

$$-16x^2 + 256, \text{ where } x \le 4$$

gives the height in feet x seconds after an object is dropped from a tall building. Find the height of the object the instant it is dropped.

87. *Group Rate* If the regular price for a room at a hotel is $120 and the manager offers a group rate that decreases this price by $2 for each room rented, then the *total* cost for a group of x people is given by

$$120x - 2x^2, \text{ where } x \le 60.$$

Find the total cost for a group of 30 people.

88. *Group Rate* Refer to Exercise 87. Find the total cost for a group of 60 people. Explain your result.

89. *Geometry* Find an expression for the perimeter of the triangle shown by adding the polynomials representing the lengths of the three sides.

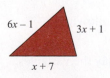

90. *Geometry* Find an expression for the perimeter of the figure shown by adding the polynomials representing the lengths of the four sides.

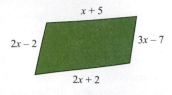

91. *Video Games* If the reaction time for a video game is given by $0.25x$ and the performance time is given by $3.75x + 8$ when a player is on level x of the game, write an expression for the total time needed to complete a task at this level.

92. *Video Games* If the reaction time for a video game is given by $0.35x + 2$ and the performance time is given by $2x^2 + 2.65x - 3$ when a player is on level x of the game, write an expression for the total time.

93. *Internet Sales* An Internet company makes a profit of $2x^2 - 3x + 6$ dollars x hours after launching a new product. If the profit in dollars from a second product is given by $x^2 + 2x - 1$, write an expression for the total profit for both products.

94. *Clearing Snow* A man can shovel $-0.1x^2 + 22x$ square feet of snow in x minutes, while his son can clear $90x$ square feet of snow in x minutes using a snow blower. Write an expression for the amount of snow they clear working together.

95. *Profit* If R represents the revenue from selling x items and C represents the cost of making x items, then the profit P for making and selling x items is given by the formula $P = R - C$. If a manufacturer finds that $R = 25x$ and $C = x^2 + 3x - 4$, write an expression for the profit.

96. *Height Difference* Two baseballs are hit at the same time. For the first few seconds, the height of the higher ball after x seconds is given by $-16x^2 + 96x + 3$ and the height of the lower ball is given by $-16x^2 + 88x + 2$. Write an expression for the difference in the heights of the two balls after x seconds.

WRITING ABOUT MATHEMATICS

97. Explain how to find the opposite of a polynomial.

98. A student says, "to add polynomials, add the first term of the first polynomial to the first term of the second polynomial, and then continue adding term by term." Explain why this will not always work.

Group Activity   Working with Real Data

Directions: Form a group of 2 to 4 people. Select someone to record the group's responses for this activity. All members of the group should work cooperatively to answer the questions. If your instructor asks for your results, each member of the group should be prepared to respond.

Global Warming From January to December 2009, the average monthly carbon dioxide levels in parts per million (ppm) measured at Mauna Loa, Hawaii are estimated by

$$-0.2x^2 + 2.7x + 381$$

where $x = 1$ corresponds to January, $x = 2$ corresponds to February, and so on. (*Source:* NOAA.)

(a) Carbon dioxide levels peak in July each year. Replace x with 7 (the 7th month) to find the maximum carbon dioxide levels in 2009.

(b) For comparison, carbon dioxide levels from January to December 2005 are estimated by

$$-0.2x^2 + 2.7x + 373.$$

Replace x with 7 to find the maximum carbon dioxide levels in July 2005.

(c) In which year were carbon dioxide levels higher, 2005 or 2009?

8.4   Multiplying and Factoring Polynomials

Multiplying Polynomials • Multiplying Polynomials Visually • Factoring Polynomials

A LOOK INTO MATH ▶

The price of a product affects the *demand* for the product. For example, very few people are willing to buy a pair of headphones for $560, but headphones priced at $19.95 are likely to sell quite well. The demand for a product is often represented mathematically by a polynomial, while a product's price is represented by a monomial. To find the revenue that the product brings to a company, we multiply the price (monomial) and the demand (polynomial). In this section, we learn how to multiply polynomial expressions.

Multiplying Polynomials

Before we can multiply two polynomials with any number of terms, we need to learn how to multiply two monomials.

NEW VOCABULARY

☐ Factor
☐ Greatest common factor (GCF)
☐ Completely factored

MULTIPLYING MONOMIALS In Section 8.3, we learned that a monomial is a number, variable, or product of numbers and variables raised to natural number powers. To multiply monomials, we often use the product rule for exponents from Section 8.1.

STUDY TIP

Are you keeping up with your assigned homework? Remember that regular and timely practice is important.

MULTIPLYING MONOMIALS IN ONE VARIABLE

To multiply monomials in one variable, do the following.

1. Multiply the coefficients.
2. Use the product rule for exponents to multiply the variables.
3. Write the above results as a product.

EXAMPLE 1   ## Multiplying monomials

Multiply.
(a) $3x^2 \cdot 7x^5$ **(b)** $(-8y^4)(9y)$ **(c)** $(-5m)(-m^3)(6m^2)$

Solution
(a) Since the product of the coefficients is $3 \cdot 7 = \mathbf{21}$ and the product of the variables is $x^2 \cdot x^5 = x^{2+5} = \mathbf{x^7}$, the product of the given monomials is $3x^2 \cdot 7x^5 = \mathbf{21x^7}$.
(b) The product of the coefficients is $-8 \cdot 9 = \mathbf{-72}$, and the product of the variables is $y^4 \cdot y = y^{4+1} = \mathbf{y^5}$. So, the product of the given monomials is $(-8y^4)(9y) = \mathbf{-72y^5}$.
(c) $(\mathbf{-5m})(\mathbf{-m^3})(\mathbf{6m^2}) = (\mathbf{-5})(\mathbf{-1})(\mathbf{6}) \cdot (\mathbf{m})(\mathbf{m^3})(\mathbf{m^2}) = \mathbf{30} \cdot \mathbf{m^{1+3+2}} = \mathbf{30m^6}$

Now Try Exercises 13, 15, 19

MULTIPLYING A MONOMIAL AND A POLYNOMIAL   To multiply a monomial and a polynomial, we use the distributive property to multiply each term of the polynomial by the monomial. Since *every term of a polynomial is a monomial*, we can use the same rules that were applied in Example 1.

EXAMPLE 2   ## Multiplying a monomial and a polynomial

Multiply.
(a) $2x(5x^2 - 7)$ **(b)** $-3y^3(4y^2 - 6y + 1)$

Solution
(a) Apply the distributive property to multiply every term in $(5x^2 - 7)$ by $\mathbf{2x}$.

$$\mathbf{2x}(5x^2 - 7) = \mathbf{2x} \cdot 5x^2 - \mathbf{2x} \cdot 7 \qquad \text{Distributive property}$$
$$= 10x^3 - 14x \qquad \text{Multiply monomials.}$$

(b) Note that the sign for the second product $\mathbf{-3y^3} \cdot (-6y)$ is $\mathbf{+}$.

$$\mathbf{-3y^3}(4y^2 - 6y + 1) = \mathbf{-3y^3} \cdot 4y^2 \mathbf{+} \mathbf{3y^3} \cdot 6y - \mathbf{3y^3} \cdot 1 \qquad \text{Distributive property}$$
$$= -12y^5 + 18y^4 - 3y^3 \qquad \text{Multiply monomials.}$$

Now Try Exercises 21, 29

MULTIPLYING POLYNOMIALS   To multiply polynomials with two or more terms, we apply the distributive property *more than once*. For example, the product $(2x + 3)(5x - 4)$ can be found by applying the distributive property a first time to get

$$(\mathbf{2x} + \mathbf{3})(5x - 4) = \mathbf{2x}(5x - 4) + \mathbf{3}(5x - 4).$$

Now, applying the distributive property to this new expression gives

$$\mathbf{2x}(5x - 4) + \mathbf{3}(5x - 4) = \mathbf{2x} \cdot 5x - \mathbf{2x} \cdot 4 + \mathbf{3} \cdot 5x - \mathbf{3} \cdot 4.$$

Note that this final expression is equal to the given product $(2x + 3)(5x - 4)$ and it could be written directly by multiplying every term in $(2x + 3)$ by every term in $(5x - 4)$.

READING CHECK

• What is the process used to multiply polynomials?

MULTIPLYING POLYNOMIALS

The product of two polynomials is found by multiplying every term in the first polynomial by every term in the second polynomial and then combining like terms.

EXAMPLE 3   **Multiplying polynomials**

Multiply.
(a) $(2x + 4)(x - 9)$ **(b)** $(2y^2 - 4)(5y + 8)$ **(c)** $(n + 2)(n^2 - 2n + 5)$

Solution
(a) Multiply every term in $(2x + 4)$ by every term in $(x - 9)$.

$$(2x + 4)(x - 9) = 2x \cdot x - 2x \cdot 9 + 4 \cdot x - 4 \cdot 9$$
$$= 2x^2 - 18x + 4x - 36 \qquad \text{Multiply monomials.}$$
$$= 2x^2 - 14x - 36 \qquad \text{Combine like terms.}$$

(b) Note that there are no like terms to combine after the monomials are multiplied.

$$(2y^2 - 4)(5y + 8) = 2y^2 \cdot 5y + 2y^2 \cdot 8 - 4 \cdot 5y - 4 \cdot 8$$
$$= 10y^3 + 16y^2 - 20y - 32 \qquad \text{Multiply monomials.}$$

(c) Multiply every term in $(n + 2)$ by every term in $(n^2 - 2n + 5)$.

$$(n + 2)(n^2 - 2n + 5) = n \cdot n^2 - n \cdot 2n + n \cdot 5 + 2 \cdot n^2 - 2 \cdot 2n + 2 \cdot 5$$
$$= n^3 - 2n^2 + 5n + 2n^2 - 4n + 10 \qquad \text{Multiply monomials.}$$
$$= n^3 + n + 10 \qquad \text{Combine like terms.}$$

Now Try Exercises 31, 33, 39

NOTE: The process used to multiply two *binomials* (two-term polynomials) is sometimes called *FOIL* because we multiply the first terms (F), outside terms (O), inside terms (I), and last terms (L). The product $(2x + 4)(x - 9)$ from Example 3(a) is found as follows.

Multiply the *First terms*:	$(2x + 4)(x - 9)$	$\rightarrow$	$2x^2$
Multiply the *Outside terms*:	$(2x + 4)(x - 9)$	$\rightarrow$	$-18x$
Multiply the *Inside terms*:	$(2x + 4)(x - 9)$	$\rightarrow$	$4x$
Multiply the *Last terms*:	$(2x + 4)(x - 9)$	$\rightarrow$	-36

The resulting monomials are added and like terms combined to get $2x^2 - 14x - 36$. It is important to remember that the *FOIL* process can only be used to multiply *two binomials*.

MULTIPLYING POLYNOMIALS VERTICALLY One way to keep all of the resulting terms organized when multiplying polynomials is to use a vertical format. To see how the vertical format works, compare the whole number product $321 \cdot 12$ with the polynomial product $(3x^2 + 2x + 1)(x + 2)$.

	$3\ 2\ 1$	$3x^2 + 2x + 1$	
	$\times\quad 1\ 2$	$\times \qquad\qquad x + 2$	
Multiply top row by 2.	$6\ 4\ 2$	$6x^2 + 4x + 2$	Multiply top row by 2.
Multiply top row by 1.	$3\ 2\ 1$	$3x^3 + 2x^2 + \ x$	Multiply top row by x.
Add like place values.	$3\ 8\ 5\ 2$	$3x^3 + 8x^2 + 5x + 2$	Combine like terms.

Like *place values* are aligned vertically when multiplying whole numbers, and like *terms* are aligned vertically when multiplying polynomials.

EXAMPLE 4   **Multiplying polynomials vertically**

Multiply $(2x - 4)(3x^2 + 6x - 5)$ using a vertical format.

Solution
Be sure to align like terms vertically.

$$
\begin{array}{r}
3x^2 + 6x - 5 \\
\times \qquad 2x - 4 \\
\hline
-12x^2 - 24x + 20 \\
6x^3 + 12x^2 - 10x \\
\hline
6x^3 \qquad\quad - 34x + 20
\end{array}
$$

Multiply top row by -4.

Multiply top row by $2x$.

Combine like terms.

The resulting product is $6x^3 - 34x + 20$.

Now Try Exercise 45

Multiplying Polynomials Visually

If the length of a rectangle is represented by a polynomial and the width is represented by a second polynomial, then the area of the rectangle represents the product of the polynomials. In other words, a visual (or geometric) process can be used to multiply two polynomials. For example, to multiply $(2x + 5)(x + 3)$, we label a rectangle as shown in Figure 8.1.

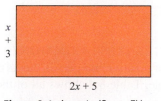

x
$+$
3

$2x + 5$

Figure 8.1 Area is $(2x + 5)(x + 3)$

The area of the rectangle in Figure 8.1 is $(2x + 5)(x + 3)$, the product of the length and width. Before combining like terms, this product results in four terms.

$$(2x + 5)(x + 3) = 2x^2 + 6x + 5x + 15$$

These four terms can be seen visually when the large rectangle is divided into four smaller rectangles with areas of $2x^2$, $6x$, $5x$, and 15, as shown in Figure 8.2.

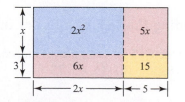

Figure 8.2 Multiplying Polynomials Visually

The total area of the large rectangle is the sum of the areas of the four smaller rectangles, $2x^2 + 6x + 5x + 15$. Combining like terms gives the product of $2x^2 + 11x + 15$.

EXAMPLE 5 **Multiplying polynomials visually**

Multiply $(x + 2)(5x + 1)$ visually.

Solution

The product $5x^2 + x + 10x + 2$ is shown in Figure 8.3. After combining like terms, the simplified product is $5x^2 + 11x + 2$.

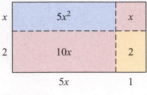

Figure 8.3

Now Try Exercise 49

Factoring Polynomials

When two or more numbers are multiplied, each number is called a factor. For example, in the equation $4 \cdot 6 = 24$, the numbers 4 and 6 are factors and we say that 24 can be *factored* into the product $4 \cdot 6$. Other ways to factor 24 include

$$2 \cdot 12, \quad 8 \cdot 3, \quad 2 \cdot 2 \cdot 6, \quad \text{and} \quad 2 \cdot 2 \cdot 2 \cdot 3.$$

When we **factor** a positive number into natural number factors, we reverse the multiplication process and write the number as a product of two or more smaller numbers. Similarly, we factor a polynomial by writing the polynomial as a product of two or more *lower degree* polynomials. For example, possible ways to factor the polynomial $12x^5 + 6x^4$ include

$$2x(6x^4 + 3x^3), \quad x^2(12x^3 + 6x^2), \quad \text{and} \quad 6x^4(2x + 1).$$

THE GREATEST COMMON FACTOR (GCF) The **greatest common factor (GCF)** of a list of positive integers is the largest integer that is a factor of every integer in the list. For example, Table 8.4 shows all of the factors of 24 and 40.

TABLE 8.4 **Factors of 24 and 40**

24	1	2	3	4	6	8	12	24
40	1	2	4	5	8	10	20	40

The integers **2**, **4**, and **8** are all *common factors* of 24 and 40, but **8** is the GCF because it is the largest of the common factors.

Because it is not always practical to list all factors of two or more numbers, we often use prime factorization to find the GCF of a list of integers, as illustrated in the next example.

EXAMPLE 6 **Finding the GCF of a list of integers**

Find the GCF of each list of integers.
(a) $18, 30$ **(b)** $15, 45, 75$

Solution

(a) Begin by finding the prime factorizations for 18 and 30.

$$18 = 2 \cdot 3 \cdot 3 \quad \text{Prime factors of 18}$$

$$30 = 2 \cdot 3 \cdot 5 \quad \text{Prime factors of 30}$$

The GCF is the product of the *common* prime factors: $2 \cdot 3 = 6$.

(b) Find the prime factorizations for 15, 45, and 75.

$$15 = \mathbf{3 \cdot 5} \qquad \text{Prime factors of 15}$$
$$45 = \mathbf{3} \cdot 3 \cdot \mathbf{5} \qquad \text{Prime factors of 45}$$
$$75 = \mathbf{3 \cdot 5} \cdot 5 \qquad \text{Prime factors of 75}$$

The GCF is $\mathbf{3 \cdot 5} = 15$.

▌ **Now Try Exercises 51, 55**

To use the process in Example 6 to find the greatest common factor of a list of terms, we must first *completely factor* each term. A term is **completely factored** when its coefficient is written as a product of prime numbers and any powers of variables are written as repeated multiplication.

EXAMPLE 7 **Finding the GCF of a list of terms**

Find the GCF of each list of terms.
(a) $4x^2, 12x^5$ **(b)** $10y^4, 30y^3, 45y^6$

Solution
(a) Find the complete factorizations for $4x^2$ and $12x^5$.

$$4x^2 = \mathbf{2 \cdot 2 \cdot x \cdot x} \qquad\qquad \text{Completely factor } 4x^2.$$
$$12x^5 = \mathbf{2 \cdot 2} \cdot 3 \cdot \mathbf{x \cdot x} \cdot x \cdot x \cdot x \quad \text{Completely factor } 12x^5.$$

The GCF is the product of the *common* factors: $\mathbf{2 \cdot 2 \cdot x \cdot x} = 4x^2$.
(b) Find the complete factorizations for $10y^4, 30y^3$, and $45y^6$.

$$10y^4 = 2 \cdot \mathbf{5 \cdot y \cdot y \cdot y} \cdot y \qquad\qquad \text{Completely factor } 10y^4.$$
$$30y^3 = 2 \cdot 3 \cdot \mathbf{5 \cdot y \cdot y \cdot y} \qquad\qquad \text{Completely factor } 30y^3.$$
$$45y^6 = 3 \cdot 3 \cdot \mathbf{5 \cdot y \cdot y \cdot y} \cdot y \cdot y \cdot y \quad \text{Completely factor } 45y^6.$$

The GCF is $\mathbf{5 \cdot y \cdot y \cdot y} = 5y^3$.

▌ **Now Try Exercises 59, 61**

READING CHECK

• What is the GCF and how is it found?

The terms in a polynomial are not separated by commas; instead, they are separated by plus signs $(+)$ and minus signs $(-)$. However, the process for finding the GCF of the terms in a polynomial is no different from the process used in Example 7.

EXAMPLE 8 **Finding the GCF of the terms in a polynomial**

Find the GCF of the terms in the polynomial $16x^4 - 12x^3 + 18x^2$.

Solution
Find the complete factorizations for $16x^4, 12x^3$, and $18x^2$.

$$16x^4 = \mathbf{2} \cdot 2 \cdot 2 \cdot 2 \cdot \mathbf{x \cdot x} \cdot x \cdot x \quad \text{Completely factor } 16x^4.$$
$$12x^3 = \mathbf{2} \cdot 2 \cdot 3 \cdot \mathbf{x \cdot x} \cdot x \qquad\quad \text{Completely factor } 12x^3.$$
$$18x^2 = \mathbf{2} \cdot 3 \cdot 3 \cdot \mathbf{x \cdot x} \qquad\qquad \text{Completely factor } 18x^2.$$

The GCF is $\mathbf{2 \cdot x \cdot x} = 2x^2$.

▌ **Now Try Exercise 65**

FACTORING THE GCF FROM A POLYNOMIAL To factor the GCF from a polynomial, write the polynomial as the product of a monomial (the GCF) and a polynomial. For example,

$$4x^3 + 60x \text{ factors as } 4x(x^2 + 15)$$

because the product $4x(x^2 + 15)$ equals $4x^3 + 60x$, and $4x$ is the GCF of $4x^3$ and $60x$. In the next example, we see how the terms of the polynomial factor $(x^2 + 15)$ are found in the complete factorizations of $4x^3$ and $60x$.

EXAMPLE 9   **Factoring the GCF from a polynomial**

Factor the GCF from the polynomial $4x^3 + 60x$.

Solution
Find the complete factorizations for $4x^3$ and $60x$.

$$4x^3 = \mathbf{2 \cdot 2} \cdot \mathbf{x} \cdot x \cdot x \quad \text{Completely factor } 4x^3.$$
$$60x = \mathbf{2 \cdot 2} \cdot 3 \cdot 5 \cdot \mathbf{x} \quad \text{Completely factor } 60x.$$

The monomial factor is the GCF, or $\mathbf{2 \cdot 2} \cdot \mathbf{x} = \mathbf{4x}$. The polynomial factor is found by adding the product of the black (non-common) factors in the factorization of $4x^3$, or $x \cdot x = \mathbf{x^2}$, and the product of the black (non-common) factors in the factorization of $60x$, or $3 \cdot 5 = \mathbf{15}$. In other words, the polynomial $4x^3 + 60x$ factors as $4x(x^2 + \mathbf{15})$.

Now Try Exercise 73

NOTE: To see if a polynomial has been factored correctly, use the distributive property to multiply the monomial factor and the polynomial factor. The result should be the given polynomial. For example, the product $4x(x^2 + 15)$ is found as follows.

$$4x(x^2 + 15) = 4x \cdot x^2 + 4x \cdot 15 \quad \text{Distributive property}$$
$$= 4x^3 + 60x \quad \text{Multiply monomials.}$$

The next example shows an organized method for factoring a polynomial.

EXAMPLE 10   **Factoring the GCF from a polynomial**

Factor the GCF from the polynomial $18x^3 - 12x^2 + 6x$.

Solution
The terms $18x^3$, $12x^2$, and $6x$ are separated by a minus sign $(-)$ and a plus sign $(+)$. The first column in Table 8.5 shows the complete factorizations for these three terms, and the second column shows the common factors with the GCF.

TABLE 8.5 Factoring $18x^3 - 12x^2 + 6x$

Complete Factorization	GCF	Remaining Factors
$18x^3 = \mathbf{2 \cdot 3 \cdot 3 \cdot x} \cdot x \cdot x$	$\mathbf{2 \cdot 3 \cdot x = 6x}$	$3 \cdot x \cdot x = 3x^2$
$12x^2 = \mathbf{2 \cdot 2 \cdot 3 \cdot x} \cdot x$	$\mathbf{2 \cdot 3 \cdot x = 6x}$	$2 \cdot x = 2x$
$6x = \mathbf{2 \cdot 3 \cdot x}$	$\mathbf{2 \cdot 3 \cdot x = 6x}$	1

The third column shows the product of the remaining (non-common) factors. The term $6x$ in the third row of the table is the GCF. Its remaining factor is $\mathbf{1}$ because we can multiply $6x$ by $\mathbf{1}$ to get the GCF. We keep the minus sign and the plus sign from the given polynomial, and $18x^3 - 12x^2 + 6x$ factors as $\mathbf{6x(3x^2 - 2x + 1)}$.

Now Try Exercise 77

8.4 Putting It All Together

CONCEPT	COMMENTS	EXAMPLES
Multiplying Monomials	1. Multiply the coefficients. 2. Use the product rule for exponents to multiply the variables. 3. Write the above results as a product.	$2y^2 \cdot 3y^4 = (2 \cdot 3) \cdot (y^2 \cdot y^4)$ $= 6 \cdot y^{4+2}$ $= 6y^6$
Multiplying a Monomial and a Polynomial	Use the distributive property to multiply each term of the polynomial by the monomial.	$2x(3x^2 - 5) = 2x \cdot 3x^2 - 2x \cdot 5$ $= 6x^3 - 10x$
Multiplying Polynomials	The product of two polynomials is found by multiplying every term in the first polynomial by every term in the second polynomial and then combining like terms.	$(2x + 3)(x - 5)$ $= 2x \cdot x - 2x \cdot 5 + 3 \cdot x - 3 \cdot 5$ $= 2x^2 - 10x + 3x - 15$ $= 2x^2 - 7x - 15$
Multiplying Polynomials Vertically	Align like terms vertically when multiplying terms.	$\begin{array}{r} 2x^2 + 3x - 1 \\ \times \quad\quad x - 4 \\ \hline -8x^2 - 12x + 4 \\ 2x^3 + 3x^2 - 1x \quad\quad \\ \hline 2x^3 - 5x^2 - 13x + 4 \end{array}$
Multiplying Polynomials Visually	If polynomials represent the length and width of a rectangle, then the area of the rectangle represents the product of the two polynomials.	$(x + 3)(4x + 1)$ $= 4x^2 + 12x + x + 3$ This simplifies to $4x^2 + 13x + 3$.
Greatest Common Factor (GCF)	The GCF of a list of positive integers is the largest integer that is a factor of every integer in the list. The greatest common factor of a list of terms is found by writing each term in completely factored form. The GCF is the product of all common factors.	The integers 16, 24, and 32 have *common* factors 2, 4, and 8. The greatest common factor (GCF) is 8. For the terms $4x^2$ and $10x$, $4x^2 = 2 \cdot 2 \cdot x \cdot x$ and $10x = 2 \cdot 5 \cdot x$. The GCF is $2x$.
Factoring the GCF from a Polynomial	To factor the GCF from a polynomial, we write the polynomial as the product of a monomial (the GCF) and a polynomial.	$4x^2 + 20x$ is factored as $4x(x + 5)$. $7y^5 + 14y^3 - 49y$ is factored as $7y(y^4 + 2y^2 - 7)$.

8.4 Exercises

MyMathLab

Math XL PRACTICE | WATCH | DOWNLOAD | READ | REVIEW

CONCEPTS AND VOCABULARY

1. When multiplying two monomials, the first step is to multiply the _____.

2. When multiplying two monomials, the second step is to use the _____ rule for exponents to multiply the variables.

3. When multiplying a monomial and a polynomial, we use the _____ property to multiply the monomial by each term of the polynomial.

4. The product of two polynomials is found by multiplying every _____ in the first polynomial by every _____ in the second polynomial and then combining like terms.

5. To multiply polynomials vertically, align the _____ vertically to aid in combining like terms.

6. When multiplying two polynomials visually, the _____ of the rectangle represents the product of the two polynomials.

7. We _____ a polynomial by writing it as a product of two or more lower degree polynomials.

8. The _____ of a list of integers is the largest integer that is a factor of every integer in the list.

9. A term is completely _____ when its coefficient is written as a product of prime numbers and any powers of variables are written as repeated multiplication.

10. To factor the GCF from a polynomial, we write the polynomial as a product of a(n) _____ (the GCF) and a polynomial.

MULTIPLYING POLYNOMIALS

Exercises 11–20: Multiply.

11. $x^2 \cdot x^3$

12. $m^3 \cdot m^5$

13. $3y^4 \cdot 6y^3$

14. $9n^2 \cdot 8n^2$

15. $(-4g^2)(7g)$

16. $(15w^5)(-2w)$

17. $(-k^2)(-6k^3)$

18. $(-5p)(-2p^3)$

19. $(4x)(-3x^3)(x^2)$

20. $(-y)(-2y^2)(4y)$

Exercises 21–30: Multiply.

21. $3y(4y^3 + 2)$

22. $2m^2(3m^2 - 5)$

23. $-3x^2(x^3 - 5x)$

24. $-w(-2w^2 - 1)$

25. $4m(-2m^6 + 5m)$

26. $9g^4(-2g^3 - 3)$

27. $7y^2(2y^2 - 3y - 8)$

28. $4a^3(2a^2 - 4a + 5)$

29. $-2n(-3n^2 + 5n - 1)$

30. $-q(2q^5 - 4q^3 + 3)$

Exercises 31–40: Multiply.

31. $(2x + 1)(3x - 4)$

32. $(5n - 8)(n - 9)$

33. $(4g - 3)(2g^2 + 5)$

34. $(6y^2 - 1)(y - 1)$

35. $(4k^2 - 2)(-3k^2 + 2)$

36. $(-w^3 + 5)(-w^3 - 2)$

37. $(b - 6)(b^2 - 3b + 2)$

38. $(n + 2)(n^2 + 3n - 2)$

39. $(3x + 3)(2x^2 - 2x - 1)$

40. $(y^2 - 3y)(2y^2 + 6y - 5)$

Exercises 41–46: Multiply using a vertical format.

41. $(4k - 5)(3k + 2)$

42. $(w^3 - 5)(-w^2 - 3)$

43. $(m - 4)(m^2 - 2m + 9)$

44. $(2n + 1)(6n^2 - 8n - 7)$

45. $(2x + 4)(5x^2 - 3x + 6)$

46. $(y^2 - 4y)(2y^2 + 8y - 7)$

Exercises 47–50: Multiply visually.

47. $(x + 3)(x + 8)$

48. $(w + 6)(w + 7)$

49. $(4m + 3)(m + 11)$

50. $(9g + 1)(3g + 2)$

FACTORING POLYNOMIALS

Exercises 51–56: Find the GCF of the list of integers.

51. 10, 14

52. 12, 30

53. 10, 50

54. 16, 48

55. 9, 36, 54

56. 7, 28, 49

Exercises 57–62: Find the GCF of the list of terms.

57. $6x^2$, $15x$

58. $10w^4$, $25w^2$

59. $9y^3$, $36y^4$

60. $11n^3$, $55n^2$

61. $12m^4$, $15m^3$, $18m^2$

62. $5x^3$, $15x^2$, $50x^5$

Exercises 63–68: Find the GCF of the terms in the given polynomial.

63. $3x - 15$

64. $14y^2 + 49$

65. $9n^5 - 27n^3 + 6n^2$

66. $6g^3 - 5g^2 - 10g$

67. $8a^4 - 16a^3 + 4a^2$

68. $12x^7 - 9x^4 + 9x$

Exercises 69–78: Factor the GCF from the polynomial.

69. $7x - 21$

70. $5w^2 + 3w$

71. $15m^3 + 5m^2$

72. $24g^3 - 12g$

73. $48a^4 - 32a$

74. $30b^5 + 48b^3$

75. $36q^5 - 20q^4 + 16q^2$

76. $15p^4 + 40p^2 + 25p$

77. $81x^4 - 45x^3 - 9x^2$

78. $8y^6 - 24y^4 + 8y^3$

APPLICATIONS

79. *Price and Demand* When the price of a video game is p dollars, where $p \geq 10$, the demand in thousands of games is given by the polynomial $25 - \frac{1}{4}p$.
(a) Find the demand when $p = \$16$.
(b) Find the demand when $p = \$40$.
(c) A polynomial expression for the revenue from selling the video games is found by multiplying the price p and the demand $25 - \frac{1}{4}p$. Write an expression for the revenue.
(d) Find the revenue when $p = \$20$.

80. *Price and Demand* When the price of a DVD is p dollars, where $p \geq 6$, the demand in millions of DVDs is given by the polynomial $10 - \frac{1}{3}p$.
(a) Find the demand when $p = \$9$.
(b) Find the demand when $p = \$15$.

(c) A polynomial expression for the revenue from selling the DVDs is found by multiplying the price p and the demand $10 - \frac{1}{3}p$. Write an expression for the revenue.
(d) Find the revenue when $p = \$12$.

81. *Numbers* Consecutive numbers are numbers that differ by 1, such as 7 and 8. If x represents a number, then $x + 1$ is the next consecutive number. Write an expression for the product of two consecutive numbers. Let x be the smaller number.

82. *Numbers* Refer to the previous exercise. Write an expression for the product of *three* consecutive numbers. Let x be the smallest number.

83. *Geometry* If the area of the following rectangle is $9x^2 + 12x$ and the width is $3x$, then the length is the polynomial factor found when $3x$ is factored from the area. Find the length.

84. *Geometry* Refer to the previous exercise. If the area of the following rectangle is $7x^3 + 49x^2$ and the length is $7x^2$, find the width.

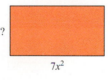

WRITING ABOUT MATHEMATICS

85. If a polynomial with m terms is multiplied by a polynomial with n terms, explain how to find the number of terms in the product before terms are combined. Give two examples.

86. A student factors the polynomial $3x^2 + 6x - 12$ as the product $3x(x + 2 - 4)$. How do you know that the student has made a mistake?

Checking Basic Concepts

1. Determine whether each given expression is a polynomial. If it is, state the number of terms and give its degree.

   (a) $2x^2 - \dfrac{7}{x}$ (b) $3 - 5x^4$

2. Evaluate $-2x + x^2$ for $x = -3$.

3. Add.
   (a) $(8m^2 + 2) + (m - 7)$
   (b) $(1 + 6x^2) + (11x - 2x^2)$

4. Write the opposite of $-5x^3 + 2x^2 - 7x + 12$.

5. Subtract.
   (a) $(12y^2 - 4) - (11y + 3)$
   (b) $(n^2 - 3n) - (-2n^2 - 3n + 5)$

6. Multiply.
   (a) $(5w^3)(-w)$ (b) $2x^2(x - 5)$
   (c) $(5k - 2)(-3k^2 + 1)$

7. Find the GCF of the integers 18, 30, and 42.

8. Find the GCF of the terms $14x^2$ and $21x$.

9. Factor the GCF from each polynomial.
   (a) $50b^4 + 75b^3$
   (b) $16p^4 + 64p^3 + 48p^2$

10. *Bird Population* The polynomial expression
$$x^3 - 40x^2 + 660x + 5200$$
gives a bird population x years after the start of a study. Find the bird population after 3 years.

CHAPTER 8   Summary

SECTION 8.1 ■ RULES FOR EXPONENTS

Exponent

A natural number exponent indicates the number of times that the base is used as a repeated factor.

Examples: $7^4 = 7 \cdot 7 \cdot 7 \cdot 7$ and $(-8)^3 = (-8) \cdot (-8) \cdot (-8)$

Rules for Exponents

For any real number a and natural number exponents m and n,

Name of Rule	Rule	Example
The Product Rule	$a^m \cdot a^n = a^{m+n}$	$5^4 \cdot 5^3 = 5^{4+3} = 5^7$
The Power Rule	$(a^m)^n = a^{m \cdot n}$	$(y^2)^5 = y^{2 \cdot 5} = y^{10}$
Power of a Product Rule	$(ab)^n = a^n b^n$	$(pq)^7 = p^7 q^7$

SECTION 8.2 ■ NEGATIVE EXPONENTS AND SCIENTIFIC NOTATION

Rules for Integer Exponents

For any nonzero real number base a and integer exponents m and n,

Name of Rule	Rule	Example
The Quotient Rule	$\dfrac{a^m}{a^n} = a^{m-n}$	$\dfrac{y^6}{y^2} = y^{6-2} = y^4$
Zero Power Rule	$a^0 = 1$	$(-13)^0 = 1$
Negative Exponents	$a^{-n} = \dfrac{1}{a^n}$	$x^{-8} = \dfrac{1}{x^8}$

Scientific Notation

A positive real number a is in scientific notation when it is written as
$$b \times 10^n,$$
where $b \geq 1$ and $b < 10$, and n is an integer.

Examples: 2.8×10^{11} and 6.15×10^{-4}

Converting Scientific Notation to Standard Form	To write $b \times 10^n$ in standard form, move the decimal point in b • $\|n\|$ places to the right if n is positive, or • $\|n\|$ places to the left if n is negative. **Example:** To write 5.8×10^{-3} in standard form, move the decimal point in 5.8 three places to the left, to get 0.0058. To write 7.2×10^6 in standard form, move the decimal point in 7.2 six places to the right, to get 7,200,000.
Converting Standard Form to Scientific Notation	If a is a positive number expressed as a decimal, **1.** Move the decimal point in a until it becomes a number b that is greater than or equal to 1 but less than 10. **2.** Let the whole number n be the number of places the decimal point was moved. **3.** Write a in scientific notation as follows. • If $a \geq 10$, then $a = b \times 10^n$. • If $a < 1$, then $a = b \times 10^{-n}$. • If $a \geq 1$ and $a < 10$, then $a = a \times 10^0$. **Example:** To write 23,000 in scientific notation, move the decimal point in 23,000 four places to the left, to get 2.3. Since $23{,}000 \geq 10$, the scientific notation is 2.3×10^4.

SECTION 8.3 ■ ADDING AND SUBTRACTING POLYNOMIALS

Monomials	A monomial is a number, a variable, or a product of numbers and variables raised to natural number powers. **Examples:** 7, $\quad m^2$, $\quad -25x^9$, $\quad$ and $\quad \frac{2}{3}y^3$
Degree and Coefficient of a Monomial	The *degree* of a monomial with one variable is the exponent on that variable. If the variable in a monomial has no exponent, its degree is 1. If a monomial has no variable, its degree is 0. The number in a monomial is called its *coefficient*. **Example:** $-6x^7$ has degree 7 and coefficient -6.
Polynomials	A polynomial is a sum of one or more monomials. **Examples:** $-4x^3 + 6x^2 - 3x + 1$, $\quad 8m^2 + 9$, $\quad$ and $\quad 5n^2 + 9n - 3$
Degree of a Polynomial	The degree of a polynomial is the degree of the term (or monomial) with the highest degree. **Examples:** The degree of $7x^3 - 2x$ is 3 and the degree of $x^4 - 2x^3 - 2$ is 4.
Rearranging Terms of a Polynomial	Using the commutative and associative properties of addition, we can rearrange the terms of a polynomial in any order. Make sure that the appropriate sign stays with its corresponding term. **Example:** The polynomial $-5x^2 - 7x + 2$ may be written as $-7x + 2 - 5x^2$ or as $2 - 7x - 5x^2$.
Adding Polynomials	To add polynomials, combine like terms. **Example:** $(4x^2 + 5x - 1) + (x^2 - 2x) = 4x^2 + 5x - 1 + x^2 - 2x$ $$= (4x^2 + x^2) + (5x - 2x) - 1$$ $$= 5x^2 + 3x - 1$$
The Opposite of a Polynomial	Change the sign of each term in the polynomial to its opposite. **Example:** The opposite of $-9x^2 + 6x - 8$ is $9x^2 - 6x + 8$.

Subtracting Polynomials

To subtract one polynomial from another, add the opposite of the second polynomial to the first polynomial.

Example: $(9n^2 + 8) - (5n^2 + 2n) = (9n^2 + 8) + (-5n^2 - 2n)$

$$= 9n^2 + 8 + (-5n^2) - 2n$$
$$= (9n^2 - 5n^2) - 2n + 8$$
$$= 4n^2 - 2n + 8$$

SECTION 8.4 ■ MULTIPLYING AND FACTORING POLYNOMIALS

Multiplying Monomials

To multiply monomials in one variable, do the following.

1. Multiply the coefficients.
2. Use the product rule for exponents to multiply the variables.
3. Write the above results as a product.

Example: $5w^2 \cdot 8w^3 = (5 \cdot 8) \cdot (w^2 \cdot w^3) = 40 \cdot w^{2+3} = 40w^5$

Multiplying a Monomial and a Polynomial

Use the distributive property to multiply each term of the polynomial by the monomial.

Example: $3x(2x^2 + 4) = 3x \cdot 2x^2 + 3x \cdot 4 = 6x^3 + 12x$

Multiplying Polynomials

The product of two polynomials is found by multiplying every term in the first polynomial by every term in the second polynomial and then combining like terms.

Example: $(3x + 8)(x - 1) = 3x \cdot x - 3x \cdot 1 + 8 \cdot x - 8 \cdot 1$

$$= 3x^2 - 3x + 8x - 8$$
$$= 3x^2 + 5x - 8$$

Multiplying Polynomials Vertically

Align like terms vertically when multiplying terms.

Example:

$$
\begin{array}{r}
7x^2 + 2x - 5 \\
\times \qquad\quad x - 3 \\
\hline
-21x^2 - 6x + 15 \\
7x^3 + 2x^2 - 5x \qquad\quad \\
\hline
7x^3 - 19x^2 - 11x + 15
\end{array}
$$

Multiplying Polynomials Visually

If polynomials represent the length and width of a rectangle, then the area of the rectangle represents the product of the two polynomials.

Example: $(x + 3)(5x + 2) = 5x^2 + 15x + 2x + 6 = 5x^2 + 17x + 6$

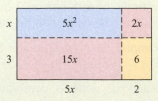

Greatest Common Factor (GCF)

The greatest common factor of a list of positive integers is the largest integer that is a factor of every integer in the list.

Examples: 36, 48, and 60 have *common* factors 2, 3, 4, 6, and 12. The greatest common factor (GCF) is 12.

Because the terms $6x^2$ and $14x$ factor completely as $6x^2 = 2 \cdot 3 \cdot x \cdot x$ and $14x = 2 \cdot 7 \cdot x$, the GCF is $2x$.

Factoring the GCF from a Polynomial	To factor the GCF from a polynomial, we write the polynomial as the product of a monomial (the GCF) and a polynomial. **Example:** $6x^2 + 24x$ is factored as $6x(x + 4)$. $9y^4 + 18y^3 - 36y^2$ is factored as $9y^2(y^2 + 2y - 4)$.

CHAPTER 8   Review Exercises

SECTION 8.1

Exercises 1 and 2: Write the given exponential expression as repeated multiplication.

1. 17^5

2. $(xy)^2$

Exercises 3–10: Simplify.

3. $x^2 \cdot x \cdot x^6$

4. $5n^8 \cdot (-n^2)$

5. $(y^4)^2$

6. $q \cdot (q^3)^2$

7. $(3w^4)^2$

8. $(-2p^4)^3$

9. $(-a^3b^4)^3$

10. $(xy^3)^2(2x^2y)^2$

SECTION 8.2

Exercises 11–20: Simplify the given expression using the definition of negative exponents. Assume that all variables represent nonzero numbers.

11. $\dfrac{x^9}{x^4}$

12. $\dfrac{20y^7}{4y^5}$

13. $\dfrac{-27w^5}{9w}$

14. $\dfrac{60k^{10}}{-12k^7}$

15. $(-19)^0$

16. $22y^0$

17. 8^{-2}

18. y^{-3}

19. a^{-1}

20. $-24n^{-6}$

Exercises 21 and 22: State whether the given number is in scientific notation.

21. 2.8×5^{-2}

22. 2.45×10^3

Exercises 23 and 24: Write the number in standard form.

23. 4.99×10^8

24. 9.2×10^{-6}

Exercises 25 and 26: Write the number in scientific notation.

25. $39{,}000{,}000{,}000$

26. 0.0000487

Exercises 27 and 28: Multiply or divide as indicated. Write answers in standard form and in scientific notation.

27. $(2.5 \times 10^{-3})(4 \times 10^6)$

28. $\dfrac{8 \times 10^5}{2 \times 10^{-2}}$

SECTION 8.3

Exercises 29 and 30: Determine if the expression is a polynomial. If so, state the number of terms in the polynomial and give its degree.

29. $6x^4 - 2x^3 + 5x$

30. $\dfrac{25x}{x^2 + 1}$

Exercises 31 and 32: Evaluate the polynomial for the given value of x.

31. $4x^3 + 5x - 12$,   for $x = 2$

32. $-x^3 + 4x - 1$,   for $x = -1$

Exercises 33 and 34: Write the polynomial so that its terms are in descending order.

33. $7x^3 - 8x + x^4$

34. $3y^2 - 4 + y - 7y^3$

Exercises 35–38: Add.

35. $(5n - 2) + (5n + 2)$

36. $(7x - 8) + (2x - 3)$

37. $(3k^2 - k) + (6k^2 + 3k + 1)$

38. $(14m^3 + 5m^2 + 3) + (-11m^3 + m - 7)$

Exercises 39 and 40: Add using a vertical format.

39. $(x^2 - 9x + 10) + (x^2 + 9x - 3)$

40. $(8w^2 - w + 14) + (-4w^2 - w - 2)$

Exercises 41 and 42: Write the opposite of the polynomial.

41. $15h^2 + 9h - 3$

42. $2b^3 - b^2 + b + 4$

Exercises 43–46: Subtract.

43. $(10a - 5) - (3a + 1)$ **44.** $(x + 19) - (x - 2)$

45. $(6k^2 - 3k - 2) - (5k^2 - 9)$

46. $(2q^3 + q + 9) - (-4q^3 + q^2 - 3)$

Exercises 47 and 48: Subtract using a vertical format.

47. $(n^2 - 3n) - (-2n^2 - 8n + 5)$

48. $(w^2 - 2w + 18) - (-7w^2 + 3w - 4)$

SECTION 8.4

Exercises 49–54: Multiply.

49. $7n^5 \cdot 8n^2$ **50.** $(2x)(-3x^3)(x^4)$

51. $6g^2(-3g^3 + 5)$ **52.** $2a^3(7a^2 - 3a + 1)$

53. $(2k^2 - 2)(9k^2 + 8)$

54. $(y^2 - 4y)(2y^2 + 8y - 3)$

55. Find the product $(2x + 3)(4x^2 - 6x + 1)$ using a vertical format.

56. Multiply $(7m + 3)(5m + 4)$ visually.

Exercises 57 and 58: Find the GCF of the list of integers.

57. $20, 44$ **58.** $18, 30, 54$

Exercises 59 and 60: Find the GCF of the list of terms.

59. $24x^2, 18x^3$ **60.** $6b^2, 22b^3, 34b^4$

Exercises 61–64: Factor the GCF from the polynomial.

61. $12m^4 + 6m^3$ **62.** $20g^2 - 15g$

63. $14x^5 - 35x^3 - 49x^2$ **64.** $30y^5 - 48y^4 + 6y^3$

APPLICATIONS

65. *Area* Write an expression for the area of the figure.

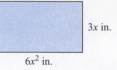

$3x$ in.

$6x^2$ in.

66. *Age in Seconds* A 20-year-old person finds that she is about 631,000,000 seconds old. Write this number in scientific notation.

67. *Height of a Baseball* For the first few seconds after a baseball is hit, the polynomial

$$-16x^2 + 72x + 3$$

gives the height of the baseball in feet after x seconds. Find the height of the baseball after 2 seconds and after 3 seconds.

68. *Height Difference* Two baseballs are hit at the same time. For the first few seconds, the height of the higher ball x seconds after it is hit is given by $-16x^2 + 90x + 3$, and the height of the lower ball is given by $-16x^2 + 78x + 1$. Write an expression for the difference in the heights of the two balls after x seconds.

69. *Geometry* If the area of the following rectangle is $6x^2 + 20x$ and the width is $2x$, find the length.

$2x$

?

70. *Group Rate* If the regular price for a room at a hotel is \$160 and the manager offers a group rate that decreases this price by \$3 for each room rented, then the *total* cost for a group of x people is given by

$$160x - 3x^2, \text{ where } x \le 26.$$

Find the total cost for a group of 20 people.

CHAPTER 8   Test

1. Simplify each expression using the definition of negative exponents. Assume that all variables represent nonzero numbers.

(a) $(-3p^3)^2$ (b) $(q^4)^2 \cdot q$

(c) $17n^{-3}$ (d) $(3x)^0$

(e) $\dfrac{28y^9}{7y^6}$

2. Write 8.3×10^{-5} in standard form.

3. Write 37,500 in scientific notation.

4. Multiply or divide as indicated. Write your answer in standard form and in scientific notation.

(a) $\dfrac{6.9 \times 10^3}{3 \times 10^{-4}}$ (b) $(1.5 \times 10^{-9})(5 \times 10^5)$

5. Evaluate $-2x^2 - 3x + 5$ for $x = 3$.

6. Write the polynomial $5x^3 - 2x^4 - 3x$ so that its terms are in descending order.

7. Add or subtract as indicated.
   (a) $(5k^2 - 7k + 4) + (2k^2 - k)$
   (b) $(8p^2 - 6) - (3p^2 - 5p + 2)$

8. Multiply.
   (a) $3b^3(5b^2 - 6b + 2)$
   (b) $(2m - 5)(7m + 3)$
   (c) $(y - 4)(5y^2 + 2y - 1)$

9. Find the GCF.
   (a) $10, 35, 50$ (b) $36x^2, 18x$

10. Factor the GCF from each polynomial.
    (a) $12x^3 - 6x^2 - 18x$
    (b) $16w^7 - 28w^3$

11. *Area* Write an expression for the area of the figure.

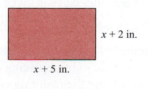

$x + 2$ in.

$x + 5$ in.

12. *Heart Rate* For the first few minutes after an athlete stops exercising, the polynomial
$$2x^2 - 35x + 180$$
gives the heart rate in beats per minute after x minutes.
   (a) What is the heart rate when the athlete first stops exercising ($x = 0$)?
   (b) What is the heart rate 3 minutes after the athlete stops exercising?

13. *Profit* If R represents the revenue from selling x items and C represents the cost of making x items, then the profit P for making and selling x items is given by the formula $P = R - C$. If a manufacturer finds that $R = 30x$ and $C = x^2 + 6x - 1$, write an expression for the profit.

14. *Geometry* If the area of the following rectangle is $8x^2 + 20x$ and the width is $4x$, find the length.

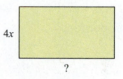

$4x$

?

15. *Thickness* A thin metal foil is 0.0043 inch thick. Write this number in scientific notation.

CHAPTERS 1–8 Cumulative Review Exercises

1. Identify the digit in the hundred-thousands place in the number 4,612,350,798.

2. Write $9 \cdot 9 \cdot 9 \cdot 9 \cdot 9$ using exponential notation.

3. Estimate the sum $698 + 401 + 197$ by rounding each value to the nearest hundred.

4. Place the correct symbol, $<, >$, or $=$, in the blank.
$$-|18| \text{\_\_\_\_} 18.$$

5. Evaluate $[x + (5 - y)^2] \div x$ for $x = 3, y = 2$.

6. Is 9 a solution to $4 + b \div 3 = -7$?

7. Complete the table and solve $-3x + 5 = -1$.

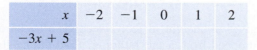

x	-2	-1	0	1	2
$-3x + 5$					

8. Translate the following sentence to an equation using x as the variable. Do not solve the equation.

   Eleven equals a number increased by 4.

9. Solve $-9 + 2x - 4 = 6x - 5$.

10. Subtracting 5 from the product of 2 and a number results in 3 more than the number. Find the number.

11. Solve the linear equation $-2x - 3 = -3$ visually.

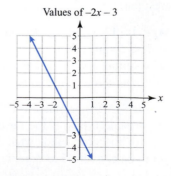

Values of $-2x - 3$

12. Give both the improper fraction and the mixed number represented by the shading.

Exercises 13 and 14: Perform the indicated arithmetic.

13. $\dfrac{11}{10} - \dfrac{1}{4}$

14. $\dfrac{3}{16} \cdot \dfrac{8}{9}$

15. Write $\dfrac{7}{20}$ as a decimal.

16. Place the correct symbol, $<$ or $>$, in the blank.

$$13.0263 \underline{\hspace{1cm}} 13.0623$$

17. Solve the equation $10.2(y - 1) = 22.44$.

18. If a car travels 171 miles in 3 hours, write its speed as a unit rate.

19. Solve the proportion.

$$\frac{6}{x} = \frac{-3}{10}$$

20. Convert 180 hours to days.

21. What percent of 120 is 15?

22. What number is 15% of 700?

23. A card is randomly selected from a standard deck. Find the probability that the card is a red card with the number 2, 5, or 8. Express your answer as a fraction in simplest form.

24. Write 6.9×10^{-4} in standard form.

Exercises 25 and 26: Multiply.

25. $3x^2(x + 4)$

26. $(4y - 1)(5y^2 + 7)$

27. Find the GCF of $48x^2$ and $12x$.

28. Factor the GCF from $14y^6 - 28y^3$.

29. *Estimating Time* An athlete stretches for 12 minutes, jogs for 58 minutes, and then walks for 19 minutes. Estimate the total time for this workout by rounding to the nearest ten.

30. *Finding Age* In 13 years, a person's age will be 9 years less than double his current age. Write an equation whose solution gives the person's current age. Use x for your variable. Do not solve the equation.

31. *Going Organic* A farmer plans to convert $\frac{13}{15}$ of his crop land to organic crops. If the farmer has already converted $\frac{8}{15}$ of his crop land, what fraction of his land has yet to be converted to organic crops?

32. *Music Sales* The total music sales S in thousands of dollars for a small band during year x can be computed using the formula

$$S = 7(x - 2005) + 8.$$

Find the year when the music sales were \$36,000 by solving the equation $36 = 7(x - 2005) + 8$ visually using the graph below.

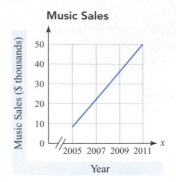

Music Sales

33. *Text Messaging* Suppose that the cost C of sending or receiving x text messages is given by the formula

$$C = 0.05(x - 400) + 10.5, \text{ where } x > 400.$$

Find the number of text messages that correspond to total charges of \$18.50 by replacing C in the formula with 18.5 and solving the resulting equation.

34. *Tree Height* A tree casts a shadow that is 55 feet long, while a nearby tree that is 12 feet tall casts a shadow that is 5 feet long. See the accompanying figure. How tall is the larger tree?

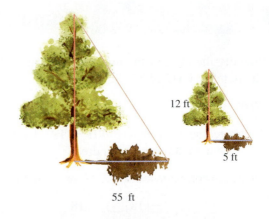

35. *HD Video* If the purchase price of an HD video recorder is \$178 and the sales tax rate is 4.5%, find the sales tax and the total price.

36. *Falling Object* The polynomial expression

$$-16x^2 + 400, \text{ where } x \leq 5$$

gives the height in feet x seconds after an object is dropped from a tall building. Find the height of the object 2 seconds after it is dropped.

9 Introduction to Graphing

We all have ability. The difference is how we use it.

—STEVIE WONDER

In colder climates, the amount of energy needed to heat a home can be reduced significantly by installing a geothermal heat pump. The following graph shows the average annual energy cost savings that a geothermal heat exchange system generates for a home in a northern climate, compared to a traditional natural gas or electric heating system.

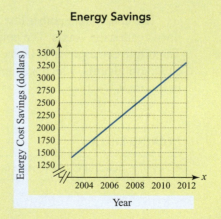

Energy Savings

Graphs help us visualize all kinds of real-world information. In this chapter, we use the *rectangular coordinate system* to graph linear equations.

Source: U.S. Department of Energy, *Energy Efficiency and Renewable Energy.*

9.1   The Rectangular Coordinate System

The *xy*-plane ● Plotting Points ● Scatterplots and Line Graphs

As Internet connection speeds have improved, the amount of visual information displayed on Web pages has increased dramatically. One full page of computer graphics contains about 100 times as much information as a page of printed text. In math, we visualize data by plotting points to make *scatterplots* and *line graphs*. In this section, we discuss the rectangular coordinate system and practice basic point-plotting skills.

NEW VOCABULARY

☐ Plane
☐ Rectangular coordinate system, or *xy*-plane
☐ *x*- and *y*-axis
☐ Origin
☐ Quadrants
☐ Coordinates
☐ Ordered pair
☐ *x*- and *y*-coordinates
☐ Plotting
☐ Scatterplot
☐ Line graph

The *xy*-plane

A **plane** is a flat surface that continues without end. Examples of portions of a plane include a wall in your classroom, the playing surface of a tennis court, and a flat piece of paper. In math, we use a plane called the **rectangular coordinate system**, or ***xy*-plane**, to graph data. The *xy*-plane is shown in Figure 9.1.

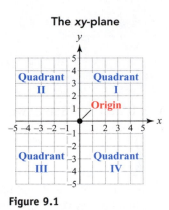

Figure 9.1

The horizontal number line in the *xy*-plane is called the ***x*-axis**, and the vertical number line is called the ***y*-axis**. The point where the two number lines intersect is called the **origin**. The origin corresponds to 0 on each axis. The two axes divide the *xy*-plane into four regions called **quadrants**, which are numbered with the Roman numerals I, II, III, and IV, as shown in Figure 9.1.

READING CHECK

• Are the quadrants numbered in a clockwise or counter-clockwise manner?

STUDY TIP

As you approach the end of the semester, remember that it is important to keep a positive attitude. Look back at your progress in this course and make special note of all of your accomplishments.

EXAMPLE 1 **Determining the location of graphed points**

Give the quadrant containing each point in Figure 9.2. If a point is on an axis, name the axis.

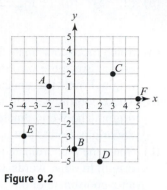

Figure 9.2

Solution

Point *A* is located in the upper-left region, which is quadrant II. Point *B* is on the vertical axis, or the *y*-axis. Point *C* is in the upper-right region, which is quadrant I. Similarly, point *D* is in quadrant IV, and point *E* is in quadrant III. Finally, point F is on the *x*-axis.

Now Try Exercise 9

Plotting Points

▶ **REAL-WORLD CONNECTION** The streets and avenues of Salt Lake City, Utah are laid out in a rectangular coordinate system with Temple Square at the origin. To find an address such as 300 East 400 South, we start at Temple Square, travel 3 city-blocks East, and then go 4 city-blocks South, as shown in Figure 9.3.

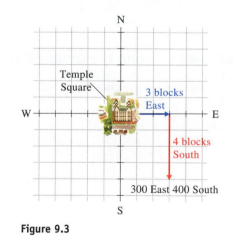

Figure 9.3

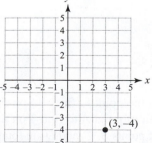

Figure 9.4

We can find the "address" of any point in the *xy*-plane in a similar manner, using two numbers called **coordinates** instead of using directions. The coordinates of a point are written in parentheses as an **ordered pair**. For example, the ordered pair $(3, -4)$ is similar to the address 300 East 400 South. Its location in the *xy*-plane is shown in Figure 9.4.

The first number in an ordered pair is called the **x-coordinate**, and the second number is called the **y-coordinate**. Points with negative *x*-coordinates are located to the left of the origin, and points with positive *x*-coordinates appear to the right of the origin. Likewise, points with negative *y*-coordinates appear below the origin, and points with positive *y*-coordinates are located above the origin.

EXAMPLE 2   **Determining the location of an ordered pair**

Give the quadrant containing each point. If a point is on an axis, name the axis.
(a) $(2, -5)$ **(b)** $(-13, -22)$ **(c)** $(0, 37)$

Solution
(a) Since the x-coordinate 2 is positive, the point is located horizontally to the right of the origin. Since the y-coordinate -5 is negative, the point is located vertically below the origin. Points that appear to the right and below the origin are in quadrant IV.
(b) The x-coordinate -13 is negative, so the point is located horizontally to the left of the origin. The y-coordinate -22 is also negative, so the point is located vertically below the origin. Points that appear to the left and below the origin are in quadrant III.
(c) Since the x-coordinate is 0, the point is located neither to the right nor to the left of the origin. In other words, the point is on the y-axis. Note that *every* point on the y-axis has an x-coordinate of 0.

Now Try Exercises 15, 19, 21

The coordinates of any point in the xy-plane can be found as follows.

WRITING THE COORDINATES OF A POINT IN THE *XY*-PLANE

The x-coordinate of a point is the number on the x-axis that is aligned directly above or below the point.

The y-coordinate of a point is the number on the y-axis that is aligned directly to the right or left of the point.

The coordinates of a point are written as the ordered pair (x-coordinate, y-coordinate).

EXAMPLE 3   **Determining the coordinates of graphed points**

Determine the coordinates of each point shown in Figure 9.5.

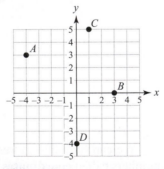

Figure 9.5

Solution
The x-coordinate of point A is found by looking directly below the point to find -4 on the x-axis. The y-coordinate is found by looking directly to the right of point A to find 3 on the y-axis. The coordinates of point A are $(-4, 3)$. The coordinates of points B, C, and D are found in a similar way. The coordinates of point B are $(3, 0)$, the coordinates of point C are $(1, 5)$, and the coordinates of point D are $(0, -4)$.

Now Try Exercise 25

When we use a point's coordinates to graph it in the *xy*-plane, we are **plotting** the point. The following procedure can be used to plot a point in the *xy*-plane.

PLOTTING A POINT IN THE *XY*-PLANE

1. Begin at the origin and move left or right along the *x*-axis to the position given by the *x*-coordinate (the first number in the ordered pair).
2. From that position on the *x*-axis, move upward or downward the amount given by the *y*-coordinate (the second number in the ordered pair).
3. Plot a point by making a dot at the final position found in Step 2.

READING CHECK

- Which coordinate places a point to the left or right of the origin?
- Which coordinate places a point above or below the origin?

EXAMPLE 4   **Plotting points in the *xy*-plane**

Plot each point in the *xy*-plane. Label each point with its coordinates.
(a) $(3, 1)$ **(b)** $(-2, -3)$ **(c)** $(0, 4)$

Solution
(a) To plot $(3, 1)$, start at the origin and move **3 units to the right**. From that position, move **1 unit upward**. Plot a point and label it as shown in Figure 9.6a.
(b) To plot $(-2, -3)$, starting at the origin, move **2 units to the left**. From that position, move **3 units downward**. Plot a point and label it as shown in Figure 9.6b.
(c) To plot $(0, 4)$, start at the origin, but do not move left or right because the *x*-coordinate is 0. Next, move **4 units upward**. Plot a point and label it as shown in Figure 9.6c.

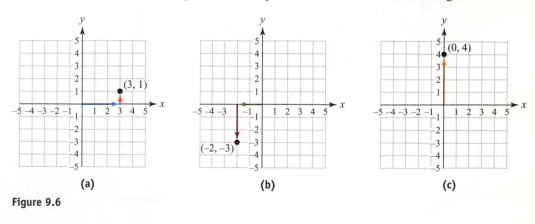

(a)   (b)   (c)

Figure 9.6

❚ **Now Try Exercise 31**

Scatterplots and Line Graphs

When points are plotted in the *xy*-plane, the resulting graph is called a **scatterplot**. By graphing data in a scatterplot, we can visualize trends in the data. Typically, the points in a scatterplot are not labeled, as shown in the next example.

EXAMPLE 5 **Making a scatterplot**

Make a scatterplot of the points $(-3, -2)$, $(-2, 1)$, $(0, 3)$, $(2, 2)$, and $(4, -3)$.

Solution
The points are plotted in Figure 9.7.

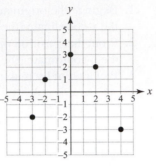

Figure 9.7

Now Try Exercise 37

READING CHECK

• How do scatterplots and line graphs differ?

To more easily visualize changes in data, consecutive data points can be connected with line segments to make a **line graph**. When making a line graph, it is important to plot all of the given data points before connecting them with line segments. The data points should be connected *from left to right*, even if data are given "out of order."

EXAMPLE 6 **Making a line graph**

Make a line graph of the points $(0, 0)$, $(-2, -2)$, $(-4, -3)$, $(2, 1)$, and $(4, 3)$.

Solution
The line graph is shown in Figure 9.8.

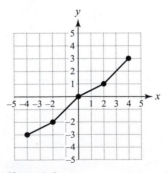

Figure 9.8

Now Try Exercise 41

▶ **REAL-WORLD CONNECTION** Scatterplots and line graphs are often used to visualize real-world data. Many times, only quadrant I is needed to make a scatterplot or line graph of real-word data because both the x-values and y-values are positive numbers. Remember, to visualize real-world data, it may be necessary to adjust the scale on the axes, as illustrated in the next example.

EXAMPLE 7   **Graphing real-world data**

Table 9.1 shows the approximate number of Facebook users in millions recorded in July of selected years. Make a scatterplot and a line graph of the data. Label each axis appropriately.

TABLE 9.1 Active Facebook Users (Millions)

Year (x)	2006	2007	2008	2009
Users (y)	10	30	80	250

(*Source:* Facebook.)

Solution

We use the ordered pair (**2006**, **10**) to indicate that there were **10** million users in **2006**. The other ordered pairs from the table are (2007, 30), (2008, 80), and (2009, 250). A scatterplot is shown in Figure 9.9, and a line graph is shown in Figure 9.10.

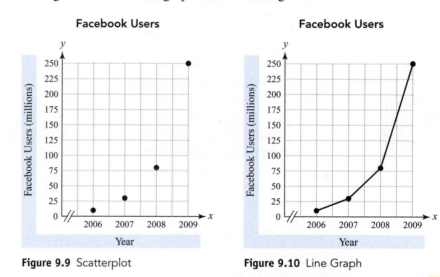

Figure 9.9 Scatterplot   **Figure 9.10** Line Graph

Now Try Exercise 45

9.1   Putting It All Together

CONCEPT	COMMENTS	EXAMPLES
The *xy*-plane (Rectangular Coordinate System)	The *xy*-plane or rectangular coordinate system is used to graph data. The horizontal axis is called the *x-axis*, and the vertical axis is called the *y-axis*. The axes intersect at the *origin* and divide the plane into four *quadrants*: I, II, III, and IV.	*(graph showing Quadrants I, II, III, IV with Origin)*
Ordered Pairs and Coordinates	The *coordinates* of a point indicate its location in the *xy*-plane. A point's coordinates are written as an *ordered pair*: (*x-coordinate, y-coordinate*).	The point given by the ordered pair (1, 3) has *x*-coordinate 1 and *y*-coordinate 3. It is located in quadrant I of the *xy*-plane.

continued on next page

continued from previous page

CONCEPT	COMMENTS	EXAMPLES
The Coordinates of a Graphed Point	The x-coordinate of a point is the number on the x-axis that is aligned directly above or below the point. The y-coordinate is the number on the y-axis that is aligned directly to the right or left of the point. The coordinates of a point are written as the ordered pair (x-coordinate, y-coordinate).	The point shown in the graph is (**3**, **4**).
Plotting Points in the xy-plane	1. From the origin, move left or right along the x-axis to the position given by the x-coordinate. 2. From that position, move upward or downward the amount given by the y-coordinate. 3. Make a dot at this final position.	$(4, 4)$ and $(-2, -3)$ are plotted below.
Scatterplot	When points are plotted in the xy-plane, the resulting graph is called a *scatterplot*.	
Line Graph	Consecutive data points can be connected with line segments to make a *line graph*. Line graphs are useful for observing trends in real-world data.	

9.1   Exercises

CONCEPTS AND VOCABULARY

1. Another name for the rectangular coordinate system is the _____.

2. In the xy-plane, the horizontal axis is the _____-axis and the vertical axis is the _____-axis.

3. The x- and y-axis intersect at the _____.

4. The x- and y-axis divide the xy-plane into four _____.

5. The "address" of a point in the xy-plane is given by its _____, written as a(n) _____.

6. The first number in an ordered pair is called the _____-coordinate, and the second number is called the _____-coordinate.

7. When points are plotted in the *xy*-plane, the resulting graph is called a(n) _____.

8. To make a(n) _____, connect consecutive data points with line segments.

THE XY-PLANE

Exercises 9–12: Give the quadrant containing each point in the given graph. If a point is on an axis, name the axis.

9.

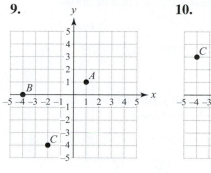

10.

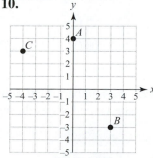

11.

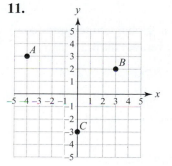

12.

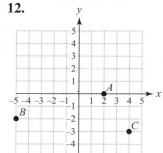

PLOTTING POINTS

Exercises 13–24: Give the quadrant containing the point. If the point is on an axis, name the axis.

13. $(-4, 5)$

14. $(2, 2)$

15. $(-4, -8)$

16. $(-23, 0)$

17. $(18, 0)$

18. $(0, 60)$

19. $(6, -11)$

20. $(-21, -20)$

21. $(0, -91)$

22. $(70, -49)$

23. $(10, 15)$

24. $(-9, 35)$

Exercises 25–30: Determine the coordinates of each point in the given graph.

25.

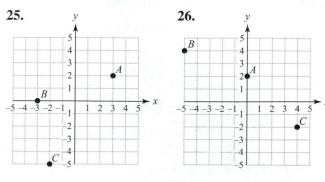

26.

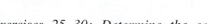

27.

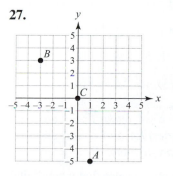

28.

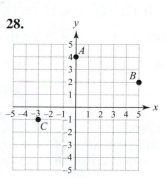

29.

30.

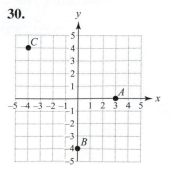

Exercises 31–36: Plot the points in the same xy-plane. Label each point with its coordinates.

31. $(0, -4)$, $(3, 5)$, $(-2, -4)$, and $(3, 0)$

32. $(2, 4)$, $(5, -3)$, $(0, 2)$, and $(-4, 1)$

33. $(-5, 0)$, $(1, 1)$, $(3, -3)$, and $(-2, -5)$

34. $(4, -1)$, $(-2, -3)$, $(-1, 4)$, and $(3, 4)$

35. $(-2, -2)$, $(3, 1)$, $(5, -4)$, and $(-2, 0)$

36. $(4, 4)$, $(0, 0)$, $(1, 3)$, and $(-3, -4)$

SCATTERPLOTS AND LINE GRAPHS

Exercises 37–40: Make a scatterplot of the points.

37. $(-4, 4)$, $(-2, 2)$, $(0, 0)$, $(2, -2)$, and $(4, -4)$

38. $(-3, -4)$, $(-1, -3)$, $(1, -2)$, $(3, -1)$, and $(5, 0)$

39. $(2, 3)$, $(-4, 3)$, $(0, 3)$, $(-1, 3)$, and $(5, 3)$

40. $(-2, -3)$, $(-3, -3)$, $(0, -1)$, $(1, 2)$, and $(4, 5)$

Exercises 41–44: Make a line graph of the points.

41. $(2, -2)$, $(-4, 1)$, $(0, 1)$, $(-2, -4)$, and $(3, 2)$

42. $(3, 1)$, $(-1, -2)$, $(0, 0)$, $(4, -1)$, and $(-5, 2)$

43. $(-4, 3)$, $(0, 1)$, $(2, 4)$, $(-1, 3)$, and $(5, 4)$

44. $(1, 0)$, $(-5, -1)$, $(-3, -3)$, $(2, 2)$, and $(4, 1)$

Exercises 45–50: **Graphing Real Data** *The table contains real data. Make a scatterplot and a line graph of the data. Label each axis appropriately.*

45. The percentage y of total music sales that were digital downloads during year x

Year (x)	2004	2005	2006	2007	2008
Percent (y)	2	5	11	15	20

(*Source:* IFPI.)

46. The number y of billionaires during year x

Year (x)	2005	2006	2007	2008	2009
People (y)	691	793	946	1125	793

(*Source:* Forbes.)

47. Annual carbon dioxide emissions y in tons for a car with an average mileage of x miles per gallon

MPG (x)	25	30	35	40	45
CO_2 (y)	7.3	6.1	5.3	4.6	4.1

(*Source:* U.S. Department of Energy.)

48. The approximate number of new apps y submitted to Apple's App Store during month x of 2008

Month (x)	8	9	10	11	12
Apps (y)	1500	3000	2700	3100	4200

(*Source:* Apple.)

49. The number of children y adopted from Ethiopia into the United States during year x

Year (x)	2004	2005	2006	2007	2008
Children (y)	289	447	732	1255	1725

(*Source:* U.S. Department of State.)

50. The predicted increase y in global temperature (compared to the 2000 global average) during year x

Year (x)	2020	2040	2060	2080	2100
°C (y)	0.6	1.0	2.1	3.6	4.7

(*Source:* National Institute for Environmental Studies.)

WRITING ABOUT MATHEMATICS

51. How can the signs of a point's x- and y-coordinates be used to determine the quadrant containing the point? Make a chart or table to visualize your answer.

52. Will a line segment connecting two points in a line graph ever intersect a different line segment in the same line graph? Explain.

9.2 Graphing Linear Equations in Two Variables

**Linear Equations in Two Variables • Tables of Solutions •
Graphing Linear Equations in Two Variables • Horizontal and Vertical Lines**

A LOOK INTO MATH ▶

Newspapers and magazines often display tables and graphs to make information easier to understand. Although a formula such as $T = 4.3x + 36.5$ could be used to describe the high temperatures for the next several days, a graph of high temperatures would quickly show that a steady increase in high temperature can be expected. In this section, we discuss the skills needed to graph linear equations in two variables.

Linear Equations in Two Variables

NEW VOCABULARY

☐ Linear equation in two variables
☐ Graph

An equation containing two variables is a *linear equation in two variables* if it can be written as follows.

STUDY TIP

It's never too late to join a study group. It is often very helpful to see how your classmates study and practice math.

> **LINEAR EQUATION IN TWO VARIABLES**
>
> A **linear equation in two variables** is an equation that can be written in the form
> $$Ax + By = C,$$
> where A, B, and C are constants (numbers) and A and B are not *both* equal to 0.

Examples of linear equations in two variables include

$$3x + 2y = 12, \quad y = 9 - 4x, \quad x = 8, \quad \text{and} \quad 5y = 11.$$

In Section 3.4, we were given three rules that show when an equation in one variable is *not* linear. Similar rules apply for equations in two variables. An equation in two variables *cannot* be written in the form $Ax + By = C$ (and is not a linear equation) if, after clearing parentheses and combining like terms, any of the following are true:

1. Either variable has an exponent other than 1.
2. Either variable is in a denominator.
3. Either variable is under the symbol $\sqrt{}$ or within absolute value symbols.

EXAMPLE 1 **Determining if equations are linear**

Determine if each equation is linear. If so, give values for A, B, and C.

(a) $6y + 5 = 9x$ **(b)** $y = 2x^2 - 5$ **(c)** $\dfrac{3}{y} - 5x = 9$ **(d)** $y = -2$

Solution

(a) The equation can be written as follows:

$$
\begin{array}{ll}
6y + 5 = 9x & \text{Given equation} \\
-9x + 6y + 5 = -9x + 9x & \text{Add } -9x \text{ to each side.} \\
-9x + 6y + 5 - 5 = 0 - 5 & \text{Subtract 5 from each side.} \\
-9x + 6y = -5 & \text{Simplify.}
\end{array}
$$

The equation $6y + 5 = 9x$ is linear because it can be written in the form $Ax + By = C$, with $A = -9$, $B = 6$, and $C = -5$.

(b) The equation $y = 2x^2 - 5$ is *not* linear. The variable x has an exponent other than 1.

(c) The equation $\frac{3}{y} - 5x = 9$ is *not* linear. The variable y is in a denominator.

(d) The equation $y = -2$ is linear because it can be written as $0x + 1y = -2$, which has the form $Ax + By = C$, with $A = 0, B = 1,$ and $C = -2$.

■ **Now Try Exercises 11, 13, 15, 17**

READING CHECK

• How many solutions does a linear equation in two variables have?

Although linear equations in one variable have exactly one solution, linear equations in two variables have *an infinite number* of solutions, each of which can be expressed as an ordered pair. For example, the ordered pair $(1, 5)$ is a solution to the equation $x + y = 6$. Other solutions include $(3, 3), (6, 0),$ and $(10, -4)$.

NOTE: An ordered pair is a solution to the equation $x + y = 6$ if it contains real numbers that sum to 6. There are infinitely many such ordered pairs.

To determine whether an ordered pair is a solution to a given linear equation in two variables, use the following procedure.

DETERMINING IF AN ORDERED PAIR IS A SOLUTION

If each variable in a linear equation in two variables is replaced with its corresponding value from an ordered pair and the resulting equation is true, then the ordered pair is a solution. Otherwise, the ordered pair is not a solution.

NOTE: Remember that the first number in an ordered pair corresponds to the variable x and the second number corresponds to the variable y.

EXAMPLE 2 **Determining if ordered pairs are solutions**

Answer each question.
(a) Is $(-3, 2)$ a solution to $2x + 3y = 0$?
(b) Is $(0, 5)$ a solution to $y = 2x + 1$?
(c) Is $(-10, 4)$ a solution to $y = 4$?

Solution
(a) Replace x with -3 and y with 2 in the given equation. Recall that a question mark is written over the equal sign because we are *checking* a possible solution.

$$2x + 3y = 0 \qquad \text{Given equation}$$
$$2(-3) + 3(2) \stackrel{?}{=} 0 \qquad \text{Replace } x \text{ with } -3 \text{ and } y \text{ with } 2.$$
$$-6 + 6 \stackrel{?}{=} 0 \qquad \text{Multiply.}$$
$$0 = 0 \checkmark \qquad \text{A true equation}$$

The ordered pair is a solution.

(b) Replace x with 0 and y with 5 in the given equation.

$$y = 2x + 1 \qquad \text{Given equation}$$
$$5 \stackrel{?}{=} 2(0) + 1 \qquad \text{Replace } x \text{ with } 0 \text{ and } y \text{ with } 5.$$
$$5 \stackrel{?}{=} 0 + 1 \qquad \text{Multiply.}$$
$$5 \stackrel{?}{=} 1 \; ✗ \qquad \text{A false equation}$$

The ordered pair is *not* a solution.

(c) Since the equation does not contain the variable x, we need only to replace y with **4**.

$$y = 4 \qquad \text{Given equation}$$

$$\mathbf{4} = 4 \ \checkmark \qquad \text{A true equation}$$

The ordered pair is a solution. In fact, *any* ordered pair with a y-coordinate of 4 is a solution to this equation.

Now Try Exercises 21, 25, 29

Tables of Solutions

Tables can be used to list solutions to a linear equation in two variables. For example, five solutions to the equation $x + y = 5$ are listed in Table 9.2.

TABLE 9.2 Solutions to $x + y = 5$

x	-2	-1	0	1	2
y	7	6	5	4	3

All linear equations in two variables have infinitely many solutions, so it is impossible to list all of the solutions in a table. However, having a table that lists a few solutions is quite helpful when graphing a linear equation. In the next example, we complete a table of solutions for given values of the variable x.

EXAMPLE 3   **Completing a table of solutions**

Complete Table 9.3 for the equation $y = 2x - 5$.

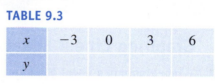

TABLE 9.3

x	-3	0	3	6
y				

Solution
For each given x-value, find the corresponding y-value. For example, when $x = -3$, the given equation $y = 2x - 5$ implies that $y = 2(\mathbf{-3}) - 5 = -6 - 5 = -11$. Similarly, if $x = \mathbf{0}$, then $y = 2(\mathbf{0}) - 5 = 0 - 5 = -5$. The other y-values are found in the same way. Filling in the y-values gives Table 9.4.

TABLE 9.4

x	-3	0	3	6
y	-11	-5	1	7

Now Try Exercise 31

When finding solutions to linear equations in two variables, *any* value of *either* variable can be used to complete a table of solutions, as described in the following procedure.

FINDING SOLUTIONS

To find a solution to a linear equation in two variables, do the following.

1. Replace *either* variable with *any* value.
2. Solve the resulting equation to find the value of the other variable.

EXAMPLE 4 **Completing a table of solutions**

Complete Table 9.5 for the equation $2x - 3y = 9$.

TABLE 9.5

x	0			9
y		-1	1	

Solution
Use each given value as a replacement for its corresponding variable in the given equation and solve for the other variable.

When $x = \mathbf{0}$,	When $y = \mathbf{-1}$,	When $y = \mathbf{1}$,	When $x = \mathbf{9}$,
$2(\mathbf{0}) - 3y = 9$	$2x - 3(\mathbf{-1}) = 9$	$2x - 3(\mathbf{1}) = 9$	$2(\mathbf{9}) - 3y = 9$
$0 - 3y = 9$	$2x + 3 = 9$	$2x - 3 = 9$	$18 - 3y = 9$
$-3y = 9$	$2x = 6$	$2x = 12$	$-3y = -9$
$y = -3$	$x = 3$	$x = 6$	$y = 3$

Filling in these values gives Table 9.6.

TABLE 9.6

x	0	3	6	9
y	-3	-1	1	3

Now Try Exercise 37

Graphing Linear Equations in Two Variables

When graphing a linear equation in two variables, we can think of the x- and y-values in a corresponding table of solutions as ordered pairs, which can be plotted on the xy-plane. To see the ordered pairs more easily, we use a vertical table rather than a horizontal one. For example, Table 9.7 is a horizontal table showing solutions to the equation $y = 2x$.

TABLE 9.7 Solutions to $y = 2x$

x	-1	0	1	2
y	-2	0	2	4

The ordered pairs are seen more easily by making the table shown in Table 9.8.

TABLE 9.8 Solutions to $y = 2x$

x	y	Ordered Pair
-1	-2	$(-1, -2)$
0	0	$(0, 0)$
1	2	$(1, 2)$
2	4	$(2, 4)$

Although the ordered pairs in Table 9.8 represent only four of the infinitely many solutions to the equation $y = 2x$, they can be plotted in the xy-plane so that a graph representing *all* of the solutions can be seen. Notice that the points $(-1, -2)$, $(0, 0)$, $(1, 2)$, and $(2, 4)$ all lie on the straight line shown in Figure 9.11.

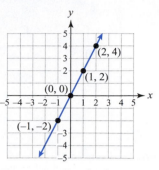

Figure 9.11 Graph of $y = 2x$

The line drawn through the plotted points in Figure 9.11 helps us visualize *every* solution to the linear equation $y = 2x$, because *every* point on the line represents a solution. The arrows at each end of the line indicate that it continues in both directions. We say that the line is the **graph** of the equation $y = 2x$.

NOTE: Every **line**ar equation in two variables has a graph that is a straight **line**.

READING CHECK

• What does the graph of a linear equation in two variables look like?

GRAPHING A LINEAR EQUATION IN TWO VARIABLES

To graph a linear equation in two variables, do the following.

1. Make a vertical table of solutions containing at least 2 ordered-pair solutions.
2. Plot the ordered-pair solutions in the *xy*-plane.
3. Draw a straight line through the plotted points.

NOTE: Only two plotted points are needed to graph a line. However, if an error was made when finding one of the points, the graph will be incorrect. To avoid such graphing errors, it is good practice to find and plot a third point. All three points must lie on a single line.

EXAMPLE 5   **Graphing a linear equation in two variables**

Graph the equation $y = 3x - 1$.

Solution
Start by selecting three values for x. It is convenient to choose integer values such as $x = -1$, 0, and 2. Using these chosen x-values, make a vertical table of solutions by finding corresponding y-values. Each y-value is found by tripling its corresponding x-value and subtracting 1. For example, when $x = -1$ the equation gives $y = 3(-1) - 1 = -3 - 1 = -4$. Three ordered pairs are shown in Table 9.9 and are plotted in Figure 9.12 on the next page. A line representing the equation $y = 3x - 1$ is graphed.

TABLE 9.9 Solutions to $y = 3x - 1$

x	y	Ordered Pair
-1	-4	$(-1, -4)$
0	-1	$(0, -1)$
2	5	$(2, 5)$

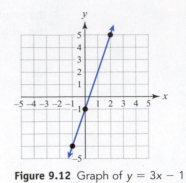

Figure 9.12 Graph of $y = 3x - 1$

■ **Now Try Exercise 55**

EXAMPLE 6   **Graphing a linear equation in two variables**

Graph the equation $2y - 3x = 6$.

Solution
Recall that solutions can be found by selecting values for *either* variable and then finding the value of the other variable. Using 0 as a replacement value—once for each variable—will often give two ordered-pair solutions that are fairly easy to find. For this equation, we will use $x = 0$, $y = 0$, and $y = -3$.

When $x = \mathbf{0}$,	When $y = \mathbf{0}$,	When $y = \mathbf{-3}$,
$2y - 3(\mathbf{0}) = 6$	$2(\mathbf{0}) - 3x = 6$	$2(\mathbf{-3}) - 3x = 6$
$2y - 0 = 6$	$0 - 3x = 6$	$-6 - 3x = 6$
$2y = 6$	$-3x = 6$	$-3x = 12$
$y = 3$	$x = -2$	$x = -4$

These three ordered pairs are shown in Table 9.10 and are plotted in Figure 9.13, where a line representing the equation $2y - 3x = 6$ is graphed.

TABLE 9.10 Solutions to $2y - 3x = 6$

x	y	Ordered Pair
0	3	$(0, 3)$
-2	0	$(-2, 0)$
-4	-3	$(-4, -3)$

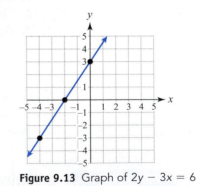

Figure 9.13 Graph of $2y - 3x = 6$

■ **Now Try Exercise 65**

Horizontal and Vertical Lines

In Example 1, we saw that the equation $y = -2$ is linear because it can be written in the form $Ax + By = C$ as $0x + 1y = -2$. Similarly, an equation such as $x = 4$ is linear because it can be written as $1x + 0y = 4$. When a linear equation has only one variable, its graph is a special kind of line.

EXAMPLE 7

Graphing a linear equation containing only the variable *y*

Graph the equation $y = -2$.

Solution

The equation $y = -2$ can be written as $0x + 1y = -2$. In this equation, *any* replacement value for x gives a y-value of -2. If x is -3, the corresponding y-value is -2. If x is 0, the y-value is -2. If x is 4, the y-value is -2. These ordered pairs are shown in Table 9.11 and are plotted in Figure 9.14, where the resulting graph is a *horizontal line*.

TABLE 9.11

x	*y*	Ordered Pair
-3	-2	$(-3, -2)$
0	-2	$(0, -2)$
4	-2	$(4, -2)$

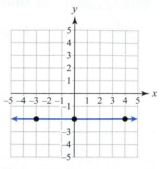

Figure 9.14 Graph of $y = -2$

Now Try Exercise 63

EXAMPLE 8

Graphing a linear equation containing only the variable *x*

Graph the equation $x = 4$.

Solution

The equation $x = 4$ can be written as $1x + 0y = 4$. In this equation, *any* replacement value for y gives an x-value of 4. If y is -2, the corresponding x-value is 4. If y is 1, the x-value is 4. If y is 5, the x-value is 4. These ordered pairs are shown in Table 9.12 and are plotted in Figure 9.15, where the resulting graph is a *vertical line*.

TABLE 9.12

x	*y*	Ordered Pair
4	-2	$(4, -2)$
4	1	$(4, 1)$
4	5	$(4, 5)$

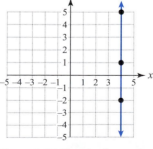

Figure 9.15 Graph of $x = 4$

Now Try Exercise 47

From Examples 7 and 8, we see that the graph of $y = -2$ is a horizontal line that crosses the y-axis at -2 and the graph of $x = 4$ is a vertical line that crosses the x-axis at 4.

> ### HORIZONTAL AND VERTICAL LINES
>
> If b is a constant (number), then the graph of the equation $y = b$ is a *horizontal* line that crosses the y-axis at b.
>
> If k is a constant (number), then the graph of the equation $x = k$ is a *vertical* line that crosses the x-axis at k.

EXAMPLE 9 | **Graphing horizontal and vertical lines**

Graph the equations $x = -2$ and $y = 5$ in the same xy-plane.

Solution

The graph of the equation $x = -2$ is a vertical line that crosses the x-axis at -2, and the graph of the equation $y = 5$ is a horizontal line that crosses the y-axis at 5. See Figure 9.16.

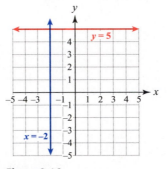

Figure 9.16

▌ **Now Try Exercise 75**

EXAMPLE 10 | **Solving for a single variable before graphing**

Graph the equation $x + 2 = 5$.

Solution

First, solve the equation for x to write it in the form $x = k$.

$$x + 2 = 5 \qquad \text{Given equation}$$
$$x + 2 - \mathbf{2} = 5 - \mathbf{2} \qquad \text{Subtract 2 from each side.}$$
$$x = 3 \qquad \text{Simplify.}$$

The graph of $x = 3$ is a vertical line that crosses the x-axis at 3, as shown in Figure 9.17.

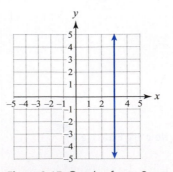

Figure 9.17 Graph of $x = 3$

▌ **Now Try Exercise 53**

9.2 Putting It All Together

CONCEPT	COMMENTS	EXAMPLES
Linear Equations in Two Variables	A linear equation in two variables is an equation that can be written in the form $$Ax + By = C,$$ where A, B, and C are constants (numbers) and A and B are not both equal to 0.	$-2x + 5y = 11$, $3x = 2y$, $y = 3x - 4$, and $x = 9$
Ordered-Pair Solutions	If each variable in a linear equation in two variables is replaced with its corresponding value from an ordered pair and the resulting equation is true, then the ordered pair is a solution. Otherwise, the ordered pair is not a solution.	The ordered pair $(3, 2)$ is a solution to the equation $2x - y = 4$ because $2(3) - 2 = 4$ is a true equation.
Tables of Solutions	Tables can be used to list solutions to a linear equation in two variables.	The table shows solutions to $y = 3x$. <table><tr><td>x</td><td>-1</td><td>0</td><td>1</td></tr><tr><td>y</td><td>-3</td><td>0</td><td>3</td></tr></table>
Finding Solutions	1. Replace *either* variable with *any* value. 2. Solve the resulting equation to find the value of the other variable.	For the linear equation $y = 5x + 1$, if we choose 0 for x, the y-value is $y = 5(0) + 1 = 0 + 1 = 1$. So the ordered pair $(0, 1)$ gives a solution.
Graphing a Linear Equation	1. Make a vertical table of solutions containing at least 2 ordered-pair solutions. 2. Plot the ordered-pair solutions in the xy-plane. 3. Draw a straight line through the plotted points.	A table of solutions for $y = 2x$: <table><tr><th>x</th><th>y</th><th>Ordered Pair</th></tr><tr><td>-2</td><td>-4</td><td>$(-2, -4)$</td></tr><tr><td>0</td><td>0</td><td>$(0, 0)$</td></tr><tr><td>1</td><td>2</td><td>$(1, 2)$</td></tr></table> The graph of $y = 2x$:

continued on next page

continued from previous page

CONCEPT	COMMENTS	EXAMPLES
Horizontal Line	If b is a constant (number), then the graph of $y = b$ is a *horizontal* line that crosses the y-axis at b.	The graph of $y = -3$ is a horizontal line crossing the y-axis at -3.
Vertical Line	If k is a constant (number), then the graph of $x = k$ is a *vertical* line that crosses the x-axis at k.	The graph of $x = -3$ is a vertical line crossing the x-axis at -3.

9.2 Exercises

MyMathLab

Math XL PRACTICE · WATCH · DOWNLOAD · READ · REVIEW

CONCEPTS AND VOCABULARY

1. If an equation can be written as $Ax + By = C$, where A, B, and C are constants and A and B are not both 0, it is a(n) _____ equation in two variables.

2. Every linear equation in two variables has a(n) _____ number of solutions.

3. A solution to a linear equation in two variables is written as a(n) _____.

4. One way to write some of the solutions to a linear equation in two variables is to list them in a(n) _____ of solutions.

5. To find a solution to a linear equation in two variables, start by replacing _____ variable with *any* value.

6. Every linear equation in two variables has a graph that is a straight _____.

7. If b is a constant (number), then the graph of $y = b$ is a(n) _____ line that crosses the y-axis at b.

8. If k is a constant (number), then the graph of $x = k$ is a(n) _____ line that crosses the x-axis at k.

LINEAR EQUATIONS IN TWO VARIABLES

Exercises 9–20: Determine whether the given equation is linear. If so, give values for A, B, and C for writing the equation in the form Ax + By = C.

9. $4x + 3y = 1$

10. $-3x + 2y = 8$

11. $2x - 7 = 5y$

12. $\dfrac{4}{x} - \dfrac{3}{y} = 10$

13. $2x + 5y^3 = 8$

14. $y = -8$

15. $x = -5$

16. $4y - 6 = -x$

17. $|-2x - y| = 12$ **18.** $5x^2 + 3y^2 = 4^2$

19. $3\sqrt{x} + 5\sqrt{y} = 9$ **20.** $|x| = |y|$

Exercises 21–30: Determine if the given ordered pair is a solution to the given equation.

21. $(1, -2)$, $2x + 4y = 8$

22. $(2, 19)$, $y = 13x - 7$

23. $(12, -1)$, $x = 9 - 3y$

24. $(2, -1)$, $7y + x = 15$

25. $(0, -3)$, $y = 7x - 3$

26. $(-10, 11)$, $y = 11$

27. $(-11, -8)$, $x = 5$

28. $(2, 8)$, $x = 5y - 2$

29. $(-17, 6)$, $y = 6$

30. $(3, -7)$, $-2y - 3x = 5$

TABLES OF SOLUTIONS

Exercises 31–42: Complete the table of solutions.

31. $y = 5x - 2$

x	-2	-1	0	1
y				

32. $y = 2x + 7$

x	-4	-2	0	2
y				

33. $-2x + 3y = 6$

x	-6	-3	0	3
y				

34. $-x - 4y = 7$

x	-7	-3	1	5
y				

35. $y = 2$

x	-23	-9	0	6
y				

36. $y = 0$

x	-8	-4	0	4
y				

37. $2x + 5y = 10$

x	-10		0	
y		4		0

38. $3x - 4y = 12$

x	-4			8
y		-3	0	

39. $y = -2x + 4$

x			2	4
y	8	4		

40. $y = 5x - 11$

x	-3	-1		
y			-6	4

41. $x = 3y - 6$

x	-9			0
y		0	1	

42. $x = -y + 4$

x	2		0	
y		3		5

GRAPHING LINEAR EQUATIONS

Exercises 43–72: Graph the given equation.

43. $y = x + 1$ **44.** $y = x - 2$

45. $x + 2y = -4$ **46.** $2x - y = 4$

47. $x = -3$ **48.** $y = 1$

49. $y = \dfrac{1}{2}x + 2$ **50.** $y = -\dfrac{1}{3}x + 4$

51. $x - y = 4$ **52.** $x + y = 3$

53. $x + 7 = 5$ **54.** $y - 4 = -1$

55. $y = 2x - 2$ **56.** $y = 3x + 1$

57. $2x - 3y = 6$ **58.** $2y + 3x = -6$

59. $y = 0$ **60.** $x = 2$

61. $-3x + 2y = -8$ **62.** $4x + 5y = 10$

63. $y = -4$ **64.** $x = 3$

65. $4x - 3y = 12$ **66.** $2x + y = -4$

67. $x = 0$ **68.** $y = -5$

69. $y = -3x + 1$ **70.** $y = -x - 3$

71. $y - 4 = 0$ **72.** $x + 4 = 0$

Exercises 73–76: Graph the given pair of equations in the same xy-plane.

73. $y = -3$ and $x = 5$ **74.** $x = -1$ and $y = 4$

75. $x = -5$ and $y = 2$ **76.** $x = 0$ and $y = 0$

WRITING ABOUT MATHEMATICS

77. When making a table of solutions for the equation

$$y = \frac{2}{5}x - 3,$$

what kinds of x-values should be chosen to be sure that the resulting y-values are not fractions? Explain.

78. Explain how you can determine whether an equation has a graph that is a vertical or horizontal line. Give an example of an equation of each type.

SECTIONS 9.1 and 9.2   Checking Basic Concepts

1. Give the quadrant containing each point. If the point is on an axis, name the axis.
   (a) $(-2, 5)$ **(b)** $(0, -4)$

2. Determine the coordinates of each point in the following graph.

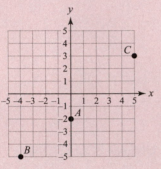

3. Make a scatterplot of the given points.
   $(-3, 4), (-2, 1), (0, -1), (2, 3),$ and $(4, 5)$

4. Make a line graph of the given points.
   $(-4, 3), (5, 2), (0, -3), (-2, 1),$ and $(1, 1)$

5. Determine whether the given equation is linear. If so, give values for A, B, and C.
   (a) $5x + 4 = 2y$ **(b)** $x^2 - 9y^2 = 12$

6. Is $(-1, -2)$ a solution to $2x - 3y = 4$?

7. Complete the table of solutions for $x + 3y = 7$.

x	-2			7
y		2	1	

8. Graph each equation.
   (a) $y = -2x - 3$ **(b)** $x = -4$
   (c) $2x - 4y = 6$ **(d)** $y = 3$

9.3   Graphical Solutions to Linear Equations

Graphical Solutions • Solving Linear Equations Graphically •
Solving Linear Equations Algebraically, Numerically, and Graphically

A LOOK INTO MATH ▶ If a store owner sets the price of a product too low, then the product is likely to sell quickly but the profit may be low. If the price is set too high, then the item may not sell well and the profit may, again, be low. To find the price that gives the highest profit, graphs of equations can be used. In math, it is often possible to solve equations graphically. In this section, we discuss the steps used to find *graphical solutions* to linear equations.

Graphical Solutions

Throughout this text, graphs have been provided for solving equations *visually*. A solution that is found by looking at a graph is called a **graphical solution**. For example, if we want to find a graphical solution to the equation $x + 2 = 5$, a graph that shows values of the left side of the equation can be used. Such a graph is shown in Figure 9.18, where the solution appears to be 3. To see that **3** is correct, we check it in the given equation.

$$x + 2 = 5 \qquad \text{Given equation}$$
$$3 + 2 \stackrel{?}{=} 5 \qquad \text{Replace } x \text{ with 3.}$$
$$5 = 5 \checkmark \qquad \text{The solution checks.}$$

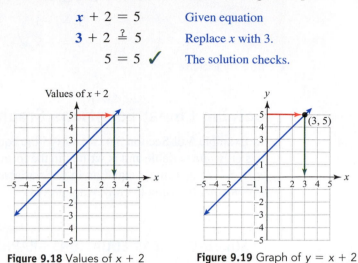

Figure 9.18 Values of x + 2   **Figure 9.19** Graph of y = x + 2

The same graphical solution is found using the graph in Figure 9.19, where the equation $y = x + 2$ is graphed. Since y equals the left side of the equation $x + 2 = 5$, the y-axis in Figure 9.19 represents the same values as the vertical axis in Figure 9.18. The point $(3, 5)$ on the graph of $y = x + 2$ tells us that the value of y is 5 when x equals 3.

EXAMPLE 1   **Finding a graphical solution**

Use the graph of $y = x - 3$ in Figure 9.20 to find a graphical solution to $x - 3 = -5$.

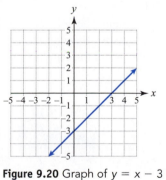

Figure 9.20 Graph of y = x − 3

Solution
The graphical solution is shown in Figure 9.21. The y-value is -5 when the x-value is -2.

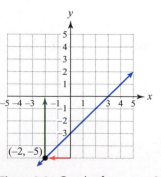

Figure 9.21 Graph of y = x − 3

The solution appears to be -2, which is checked as follows.

$$x - 3 = -5 \qquad \text{Given equation}$$
$$-2 - 3 \stackrel{?}{=} -5 \qquad \text{Replace } x \text{ with } -2.$$
$$-5 = -5 \checkmark \qquad \text{The solution checks.}$$

Now Try Exercise 5

Solving Linear Equations Graphically

In Example 5 of Section 3.4, we solved the equation $-2x - 5 = 3$ *visually*. This equation has only *one* variable and is written in the form $ax + b = c$, where $a = -2$, $b = -5$, and $c = 3$. When a linear equation in one variable has the form $ax + b = c$, we can solve it *graphically* by using the following process.

SOLVING LINEAR EQUATIONS GRAPHICALLY

To solve the linear equation $ax + b = c$ graphically, do the following.

1. Let y equal the left side of the equation and then graph $y = ax + b$.
2. Use the graph to solve $ax + b = c$ visually.
3. Check the solution in the given equation.

EXAMPLE 2   Solving a linear equation graphically

Solve the equation $2x - 3 = 9$ graphically.

Solution
Let y equal the left side of the equation and then graph $y = 2x - 3$. Three ordered-pair solutions for this equation are shown in Table 9.13. The ordered pairs are plotted, and a line representing the equation $y = 2x - 3$ is graphed in Figure 9.22.

TABLE 9.13

x	y	Ordered Pair
-2	-7	$(-2, -7)$
0	-3	$(0, -3)$
3	3	$(3, 3)$

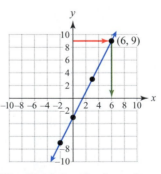

Figure 9.22 Graph of $y = 2x - 3$

From the graph in Figure 9.22, the solution appears to be 6, which is checked as follows.

$$2x - 3 = 9 \qquad \text{Given equation}$$
$$2(\mathbf{6}) - 3 \stackrel{?}{=} 9 \qquad \text{Replace } x \text{ with 6.}$$
$$12 - 3 \stackrel{?}{=} 9 \qquad \text{Simplify.}$$
$$9 = 9 \checkmark \qquad \text{The solution checks.}$$

The solution to the equation $2x - 3 = 9$ is 6.

Now Try Exercise 13

EXAMPLE 3 **Solving a linear equation graphically**

Solve the equation $-x + 6 = -4$ graphically.

Solution
Let y equal the left side of the equation and then graph $y = -x + 6$. Three ordered-pair solutions for this equation are shown in Table 9.14. The ordered pairs are plotted, and a line representing the equation $y = -x + 6$ is graphed in Figure 9.23.

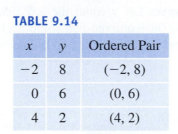

TABLE 9.14

x	y	Ordered Pair
-2	8	$(-2, 8)$
0	6	$(0, 6)$
4	2	$(4, 2)$

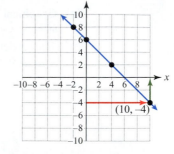

Figure 9.23 Graph of $y = -x + 6$

From the graph in Figure 9.23, the solution appears to be 10, which is checked as follows.

$$-x + 6 = -4 \qquad \text{Given equation}$$
$$-10 + 6 \overset{?}{=} -4 \qquad \text{Replace } x \text{ with 10.}$$
$$-4 = -4 \;\checkmark \qquad \text{The solution checks.}$$

The solution to the equation $-x + 6 = -4$ is 10.

▌ **Now Try Exercise 15**

Solving Linear Equations Algebraically, Numerically, and Graphically

We can now solve linear equations algebraically, numerically, and graphically. Remember that no matter which of the three methods we use, the solution should be the same.

EXAMPLE 4 **Solving an equation algebraically, numerically, and graphically**

Solve the equation $-\dfrac{1}{2}x + 5 = 8$

(a) algebraically, **(b)** numerically, and **(c)** graphically.

Solution
(a) Algebraic Solution:

$$-\frac{1}{2}x + 5 = 8 \qquad \text{Given equation}$$

$$-\frac{1}{2}x + 5 - 5 = 8 - 5 \qquad \text{Subtract 5 from each side.}$$

$$-\frac{1}{2}x = 3 \qquad \text{Simplify.}$$

$$-2 \cdot \left(-\frac{1}{2}x\right) = -2 \cdot 3 \qquad \text{Multiply each side by } -2$$

$$x = -6 \qquad \text{Simplify.}$$

The solution is -6.

(b) Numerical Solution: Since the expression $-\frac{1}{2}x + 5$ will evaluate to a whole number only for even values of x, we make our table as shown in Table 9.15, where we see that $-\frac{1}{2}x + 5 = \mathbf{8}$ when x is $\mathbf{-6}$. The solution is -6.

TABLE 9.15

x	-10	-8	$\mathbf{-6}$	-4	-2	0	2	4
$-\frac{1}{2}x + 5$	10	9	$\mathbf{8}$	7	6	5	4	3

(c) Graphical Solution: Let y equal the left side of the given equation and then graph $y = -\frac{1}{2}x + 5$. Three ordered-pair solutions for this equation are shown in Table 9.16. The ordered pairs are plotted, and a line representing the equation is graphed in Figure 9.24.

TABLE 9.16

x	y	Ordered Pair
-8	9	$(-8, 9)$
0	5	$(0, 5)$
6	2	$(6, 2)$

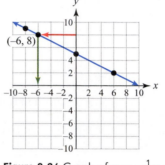

Figure 9.24 Graph of $y = -\frac{1}{2}x + 5$

From Figure 9.24, the solution appears to be -6, which is checked as follows.

READING CHECK

• How do algebraic, numerical, and graphical solutions to linear equations compare?

$$-\frac{1}{2}x + 5 = 8 \qquad \text{Given equation}$$

$$-\frac{1}{2}(\mathbf{-6}) + 5 \stackrel{?}{=} 8 \qquad \text{Replace } x \text{ with } -6.$$

$$3 + 5 \stackrel{?}{=} 8 \qquad \text{Simplify.}$$

$$8 = 8 \ \checkmark \qquad \text{The solution checks.}$$

The algebraic, numerical, and graphical solutions are the same.

Now Try Exercise 33

9.3   Putting It All Together

CONCEPT	COMMENTS	EXAMPLES
Graphical Solution	A solution that is found visually from a graph is referred to as a graphical solution.	The solution to $2x - 1 = 5$ is 3. 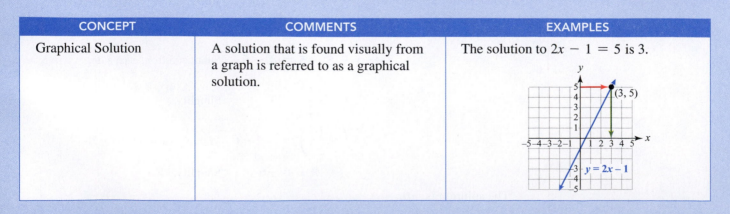

CONCEPT	COMMENTS	EXAMPLES
Solving a Linear Equation Graphically	To solve the equation $ax + b = c$ graphically, do the following. 1. Let y equal the left side of the equation and then graph $y = ax + b$. 2. Use the graph to solve $ax + b = c$ visually. 3. Check the solution in the given equation.	To solve the equation $2x + 1 = 9$, graph the equation $y = 2x + 1$. Three ordered-pair solutions are shown in the following table.

<table>
<tr><th>x</th><th>y</th><th>Ordered Pair</th></tr>
<tr><td>-2</td><td>-3</td><td>$(-2, -3)$</td></tr>
<tr><td>0</td><td>1</td><td>$(0, 1)$</td></tr>
<tr><td>3</td><td>7</td><td>$(3, 7)$</td></tr>
</table>

The ordered pairs are plotted, and a line representing $y = 2x + 1$ is graphed in the following figure.

The solution is 4, which checks in the given equation: $2(4) + 1 = 9$.

9.3 Exercises

MyMathLab   Math XL PRACTICE   WATCH   DOWNLOAD   READ   REVIEW

CONCEPTS AND VOCABULARY

1. A solution that is found by looking at a graph is called a(n) _____ solution.

2. To solve the equation $ax + b = c$ graphically, start by graphing the equation _____.

3. To be sure that a graphical solution is correct, _____ it in the given equation.

4. Whether we solve an equation algebraically, numerically, or graphically, the solution is the _____.

GRAPHICAL SOLUTIONS

Exercises 5–10: Use the graph shown to find a graphical solution to the given equation.

5. $2x - 3 = -1$

6. $-3x - 4 = 2$

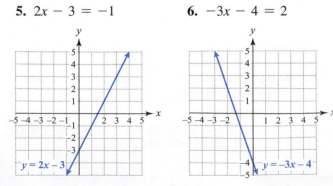

7. $-x - 2 = -4$ **8.** $3x + 2 = 2$

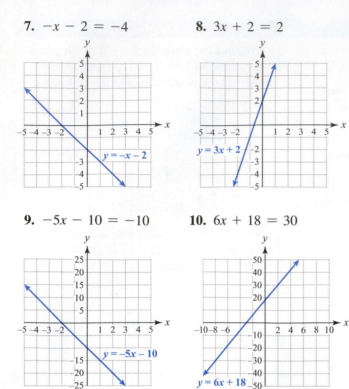

9. $-5x - 10 = -10$ **10.** $6x + 18 = 30$

SOLVING LINEAR EQUATIONS GRAPHICALLY

Exercises 11–26: Solve the equation graphically.

11. $x - 4 = -7$ **12.** $-x + 6 = 5$

13. $-2x - 3 = 5$ **14.** $2x + 5 = 9$

15. $-3x + 8 = -1$ **16.** $3x - 8 = 4$

17. $2x - 7 = -7$ **18.** $-2x + 3 = 3$

19. $\frac{1}{2}x - 2 = -5$ **20.** $-\frac{1}{2}x + 9 = 5$

21. $-\frac{1}{3}x + 9 = 6$ **22.** $\frac{1}{4}x + 7 = 8$

23. $1.5x - 4 = 5$ **24.** $-0.5x + 1 = 0$

25. $0.25x - 2 = 0$ **26.** $0.5x - 4 = -5$

SOLVING LINEAR EQUATIONS

Exercises 27–36: Solve the equation
    (a) algebraically,
    (b) numerically, and
    (c) graphically.

27. $-x - 7 = -6$ **28.** $x + 8 = 2$

29. $2x + 7 = -9$ **30.** $-3x + 14 = 2$

31. $4x - 3 = 5$ **32.** $3x - 1 = 8$

33. $\frac{1}{3}x + 4 = 6$ **34.** $\frac{1}{2}x - 12 = -8$

35. $-0.5x + 7 = 6$ **36.** $0.25x - 6 = -7$

WRITING ABOUT MATHEMATICS

37. Explain why graphical solutions should always be checked in the given equation.

38. When a student solves a linear equation algebraically, numerically, and graphically, he gets three different answers. What can be said about the student's work? Why?

Group Activity   Working with Real Data

Directions: Form a group of 2 to 4 people. Select someone to record the group's responses for this activity. All members of the group should work cooperatively to answer the questions. If your instructor asks for your results, each member of the group should be prepared to respond.

The Value of an Education Research has shown that a person's level of education often affects his or her earning potential. The ordered pairs

$$(0, 27.2), (1, 32.8), (2, 36.3), \text{ and } (4, 45.8)$$

have the form (x, y), where x represents the number of years of education beyond high school and y represents the corresponding median annual salary, in thousands of dollars, for full-time working women. (*Source: Digest of Educational Statistics, 2008.*)

(a) What is the median salary for women with a high school diploma (0 years of college)?

(b) How much more is the median salary for women with a 4-year bachelor's degree when compared to women with a 2-year associate's degree?

(c) Make a scatterplot of the ordered pairs.

(d) Use your scatterplot to estimate the median annual salary for women who have completed 3 years of college.

(e) The linear equation

$$y = 4.5x + 27.6$$

can be used to model this situation. Use the equation to estimate the median salary for women who have completed 3 years of college.

9.4 Solving Applications Using Graphs

Data Represented in Pictographs • Representing Linear Data Graphically • Solving Applications Graphically

A LOOK INTO MATH ▶ Every year, billions of dollars are spent on Internet advertising. To market their products and services more effectively, advertisers hire marketing firms that analyze huge amounts of data in an effort to predict future consumer trends. Sometimes, a *linear trend* is found. When this happens, graphs of lines can be used to analyze the linear data so that estimates of past and future data values can be made. (*Source: Newsweek.*)

NEW VOCABULARY

☐ Pictograph
☐ Key

Data Represented in Pictographs

Newspapers and magazines commonly display information in graphs that use a picture or symbol to represent a particular quantity. This type of graph is called a **pictograph**. Like bar graphs, pictographs can be used to easily compare data. Every pictograph contains a **key** that gives the meaning of the symbol in the graph. Figure 9.25 shows a pictograph that gives the number of iPhones sold during selected quarters of 2009 and 2010. (*Source: Apple.*)

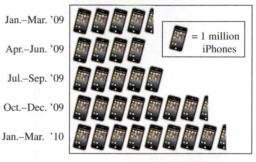

iPhone Sales by Quarter

Figure 9.25

READING CHECK

• What does a pictograph key tell us?

The pictograph key shows that each iPhone symbol represents sales of 1,000,000 phones. For example, for the third quarter of 2009, from July through September, the pictograph displays **5** iPhone symbols, indicating that **5,000,000** iPhones were sold that quarter.

EXAMPLE 1 Reading a pictograph

The pictograph in Figure 9.26 shows the number of college graduates with more than $40,000 in student loan debt during selected years. Find the number of college graduates with more than $40,000 in student loan debt in 2004 and 2008. (*Source:* U.S. Department of Education.)

Students Owing More Than $40,000

Figure 9.26

Solution

Student loan debt: 2004 Since each student symbol represents **15,000** students, and there are **7** symbols shown for 2004, the total number of students with more than $40,000 in student loan debt in 2004 is **15,000** · **7** = 105,000.

Student loan debt: 2008 There are $13\frac{1}{2}$ student symbols shown for 2008. Write $13\frac{1}{2}$ as 13.5 and multiply by 15,000. In 2008, there were 15,000 · 13.5 = 202,500 students with more than $40,000 in student loan debt.

Now Try Exercise 3

Representing Linear Data Graphically

When data are linear, a graphical representation can be made by plotting two data points in the *xy*-plane and then drawing a line that passes through them.

▶ **REAL-WORLD CONNECTION** In 2000, the average price of a new car was $20,400. For the next ten years, this average followed a linear trend until it reached $29,300 in 2010. If we let *x* represent the year and *y* represent the average price, we can graph a line that passes through the points (2000, 20400) and (2010, 29300). This graph can help us analyze average price data for years other than 2000 and 2010. (*Source:* U.S. Department of Commerce.)

Figure 9.27 shows a line passing through the points (2000, 20400) and (2010, 29300). Every point on this graph represents an ordered pair of the form (year, price).

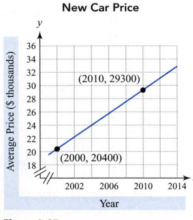

Figure 9.27

The graph can be used to determine other information. For example, we can estimate graphically the year when the average price of a new car was $27,520. Figure 9.28 shows that the average price was $27,520 in 2008.

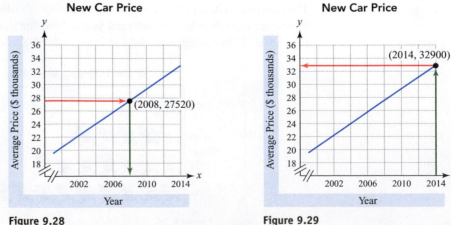

Figure 9.28

Figure 9.29

The graph can also be used to predict the average price of a new car in 2014. Figure 9.29 shows an average price of about $32,900 in 2014.

EXAMPLE 2 | **Analyzing linear data graphically**

The cost of taking 4 credits at a two-year college is $500, and the cost of taking 12 credits is $1500. Let x represent the number of credits, and let y represent the cost.
(a) Write two ordered pairs for the given information.
(b) Plot the ordered pairs in the xy-plane and draw a line passing through them.
(c) Use the graph to estimate the cost of taking 8 credits.
(d) If the cost is $1750, about how many credits does this purchase?

Solution
(a) The ordered pairs have the form (credits, cost). They are (4, 500) and (12, 1500).
(b) Figure 9.30 shows the graph of a line passing through (4, 500) and (12, 1500).

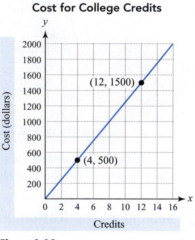

Figure 9.30

(c) The cost of 8 credits is $1000, as shown in Figure 9.31.

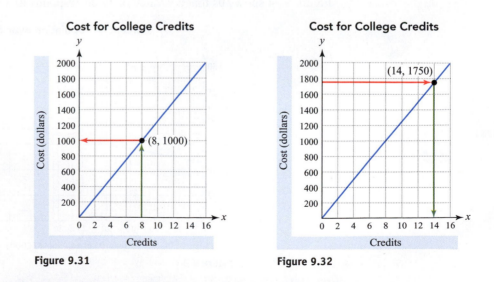

Figure 9.31   **Figure 9.32**

(d) Figure 9.32 shows that $1750 purchases 14 credits.

Now Try Exercise 7

Solving Applications Graphically

Using computer graphing software or graphing calculators, we can solve many real-world applications graphically.

▶ **REAL-WORLD CONNECTION** Let x represents a number of years after 2000. For example, $x = 20$ represents 2020 and $x = 34$ represents 2034. If we use y to represent the projected percentage of the U.S. population that will be over the age of 65 during year x, then the linear equation $y = 0.18x + 11.4$ can be used to estimate this percentage y for a given year x. Figure 9.33 shows a computer-generated graph of $y = 0.18x + 11.4$. (*Source:* U.S. Census Bureau.)

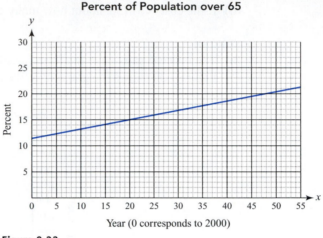

Percent of Population over 65

Year (0 corresponds to 2000)

Figure 9.33

In the next example, we use Figure 9.33 to answer a question about the percentage of the U.S. population that is over the age of 65.

EXAMPLE 3 **Solving an application graphically**

Use Figure 9.33 to estimate the year when 17% of the population will be over the age of 65.

Solution
Figure 9.34 shows us that a y-value of 17 corresponds to an x-value of 31.

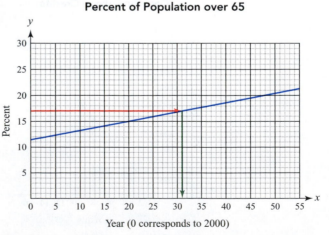

Percent of Population over 65

Year (0 corresponds to 2000)

Figure 9.34

By replacing x with 31 in the given equation, we find that $y = 0.18(31) + 11.4 = 16.98$, which is very close to 17%. Since an x-value of 31 corresponds to the year 2031, we find that 17% of the population will be over the age of 65 in 2031.

Now Try Exercise 11

When a graph represents real-world data, special meaning is often associated with the points where the graph crosses (or touches) each axis. These meanings are discussed in the next Real-World Connection and Example 4.

▶ **REAL-WORLD CONNECTION** Suppose that a person drives home after visiting a friend who lives 180 miles away. As time passes, the distance between the driver and her home decreases. If the driver travels at a constant speed, a linear graph can be used to represent the trip. Figure 9.35 shows such a graph when the speed is 60 miles per hour.

Distance from Home

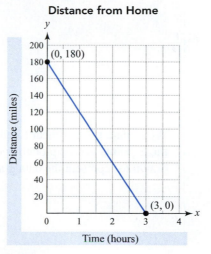

Figure 9.35

The graph in Figure 9.35 touches the *y*-axis at (0, 180) and the *x*-axis at (3, 0). Since each point on the graph has the form (time, distance), the point (0, 180) tells us that the *initial* distance, when the time is $x = 0$, is $y = 180$ miles. Similarly, the point (3, 0) tells us that after $x = 3$ hours, the driver is home. The distance between the driver and home is $y = 0$.

EXAMPLE 4 **Analyzing a graph that represents linear data**

The graph in Figure 9.36 shows the distance of a driver from home.
(a) How far away from home is the driver initially?
(b) How many hours pass before the driver is home?

Distance from Home

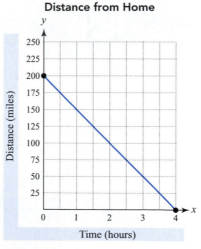

Figure 9.36

Solution
(a) The graph touches the *y*-axis at (0, **200**). The driver is initially **200** miles from home.
(b) The graph touches the *x*-axis at (**4**, 0). The driver is home after **4** hours.

Now Try Exercise 17

9.4   **Putting It All Together**

CONCEPT	COMMENTS	EXAMPLES
Pictographs	A pictograph is a graph that uses a picture or symbol to represent a particular quantity.	The following pictograph shows that 3000 boats sold in 2012 and 5500 boats sold in 2014. 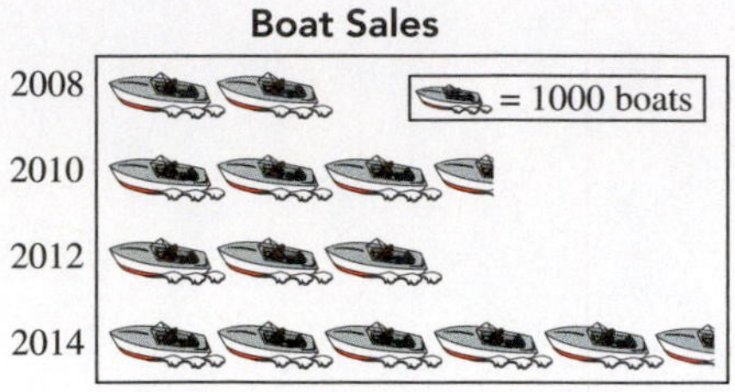
Representing Linear Data Graphically	When data have a linear trend, two or more points can be plotted and a line drawn to represent the data graphically.	If x and y represent time and distance, respectively, then the ordered pairs (2, 40) and (7, 140) can be used to represent distances of 40 miles after 2 hours and 140 miles after 7 hours. 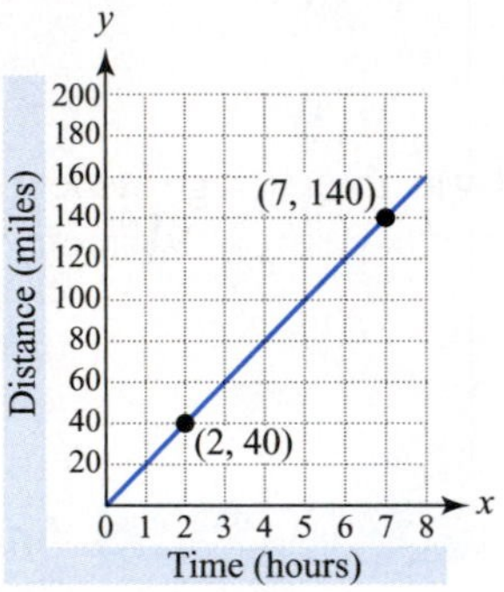
Solving Applications Graphically	Linear graphs can be used to find solutions to problems involving real-world data.	The following graph can be used to find how long it will take to drive 120 miles at 30 miles per hour. 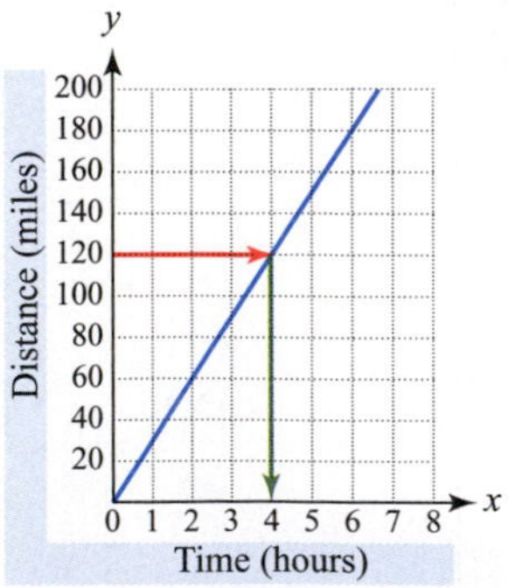

CONCEPTS AND VOCABULARY

1. When plotted data points align along a graphed line, we say that the data follow a(n) _____ trend.

2. The initial value for linear data is often found where the line crosses or touches the _____-axis.

REPRESENTING DATA IN PICTOGRAPHS

Exercises 3 and 4: The following pictograph shows the average student loan debt for 2008 college graduates from selected states. (*Source:* U.S. Department of Education.)

Average Student Loan Debt

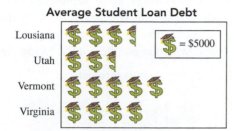

3. What was the average student loan debt for 2008 college graduates from Vermont? From Louisiana?

4. What is the difference between the average student loan debt for 2008 college graduates from Virginia and Utah?

Exercises 5 and 6: The following pictograph shows the number of earthquakes measuring 7.0 to 7.9 in magnitude during selected years. (*Source:* U.S. Geological Survey.)

Earthquakes: 7.0–7.9 Magnitude

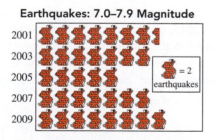

5. How many earthquakes measuring 7.0 to 7.9 in magnitude occurred in 2001? In 2005?

6. How many more earthquakes measuring 7.0 to 7.9 in magnitude occurred in 2009 compared to 2005?

REPRESENTING LINEAR DATA GRAPHICALLY

7. *Movie Downloads* A student downloads 4 movies for $12 and 7 movies for $21. Let x and y represent the number of downloads and the cost, respectively.

(a) Write two ordered pairs that correspond to the given information.

(b) Plot the ordered pairs in the xy-plane and draw a line through them.

(c) Use the graph to estimate the cost for downloading 6 movies.

(d) If the cost is $9, estimate the number of movie downloads.

8. *Facebook Friends* After using Facebook for 4 months, a girl has 24 Facebook friends. At the end of 12 months, she has 72 Facebook friends. Let x and y represent the number of months and the number of friends, respectively.

(a) Write two ordered pairs that correspond to the given information.

(b) Plot the ordered pairs in the xy-plane and draw a line through them.

(c) Use the graph to estimate the number of friends after 5 months.

(d) If the girl has 48 friends, estimate the number of months she has been using Facebook.

9. *File Backups* When a computer begins a hard drive backup, it has 120 gigabytes of information to save. After 8 minutes, the backup is complete. Let x and y represent minutes from the start of the backup and the remaining number of gigabytes, respectively.

(a) Write two ordered pairs that correspond to the given information.

(b) Plot the ordered pairs in the xy-plane and draw a line through them.

(c) Use the graph to estimate the number of gigabytes *completed* in 2 minutes.

(d) If 30 gigabytes remain, how many minutes have passed since the backup began?

10. *Math Homework* A student has 15 linear equations to graph for homework. The student completes the homework in 45 minutes. Let x and y represent the minutes since homework was started and the remaining number of equations to be graphed, respectively.

(a) Write two ordered pairs that correspond to the given information.

(b) Plot the ordered pairs in the xy-plane and draw a line through them.

(c) Use the graph to estimate the remaining number of linear equations to be graphed after 9 minutes.

(d) If 3 equations remain to be graphed, how many minutes have passed since homework began?

SOLVING APPLICATIONS GRAPHICALLY

11. *Wyoming Tuition* If x represents years after 2000 and y represents the average tuition at Wyoming public 2-year colleges during year x, then the linear equation $y = 70.7x + 1344$ can be used to estimate the tuition y for a given year x. The following figure shows the graph of $y = 70.7x + 1344$. (*Source:* SSTI *Weekly Digest.*)

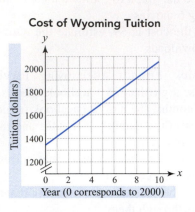

Cost of Wyoming Tuition

Estimate the year when the tuition was $2051.

12. *Distance to Lightning* If x represents the number of seconds between seeing a bolt of lightning and hearing the related thunder, then the distance y in miles between the observer and the lightning bolt is approximated by the equation $y = 0.22x$, which is graphed in the following figure.

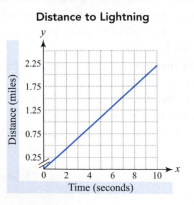

Distance to Lightning

Estimate the distance between the observer and the lightning bolt if 8 seconds pass between seeing the lightning and hearing the related thunder.

13. *Crutch Length* If x represents the height in inches of a person with a leg injury, then a reasonable crutch length y in inches is approximated by the equation $y = 0.72x + 2$, which is graphed in the following figure. (*Source:* American Physical Therapy Association.)

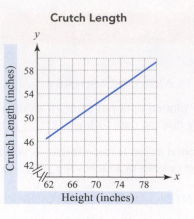

Crutch Length

Estimate the height of a person using crutches that are 56 inches long.

14. *Investment Earnings* If x represents an amount of money invested at 6% per year, then the simple interest y after one year is given by the equation $y = 0.06x$, which is graphed in the following figure.

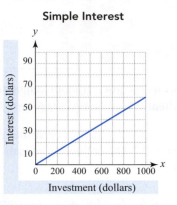

Simple Interest

Find the amount of money invested if the simple interest is $45.

15. *Cigarette Consumption* If x represents a year from 1975 to 2015, then the number of cigarettes y in billions consumed in the United States is approximated by the equation $y = -10.33x + 21{,}087$, which is graphed in the following figure. (*Source:* The Tobacco Outlook Report.)

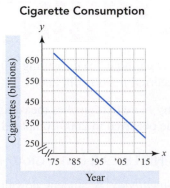

Cigarette Consumption

Estimate the year when 375 billion cigarettes were consumed in the United States.

16. *Alcohol Consumption* If x represents a year from 2000 to 2008, then the annual *average* number of gallons y of pure alcohol consumed by each person aged 14 years and older in the United States can be approximated by the equation $y = 0.015x - 27.825$, which is graphed in the following figure. (*Source:* National Institutes of Health.)

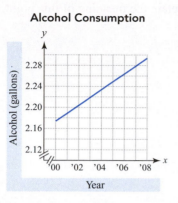

Alcohol Consumption

Estimate the year when average alcohol consumption reached 2.28 gallons per person.

17. *Driving Distance* The following graph shows the distance that a driver is from home.
(a) How far is the driver from home, initially?
(b) How many hours pass before the driver is home?

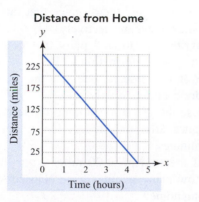

Distance from Home

18. *Running Distance* The following graph shows the distance that a runner is from home.
(a) How far is the runner from home, initially?
(b) How many minutes pass before the runner is home?

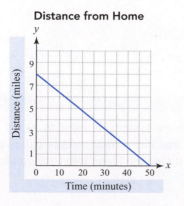

Distance from Home

19. *Flying Distance* The following graph shows the distance that an airplane is from an airport. Is the airplane flying toward the airport or away from it?

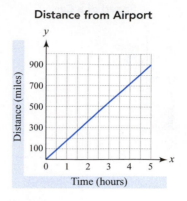

Distance from Airport

20. *Altitude* The following graph shows the altitude of a hiker on a mountain. Is the hiker moving up the mountain or down?

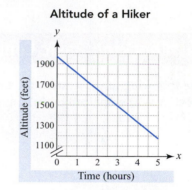

Altitude of a Hiker

21. *Driving Distance* The following graph shows the distance that a driver is from home. Give an explanation for this situation.

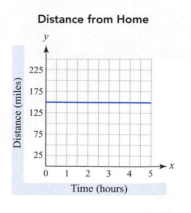

Distance from Home

22. *Driving Speed* The following graph shows the speed of a car. What can be said about the speed of the car over this time period?

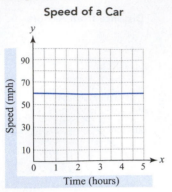

Speed of a Car

WRITING ABOUT MATHEMATICS

23. Give an example of real-world data that have a linear trend. Describe a graph representing these data.

24. A line is graphed in an *xy*-plane where the *x*-axis is labeled "Cookies (dozens)" and the *y*-axis is labeled "Sugar (cups)." If the line passes through the point $(6, 4)$, explain the meaning of this point.

Checking Basic Concepts

1. Use the following graph to solve the linear equation $3x + 3 = 0$ graphically.

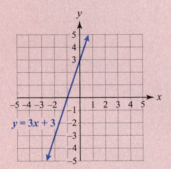

$y = 3x + 3$

2. Solve each linear equation graphically.
   (a) $x + 2 = 3$ **(b)** $-2x - 1 = 5$

3. Solve the linear equation $-x + 8 = 6$ algebraically, numerically, and graphically.

4. *Travel Time* A student has an 80-mile commute to college and takes 96 minutes to make the trip. Let *x* and *y* represent the number of minutes since leaving home and the distance that the student is from the college, respectively.
   (a) Write two ordered pairs that correspond to the given information.
   (b) Plot the ordered pairs in the *xy*-plane and draw a line through them.
   (c) Use the graph to estimate the remaining distance after the student has traveled for 24 minutes.
   (d) If the student is 20 miles from the college, how many minutes have elapsed since leaving home?

CHAPTER 9   Summary

SECTION 9.1 ■ THE RECTANGULAR COORDINATE SYSTEM

The xy-plane (Rectangular Coordinate System)

The *xy*-plane, or rectangular coordinate system, is used to graph data. The horizontal axis is called the *x-axis*, and the vertical axis is called the *y-axis*. The axes intersect at the *origin* and divide the plane into four *quadrants*: I, II, III, and IV.

Example:

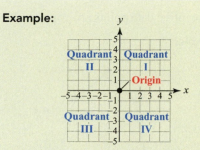

Ordered Pairs and Coordinates

The *coordinates* of a point indicate its location in the *xy*-plane. A point's coordinates are written as an *ordered pair* with the *x-coordinate* first and the *y-coordinate* second.

Example: The point given by the ordered pair (2, −5) has *x*-coordinate 2 and *y*-coordinate −5. It is located in quadrant IV of the *xy*-plane.

Coordinates of a Graphed Point

The *x*-coordinate of a point in the *xy*-plane is the number on the *x*-axis that is aligned directly above or below the point. The *y*-coordinate is the number on the *y*-axis that is aligned directly to the right or left of the point. The coordinates of a point are written as the ordered pair (*x*-coordinate, *y*-coordinate).

Example: The point shown in the graph is (**− 4, 2**).

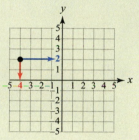

Plotting Points in the xy-plane

1. From the origin, move left or right along the *x*-axis to the position given by the *x*-coordinate.
2. From that position, move upward or downward the amount given by the *y*-coordinate.
3. Make a dot at this final position.

Example: (−2, 3) and (5, −4) are plotted in the graph.

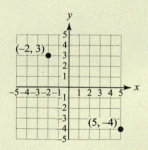

Scatterplot

When points are plotted in the *xy*-plane, the resulting graph is called a scatterplot.

Example:

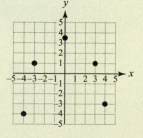

Line Graph	Consecutive data points are connected with line segments to make a line graph.

Example:

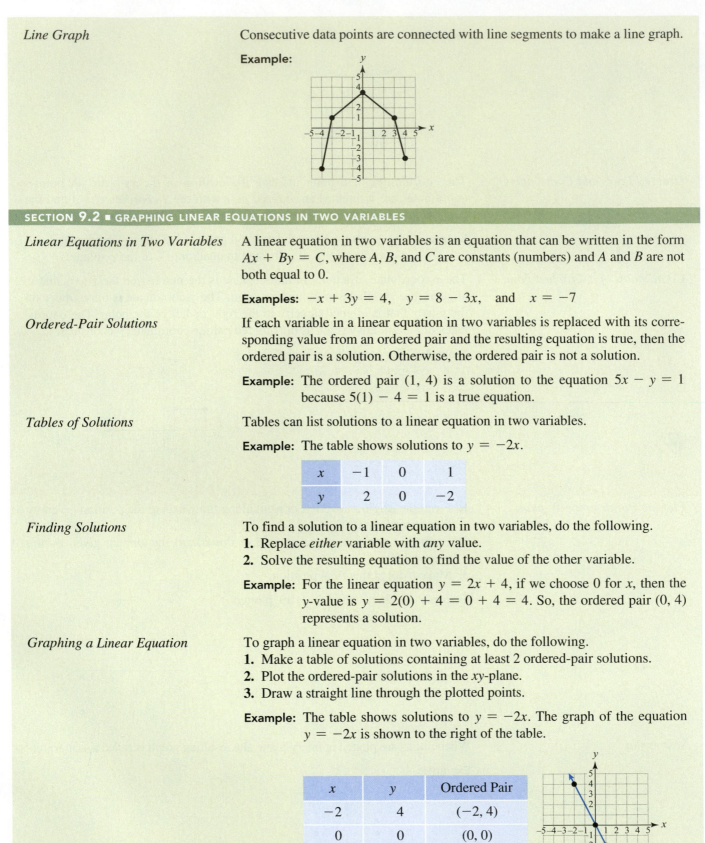

Linear Equations in Two Variables	A linear equation in two variables is an equation that can be written in the form $Ax + By = C$, where A, B, and C are constants (numbers) and A and B are not both equal to 0.

Examples: $-x + 3y = 4$, $y = 8 - 3x$,   and $x = -7$

Ordered-Pair Solutions	If each variable in a linear equation in two variables is replaced with its corresponding value from an ordered pair and the resulting equation is true, then the ordered pair is a solution. Otherwise, the ordered pair is not a solution.

Example: The ordered pair $(1, 4)$ is a solution to the equation $5x - y = 1$ because $5(1) - 4 = 1$ is a true equation.

Tables of Solutions	Tables can list solutions to a linear equation in two variables.

Example: The table shows solutions to $y = -2x$.

x	-1	0	1
y	2	0	-2

Finding Solutions	To find a solution to a linear equation in two variables, do the following. 1. Replace *either* variable with *any* value. 2. Solve the resulting equation to find the value of the other variable.

Example: For the linear equation $y = 2x + 4$, if we choose 0 for x, then the y-value is $y = 2(0) + 4 = 0 + 4 = 4$. So, the ordered pair $(0, 4)$ represents a solution.

Graphing a Linear Equation	To graph a linear equation in two variables, do the following. 1. Make a table of solutions containing at least 2 ordered-pair solutions. 2. Plot the ordered-pair solutions in the *xy*-plane. 3. Draw a straight line through the plotted points.

Example: The table shows solutions to $y = -2x$. The graph of the equation $y = -2x$ is shown to the right of the table.

x	y	Ordered Pair
-2	4	$(-2, 4)$
0	0	$(0, 0)$
1	-2	$(1, -2)$

Horizontal Line	If b is a constant (number), then the graph of the equation $y = b$ is a *horizontal* line that crosses the y-axis at b.

Example: The graph of $y = 3$ is a horizontal line crossing the y-axis at 3.

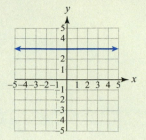

Vertical Line	If k is a constant (number), then the graph of the equation $x = k$ is a *vertical* line that crosses the x-axis at k.

Example: The graph of $x = 3$ is a vertical line crossing the x-axis at 3.

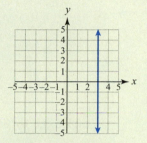

SECTION 9.3 ■ GRAPHICAL SOLUTIONS TO LINEAR EQUATIONS

Graphical Solution	A solution that is found visually from a graph is called a graphical solution.

Example: Since $x = 2$ when $y = 3$, the solution to $2x - 1 = 3$ is 2.

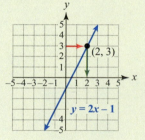

Solving a Linear Equation Graphically	To solve the linear equation $ax + b = c$ graphically, do the following. **1.** Graph $y = ax + b$ in the xy-plane. **2.** Use the graph to solve $ax + b = c$ visually. **3.** Check the solution in the given equation.

Example: To solve $2x + 2 = 8$, we graph the equation $y = 2x + 2$. Three ordered-pair solutions are shown in the table on the next page. The ordered pairs are plotted in the xy-plane, and a line representing the equation $y = 2x + 2$ is drawn through them.

x	y	Ordered Pair
−2	−2	(−2, −2)
0	2	(0, 2)
2	6	(2, 6)

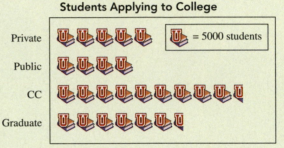

The solution is 3, which checks in the equation: 2(3) + 2 = 8.

SECTION 9.4 ■ SOLVING APPLICATIONS USING GRAPHS

Pictograph

A pictograph is a graph that uses a picture or symbol to represent a particular quantity.

Example: The following pictograph shows that 25,000 students applied to private colleges and 47,500 applied to community colleges.

Students Applying to College

Private	🎓🎓🎓🎓🎓 🎓 = 5000 students
Public	🎓🎓🎓🎓
CC	🎓🎓🎓🎓🎓🎓🎓🎓🎓
Graduate	🎓🎓🎓🎓🎓🎓🎓

Representing Linear Data Graphically

When data have a linear trend, two or more points can be plotted and a line drawn to represent the data graphically.

Example: If x and y represent time and distance, respectively, then the ordered pairs (3, 60) and (6, 120) represent distances of 60 miles after 3 hours and 120 miles after 6 hours.

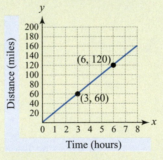

Solving Applications Graphically

Linear graphs can be used to find solutions to problems with real-world data.

Example: The following graph can be used to find out how long it will take to drive 100 miles at 25 miles per hour.

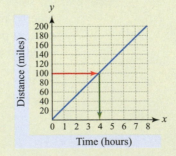

CHAPTER 9   Review Exercises

SECTION 9.1

1. Give the quadrant containing each point in the given graph. If a point is on an axis, name the axis.

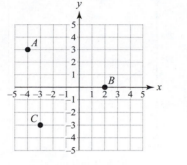

2. Write the coordinates of each point in the graph.

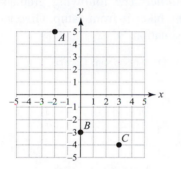

Exercises 3 and 4: Give the quadrant containing the point. If the point is on an axis, name the axis.

3. $(-4, -5)$

4. $(-3, 0)$

5. Make a scatterplot of the given points.
   $(-4, -2)$, $(-2, 1)$, $(0, 4)$, $(1, 2)$, and $(3, -2)$

6. Make a line graph of the given points.
   $(4, 1)$, $(-2, -2)$, $(0, 0)$, $(-3, -4)$, and $(1, 3)$

SECTION 9.2

Exercises 7 and 8: Determine whether the given equation is linear. If so, give values for A, B, and C for writing the equation in the form $Ax + By = C$.

7. $y = -2$ 8. $x - 4y^3 = 9$

9. Is $(3, 0)$ a solution to $4y + x = 12$?

10. Is $(1, -3)$ a solution to $y = 4x - 7$?

Exercises 11 and 12: Complete the table of solutions.

11. $-x - 3y = 8$

x	-5	-2	1	4
y				

12. $y = 2x - 5$

x	-2			4
y		-5	-1	

Exercises 13–18: Graph the given equation.

13. $y = 3x - 2$

14. $-2y + 3x = 6$

15. $y = -\dfrac{1}{3}x + 1$

16. $x = 4$

17. $y = -2$

18. $x - y = 3$

SECTION 9.3

Exercises 19 and 20: Use the graph to find a graphical solution to the given equation.

19. $-3x - 4 = -4$ 20. $2x - 3 = 1$

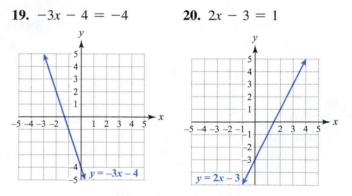

Exercises 21 and 22: Solve the equation graphically.

21. $\dfrac{1}{2}x - 3 = 0$

22. $2x + 3 = 7$

Exercises 23 and 24: Solve the equation
   (a) algebraically,
   (b) numerically, and
   (c) graphically.

23. $-3x - 4 = 2$

24. $\dfrac{1}{3}x - 9 = -6$

SECTION 9.4

Exercises 25 and 26: The following pictograph shows the projected number of job openings from 2008 to 2018 for selected occupations. (*Source:* College Board.)

Job Openings 2008 to 2018

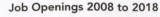
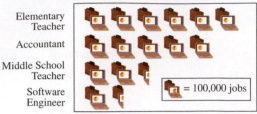

25. How many job opening are there for accountants? For middle school teachers?

26. How many more job openings are there for elementary teachers compared to software engineers?

27. *Investment* An investment of $200 earns $10 in simple interest, and an investment of $700 earns $35 in simple interest. Let *x* and *y* represent the investment amount and the simple interest, respectively.
    (a) Write two ordered pairs that correspond to the given information.
    (b) Plot the ordered pairs in the *xy*-plane and draw a line through them.
    (c) Use the graph to estimate the interest on a $500 investment.
    (d) If the simple interest is $30, how much is the investment?

28. *Children and HIV/AIDS* If *x* represents a year after 2003, then the number of children worldwide living with HIV/AIDS (in millions) can be estimated by the equation $y = 0.7x - 1400.3$, which is graphed in the following figure. (*Source:* World Health Organization.)

Children with HIV/AIDS

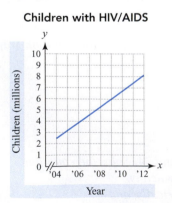

Estimate the year when there were 6 million children living with HIV/AIDS.

29. *Driving Distance* The following graph shows the distance that a driver is from home.
    (a) How far is the driver from home, initially?
    (b) How many hours pass before the driver has reached home?

Distance from Home

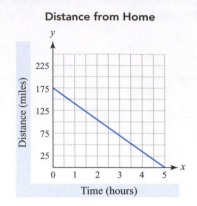

30. *Walking Distance* The following graph shows the distance that a hiker is from camp. Give an explanation for this situation.

Distance from Camp

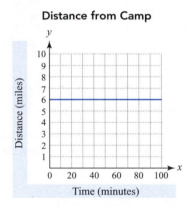

CHAPTER 9 Test

1. For each point in the graph, write the coordinates of the point and give the quadrant containing the point. If any point is on an axis, name the axis.

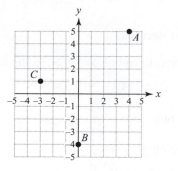

2. Name the quadrant containing the point $(7, -1)$.

3. Make a line graph of the given points.

$(3, -1)$, $(-2, 0)$, $(0, 3)$, $(-4, -3)$, and $(4, -4)$

4. Is the equation $x - y = -5$ linear? If so, give values for A, B, and C for writing the equation in the form $Ax + By = C$.

5. Is $(2, 3)$ a solution to $x = 10 - 4y$?

6. Complete the table of solutions for $y = -4x + 9$.

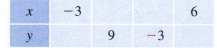

x	-3			6
y		9	-3	

Exercises 7–10: Graph the given equation.

7. $y = -x - 3$

8. $-x + 2y = 4$

9. $y = 3$

10. $x = 1$

11. Use the following graph to find a graphical solution to the equation $5x + 10 = 20$.

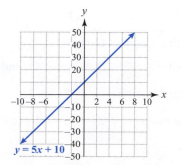

12. Solve the equation $-2x + 3 = 7$ graphically.

13. *Distance to Lightning* If x represents the number of seconds between seeing a bolt of lightning and hearing the related thunder, then the distance y in miles between the observer and the lightning bolt is approximated by the equation $y = 0.22x$, which is graphed in the following figure.

Distance to Lightning

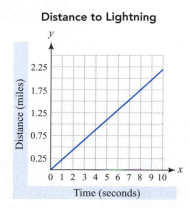

Estimate the time between seeing a bolt of lightning and hearing the related thunder if the lightning is 1.54 miles away from the observer.

14. *Height of a Balloon* The following graph shows the height of a hot air balloon.
  (a) How high is the balloon, initially?
  (b) How many minutes pass before the balloon is on the ground?

Height of a Balloon

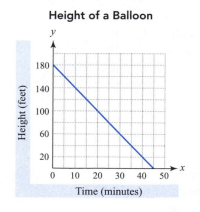

CHAPTERS 1–9 Cumulative Review Exercises

1. Identify the digit in the hundred-thousands place in the number 8,514,293,057.

2. Round 73,988 to the nearest thousand.

3. Approximate $\sqrt{65}$ to the nearest whole number.

4. Evaluate $4 \cdot 7 - 48 \div 8$.

5. Evaluate $y + (1 - x)^2 \div 4y$ for $x = 7, y = -1$.

6. Is -12 a solution to $6 + b \div 4 = 3$?

7. Complete the table and solve $3x - 5 = -2$.

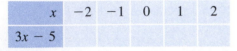

x	-2	-1	0	1	2
$3x - 5$					

8. Simplify the expression $-5w - (2w + 3)$.

9. Translate the following sentence to an equation using x as the variable. Do not solve the equation.

    Four equals a number decreased by 8.

10. Is the equation $-x + 7 = 1$ linear?

11. Solve the equation $-3y + (-8) = y$.

12. Solve the linear equation $-2x - 1 = 3$ visually. Check your answer.

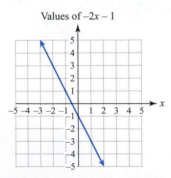

Values of $-2x - 1$

13. Give both the improper fraction and the mixed number represented by the shading.

14. Solve the following equation algebraically.

$$\frac{5}{10}y + \frac{3}{4} = \frac{7}{2}$$

Exercises 15 and 16: Perform the indicated arithmetic.

15. $-\dfrac{8}{9} \div \dfrac{24}{27}$

16. $-\dfrac{1}{5} + \dfrac{1}{7} \cdot \dfrac{7}{10}$

17. Write $\frac{7}{20}$ as a decimal.

18. Place the correct symbol, $<$ or $>$, in the blank.

$$21.0214 \underline{\quad} 21.0241$$

19. Multiply: $0.6(6.15)$.

20. Solve the equation $2.1(y - 2) = 5.25$.

21. If a car travels 224 miles in 4 hours, write its speed as a unit rate.

22. Solve the following proportion.

$$\frac{18}{x} = \frac{-3}{7}$$

23. Convert 0.051 square dekameters to square meters.

24. Convert 14 days to hours.

25. Write 193% as a decimal.

26. What percent of 120 is 48?

27. What number is 35% of 760?

28. A card is randomly selected from a standard deck. Find the probability that the card is a red card with the number 5 or 7. Express your answer as a fraction in simplest form.

29. Write 4.7×10^{-3} in standard form.

30. Subtract: $(7k^2 - k - 4) - (5k^2 - 9)$.

31. Find the GCF of $54x^2$ and $24x$.

32. Factor: $10y^5 - 25y^3$.

33. Make a line graph of the points.

    $(3, 1), (-1, -2), (0, 3), (-4, 4)$, and $(2, 3)$

34. Graph the equation: $-y + 2x = 4$.

35. Solve the equation $-2y + 6 = 4$ graphically.

36. Use the following graph to find a graphical solution to the equation $5x + 10 = 40$.

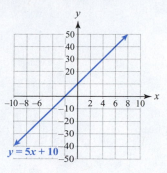

37. *Geometry* Find the perimeter of the figure.

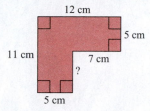

38. *Converting Currency* At one time, the formula for converting E euros (€) to D dollars was

$$D = \frac{29}{20}E.$$

Use the formula to find the number of dollars that could be purchased for €100.

39. *Target Heart Rate* For general health and weight loss, a person who is x years old should maintain a minimum target heart rate of T beats per minute (bpm) during extended exercise, where

$$T = -0.5x + 120.$$

Determine the minimum target heart rate for a person who is 48 years old.

40. *Speed Limits* An American drives 55 miles per hour after crossing the border into Canada. If the posted speed limit is 80 kilometers per hour, is the driver exceeding the speed limit?

41. *The Human Body* An adult female's body is about 55% water. If a woman's body has 66 pounds of water, what is her total body weight?

42. *Geometry* If the area of the following rectangle is $12x^2 + 28x$ and the width is $4x$, then the length of the rectangle is the polynomial factor found when $4x$ is factored from the area. Find the length.

10 Geometry

We can do anything we want to do if we stick to it long enough.
— HELEN KELLER

One of the most noted landmark buildings ever constructed is the Beijing National Stadium, the venue for the 2008 Olympic Games. Because of its seemingly random web of twisting steel, the building is commonly called the *Bird's Nest*. Its 110,000 tons of steel make the Bird's Nest the largest steel structure in the world.

From a distance, the outer surface of the stadium appears smooth and simple, but the *geometry* used in its design is very complex.

Without geometry, building structures as remarkable as the Bird's Nest or as simple as a backyard shed would be impossible. In this chapter, we discuss many of the basic concepts in the branch of mathematics known as geometry.

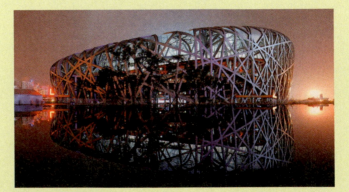

Source: Design Build Network, *Beijing National Stadium, China.*

10.1   Plane Geometry: Points, Lines, and Angles

Geometric Terms and Concepts ● Classifying Angles ● Parallel, Intersecting, and Perpendicular Lines ● Parallel Lines Cut by a Transversal

A LOOK INTO MATH ▶

Developed in 1984, Tetris is one of the most popular video games of all time. It is available for nearly every video game console and computer operating system and can be played on graphing calculators, mobile phones, and iPods. When writing computer code for Tetris (or any other video game), programmers make use of many geometric concepts. In this section, we discuss plane geometry.

NEW VOCABULARY

- ☐ Plane
- ☐ Plane geometry
- ☐ Vertex
- ☐ Side (of an angle)
- ☐ Degree
- ☐ Right angle
- ☐ Straight angle
- ☐ Acute angle
- ☐ Obtuse angle
- ☐ Congruent angles
- ☐ Complementary angles
- ☐ Supplementary angles
- ☐ Parallel lines
- ☐ Intersecting lines
- ☐ Perpendicular
- ☐ Vertical angles
- ☐ Adjacent angles
- ☐ Transversal
- ☐ Corresponding angles
- ☐ Alternate interior angles
- ☐ Alternate exterior angles

Geometric Terms and Concepts

Recall from Section 9.1 that a **plane** is a flat surface that continues without end. When a child draws on a sidewalk with chalk, the sidewalk represents (a portion of) a plane. Other plane surfaces include a white board in a classroom and a flat sheet of paper. A plane is *two-dimensional*—it has length and width but no height.

Geometric figures with two or fewer dimensions are the focus of **plane geometry**. Table 10.1 shows descriptions and examples of terms used in plane geometry.

TABLE 10.1 Terms Used in Plane Geometry

Term	Description	Example(s)	Notation
Point	A location in space having no length, width, or height	• P	Point P
Line	A straight figure representing a set of points extending forever in two directions	A B m	Line AB or $\overleftrightarrow{AB}$ Line m
Line Segment	A straight figure representing a set of points extending between two endpoints	A B	Segment AB or $\overline{AB}$
Ray	A straight figure representing a set of points extending forever in one direction	A B	Ray AB or $\overrightarrow{AB}$
Angle	A figure formed by two rays with a common endpoint	A B x C	$\angle ABC$ or $\angle CBA$ or $\angle B$ or $\angle x$

STUDY TIP

The terms listed in Table 10.1 are the "building blocks" of plane geometry. Spend extra time learning the meanings of these terms.

EXAMPLE 1 | **Identifying geometric figures**

Identify the figure and name it using the labels shown.

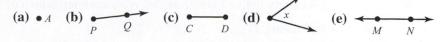

(a) • A (b) P → Q (c) • C • D (d) x (e) M N

Solution

(a) Point *A* is a dot representing a location.

(b) This figure is straight, has one endpoint, and extends in one direction. It is a ray. To name this ray, we write ray *PQ* or $\overrightarrow{PQ}$.

(c) The figure contains all points extending between two endpoints. It is a line segment. To name this line segment, we write segment *CD* or $\overline{CD}$.

(d) An angle is formed by two rays with a common endpoint. It is ∠*x*.

(e) A line is straight and extends in two directions. This line is line *MN* or $\overleftrightarrow{MN}$.

Now Try Exercises 23, 25, 27, 29, 31

When two rays form an angle, the common endpoint is the angle's **vertex** and the rays are the angle's **sides**. Some angles can be named using only the vertex letter. However, when three letters are used to name an angle, the vertex letter should always be in the middle. For example, the angle in Figure 10.1 can be named as

∠*Y*, ∠*XYZ*, or ∠*ZYX*.

Vertex

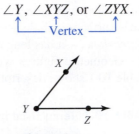

Figure 10.1

In the next example, we analyze the angles in a figure and discuss why it is not always appropriate to name an angle using only its vertex letter.

EXAMPLE 2   **Analyzing angles in a figure**

Refer to Figure 10.2.

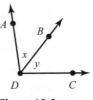

Figure 10.2

(a) Name the vertex of ∠*x*.

(b) Name the two sides of ∠*y*.

(c) Use three-letter naming to name three *different* angles that have vertex *D*.

(d) Explain why it is not appropriate to name an angle as ∠*D*.

Solution

(a) The common endpoint of the rays that form ∠*x* is the vertex. It is *D*.

(b) The two sides of ∠*y* are the rays that form the angle. They are $\overrightarrow{DB}$ and $\overrightarrow{DC}$.

(c) The angles are ∠*ADB*, ∠*BDC*, and ∠*ADC*.

(d) Because there are three different angles with vertex D, naming an angle as ∠*D* would not clearly define a single angle. We cannot tell which of the angles is being named.

Now Try Exercise 33

Classifying Angles

Angles are often measured in *degrees*. There are 360 degrees in one revolution, as shown in Figure 10.3, so a **degree** is a measure representing $\frac{1}{360}$ of a revolution.

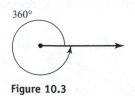

Figure 10.3

RIGHT, STRAIGHT, ACUTE, AND OBTUSE ANGLES An angle that measures 90° is called a **right angle**. A right angle is $\frac{1}{4}$ of a revolution. A small square near its vertex, as shown in Figure 10.4, tells us that the angle is a right angle and measures 90°.

Figure 10.4 Right Angle   **Figure 10.5** Straight Angle

READING CHECK

• What is the difference between an acute angle and an obtuse angle?

An angle that measures 180° is called a **straight angle**. A straight angle is $\frac{1}{2}$ of a revolution, as shown in Figure 10.5.

An angle that measures between 0° and 90° is called an **acute angle**, while an angle that measures between 90° and 180° is called an **obtuse angle**. Examples of acute and obtuse angles are shown in Figures 10.6 and 10.7, respectively.

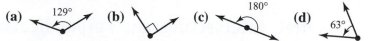

Figure 10.6 Acute Angles   **Figure 10.7** Obtuse Angles

EXAMPLE 3   **Classifying angles as acute, right, obtuse, or straight**

Classify each angle as acute, right, obtuse, or straight.

(a) 129°   (b)   (c) 180°   (d) 63°

Solution
(a) The measure of the angle is between 90° and 180°. It is an obtuse angle.
(b) The square near the vertex indicates that the angle is a right angle.
(c) The angle forms a straight line and has measure 180°. It is a straight angle.
(d) The measure of the angle is between 0° and 90°. It is an acute angle.

Now Try Exercises 35, 37, 39, 41

CONGRUENT, SUPPLEMENTARY, AND COMPLEMENTARY ANGLES There are three special ways to classify two angles considered together.

1. **Congruent angles**: Two angles whose measures are equal.
2. **Complementary angles**: Two angles whose measures sum to 90°.
3. **Supplementary angles**: Two angles whose measures sum to 180°.

NOTE: If two angles are complementary, then each angle is the **complement** of the other. If two angles are supplementary, then each angle is the **supplement** of the other.

Figures 10.8, 10.9, and 10.10 show examples of congruent, complementary, and supplementary angles, respectively.

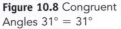

Figure 10.8 Congruent
Angles 31° = 31°

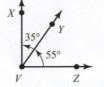

Figure 10.9 Complementary
Angles 35° + 55° = 90°

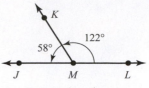

Figure 10.10 Supplementary
Angles 58° + 122° = 180°

NOTE: The symbol ($\cong$) is used to show that two angles are congruent. For the angles shown in Figure 10.8, we write $\angle ADB \cong \angle BDC$.

EXAMPLE 4 **Finding congruent, complementary, and supplementary angles**

Find the measure of an angle that is
(a) the complement of an angle with measure 71°.
(b) the supplement of an angle with measure 32°.
(c) congruent to an angle with measure 49°.

Solution
(a) Two angles are complements of each other if the sum of their angle measures is 90°. If one angle measures 71°, then the other must measure **19°** because

$$71° + \mathbf{19°} = 90°.$$

(b) The supplement of an angle with measure 32° is an angle with measure **148°** because

$$32° + \mathbf{148°} = 180°.$$

(c) Congruent angles have equal measure, so an angle that is congruent to an angle with measure 49° also has measure 49°.

Now Try Exercises 43, 45, 47

Parallel, Intersecting, and Perpendicular Lines

If two lines in a plane never touch, they are **parallel lines**. We can indicate that the lines in Figure 10.11 are parallel by using the symbol ($\parallel$), as in $m \parallel n$. When two lines meet, they are called **intersecting lines**. The lines in Figure 10.12 intersect at the point I.

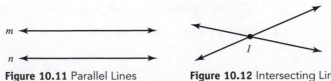

Figure 10.11 Parallel Lines   **Figure 10.12** Intersecting Lines

When two lines intersect to form right angles, we say that the lines are **perpendicular**. We can indicate that the intersecting lines in Figure 10.13 are perpendicular by using the symbol ($\perp$), as in $g \perp h$.

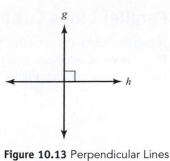

Figure 10.13 Perpendicular Lines

READING CHECK

- What symbol indicates that two lines are parallel?
- What symbol indicates that two lines are perpendicular?

Two intersecting lines that are not perpendicular form two *pairs* of congruent angles and four *pairs* of supplementary angles. Two angles in a congruent pair are called **vertical angles**, and two angles in a supplementary pair are called **adjacent angles**. The intersecting lines in Figure 10.14 form the following pairs of vertical and adjacent angles.

Vertical Angles	**Adjacent Angles**
$\angle w$ and $\angle y$	$\angle w$ and $\angle x$
$\angle x$ and $\angle z$	$\angle x$ and $\angle y$
	$\angle y$ and $\angle z$
	$\angle z$ and $\angle w$

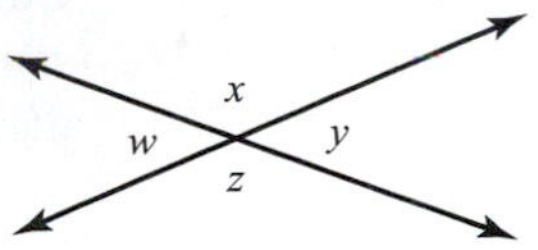

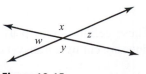

Figure 10.14 Intersecting Lines

We can use mathematical notation to say that an angle has a specified measure. For example, if $\angle b$ has measure 79°, we can write $m\angle b = 79°$. The m in front of the angle symbol tells us that we are talking about the *measure of the angle*.

In the next example, this notation is used in finding the measures of angles formed by intersecting lines.

EXAMPLE 5 **Finding angle measures**

If $m\angle w = 36°$ in Figure 10.15, find the measures of $\angle x$, $\angle y$, and $\angle z$.

Figure 10.15

Solution
Since $\angle x$ and $\angle w$ are adjacent angles, they are supplementary. So, $m\angle x = \mathbf{144°}$ because $36° + \mathbf{144°} = 180°$. Similarly, since $\angle y$ and $\angle w$ are also adjacent angles, $m\angle y = \mathbf{144°}$. Since $\angle w$ and $\angle z$ are vertical angles, they have the same measure. That is, $m\angle z = 36°$.

Now Try Exercise 49

NOTE: After determining that $m\angle x = 144°$ in Example 5, we could have determined that $m\angle y = 144°$ because $\angle x$ and $\angle y$ are vertical angles.

Parallel Lines Cut by a Transversal

A line that intersects two other lines in the same plane (at different points) is called a **transversal**. If a transversal intersects two *parallel* lines, then the eight angles formed have special relationships with one another. Figure 10.16 shows parallel lines *m* and *n* "cut" by the transversal *t*.

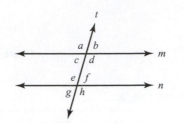

Figure 10.16 Parallel Lines Cut by a Transversal

In order to discuss the relationships among the angles formed when two parallel lines are cut by a transversal, we first define the special angle pairs in Table 10.2.

TABLE 10.2 **Key Definitions When Parallel Lines Are Cut by a Transversal**

Definition	Angle Pairs and Illustration	
Corresponding angles are angles in the same relative position with respect to the parallel lines.	$\angle a$ and $\angle e$ $\angle b$ and $\angle f$ $\angle c$ and $\angle g$ $\angle d$ and $\angle h$	
Alternate interior angles are angles that are between the parallel lines, on opposite sides of the transversal, and are *not* adjacent angles.	$\angle c$ and $\angle f$ $\angle d$ and $\angle e$	
Alternate exterior angles are angles that are outside of the parallel lines, on opposite sides of the transversal, and are *not* adjacent angles.	$\angle a$ and $\angle h$ $\angle b$ and $\angle g$	

These definitions help us to define the following properties of angles when two parallel lines are cut by a transversal.

TWO PARALLEL LINES CUT BY A TRANSVERSAL

When two parallel lines are cut by a transversal, the following properties apply.

1. Vertical angles are congruent.
2. Corresponding angles are congruent.
3. Alternate interior angles are congruent.
4. Alternate exterior angles are congruent.

Two angles in any other angle pair are supplementary.

EXAMPLE 6 **Finding angle measures**

If $p\|q$ and $m\angle h = 78°$ in Figure 10.17, find the measures of $\angle a$, $\angle c$, and $\angle f$.

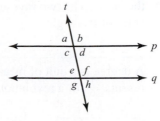

Figure 10.17

Solution

Since $\angle h$ and $\angle a$ are alternate exterior angles, they are congruent. So, $m\angle a = 78°$. Since $\angle c$ and $\angle a$ are adjacent angles, the sum of their measures is 180°. That is, $m\angle c = \mathbf{102°}$ because $78° + \mathbf{102°} = 180°$. Finally, $\angle c$ and $\angle f$ are congruent because they are alternate interior angles. So, $m\angle f = 102°$.

Now Try Exercise 65

READING CHECK

• When parallel lines are cut by a transversal, how many angle measures result?

NOTE: In Example 6, each of the angles has a measure of either 78° or 102°. In general, when a transversal cuts two parallel lines, all angles have one of two measures and the two measures sum to 180°.

10.1 Putting It All Together

CONCEPT	COMMENTS	EXAMPLES
Point	A point is a location in space having no length, width, or height.	• P
Line	A line is a straight figure representing a set of points extending forever in two directions.	A B
Line Segment	A line segment is a straight figure representing a set of points extending between two endpoints.	A B
Ray	A ray is a straight figure representing a set of points extending forever in one direction.	A B
Angle	An angle is a figure formed by two rays with a common endpoint.	A x B C

continued on next page

continued from previous page

CONCEPT	COMMENTS	EXAMPLES
Vertex and Sides of an Angle	The common endpoint of the rays that form an angle is called the vertex of the angle. The two rays are called the sides of the angle.	The vertex is B, and the sides are $\overrightarrow{BA}$ and $\overrightarrow{BC}$.
Degree	A degree is a unit of measure that represents $\frac{1}{360}$ of a revolution.	360°
Right Angle	A right angle has measure 90°. A small square near the vertex indicates a right angle.	90°
Straight Angle	A straight angle has measure 180°.	180°
Acute Angle	An acute angle measures between 0° and 90°.	82°
Obtuse Angle	An obtuse angle measures between 90° and 180°.	159°
Congruent Angles	Angles with equal measures are congruent.	29° 29°
Complementary Angles	The sum of the measures of two complementary angles is 90°. Each angle is the complement of the other.	37° 53°
Supplementary Angles	The sum of the measures of two supplementary angles is 180°. Each angle is the supplement of the other.	56° 124°
Parallel and Intersecting Lines	Parallel lines never touch. We indicate that two lines are parallel by using the symbol ($\parallel$). When two lines meet, they are intersecting lines.	**Parallel Lines** **Intersecting Lines** m n I
Perpendicular Lines	Two lines are perpendicular if they intersect to form right angles. We use the symbol ($\perp$) to represent perpendicular lines.	
Vertical and Adjacent Angles	Two intersecting lines that are not perpendicular form two pairs of congruent angles called vertical angles and four pairs of supplementary angles called adjacent angles.	x w y z **Vertical** **Adjacent** $\angle w$ and $\angle y$ $\angle w$ and $\angle x$ $\angle x$ and $\angle z$ $\angle x$ and $\angle y$ $\angle y$ and $\angle z$ $\angle z$ and $\angle w$

CONCEPT	COMMENTS	EXAMPLES
Parallel Lines Cut by a Transversal	When two parallel lines are cut by a transversal, the following properties apply. 1. Vertical angles are congruent. 2. Corresponding angles are congruent. 3. Alternate interior angles are congruent. 4. Alternate exterior angles are congruent. Any other angle pairs are supplementary.	 **Corresponding** **Alternate Interior** $\angle a$ and $\angle e$ $\angle c$ and $\angle f$ $\angle b$ and $\angle f$ $\angle d$ and $\angle e$ $\angle c$ and $\angle g$ **Alternate Exterior** $\angle d$ and $\angle h$ $\angle a$ and $\angle h$ $\angle b$ and $\angle g$

10.1 Exercises

CONCEPTS AND VOCABULARY

1. A position in space having no length, width, or height is called a(n) _____.

2. A(n) _____ is a straight figure representing a set of points extending forever in two directions.

3. A(n) _____ is a straight figure representing a set of points extending between two endpoints.

4. A(n) _____ is a straight figure representing a set of points extending forever in one direction.

5. A figure formed by two rays that share a common endpoint is called a(n) _____.

6. The shared endpoint of the rays that form an angle is called the _____ of the angle.

7. An angle that measures 90° is a(n) _____ angle.

8. A(n) _____ angle measures 180°.

9. An angle that measures between 0° and 90° is called a(n) _____ angle.

10. An angle that measures between 90° and 180° is a(n) _____ angle.

11. Two angles with the same measure are _____ angles.

12. Two angles whose measures sum to 90° are called _____ angles.

13. Two angles whose measures sum to 180° are called _____ angles.

14. If two lines in a plane never touch, they are called _____ lines.

15. If two lines in a plane meet at a point, they are called _____ lines.

16. A pair of congruent angles formed when two lines intersect are called _____ angles.

17. A pair of supplementary angles formed when two lines intersect are called _____ angles.

18. Two lines are _____ if their intersection forms right angles.

19. When parallel lines are cut by a transversal, _____ angles are in the same position with respect to the parallel lines.

20. When parallel lines are cut by a transversal, _____ angles are between the parallel lines, on opposite sides of the transversal, and are not adjacent angles.

21. When parallel lines are cut by a transversal, _____ angles are outside of the parallel lines, on opposite sides of the transversal, and are not adjacent angles.

22. When parallel lines are cut by a transversal, any two angles that are *not* vertical, corresponding, alternate interior, or alternate exterior angles are _____.

GEOMETRIC TERMS AND CONCEPTS

Exercises 23–32: Identify the figure and name it using the labels shown.

23.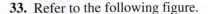
A B

24. • D

25.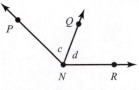
m

26. •————•
R S

27. •————•——→
C D

28. ←——•————•——→
X Y

29. • K

30. •————•——→
G H

31. •————•
P Q

32.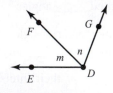
y

33. Refer to the following figure.

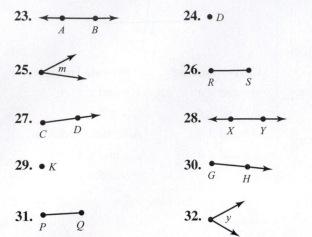

(a) Name the vertex of ∠c.
(b) Name the two sides of ∠d.
(c) Use three-letter naming to name three *different* angles that have vertex N.
(d) Explain why it is not appropriate to name an angle as ∠N.

34. Refer to the following figure.

(a) Name the vertex of ∠n.
(b) Name the two sides of ∠m.
(c) Use three-letter naming to name three *different* angles that have vertex D.
(d) Explain why it is not appropriate to name an angle as ∠D.

CLASSIFYING ANGLES

Exercises 35–42: Classify the angle as acute, right, obtuse, or straight.

35.

36.

37.

38.

39.

40.

41.

42.

Exercises 43–48: Find the measure of an angle that has the specified description.

43. The supplement of an angle with measure 14°

44. The complement of an angle with measure 43°

45. Congruent to an angle with measure 109°

46. The supplement of an angle with measure 154°

47. The complement of an angle with measure 9°

48. Congruent to an angle with measure 13°

INTERSECTING LINES

49. If $m\angle c = 49°$ in the following figure, find the measures of ∠a, ∠b, and ∠d.

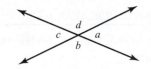

50. If $m\angle y = 158°$ in the following figure, find the measures of ∠x, ∠w, and ∠z.

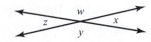

51. If $m\angle k = 84°$ in the following figure, find the measures of ∠j, ∠l, and ∠m.

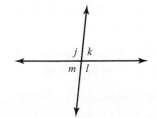

52. If $m\angle f = 102°$ in the following figure, find the measures of ∠c, ∠d, and ∠e.

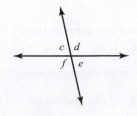

PARALLEL LINES CUT BY A TRANSVERSAL

Exercises 53–64: In the following figure, m∥n. Determine whether the given angles are congruent or supplementary.

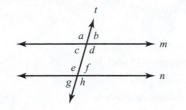

53. ∠d and ∠a

54. ∠e and ∠g

55. ∠c and ∠f

56. ∠a and ∠e

57. ∠g and ∠h

58. ∠b and ∠g

59. ∠f and ∠a

60. ∠e and ∠d

61. ∠c and ∠g

62. ∠b and ∠h

63. ∠h and ∠a

64. ∠e and ∠h

65. If $m \| n$ and $m\angle c = 119°$ in the following figure, find the measures of ∠b, ∠g, and ∠h.

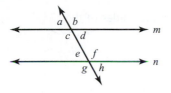

66. If $m \| n$ and $m\angle g = 63°$ in the following figure, find the measures of ∠a, ∠b, and ∠c.

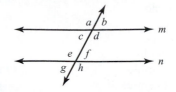

67. If $m \| n$ and $m\angle e = 92°$ in the following figure, find the measures of ∠a, ∠c, and ∠h.

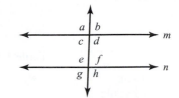

68. If $m \| n$ and $m\angle b = 142°$ in the following figure, find the measures of ∠d, ∠e, and ∠g.

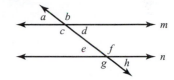

WRITING ABOUT MATHEMATICS

69. Suppose that two angles are supplementary. Can both of the angles be acute? Explain.

70. If two parallel lines are cut by a transversal that is perpendicular to one of the lines, explain how you know that the transversal is also perpendicular to the other line.

10.2   Triangles

Classifying Triangles • The Sum of the Angle Measures • Congruent Triangles

A LOOK INTO MATH ▶

The figure in the margin appears to show a simple pattern of small red and yellow triangles. However, a closer look will reveal that there are many triangles of different sizes "hidden" within the figure. Can you find all 48 triangles in the figure? It may take several tries to find them all. In this section, we discuss different kinds of triangles and their properties.

Classifying Triangles

We can classify a triangle as *acute*, *obtuse*, or *right* by looking at its angles. We can classify a triangle as *scalene*, *isosceles*, or *equilateral* by looking at its sides. Table 10.3 on the next page shows three types of triangles that are classified by looking at angles.

NEW VOCABULARY

☐ Acute triangle
☐ Obtuse triangle
☐ Right triangle
☐ Scalene triangle
☐ Isosceles triangle
☐ Equilateral triangle
☐ Congruent triangles
☐ Angle-side-angle (ASA)
☐ Side-angle-side (SAS)
☐ Side-side-side (SSS)

STUDY TIP

The end of the semester can be a hectic time. Be sure that you plan your schedule to make extra time for your studies.

TABLE 10.3 Classifying Triangles by Looking at the Angles

Type of Triangle	Description	Example
Acute	A triangle in which every angle measures less than 90°	
Obtuse	A triangle in which one angle measures between 90° and 180°	
Right	A triangle in which one angle measures exactly 90°	

Table 10.4 shows three types of triangles that are classified by looking at sides.

TABLE 10.4 Classifying Triangles by Looking at the Sides

Type of Triangle	Description	Example
Scalene	A triangle with no sides of the same length	
Isosceles	A triangle with at least two sides of the same length	
Equilateral	A triangle with three sides of the same length	

NOTE: *Every* triangle must be one of the types in Table 10.3 and, at the same time, must be one of the types in Table 10.4. For example, one triangle might be both acute and scalene, while a different triangle might be both right and isosceles.

EXAMPLE 1 **Classifying triangles as acute, obtuse, or right**

Classify each triangle as acute, obtuse, or right.

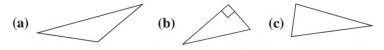

Solution
(a) Since one angle of the triangle measures between 90° and 180°, it is an obtuse triangle.
(b) The triangle has one angle that measures exactly 90°, so it is a right triangle.
(c) Every angle in the triangle measures less than 90°, so it is an acute triangle.

Now Try Exercises 13, 17, 19

EXAMPLE 2 **Classifying triangles as scalene, isosceles, or equilateral**

Classify each triangle as scalene, isosceles, or equilateral.

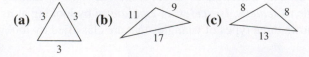

Solution

(a) Since the triangle has three sides of the same length, it is an equilateral triangle.

(b) The triangle has no sides of the same length, so it is a scalene triangle.

(c) Since the triangle has two sides of the same length, it is an isosceles triangle.

▍ Now Try Exercises 25, 29, 31

The Sum of the Angle Measures

No matter what type of triangle is being considered, one property is always true—the sum of the measures of the angles is 180°. To show that this property is true, we position the triangle between two parallel lines as shown in Figure 10.18, where $m\|n$ and the three angles of the triangle are $\angle 1$, $\angle 2$, and $\angle 3$.

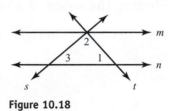

Figure 10.18

Since transversal t forms one side of the triangle, the measure of $\angle 1$ is equal to that of its corresponding angle, $\angle \mathbf{1}$. Since transversal s forms another side of the triangle, the measure of $\angle 3$ is equal to that of its corresponding angle, $\angle \mathbf{3}$. Finally, transversals s and t intersect to form vertical angles with the measure of $\angle 2$ equal to that of $\angle \mathbf{2}$. See Figure 10.19.

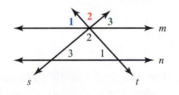

Figure 10.19

READING CHECK

• What is the sum of the angle measures for any triangle?

Together, $\angle \mathbf{1}$, $\angle \mathbf{2}$, and $\angle \mathbf{3}$ form a straight angle with measure 180°. Since the measures of $\angle 1$, $\angle 2$, and $\angle 3$ are equal to the measures of $\angle \mathbf{1}$, $\angle \mathbf{2}$, and $\angle \mathbf{3}$, respectively, the three angles of the triangle also sum to 180°.

> **THE SUM OF THE ANGLE MEASURES**
>
> For any triangle, the sum of the measures of the three angles is 180°.

EXAMPLE 3 **Finding a missing angle measure in a triangle**

Find the measure of $\angle x$ in Figure 10.20.

Figure 10.20

Figure 10.20 (Repeated)

Solution

Figure 10.20 is repeated in the margin. For simplicity, the degree symbol can be left out during the computation.

$$38 + 67 + x = 180 \qquad \text{Angle measures sum to 180.}$$
$$105 + x = 180 \qquad \text{Add.}$$
$$105 + x - \mathbf{105} = 180 - \mathbf{105} \qquad \text{Subtract 105 from each side.}$$
$$x = 75 \qquad \text{Simplify.}$$

The measure of $\angle x$ is 75°.

Now Try Exercise 37

EXAMPLE 4 **Finding the value of a variable in a triangle**

Find the value of x in Figure 10.21. Assume that all angle measures are given in degrees.

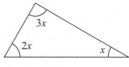

Figure 10.21

Solution

$$x + 2x + 3x = 180 \qquad \text{Angle measures sum to 180.}$$
$$6x = 180 \qquad \text{Combine like terms.}$$
$$\frac{6x}{\mathbf{6}} = \frac{180}{\mathbf{6}} \qquad \text{Divide each side by 6.}$$
$$x = 30 \qquad \text{Simplify.}$$

The value of x is 30°. We can check this by finding the sum of the measures of the three angles. Here, $x = 30°$, $2x = 60°$, and $3x = 90°$, so the sum is $30° + 60° + 90° = 180°$.

Now Try Exercise 45

Congruent Triangles

In Section 10.1, we learned that two angles are *congruent angles* if they have the same measure. Two triangles are **congruent triangles** if they have exactly the same shape and size. In other words, two triangles are congruent if the corresponding angles are congruent and the lengths of the corresponding sides are equal. Figure 10.22 shows congruent triangles.

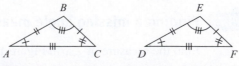

Figure 10.22 Congruent Triangles

To show that the corresponding sides or angles of two triangles have the same measure, we use single, double, or triple hash marks, as shown in Figure 10.22.

When comparing two triangles, we do not need to know the measures of every side and angle to know if the triangles are congruent. Any of the following properties can be used.

CONGRUENT TRIANGLE PROPERTIES

The Angle-Side-Angle Property (ASA)

If two angles and the included side of one triangle are congruent to two angles and the included side of another triangle, then the triangles are congruent.

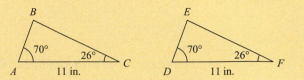

The Side-Angle-Side Property (SAS)

If two sides and the included angle of one triangle are congruent to two sides and the included angle of another triangle, then the triangles are congruent.

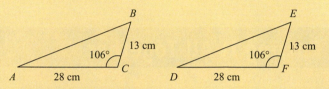

The Side-Side-Side Property (SSS)

If three sides of one triangle are congruent to three sides of another triangle, then the triangles are congruent.

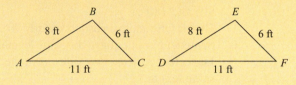

NOTE: Even though SSS can be used to show that two triangles are congruent, there is **not** an AAA property. Two triangles of different sizes can have three pairs of congruent angles, as shown in Figure 10.23.

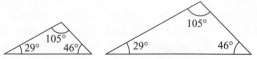

Figure 10.23 Triangles That Are Not Congruent

EXAMPLE 5

Stating the property that shows triangles are congruent

For each pair of triangles, state the property that shows the triangles are congruent.

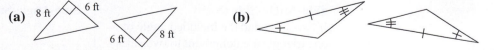

(a) 8 ft 6 ft 6 ft 8 ft

(b)

Solution
(a) Two sides and the included angle of the first triangle are congruent to two sides and the included angle of the second triangle. The triangles are congruent by SAS.
(b) Since two angles and the included side of the first triangle are congruent to two angles and the included side of the second triangle, the triangles are congruent by ASA.

Now Try Exercises 53, 55

READING CHECK

• Describe the three ways that are used to determine when two triangles are congruent.

EXAMPLE 6   **Determining whether triangles are congruent**

Determine whether the triangles in each pair are congruent. If so, state the property that shows that the triangles are congruent.

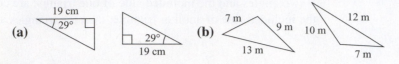

(a)

(b)

Solution

(a) Each triangle has a 19-centimeter side included between a 90° angle and a 29° angle. The two triangles are congruent by ASA.

(b) The two triangles are not congruent because the corresponding sides do not have equal measures.

Now Try Exercises 59, 61

10.2   Putting It All Together

CONCEPT	COMMENTS	EXAMPLES
Types of Triangles	**Acute:** Every angle measures less than 90°. **Obtuse:** One angle measures between 90° and 180°. **Right:** One angle measures exactly 90°. **Scalene:** No sides have the same length. **Isosceles:** At least two sides have the same length. **Equilateral:** Three sides have the same length.	Acute Obtuse Right Scalene Isosceles Equilateral
The Sum of the Angle Measures	For any triangle, the sum of the measures of the three angles is 180°.	$37° + 68° + 75° = 180°$ 75° 37° 68°
Congruent Triangles	Any of the following properties can be used to determine whether two triangles are congruent. **Angle-Side-Angle (ASA)** If two angles and the included side of one triangle are congruent to two angles and the included side of another triangle, then the triangles are congruent. **Side-Angle-Side (SAS)** If two sides and the included angle of one triangle are congruent to two sides and the included angle of another triangle, then the triangles are congruent. **Side-Side-Side (SSS)** If three sides of one triangle are congruent to three sides of another triangle, then the triangles are congruent.	**Angle-Side-Angle (ASA)** **Side-Angle-Side (SAS)** **Side-Side-Side (SSS)**

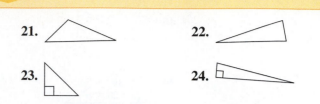

10.2 Exercises

CONCEPTS AND VOCABULARY

1. If every angle of a triangle measures less than 90°, then the triangle is a(n) _____ triangle.

2. If one angle of a triangle measures between 90° and 180°, then the triangle is a(n) _____ triangle.

3. If one angle of a triangle measures exactly 90°, then the triangle is a(n) _____ triangle.

4. A(n) _____ triangle has no sides of the same length.

5. A(n) _____ triangle has two sides of the same length.

6. A(n) _____ triangle has three sides of the same length.

7. For any triangle, the sum of the measures of the three angles is _____.

8. Two triangles are _____ triangles if they have exactly the same shape and size.

9. According to the _____ property, if two angles and the included side of one triangle are congruent to two angles and the included side of another triangle, then the triangles are congruent.

10. According to the _____ property, if two sides and the included angle of one triangle are congruent to two sides and the included angle of another triangle, then the triangles are congruent.

11. According to the _____ property, if three sides of one triangle are congruent to three sides of another triangle, then the two triangles are congruent.

12. There is not a(n) _____ property to determine if two triangles are congruent because triangles of different sizes can have three pairs of congruent angles.

CLASSIFYING TRIANGLES

Exercises 13–24: Classify the triangle as acute, obtuse, or right.

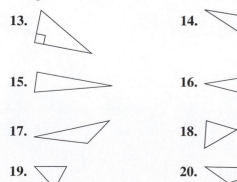

THE SUM OF THE ANGLE MEASURES

Exercises 37–44: Find the measure of ∠x in the triangle.

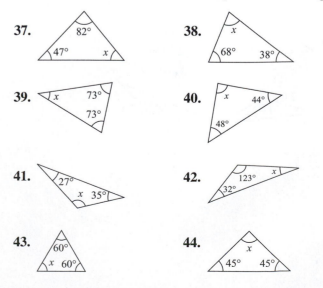

Exercises 25–36: Classify the triangle as scalene, isosceles, or equilateral.

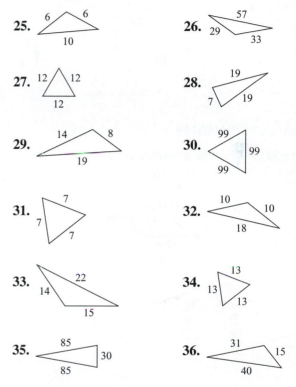

Exercises 45–52: Find the value of x. Assume that all angle measures are given in degrees.

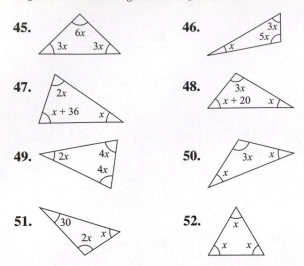

45.

46.

47.

48.

49.

50.

51.

52.

CONGRUENT TRIANGLES

Exercises 53–58: For the triangle pair, state the property that shows the triangles are congruent.

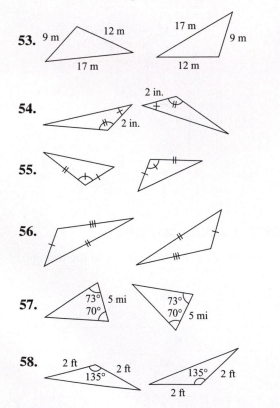

53.

54.

55.

56.

57.

58.

Exercises 59–64: Determine whether the triangles in the given pair are congruent. If so, state the property that shows that the triangles are congruent.

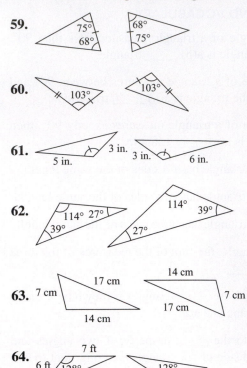

59.

60.

61.

62.

63.

64.

WRITING ABOUT MATHEMATICS

65. Can a right triangle also be obtuse? Explain.

66. For every triangle, the sum of the lengths of any two sides must be greater than the length of the third side. Use this fact to explain why a triangle with side lengths of 2, 7, and 4 inches cannot exist.

67. Can an equilateral triangle also be right? Explain.

68. If two angles and a non-included side of one triangle are congruent to two angles and a non-included side of a second triangle, then the two triangles are congruent. This property is known as AAS. Explain why this property is equivalent to the ASA property.

Checking Basic Concepts

1. Identify each of the given figures and name it using the labels shown.

    (a) 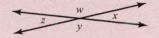   (b)

2. Classify each of the given angles as acute, right, obtuse, or straight.

    (a)   (b)

3. Find the measure of the complement of an angle with measure 19°.

4. Find the measure of the supplement of an angle with measure 114°.

5. If $m\angle y = 161°$ in the following figure, find the measures of $\angle x$, $\angle w$, and $\angle z$.

    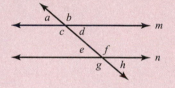

6. If $m \| n$ and $m\angle b = 139°$ in the following figure, find the measures of $\angle d$, $\angle e$, and $\angle g$.

7. Classify each triangle as acute, obtuse, or right.

    (a)   (b)

8. Classify each of the given triangles as scalene, isosceles, or equilateral.

    (a)   (b)

9. Find the measure of $\angle x$ in the triangle.

10. Determine whether the triangles in the given pair are congruent. If so, state the property that shows that the triangles are congruent.

10.3 Polygons and Circles

Polygons • Regular Polygons • Quadrilaterals • Circles

A LOOK INTO MATH ▶

Even though geometric shapes occur often in math, construction, and architecture, many remarkable examples can be found in nature. For example, it has taken hundreds of years for earth-covered mounds of ice called *pingos* to form in the Arctic National Wildlife Refuge. The annual thawing and freezing that occurs near a pingo can cause a geometric pattern to form. After studying this section, you will understand why arctic scientists call this pattern *polygons*. (*Source*: U.S. Fish and Wildlife Service.)

Polygons

A **polygon** is a closed plane figure determined by three or more line segments. Each line segment is called a **side** of the polygon, where two sides never touch except at a common endpoint called a **vertex**. Figure 10.24 on the next page shows several examples of polygons.

NEW VOCABULARY

- ☐ Polygon
- ☐ Side (of a polygon)
- ☐ Vertex
- ☐ Triangle, quadrilateral, pentagon, hexagon, heptagon, octagon
- ☐ Regular polygon
- ☐ Square, rectangle, parallelogram, trapezoid, rhombus, kite
- ☐ Circle
- ☐ Center (of a circle)
- ☐ Radius
- ☐ Diameter

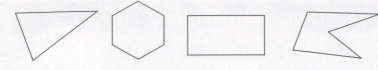

Figure 10.24 Examples of Polygons

PROPERTIES OF POLYGONS

Every polygon has the following three properties.

1. All sides must be *straight* line segments.
2. Sides cannot intersect, except at endpoints.
3. The figure must be closed.

EXAMPLE 1 **Explaining why figures are not polygons**

Explain why each figure is not a polygon.

(a)   (b)    (c)

Solution
(a) The figure is not closed.
(b) One side is curved.
(c) The sides intersect.

Now Try Exercises 11, 13, 15

Every polygon is named by the number of its sides. Table 10.5 lists the names of polygons with 3 to 8 sides.

TABLE 10.5 Names of Polygons

Number of Sides	Polygon Name	Example
3	**Triangle**	
4	**Quadrilateral**	
5	**Pentagon**	
6	**Hexagon**	
7	**Heptagon**	
8	**Octagon**	

STUDY TIP

Does your school use a final exam schedule? Do you know the time and location of your final exam? Make sure that you know the answers to these important questions.

EXAMPLE 2   **Determining the name of a polygon**

Determine the name of each polygon.

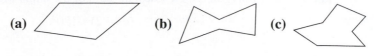

(a)   **(b)**   **(c)**

Solution
(a) The polygon has four sides, so it is a quadrilateral.
(b) Since the polygon has six sides, it is a hexagon.
(c) A polygon with seven sides is a heptagon.

▌ **Now Try Exercises 17, 23, 27**

Regular Polygons

When all sides of a polygon have equal length and all angles have equal measure, the polygon is a **regular polygon**. Figure 10.25 shows a regular quadrilateral (square), a regular hexagon, and a regular octagon.

Square   Hexagon   Octagon

Figure 10.25 Examples of Regular Polygons

The sum of the angles of any regular polygon can be found using the fact that the angle measures of any triangle add up to 180°. For example, the regular pentagon in Figure 10.26 is divided into 3 triangles.

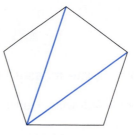

Figure 10.26

The sum of the angle measures of the pentagon is equal to the sum of the angle measures of 3 triangles. Since we know that the sum of the angle measures of a triangle is equal to 180°, we know that the sum of the angle measures of a pentagon is $3 \cdot 180° = 540°$.

We can find the sum of the angle measures of *any* regular polygon by simply counting the number of triangles inside and multiplying by 180°. The regular polygons in Figure 10.27 show that the number of triangles is always 2 less than the number of sides.

READING CHECK

• Explain how triangles are used to find the sum of the angle measures in a regular polygon.

Figure 10.27 Always 2 Fewer Triangles Than Sides

This discussion leads to the following rule for finding the sum of the angle measures of any regular polygon.

THE SUM OF THE ANGLE MEASURES OF A REGULAR POLYGON

For a regular polygon with n sides, the sum of the angle measures is

$$(n - 2) \cdot 180°.$$

EXAMPLE 3 **Finding the sum of the angle measures of a regular polygon**

Find the sum of the angle measures for the regular polygon shown.

Solution
Since $n = 6$, the sum of the angle measures is $(6 - 2) \cdot 180° = 4 \cdot 180° = 720°$.

Now Try Exercise 33

Because all angles of a regular polygon have equal measure, we can find the measure of a *single* angle by dividing the angle measure sum by the number of angles. For a regular polygon, the number of angles equals the number of sides and the following rule can be used to find the measure of one angle of a regular polygon.

THE MEASURE OF ONE ANGLE OF A REGULAR POLYGON

For a regular polygon with n sides, the measure of one angle is

$$\frac{(n - 2) \cdot 180°}{n}.$$

EXAMPLE 4 **Finding the measure of one angle of a regular polygon**

For the following regular polygon, find the measure of one angle.

Solution
Since $n = 5$, the measure of one angle is

$$\frac{(5 - 2) \cdot 180°}{5} = \frac{3 \cdot 180°}{5} = \frac{540°}{5} = 108°.$$

Now Try Exercise 37

Quadrilaterals

Recall that a quadrilateral is a 4-sided polygon. In the last section, we learned that there are several special classifications for triangles. The same is true for quadrilaterals. Table 10.6 lists different types of quadrilaterals.

TABLE 10.6 Types of Quadrilaterals

Quadrilateral	Defining Feature	Example
Square	A regular quadrilateral	
Rectangle	All angles measure 90°	
Parallelogram	Two pairs of parallel sides	
Trapezoid	One pair of parallel sides	
Rhombus	Four sides of equal length	
Kite	Two pairs of adjacent sides equal in length	

NOTE: Small arrowheads are used along the sides of the parallelogram and trapezoid shown in Table 10.6 to indicate which sides are parallel. However, such arrowheads are not commonly shown on a square, rectangle, or rhombus, which can each be classified as a parallelogram.

EXAMPLE 5   **Naming quadrilaterals**

Name each quadrilateral using every classification that applies.

(a) 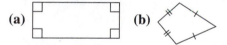   **(b)**

Solution
(a) Since all angles measure 90°, the quadrilateral is a rectangle. It is also a parallelogram because it has two pairs of parallel sides.
(b) The quadrilateral has two pairs of adjacent sides that are equal in length, so it is a kite.

Now Try Exercises 41, 43

Circles

A **circle** is a closed plane figure consisting of all points that are equally distant from a single point called the **center** of the circle. A line segment with one endpoint on the circle and the other endpoint at the center is called a **radius**. A line segment that contains the center and has both endpoints on the circle is called a **diameter**. Figure 10.28 shows these terms labeled on a circle.

Figure 10.28 A Circle and Related Terms

NOTE: Not only are the terms radius and diameter used to name the segments that they represent, but they are also used to refer to the length of those segments. For example, we may say that the radius is 5 inches, or the diameter is 13 feet.

The radius and diameter of a circle are related in the following way.

THE RADIUS AND DIAMETER OF A CIRCLE

If r represents the radius and d represents the diameter of a circle, then

$$d = 2r \quad \text{and} \quad r = \frac{d}{2}.$$

READING CHECK

• How are the radius and diameter of a circle related?

EXAMPLE 6 **Finding the radius when given the diameter**

Find the radius of the following circle.

16 in.

Solution
The radius is half the diameter, or

$$r = \frac{d}{2} = \frac{16}{2} = 8 \text{ in.}$$

▎ **Now Try Exercise 51**

EXAMPLE 7 **Finding the diameter when given the radius**

Find the diameter of the following circle.

1.5 m

Solution
The diameter is twice the radius, or

$$d = 2r = 2(1.5) = 3 \text{ m.}$$

▎ **Now Try Exercise 59**

10.3   Putting It All Together

CONCEPT	COMMENTS	EXAMPLES
Polygon	A polygon is a closed plane figure determined by three or more line segments. To be a polygon, the following must be true. **1.** All sides must be *straight* line segments. **2.** Sides cannot intersect, except at endpoints. **3.** The figure must be closed.	
Names of Polygons	A polygon is named by the number of its sides. 3 sides: **Triangle** 4 sides: **Quadrilateral** 5 sides: **Pentagon** 6 sides: **Hexagon** 7 sides: **Heptagon** 8 sides: **Octagon**	Triangle Quadrilateral Pentagon Hexagon Heptagon Octagon
Regular Polygon	In a regular polygon, all sides have equal length and all angles have equal measure.	
Angle Measures for a Regular Polygon	For a regular polygon with n sides, $$(n - 2) \cdot 180°$$ gives the sum of the angle measures, and $$\frac{(n - 2) \cdot 180°}{n}$$ gives the measure of one angle.	For a regular polygon with 6 sides, the sum of the angle measures is $$(6 - 2) \cdot 180° = 720°.$$ The measure of one angle is $$\frac{(6 - 2) \cdot 180°}{6} = \frac{720°}{6} = 120°.$$
Quadrilaterals	There are six kinds of quadrilaterals. **Square:** A regular quadrilateral **Rectangle:** All angles measure 90° **Parallelogram:** Two pairs of parallel sides **Trapezoid:** One pair of parallel sides **Rhombus:** Four sides of equal length **Kite:** Two pairs of adjacent sides equal in length	Square Rectangle Parallelogram Trapezoid Rhombus Kite
Circle	A circle is a closed plane figure that consists of all points that are equally distant from a central point (center). A radius is a line segment with one endpoint on the circle and the other endpoint at the center. A diameter is a line segment that contains the center and has both endpoints on the circle.	Radius r Diameter d Center $$d = 2r \quad \text{and} \quad r = \frac{d}{2}$$

10.3   Exercises

MyMathLab   Math XL PRACTICE   WATCH   DOWNLOAD   READ   REVIEW

CONCEPTS AND VOCABULARY

1. A(n) _____ is a closed plane figure determined by three or more line segments.

2. A polygon with 6 sides is called a(n) _____.

3. A polygon with 7 sides is called a(n) _____.

4. In a(n) _____ polygon, all sides have the same length and all angles have the same measure.

5. For a regular polygon with n sides, the sum of the angle measures is given by _____.

6. For a regular polygon with n sides, the measure of one angle is given by _____.

7. A(n) _____ is a closed plane figure consisting of all points that are equally distant from a single point called the _____.

8. A line segment with one endpoint on a circle and the other endpoint at the center is a(n) _____.

9. A line segment that contains the center and has both endpoints on the circle is a(n) _____.

10. If r represents the radius and d represents the diameter of a circle, then $d =$ _____ and $r =$ _____.

POLYGONS

Exercises 11–16: Explain why the figure is not a polygon.

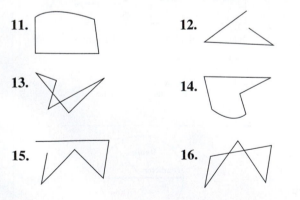

11.   12.

13.   14.

15.   16.

Exercises 17–28: Determine the name of the polygon.

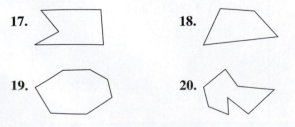

17.   18.

19.   20.

21.   22.

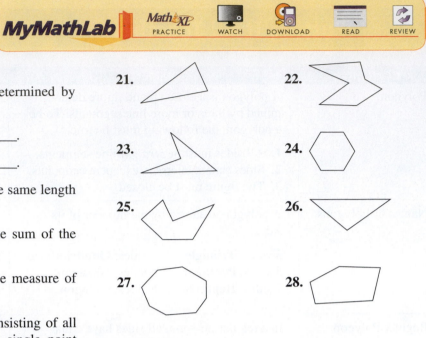

23.   24.

25.   26.

27.   28.

REGULAR POLYGONS

Exercises 29–34: Find the sum of the angle measures for the given regular polygon.

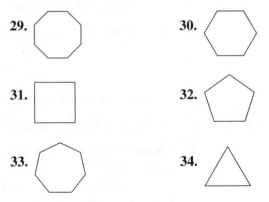

29.   30.

31.   32.

33.   34.

Exercises 35–40: For the given regular polygon, find the measure of one angle.

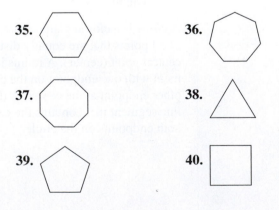

35.   36.

37.   38.

39.   40.

QUADRILATERALS

Exercises 41–50: Name the quadrilateral using every classification that applies.

41.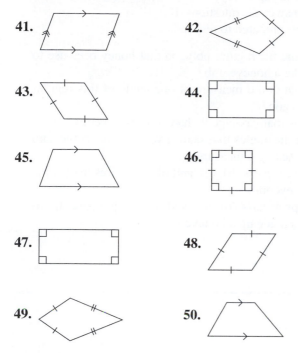

42.

43.

44.

45.

46.

47.

48.

49.

50.

CIRCLES

Exercises 51–56: Find the radius of the given circle.

51. 18 m

52. 7 in.

53. $\frac{4}{9}$ ft

54. 216 mi

55. 0.3 cm

56. 7.5 km

Exercises 57–62: Find the diameter of the given circle.

57. 8 mi

58. 12 cm

59. 4.5 ft

60. 1.7 in.

61. $\frac{5}{2}$ m

62. $\frac{7}{4}$ ft

APPLICATIONS

63. *Traffic Sign* A stop sign has the shape of a regular octagon. Find the measure of one of its angles.

64. *Wall Clock* A circular clock has a 5.5-inch radius. Find the diameter of the clock.

65. *National Defense* The U.S. Department of Defense is housed in a building called the Pentagon. The building has the shape of a regular five-sided polygon. Find the sum of the angles of the Pentagon.

66. *Beverage Can* The top of a soda can is a circle with a diameter of 5.4 centimeters. Find the radius of the top of the can.

WRITING ABOUT MATHEMATICS

67. How are the radius and diameter of a circle related?

68. Give three examples of real-world things with polygon shapes.

69. A student draws a rhombus with four right angles. What are some other quadrilateral names that describe what the student has drawn? Explain.

70. Explain why it is necessary to have a *regular* polygon when finding the measure of one angle.

Group Activity | Working with Real Data

Directions: Form a group of 2 to 4 people. Select someone to record the group's responses for this activity. All members of the group should work cooperatively to answer the questions. If your instructor asks for your results, each member of the group should be prepared to respond.

The Honeycomb Tessellation Honey bees can construct a honeycomb that fits regular polygons together perfectly to make the pattern shown.

A tiling pattern made of non-overlapping polygons with no gaps is called a *tessellation*.

(a) Name the regular polygon that honey bees use to make a honeycomb.

(b) What is the measure of one angle of this regular polygon?

(c) How many polygons share any given vertex?

(d) For the angles that share a vertex, what is the sum of the angle measures?

(e) Repeat parts (b), (c), and (d) if squares made up a honeycomb.

(f) Repeat parts (b), (c), and (d) if equilateral triangles made up a honeycomb.

10.4 | Perimeter and Circumference

Perimeter of a Polygon • Circumference of a Circle • Composite Figures

A LOOK INTO MATH ▶

The Spaceship Earth attraction at Walt Disney World's Epcot Theme Park is among the most recognizable structures in the world. Architects used 11,324 triangular panels to create the outer surface of the ball-shaped building. When it rains, water is channeled through one-inch gaps along the *perimeter* of the triangles and is collected in a gutter system. In this section, we discuss how to measure the perimeter of a polygon and a related measure for circles called the circumference. (*Source:* Walt Disney Company.)

Perimeter of a Polygon

NEW VOCABULARY

☐ Perimeter
☐ Circumference
☐ Composite figure

The **perimeter** of a polygon is the distance around the polygon. We can find the perimeter of a polygon by adding the lengths of its sides.

THE PERIMETER OF A POLYGON

To find the perimeter of a polygon, add the lengths of all sides of the polygon.

▶ **REAL-WORLD CONNECTION** City planners often include *green space* in new city development projects. To determine the amount of material needed to create a walkway around a city park, the perimeter of the park is measured. The perimeter of the rectangular park in Figure 10.29 is 780 feet because $200 + 190 + 200 + 190 = 780$.

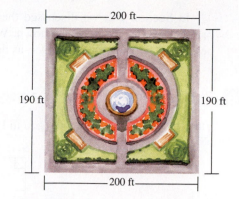

Figure 10.29 Perimeter of a Rectangular Park

EXAMPLE 1   **Finding the perimeter of polygons**

Find the perimeter of each polygon.

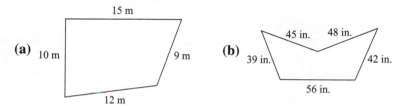

Solution
(a) The perimeter is $10 + 15 + 9 + 12 = 46$ meters.
(b) The perimeter is $39 + 45 + 48 + 42 + 56 = 230$ inches.

Now Try Exercises 9, 11

We use multiplication rather than addition to find the perimeter of a *regular* polygon because all sides of a regular polygon have the same measure.

THE PERIMETER OF A REGULAR POLYGON

To find the perimeter of a regular polygon, multiply the number of sides by the length of one side.

EXAMPLE 2   **Finding the perimeter of a regular polygon**

Find the perimeter of the regular polygon in Figure 10.30.

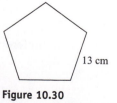

Figure 10.30

Solution
A regular pentagon has 5 sides with the same length. The perimeter is $5 \cdot 13 = 65$ cm.

Now Try Exercise 15

In Section 1.2, we discussed the perimeter of a region in which the measure of one or more of the sides was unknown. When this happens, we must find any missing lengths before we can find the perimeter, as demonstrated in the next example.

| EXAMPLE 3 | **Finding the perimeter of a polygon with missing measures** |

Find the perimeter of the polygon in Figure 10.31.

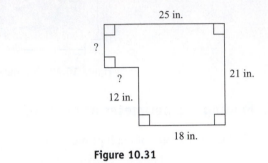

Figure 10.31

Solution
The length of the missing sides can be found by subtraction, as shown in Figure 10.32.

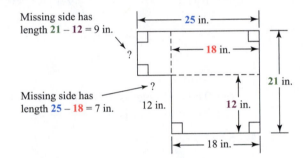

Figure 10.32 Finding the Lengths of Missing Sides

The perimeter is $25 + 21 + 18 + 12 + 7 + 9 = 92$ in.

Now Try Exercise 21

Circumference of a Circle

The perimeter of a circle is called the **circumference** of the circle. A formula involving the number π (pi) is needed to find the circumference of a circle. Recall from Section 5.4 that π is an irrational number whose decimal representation neither terminates nor repeats.

$$\pi \approx 3.1415926536$$

The ratio of the circumference to the diameter of *any* circle is always equal to the number π. In other words,

$$\frac{\textbf{Circumference}}{\textbf{diameter}} = \pi.$$

Using C for circumference and d for diameter, we have

$$\frac{C}{d} = \pi.$$

Multiplying each side of this equation by d gives the formula

$$C = \pi \cdot d.$$

Since diameter is twice the radius, the circumference formula can be written in two ways.

> **THE CIRCUMFERENCE OF A CIRCLE**
>
> The circumference C of a circle with radius r and diameter d is found using
>
> $$C = \pi d \quad \text{or} \quad C = 2\pi r,$$
>
> where $\pi \approx 3.14$ or $\pi \approx \frac{22}{7}$.

EXAMPLE 4   **Finding the circumference of a circle using the diameter**

Refer to the circle in Figure 10.33.

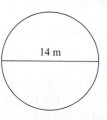

14 m

Figure 10.33

(a) Find the exact circumference of the circle.
(b) Approximate the circumference using 3.14 as an approximation for π.

Solution
(a) Using the formula $C = \pi d$ gives $C = \pi \cdot 14 = 14\pi$ m.
(b) Using $\pi \approx 3.14$ gives $C \approx 14 \cdot 3.14 = 43.96$ m.

▎**Now Try Exercise 27**

EXAMPLE 5   **Finding the circumference of a circle using the radius**

Refer to the circle in Figure 10.34.

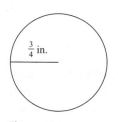

$\frac{3}{4}$ in.

Figure 10.34

(a) Find the exact circumference of the circle.
(b) Approximate the circumference, using $\frac{22}{7}$ as an approximation for π.

Solution
(a) Using the formula $C = 2\pi r$ gives $C = 2 \cdot \pi \cdot \frac{3}{4} = \frac{3}{2}\pi$ in.
(b) Using $\pi \approx \frac{22}{7}$ gives $C \approx \frac{3}{2} \cdot \frac{22}{7} = \frac{33}{7}$ in.

▎**Now Try Exercise 37**

Composite Figures

An enclosed geometric region made up of polygons and semicircles (half circles) is called a **composite figure**. Examples of composite figures can be found in everyday situations.

▶ **REAL-WORLD CONNECTION** On a basketball court, the *key* is an enclosed region on the floor at each end of the court. The key is a composite figure made up of a rectangle and a semi-circle. In the next example, we find the perimeter of the key in Figure 10.35.

READING CHECK

• How do we find the perimeter of a composite figure?

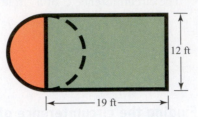

Figure 10.35 The Key on a Basketball Court

EXAMPLE 6   **Finding the perimeter of a basketball court key**

Find the perimeter of the key in Figure 10.35. Use 3.14 as an approximation for π.

Solution
The key is enclosed by three line segments and a semicircle. The segments have lengths 19, 12, and 19 feet. The length of the semicircle is one-half the circumference of a circle with a 12-foot diameter. Since the circumference of the circle is $C = \pi \cdot 12 \approx 3.14 \cdot 12 = 37.68$ feet, the length of the semicircle is approximately $0.5 \cdot 37.68 = 18.84$ feet. The total perimeter is approximately $19 + 12 + 19 + 18.84 = 68.84$ feet.

Now Try Exercise 53

10.4   Putting It All Together

CONCEPT	COMMENTS	EXAMPLES
Perimeter of a Polygon	To find the perimeter of polygon, add the lengths of all sides of the polygon.	15 cm, 8 cm, 11 cm $15 + 11 + 8 = 34$ cm
Perimeter of a Regular Polygon	To find the perimeter of a regular polygon, multiply the number of sides by the length of one side.	7 ft $3 \cdot 7 = 21$ ft
Circumference of a Circle	The circumference C of a circle with radius r and diameter d is found using $C = \pi d$ or $C = 2\pi r$, where $\pi \approx 3.14$ or $\pi \approx \frac{22}{7}$.	4 mi $C = 4\pi \approx 4 \cdot 3.14 = 12.56$ mi $\frac{7}{11}$ m $C = 2 \cdot \frac{7}{11}\pi \approx 2 \cdot \frac{7}{11} \cdot \frac{22}{7} = 4$ m

10.4 Exercises

MyMathLab · Math XL PRACTICE · WATCH · DOWNLOAD · READ · REVIEW

CONCEPTS AND VOCABULARY

1. The distance around a polygon is called the _____ of the polygon.

2. To find the perimeter of a polygon, add the lengths of all _____ of the polygon.

3. The perimeter of a circle is its _____.

4. To find the circumference of a circle with a given diameter, use the formula _____.

5. To find the circumference of a circle with a given radius, use the formula _____.

6. An enclosed geometric region made up of polygons and semicircles is called a(n) _____ figure.

PERIMETER OF A POLYGON

Exercises 7–14: Find the perimeter of the polygon.

7. 12 cm, 9 cm, 14 cm

8. 8 cm, 8 cm, 12 cm

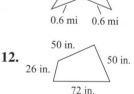

9. 4.5 ft, 3.7 ft, 3.0 ft, 6.1 ft

10. 0.8 mi, 0.5 mi, 1.1 mi, 0.6 mi, 0.6 mi

11. 6 km, 2 km, 6 km, 5 km, 4 km

12. 50 in., 26 in., 50 in., 72 in.

13. $\frac{1}{2}$ m, $\frac{7}{8}$ m, 1 m

14. $\frac{9}{4}$ in., $\frac{9}{4}$ in., $\frac{5}{2}$ in.

Exercises 15–20: Find the perimeter of the regular polygon.

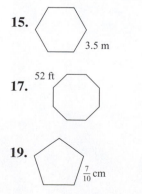

15. 3.5 m

16. 13 in.

17. 52 ft

18. 16.4 mi

19. $\frac{7}{10}$ cm

20. $\frac{9}{2}$ ft

Exercises 21–26: Find the perimeter of the polygon.

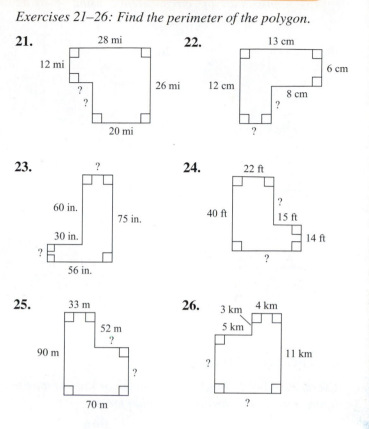

21. 28 mi, 12 mi, 26 mi, ?, ?, 20 mi

22. 13 cm, 6 cm, 12 cm, 8 cm, ?, ?

23. ?, 60 in., 75 in., 30 in., ?, 56 in.

24. 22 ft, ?, 40 ft, 15 ft, 14 ft, ?

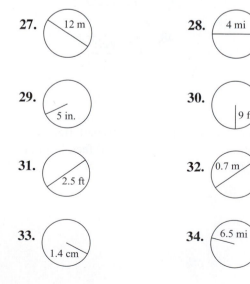

25. 33 m, 52 m, ?, 90 m, ?, 70 m

26. 3 km, 4 km, 5 km, 11 km, ?, ?

CIRCUMFERENCE OF A CIRCLE

Exercises 27–34: Do the following.

(a) Find the exact circumference of the circle.
(b) Approximate the circumference, using 3.14 as an approximation for π.

27. 12 m

28. 4 mi

29. 5 in.

30. 9 ft

31. 2.5 ft

32. 0.7 m

33. 1.4 cm

34. 6.5 mi

Exercises 35–42: Do the following.

(a) *Find the exact circumference of the circle.*
(b) *Approximate the circumference, using $\frac{22}{7}$ as an approximation for π.*

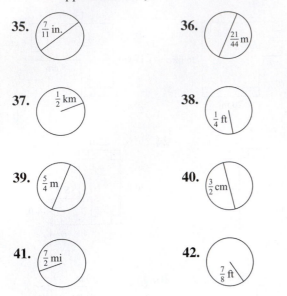

35. $\frac{7}{11}$ in.

36. $\frac{21}{44}$ m

37. $\frac{1}{2}$ km

38. $\frac{1}{4}$ ft

39. $\frac{5}{4}$ m

40. $\frac{3}{2}$ cm

41. $\frac{7}{2}$ mi

42. $\frac{7}{8}$ ft

COMPOSITE FIGURES

Exercises 43–48: Find the perimeter of the composite figure. Use 3.14 as an approximation for π.

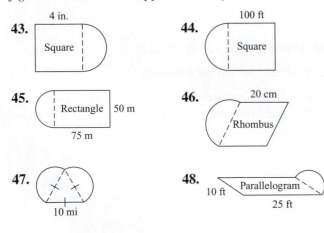

43.
4 in.
Square

44.
100 ft
Square

45.
Rectangle   50 m
75 m

46.
20 cm
Rhombus

47.
x   x
10 mi

48.
Parallelogram
10 ft   25 ft

APPLICATIONS

49. *Football Field* An official NFL football field is a rectangle that measures 160 feet by 360 feet. Find the perimeter of an official NFL football field.

50. *National Defense* The U.S. Department of Defense building has the shape of a regular pentagon with each wall measuring 921 feet. Find the perimeter of the Pentagon building.

51. *Bermuda Triangle* A triangular region with vertices at Bermuda, Puerto Rico, and southern Florida is known as the Bermuda Triangle. It is roughly an equilateral triangle measuring 1580 km on each side. Find the perimeter of the Bermuda Triangle.

52. *Olympic Shotput* An Olympic shotput ring is a circle with a diameter of 7 feet. Each athlete must stay within the ring during a throw. Find the circumference of an Olympic shotput ring, using 3.14 as an approximation for π.

53. *Track and Field* A running track is constructed with the dimensions shown in the following figure. Find the perimeter of the track, using 3.14 as an approximation for π. Round to the nearest meter.

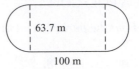

63.7 m
100 m

54. *Raceways* A racing track is constructed with the dimensions shown in the following figure. Find the perimeter of the track, using 3.14 as an approximation for π. Round to the nearest tenth of a mile.

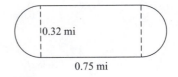

0.32 mi
0.75 mi

WRITING ABOUT MATHEMATICS

55. Which number, 3.14 or $\frac{22}{7}$, is a more accurate approximation for π? Explain your reasoning.

56. If the perimeter of a regular polygon is known, explain how you can find the length of one side.

1. Name each polygon.

(a) 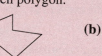   **(b)**

2. Find the sum of the angle measures for each regular polygon.

(a)    **(b)**

3. For each regular polygon, find the measure of one angle.

(a)    **(b)**

4. Name each quadrilateral, using every classification that applies.

(a)    **(b)**

5. Find the radius of the circle shown.

42 m

6. Find the diameter of the circle shown.

5.7 in.

7. Find the perimeter of the following polygon.

3 in.   3 in.
$\frac{9}{2}$ in.

8. Find the perimeter of the following regular polygon.

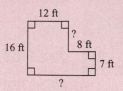
4.5 cm

9. Find the perimeter of the polygon.

12 ft
?
16 ft   8 ft
7 ft
?

10. Approximate the circumference, using 3.14 as an approximation for π.

1.5 m

11. Approximate the circumference, using $\frac{22}{7}$ as an approximation for π.

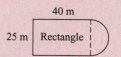
$\frac{4}{11}$ ft

12. Approximate the perimeter of the given composite figure, using 3.14 as an approximation for π.

40 m
25 m | Rectangle

13. *Bicycle Tire* A bicycle tire has a 9.5-inch radius. Find the diameter of the tire.

14. *Flower Garden* A flower garden is constructed in the shape of a rhombus, measuring 16 feet on each side. Find the perimeter of the garden.

10.5   Area, Volume, and Surface Area

Area of Plane Figures • Volume and Surface Area of Geometric Solids

A LOOK INTO MATH ▶   When sand is poured onto a flat surface, it naturally forms a cone-shaped pile. To find the amount of sand in the pile, we use a volume formula for a *right, circular cone*. In this section, we discuss area formulas for common two-dimensional geometric figures as well as volume and surface area formulas for common three-dimensional geometric solids.

Area of Plane Figures

As we first discussed in Section 1.3, the *area* of a region is computed by finding the number of square units that are needed to cover the region. In this section, we are interested in regions that are common polygons or circles.

NEW VOCABULARY

☐ Volume
☐ Surface area
☐ Cube
☐ Rectangular prism
☐ Circular cylinder
☐ Cone
☐ Square-based pyramid
☐ Sphere

AREA OF COMMON POLYGONS   Earlier in this text, area formulas were presented for squares, rectangles, and triangles. For convenience, these formulas are provided again in Figure 10.36.

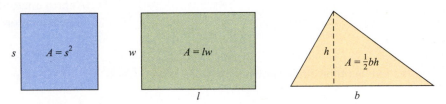

Figure 10.36 Area Formulas for Squares, Rectangles, and Triangles

STUDY TIP

Organize your completed notes for this course and keep them ready as an aid for your studies in your next math course.

EXAMPLE 1   **Finding the area of a square, rectangle, or triangle**

Find the area of each figure.

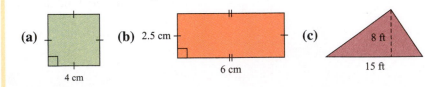

Solution
(a) The area of the square is $A = s^2 = 4^2 = 16 \text{ cm}^2$.
(b) The area of the rectangle is $A = lw = 6(2.5) = 15 \text{ cm}^2$.
(c) The area of the triangle is $A = \frac{1}{2}bh = \frac{1}{2}(15)(8) = 60 \text{ ft}^2$.

Now Try Exercises 7, 9, 11

The area formula for a parallelogram can be found by "cutting" a right triangle off one end of the parallelogram and connecting it to the other end, as shown in Figure 10.37.

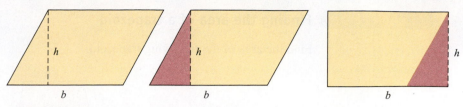

Figure 10.37 Area of the Parallelogram Equals Area of the Rectangle

Rearranging the parallelogram in this way results in a rectangle. Since the area of the original parallelogram is equal to the area of the rectangle, the area formula for a parallelogram is $A = bh$, where b is the length of the base and h is the height.

AREA OF A PARALLELOGRAM

The area A of a parallelogram with base b and height h is

$$A = bh.$$

EXAMPLE 2 | **Finding the area of a parallelogram**

Find the area of the following parallelogram.

12 mi

20 mi

Solution
The area of the parallelogram is $A = bh = 20(12) = 240$ mi^2.

▌ **Now Try Exercise 13**

The area formula for a trapezoid can be found by connecting the trapezoid to a second, identical but rotated, trapezoid to form a parallelogram, as shown in Figure 10.38.

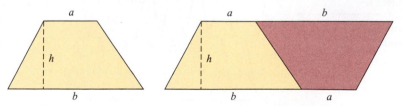

Figure 10.38 Area of the Trapezoid Is Half the Area of the Parallelogram

The area of the trapezoid is equal to half the area of the resulting parallelogram. Since the base of the parallelogram in Figure 10.38 is the sum $a + b$, the area formula for a trapezoid is $A = \frac{1}{2}(a + b)h$, where $a + b$ is the sum of the lengths of the trapezoid's parallel sides and h is the height.

AREA OF A TRAPEZOID

The area A of a trapezoid with parallel sides of lengths a and b and height h is

$$A = \frac{1}{2}(a + b)h.$$

EXAMPLE 3   **Finding the area of a trapezoid**

Find the area of the following trapezoid.

Solution

The area of the trapezoid is $A = \frac{1}{2}(a + b)h = \frac{1}{2}(5.8 + 8.2) \cdot 6 = 42 \text{ ft}^2$.

Now Try Exercise 15

AREA OF A CIRCLE To find the area formula for a circle, we will take the circle apart in a way that allows us to reassemble it as a figure whose shape closely resembles a parallelogram. First, divide the circle as shown in Figure 10.39.

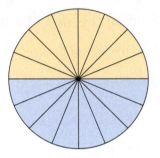

Figure 10.39 Slicing a Circle

Both semicircles (or halves of the circle) can be pulled apart to form sawtooth patterns that fit together to form a shape resembling a parallelogram, as shown in Figure 10.40.

READING CHECK

• After transforming a circle into a parallelogram shape, why is the base equal to πr?

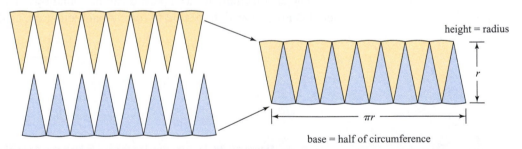

Figure 10.40 Transforming a Circle to a Parallelogram Shape

Since the area of the parallelogram equals the base πr times the height r, we approximate the area of the circle as $A = \pi r \cdot r$, which simplifies to $A = \pi r^2$. This is the formula for the area of a circle.

AREA OF A CIRCLE

The area A of a circle with radius r is found using

$$A = \pi r^2,$$

where $\pi \approx 3.14$ or $\pi \approx \frac{22}{7}$.

EXAMPLE 4   **Finding the area of circles**

Find the approximate area of each circle, using 3.14 for π.

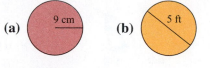

(a)   9 cm   (b)   5 ft

Solution

(a) The area of the circle is $A = \pi r^2 \approx 3.14(9^2) = 254.34$ cm^2.

(b) The radius of the circle is half the diameter, or $0.5(5) = 2.5$ feet. So, the area of the circle is $A = \pi r^2 \approx 3.14(2.5)^2 = 19.625$ ft^2.

Now Try Exercises 17, 19

Volume and Surface Area of Geometric Solids

For geometric solids (three-dimensional figures), we are interested in finding *volume* and *surface area*. As noted in Section 6.3, **volume** is a measure of the amount of a substance needed to fill a defined space. Volume is measured in *cubic units*. Figure 10.41 shows a cubic centimeter and a cubic inch.

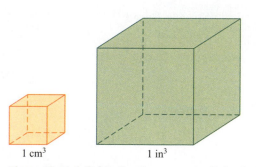

1 cm^3 1 in^3

Figure 10.41 A Cubic Centimeter and a Cubic Inch

Surface area measures the amount of exposed area for a given solid. For example, a simple die from a board game has the shape of a cube with six square surfaces called *faces*, as shown in Figure 10.42.

One of six faces

Figure 10.42 A Die with Six Faces

The surface area of a die is the sum of the areas of the six exposed faces. This discussion suggests that surface area is found by adding the areas of the exposed faces or surfaces.

Several common geometric solids are shown in Table 10.7 on the next page, along with formulas for the volume and surface area. In the table, we use V to represent volume, S to represent surface area, and B to represent the *area of the base* of a solid.

TABLE 10.7 Volume and Surface Area Formulas for Common Solids

Solid		Formulas
Cube		$V = s^3$ (or $V = Bh$), $S = 6s^2$ (or $S = 6B$), where s = side length, $B = s^2$, and $h = s$ = height
Rectangular Prism		$V = lwh$ (or $V = Bh$), $S = 2lw + 2lh + 2wh$, where l = length, w = width, h = height, and $B = lw$
Circular Cylinder		$V = \pi r^2 h$ (or $V = Bh$), $S = 2\pi rh + 2\pi r^2$, where r = radius, h = height, and $B = \pi r^2$
Cone		$V = \frac{1}{3}\pi r^2 h$ (or $V = \frac{1}{3}Bh$), $S = \pi r\sqrt{r^2 + h^2} + \pi r^2$, where r = radius, h = height, and $B = \pi r^2$
Square-Based Pyramid		$V = \frac{1}{3}s^2 h$ (or $V = \frac{1}{3}Bh$), $S = B + 2sl$, where s = side, h = height, $B = s^2$, and l = slant height
Sphere		$V = \frac{4}{3}\pi r^3$, $S = 4\pi r^2$, where r = radius

READING CHECK

• What does the variable B represent when finding the volume of geometric solids?

EXAMPLE 5 **Finding the volume of geometric solids**

Find the volume of each geometric solid. If needed, use 3.14 as an approximation for π and round answers to the nearest tenth.
(a) Circular cylinder: $r = 3$ ft and $h = 12$ ft
(b) Rectangular prism: $l = 14$ m, $w = 10$ m, and $h = 3$ m

Solution
(a) $V = \pi r^2 h = (3.14)(3^2)(12) \approx 339.1$ ft^3
(b) $V = lwh = (14)(10)(3) = 420$ m^3

Now Try Exercises 27, 29

EXAMPLE 6 **Finding the surface area of geometric solids**

Find the surface area of each geometric solid. If needed, use 3.14 as an approximation for π and round answers to the nearest tenth.
(a) Square-based pyramid: $s = 6$ ft, $h = 4$ ft and $l = 5$ ft
(b) Sphere: $r = 6$ cm

Solution

(a) $S = B + 2sl = 6^2 + 2(6)(5) = 96 \text{ ft}^2$

(b) $S = 4\pi r^2 \approx 4(3.14)(6^2) \approx 452.2 \text{ cm}^2$

Now Try Exercises 45, 47

10.5 Putting It All Together

CONCEPT	COMMENTS	EXAMPLES
Area of Plane Figures	**Square:** $A = s^2$, (s = side) **Rectangle:** $A = lw$, (l = length; w = width) **Triangle:** $A = \frac{1}{2}bh$, (b = base; h = height) **Parallelogram:** $A = bh$, (b = base; h = height) **Trapezoid:** $A = \frac{1}{2}(a + b)h$, (h = height; a and b are lengths of parallel sides) **Circle:** $A = \pi r^2$, (r = radius)	The area of a rectangle with $l = 9$ ft and $w = 5$ ft is $A = 9(5) = 45 \text{ ft}^2$. The area of a circle with $r = 7$ cm is $A \approx 3.14(7^2) \approx 153.9 \text{ cm}^2$.
Volume and Surface Area of Geometric Solids	A full summary of the volume and surface area formulas for the cube, rectangular prism, circular cylinder, cone, square-based pyramid, and sphere can be found in Table 10.7 on page 626.	The volume of a sphere with $r = 2$ m is $V \approx \left(\frac{4}{3}\right)(3.14)(2^3) \approx 33.5 \text{ m}^3$. The surface area of a cube with $s = 9$ mi is $S = 6(9^2) = 486 \text{ mi}^2$.

10.5 Exercises

MyMathLab | Math XL PRACTICE | WATCH | DOWNLOAD | READ | REVIEW

CONCEPTS AND VOCABULARY

1. When finding the area of a plane figure, the units for the result are (square/cubic) units.

2. When finding the volume of a geometric solid, the units for the result are (square/cubic) units.

3. When finding the surface area of a geometric solid, the units for the result are (square/cubic) units.

4. (True or False?) When finding the area of a circle, the units for the result are *circular* units.

5. When we find the amount of a substance needed to fill a defined space, we are finding _____.

6. When we find the amount of exposed area on a solid, we are finding _____.

AREA OF PLANE FIGURES

Exercises 7–16: Find the area of the polygon.

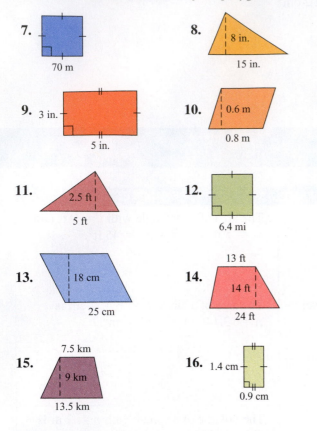

7.

70 m

8.

8 in.

15 in.

9. 3 in.

5 in.

10. 0.6 m

0.8 m

11. 2.5 ft

5 ft

12.

6.4 mi

13. 18 cm

25 cm

14. 13 ft

14 ft

24 ft

15. 7.5 km

9 km

13.5 km

16. 1.4 cm

0.9 cm

Exercises 17–24: Determine the approximate area of the given circle, using 3.14 for π. Round your answers to the nearest tenth.

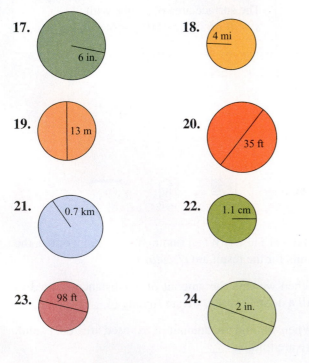

17.

6 in.

18.

4 mi

19.

13 m

20.

35 ft

21.

0.7 km

22.

1.1 cm

23.

98 ft

24.

2 in.

VOLUME AND SURFACE AREA

Exercises 25–36: Find the volume of the geometric solid. If needed, use 3.14 as an approximation for π and round answers to the nearest tenth.

25. Cube: $s = 7$ in.

26. Cube: $s = 30$ m

27. Rectangular prism: $l = 7$ ft, $w = 3$ ft, $h = 2$ ft

28. Rectangular prism: $l = 16$ mi, $w = 5$ mi, $h = 4$ mi

29. Circular cylinder: $r = 8$ cm, $h = 15$ cm

30. Circular cylinder: $r = 12$ mm, $h = 56$ mm

31. Cone: $r = 3$ in., $h = 4$ in.

32. Cone: $r = 14$ m, $h = 50$ m

33. Square-based pyramid: $s = 9$ ft, $h = 10$ ft

34. Square-based pyramid: $s = 4$ in., $h = 6$ in.

35. Sphere: $r = 2$ cm

36. Sphere: $r = 10$ m

Exercises 37–48: Find the surface area of the geometric solid. If needed, use 3.14 as an approximation for π and round answers to the nearest tenth.

37. Cube: $s = 6$ in.

38. Cube: $s = 56$ m

39. Rectangular prism: $l = 8$ ft, $w = 4$ ft, $h = 3$ ft

40. Rectangular prism: $l = 10$ mi, $w = 3$ mi, $h = 1$ mi

41. Circular cylinder: $r = 4$ cm, $h = 9$ cm

42. Circular cylinder: $r = 20$ mm, $h = 80$ mm

43. Cone: $r = 5$ in., $h = 10$ in.

44. Cone: $r = 7$ m, $h = 4$ m

45. Square-based pyramid: $s = 3$ ft, $l = 8$ ft

46. Square-based pyramid: $s = 12$ in., $l = 9$ in.

47. Sphere: $r = 7$ cm

48. Sphere: $r = 25$ m

APPLICATIONS

49. *Pile of Sand* A pile of sand has the shape of a cone with radius 6 feet and height 5 feet. Find the volume of sand in the pile. Use 3.14 as an approximation for π and round to the nearest tenth.

50. *Great Pyramid* The Great Pyramid of Giza has a square base measuring about 756 feet. Approximate the volume of the Great Pyramid if the height is about 480 feet. (*Source: National Geographic.*)

51. *Basketball* A basketball has a 4.7-inch radius. Find the surface area of the ball. Use 3.14 as an approximation for π and round to the nearest tenth.

52. *Basketball Hoop* A basketball hoop has an 18-inch diameter. Find the area enclosed by the hoop. Use 3.14 as an approximation for π and round to the nearest tenth.

53. *Heavy Lifting* A hydraulic lift has the shape of a parallelogram. Find the area of the parallelogram if the base measures 3.5 feet and the lift is extended to a height of 2 feet.

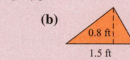

2 feet

3.5 feet

54. *Geometric Garden* A garden is constructed in the shape of a trapezoid. If the parallel sides are 20 feet apart and measure 18 feet and 32 feet, find the area of the garden.

55. *Paint Can* A paint can has the shape of a cylinder with a 6-inch diameter and a 7-inch height. Find the surface area of the can. Use 3.14 as an approximation for π and round to the nearest tenth.

56. *Water Cup* How much water fits in a cone-shaped cup with a 4-centimeter radius and an 8-centimeter height? Use 3.14 as an approximation for π and round to the nearest cubic centimeter.

WRITING ABOUT MATHEMATICS

57. Half of a sphere is called a semi-sphere. Write the formulas for the volume and surface area of a semi-sphere. Explain how you arrived at your answers.

58. If you knew the distance around Earth's equator, explain how you could compute Earth's radius.

59. Suppose a box (rectangular prism) has no top. Write a formula for the surface area for the box. Explain how you arrived at your answer.

60. Explain how you could find the volume of wood that remains after a circular hole is drilled through a wooden cube.

SECTION 10.5   Checking Basic Concepts

1. Find the area of each polygon.

(a) 4 m / 5 m

(b) 0.8 ft / 1.5 ft

2. Find the area of each circle, using 3.14 as an approximation for π. Round your answers to the nearest tenth.

(a) 15 ft

(b) 24 in.

3. Find the volume of each geometric solid. If needed, use 3.14 as an approximation for π and round answers to the nearest tenth.
 (a) Cube: $s = 14$ km
 (b) Cone: $r = 6$ ft, $h = 15$ ft

4. Find the surface area of each geometric solid. If needed, use 3.14 as an approximation for π and round answers to the nearest tenth.
 (a) Circular cylinder: $r = 1$ m, $h = 3$ m
 (b) Sphere: $r = 60$ cm

CHAPTER 10 Summary

Point	A point is a location in space having no length, width, or height. **Example:** • *P*
Line	A line is a straight figure representing a set of points extending forever in two directions. **Example:** ◄—•——•—► *A* *B*
Line Segment	A line segment is a straight figure representing a set of points extending between two endpoints. **Example:** •————• *A* *B*
Ray	A ray is a straight figure representing a set of points extending forever in one direction. **Example:** •————► *A* *B*
Angle	An angle is a figure formed by two rays with a common endpoint. The common endpoint is called the *vertex*, and the two rays are called the *sides* of the angle. **Example:** The vertex is *B* and the sides are $\overrightarrow{BA}$ and $\overrightarrow{BC}$.

$$
\begin{array}{c}
A \\
B \;\; x \\
C
\end{array}
$$

Types of Angles	A *right* angle measures 90°. A *straight* angle measures 180°. An *acute* angle measures between 0° and 90°. An *obtuse* angle measures between 90° and 180°. **Examples:**

90°	180°	77°	154°
Right Angle	Straight Angle	Acute Angle	Obtuse Angle

Pairs of Angles	Angles with equal measures are *congruent*. The sum of the measures of two *complementary* angles is 90°. The sum of the measures of two *supplementary* angles is 180°. **Examples:**

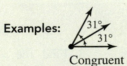

31° / 31°	39° / 51°	59° / 121°
Congruent Angles	Complementary Angles	Supplementary Angles

Perpendicular Lines	Two intersecting lines are *perpendicular* if they form right angles. **Example:**

Vertical and Adjacent Angles	Two intersecting lines that are not perpendicular form two pairs of congruent angles called vertical angles and four pairs of supplementary angles called adjacent angles.

Example:

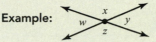

Vertical: ∠w and ∠y, ∠x and ∠z
Adjacent: ∠w and ∠x, ∠x and ∠y, ∠y and ∠z, ∠z and ∠w

Parallel Lines Cut by a Transversal

When two parallel lines are cut by a transversal, the following properties apply.

1. Vertical angles are congruent.
2. Corresponding angles are congruent.
3. Alternate interior angles are congruent.
4. Alternate exterior angles are congruent.

Any other angle pairs are supplementary.

Example:

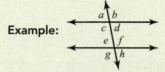

Corresponding: ∠a and ∠e, ∠b and ∠f, ∠c and ∠g, ∠d and ∠h
Alternate Interior: ∠c and ∠f, ∠d and ∠e
Alternate Exterior: ∠a and ∠h, ∠b and ∠g

SECTION 10.2 ■ TRIANGLES

Types of Triangles

Acute: Every angle measures less than 90°.
Obtuse: One angle measures between 90° and 180°.
Right: One angle measures exactly 90°.
Scalene: No sides have the same length.
Isosceles: At least two sides have the same length.
Equilateral: Three sides have the same length.

Examples:

Acute   Obtuse   Right

Scalene   Isosceles   Equilateral

The Sum of the Angle Measures

For any triangle, the sum of the measures of the three angles is 180°.

Example: 38° + 66° + 76° = 180°

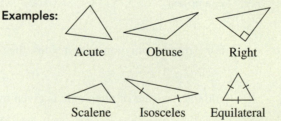

Congruent Triangles

Any of the following properties can be used to determine whether two triangles are congruent. (See examples on the next page.)
Angle-Side-Angle (ASA): If two angles and the included side of one triangle are congruent to two angles and the included side of another triangle, then the triangles are congruent.
Side-Angle-Side (SAS): If two sides and the included angle of one triangle are congruent to two sides and the included angle of another triangle, then the triangles are congruent.
Side-Side-Side (SSS): If three sides of one triangle are congruent to three sides of another triangle, then the triangles are congruent.

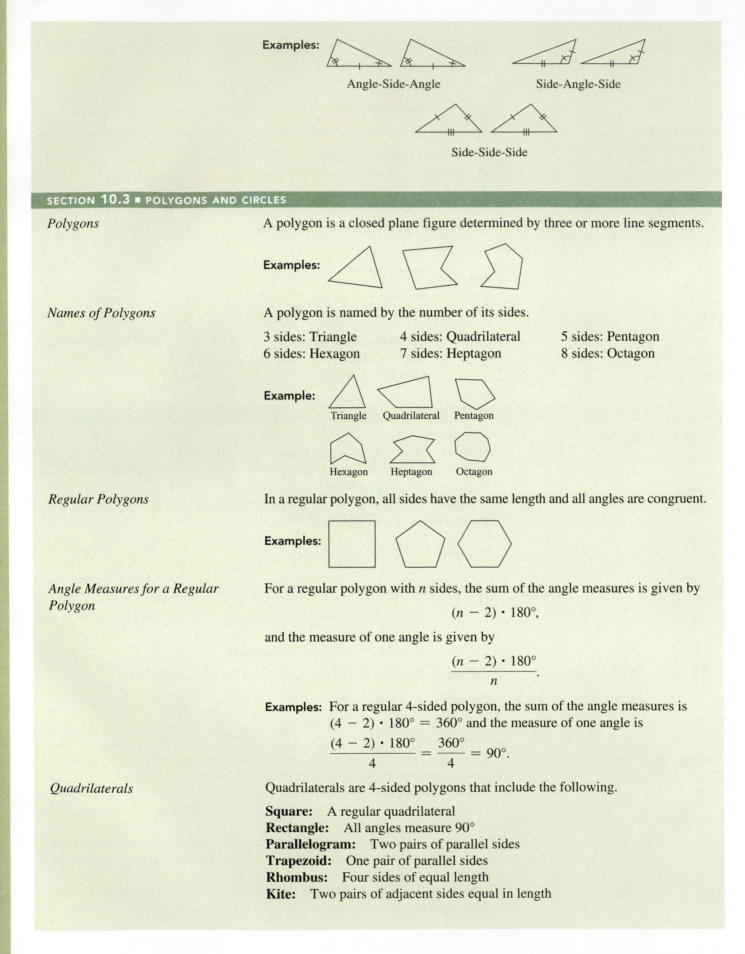

Examples:

Angle-Side-Angle   Side-Angle-Side

Side-Side-Side

SECTION **10.3** ■ POLYGONS AND CIRCLES

Polygons

A polygon is a closed plane figure determined by three or more line segments.

Examples:

Names of Polygons

A polygon is named by the number of its sides.

3 sides: Triangle	4 sides: Quadrilateral	5 sides: Pentagon
6 sides: Hexagon	7 sides: Heptagon	8 sides: Octagon

Example:

Triangle   Quadrilateral   Pentagon

Hexagon   Heptagon   Octagon

Regular Polygons

In a regular polygon, all sides have the same length and all angles are congruent.

Examples:

Angle Measures for a Regular Polygon

For a regular polygon with n sides, the sum of the angle measures is given by

$$(n - 2) \cdot 180°,$$

and the measure of one angle is given by

$$\frac{(n - 2) \cdot 180°}{n}.$$

Examples: For a regular 4-sided polygon, the sum of the angle measures is $(4 - 2) \cdot 180° = 360°$ and the measure of one angle is

$$\frac{(4 - 2) \cdot 180°}{4} = \frac{360°}{4} = 90°.$$

Quadrilaterals

Quadrilaterals are 4-sided polygons that include the following.

Square: A regular quadrilateral
Rectangle: All angles measure 90°
Parallelogram: Two pairs of parallel sides
Trapezoid: One pair of parallel sides
Rhombus: Four sides of equal length
Kite: Two pairs of adjacent sides equal in length

Examples: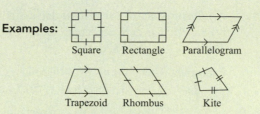

Square Rectangle Parallelogram

Trapezoid Rhombus Kite

Circles

A circle is a closed plane figure that consists of all points that are equally distant from a central point (called the center of the circle). A radius is a line segment with one endpoint on the circle and the other endpoint at the center. A diameter is a line segment that contains the center and has both endpoints on the circle.

Example: $d = 2r$ and $r = \dfrac{d}{2}$

SECTION 10.4 ■ PERIMETER AND CIRCUMFERENCE

Perimeter of a Polygon

To find the perimeter of a polygon, add the lengths of all sides of the polygon.

Example: $9 + 14 + 17 = 40$ cm

9 cm 14 cm 17 cm

Perimeter of a Regular Polygon

To find the perimeter of a regular polygon, multiply the number of sides by the length of one side.

Example: $4 \cdot 7 = 28$ ft

7 ft

Circumference of a Circle

The circumference C of a circle with radius r and diameter d is found using $C = \pi d$ or $C = 2\pi r$, where $\pi \approx 3.14$ or $\pi \approx \frac{22}{7}$.

Example: $C = 16\pi \approx 16 \cdot 3.14 = 50.24$ cm

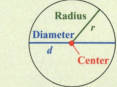

16 cm

SECTION 10.5 ■ AREA, VOLUME, AND SURFACE AREA

Area of Plane Figures

The following formulas can be used to find the area of common plane figures.

Square: $A = s^2$, (s = side)
Rectangle: $A = lw$, (l = length; w = width)
Triangle: $A = \frac{1}{2}bh$, (b = base; h = height)
Parallelogram: $A = bh$, (b = base; h = height)
Trapezoid: $A = \frac{1}{2}(a + b)h$, (h = height; a and b are lengths of parallel sides)
Circle: $A = \pi r^2$, (r = radius)

Examples: The area of a parallelogram with $b = 10$ ft and $h = 13$ ft is
$A = 10(13) = 130$ ft^2.

The area of a circle with $r = 2$ cm is $A \approx 3.14(2^2) = 12.56$ cm^2.

Volume and Surface Area of Plane Figures	A full summary of the volume and surface area formulas for the cube, rectangular prism, circular cylinder, cone, square-based pyramid, and sphere can be found in Table 10.7 on page 626.

Example: The volume of a circular cylinder with $r = 2$ m and $h = 4$ m is
$$V \approx (3.14)(2^2)(4) = 50.24 \text{ m}^3.$$

The surface area of a rectangular prism with $l = 8$ ft, $w = 5$ ft, and $h = 3$ ft is $S \approx 2(8)(5) + 2(8)(3) + 2(5)(3) = 158 \text{ ft}^2.$

CHAPTER 10   Review Exercises

SECTION 10.1

Exercises 1–4: Identify the figure and name it using the labels shown.

1.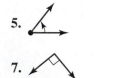
$\quad X \quad Y$

2.
$\quad P \quad\quad Q$

3. $\bullet\!\!-\!\!-\!\!-\!\!\bullet$
$\quad A \quad\quad B$

4. $\bullet\ T$

Exercises 5–8: Classify the angle as acute, right, obtuse, or straight.

5. **6.**

7. **8.**

Exercises 9 and 10: Find the measure of an angle that has the specified description.

9. The complement of an angle with measure $24°$

10. The supplement of an angle with measure $33°$

11. If $m\angle d = 104°$ in the following figure, find the measures of $\angle c$, $\angle f$, and $\angle e$.

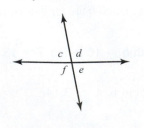

12. If $m \| n$ and $m\angle f = 65°$ in the following figure, find the measures of $\angle a$, $\angle b$, and $\angle c$.

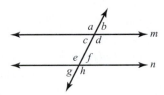

Exercises 13–16: In the following figure, $m \| n$. Determine whether the given angles are congruent or supplementary.

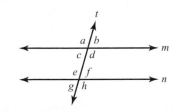

13. $\angle b$ and $\angle e$ **14.** $\angle a$ and $\angle h$

15. $\angle b$ and $\angle f$ **16.** $\angle c$ and $\angle h$

SECTION 10.2

Exercises 17–20: Classify the given triangle as acute, obtuse, or right.

17.   **18.**

19.   **20.**

Exercises 21–24: Classify the given triangle as scalene, isosceles, or equilateral.

21.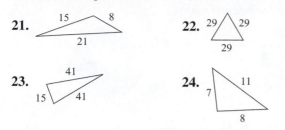

22. 29 / 29 / 29

23.

24. 11 / 7 / 8

Exercises 25 and 26: Find the measure of ∠x in the triangle.

25. 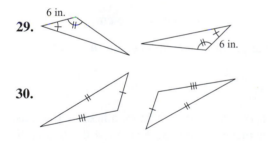 x, 46°, 50°

26. x, 43°, 43°

Exercises 27 and 28: Find the value of x. Assume that all angle measures are given in degrees.

27. 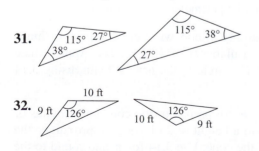 6x, 2x, 2x

28. 2x, x + 20, x

Exercises 29 and 30: For the triangle pair, state the property that shows the triangles are congruent.

29. 6 in. ... 6 in.

30.

Exercises 31 and 32: Determine whether the triangles in the given pair are congruent. If so, state the property that shows the triangles are congruent.

31. 115° 27° / 38° ... 115° 38° / 27°

32. 10 ft / 9 ft 126° ... 126° 10 ft 9 ft

SECTION 10.3

Exercises 33–36: Determine the name of the polygon.

33.

34.

35.

36.

Exercises 37 and 38: Find the sum of the angle measures for the given regular polygon.

37.

38.

Exercises 39 and 40: For the given regular polygon, find the measure of one angle.

39.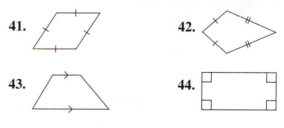

40.

Exercises 41–44: Name the quadrilateral using every classification that applies.

41.

42.

43.

44.

Exercises 45 and 46: Find the radius of the given circle.

45. 28 m

46. $\frac{8}{9}$ ft

Exercises 47 and 48: Find the diameter of the given circle.

47. 5.3 in.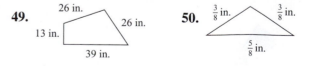

48. 0.5 cm

SECTION 10.4

Exercises 49 and 50: Find the perimeter of the polygon.

49. 26 in. / 26 in. / 13 in. / 39 in.

50. $\frac{3}{8}$ in. / $\frac{3}{8}$ in. / $\frac{5}{8}$ in.

Exercises 51 and 52: Find the perimeter of the given regular polygon.

51.  $\frac{4}{5}$ m

52. 7.5 in.

Exercises 53 and 54: Do the following.

(a) *Find the exact circumference of the circle.*
(b) *Approximate the circumference, using 3.14 as an approximation for π.*

53. 5 ft

54. 6.5 m

Exercises 55 and 56: Do the following.

(a) *Find the exact circumference of the circle.*
(b) *Approximate the circumference, using $\frac{22}{7}$ as an approximation for π.*

55. $\frac{3}{2}$ km

56. $\frac{14}{11}$ cm

57. Find the perimeter of the polygon.

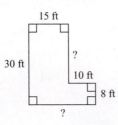
15 ft
30 ft
?
10 ft
8 ft
?

58. Find the perimeter of the composite figure. Use 3.14 as an approximation for π.

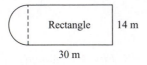
Rectangle   14 m
30 m

SECTION 10.5

Exercises 59–62: Find the area of the given polygon.

59. 4.6 m
5.5 m

60. 4 ft
5 ft
11 ft

61. 7 ft
14 ft

62. 6 cm
2.5 cm

Exercises 63 and 64: Find the approximate area of the circle using 3.14 as an approximation for π. Round your answers to the nearest tenth.

63. 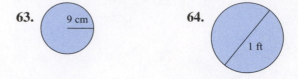 9 cm

64. 1 ft

Exercises 65–68: Find the volume of the geometric solid. If needed, use 3.14 as an approximation for π and round answers to the nearest tenth.

65. Cube: $s = 5$ ft

66. Sphere: $r = 9$ cm

67. Circular cylinder: $r = 2$ in., $h = 4$ in.

68. Cone: $r = 7$ m, $h = 5$ m

Exercises 69–72: Find the surface area of the geometric solid. If needed, use 3.14 as an approximation for π and round answers to the nearest tenth.

69. Square-based pyramid: $s = 9$ in., $l = 7$ in.

70. Cone: $r = 10$ m, $h = 3$ m

71. Circular cylinder: $r = 2$ cm, $h = 6$ cm

72. Rectangular prism: $l = 9$ ft, $w = 5$ ft, $h = 2$ ft

APPLICATIONS

73. *Hockey Rink* The center circle on a hockey rink has a diameter that measures 30 feet. Find the radius of the center circle on a hockey rink.

74. *Stop Sign* A standard stop sign has the shape of a regular octagon with one side measuring about 16.24 inches. Find the perimeter of a stop sign.

75. *Sumo Wrestling* Sumo wrestlers compete within a circle with a diameter of 4.55 meters. Approximate the area of the circle to the nearest tenth, using 3.14 for π.

76. *Traffic Cone* An orange traffic cone has a radius of 5 inches and a height of 24 inches. Approximate the volume of the cone. Use 3.14 for π and round to the nearest whole number.

CHAPTER 10   Test

1. Name the figure using the labels shown.

2. Classify the angle as acute, right, obtuse, or straight.

3. Find the measure of the complement of a 71° angle.

4. Find the measure of the supplement of a 118° angle.

5. If $m \parallel n$ and $m\angle c = 72°$ in the following figure, find the measures of $\angle a$, $\angle f$, and $\angle h$.

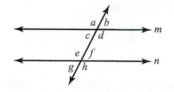

6. Classify the given triangle as acute, obtuse, or right.

7. Classify the given triangle as scalene, isosceles, or equilateral.

8. Find the measure of $\angle x$ in the triangle.

9. Determine whether the triangles in the given pair are congruent. If so, state the property that shows the triangles are congruent.

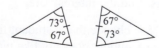

10. Determine the name of the polygon.

11. For the regular polygon shown, do the following.

(a) Find the sum of the angle measures.
(b) Find the measure of one angle.

12. Name the given quadrilateral using every classification that applies.

13. Find the diameter of the given circle.

14. Find the perimeter of the polygon.

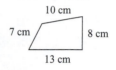

15. For the circle shown, do the following.
(a) Find the exact circumference of the circle.
(b) Approximate the circumference, using 3.14 for π.

16. Find the perimeter of the polygon.

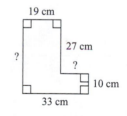

17. Find the area of each polygon.

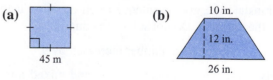

18. Find the approximate area of the circle, using 3.14 for π. Round your answer to the nearest tenth.

19. Find the volume of the geometric solid. If needed, use 3.14 as an approximation for π and round answers to the nearest tenth.
(a) Sphere: $r = 2$ in.
(b) Rectangular prism: $l = 14$ ft, $w = 6$ ft, $h = 4$ ft

20. Find the surface area of the geometric solid. If needed, use 3.14 as an approximation for π and round answers to the nearest tenth.
(a) Cube: $s = 16$ cm
(b) Circular cylinder: $r = 3$ ft, $h = 8$ ft

CHAPTERS 1–10   Cumulative Review Exercises

1. Write 18,020 in expanded form.

2. Write $9 \cdot 9 \cdot 9 \cdot 9$ using exponential notation.

3. Round 73,987 to the nearest thousand.

4. Evaluate $4 \cdot 8 - 50 \div 10$.

5. Graph the integers $-4, 0,$ and 3 on a number line.

6. Evaluate $6 - 8 + (-3) - (-9)$.

7. Evaluate $25 - 3^2 \div (4 - 7)$.

8. Complete the table and solve $5x - 2 = 3$.

x	-2	-1	0	1	2
$5x - 2$					

9. Is -2 a solution to the equation $-4x + 3 = 11$?

10. Solve the equation $-3y + (-8) = y$.

11. Solve the linear equation $-2x - 3 = -5$ visually.

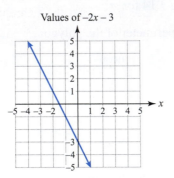

Values of $-2x - 3$

12. Translate the given sentence to an equation using x as the variable. Do not solve the equation.

    Four equals a number increased by 9.

13. Give both the improper fraction and mixed number represented by the shading.

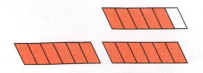

14. Determine whether the given whole number is prime, composite or neither.
    (a) 39   (b) 1
    (c) 17   (d) 75

Exercises 15 and 16: Perform the indicated arithmetic.

15. $-\dfrac{8}{7} \div \dfrac{4}{35}$

16. $-\dfrac{5}{6} + \dfrac{1}{3} \cdot \dfrac{2}{5}$

17. Approximate the length of the unknown side of the right triangle to 1 decimal place.

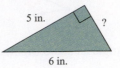

5 in.   ?

6 in.

18. Round 841.5706 to the nearest hundredth.

19. Solve the equation $2.1y = 6 - 0.4y$.

20. For the numbers in the given list, identify any of the following types of numbers:
    (a) natural numbers   (b) whole numbers
    (c) integers   (d) rational numbers
    (e) irrational numbers

    $$8, -\frac{1}{5}, \sqrt{9}, 0, \pi$$

21. The triangles are similar. Find the value of x.

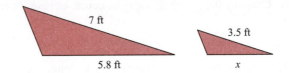

7 ft   3.5 ft

5.8 ft x

22. A 16-ounce box of crackers sells for $3.68. Find the unit price of the crackers.

23. Convert $7\frac{1}{2}$ gallons to pints.

24. Convert 7 inches per second to feet per hour.

25. Translate the following percent problem to a basic percent equation. Do not solve the equation.

    What percent of 80 is 16?

26. If an investment of $1200 earns $144 simple interest after 3 years, find the interest rate.

27. There is a 78% chance that a student likes pie. What is the chance a student does not like pie?

28. 14.5% of what number is 116?

29. Simplify each expression, writing answers with positive exponents. Assume that all variables represent nonzero numbers.

    (a) $(-5w^4)^2$ (b) $\dfrac{45y^8}{9y^5}$

30. Write 617,000 in scientific notation.

31. Add $(3x^2 - 9x + 1) + (2x^2 - x)$.

32. Multiply.
(a) $2a^3(7a^2 - 3a + 5)$
(b) $(3m - 2)(7m - 8)$

33. For each point in the graph, write the coordinates of the point and name its quadrant. If any point is on an axis, name the axis.

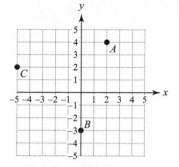

34. Complete the table of solutions for $y = -3x + 7$.

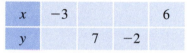

x	-3		6
y		7	-2

35. Use the following graph to find a graphical solution to the equation $5x + 10 = 40$.

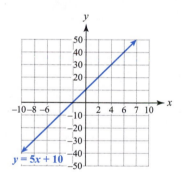

$y = 5x + 10$

36. Solve the equation $-x + 3 = 2$ graphically.

37. If $m\|n$ and $m\angle g = 71°$ in the following figure, find the measures of $\angle a$, $\angle b$, and $\angle h$.

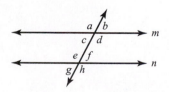

38. Find the measure of $\angle x$ in the triangle.

39. For the regular polygon shown, do the following.

(a) Find the sum of the angle measures.
(b) Find the measure of one angle.

40. Find the surface area of the sphere with radius 6 cm. Use 3.14 as an approximation for π and round your answer to the nearest tenth.

41. *Gardening* A garden is built in the shape of a square with an area of 330 square feet. Estimate the length of one side of the garden to the nearest foot.

42. *Finding Age* In 17 years, a person's age will be 6 years less than double his current age. Write an equation whose solution gives the person's current age. Use x for your variable. Do not solve the equation.

43. *Converting Temperature* The formula

$$C = \frac{(F - 32)}{1.8}$$

gives the relationship between C degrees Celsius and F degrees Fahrenheit. Find the Fahrenheit temperature that is equivalent to 26.5°C.

44. *Smart Phone* If the purchase price of a smart phone is $178 and the sales tax rate is 6.5%, find the sales tax and the total price.

45. *Driving Distance* The following graph shows the distance of a driver from home.
(a) How far is the driver from home, initially?
(b) How many hours pass before the driver is home?

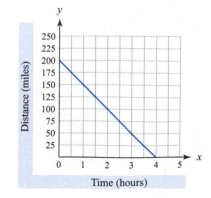

46. *Height of a Baseball* The polynomial expression

$$-16x^2 + 88x + 2$$

gives the height of a baseball in feet x seconds after it is hit. Find the height of the baseball 2 seconds and 4 seconds after it is hit.

Appendix: Using A Calculator

Overview of the Appendix

This appendix offers instruction for performing arithmetic on common scientific calculators as well as the TI-83, TI-83 Plus, and TI-84 Plus graphing calculators. For more detailed instructions on the use of a specific calculator, students are advised to consult the user manual provided by the manufacturer.

Operations on Whole Numbers

ADDING WHOLE NUMBERS To find a sum such as $328 + 4169$ on a calculator, use the following keystrokes.

Scientific Calculator: $\boxed{3}\,\boxed{2}\,\boxed{8}\,\boxed{+}\,\boxed{4}\,\boxed{1}\,\boxed{6}\,\boxed{9}\,\boxed{=}$

The result shown in the display will be $\boxed{\qquad 4497}$.

Graphing Calculator: $\boxed{3}\,\boxed{2}\,\boxed{8}\,\boxed{+}\,\boxed{4}\,\boxed{1}\,\boxed{6}\,\boxed{9}\,\boxed{\text{Enter}}$

The result is shown in the first and second lines of Figure A.1.

```
328+4169
           4497
4039-372
           3667
```

Figure A.1

SUBTRACTING WHOLE NUMBERS A difference such as $4039 - 372$ can be found on a calculator by using the following keystrokes.

Scientific Calculator: $\boxed{4}\,\boxed{0}\,\boxed{3}\,\boxed{9}\,\boxed{-}\,\boxed{3}\,\boxed{7}\,\boxed{2}\,\boxed{=}$

The result shown in the display will be $\boxed{\qquad 3667}$.

Graphing Calculator: $\boxed{4}\,\boxed{0}\,\boxed{3}\,\boxed{9}\,\boxed{-}\,\boxed{3}\,\boxed{7}\,\boxed{2}\,\boxed{\text{Enter}}$

The result is shown in the third and fourth lines of Figure A.1.

MULTIPLYING WHOLE NUMBERS The following keystrokes can be used to find a product such as 386×73 on a calculator.

Scientific Calculator: $\boxed{3}\,\boxed{8}\,\boxed{6}\,\boxed{\times}\,\boxed{7}\,\boxed{3}\,\boxed{=}$

The result shown in the display will be $\boxed{\qquad 28178}$.

Graphing Calculator: $\boxed{3}\,\boxed{8}\,\boxed{6}\,\boxed{\times}\,\boxed{7}\,\boxed{3}\,\boxed{\text{Enter}}$

The result is shown in the first and second lines of Figure A.2.

```
386*73
          28178
1248/8
            156
```

Figure A.2

DIVIDING WHOLE NUMBERS To find a quotient such as $1248 \div 8$ on a calculator, use the following keystrokes.

Scientific Calculator: $\boxed{1}\,\boxed{2}\,\boxed{4}\,\boxed{8}\,\boxed{\div}\,\boxed{8}\,\boxed{=}$

The result shown in the display will be $\boxed{\qquad 156}$.

Graphing Calculator: $\boxed{1}\,\boxed{2}\,\boxed{4}\,\boxed{8}\,\boxed{\div}\,\boxed{8}\,\boxed{\text{Enter}}$

The result is shown in the third and fourth lines of Figure A.2.

EVALUATING EXPONENTS On a scientific calculator, an exponent is entered using the $\boxed{y^x}$ key. On a graphing calculator, an exponent is entered using the $\boxed{\wedge}$ key. To evaluate an exponential expression such as 5^3 on a calculator, use the following keystrokes.

Scientific Calculator: $\boxed{5}$ $\boxed{y^x}$ $\boxed{3}$ $\boxed{=}$

The result shown in the display will be $\boxed{\qquad 125}$.

Graphing Calculator: $\boxed{5}$ $\boxed{\wedge}$ $\boxed{3}$ $\boxed{\text{Enter}}$

The result is shown in the first and second lines of Figure A.3.

```
5^3
              125
100/(8−3)
               20
```

Figure A.3

USING PARENTHESES An expression such as $100 \div (8 - 3)$ can be evaluated on a calculator by using the following keystrokes.

Scientific Calculator: $\boxed{1}$ $\boxed{0}$ $\boxed{0}$ $\boxed{\div}$ $\boxed{(}$ $\boxed{8}$ $\boxed{-}$ $\boxed{3}$ $\boxed{)}$ $\boxed{=}$

The result shown in the display will be $\boxed{\qquad 20}$.

Graphing Calculator: $\boxed{1}$ $\boxed{0}$ $\boxed{0}$ $\boxed{\div}$ $\boxed{(}$ $\boxed{8}$ $\boxed{-}$ $\boxed{3}$ $\boxed{)}$ $\boxed{\text{Enter}}$

The result is shown in the third and fourth lines of Figure A.3.

Operations on Integers

ENTERING NEGATIVE NUMBERS On a scientific calculator, a negative integer is entered using the $\boxed{+/-}$ key. On a graphing calculator, a negative integer is entered using the $\boxed{(-)}$ key. For example, -7 is entered in a calculator as follows.

Scientific Calculator: $\boxed{7}$ $\boxed{+/-}$

The result shown in the display will be $\boxed{\qquad -7}$.

```
−7
```

Figure A.4

Graphing Calculator: $\boxed{(-)}$ $\boxed{7}$

The result is shown in Figure A.4.

FINDING ABSOLUTE VALUE On some scientific calculators, an absolute value can be found by pressing the $\boxed{\text{abs}}$ key or a key with "abs" displayed directly above it. On a graphing calculator, absolute value is found under the "NUM" menu after pressing the $\boxed{\text{MATH}}$ key. For example, to find $|-3|$ with a calculator, use the following keystrokes.

Scientific Calculator: $\boxed{\text{abs}}$ $\boxed{3}$ $\boxed{+/-}$ $\boxed{=}$ or $\boxed{\text{2nd}}$ $\boxed{\text{[abs]}}$ $\boxed{3}$ $\boxed{+/-}$ $\boxed{=}$

The result shown in the display will be $\boxed{\qquad 3}$.

```
abs(−3)
                3
```

Figure A.5

Graphing Calculator: $\boxed{\text{MATH}}$ $\boxed{\blacktriangleright}$ $\boxed{1}$ $\boxed{(-)}$ $\boxed{3}$ $\boxed{)}$ $\boxed{\text{Enter}}$

The result is shown in Figure A.5.

ADDING INTEGERS The following calculator keystrokes can be used to find a sum such as $14 + (-9)$.

Scientific Calculator: $\boxed{1}$ $\boxed{4}$ $\boxed{+}$ $\boxed{9}$ $\boxed{+/-}$ $\boxed{=}$

The result shown in the display will be $\boxed{\qquad 5}$.

```
14+(−9)
                5
−4−(−11)
                7
5*(−19)
              −95
```

Figure A.6

Graphing Calculator: $\boxed{1}$ $\boxed{4}$ $\boxed{+}$ $\boxed{(}$ $\boxed{(-)}$ $\boxed{9}$ $\boxed{)}$ $\boxed{\text{Enter}}$

The result is shown in the first and second lines of Figure A.6.

SUBTRACTING INTEGERS To find a difference such as $-4 - (-11)$ on a calculator, use the following keystrokes.

Scientific Calculator: [4] [+/−] [−] [1] [1] [+/−] [=]

The result shown in the display will be [7].

Graphing Calculator: [(−)] [4] [−] [(] [(−)] [1] [1] [)] [Enter]

The result is shown in the third and fourth lines of Figure A.6.

MULTIPLYING INTEGERS A product such as $5 \cdot (-19)$ can be found on a calculator by using the following keystrokes.

Scientific Calculator: [5] [×] [1] [9] [+/−] [=]

The result shown in the display will be [−95].

Graphing Calculator: [5] [×] [(] [(−)] [1] [9] [)] [Enter]

The result is shown in the fifth and sixth lines of Figure A.6.

EVALUATING EXPONENTS ON INTEGERS When using a calculator to evaluate exponential expressions such as -2^4 and $(-2)^4$, it is important to apply the order of operations agreement correctly. To evaluate the expression -2^4 on a calculator, use the following keystrokes.

Scientific Calculator: [2] [y^x] [4] [=] [+/−]

The result shown in the display will be [−16].

Graphing Calculator: [(−)] [2] [∧] [4] [Enter]

The result is shown in the first and second lines of Figure A.7. To evaluate the expression $(-2)^4$ on a calculator, use the following keystrokes.

Scientific Calculator: [2] [+/−] [y^x] [4] [=]

The result shown in the display will be [16].

Graphing Calculator: [(] [(−)] [2] [)] [∧] [4] [Enter]

The result is shown in the third and fourth lines of Figure A.7.

DIVIDING INTEGERS The following keystrokes can be used to find a quotient such as $-144 \div 6$ on a calculator.

Scientific Calculator: [1] [4] [4] [+/−] [÷] [6] [=]

The result shown in the display will be [−24].

Graphing Calculator: [(−)] [1] [4] [4] [÷] [6] [Enter]

The result is shown in the first and second lines of Figure A.8.

FINDING A SQUARE ROOT To find a square root such as $\sqrt{196}$ on a calculator, use the following keystrokes.

Scientific Calculator: [1] [9] [6] [√]

The result shown in the display will be [14].

Graphing Calculator: [2nd] [x^2 [√]] [1] [9] [6] [)] [Enter]

The result is shown in the third and fourth lines of Figure A.8.

```
14+(-9)
                    5
-4-(-11)
                    7
5*(-19)
                   -95
```

Figure A.6 (Repeated)

```
-2^4
                  -16
(-2)^4
                   16
```

Figure A.7

```
-144/6
                  -24
√(196)
                   14
```

Figure A.8

Operations on Fractions

SIMPLIFYING FRACTIONS On some scientific calculators, a fraction can be simplified using the $\boxed{a^{b/c}}$ key. Such calculators often display the symbol ($\lrcorner$) to indicate a fraction bar. On a graphing calculator, a fraction feature is found under the "MATH" menu after pressing the $\boxed{MATH}$ key. For example, the fraction $\frac{84}{120}$ can be simplified to $\frac{7}{10}$ with a calculator by using the following keystrokes.

Scientific Calculator: $\boxed{8}\ \boxed{4}\ \boxed{a^{b/c}}\ \boxed{1}\ \boxed{2}\ \boxed{0}\ \boxed{=}$

The result shown in the display will be $\boxed{\qquad 7\lrcorner10\qquad}$.

Graphing Calculator: $\boxed{8}\ \boxed{4}\ \boxed{\div}\ \boxed{1}\ \boxed{2}\ \boxed{0}\ \boxed{MATH}\ \boxed{1}\ \boxed{Enter}$

The result is shown in the first and second lines of Figure A.9.

```
84/120►Frac
             7/10
(-1/6)*(5/8)►Fra
c
            -5/48
```

Figure A.9

MULTIPLYING FRACTIONS A product such as $-\frac{1}{6} \cdot \frac{5}{8}$ can be found on a calculator by using the following keystrokes.

Scientific Calculator: $\boxed{1}\ \boxed{a^{b/c}}\ \boxed{6}\ \boxed{+/-}\ \boxed{\times}\ \boxed{5}\ \boxed{a^{b/c}}\ \boxed{8}\ \boxed{=}$

The result shown in the display will be $\boxed{\qquad -5\lrcorner48\qquad}$.

Graphing Calculator: $\boxed{(}\ \boxed{(-)}\ \boxed{1}\ \boxed{\div}\ \boxed{6}\ \boxed{)}\ \boxed{\times}\ \boxed{(}\ \boxed{5}\ \boxed{\div}\ \boxed{8}\ \boxed{)}\ \boxed{MATH}\ \boxed{1}\ \boxed{Enter}$

The result is shown in the third, fourth, and fifth lines of Figure A.9.

EVALUATING EXPONENTS ON FRACTIONS On some scientific calculators, the keys $\boxed{2nd}$ and $\boxed{F \blacktriangleleft \blacktriangleright D}$ are used to express an answer in fraction form. The following keystrokes can be used when evaluating an exponential expression such as $\left(-\frac{2}{3}\right)^3$ with a calculator.

Scientific Calculator: $\boxed{2}\ \boxed{a^{b/c}}\ \boxed{3}\ \boxed{+/-}\ \boxed{y^x}\ \boxed{3}\ \boxed{=}\ \boxed{2nd}\ \boxed{F \blacktriangleleft \blacktriangleright D}$

The result shown in the display will be $\boxed{\qquad -8\lrcorner27\qquad}$.

Graphing Calculator: $\boxed{(}\ \boxed{(-)}\ \boxed{2}\ \boxed{\div}\ \boxed{3}\ \boxed{)}\ \boxed{\wedge}\ \boxed{3}\ \boxed{MATH}\ \boxed{1}\ \boxed{Enter}$

The result is shown in the first and second lines of Figure A.10.

```
(-2/3)^3►Frac
            -8/27
√(45/80)►Frac
              3/4
```

Figure A.10

FINDING A SQUARE ROOT OF A FRACTION To find a square root such as $\sqrt{\frac{45}{80}}$ on a calculator, use the following keystrokes.

Scientific Calculator: $\boxed{4}\ \boxed{5}\ \boxed{a^{b/c}}\ \boxed{8}\ \boxed{0}\ \boxed{\sqrt{}}\ \boxed{2nd}\ \boxed{F \blacktriangleleft \blacktriangleright D}$

The result shown in the display will be $\boxed{\qquad 3\lrcorner4\qquad}$.

Graphing Calculator: $\boxed{2nd}\ \boxed{x^2\ [\sqrt{}]}\ \boxed{4}\ \boxed{5}\ \boxed{\div}\ \boxed{8}\ \boxed{0}\ \boxed{)}\ \boxed{MATH}\ \boxed{1}\ \boxed{Enter}$

The result is shown in the third and fourth lines of Figure A.10.

DIVIDING FRACTIONS A quotient such as $-\frac{5}{6} \div \frac{5}{4}$ can be found on a calculator by using the following keystrokes.

Scientific Calculator: $\boxed{5}\ \boxed{a^{b/c}}\ \boxed{6}\ \boxed{+/-}\ \boxed{\div}\ \boxed{5}\ \boxed{a^{b/c}}\ \boxed{4}\ \boxed{=}$

The result shown in the display will be $\boxed{\qquad -2\lrcorner3\qquad}$.

Graphing Calculator: $\boxed{(}\ \boxed{(-)}\ \boxed{5}\ \boxed{\div}\ \boxed{6}\ \boxed{)}\ \boxed{\div}\ \boxed{(}\ \boxed{5}\ \boxed{\div}\ \boxed{4}\ \boxed{)}\ \boxed{MATH}\ \boxed{1}\ \boxed{Enter}$

The result is shown in the first, second, and third lines of Figure A.11.

```
(-5/6)/(5/4)►Fra
c
            -2/3
(5/6)+(1/8)-(2/3
)►Frac
            7/24
```

Figure A.11

```
(-5/6)/(5/4)►Fra
c
            -2/3
(5/6)+(1/8)-(2/3
)►Frac
            7/24
```

Figure A.11 (Repeated)

ADDING OR SUBTRACTING FRACTIONS To evaluate expressions involving sums and differences such as $\frac{5}{6} + \frac{1}{8} - \frac{2}{3}$ with a calculator, use the following keystrokes.

Scientific Calculator: [5] [abᶜ] [6] [+] [1] [abᶜ] [8] [−] [2] [abᶜ] [3] [=]

The result shown in the display will be [7⌐24].

Graphing Calculator: [(] [5] [÷] [6] [)] [+] [(] [1] [÷] [8] [)] [−] [(] [2] [÷] [3] [)]
[MATH] [1] [Enter]

The result is shown in the fourth, fifth, and sixth lines of Figure A.11.

Operations on Decimals

ADDING OR SUBTRACTING DECIMALS To evaluate expressions involving sums and differences such as $4.3 + 6.7 - (-1.2)$ with a calculator, use the following keystrokes.

Scientific Calculator: [4] [.] [3] [+] [6] [.] [7] [−] [1] [.] [2] [+/−] [=]

The result shown in the display will be [12.2].

Graphing Calculator: [4] [.] [3] [+] [6] [.] [7] [−] [(] [(−)] [1] [.] [2] [)] [Enter]

The result is shown in the first and second lines of Figure A.12.

```
4.3+6.7-(-1.2)
            12.2
(-8.2)(-0.51)
            4.182
```

Figure A.12

MULTIPLYING DECIMALS The following keystrokes can be used to find a product such as $(-8.2)(-0.51)$ on a calculator.

Scientific Calculator: [8] [.] [2] [+/−] [×] [0] [.] [5] [1] [+/−] [=]

The result shown in the display will be [4.182].

Graphing Calculator: [(] [(−)] [8] [.] [2] [)] [(] [(−)] [0] [.] [5] [1] [)] [Enter]

The result is shown in the third and fourth lines of Figure A.12.

DIVIDING DECIMALS A quotient such as $3.72 \div 5$ can be found on a calculator by using the following keystrokes.

Scientific Calculator: [3] [.] [7] [2] [÷] [5] [=]

The result shown in the display will be [0.744].

Graphing Calculator: [3] [.] [7] [2] [÷] [5] [Enter]

The result is shown in the first and second lines of Figure A.13.

ESTIMATING A SQUARE ROOT To estimate a square root such as $\sqrt{11}$ on a calculator, use the following keystrokes.

Scientific Calculator: [1] [1] [√]

The result shown in the display will be [3.31662479].

```
3.72/5
            .744
√(11)
     3.31662479
```

Figure A.13

Graphing Calculator: [2nd] [x² [√]] [1] [1] [)] [Enter]

The result is shown in the third and fourth lines of Figure A.13.

Financial Math

USING THE SIMPLE INTEREST FORMULA The formula $I = Prt$ can be used to find simple interest. For example, when $P = \$9520$, $r = 3\%$, and $t = 5$ years, the following keystrokes can be used to find the simple interest.

Scientific Calculator: ⑨ ⑤ ② ⓪ ✕ ⓪ ⋅ ⓪ ③ ✕ ⑤ ═

The result shown in the display will be ⌑ 1428 .

Graphing Calculator: ⑨ ⑤ ② ⓪ ✕ ⓪ ⋅ ⓪ ③ ✕ ⑤ Enter

The result is shown in the first and second lines of Figure A.14.

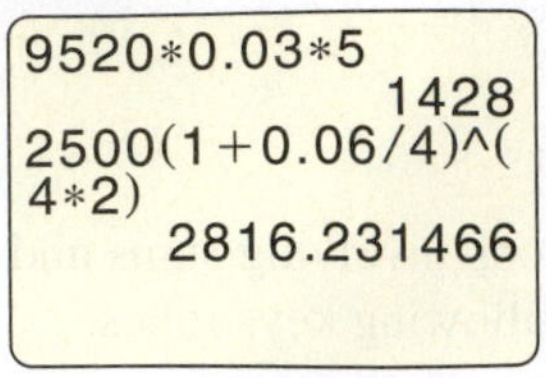

USING THE COMPOUND INTEREST FORMULA The final amount in an account that pays compound interest is given by the formula

$$A = P\left(1 + \frac{r}{n}\right)^{nt}.$$

```
9520*0.03*5
           1428
2500(1+0.06/4)^(
4*2)
      2816.231466
```

Figure A.14

It is important to use parentheses correctly when working with this formula. For example, when $P = \$2500$, $r = 6\%$, $n = 4$, and $t = 2$ years, the following keystrokes can be used to find the final amount in the account.

Scientific Calculator: ② ⑤ ⓪ ⓪ ✕ ⟮ ① ➕ ⓪ ⋅ ⓪ ⑥ ➗ ④ ⟯ y^x
⟮ ④ ✕ ② ⟯ ═

The result shown in the display will be ⌑ 2816.231467 .

Graphing Calculator: ② ⑤ ⓪ ⓪ ⟮ ① ➕ ⓪ ⋅ ⓪ ⑥ ➗ ④ ⟯ ⌃
⟮ ④ ✕ ② ⟯ Enter

The result is shown in the third, fourth, and fifth lines of Figure A.14.

Scientific Notation

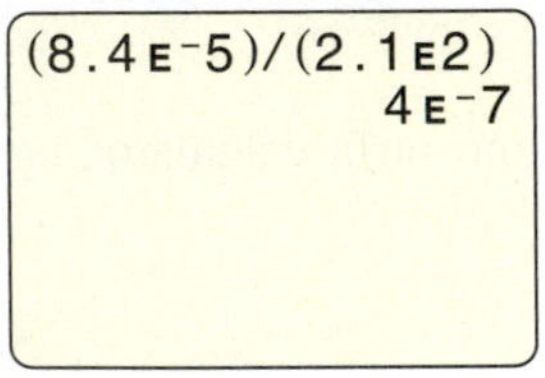

```
(8.4E-5)/(2.1E2)
            4E-7
```

Figure A.15

USING SCIENTIFIC NOTATION On some scientific calculators, scientific notation can be entered into the calculator by pressing the EE key. On a graphing calculator, the EE feature is accessed by pressing the 2nd and , keys. For example, to find a quotient such as $(8.4 \times 10^{-5}) \div (2.1 \times 10^2)$ with a calculator, use the following keystrokes.

Scientific Calculator: ⑧ ⋅ ④ EE ⑤ +/– ➗ ② ⋅ ① EE ② ═

The result shown in the display will be ⌑ 0.0000004 .

Graphing Calculator: ⟮ ⑧ ⋅ ④ 2nd ▸[EE] (–) ⑤ ⟯ ➗ ⟮ ② ⋅ ①
2nd ▸[EE] ② ⟯ Enter

The result is shown (in scientific notation) in Figure A.15.

Answers to Selected Exercises

1 Whole Numbers

SECTION 1.1 (pp. 11–14)

1. counting **2.** 0 **3.** periods **4.** standard **5.** place value
6. trillions **7.** word **8.** expanded **9.** graph **10.** bar graph
and line graph **11.** thousands **13.** ten-millions
15. hundreds **17.** ones **19.** hundred-millions **21.** 7
23. 8 **25.** 3 **27.** 2 **29.** 4 **31.** Four hundred seventy-
two thousand, five hundred **33.** Ninety-three thousand,
two hundred six **35.** One thousand, six hundred fifty-one
37. 2055 **39.** 599,616,423
41. 2,000,000 + 500,000 + 10,000 + 30 + 6
43. 600 + 20 + 9 **45.** 600,000 + 3000 + 100 + 30 + 8
47. 39,410,000 **49.** 83,000,600,012 **51.** 342,563
53. 7,905,377

55. [number line 0 to 5]; **(a)** < **(b)** > **(c)** >

57. [number line 0 to 25, marks at 4, 8, 11, 22]; **(a)** > **(b)** < **(c)** <

59. [number line 0 to 100, marks at 10, 24, 64, 86]; **(a)** > **(b)** > **(c)** >

61. > **63.** < **65.** > **67.** < **69.** < **71.** India
73. South Korea **75.** Mackenzie **77.** 2575 km
79. 2007 **81.** $423,000,000 **83.** Minimum wage
increased. **85.** 1990 to 2010 **87.** 2006 **89.** Males
91. 2006 **93.** Public **95.** 1000 + 100 + 20 + 4
97. 34,359,738,378 **99.** One hundred thirty-five billion
101. The truck driver

SECTION 1.2 (pp. 25–28)

1. addends **2.** sum **3.** Yes **4.** commutative
5. associative **6.** identity **7.** addition
8. minuend; subtrahend **9.** difference **10.** No
11. identity **12.** subtraction **13.** solution **14.** solutions
15. 28 **17.** 599 **19.** 885 **21.** 7868 **23.** 7872
25. 27,064 **27.** 81,827 **29.** 1,064,159 **31.** 8967
33. 88,051 **35.** 66 **37.** 70 **39.** 13 **41.** 431
43. 1422 **45.** 151 **47.** 2316 **49.** 32,905 **51.** 31,433
53. 101,027 **55.** 2618 **57.** 17,685 **59.** 10,876
61. 22 + 57; $79 **63.** 793 − 54; 739 photos **65.** 62 − 19;
43 eggs **67.** 1200 + 300; 1500 patients **69.** 645 − 3;
642 DVDs **71.** 39 + 71; 110 Web pages **73.** 3 **75.** 19
77. 24 **79.** 151 **81.** 26 **83.** 532 **85.** 56 ft **87.** 35 cm
89. 54 mi **91.** 78 in. **93.** 43 in. **95.** 207 **97.** 6
99. 17 **101.** 1075 pounds **103.** 2008 **105.** 6,000,000

CHECKING BASIC CONCEPTS 1.1 & 1.2 (p. 28)

1. (a) ten-thousands **(b)** hundreds
2. 70,000 + 4000 + 200 + 90 + 3 **3.** 48,239,610
4. [number line 0 to 5]
5. (a) > **(b)** < **6. (a)** 4317 **(b)** 210,530
7. (a) 8659 **(b)** 600,884
8. (a) 97 − 45; 52 **(b)** 106 + 73; 179
9. (a) 5 **(b)** 29 **10.** 108 cm

SECTION 1.3 (pp. 41–43)

1. addition **2.** factors **3.** product **4.** commutative
5. associative **6.** identity **7.** zero **8.** distributive
9. multiplication **10.** subtraction **11.** dividend; divisor
12. quotient **13.** identity **14.** 0; undefined
15. long division **16.** division **17.** 1 square unit **18.** area
19. $5 \cdot 6 + 5 \cdot 9$ **21.** $4 \cdot 8 - 4 \cdot 1$ **23.** $6 \cdot 3 - 2 \cdot 3$
25. 7 **27.** 0 **29.** 54 **31.** 336 **33.** 1812 **35.** 1704
37. 2408 **39.** 51,625 **41.** 283,504 **43.** 117,066
45. 21,000 **47.** 680,000 **49.** 1,500,000 **51.** 9 **53.** 1
55. 11 **57.** Undefined **59.** 0 **61.** 1 **63.** 12 **65.** 961 r2
67. 80 r7 **69.** 200 **71.** Undefined **73.** 126 r67 **75.** 65
77. 62 r735 **79.** 14 · 3; 42 square feet **81.** 5 · 15; $75
83. 126 ÷ 7; 18 miles per gallon **85.** 75 ÷ 15; 5 days
87. 3 **89.** 8 **91.** 65 **93.** 8 **95.** 40 square inches
97. 289 square miles **99.** 8100 square feet **101.** 3
103. 5000 **105.** 0 calories **107.** 240,000 pixels **109.** 250
111. (a) 120 feet **(b)** 360 feet **113.** 13 **115.** 9; $3

SECTION 1.4 (pp. 53–55)

1. arithmetic **2.** algebra **3.** exponential notation
4. 4; 7 **5.** 2 **6.** 3 **7.** 9 **8.** 10^7 **9.** variable
10. algebraic expression **11.** equation **12.** formula
13. evaluate **14.** variable **15.** expression **16.** equation
17. 8^3 **19.** 2^5 **21.** $2^3 \cdot 5^2$ **23.** $5^3 \cdot 7^3$ **25.** 7^2 **27.** 4^9
29. 2^3 **31.** 3^5 **33.** 81 **35.** 32 **37.** 256 **39.** 216
41. 1000 **43.** 8,000,000 **45.** 3000 **47.** $A - 5$
49. $6 \cdot G$ **51.** $S + 10$ **53.** $P \div 2$ **55.** 20 **57.** 0 **59.** 65
61. 46 **63.** 125 **65.** 10 **67.** 4 **69.** 64 **71.** 16 inches
73. 1500 square inches **75.** 69 yards **77.** $6M$, where M
represents monthly income **79.** $p \div 3$, where p represents
the number of pizza slices **81.** $a + h$, where a represents
age and h represents heart rate **83.** Yes **85.** No **87.** No
89. 8 **91.** 3 **93.** 12 **95.** 8 **97.** 10 feet **99.** 40 inches
101. (a) $C = 6p$ **(b)** $C = 8d$ **(c)** painting: 1440; clearing:
1920 **103.** 15,756 **105.** 3; $1 **107.** $20 **109.** 3 **111.** 13

CHECKING BASIC CONCEPTS 1.3 & 1.4 (p. 55)

1. (a) 286 **(b)** 0 **(c)** 207 **(d)** 64,668 **2. (a)** 35
(b) 140 r5 **(c)** Undefined **(d)** 171 **3. (a)** 12 **(b)** 72
4. 120,000 **5. (a)** 7^4 **(b)** $2^2 \cdot 8^2$ **6. (a)** 40 **(b)** 300,000
7. (a) 13 **(b)** 60 **8.** Yes **9. (a)** 8 **(b)** 12 **10.** 44 inches
11. 7 **12.** 800 square feet

SECTION 1.5 (pp. 63–64)

1. rounding **2.** right **3.** estimation **4.** approximation
5. highest **6.** perfect **7.** square root **8.** radical; radicand
9. 6 **10.** is approximately equal to **11.** 700 **13.** 59,000
15. 80 **17.** 900 **19.** 380,000 **21.** 54,200 **23.** 780
25. 30,000,000 **27.** 3000 **29.** 90,000,000 **31.** 70
33. 20,000 **35.** 1500 **37.** 400 **39.** 4,900,000 **41.** 20
43. 5 **45.** 11 **47.** 19 **49.** 25 **51.** 3 **53.** 13 **55.** 16
57. 28 **59.** 2000 hours **61.** 180 years **63.** 9000
65. 12,000 feet **67.** 900 **69.** 41 feet **71.** 220 million
73. 2003

SECTION 1.6 (pp. 69–71)

1. order of operations **2.** Parentheses; radical **3.** before
4. subtraction **5.** the quantity **6.** addition **7.** 24 **9.** 16
11. 21 **13.** 14 **15.** 73 **17.** 25 **19.** 30 **21.** 19 **23.** 180
25. 256 **27.** 86 **29.** 0 **31.** 4 **33.** 10 **35.** 9 **37.** 67
39. 8 **41.** 12 **43.** 22 **45.** $(14 - 12) \cdot 5 - 10$
47. $(36 - 6^2) \div (5 - 1)$ **49.** $(32 \div 4^2 - 2) \cdot 9$ **51.** 8
53. 7 **55.** 41 **57.** 1 **59.** 21 **61.** 19 **63.** 5 + 12; 17
65. 21 − 9; 12 **67.** $7^2 + 4$; 53 **69.** $(6 + 5) \cdot 9$; 99
71. $(6 + 2) \cdot 5$; 40 **73.** $2^3 \cdot 3^2$; 72 **75.** $\sqrt{16} + 9$; 13
77. $7 \cdot 3 - 2$; 19 **79.** 177 bpm **81.** 8000 watts
83. 8 thousand per acre **85.** 40°C

CHECKING BASIC CONCEPTS 1.5 & 1.6 (p. 71)

1. 45,000 **2. (a)** 1200 **(b)** 4600 **3. (a)** 9 **(b)** 13
4. (a) 5 **(b)** 14 **5. (a)** 13 **(b)** 12 **6.** 2 **7.** 9
8. $(4 + 2) \cdot 3$; 18 **9.** 8400 **10.** 20 feet

SECTION 1.7 (pp. 77–79)

1. Expression **2.** Equation **3.** term **4.** coefficient
5. like **6.** unlike **7.** Expression **9.** Equation
11. Equation **13.** Expression **15.** Like **17.** Unlike
19. Like **21.** Unlike **23.** $11x$ **25.** $7yz$ **27.** Not possible
29. $16ab$ **31.** $7x + 9$ **33.** $11y + 5$ **35.** $5a + 4$
37. $8z - 14$ **39.** $3x + 2$ **41.** $5x + 9$ **43.** $3ab + 4y$
45. (a) $7x = 4x + 15$ **(b)** 5 checks in both equations.
47. (a) $2x + 11 = 3x + 6$ **(b)** 5 checks in both equations.
49. $6x - 2x = 36$, where x is the number of inches.
51. $14 = x - 9$, where x is her score. **53.** $4x = 28$, where
x is her shoe size. **55.** $x = 3x - 8$, where x is the score.
57. $53,000 **59.** 4 gallons **61.** 14,940 births per hour
63. 12 **65.** 4 **67.** $21 **69.** 3 inches

CHECKING BASIC CONCEPTS 1.7 (p. 79)

1. (a) Equation **(b)** Expression **2. (a)** Like **(b)** Unlike
3. (a) $5pq$ **(b)** Not possible **4. (a)** $2x + 1$ **(b)** $9x + 3$
5. $x - 7 = 23$, where x is his age. **6.** 4 feet

CHAPTER 1 REVIEW (pp. 84–86)

1. hundreds **2.** hundred-thousands **3.** 1 **4.** 0
5. Forty-eight thousand, three hundred nine **6.** Thirty-seven
7. 600 + 70 + 3 **8.** 60,000 + 1000 + 4 **9.** 58,345
10. ⊢——◆——⊢——⊢——◆——⊢→ **11.** > **12.** <
     0   1   2   3   4   5
13. 35 **14.** 1125 **15.** 6005 **16.** 11,886 **17.** 766
18. 2134 **19.** 32,827 **20.** 32,708 **21.** 83 − 21; 62
22. 103 + 48; 151 **23.** 22 **24.** 11 **25.** $5 \cdot 4 + 5 \cdot 2$
26. $7 \cdot 8 - 7 \cdot 5$ **27.** 0 **28.** 99 **29.** 79,636
30. 367,392 **31.** 21 **32.** 67 **33.** 83 r6 **34.** Undefined
35. 66 ÷ 11; 6 **36.** 26 × 17; 442 **37.** 4 **38.** 9 **39.** 8^5
40. 9^3 **41.** 49 **42.** 125 **43.** 400 **44.** 900,000 **45.** 54
46. 216 **47.** 35 feet **48.** 480 square inches **49.** 5 **50.** 3
51. 9 **52.** 27 **53.** 6 **54.** 42 **55.** 160 **56.** 980,000
57. 50,000 **58.** 400,000 **59.** 1300 **60.** 1400 **61.** 16
62. 11 **63.** 6 **64.** 8 **65.** 35 **66.** 31 **67.** 23 **68.** 52
69. 8 **70.** 42 **71.** 10 **72.** 13 **73.** 31 **74.** 21
75. 9 − (2 + 6); 1 **76.** $4 \cdot 3 - 1$; 11 **77.** Expression
78. Equation **79.** Unlike **80.** Like **81.** $18x$ **82.** $10b$
83. $13mn$ **84.** Not possible **85.** $4y + 7$ **86.** $10z - 6$
87. $15a + 4$ **88.** $5y + 9$ **89.** $x - 7 = 64$, where x is
his height. **90.** $14 = x + 12$, where x is her score.
91. Stamp price increased. **92.** 15¢ **93.** 3
94. 88 square inches **95.** 5698 **96.** 80 minutes
97. 59 bpm **98.** 68 bpm **99.** 64 cm **100.** 7; $2
101. $40 **102.** 15 ft **103.** 25°C **104.** 5 **105.** $15 billion
106. 2008 **107.** 2010 **108.** $18 billion

CHAPTER 1 TEST (p. 87)

1. Ten-thousands **2.** 7000 + 300 + 40 + 1 **3.** 78,000
4. > **5.** 4341 **6.** 6026 **7.** 130,059 **8.** 514 r38
9. 3^4 **10.** 200 **11.** 13 **12.** 12 **13.** 19 **14.** 5 **15.** 12
16. 11 **17.** 16 **18.** 9 **19.** $5p^2$ **20.** Not possible **21.** 6
22. ⊢——⊢——◆——⊢——⊢——◆——→ **23.** $8x + 2$
    0   1   2   3   4   5
24. $4y + 13$ **25.** 56 mi **26.** 6; $8 **27.** $1268
28. $C = 14h$

2 Integers

SECTION 2.1 (pp. 96–98)

1. positive **2.** negative **3.** opposite **4.** 0 **5.** a
6. integers **7.** origin **8.** absolute value **9.** −3
10. +17 or 17 **11.** −7 **13.** 43 **15.** 237 **17.** −93,000
19. −8 **21.** 26 **23.** 0 **25.** 23 **27.** −5 **29.** 1
31. ⊣——◆——⊢——◆——⊢——⊢——⊢——⊢——◆——⊢——⊢→
   −5 −4 −3 −2 −1   0   1   2   3   4   5

33.

35.

37. >   **39.** >   **41.** <   **43.** <   **45.** >   **47.** <   **49.** 10
51. 0   **53.** 18   **55.** 87   **57.** −2   **59.** −19   **61.** 0
63. >   **65.** >   **67.** =   **69.** <   **71.** −282   **73.** 19,340
75. (a) Romania   (b) Malta and Tonga   **77.** −$1745
79. −$200   **81.** (a) Pacific   (b) Arctic   (c) Southern
(d) Indian   **83.** −$4000   **85.** 400 videos

SECTION 2.2 (pp. 104–106)

1. absolute values   **2.** Negative   **3.** Positive   **4.** Negative
5. additive inverses or opposites   **6.** 0   **7.** right; left
8. $\frown$; $\smile$ **9.** 12   **11.** −12   **13.** 41   **15.** −42   **17.** −15
19. 23   **21.** 0   **23.** 39   **25.** −1   **27.** −66   **29.** −143
31. Commutative   **33.** Inverse   **35.** Associative
37. Identity   **39.** −4   **41.** −17   **43.** 0   **45.** 64
47. −45   **49.** −16   **51.** 13   **53.** −93   **55.** −4   **57.** 4
59. −5   **61.** 0   **63.** 4   **65.** −2   **67.** −9   **69.** 5
71. 49°F   **73.** 13 yards   **75.** −1019 feet, or 1019 feet
below ground level   **77.** −$25 million   **79.** $2489

CHECKING BASIC CONCEPTS 2.1 & 2.2 (p. 106)

1. (a) −23   (b) 16   **2.** (a) 52   (b) −9
3.

   4. (a) >   (b) >

5. (a) 17   (b) 31   **6.** (a) 8   (b) −35   (c) −21   (d) 83
7. (a) 4   (b) −6   **8.** (a) 2   (b) −5   **9.** −$420
10. 2405 feet

SECTION 2.3 (pp. 112–114)

1. opposite   **2.** (−7)   **3.** 9   **4.** step to the right
5. Step to the left   **6.** stop and change direction
7. subtraction symbol   **8.** True   **9.** 6   **11.** −5   **13.** −15
15. −42   **17.** 27   **19.** 29   **21.** −5   **23.** 8   **25.** 34
27. 52   **29.** 9   **31.** 45   **33.** −67   **35.** −2   **37.** 13
39. 4   **41.** −31   **43.** 14   **45.** −4   **47.** 3   **49.** −6
51. 0   **53.** 4   **55.** 5   **57.** −5   **59.** −2   **61.** 10
63. 7   **65.** 87°F   **67.** 14,776 feet   **69.** 15 feet
71. $189,400   **73.** $176   **75.** −10°F

SECTION 2.4 (pp. 121–123)

1. positive   **2.** negative   **3.** identity   **4.** distributive
5. positive   **6.** negative   **7.** False   **8.** True   **9.** positive
10. negative   **11.** $\sqrt{4}$ **12.** −$\sqrt{4}$ **13.** −12   **15.** 40
17. −18   **19.** 170   **21.** 0   **23.** −42   **25.** −150
27. −36   **29.** 30   **31.** −21   **33.** −90   **35.** 0   **37.** −80
39. 400   **41.** Associative   **43.** Zero   **45.** Commutative
47. Distributive   **49.** Identity   **51.** −8   **53.** 16   **55.** −81
57. −1   **59.** −27   **61.** −1,000,000   **63.** 10,000   **65.** −3
67. 5   **69.** 12   **71.** −2   **73.** Undefined   **75.** 1   **77.** 0
79. −6   **81.** −5, 5   **83.** −9, 9   **85.** No integer square roots

87. 0   **89.** 4   **91.** −6   **93.** 10   **95.** Not an integer
97. −1   **99.** −21   **101.** −10   **103.** −5   **105.** 25
107. 10   **109.** Not an integer   **111.** −44°C
113. $5 \cdot (-107) = -535$; 535 feet deep
115. $-300 \div 12 = -25$; 25 fewer prisoners
117. $5 \cdot (-29) = -145$; $145

CHECKING BASIC CONCEPTS 2.3 & 2.4 (p. 123)

1. (a) −34   (b) −14   (c) 33   (d) 15
2. (a) −8   (b) 9   **3.** (a) −8   (b) −5   **4.** 4
5. (a) −44   (b) 39   (c) −48   (d) 50   **6.** −48
7. (a) −4   (b) −5   (c) 9   (d) −25   **8.** (a) 8   (b) −4
9. $165   **10.** $3 \cdot (-23) = -69$; 69 feet deep

SECTION 2.5 (pp. 128–130)

1. order of operations   **2.** innermost   **3.** parentheses
4. average   **5.** −10   **7.** 3   **9.** 27   **11.** −1   **13.** −2
15. 9   **17.** 39   **19.** 30   **21.** 0   **23.** 8   **25.** 50
27. −13   **29.** 3   **31.** −15   **33.** −8   **35.** 30   **37.** 1
39. 16   **41.** $-20 + 10 \cdot (14 - 12)$
43. $-5^2 \div (3 + 2) + 5$ **45.** $(32 \div 4^2 - 2) \cdot 9$
47. $(16 - 4^2) \div (4 - 9)$ **49.** −1   **51.** 0   **53.** 12
55. 11   **57.** −5   **59.** −6   **61.** 29   **63.** −20°C
65. 5°F   **67.** −$1000   **69.** −6°F

SECTION 2.6 (pp. 137–139)

1. equation   **2.** solution   **3.** variable   **4.** check   **5.** Yes
7. No   **9.** No   **11.** No   **13.** No   **15.** Yes   **17.** Yes
19. No   **21.** −15   **23.** −5   **25.** 8   **27.** −5   **29.** −1
31. 64   **33.** −3   **35.** 100   **37.** 5   **39.** 7
41. Table values: 0, 1, 2, 3, 4; solution: −1
43. Table values: −1, 2, 5, 8, 11; solution: 1
45. Table values: −7, −1, 5, 11, 17; solution: 2
47. Table values: 4, 3, 2, 1, 0; solution: −6
49. −1   **51.** −2   **53.** 2   **55.** 0   **57.** 7   **59.** 3   **61.** 3
63. 2008   **65.** Table values: 120, 60, 40, 30, 24; solution:
4 gallons   **67.** 2000

CHECKING BASIC CONCEPTS 2.5 & 2.6 (p. 140)

1. (a) 23   (b) 0   (c) −14   **2.** (a) 8   (b) −11   (c) 6
3. No   **4.** Yes   **5.** (a) −5   (b) 6
6. Table values: 9, 7, 5, 3, 1; solution: −1   **7.** −58°F

CHAPTER 2 REVIEW (pp. 144–146)

1. −19   **2.** 52   **3.** 31   **4.** −2
5.

6.

7. <   **8.** >   **9.** >   **10.** <   **11.** −6   **12.** −1
13. 0   **14.** 12   **15.** >   **16.** =   **17.** Sunday
18. Increase   **19.** −1   **20.** −15   **21.** −51   **22.** 22

23. 53   **24.** -18 **25.** -46 **26.** 19   **27.** 5   **28.** -5
29. Inverse   **30.** Associative   **31.** Commutative
32. Identity   **33.** -3 **34.** 4   **35.** -4 **36.** 2   **37.** 19
38. -32 **39.** -30 **40.** -18 **41.** -17 **42.** -22
43. 7   **44.** -22 **45.** -21 **46.** -12 **47.** -49 **48.** 160
49. 3   **50.** -2 **51.** -3 **52.** 9   **53.** -3 **54.** 9   **55.** -6
56. 90   **57.** -24 **58.** 1   **59.** -15 **60.** 3   **61.** -90
62. 48   **63.** Associative   **64.** Identity   **65.** Zero
66. Distributive   **67.** -49 **68.** 49   **69.** -4
70. Not an integer   **71.** -8 **72.** -45 **73.** -5
74. Not an integer   **75.** -22 **76.** 8   **77.** 1   **78.** -4
79. 21   **80.** -5 **81.** $-10 + 5 \cdot (8 - 6)$
82. $14 - (16 - 3) - 1$ **83.** $(7 - 11) \cdot 4^2 + 64$
84. $-3^2 \div (5 - 2) + 3$ **85.** 0   **86.** 6   **87.** -20
88. -5 **89.** Yes   **90.** Yes   **91.** No   **92.** No
93. -11 **94.** -3 **95.** 36   **96.** 20
97. Table values: $-11, -8, -5, -2, 1$; solution: 2
98. Table values: 11, 9, 7, 5, 3; solution: -1
99. 6   **100.** 5   **101. (a)** Mormon   **(b)** Lutheran and Jewish
102. \$592   **103.** 2006   **104.** $-60,000 \div 8 = -7500$;
a 7500 decrease each year   **105.** $-25°C$ **106.** 2006
107. $-3°F$ **108.** $8 \cdot (-19) = -152$; \$152
109. $-\$7$ million   **110.** $3°C$ **111.** $-\$64$ **112.** \$850

CHAPTER 2 TEST (p. 147)

1.
2. 5   **3.** $>$ **4.** $=$ **5.** -5 **6.** -50 **7.** 84   **8.** -7
9. 1   **10.** -84 **11.** -121 **12.** -10 **13.** -8 **14.** -7
15. 8   **16.** -4 **17.** 1   **18.** -4 **19.** Yes   **20.** No
21. -6 **22.** 9   **23.** Table values: 13, 9, 5, 1, -3;
solution: 2   **24.** 7   **25.** \$132,900
26. Table values: 70, 51, 32, 13, -6; solution: 4 miles

CHAPTERS 1–2 CUMULATIVE REVIEW (p. 148)

1. 8   **2.** $30,000 + 2000 + 10$ **3.** 6064   **4.** 11,105
5. 6417   **6.** 81 r27   **7.** 7^3 **8.** 19   **9.** 33,000,000
10. 1500   **11.** 9   **12.** 39   **13.** $5x - 5$ **14.** $<$
15. -17 **16.** 5   **17.** -10 **18.** 100   **19.** -1 **20.** 4
21. -80 **22.** Zero property   **23.** 8   **24.** 1   **25.** No
26. Table values: $-26, -19, -12, -5, 2$; solution: 1
27. \$979   **28.** $-4°F$ **29.** 177 bpm   **30.** 40 cm

3   Algebraic Expressions and Linear Equations

SECTION 3.1 (pp. 156–157)

1. term   **2.** coefficient   **3.** unlike   **4.** like
5. commutative; associative   **6.** evaluate   **7.** $7y$ **9.** $-6m$
11. $10x^3$ **13.** $4b - 8$ **15.** 7   **17.** $8x + 16$
19. $2m - 4$ **21.** $4x + 8$ **23.** $2y - 18$
25. $3n^2 - 7n + 3$ **27.** 0   **29.** $7a$ **31.** $4x + y$
33. $-2y + 9$ **35.** $-15m - 3$ **37.** $8a - 7$
39. $-2x^2 + 3x - 1$ **41.** $10m + 13$ **43.** $y - 5$ **45.** 0

47. $4x - 12$ **49.** $-2t$ **51.** $10x$ **53.** $-40n$ **55.** $-27p$
57. 0   **59.** $-8y - 24$ **61.** $-63x + 7$ **63.** 0
65. $12a + 1$ **67.** $2b + 10$ **69.** $10x - 1$ **71.** $y + 3$
73. $3m - 1$ **75.** $11x$ **77.** $6t - 28$ **79.** $9x + 8$
81. $-7a + 38$
83. (a) $16x + 14x$ **(b)** $30x$ **(c)** 150 square feet
85. (a) $3(x + 2)$ **(b)** $3x + 6$ **(c)** 21 square units

SECTION 3.2 (pp. 162–163)

1. $-$ **2.** $\cdot$ **3.** $=$ **4.** $\div$ **5.** $+$ **6.** define
7. $2p$, where p represents the ticket price
9. $h - 8$, where h represents his height
11. $5 + g$, where g represents the number of gallons
13. $t \div 6$, where t represents the number of toys
15. $7(a + 6)$, where a represents her age
17. $9(a - 6)$, where a represents his age
19. $3n - 8$, where n represents the number
21. $x \div (-10) = 9$ **23.** $64x = -256$ **25.** $x - 3 = 12$
27. $x - 12 = x \div 2$ **29.** $2(x + 6) = -20$
31. $x - 30 = 135$ **33.** $2x - 188 = 356$
35. $x + 10 = 3x - 6$ **37.** $(x + 2) \div 2 = 13$
39. $2x + 2(x + 5) = 70$ **41.** $4x = x^3$

CHECKING BASIC CONCEPTS 3.1 & 3.2 (p. 163)

1. $-6y + 5$
2. (a) $a + 9$ **(b)** $-7y + 7$ **(c)** $-5m - 3$ **(d)** $-5x + 6$
3. $s + 5$, where s represents her score   **4.** $x - 9 = 15$
5. $x + 10,000 = 76,300$ **6.** $3x + 53 = 806$

SECTION 3.3 (pp. 170–171)

1. solution   **2.** variable   **3.** equivalent   **4.** addition
5. subtraction   **6.** multiplication   **7.** division   **8.** variable
9. Yes   **11.** No   **13.** Yes   **15.** Yes   **17.** No
19. Equivalent   **21.** Not equivalent   **23.** Equivalent
25. Not equivalent   **27.** Not equivalent   **29.** -7 **31.** 7
33. -3 **35.** 17   **37.** -49 **39.** -13 **41.** 10   **43.** 6
45. 4   **47.** 1   **49.** 6   **51.** -10 **53.** 7   **55.** 27
57. 48   **59.** -35 **61.** -17 **63.** -8 **65.** 4   **67.** -48
69. 14   **71.** 9   **73.** 60   **75.** -8 **77.** 19   **79.** -88
81. 5 inches   **83.** 3 years

SECTION 3.4 (pp. 181–183)

1. linear   **2.** $-2; 0$ **3.** 1   **4.** numerically
5. algebraically   **6.** given   **7.** Linear; $a = 3, b = 7$
9. Not linear   **11.** Linear; $a = 4, b = -9$ **13.** Not linear
15. Not linear   **17.** Linear; $a = 1, b = -1$
19. Linear; $a = 6, b = 15$ **21.** -4 **23.** -7 **25.** 3
27. 11   **29.** 5   **31.** 2   **33.** 5   **35.** -24 **37.** -3
39. -3 **41.** 4   **43.** 15   **45.** -1 **47.** -12 **49.** -6
51. 16   **53.** 0   **55.** -13 **57.** 2   **59.** 6   **61.** 7
63. -2 **65.** 17   **67.** 6   **69.** 1   **71.** -22 **73.** 8
75. Table values: $-4, -3, -2, -1, 0$; solution: 2
77. Table values: 11, 9, 7, 5, 3; solution: 0
79. Table values: $-12, -9, -6, -3, 0$; solution: 2

81. -2 **83.** 0   **85.** 1   **87.** 8
89. (a) 6 miles   **(b)** 6 miles   **(c)** 6 miles
91. (a) 2007   **(b)** 2007   **(c)** 2007

CHECKING BASIC CONCEPTS 3.3 & 3.4 (p. 184)

1. Yes   **2.** No   **3. (a)** 7   **(b)** -5 **(c)** -32 **(d)** -56
4. (a) Linear; $a = 4, b = -3$ **(b)** Not linear
(c) Not linear   **(d)** Linear; $a = -2, b = -8$
5. (a) 3   **(b)** 2   **(c)** -3 **(d)** 1
6. Table values: 9, 6, 3, 0, -3; solution: -1
7. -2 **8.** 5 inches   **9.** 4 miles

SECTION 3.5 (pp. 190–192)

1. understanding; variable   **2.** equation   **3.** solve; original
4. check   **5.** 24   **7.** -3 **9.** 3   **11.** -5 **13.** -2
15. 8   **17.** 10   **19.** -2 **21.** 165 pounds
23. 272 billion kilowatt-hours   **25.** 8   **27.** 24   **29.** 24
31. 1812 million   **33.** Length: 19 inches; width: 12 inches
35. 26 feet, 35 feet, 41 feet   **37.** 2003: 150,000; 2010: 98,000
39. Mother: 49; daughter: 17   **41.** Prius: 48 mpg; Insight:
43 mpg   **43.** 2003

CHECKING BASIC CONCEPTS 3.5 (p. 193)

1. 6   **2.** -3 **3.** 0   **4.** 2493 megawatts

CHAPTER 3 REVIEW (pp. 195–197)

1. $9y$ **2.** $-17n$ **3.** 0   **4.** $8b - 8$ **5.** $9a - 8$
6. $-4y + 5$ **7.** $-35p$ **8.** $15w - 5$ **9.** $2b + 21$
10. $-3x - 15$ **11.** $-5t - 26$ **12.** $-5m + 1$
13. $2p$, where p represents the number of pancakes
14. $3a$, where a represents her age
15. $5(s + 3)$, where s represents her score
16. $2n - 8$, where n represents the number
17. $(p + 3) \div 5$, where p represents the price
18. $3n + 4$, where n represents the number
19. $5x = 45$ **20.** $x + 7 = 14$ **21.** $9 = x - 11$
22. $x - 10 = 3x$ **23.** $2(x + 4) = -12$
24. $(x + 14) \div 3 = 7$ **25.** Yes   **26.** No   **27.** No
28. Yes   **29.** Equivalent   **30.** Equivalent
31. Not equivalent   **32.** Not equivalent   **33.** 15   **34.** 2
35. 3   **36.** -1 **37.** -6 **38.** 36   **39.** 35   **40.** -72
41. 1   **42.** -7 **43.** -45 **44.** -38
45. Linear; $a = 2, b = -6$ **46.** Linear; $a = 6, b = -10$
47. Not linear   **48.** Not linear   **49.** -7 **50.** 3
51. 4   **52.** 2   **53.** -7 **54.** -1 **55.** -13 **56.** 2
57. -2 **58.** 5   **59.** -2 **60.** 5   **61.** 12   **62.** -5
63. Table values: $-9, -7, -5, -3, -1$; solution: -1
64. Table values: 12, 9, 6, 3, 0; solution: 2   **65.** -2
66. -8 **67.** 18   **68.** -2 **69.** -3 **70.** 5   **71.** 10
72. -3 **73. (a)** $8x + 6x$ **(b)** $14x$ **(c)** 112 windows
74. $x + 23 = 3x - 7$ **75.** 16 inches   **76.** 4 miles
77. 12   **78.** Length: 22 inches; width: 15 inches
79. *Gunsmoke:* 20 yr; *Lassie:* 17 yr   **80.** 15 inches
81. $(6x - 1) + x = 13$ **82.** Mexico: 2; Sweden: 11
83. 2002   **84.** 4 inches

CHAPTER 3 TEST (p. 198)

1. $3x + 4$ **2.** $-3n - 7$ **3.** $24y$ **4.** $-2m + 13$ **5.** 14
6. $2w - 7$ **7.** $x + 8 = 3x$ **8.** $5(x + 3) = -60$ **9.** 56
10. -2 **11.** -7 **12.** -14 **13.** -180 **14.** -24
15. Linear; $a = 6, b = -13$ **16.** Not linear   **17.** 3
18. -1 **19.** -7 **20.** -9
21. Table values: 7, 4, 1, -2, -5; solution: 2
22. 4   **23.** -32 **24.** 3   **25.** $x + 15 = 139$
26. Length: 19 inches; width: 11 inches
27. AK: 20; SD: 54

CHAPTERS 1–3 CUMULATIVE REVIEW (p. 199)

1. 0   **2.** 5^4 **3.** 80,000   **4.** 8000   **5.** 11   **6.** 19   **7.** $2x + 2$
8. $=$ **9.** 5   **10.** -8 **11.** 0   **12.** -1 **13.** Yes
14. Table values: $-15, -11, -7, -3, 1$; solution: 1
15. $-5w - 3$ **16.** $7y - 9$ **17.** $10 = x + 7$
18. $3(x + 6) = -18$ **19.** -9 **20.** -9 **21.** -1 **22.** 0
23. 9   **24.** -5 **25.** 8; \$4   **26.** 56 in.   **27.** -2
28. \$200,000   **29.** $1400 + x = 1550$
30. AR: 14 million acres; CA: 25 million acres

SECTION 4.1 (pp. 210–212)

1. fraction   **2.** numerator; denominator   **3.** rational
4. proper   **5.** improper   **6.** mixed number
7. Numerator: 6; Denominator: 13   **9.** Numerator: 12;
Denominator: 5   **11.** Numerator: x; Denominator: y
13. Numerator: $3p$; Denominator: 14   **15.** $\frac{1}{6}$ **17.** $\frac{5}{8}$ **19.** $\frac{13}{16}$
21. $\frac{1}{8}$ **23.** 1   **25.** 0   **27.** -13 **29.** Undefined   **31.** 1
33. 53   **35.** $\frac{13}{4}$; $3\frac{1}{4}$ **37.** $\frac{17}{10}$; $1\frac{7}{10}$ **39.** $\frac{43}{8}$; $5\frac{3}{8}$ **41.** $5\frac{1}{2}$
43. $2\frac{5}{6}$ **45.** -4 **47.** $11\frac{3}{8}$ **49.** $-7\frac{2}{5}$ **51.** $\frac{23}{4}$ **53.** $-\frac{26}{3}$
55. $\frac{77}{8}$ **57.** $-\frac{107}{3}$ **59.** $\frac{561}{5}$
61.
63.
65.
67.
69.
71. $\frac{3}{50}$ **73.** $\frac{356}{365}$ **75.** $\frac{807}{1000}$ **77.** $\frac{39}{134}$ **79.** $\frac{35}{68}$

SECTION 4.2 (pp. 223–225)

1. factor   **2.** divisible   **3.** 5   **4.** 3   **5.** prime
6. composite   **7.** prime   **8.** factor   **9.** equivalent
10. greatest common factor   **11.** both   **12.** lowest terms
13. cross   **14.** rational   **15.** No   **17.** Yes   **19.** Yes
21. No   **23.** Yes   **25.** Yes   **27.** No   **29.** Yes
31. Yes   **33.** Prime   **35.** Composite   **37.** Neither
39. Composite   **41.** $2 \cdot 2 \cdot 2 \cdot 2$ **43.** $3 \cdot 3 \cdot 5$

45. $2 \cdot 2 \cdot 5 \cdot 7$ **47.** $3 \cdot 7 \cdot 11$ **49.** $2 \cdot 13 \cdot 17$ **51.** 21
53. 42   **55.** 5   **57.** 3   **59.** 20   **61.** 4   **63.** 2   **65.** 10
67. 24   **69.** 4   **71.** 16   **73.** $\frac{2}{3}$ **75.** $\frac{2}{5}$ **77.** $-\frac{6}{11}$ **79.** $\frac{3}{10}$
81. $-\frac{4}{9}$ **83.** $<$ **85.** $<$ **87.** $>$ **89.** $=$ **91.** Yes
93. No   **95.** Yes   **97.** $2x$ **99.** $7a^2$ **101.** 50   **103.** $8a^2c$
105. $\frac{x}{3}$ **107.** $-\frac{3}{10y}$ **109.** $\frac{1}{2}$ **111.** $-2xy$ **113.** $\frac{5x}{9y^2}$
115. $\frac{1}{3}$ **117.** 43   **119.** 1970: $\frac{2}{25}$; 2008: $\frac{7}{25}$

CHECKING BASIC CONCEPTS 4.1 & 4.2 (p. 226)

1. Numerator: 5; Denominator: 18   **2.** $\frac{11}{4}$; $2\frac{3}{4}$ **3.** $3\frac{4}{5}$
4. $\frac{45}{7}$ **5.** Yes   **6. (a)** Composite   **(b)** Prime
7. (a) $2 \cdot 5 \cdot 5$ **(b)** $3 \cdot 5 \cdot 7$ **8.** 30   **9. (a)** $\frac{5}{8}$ **(b)** $-\frac{4x}{7}$
10. $\frac{6}{25}$

SECTION 4.3 (pp. 236–238)

1. numerators; denominators   **2.** fractions   **3.** multiply
4. numerator; denominator   **5.** reciprocals
6. multiplication   **7.** $\frac{3}{20}$ **9.** $\frac{22}{27}$ **11.** $-\frac{10}{9}$ **13.** $-\frac{63}{20}$
15. $\frac{1}{24}$ **17.** 6   **19.** $-\frac{9}{2}$ **21.** $\frac{1}{5}$ **23.** $-\frac{4}{3}$ **25.** $\frac{4}{21}$ **27.** $\frac{1}{10}$
29. -3 **31.** $\frac{4}{15}$ **33.** $-\frac{15}{22}$ **35.** $\frac{11}{18}$ **37.** $\frac{1}{4}$ **39.** $\frac{5x}{2y}$
41. $\frac{8p^2}{15q^2}$ **43.** $\frac{3ab^2}{2}$ **45.** $\frac{9x^2}{2y}$ **47.** $-\frac{2u^3}{3v^2}$ **49.** $-\frac{7y}{3x}$ **51.** $\frac{1}{6x}$
53. 4   **55.** $\frac{x}{5y}$ **57.** $\frac{1}{16}$ **59.** $\frac{9}{25}$ **61.** $-\frac{27}{64}$ **63.** $\frac{64}{121}$ **65.** $\frac{3}{5}$
67. $\frac{1}{8}$ **69.** 4   **71.** $\frac{2}{3}$ **73.** $\frac{5}{3}$ **75.** -12 **77.** $\frac{1}{15}$ **79.** 1
81. $\frac{5}{3}$ **83.** $-\frac{20}{3}$ **85.** $-\frac{24}{5}$ **87.** 6   **89.** $-\frac{5}{3}$ **91.** $\frac{5}{9}$
93. $-\frac{9}{22}$ **95.** $\frac{7}{26}$ **97.** $\frac{xy}{2}$ **99.** $-9b$ **101.** $-\frac{5p}{6q}$ **103.** $9xy^2$
105. 1   **107.** $\frac{5}{6}$ **109.** -3 **111.** $\frac{70}{3}$ **113.** 81   **115.** \$480
117. $\frac{1}{3}$ cup   **119.** $\frac{1}{12}$ **121.** 20 square inches
123. $\frac{91}{2}$ square miles

SECTION 4.4 (pp. 243–245)

1. like; common   **2.** numerators   **3.** denominator
4. numerators   **5.** denominator   **6.** fractions   **7.** $\frac{3}{5}$
9. $-\frac{5}{9}$ **11.** $\frac{24}{25}$ **13.** $\frac{10}{13}$ **15.** $\frac{4}{5}$ **17.** $\frac{2}{3}$ **19.** -2 **21.** 0
23. 1   **25.** $\frac{2}{7}$ **27.** $\frac{8}{9}$ **29.** $-\frac{12}{17}$ **31.** $-\frac{5}{11}$ **33.** $\frac{1}{4}$ **35.** 0
37. $-\frac{1}{5}$ **39.** $-\frac{1}{3}$ **41.** 1   **43.** $\frac{1}{3}$ **45.** $-\frac{1}{2}$ **47.** $\frac{2}{3}$ **49.** $\frac{1}{15}$
51. $-\frac{8}{15}$ **53.** -1 **55.** $\frac{7x}{9y}$ **57.** $-\frac{4}{5m^2}$ **59.** $\frac{7x-3w}{y}$
61. $\frac{1}{d}$ **63.** $\frac{4y}{x^2}$ **65.** $-\frac{k^2}{3c}$ **67.** $3x^2$ **69.** $\frac{3}{4}$ inch   **71.** $\frac{1}{125}$
73. $\frac{13}{2500}$ **75.** $\frac{2}{5}$ **77.** $\frac{1}{5}$ **79.** $\frac{3}{5}$

CHECKING BASIC CONCEPTS 4.3 & 4.4 (p. 245)

1. (a) $\frac{4}{3}$ **(b)** -5 **(c)** $\frac{4}{5}$ **(d)** $-\frac{3}{4}$ **(e)** $\frac{1}{12y}$ **(f)** -3
(g) $-10a^2$ **(h)** $\frac{3x}{2y}$ **2. (a)** $\frac{4}{25}$ **(b)** 5   **3. (a)** $\frac{12}{19}$
(b) $\frac{1}{3}$ **(c)** $\frac{2}{7}$ **(d)** -1 **(e)** $4x^2$ **(f)** $\frac{m^2}{4n^3}$ **4.** 39

SECTION 4.5 (pp. 254–256)

1. least   **2.** divisible   **3.** listing   **4.** prime factorization
5. maximum   **6.** LCM   **7.** 20   **9.** 60   **11.** 12   **13.** 80
15. 90   **17.** 30   **19.** 180   **21.** 135   **23.** 1296   **25.** $9x$
27. $24y^3$ **29.** $24a^3b^2$ **31.** $12xyz$ **33.** 36   **35.** 28
37. 120   **39.** 60   **41.** $6xy^2$ **43.** $16mn^2$ **45.** abc

47. $\frac{9}{12}$; $\frac{2}{12}$ **49.** $\frac{22}{36}$; $\frac{27}{36}$ **51.** $\frac{15}{30}$; $\frac{25}{30}$; $\frac{27}{30}$ **53.** $\frac{11}{20}$ **55.** $-\frac{7}{40}$
57. $-\frac{13}{9}$ **59.** $\frac{1}{6}$ **61.** $\frac{2}{3}$ **63.** $-\frac{7}{6}$ **65.** 1   **67.** $\frac{3}{4}$ **69.** $\frac{5}{2}$
71. $\frac{19x}{12}$ **73.** $\frac{m^2}{3}$ **75.** $\frac{3x+20y}{48}$ **77.** $\frac{20y-3}{24y^2}$ **79.** $\frac{9x^2+25y^2}{15xy}$
81. $\frac{37}{100}$ **83.** $\frac{3}{50}$ **85.** $\frac{41}{100}$ **87.** $\frac{5}{8}$ **89.** 4 ft   **91.** $\frac{37}{18}$ in.
93. 7 yd

SECTION 4.6 (pp. 263–265)

1. $\frac{1}{2}$ **2.** less   **3.** improper fractions   **4.** estimate
5. vertically   **6.** borrow   **7.** $\frac{21}{5}$ **9.** $-\frac{52}{15}$ **11.** $\frac{89}{10}$
13. $-\frac{61}{4}$ **15.** 8   **17.** -14 **19.** 11   **21.** -10 **23.** 10
25. $39\frac{1}{2}$ **27.** $7\frac{1}{24}$ **29.** $6\frac{3}{4}$ **31.** $-33\frac{4}{9}$ **33.** $-1\frac{1}{8}$
35. $-9\frac{1}{10}$ **37.** $13\frac{9}{19}$ **39.** $4\frac{23}{35}$ **41.** $11\frac{7}{12}$ **43.** $7\frac{8}{21}$
45. $5\frac{7}{24}$ **47.** $14\frac{13}{24}$ **49.** 4   **51.** $11\frac{6}{11}$ **53.** $-13\frac{19}{21}$
55. $14\frac{7}{8}$ **57.** $-23\frac{2}{3}$ **59.** $2\frac{13}{15}$ **61.** 10 pieces; $4\frac{1}{2}$ inches
63. $5\frac{1}{3}$ cups   **65.** About 629 thousand   **67.** $2\frac{3}{8}$ inches
69. $\frac{7}{12}$ gallon   **71.** $16\frac{2}{3}$ square feet   **73.** $57\frac{19}{25}$ square inches

CHECKING BASIC CONCEPTS 4.5 & 4.6 (p. 265)

1. (a) 60   **(b)** $20x^2$ **2. (a)** 72   **(b)** $30x^2y$
3. (a) $\frac{7}{24}$ **(b)** $-\frac{1}{20}$ **(c)** $\frac{33}{40}$ **(d)** $-\frac{x}{24}$
4. (a) $-\frac{1}{2}$ **(b)** $6\frac{17}{20}$ **(c)** $30\frac{3}{4}$ **(d)** $5\frac{1}{3}$ **5.** $\frac{13}{24}$

SECTION 4.7 (pp. 270–272)

1. complex   **2.** reciprocal   **3.** I   **4.** II   **5.** $\frac{5}{14}$ **7.** 9   **9.** $\frac{16}{5}$
11. $\frac{1}{12}$ **13.** $\frac{25}{2xy}$ **15.** 12   **17.** $\frac{5}{3}$ **19.** $\frac{4}{5}$ **21.** 5   **23.** $\frac{17}{22}$
25. $\frac{16x}{21}$ **27.** $\frac{21x}{y}$ **29.** $\frac{1}{5}$ **31.** $\frac{11}{8}$ **33.** $\frac{2}{7}$ **35.** $\frac{4}{9}$ **37.** $\frac{3x}{4}$
39. 8   **41.** 3   **43.** $\frac{3}{4}$ **45.** $-\frac{17}{32}$ **47.** $\frac{1}{10}$ **49.** -11
51. $-\frac{9}{16}$ **53.** $\frac{1}{6}$ **55.** $\frac{29}{30}$ **57.** $3\frac{1}{8}$ mph   **59.** $\frac{3}{4}$ hour
61. $28°C$

SECTION 4.8 (pp. 284–286)

1. addition   **2.** multiplication   **3.** distributive
4. LCD   **5.** $\frac{13}{20}$ **7.** $-\frac{2}{15}$ **9.** $-\frac{7}{12}$ **11.** $\frac{2}{5}$ **13.** 14
15. $-\frac{15}{2}$ **17.** $\frac{9}{4}$ **19.** $-\frac{3}{10}$ **21.** 11   **23.** $\frac{2}{9}$ **25.** $-\frac{5}{12}$
27. $\frac{9}{20}$ **29.** $\frac{1}{9}$ **31.** $-\frac{6}{7}$ **33.** $\frac{1}{6}$ **35.** 2   **37.** $-\frac{11}{16}$
39. 6   **41.** $\frac{12}{7}$ **43.** $\frac{4}{5}$ **45.** $\frac{7}{20}$ **47.** $\frac{1}{3}$
49. Table values: $-\frac{8}{3}$, -2, $-\frac{4}{3}$, $-\frac{2}{3}$, 0; solution: 9
51. Table values: $\frac{23}{5}$, $\frac{19}{5}$, 3, $\frac{11}{5}$, $\frac{7}{5}$; solution: -1
53. Table values: $-\frac{7}{3}$, $-\frac{4}{3}$, $-\frac{1}{3}$, $\frac{2}{3}$, $\frac{5}{3}$; solution: 4
55. -2 **57.** 4   **59.** 1   **61.** $-\frac{11}{4}$ **63.** $\frac{21}{2}$ **65.** $-\frac{34}{3}$
67. $-\frac{7}{8}$ **69.** $-\frac{1}{45}$ **71.** $\frac{2}{5}$ **73.** $\frac{7}{4}$ feet   **75.** $\frac{10}{3}$ inches
77. $8°C$ **79.** 3 minutes   **81. (a)** 2006   **(b)** 2006
(c) 2006   **(d)** The solutions are the same.

CHECKING BASIC CONCEPTS 4.7 & 4.8 (p. 287)

1. (a) $\frac{3}{2}$ **(b)** $\frac{w}{27}$ **(c)** $\frac{11}{6}$ **(d)** $\frac{2x}{3}$ **2. (a)** $\frac{7}{24}$ **(b)** $\frac{1}{4}$
3. (a) $\frac{5}{8}$ **(b)** -30 **(c)** 6   **(d)** 1
4. Table values: 6, 5, 4, 3, 2; solution: -7 **5.** -2 **6.** $74°F$

CHAPTER 4 REVIEW (pp. 294–296)

1. (a) Numerator: 7; Denominator: 18
(b) Numerator: x; Denominator: 5
2. (a) $\frac{2}{5}$ **(b)** $\frac{7}{8}$ **3. (a)** 0 **(b)** -2 **(c)** 1 **(d)** Undefined
4. (a) $\frac{13}{10}$; $1\frac{3}{10}$ **(b)** $\frac{17}{6}$; $2\frac{5}{6}$ **5. (a)** $6\frac{1}{3}$ **(b)** $-2\frac{4}{5}$
6. (a) $-\frac{32}{5}$ **(b)** $\frac{70}{9}$
7. (a)
(b)

8. (a) No **(b)** Yes **9. (a)** Composite **(b)** Prime
(c) Neither **(d)** Composite **10. (a)** $2 \cdot 2 \cdot 2 \cdot 5$
(b) $2 \cdot 5 \cdot 11$ **11. (a)** 5 **(b)** 27 **12. (a)** 6 **(b)** 8
13. (a) $-\frac{4}{13}$ **(b)** $\frac{8}{11}$ **14. (a)** $<$ **(b)** $>$ **15. (a)** $3x$
(b) $2xy^2$ **16. (a)** $\frac{4x}{5}$ **(b)** $-3x^2$ **17.** $\frac{25}{48}$ **18.** $\frac{5}{12}$ **19.** -3
20. $\frac{8}{21}$ **21.** $\frac{3x}{10w}$ **22.** $-\frac{9y}{4x}$ **23. (a)** $-\frac{1}{64}$ **(b)** $\frac{7}{10}$
24. (a) $-\frac{9}{4}$ **(b)** 10 **25.** $\frac{2}{15}$ **26.** $-\frac{18}{5}$ **27.** $\frac{xy}{8}$ **28.** $\frac{3p^2}{8q^2}$
29. -20 **30.** $\frac{5}{24}$ **31.** 1 **32.** $\frac{4}{5}$ **33.** 2 **34.** $-\frac{5}{8}$ **35.** $\frac{y^2}{4x}$
36. 0 **37. (a)** 60 **(b)** 48 **(c)** $12x^2$ **(d)** $12mn$
38. (a) 48 **(b)** $24y^2$ **39.** $\frac{3}{18}$; $\frac{16}{18}$ **40.** $\frac{35}{42}$; $\frac{12}{42}$; $\frac{9}{42}$ **41.** $\frac{11}{12}$
42. $-\frac{17}{40}$ **43.** $\frac{2}{3}$ **44.** $\frac{4}{9}$ **45.** $\frac{7y^2}{3}$ **46.** $\frac{11}{8m}$ **47.** -4 **48.** 5
49. $13\frac{1}{12}$ **50.** $-21\frac{1}{2}$ **51.** $2\frac{14}{15}$ **52.** $5\frac{5}{9}$ **53.** $-13\frac{3}{4}$
54. $4\frac{1}{2}$ **55.** $2\frac{2}{3}$ **56.** $-8\frac{2}{9}$ **57.** $\frac{2}{3}$ **58.** $\frac{5}{2}$ **59.** $2x$ **60.** $\frac{9y}{25}$
61. $\frac{2}{3}$ **62.** $\frac{5}{4}$ **63.** $\frac{23}{24}$ **64.** -72 **65.** $\frac{3}{14}$ **66.** 2 **67.** $\frac{7}{12}$
68. $-\frac{10}{3}$ **69.** -2 **70.** -3 **71.** Table values: $-\frac{23}{5}$, $-\frac{21}{5}$,
$-\frac{19}{5}$, $-\frac{17}{5}$, -3; solution: 5 **72.** Table values: $\frac{1}{2}$, 2, $\frac{7}{2}$, 5, $\frac{13}{2}$;
solution: 3 **73.** -2 **74.** 4 **75.** $\frac{93}{100}$ **76.** Red class
77. $\frac{1}{2}$ cup **78.** $\frac{4}{15}$ **79.** $\frac{7}{10}$ **80.** $25\frac{1}{4}$ inches **81.** \$132
82. $4\frac{4}{5}$ hours

CHAPTER 4 TEST (pp. 296–297)

1. $\frac{19}{6}$; $3\frac{1}{6}$ **2.** $\frac{33}{7}$ **3.** $7\frac{1}{3}$ **4. (a)** Composite **(b)** Neither
(c) Prime **(d)** Composite **5.** $2 \cdot 2 \cdot 2 \cdot 2 \cdot 3 \cdot 5$
6. (a) 5 **(b)** 27 **7. (a)** 8 **(b)** $7x$ **8. (a)** $-\frac{3}{4}$ **(b)** $4x$
9. (a) $-\frac{3}{2}$ **(b)** $\frac{7}{20}$ **(c)** $\frac{5x}{7y}$ **(d)** $-\frac{8xy}{3}$ **10. (a)** $\frac{9}{11}$ **(b)** $\frac{25}{36}$
11. (a) $\frac{7}{12}$ **(b)** $-\frac{7}{30}$ **(c)** $-\frac{23}{15}$ **(d)** $\frac{4w}{x}$ **12. (a)** $11\frac{5}{12}$
(b) $-22\frac{2}{5}$ **13. (a)** $\frac{25}{26}$ **(b)** 1 **14. (a)** -16 **(b)** $\frac{3}{4}$ **(c)** 5
(d) $\frac{1}{2}$ **15.** Table values: $-\frac{13}{6}$, $-\frac{5}{6}$, $\frac{1}{2}$, $\frac{11}{6}$, $\frac{19}{6}$; solution: 2
16. 3 **17.** \$672 **18.** 2000

CHAPTERS 1–4 CUMULATIVE REVIEW (pp. 297–298)

1. 36,285 **2.** $15 \cdot 4$; 60 **3.** 222 r15 **4.** 5 **5.** 3700 **6.** 6
7. $3x - 1$ **8.** 4 **9.** Associative **10.** 6 **11.** -60 **12.** 1
13. Table values: $-11, -8, -5, -2, 1$; solution: 0
14. $2y + 5$ **15.** $7 = x - 13$ **16.** $2(x + 3) = -14$
17. No **18.** No **19.** -2 **20.** -3 **21.** $\frac{37}{10}$; $3\frac{7}{10}$
22. $2 \cdot 3 \cdot 3 \cdot 11$ **23.** $-\frac{3}{4}$ **24.** $\frac{1}{6}$ **25.** 0 **26.** $\frac{7}{15}$
27. $8\frac{23}{24}$ **28.** $\frac{13}{30}$ **29.** $\frac{1}{3}$ **30.** $\frac{11}{4}$ **31.** \$40 **32.** 2002
33. Length: 9 inches; width: 4 inches **34.** $\frac{2}{5}$

---

5 Decimals

SECTION 5.1 (pp. 308–309)

1. decimal **2.** decimal; decimals **3.** and **4.** 10
5. 10 **6.** unequal **7.** fifty-six hundredths
9. seven and one hundred sixteen thousandths
11. negative fifty-eight and seven tenths
13. negative two and one thousand three millionths
15. five hundred one and twelve ten-thousandths
17. One hundred twenty-nine and 68/100
19. $\frac{3}{10}$ **21.** $-\frac{1}{25}$ **23.** $\frac{17}{20}$ **25.** $-8\frac{1}{5}$ **27.** $12\frac{3}{4}$
29. $23\frac{41}{200}$ **31.** $-1\frac{7}{250}$ **33.** $6\frac{41}{80}$
35.
37.
39.
41.
43.
45. $<$ **47.** $<$ **49.** $>$ **51.** $<$ **53.** $>$ **55.** $<$ **57.** $<$
59. $>$ **61.** 0.4 **63.** 52.01 **65.** -7.0094 **67.** 9.003
69. -1.106021 **71.** 5.73829 **73.** \$4 **75.** \$143.30
77. \$20 **79.** one thousand four hundred fifty-three and
seventy-one hundredths **81.** $2\frac{9}{20}$ **83.** Usain Bolt
85. \$1.91 **87.** $\frac{4}{125}$

SECTION 5.2 (pp. 316–318)

1. estimate **2.** rounded **3.** decimal points **4.** vertically
5. $20 + 400 = 420$ **7.** $1500 - 300 = 1200$
9. $690 - 90 = 600$ **11.** $1650 + 350 = 2000$
13. $-300 + 70 = -230$ **15.** $-40 - 130 = -170$
17. Estimate: 10; Actual: 10.388
19. Estimate: 670; Actual: 672.899
21. Estimate: 830; Actual: 833.62
23. Estimate: 760; Actual: 758.259
25. Estimate: 6550; Actual: 6547.18
27. Estimate: 170; Actual: 173.48
29. Estimate: 700; Actual: 700.857
31. Estimate: 160; Actual: 162.7
33. Estimate: -200; Actual: -198.37
35. Estimate: -690; Actual: -695.61
37. Estimate: 1300; Actual: 1298.94
39. Estimate: 60; Actual: 56.32 **41.** 134.207 **43.** 800.96
45. -3154.29 **47.** -61.692 **49.** 1235.899 **51.** 5006.555
53. 2.598 **55.** 1.473 **57.** \$2.78 **59.** \$4.14 **61.** $11.03y$
63. $9.6w - 5$ **65.** $1.8n^2$ **67.** $-2.6p^2 + 13.6$
69. $1.6x + 2.7y$ **71.** 45.3 million tons **73.** \$19.52
75. 84.2 thousand **77.** 1980 **79.** 0.92 thousand gallons
81. 296.9 miles **83.** 40.2 inches **85.** 40.16 feet
87. 15.5 inches **89.** \$867.85 **91.** 3.517 hours

CHECKING BASIC CONCEPTS 5.1 & 5.2 (p. 319)

1. twenty-three and ninety-seven thousandths
2. (a) $-5\frac{3}{5}$ **(b)** $\frac{13}{25}$
3. ◄─┼─┼─┼─┼─┼─┼─♦─┼─┼─┼─►
 34.20 34.22 34.24 34.26 34.28 34.30
4. (a) $<$ **(b)** $>$ **5.** 0.278 **6.** $9
7. (a) $150 + 20 = 170$ **(b)** $7000 - 200 = 6800$
8. (a) 36.02 **(b)** 434.59 **(c)** -71.66 **(d)** 282.62
9. (a) $5.04x$ **(b)** $3.5y^2 + 2$ **10.** $8.07

SECTION 5.3 (pp. 331–333)

1. estimate **2.** sum **3.** right **4.** above **5.** right
6. 2 **7.** left **8.** denominator; numerator **9.** $2 \cdot 27 = 54$
11. $12 \div 4 = 3$ **13.** $500 \cdot 5 = 2500$ **15.** $87 \div 1 = 87$
17. 123.3 **19.** 2.94 **21.** 3.953 **23.** 15.96 **25.** 34.4344
27. -43.8 **29.** -7.02 **31.** 2.091 **33.** 124.89
35. -467.9 **37.** 410 **39.** -9.8 **41.** 34,498 **43.** 9.4
45. 2.97 **47.** -12.9 **49.** $4.\overline{6}$ **51.** $3.7\overline{3}$ **53.** 1.38
55. 14.75 **57.** 86.25 **59.** -0.376 **61.** -600
63. 1.779 **65.** -0.634 **67.** 78.94 **69.** -0.0076
71. 0.0001 **73.** 0.25 **75.** 0.375 **77.** -0.24
79. $0.\overline{3}$ **81.** $0.2\overline{6}$ **83.** -0.365 **85.** 3.2 **87.** $-9.5\overline{3}$
89. $17.\overline{6}$ **91.** 49.68 **93.** -2.3364 **95.** 42.2
97. -0.05 **99.** $3.51x + 21.6$ **101.** $-20.4y - 18.36$
103. $0.8x$ **105.** $-80w$ **107.** 578.5 calories **109.** 51 mpg
111. $4.68 **113.** $264.12 **115.** $459,334.80
117. $512.46

SECTION 5.4 (pp. 340–342)

1. irrational **2.** real **3.** real **4.** $\sqrt{36}$ **5.** Guess-and-check
6. Babylonian **7. (a)** $3, \sqrt{4}$ **(b)** $3, 0, \sqrt{4}$ **(c)** $3, 0, \sqrt{4}$
(d) $-\frac{5}{8}, 3, 0, \sqrt{4}$ **(e)** $\sqrt{5}$ **9. (a)** $\frac{9}{3}$ **(b)** $0, \frac{9}{3}$ **(c)** $0, \frac{9}{3}$
(d) $0, \frac{9}{3}, -1.\overline{2}, 6.4$ **(e)** $\sqrt{10}$ **11.** 2.2 **13.** 3.9 **15.** 9.1
17. 3.74 **19.** 6.48 **21.** 9.95 **23.** 2.45 **25.** 4.24
27. 8.83 **29.** 1.4142 **31.** 5.4772 **33.** 7.4162 **35.** 9.8
37. 5.15 **39.** 13.9 **41.** 1.25 **43.** $16.8\overline{3}$ **45.** 3 **47.** 7.18
49. 10.61 **51.** -9.1 **53.** 9 **55.** $-10.\overline{6}$ **57.** 32 feet
59. $147.95 million per year **61.** 46.28 square inches
63. 16.35 square miles **65.** 29.5 mpg **67.** 2.4 million
69. 6.93 feet

CHECKING BASIC CONCEPTS 5.3 & 5.4 (p. 343)

1. (a) 7.56 **(b)** -244.3 **(c)** 26.3 **(d)** -19.096
2. (a) 3.56 **(b)** $9.1\overline{5}$ **(c)** 0.635 **(d)** -0.4
3. (a) 0.45 **(b)** $0.4\overline{6}$ **(c)** -3.125 **(d)** $9.8\overline{3}$
4. (a) $19x + 7.5$ **(b)** $-70y$ **5.** 4.9 **6.** 4.47
7. (a) 6.7 **(b)** $1.\overline{8}$ **8.** -6.3 **9.** $1169.62 **10.** 35 feet

SECTION 5.5 (pp. 351–354)

1. Algebraic, numerical, and visual **2.** decimals
3. decimal places **4.** numerically **5.** 16.5 **7.** -12.4
9. 3.5 **11.** 4.9 **13.** 3.2 **15.** 6.8 **17.** $-0.\overline{3}$ **19.** -3.05
21. 3.7 **23.** 18.7 **25.** -40.25 **27.** -0.5 **29.** 44.8

31. Table values: -2.5, 2.3, 7.1, 11.9, 16.7; solution: 2
33. Table values: 9.5, 5.6, 1.7, -2.2, -6.1; solution: -2
35. Table values: 11, 11.7, 12.4, 13.1, 13.8; solution: 2.1
37. -5 **39.** 3 **41.** 3 **43.** 9.15 **45.** 3 **47.** 8.5
49. -1.2 **51.** 417 **53.** 6.5 feet **55.** 5.7 inches
57. 2009 **59.** 14°C **61. (a)** 2007 **(b)** 2007 **(c)** 2007
(d) The solutions are the same.

SECTION 5.6 (pp. 364–368)

1. circle **2.** radius **3.** diameter **4.** radius
5. circumference **6.** π **7.** semicircle **8.** right
9. legs **10.** hypotenuse **11.** mean **12.** median
13. mode **14.** weighted **15.** 12π; 37.68 inches
17. 50π; 157 yards **19.** 0.6π; 1.884 inches
21. 3π; 9.42 feet **23.** 25π; 78.5 square feet
25. 0.16π; 0.5024 square yards **27.** π; 3.14 square inches
29. 2.25π; 7.065 square feet **31.** 84 square inches
33. 114.24 square yards **35.** 178.46 square feet
37. 35.44 square yards **39.** 10 inches **41.** 8 yards
43. 6.6 feet **45.** 10.8 yards **47.** 69.8 **49.** 29.2 **51.** 3.4
53. 40.5 **55.** 27.65 **57.** 1 **59.** 4.7 **61.** No mode
63. Two modes: 5 and 9 **65.** 3.25 **67.** 3.8
69. 1293.68 feet **71.** 103,918.5 square feet **73.** 17 feet
75. 22.3 seasons **77.** 19 seasons **79.** 17 seasons

CHECKING BASIC CONCEPTS 5.5 & 5.6 (p. 368)

1. (a) 3.2 **(b)** 18.1 **(c)** $8.\overline{36}$
2. Table values: -6.1, -2.7, 0.7, 4.1, 7.5; solution: -2
3. 4 **4.** $C \approx 43.96$ inches; $A \approx 153.86$ square inches
5. 7 inches **6.** Mean: 7.875; Median: 7.5; Mode: 6
7. 2.625 **8.** 2010

CHAPTER 5 REVIEW (pp. 374–376)

1. seventy-six hundredths
2. negative five and two hundred six thousandths
3. $-\frac{2}{25}$ **4.** $37\frac{1}{4}$
5. ◄─┼─┼─┼─┼─┼─┼─♦─┼─┼─┼─►
 7.0 7.2 7.4 7.6 7.8 8.0
6. ◄─┼─┼─┼─┼─┼─┼─♦─┼─┼─┼─►
 -5.20 -5.18 -5.16 -5.14 -5.12 -5.10
7. $>$ **8.** $<$ **9.** -4.0083 **10.** 3591.01 **11.** $12
12. $41.81 **13.** Estimate: 250; Actual: 250.145
14. Estimate: 660; Actual: 659.98
15. Estimate: 670; Actual: 669.979
16. Estimate: 840; Actual: 840.07
17. Estimate: 290; Actual: 290.34
18. Estimate: 180; Actual: 177.8
19. $10.77y$ **20.** $145.7x^3$ **21.** $7q - 1.1$ **22.** $7.41n + 4b$
23. $3.6x + 4.7y$ **24.** $0.3a - 3.5b + 0.9$
25. $4 \cdot 30 = 120$ **26.** $200 \div 10 = 20$ **27.** -1.56
28. -1.89 **29.** 194.6 **30.** 16.3 **31.** 13.56 **32.** 6.256
33. 17.475 **34.** 92.46 **35.** 140 **36.** 6.8125 **37.** -13.8
38. 10.504 **39.** -0.44 **40.** $0.4\overline{3}$ **41.** $8.1\overline{3}$ **42.** 64.85
43. $7.7x + 17.6$ **44.** $-1.5y + 0.6$ **45.** $0.6x$ **46.** $-60n$

47. (a) 2, $\sqrt{9}$ (b) 2, 0, $\sqrt{9}$ (c) 2, 0, $\sqrt{9}$
(d) $\frac{5}{8}$, 2, 0, $\sqrt{9}$ (e) $-\sqrt{7}$ **48.** (a) 7   (b) 7   (c) 7
(d) 7, $-\frac{2}{5}$, $3.\overline{6}$, -7.4 (e) $\sqrt{5}$ **49.** 3.61   **50.** 4.58
51. 2.65   **52.** 5.29   **53.** 6.775   **54.** 1   **55.** 8.25   **56.** 3
57. 4.9   **58.** 9.5   **59.** -32.3 **60.** 9.2   **61.** 16.9   **62.** 5.6
63. 3.1   **64.** 1.25   **65.** -9.9 **66.** 5   **67.** 4.8
68. 0.125   **69.** -30.2 **70.** 2.2
71. Table values: -7.5, -1.6, 4.3, 10.2, 16.1; solution: 3
72. Table values: 9, 9.8, 10.6, 11.4, 12.2; solution: 0.6
73. -5 **74.** 3   **75.** 18π; 56.52 inches   **76.** 22π; 69.08 feet
77. 81π; 254.34 square feet   **78.** 64π; 200.96 square inches
79. 64.26 square yards   **80.** 42.74 square feet   **81.** 20 feet
82. 10.9 feet   **83.** Mean: 4.8; median: 4.7; mode: 4.7
84. Mean: 12.4; median: 12.5; mode: 15   **85.** 2.8
86. 1.25   **87.** Bacterium A   **88.** 470.4 miles
89. 189 calories   **90.** 72 feet   **91.** 94.1°F   **92.** 24 feet
93. $26.37   **94.** $\frac{38}{125}$ **95.** 104 bpm   **96.** $20.15
97. 540.08 feet   **98.** 493

CHAPTER 5 TEST (pp. 377–378)

1. $-\frac{17}{20}$ **2.** $13\frac{5}{8}$ **3.** $<$ **4.** 91.581   **5.** 130.347
6. 3708.23   **7.** 22.685   **8.** 413.8   **9.** 69.24   **10.** 48
11. $15x + 1$ **12.** $-50x$ **13.** $17.1\overline{6}$ **14.** -0.275
15. (a) 9, $\sqrt{4}$ (b) 9, $\sqrt{4}$, 0   (c) 9, $\sqrt{4}$, 0
(d) 9, $-\frac{2}{3}$, $\sqrt{4}$, 0   (e) π **16.** 4.58   **17.** 4.5   **18.** 3.1
19. 4.1   **20.** 7.5   **21.** Table values: -7.4, -6.7, -6, -5.3,
-4.6; solution: -0.5 **22.** 2   **23.** $C \approx 18.84$ inches;
$A \approx 28.26$ square inches   **24.** 6.9 inches   **25.** Mean: 11.9;
median: 12.2; mode: 12.6   **26.** 2.75   **27.** 65.5 feet
28. 62 years

CHAPTERS 1–5 CUMULATIVE REVIEW (pp. 378–379)

1. $60,000 + 1000 + 5$ **2.** 48,000   **3.** 8   **4.** 1500   **5.** 11
6. $=$ **7.** -7 **8.** Commutative   **9.** $4x + 8$ **10.** -2
11. Table values: -14, -9, -4, 1, 6; solution: -1 **12.** No
13. $x - 12 = -5$ **14.** $2(x + 4) = -10$ **15.** 2   **16.** 4
17. -2 **18.** Yes   **19.** 2   **20.** $\frac{23}{10}$; $2\frac{3}{10}$ **21.** $-\frac{3}{2}$ **22.** $\frac{11}{40}$
23. $0.2\overline{6}$ **24.** -10.7 **25.** 6.71   **26.** $<$ **27.** 14.76
28. 3.2   **29.** 64 inches   **30.** $2862   **31.** 2007   **32.** $\frac{43}{50}$
33. 2005   **34.** $54,000

6 Ratios, Proportions, and Measurement

SECTION 6.1 (pp. 387–389)

1. ratio   **2.** 1   **3.** denominator; numerator   **4.** rate
5. unit   **6.** pricing   **7.** $\frac{1}{3}$ **9.** $\frac{3}{8}$ **11.** $\frac{4}{5}$ **13.** $\frac{9}{10}$ **15.** $\frac{2}{5}$
17. $\frac{10}{3}$ **19.** $\frac{5}{4}$ **21.** $\frac{7}{6}$ **23.** $\frac{3}{1} = 3$ **25.** $\frac{0.625}{1} = 0.625$
27. $\frac{5.5}{1} = 5.5$ **29.** $\frac{1}{1} = 1$ **31.** $\frac{2 \text{ inches}}{3 \text{ hours}}$ **33.** $\frac{13 \text{ dollars}}{2 \text{ hours}}$
35. $\frac{12 \text{ seats}}{1 \text{ row}}$ **37.** $\frac{1 \text{ copier}}{17 \text{ employees}}$ **39.** $\frac{2 \text{ slices}}{1 \text{ person}}$ **41.** $8.75/hr
43. 25.6 mi/gal   **45.** 0.8 in./hr   **47.** $2.25/drink
49. 62 beats/min   **51.** $0.42/oz   **53.** $3.95/lb
55. $0.00425/lb   **57.** $\frac{5}{6}$ **59.** $\frac{2}{3}$ **61.** $\frac{18}{5}$ **63.** $\frac{19}{10}$

65. (a) $\frac{1}{3}$ (b) $\frac{11}{250}$ (c) $\frac{3}{41}$ **67.** $\frac{3}{197}$ **69.** 0.72; yes
71. University: 43.25; community college: 32.5; there
are 32.5 students for each instructor.   **73.** Receptionist:
66 words/min; office manager: 63.5 words/min; the receptionist   **75.** Large: $0.22/oz; small: $0.205/oz; the
small jar   **77.** Generic ($0.256/pill vs. $0.365/pill)
79. (a) Beets: $0.14/oz; mints: $0.35/oz   (b) We are not
comparing size options for the same product.

SECTION 6.2 (pp. 397–400)

1. proportion   **2.** $\frac{a}{b} = \frac{c}{d}$ **3.** cross products   **4.** equal
5. similar   **6.** proportional   **7.** Yes   **9.** No   **11.** Yes
13. Yes   **15.** No   **17.** Yes   **19.** No   **21.** No   **23.** 35
25. 6.3   **27.** $\frac{1}{3}$ **29.** 1   **31.** -12 **33.** 19.2   **35.** 3.6
37. 6.9   **39.** $\frac{2}{3}$ **41.** $-\frac{3}{5}$ **43.** $-\frac{3}{4}$ **45.** 14 inches
47. 1.6 yards   **49.** $\frac{1}{4}$ inch   **51.** 140 inches   **53.** 14 days
55. 12 inches   **57.** $1.14   **59.** 4   **61.** 288   **63.** $2\frac{1}{2}$
65. 1750   **67.** 10,500   **69.** 240   **71.** Niger: 7200;
China: 900   **73.** 25 meters   **75.** 143 feet   **77.** 0.6 inch

CHECKING BASIC CONCEPTS 6.1 & 6.2 (p. 400)

1. (a) $\frac{4}{9}$ (b) $\frac{16}{3}$ **2.** (a) $\frac{2 \text{ inches}}{5 \text{ hours}}$ (b) $\frac{291 \text{ miles}}{4 \text{ hours}}$
3. (a) $9.25/hr   (b) 1.5 in./hr   **4.** (a) $2.95/lb   (b) $0.18/oz
5. (a) Yes   (b) No   **6.** (a) -30 (b) 1   **7.** 3 inches
8. Large: $0.07/oz; small: $0.095/oz; the large drink
9. $1\frac{2}{3}$ **10.** 84 feet

SECTION 6.3 (pp. 406–408)

1. length   **2.** unit   **3.** denominator   **4.** numerator   **5.** area
6. capacity   **7.** volume   **8.** capacity; volume   **9.** weight
10. ounce   **11.** 4 ft   **13.** $\frac{1}{2}$ or 0.5 mi   **15.** 24 in.   **17.** 29 yd
19. 6336 ft   **21.** 1584 in.   **23.** $\frac{1}{12}$ or $0.08\overline{3}$ ft   **25.** 1440 in^2
27. 30,976 yd^2 **29.** 36 yd^2 **31.** $\frac{1}{100}$ or 0.01 mi^2 **33.** 6 c
35. 1 pt   **37.** 41 qt   **39.** 13 pt   **41.** $6\frac{1}{4}$ or 6.25 pt   **43.** 6 pt
45. $3\frac{3}{4}$ or 3.75 gal   **47.** 1 c   **49.** 80 c   **51.** 25 qt
53. $27\frac{1}{2}$ or 27.5 lb   **55.** 122 oz   **57.** 8 oz   **59.** 20 lb
61. 55 T   **63.** $11\frac{1}{4}$ or 11.25 T   **65.** $1\frac{1}{4}$ or 1.25 pt
67. 110 yd   **69.** 11 stones   **71.** (a) 225 ft^2 (b) 25 yd^2
(c) $800   **73.** 16 servings   **75.** 56; 1 foot   **77.** 500

SECTION 6.4 (pp. 415–417)

1. 1000   **2.** deci   **3.** meter   **4.** Decimeter   **5.** liter
6. mL   **7.** gram   **8.** Hectogram   **9.** kiloliter
11. centimeter   **13.** dekameter   **15.** milligram   **17.** 34 dm
19. 1210 hm   **21.** 4.5 dm   **23.** 0.01459 km   **25.** 60 cm
27. 2.5 dm   **29.** 40 cm^2 **31.** 0.07 km^2 **33.** 0.012 hm^2
35. 2500 m^2 **37.** 51,000 dl   **39.** 9000 L   **41.** 130 hl
43. 9 mL   **45.** 120,000 dl   **47.** 0.0005 dl   **49.** 0.0038 hg
51. 0.095 dg   **53.** 4.5 g   **55.** 570 cg   **57.** 8.793 kg
59. 0.55 cg   **61.** 0.5 L   **63.** 23 kg   **65.** (a) 1250 m^2
(b) 12.5 dam^2 **67.** 5.5 t   **69.** 25 servings   **71.** 125   **73.** 75

CHECKING BASIC CONCEPTS 6.3 & 6.4 (p. 417)

1. (a) 6 yd   **(b)** 12 in.   **2.** 1152 in^2 **3. (a)** 36 oz
(b) $5\frac{1}{2}$ or 5.5 gal   **4. (a)** 4 lb   **(b)** 23 lb   **5. (a)** 4.7 m
(b) 0.3 hm   **6.** 30,000 cm^2 **7. (a)** 5800 mL   **(b)** 3.25 dal
8. (a) 83 kg   **(b)** 0.62 cg   **9.** 24 servings   **10.** 50

SECTION 6.5 (pp. 424–427)

1. 2.54   **2.** 1.09   **3.** 0.62   **4.** 1.06   **5.** 29.57   **6.** 2.2
7. 28.35   **8.** Fahrenheit; Celsius   **9.** 43.18 cm   **11.** 220.18 m
13. 135.48 km   **15. (a)** 63.43 yd   **(b)** 63.22 yd
17. (a) 2.79 mi   **(b)** 2.79 mi   **19.** 8.26 m   **21.** 111.55 ft
23. 9.84 in.   **25.** 4.84 km   **27.** 9.81 ft   **29.** 473.12 mL
31. 52.83 L   **33.** 19.88 gal   **35.** 7.55 L   **37.** 4.04 c
39. 14.53 gal   **41.** 14.81 oz   **43.** 134.64 lb   **45.** 3827.25 mg
47. 91.52 oz   **49.** 659.09 g   **51.** 2909.09 mg   **53.** 25°C
55. 37.4°F   **57.** $-26.\overline{1}$°C   **59.** 212°F   **61.** 36.9°C
63. 101.9°F   **65.** 37.0°C   **67.** 68°F (actual: 68°F)
69. -23°C (actual: $-23.\overline{3}$°C)   **71.** 10°C (actual: 10°C)
73. 591 mL   **75.** 50 lb   **77.** 115.7°F
79. Yes (about 60.007 ft)   **81.** 3 L   **83.** No (about 18.46 ft)
85. 3.6 in. by 1.5 in.   **87.** -140°C   **89.** 3.9 kg

SECTION 6.6 (pp. 430–431)

1. days   **2.** 60; 60   **3.** speed   **4.** distance; time
5. 315 min   **7.** 1.5 hr   **9.** 14,760 sec   **11.** 420 hr
13. 1.15 d   **15.** 0.5 yr   **17.** 102 ft/sec   **19.** 12 m/d
21. 0.2 m/sec   **23.** 22.4 in./wk   **25.** 44 ft/sec
27. 420,000 m/hr   **29.** 4000 yd/hr   **31.** 10.8 km/hr
33. 36 ft/sec   **35.** 57 mi/hr   **37.** 180 mi/hr   **39.** 70 mi/hr
41. No (48.4 km/hr)

CHECKING BASIC CONCEPTS 6.5 & 6.6 (p. 431)

1. (a) 6.35 m   **(b)** 1636.8 ft   **(c)** 1182.8 mL   **(d)** 3.18 pt
(e) 2.75 lb   **(f)** 20 kg   **2. (a)** 23°F   **(b)** 70°C   **3.** 90 cm/sec
4. 18 in./d   **5.** 2.9 lb   **6.** No (49.6 mi/hr)

CHAPTER 6 REVIEW (pp. 435–437)

1. $\frac{1}{4}$ **2.** $\frac{4}{11}$ **3.** $\frac{1}{6}$ **4.** $\frac{8}{25}$ **5.** $\frac{0.8}{1}=0.8$ **6.** $\frac{4.\overline{6}}{1}=4.\overline{6}$
7. $\frac{3\ \text{miles}}{16\ \text{minutes}}$ **8.** $\frac{1\ \text{laptop}}{4\ \text{students}}$ **9.** \$7.35/hr   **10.** 74 beats/min
11. \$0.143/lb   **12.** \$0.49/L   **13.** Yes   **14.** No   **15.** No
16. Yes   **17.** -16 **18.** -1 **19.** 3   **20.** $-\frac{7}{6}$ **21.** 7 inches
22. 5 inches   **23.** 150 in.   **24.** 9504 ft   **25.** $\frac{3}{5}$ or 0.6 mi
26. 4 yd   **27.** 30 yd^2 **28.** 1,115,136 ft^2 **29.** 56 qt
30. 3.5 qt   **31.** 17 c   **32.** 44 c   **33.** 120 oz   **34.** 24.5 lb
35. 8 lb   **36.** 41 T   **37.** milligram   **38.** dekaliter
39. 75 cm   **40.** 0.114 dam   **41.** 126 hm   **42.** 1.9 dm
43. 7.8 m^2 **44.** 850 m^2 **45.** 525.3 kl   **46.** 7.5 mL
47. 7.69 dal   **48.** 0.001 dl   **49.** 8.7 dg   **50.** 28 kg
51. 77 cg   **52.** 2.1 g   **53. (a)** 7.20 km   **(b)** 7.19 km
54. (a) 1.55 mi   **(b)** 1.55 mi   **55.** 5.59 m   **56.** 272.80 yd
57. 11.29 km   **58.** 274.32 cm   **59.** 26.5 gal   **60.** 94.62 mL
61. 38.68 L   **62.** 10.40 gal   **63.** 14.33 lb   **64.** 200,000 mg
65. 0.13 oz   **66.** 477.27 g   **67.** 48.2°F   **68.** -25°F
69. 36.8°C   **70.** 98.5°F   **71.** -35°C (actual: $-34.\overline{4}$°C)

72. 77°F (actual: 77°F)   **73.** 72 hr   **74.** 576 min   **75.** 4.7 hr
76. 2.5 wk   **77.** 7500 m/hr   **78.** 2.5 ft/wk   **79.** 40 in./hr
80. 480,000 m/hr   **81.** 121 km/hr   **82.** 392 ft/sec
83. University: 51.6; college: 38.2; there are 38.2 students
for each instructor.   **84.** 8 students   **85.** 337; 80 in.
86. 20 servings   **87.** 233 lb   **88.** Yes (56.45 km/hr)
89. 117 ft   **90.** Large: \$0.195/oz; small: \$0.205/oz; the
large coffee drink   **91.** 22,700 g   **92.** More   **93.** 50°C
94. -297.4°F

CHAPTER 6 TEST (p. 438)

1. $\frac{1}{3}$ **2.** 0.125 mi/min   **3.** \$0.215/oz   **4.** No   **5.** 27
6. $-\frac{1}{42}$ **7.** 2.1 ft   **8.** 234 in.   **9.** 3.75 ft^2 **10.** 68 pt
11. 17.5 T   **12.** 0.0154 m   **13.** 0.54 km^2 **14.** 75 mL
15. 2.4 hg   **16.** 0.32 km   **17.** 1.82 c   **18.** 1136.36 g
19. 59°F   **20.** 21°C (actual: 21.$\overline{1}$°C)   **21.** 28.5 d
22. 259,200 sec   **23.** 300 ft/hr   **24.** 3226 m/d
25. Large: \$0.13/oz; small: \$0.148/oz; the large fruit smoothie
26. 8   **27.** Tuesday

CHAPTERS 1–6 CUMULATIVE REVIEW (pp. 439–440)

1. Thirty-four thousand two hundred six   **2.** Unlike
3.

4. 730,000   **5.** <   **6.** -16 **7.** -60 **8.** 1   **9.** $4x$
10. 4   **11.** Table values: $-9, -7, -5, -3, -1$; solution: 2
12. 9   **13.** $\frac{23}{10}$; $2\frac{3}{10}$ **14. (a)** Yes   **(b)** No   **15.** $\frac{2}{9}$ **16.** $\frac{7}{10}$
17. 594.03   **18.** 64   **19.** 100π; 314 square feet   **20.** 3.00
21. 58 mi/hr   **22.** -20 **23.** 80 bpm   **24.** 2010   **25.** 4 in.
26. $29\frac{3}{4}$ in.   **27.** 739   **28.** 27 in.

7   Percents

SECTION 7.1 (pp. 447–450)

1. percent   **2.** %   **3.** $\frac{x}{100}$ **4.** $0.01x$ **5.** left   **6.** 100%
7. right   **8.** circle graph   **9.** $\frac{7}{25}$ **11.** $\frac{3}{80}$ **13.** $\frac{11}{20}$
15. $\frac{1}{25}$ **17.** $\frac{3}{40}$ **19.** $\frac{29}{25}$ or $1\frac{4}{25}$ **21.** $\frac{33}{400}$ **23.** $\frac{1}{3}$
25. 0.58   **27.** 0.095   **29.** 1.73   **31.** 0.06   **33.** 0.0025
35. 0.003   **37.** 1.16   **39.** 0.084   **41.** 70%   **43.** 22.5%
45. 550%   **47.** 18%   **49.** 66.$\overline{6}$%   **51.** 58.$\overline{3}$%   **53.** 125%
55. 81%   **57.** 1%   **59.** 160%   **61.** 7.2%   **63.** 299%
65. 0.5%   **67.** 4.01%
69.

Percent	Decimal	Fraction
80%	0.8	$\frac{4}{5}$
30%	0.3	$\frac{3}{10}$
65%	0.65	$\frac{13}{20}$
22.5%	0.225	$\frac{9}{40}$
260%	2.6	$\frac{13}{5}$ or $2\frac{3}{5}$
40%	0.4	$\frac{2}{5}$

71. (a) Recycling (b) $\frac{1}{4}$ (c) 0.33 **73.** (a) 10% (b) $\frac{8}{25}$ (c) 0.05 **75.** $2.\overline{3}\%$ **77.** $\frac{3}{5}$ **79.** 110%; Sweden has more cell phones than people. **81.** 0.236 **83.** $\frac{13}{25}$ **85.** 60%

SECTION 7.2 (pp. 455–456)

1. whole; part **2.** of **3.** multiply; equals
4. percent · whole = part **5.** problem **6.** decimal
7. 15% of 120 is 18 **9.** 150% of 42 is 63
11. 2.5% of 760 is 19 **13.** 80% of 5 is 4
15. 85% of 20 is 17 **17.** $0.18 \cdot 40 = 7.2$
19. $1.4 \cdot 35 = 49$ **21.** $0.64 \cdot 75 = 48$
23. $0.125 \cdot 24 = 3$ **25.** $0.625 \cdot 24 = 15$
27. $0.68 \cdot x = 17$ **29.** $x \cdot 95 = 39.9$ **31.** $1.04 \cdot 70 = x$
33. $x \cdot 60 = 48$ **35.** $x \cdot 9 = 3$ **37.** 3 **39.** $33.\overline{3}\%$
41. 6 **43.** 748.6 **45.** 60 **47.** 150 **49.** $66.\overline{6}\%$ **51.** 3550
53. 17.5% **55.** 27.5 **57.** 20% **59.** 76

CHECKING BASIC CONCEPTS 7.1 & 7.2 (p. 456)

1. (a) $\frac{19}{25}$ (b) $\frac{1}{15}$ (c) $\frac{12}{125}$ **2.** (a) 0.065 (b) 2.14 (c) 0.028
3. (a) 12% (b) 350% (c) 87.5% **4.** (a) 29% (b) 7.3%
(c) 297% **5.** (a) 350% of 14 is 49 (b) 44% of 25 is 11
6. (a) $1.8 \cdot 30 = 54$ (b) $0.25 \cdot 36 = 9$
7. (a) $0.75 \cdot x = 64$ (b) $x \cdot 25 = 5$ **8.** (a) 20%
(b) 45 (c) 130.8 **9.** $\frac{23}{50}$ **10.** (a) Upper class
(b) $\frac{8}{25}$ (c) 0.2

SECTION 7.3 (p. 462)

1. proportion **2.** equal **3.** percent **4.** $\frac{percent}{100} = \frac{part}{whole}$
5. $\frac{24}{100} = \frac{16.8}{70}$ **7.** $\frac{180}{100} = \frac{45}{25}$ **9.** $\frac{40}{100} = \frac{34}{85}$ **11.** $\frac{12.5}{100} = \frac{8}{64}$
13. $\frac{62.5}{100} = \frac{25}{40}$ **15.** $\frac{48}{100} = \frac{19}{x}$ **17.** $\frac{x}{100} = \frac{38.4}{114}$ **19.** $\frac{x}{100} = \frac{24}{80}$
21. $\frac{42.5}{100} = \frac{39}{x}$ **23.** $\frac{99}{100} = \frac{297}{x}$ **25.** 18 **27.** $66.\overline{6}\%$
29. 400% **31.** 27 **33.** 40 **35.** 25 **37.** 130 **39.** 35
41. $33.\overline{3}\%$ **43.** 32.5 **45.** 20% **47.** 23.5%

SECTION 7.4 (pp. 470–472)

1. *percent* **2.** sales **3.** total price **4.** discount
5. sale price **6.** commission **7.** gross pay **8.** withholdings
9. net pay **10.** increase; decrease **11.** $279 **13.** 720
15. $1.9345 million **17.** 185 pounds **19.** 20,000
21. $70,000 **23.** 7% **25.** 65% **27.** $26.\overline{6}\%$
29. $4.16; $68.16 **31.** $11.16; $159.96 **33.** 4%
35. $160 **37.** $140; $420 **39.** $53.55; $22.95 **41.** 18%
43. $35 **45.** $11,400 **47.** $4370 **49.** $8700 **51.** 6%
53. $177.60; $562.40 **55.** 12.5% **57.** $3900
59. 20% increase **61.** 58% increase **63.** 12% decrease

CHECKING BASIC CONCEPTS 7.3 & 7.4 (p. 473)

1. (a) $\frac{64}{100} = \frac{144}{x}$ (b) $\frac{x}{100} = \frac{80}{60}$ (c) $\frac{99}{100} = \frac{x}{70}$
2. (a) 200% (b) 15 (c) 25 (d) $16.\overline{6}\%$
3. 15 milligrams **4.** 3800 **5.** $7.15; $137.15
6. $12.16; $3.04 **7.** $5112 **8.** 2% decrease

SECTION 7.5 (pp. 479–480)

1. principal **2.** interest **3.** interest rate **4.** simple
5. annual **6.** decimal; years **7.** principal; interest
8. compound **9.** annual percentage rate **10.** decimal; years
11. $32 **13.** $21 **15.** $520 **17.** $131.25 **19.** $1
21. $14,000 **23.** $729.60 **25.** $2320 **27.** $922.50
29. 4 years **31.** 6% **33.** $7000 **35.** $1856.03
37. $1348.35 **39.** $275.38 **41.** $6087.60
43. $111,177.59

SECTION 7.6 (pp. 486–487)

1. experiment **2.** outcome **3.** event **4.** probability
5. percent chance **6.** impossible event **7.** certain event
8. complement **9.** True, false **11.** 1, 2, 3, 4, 5, 6, 7, 8
13. Monday, Tuesday, Wednesday, Thursday, Friday
15. Certain; 1 **17.** Impossible; 0 **19.** Certain; 1
21. 50% **23.** 25% **25.** 40% **27.** $\frac{1}{3}$ **29.** $\frac{1}{2}$ **31.** $\frac{3}{5}$
33. $\frac{1}{52}$ **35.** $\frac{3}{26}$ **37.** $\frac{1}{2}$ **39.** $\frac{1}{12}$ **41.** $\frac{1}{3}$ **43.** 20%
45. $\frac{3}{10}$ **47.** 40%

CHECKING BASIC CONCEPTS 7.5 & 7.6 (p. 487)

1. $2.50 **2.** $1227 **3.** 9 months **4.** $25,776.80 **5.** 0
6. $\frac{3}{52}$ **7.** 80% **8.** $\frac{2}{11}$

CHAPTER 7 REVIEW (pp. 491–494)

1. $\frac{17}{25}$ **2.** $\frac{9}{5}$ or $1\frac{4}{5}$ **3.** $\frac{1}{16}$ **4.** $\frac{9}{125}$ **5.** 0.29 **6.** 0.004
7. 6.25 **8.** 0.3575 **9.** 68% **10.** 320% **11.** $77.\overline{7}\%$
12. 31.25% **13.** 16% **14.** 4.9% **15.** 702%
16. 3.05% **17.** Tuition **18.** $\frac{3}{25}$ **19.** 130% of 40 is 52
20. 75% of 12 is 9 **21.** $0.775 \cdot 80 = 62$
22. $0.50 \cdot 184 = 92$ **23.** $0.75 \cdot x = 46.5$
24. $x \cdot 72 = 48$ **25.** 13.6 **26.** 35 **27.** $33.\overline{3}\%$
28. 60% **29.** $\frac{240}{100} = \frac{96}{40}$ **30.** $\frac{80}{100} = \frac{28}{35}$ **31.** $\frac{165}{100} = \frac{x}{20}$
32. $\frac{x}{100} = \frac{3}{9}$ **33.** 7.8 **34.** $66.\overline{6}\%$ **35.** 50% **36.** 4400
37. 2.8 grams **38.** 3120 **39.** 4% **40.** 35%
41. $10.89; $208.89 **42.** 4% **43.** $76; $304 **44.** 17%
45. $187.50 **46.** $555 **47.** 12% **48.** $196.80; $623.20
49. $22.\overline{2}\%$ **50.** $2280 **51.** $33.\overline{3}\%$ increase
52. 20% decrease **53.** $7.20 **54.** $9276 **55.** $2520
56. $26,880 **57.** $6660 **58.** $1245 **59.** $3900
60. 9 months **61.** $114,840.86 **62.** $8933.82
63. $58,533.43 **64.** $1512.22 **65.** Red, black, blue
66. 2, 3, 5, 7, 11, 13, 17, 19 **67.** Certain; 1 **68.** Impossible; 0
69. 25% **70.** 37.5% **71.** $\frac{1}{3}$ **72.** $\frac{4}{5}$ **73.** $\frac{1}{13}$ **74.** $\frac{5}{26}$
75. 25% **76.** $\frac{3}{11}$ **77.** 66% **78.** $\frac{2}{5}$ **79.** 0.428 **80.** 62.5%

CHAPTER 7 TEST (pp. 494–495)

1. $\frac{11}{20}$ **2.** $\frac{13}{5}$ or $2\frac{3}{5}$ **3.** 95% **4.** 7.8% **5.** $x \cdot 160 = 32$
6. $\frac{12}{100} = \frac{x}{90}$ **7.** 140 **8.** $33.\overline{3}\%$ **9.** 95% **10.** 800
11. $31.50 **12.** $26,000 **13.** 3% **14.** $2391.24
15. 2, 4, 6, 8, 10 **16.** Impossible; 0 **17.** $\frac{1}{2}$ **18.** $16.\overline{6}\%$

19. $\frac{2}{13}$ **20.** 62%   **21.** 140 pounds   **22.** $64,000
23. $17.45; $366.45   **24.** $11.55; $7.70   **25.** $4620
26. 15% increase

CHAPTERS 1–7 CUMULATIVE REVIEW (pp. 495–496)

1. 46,391   **2.** 658,000   **3.** 11   **4.** $9w - 10$ **5.** -21
6. -72 **7.** -2 **8.** 1   **9.** $-7t - 8$ **10.** $2(x + 5) = -14$
11. -16 **12.** Table values: 12, 9, 6, 3, 0; solution: 1
13. $3 \cdot 5 \cdot 7$ **14.** $-\frac{1}{14}$ **15.** $6\frac{1}{3}$ **16.** $\frac{47}{48}$ **17.** 468.05
18. (a) 9   (b) 9   (c) 9   (d) $9, -\frac{1}{5}, 4.\overline{6}, -2.1$ (e) $\sqrt{7}$
19. 6   **20.** 49π; 153.86 square feet   **21.** 13,728 ft
22. 8.9 mL   **23.** $-35°C$ **24.** 300,000 m/hr   **25.** 1.37
26. $33.\overline{3}\%$ **27.** 126   **28.** $\frac{3}{26}$ **29.** 177 bpm   **30.** 2008
31. 3 miles   **32.** $26\frac{1}{4}$ inches   **33.** 521
34. Large: $0.175/oz; small: $0.185/oz; the large fruit drink
35. $9.50; $28.50   **36.** 12% decrease

8   Exponents and Polynomials

SECTION 8.1 (pp. 502–503)

1. x; 6   **2.** a^{m+n} **3.** $a^{m \cdot n}$ **4.** $a^n b^n$ **5.** add
6. multiply   **7.** $8 \cdot 8 \cdot 8$ **9.** $y \cdot y \cdot y \cdot y \cdot y \cdot y$
11. $(-7) \cdot (-7) \cdot (-7)$ **13.** $-(3 \cdot 3)$
15. $(-2x) \cdot (-2x) \cdot (-2x) \cdot (-2x)$ **17.** x^5 **19.** m^{13}
21. w^8 **23.** x^{11} **25.** a^{14} **27.** $28x^5$ **29.** $54y^6$
31. $-18m^{11}$ **33.** $6x^6$ **35.** x^{10} **37.** n^{27} **39.** m^{14}
41. q^{11} **43.** $9x^2$ **45.** $-64y^{12}$ **47.** x^6y^2 **49.** $v^{12}w^8$
51. $16x^8y^{12}$ **53.** $-27a^{15}b^9$ **55.** $81x^{16}y^6$ **57.** $200p^8q^9$
59. $-72a^{14}b^{12}$ **61.** $4x^5$ square inches
63. $3x^3$ square centimeters

SECTION 8.2 (pp. 510–511)

1. a^{m-n} **2.** 1   **3.** $\frac{1}{a^n}$ **4.** scientific notation   **5.** positive
6. negative   **7.** $3^3 = 27$ **9.** w^7 **11.** $-a^3$ **13.** x^4
15. $5x$ **17.** $2w^3$ **19.** $-3b^6$ **21.** 1   **23.** 1   **25.** -1
27. 32   **29.** -14 **31.** $\frac{1}{36}$ **33.** $\frac{1}{8}$ **35.** $\frac{1}{x^5}$ **37.** $\frac{7}{a^3}$
39. $-\frac{24}{n^6}$ **41.** No   **43.** Yes   **45.** No   **47.** Yes
49. 6,430,000,000   **51.** 0.00059   **53.** 41,100   **55.** 3
57. 8.64×10^4 **59.** 6.8×10^{-9} **61.** 5.9472×10^{24}
63. 3,600,000; 3.6×10^6 **65.** 200,000; 2×10^5
67. 0.0122; 1.22×10^{-2} **69.** 0.000025; 2.5×10^{-5}
71. 8.6×10^9 **73.** 8.6×10^{10}
75. (a) 3.1×10^8; 1.302×10^{13} (b) $42,000

CHECKING BASIC CONCEPTS 8.1 & 8.2 (p. 512)

1. (a) y^{10} (b) $-18m^6$ **2.** (a) x^{12} (b) b^{11} (c) $8x^{15}$
(d) n^{12} (e) $49x^4y^6$ (f) $-72a^{11}b^8$
3. (a) w^8 (b) $-8n^4$ (c) 1   (d) $\frac{16}{m^3}$
4. (a) 0.0000028   (b) 518,000
5. (a) 3.4×10^7 (b) 9.5×10^{-6}
6. (a) 50,000; 5×10^4 (b) 0.00014; 1.4×10^{-4}

SECTION 8.3 (pp. 519–521)

1. monomial   **2.** degree   **3.** coefficient   **4.** 1; -1
5. polynomial   **6.** term   **7.** descending   **8.** sign
9. opposite   **10.** like   **11.** Yes; 2; 5   **13.** No
15. Yes; 5; 4   **17.** Yes; 1; 1   **19.** 7   **21.** 15   **23.** 24
25. 49   **27.** $2x + 3$ **29.** $-3n^2 + 2n + 4$
31. $3x^3 + x^2 - 5x$ **33.** $4n^3 - 3n^2 + n - 2$
35. $10y + 6$ **37.** 1   **39.** $a^2 + 4a - 1$
41. $-x^2 + 5x + 7$ **43.** $7w^3 + 3w^2 + 4w - 8$
45. $3x^3 - 2x^2 - 5x + 7$ **47.** 0   **49.** $a - 8$
51. $5x + 7$ **53.** $8w^2 - 4w + 9$ **55.** $4x + 13$
57. $-2m^2 - 3m + 19$ **59.** $y^3 - 3y^2 - 2y + 6$
61. $-165n^2 + 12n + 17$ **63.** $5a - 10$ **65.** 8
67. $6n^2 + 4n - 14$ **69.** $-9x^2 + 3x - 19$
71. $-3m^3 + 3m^2 - 4m - 8$ **73.** $-9k^2 + 2k - 13$
75. $-7g - 10$ **77.** $-5x + 4$ **79.** $14w^2 - 4w + 18$
81. 1 second: 74 ft; 3 seconds: 122 ft   **83.** 60 bpm
85. 0 ft; The object is on the ground.   **87.** $1800
89. $10x + 7$ **91.** $4x + 8$
93. $3x^2 - x + 5$ **95.** $-x^2 + 22x + 4$

SECTION 8.4 (pp. 530–531)

1. coefficients   **2.** product   **3.** distributive   **4.** term; term
5. like terms   **6.** area   **7.** factor   **8.** GCF   **9.** factored
10. monomial   **11.** x^5 **13.** $18y^7$ **15.** $-28g^3$ **17.** $6k^5$
19. $-12x^6$ **21.** $12y^4 + 6y$ **23.** $-3x^5 + 15x^3$
25. $-8m^7 + 20m^2$ **27.** $14y^4 - 21y^3 - 56y^2$
29. $6n^3 - 10n^2 + 2n$ **31.** $6x^2 - 5x - 4$
33. $8g^3 - 6g^2 + 20g - 15$ **35.** $-12k^4 + 14k^2 - 4$
37. $b^3 - 9b^2 + 20b - 12$ **39.** $6x^3 - 9x - 3$
41. $12k^2 - 7k - 10$ **43.** $m^3 - 6m^2 + 17m - 36$
45. $10x^3 + 14x^2 + 24$ **47.** $x^2 + 11x + 24$
49. $4m^2 + 47m + 33$ **51.** 2   **53.** 10   **55.** 9   **57.** $3x$
59. $9y^3$ **61.** $3m^2$ **63.** 3   **65.** $3n^2$ **67.** $4a^2$
69. $7(x - 3)$ **71.** $5m^2(3m + 1)$ **73.** $16a(3a^3 - 2)$
75. $4q^2(9q^3 - 5q^2 + 4)$ **77.** $9x^2(9x^2 - 5x - 1)$
79. (a) 21 thousand games   (b) 15 thousand games
(c) $25p - \frac{1}{4}p^2$ (d) $400 thousand
81. $x(x + 1)$ or $x^2 + x$ **83.** $3x + 4$

CHECKING BASIC CONCEPTS 8.3 & 8.4 (p. 532)

1. (a) No   (b) Yes; 2; 4   **2.** 15   **3.** (a) $8m^2 + m - 5$
(b) $4x^2 + 11x + 1$ **4.** $5x^3 - 2x^2 + 7x - 12$
5. (a) $12y^2 - 11y - 7$ (b) $3n^2 - 5$
6. (a) $-5w^4$ (b) $2x^3 - 10x^2$ (c) $-15k^3 + 6k^2 + 5k - 2$
7. 6   **8.** $7x$ **9.** (a) $25b^3(2b + 3)$ (b) $16p^2(p^2 + 4p + 3)$
10. 6847

CHAPTER 8 REVIEW (pp. 535–536)

1. $17 \cdot 17 \cdot 17 \cdot 17 \cdot 17$ **2.** $(xy) \cdot (xy)$ **3.** x^9 **4.** $-5n^{10}$
5. y^8 **6.** q^7 **7.** $9w^8$ **8.** $-8p^{12}$ **9.** $-a^9b^{12}$ **10.** $4x^6y^8$
11. x^5 **12.** $5y^2$ **13.** $-3w^4$ **14.** $-5k^3$ **15.** 1   **16.** 22
17. $\frac{1}{64}$ **18.** $\frac{1}{y^3}$ **19.** $\frac{1}{a}$ **20.** $-\frac{24}{n^6}$ **21.** No   **22.** Yes
23. 499,000,000   **24.** 0.0000092   **25.** 3.9×10^{10}

26. 4.87×10^{-5} **27.** $10,000; 1 \times 10^4$
28. $40,000,000; 4 \times 10^7$ **29.** Yes; 3; 4   **30.** No   **31.** 30
32. -4 **33.** $x^4 + 7x^3 - 8x$ **34.** $-7y^3 + 3y^2 + y - 4$
35. $10n$ **36.** $9x - 11$ **37.** $9k^2 + 2k + 1$
38. $3m^3 + 5m^2 + m - 4$ **39.** $2x^2 + 7$
40. $4w^2 - 2w + 12$ **41.** $-15h^2 - 9h + 3$
42. $-2b^3 + b^2 - b - 4$ **43.** $7a - 6$ **44.** 21
45. $k^2 - 3k + 7$ **46.** $6q^3 - q^2 + q + 12$
47. $3n^2 + 5n - 5$ **48.** $8w^2 - 5w + 22$ **49.** $56n^7$
50. $-6x^8$ **51.** $-18g^5 + 30g^2$ **52.** $14a^5 - 6a^4 + 2a^3$
53. $18k^4 - 2k^2 - 16$ **54.** $2y^4 - 35y^2 + 12y$
55. $8x^3 - 16x + 3$ **56.** $35m^2 + 43m + 12$ **57.** 4
58. 6   **59.** $6x^2$ **60.** $2b^2$ **61.** $6m^3(2m + 1)$
62. $5g(4g - 3)$ **63.** $7x^2(2x^3 - 5x - 7)$
64. $6y^3(5y^2 - 8y + 1)$ **65.** $18x^3$ square inches
66. 6.31×10^8 **67.** 2 seconds: 83 ft; 3 seconds: 75 ft
68. $12x + 2$ **69.** $3x + 10$ **70.** \$2000

CHAPTER 8 TEST (pp. 536–537)

1. (a) $9p^6$ **(b)** q^9 **(c)** $\frac{17}{n^3}$ **(d)** 1   **(e)** $4y^3$
2. 0.000083   **3.** 3.75×10^4
4. (a) $23,000,000; 2.3 \times 10^7$ **(b)** $0.00075; 7.5 \times 10^{-4}$
5. -22 **6.** $-2x^4 + 5x^3 - 3x$
7. (a) $7k^2 - 8k + 4$ **(b)** $5p^2 + 5p - 8$
8. (a) $15b^5 - 18b^4 + 6b^3$ **(b)** $14m^2 - 29m - 15$
(c) $5y^3 - 18y^2 - 9y + 4$ **9. (a)** 5   **(b)** $18x$
10. (a) $6x(2x^2 - x - 3)$ **(b)** $4w^3(4w^4 - 7)$
11. $x^2 + 7x + 10$ square inches
12. (a) 180 bpm   **(b)** 93 bpm   **13.** $-x^2 + 24x + 1$
14. $2x + 5$ **15.** 4.3×10^{-3}

CHAPTERS 1–8 CUMULATIVE REVIEW (pp. 537–538)

1. 3   **2.** 9^5 **3.** 1300   **4.** $<$ **5.** 4   **6.** No
7. Table values: 11, 8, 5, 2, -1; solution: 2
8. $11 = x + 4$ **9.** -2 **10.** 8   **11.** 0   **12.** $\frac{27}{10}; 2\frac{7}{10}$
13. $\frac{17}{20}$ **14.** $\frac{1}{6}$ **15.** 0.35   **16.** $<$ **17.** 3.2   **18.** 57 mi/hr
19. -20 **20.** 7.5 d   **21.** 12.5%   **22.** 105   **23.** $\frac{3}{26}$
24. 0.00069   **25.** $3x^3 + 12x^2$ **26.** $20y^3 - 5y^2 + 28y - 7$
27. $12x$ **28.** $14y^3(y^3 - 2)$ **29.** 90 minutes
30. $x + 13 = 2x - 9$ **31.** $\frac{1}{3}$ **32.** 2009   **33.** 560
34. 132 feet   **35.** \$8.01; \$186.01   **36.** 336 ft

9   Introduction to Graphing

SECTION 9.1 (pp. 546–548)

1. xy-plane   **2.** $x; y$ **3.** origin   **4.** quadrants
5. coordinates; ordered pair   **6.** $x; y$ **7.** scatterplot
8. line graph   **9.** A: QI; B: x-axis; C: QIII
11. A: QII; B: QI; C: y-axis   **13.** QII   **15.** QIII   **17.** x-axis
19. QIV   **21.** y-axis   **23.** QI

25. A: $(3, 2)$; B: $(-3, 0)$; C: $(-2, -5)$
27. A: $(1, -5)$; B: $(-3, 3)$; C: $(0, 0)$
29. A: $(5, -3)$; B: $(0, 1)$; C: $(-4, -2)$
31.

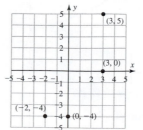

33.

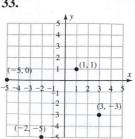

35.

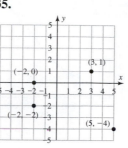

37.

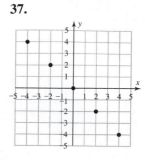

39.

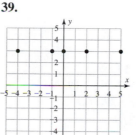

41.

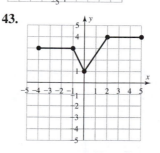

43.

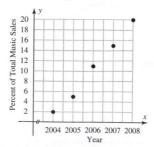

45. Scatterplot   Line graph

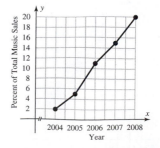

47. Scatterplot   Line graph

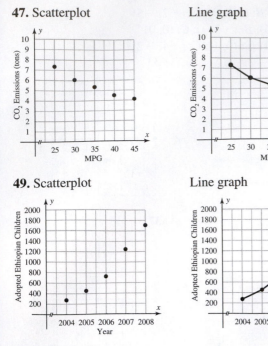

49. Scatterplot   Line graph

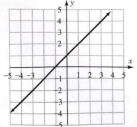
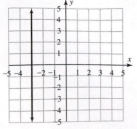

SECTION 9.2 (pp. 558–560)

1. linear   **2.** infinite   **3.** ordered pair   **4.** table
5. either   **6.** line   **7.** horizontal   **8.** vertical
9. Yes, $A = 4, B = 3, C = 1$
11. Yes, $A = 2, B = -5, C = 7$ **13.** No
15. Yes, $A = 1, B = 0, C = -5$ **17.** No   **19.** No
21. No   **23.** Yes   **25.** Yes   **27.** No   **29.** Yes
31. Table values: $-12, -7, -2, 3$
33. Table values: $-2, 0, 2, 4$
35. Table values: $2, 2, 2, 2$
37. Table values: $6, -5, 2, 5$
39. Table values: $-2, 0, 0, -4$
41. Table values: $-1, -6, -3, 2$

43.   **45.**

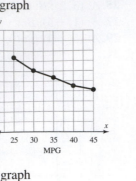
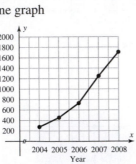

47.   **49.**

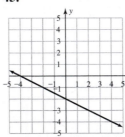
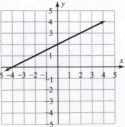

51.
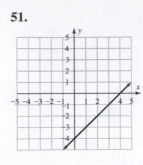

53.
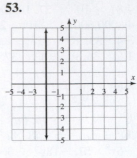

55.
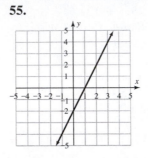

57.
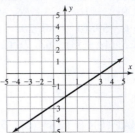

59.
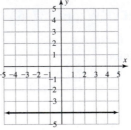

61.
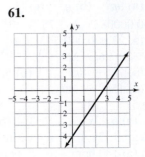

63.
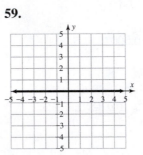

65.

67.
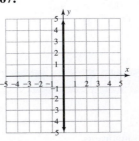

69.
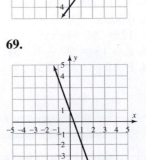

71.

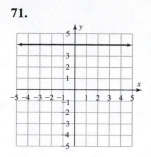

73.

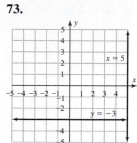

75.

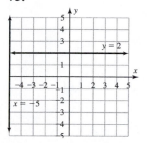

CHECKING BASIC CONCEPTS 9.1 & 9.2 (p. 560)

1. (a) QII   **(b)** y-axis   **2.** A: $(0, -2)$; B: $(-4, -5)$; C: $(5, 3)$

3.

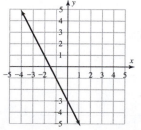

4.

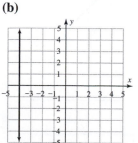

5. (a) Yes, $A = 5, B = -2, C = -4$ **(b)** No
6. Yes   **7.** Table values: 3, 1, 4, 0
8. (a)   **(b)**

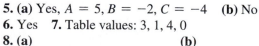

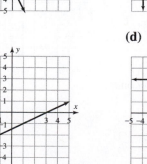

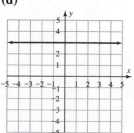

SECTION 9.3 (pp. 565–566)

1. graphical   **2.** $y = ax + b$ **3.** check   **4.** same
5. 1   **7.** 2   **9.** 0   **11.** -3 **13.** -4 **15.** 3   **17.** 0
19. -6 **21.** 9   **23.** 6   **25.** 8
27. (a) -1 **(b)** -1 **(c)** -1 **29. (a)** -8 **(b)** -8 **(c)** -8
31. (a) 2   **(b)** 2   **(c)** 2   **33. (a)** 6   **(b)** 6   **(c)** 6
35. (a) 2   **(b)** 2   **(c)** 2

SECTION 9.4 (pp. 573–576)

1. linear   **2.** y **3.** Vermont: \$25,000; Louisiana: \$17,500
5. 2001: 15; 2005: 10
7. (a) $(4, 12), (7, 21)$ **(b)**

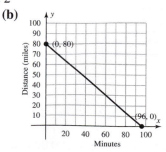

(c) \$18   **(d)** 3
9. (a) $(0, 120), (8, 0)$ **(b)**

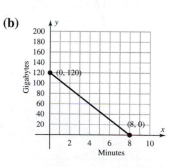

(c) 30   **(d)** 6
11. 2010   **13.** 75 inches (or 6 feet 3 inches)   **15.** 2005
17. (a) 250 mi   **(b)** 4.5 h   **19.** Away
21. The car is parked.

CHECKING BASIC CONCEPTS 9.3 & 9.4 (p. 576)

1. -1 **2. (a)** 1   **(b)** -3 **3.** 2
4. (a) $(0, 80), (96, 0)$ **(b)**

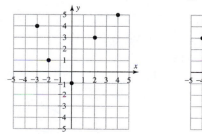

(c) 60 miles   **(d)** 72 min

CHAPTER 9 REVIEW (pp. 581–582)

1. *A*: QII; *B*: *x*-axis; *C*: QIII
2. *A*: $(-2, 5)$; *B*: $(0, -3)$; *C*: $(3, -4)$
3. QIII   4. *x*-axis
5.   6.

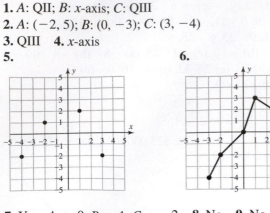   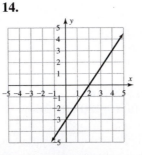

7. Yes; $A = 0, B = 1, C = -2$ 8. No   9. No   10. Yes
11. Table values: $-1, -2, -3, -4$
12. Table values: $-9, 0, 2, 3$
13.   14.

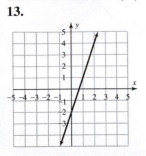   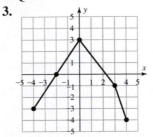

15.   16.

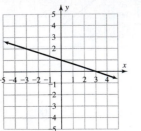   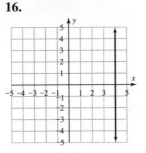

17.   18.

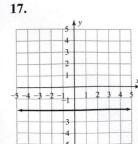   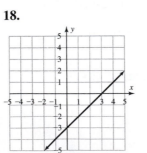

19. 0   20. 2   21. 6   22. 2   23. (a) -2 (b) -2 (c) -2
24. (a) 9   (b) 9   (c) 9
25. Accountants: 500,000; middle school teachers: 250,000
26. 450,000

27. (a) $(200, 10), (700, 35)$ (b)

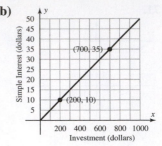

(c) $25   (d) $600
28. 2009   29. (a) 175 mi   (b) 5 h
30. The hiker has stopped.

CHAPTER 9 TEST (p. 583)

1. *A*: $(4, 5)$, QI; *B*: $(0, -4)$, *y*-axis; *C*: $(-3, 1)$, QII
2. QIV
3.

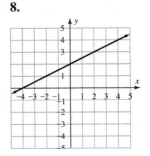

4. Yes; $A = 1, B = -1, C = -5$ 5. No
6. Table values: $21, 0, 3, -15$
7.   8.

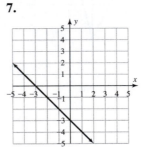   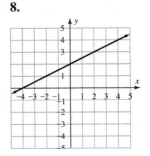

9.   10.

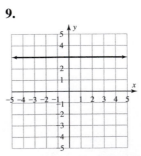   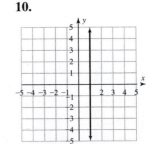

11. 2   12. -2 13. 7 seconds
14. (a) 180 ft   (b) 45 min

CHAPTERS 1–9 CUMULATIVE REVIEW (pp. 584–585)

1. 2   **2.** 74,000   **3.** 8   **4.** 22   **5.** -10 **6.** Yes
7. Table values: $-11, -8, -5, -2, 1$; solution: 1
8. $-7w - 3$ **9.** $4 = x - 8$ **10.** Yes   **11.** -2 **12.** -2
13. $\frac{13}{6}$; $2\frac{1}{6}$ **14.** $\frac{11}{2}$ **15.** -1 **16.** $-\frac{1}{10}$ **17.** 0.35   **18.** <
19. 3.69   **20.** 4.5   **21.** 56 mi/hr   **22.** -42 **23.** 5.1 m²
24. 336 hr   **25.** 1.93   **26.** 40%   **27.** 266   **28.** $\frac{1}{13}$
29. 0.0047   **30.** $2k^2 - k + 5$ **31.** $6x$ **32.** $5y^3(2y^2 - 5)$
33.   **34.**

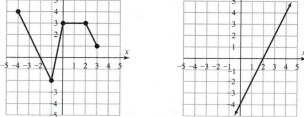

35. 1   **36.** 6   **37.** 46 cm   **38.** $145   **39.** 96 bpm
40. Yes (88.7 km/hr)   **41.** 120 pounds   **42.** $3x + 7$

10 Geometry

SECTION 10.1 (pp. 595–597)

1. point   **2.** line   **3.** line segment   **4.** ray   **5.** angle
6. vertex   **7.** right   **8.** straight   **9.** acute
10. obtuse   **11.** congruent   **12.** complementary
13. supplementary   **14.** parallel   **15.** intersecting
16. vertical   **17.** adjacent   **18.** perpendicular
19. corresponding   **20.** alternate interior   **21.** alternate
exterior   **22.** supplementary   **23.** Line AB, or $\overleftrightarrow{AB}$
25. $\angle m$ **27.** Ray CD, or $\overrightarrow{CD}$ **29.** Point K
31. Line segment PQ, or $\overline{PQ}$ **33.** (a) N (b) $\overrightarrow{NQ}$ and $\overrightarrow{NR}$
(c) $\angle PNQ$, $\angle QNR$, and $\angle PNR$ (d) The naming does not
clearly define a single angle.   **35.** Right   **37.** Acute
39. Obtuse   **41.** Straight   **43.** 166°   **45.** 109°   **47.** 81°
49. $m\angle a = 49°$; $m\angle b = 131°$; $m\angle d = 131°$
51. $m\angle j = 96°$; $m\angle l = 96°$; $m\angle m = 84°$
53. Congruent   **55.** Congruent   **57.** Supplementary
59. Supplementary   **61.** Congruent   **63.** Congruent
65. $m\angle b = 119°$; $m\angle g = 119°$; $m\angle h = 61°$
67. $m\angle a = 92°$; $m\angle c = 88°$; $m\angle h = 92°$

SECTION 10.2 (pp. 603–604)

1. acute   **2.** obtuse   **3.** right   **4.** scalene   **5.** isosceles
6. equilateral   **7.** 180°   **8.** congruent   **9.** ASA   **10.** SAS
11. SSS   **12.** AAA   **13.** Right   **15.** Acute   **17.** Obtuse
19. Acute   **21.** Obtuse   **23.** Right   **25.** Isosceles
27. Equilateral   **29.** Scalene   **31.** Equilateral   **33.** Scalene
35. Isosceles   **37.** 51°   **39.** 34°   **41.** 118°   **43.** 60°
45. 15°   **47.** 36°   **49.** 18°   **51.** 50°   **53.** SSS   **55.** SAS
57. ASA   **59.** Congruent; ASA   **61.** Not congruent
63. Congruent; SSS

CHECKING BASIC CONCEPTS 10.1 & 10.2 (p. 605)

1. (a) $\angle n$ (b) Ray AB, or $\overrightarrow{AB}$ **2.** (a) Right   (b) Acute
3. 71°   **4.** 66°   **5.** $m\angle x = 19°$; $m\angle w = 161°$; $m\angle z = 19°$
6. $m\angle d = 41°$; $m\angle e = 41°$; $m\angle g = 139°$
7. (a) Obtuse   (b) Right   **8.** (a) Equilateral   (b) Scalene
9. 25°   **10.** Congruent; SAS

SECTION 10.3 (pp. 612–613)

1. polygon   **2.** hexagon   **3.** heptagon   **4.** regular
5. $(n - 2) \cdot 180°$ **6.** $\frac{(n - 2) \cdot 180°}{n}$ **7.** circle; center
8. radius   **9.** diameter   **10.** $2r$; $\frac{d}{2}$ **11.** It has a curved side.
13. The sides intersect.   **15.** It is not closed.   **17.** Pentagon
19. Heptagon   **21.** Triangle   **23.** Quadrilateral
25. Hexagon   **27.** Octagon   **29.** 1080°   **31.** 360°
33. 900°   **35.** 120°   **37.** 135°   **39.** 108°
41. Parallelogram   **43.** Rhombus, parallelogram
45. Trapezoid   **47.** Rectangle, parallelogram   **49.** Kite
51. 9 m   **53.** $\frac{2}{9}$ ft   **55.** 0.15 cm   **57.** 16 mi   **59.** 9 ft
61. 5 m   **63.** 135°   **65.** 540°

SECTION 10.4 (pp. 619–620)

1. perimeter   **2.** sides   **3.** circumference   **4.** $C = \pi d$
5. $C = 2\pi r$ **6.** composite   **7.** 35 cm   **9.** 17.3 ft
11. 23 km   **13.** $2\frac{3}{8}$ or $\frac{19}{8}$ m   **15.** 21 m   **17.** 416 ft
19. $\frac{7}{2}$ cm   **21.** 108 mi   **23.** 262 in.   **25.** 320 m
27. (a) 12π m   (b) 37.68 m   **29.** (a) 10π in.   (b) 31.4 in.
31. (a) 2.5π ft   (b) 7.85 ft   **33.** (a) 2.8π cm   (b) 8.792 cm
35. (a) $\frac{7}{11}\pi$ in.   (b) 2 in.   **37.** (a) π km   (b) $\frac{22}{7}$ km
39. (a) $\frac{5}{4}\pi$ m   (b) $\frac{55}{14}$ m   **41.** (a) 7π mi   (b) 22 mi
43. 18.28 in.   **45.** 278.5 m   **47.** 41.4 mi   **49.** 1040 feet
51. 4740 km   **53.** 400 m

CHECKING BASIC CONCEPTS 10.3 & 10.4 (p. 621)

1. (a) Octagon   (b) Quadrilateral   **2.** (a) 540°   (b) 900°
3. (a) 60°   (b) 120°   **4.** (a) Kite   (b) Rhombus,
parallelogram   **5.** 21 m   **6.** 11.4 in.   **7.** $10\frac{1}{2}$ or $\frac{21}{2}$ in.
8. 31.5 cm   **9.** 72 ft   **10.** 9.42 m   **11.** $\frac{8}{7}$ ft   **12.** 144.25 m
13. 19 in.   **14.** 64 ft

SECTION 10.5 (pp. 627–629)

1. square   **2.** cubic   **3.** square   **4.** False   **5.** volume
6. surface area   **7.** 4900 m²   **9.** 15 in²   **11.** 6.25 ft²
13. 450 cm²   **15.** 94.5 km²   **17.** 113.0 in²   **19.** 132.7 m²
21. 1.5 km²   **23.** 7539.1 ft²   **25.** 343 in³   **27.** 42 ft³
29. 3014.4 cm³   **31.** 37.7 in³   **33.** 270 ft³   **35.** 33.5 cm³
37. 216 in²   **39.** 136 ft²   **41.** 326.6 cm²   **43.** 254.0 in²
45. 57 ft²   **47.** 615.4 cm²   **49.** 188.4 ft²   **51.** 277.5 in²
53. 7 ft²   **55.** 188.4 in²

CHECKING BASIC CONCEPTS 10.5 (p. 629)

1. (a) 20 m²   (b) 0.6 ft²   **2.** (a) 706.5 ft²   (b) 452.2 in²
3. (a) 2744 km³   (b) 565.2 ft³
4. (a) 25.1 m²   (b) 45,216 cm²

CHAPTER 10 REVIEW (pp. 634–636)

1. Line XY, or $\overleftrightarrow{XY}$ **2.** Ray PQ, or $\overrightarrow{PQ}$

3. Line segment AB, or $\overline{AB}$ **4.** Point T **5.** Acute

6. Obtuse   **7.** Right   **8.** Straight   **9.** 66°   **10.** 147°

11. $m\angle c = 76°$; $m\angle f = 104°$; $m\angle e = 76°$

12. $m\angle a = 115°$; $m\angle b = 65°$; $m\angle c = 65°$

13. Supplementary   **14.** Congruent   **15.** Congruent

16. Supplementary   **17.** Acute   **18.** Obtuse   **19.** Right

20. Obtuse   **21.** Scalene   **22.** Equilateral   **23.** Isosceles

24. Scalene   **25.** 84°   **26.** 94°   **27.** 18°   **28.** 40°   **29.** ASA

30. SSS   **31.** Not congruent   **32.** Congruent; SAS

33. Quadrilateral   **34.** Hexagon   **35.** Pentagon

36. Octagon   **37.** 720°   **38.** 540°   **39.** 60°   **40.** 135°

41. Rhombus, parallelogram   **42.** Kite   **43.** Trapezoid

44. Rectangle, parallelogram   **45.** 14 m   **46.** $\frac{4}{9}$ ft

47. 10.6 in.   **48.** 1 cm   **49.** 104 in.   **50.** $\frac{11}{8}$ in.

51. 4 m   **52.** 52.5 in.   **53. (a)** 5π ft   **(b)** 15.7 ft

54. (a) 13π m   **(b)** 40.82 m   **55. (a)** 3π km   **(b)** $\frac{66}{7}$ km

56. (a) $\frac{14}{11}\pi$ cm   **(b)** 4 cm   **57.** 110 ft   **58.** 95.98 m

59. 25.3 m²   **60.** 37.5 ft²   **61.** 49 ft²   **62.** 15 cm²

63. 254.3 cm²   **64.** 0.8 ft²   **65.** 125 ft³   **66.** 3052.1 cm³

67. 50.2 in³   **68.** 256.4 m³   **69.** 207 in²   **70.** 641.8 m²

71. 100.5 cm²   **72.** 146 ft²   **73.** 15 ft   **74.** 129.92 in.

75. 16.3 m²   **76.** 628 in³

CHAPTERS 10 TEST (p. 637)

1. $\angle ABC$ or $\angle B$ **2.** Obtuse   **3.** 19°   **4.** 62°

5. $m\angle a = 108°$; $m\angle f = 72°$; $m\angle h = 108°$ **6.** Acute

7. Isosceles   **8.** 99°   **9.** Congruent; ASA   **10.** Pentagon

11. (a) 1080°   **(b)** 135°   **12.** Parallelogram   **13.** $\frac{4}{9}$ ft

14. 38 cm   **15. (a)** 40π ft   **(b)** 125.6 ft   **16.** 140 cm

17. (a) 2025 m²   **(b)** 216 in²   **18.** 28.3 ft²

19. (a) 33.5 in³   **(b)** 336 ft³

20. (a) 1536 cm²   **(b)** 207.2 ft²

CHAPTERS 1–10 CUMULATIVE REVIEW (pp. 638–639)

1. $10,000 + 8000 + 20$ **2.** 9^4 **3.** 74,000   **4.** 27

5. ← + ◆ + + + + ◆ + + + + →
    $-5\ -4\ -3\ -2\ -1\ \ 0\ \ 1\ \ 2\ \ 3\ \ 4\ \ 5$ **6.** 4   **7.** 28

8. Table values: $-12, -7, -2, 3, 8$; solution: 1   **9.** Yes

10. -2 **11.** 1   **12.** $4 = x + 9$ **13.** $\frac{17}{6}$; $2\frac{5}{6}$

14. (a) Composite   **(b)** Neither   **(c)** Prime   **(d)** Composite

15. -10 **16.** $-\frac{7}{10}$ **17.** 3.3 in.   **18.** 841.57   **19.** 2.4

20. (a) $8, \sqrt{9}$ **(b)** $8, \sqrt{9}, 0$ **(c)** $8, \sqrt{9}, 0$

(d) $8, -\frac{1}{5}, \sqrt{9}, 0$ **(e)** π **21.** 2.9 ft   **22.** \$0.23/oz

23. 60 pt   **24.** 2100 ft/hr   **25.** $x \cdot 80 = 16$ **26.** 4%

27. 22%   **28.** 800   **29. (a)** $25w^8$ **(b)** $5y^3$

30. 6.17×10^5 **31.** $5x^2 - 10x + 1$

32. (a) $14a^5 - 6a^4 + 10a^3$ **(b)** $21m^2 - 38m + 16$

33. A: $(2, 4)$, QI; B: $(0, -3)$, y-axis; C: $(-5, 2)$, QII

34. Table values: $16, 0, 3, -11$ **35.** 6   **36.** 1

37. $m\angle a = 109°$; $m\angle b = 71°$; $m\angle h = 109°$

38. 105°   **39. (a)** 1080°   **(b)** 135°   **40.** 452.2 cm²

41. 18 feet   **42.** $x + 17 = 2x - 6$ **43.** 79.7°F

44. \$11.57; \$189.57   **45. (a)** 200 mi   **(b)** 4 h

46. 114 ft; 98 ft

Glossary

absolute value A real number a, written $|a|$, is equal to its distance from the origin on the number line.

acute angle An angle that measures between 0° and 90°.

acute triangle A triangle with only acute angles.

addends In an addition problem, the two numbers that are added.

addition property of equality If a, b, and c are real numbers, then $a = b$ is equivalent to $a + c = b + c$.

additive inverse (opposite) The additive inverse or opposite of a number a is $-a$.

adjacent angles Supplementary pairs of angles formed by intersecting lines.

algebra A generalization of arithmetic in which letters representing numbers are combined according to the rules of arithmetic.

algebraic expression An expression consisting of numbers, variables, operation symbols such as $+$, $-$, $\times$, and $\div$, and grouping symbols such as parentheses.

alternate exterior angles When two parallel lines are cut by a transversal, alternate exterior angles are outside of the parallel lines on opposite sides of the transversal, and are *not* adjacent angles.

alternate interior angles When two parallel lines are cut by a transversal, alternate interior angles are between the parallel lines on opposite sides of the transversal, and are *not* adjacent angles.

angle-side-angle (ASA) Two triangles are congruent if two angles and the included side of one triangle are congruent to two angles and the included side of the other triangle.

annual interest rate An interest rate that is applied once per year.

annual percentage rate (APR) The interest rate usually used in the computation of compound interest.

approximately equal The symbol $\approx$ indicates that two quantities are nearly equal.

approximation An answer found by estimating that is usually not exactly accurate.

area The number of square units that are needed to cover a region.

arithmetic The branch of mathematics that combines numbers using addition, subtraction, multiplication, and division.

arithmetic mean (mean) The arithmetic mean of a list of numbers is the sum of the numbers divided by the number of numbers in the list.

associative property of addition For any real numbers a, b, and c, $(a + b) + c = a + (b + c)$.

associative property of multiplication For any real numbers a, b, and c, $(a \cdot b) \cdot c = a \cdot (b \cdot c)$.

average The result of adding up the numbers of a list and then dividing the sum by the number of numbers in the list.

bar graph A graph with rectangular bars, where the length of each bar represents a data value.

base The value of b in the expression b^n.

base metric units Meters, grams, and liters.

bases (of a trapezoid) The parallel sides of a trapezoid.

basic percent equation An equation that is written in the form $percent \cdot whole = part$.

basic percent statement form A percent statement written in the form *a percent of the whole is a part*.

basic principle of fractions When simplifying fractions, the principle which states $\dfrac{a \cdot c}{b \cdot c} = \dfrac{a}{b}$.

capacity A measure of the amount of a substance that a container can hold.

Celsius Units used in the metric system to measure temperature.

center of a circle The point in the interior of a circle that is the same distance from every point on the circle.

certain event An event that is certain to occur.

circle A collection of points that are all the same distance from a central point.

circle graph A circle with shaded regions that visually represent percentages.

circumference The perimeter of a circle.

coefficient The number that appears in a term.

coefficient of a monomial The number in a monomial.

commission A percent of total sales, usually paid to the salesperson.

common denominator A denominator that is the same in two or more fractions.

common multiple A number that two or more numbers will divide into evenly.

commutative property of addition For any real numbers a and b, $a + b = b + a$.

commutative property of multiplication For any real numbers a and b, $a \cdot b = b \cdot a$.

complement of an event All outcomes of an experiment that are *not* part of the given event.

complementary angles Two angles whose measures sum to 90°.

completely factored When the coefficient of a term is written as the product of prime numbers and any powers of variables are written as repeated multiplications.

complex fraction A rational expression that contains fractions in its numerator, denominator, or both.

composite number A natural number greater than 1 that is not a prime number.

composite region A region that consists of more than one geometric shape.

compound interest Interest that is based on both the original principal and any earned interest.

congruent angles Angles with the same measure.

congruent triangles Triangles in which the measures of corresponding angles are equal and corresponding sides have the same length.

coordinates The numbers in an ordered pair.

corresponding angles When two parallel lines are cut by a transversal, corresponding angles are in the same relative position with respect to the parallel lines.

cross product For the equation $a/b = c/d$ the cross products are ad and bc.

decimal notation Notation used to represent a number with an integer part and a fraction part separated by a decimal point.

decimal number A number written in decimal notation.

decimal point The dot in a decimal number that separates the integer part and the fraction part.

defining a variable Specifically stating what a variable represents.

degree A degree (°) is 1/360 of a revolution.

degree of a monomial The sum of the exponents of the variables.

degree of a polynomial The degree of the term (or monomial) with greatest degree.

denominator The bottom number in a fraction.

descending order A polynomial in one variable is written in descending order if the exponents on the variable decrease as the terms are written from left to right.

diameter The distance across a circle on a straight line through its center.

difference The answer to a subtraction problem.

discount A percent of the original price that is subtracted from the original price.

distributive properties For any real numbers a, b, and c, $a(b + c) = ab + ac$ and $a(b - c) = ab - ac$.

dividend In a division problem, the number being divided.

divisible One whole number is divisible by a second whole number if their quotient has remainder 0.

divisor In a division problem, the number being divided *into* another.

equation A mathematical statement that two algebraic expressions are equal.

equilateral triangle A triangle with three sides of the same length.

equivalent equations Equations that have the same solution set.

equivalent fractions Fractions that name the same number.

estimation A rough calculation used to find a reasonably accurate answer.

evaluate To find the value of an expression by replacing any variables with given values.

event Any outcome or group of outcomes of an experiment.

expanded form of a whole number A whole number written as a sum of the numbers represented by each digit in the whole number.

experiment An activity with an observable result.

exponent The value of n in the expression b^n.

exponential expression An expression that has an exponent.

exponential notation A notation involving exponents that can be used to represent repeated multiplication.

factor tree A visual diagram used to find the prime factorization of a number.

factoring The process of writing a polynomial as a product of lower degree polynomials.

factors In a multiplication problem, the two numbers multiplied.

Fahrenheit Units used in the American system to measure temperature.

FOIL A method for multiplying two binomials $(A + B)$ and $(C + D)$. Multiply First terms AC, Outside terms AD, Inside terms BC, and Last terms BD; then combine like terms.

formula A special type of equation used to calculate one quantity from given values of other quantities.

fraction A number that can be used to describe a portion of a whole.

graph of a linear equation A straight line in the xy-plane.

graph of a whole number A dot placed on a number line at a whole number's position.

graphical solution A solution to an equation obtained by graphing.

greater than If a real number b is located to the right of a real number a on the number line, we say that b is greater than a, and write $b > a$.

greater than or equal to If a real number a is greater than or equal to b, denoted $a \geq b$, then either $a > b$ or $a = b$ is true.

greatest common factor (GCF) for numbers The largest number that divides evenly into two or more given numbers.

greatest common factor (GCF) for polynomials The term with the greatest degree and greatest integer coefficient that is a factor of all terms in the polynomial.

gross pay Salary or hourly wages before withholdings have been subtracted.

heptagon A polygon with seven sides.

hexagon A polygon with six sides.

hypotenuse The longest side of a right triangle.

identity properties of division If $a \neq 0$, then $a/a = 1$; for any number a, $a/1 = a$.

identity property of multiplication If any number a is multiplied by 1, the result is a, that is, $a \cdot 1 = 1 \cdot a = a$.

impossible event An event that cannot possibly occur.

improper fraction A fraction whose numerator is greater than or equal to its denominator.

inequality When the equals sign in an equation is replaced with any one of the symbols $<$, $\leq$, $>$, or $\geq$.

integers A set of numbers including natural numbers, their opposites, and 0, or $\ldots, -3, -2, -1, 0, 1, 2, 3, \ldots$.

interest A fee for the use of someone's money, which is usually a percent of the principal.

interest rate The percent used to calculate the interest on an investment or loan.

intersecting lines Two lines that cross at a point.

inverse property for addition For any number a, $a + (-a) = 0$.

irrational numbers Real numbers that cannot be expressed as fractions, such as π or $\sqrt{2}$.

isosceles triangle A triangle with at least two sides of the same length.

key (on a pictograph) The portion of a pictograph that gives the meaning of one picture or symbol in the graph.

kite A quadrilateral with two pairs of adjacent sides that are equal in length.

leading coefficient In a polynomial of one variable, the coefficient of the monomial with greatest degree.

least common denominator (LCD) The common denominator with the fewest factors.

least common multiple (LCM) The smallest number that two or more numbers will divide into evenly.

legs of a right triangle The two shorter sides of a right triangle.

less than If a real number a is located to the left of a real number b on the number line, we say that a is less than b and write $a < b$.

less than or equal to If a real number a is less than or equal to b, denoted $a \le b$, then either $a < b$ or $a = b$ is true.

like fractions Fractions with the same denominator.

like terms Two terms that contain the same variables raised to the same powers.

line graph The resulting graph when consecutive data points in a scatterplot are connected with straight line segments.

linear equation in one variable An equation that can be written in the form $ax + b = 0$, where $a \ne 0$.

linear equation in two variables An equation that can be written in the form $Ax + By = C$, where A, B, and C are fixed numbers and A and B are not both equal to 0.

listing method A method for finding the LCM that involves listing multiples of the given numbers.

lowest terms A fraction is in lowest terms if its numerator and denominator have no factors in common.

mass A measure of the amount of matter in an object.

mean (arithmetic mean) The mean of a list of numbers is the sum of the numbers divided by the number of numbers in the list.

measures of central tendency Measures of the location of the "middle" of a list of numbers.

median In an ordered list of numbers, the median is the middle number in a list with an odd number of values, or it is the mean of the two middle numbers in a list with an even number of values.

minuend The number a in the difference $a - b$.

mixed number An integer written with a proper fraction.

mode The value that occurs most often in a list of numbers. A data set can have more than one mode or no mode.

monomial A number, a variable, or a product of numbers and variables raised to natural number powers.

multiplication property of equality If a, b, and c are real numbers with $c \ne 0$, then $a = b$ is equivalent to $ac = bc$.

multiplicative identity The number 1.

multiplicative inverse (reciprocal) The multiplicative inverse of a nonzero number a is $1/a$.

natural numbers The set of numbers 1, 2, 3, 4, 5, 6, ...

negative number A number that is less than zero.

negative square root The negative square root is denoted $-\sqrt{a}$.

net pay The amount remaining after withholdings have been subtracted from gross pay.

number line A horizontal line marked with evenly spaced tick marks that is often used to graph numbers.

numerator The top number in a fraction.

numerical solution A solution often obtained by using a table of values.

obtuse angle An angle that measures between 90° and 180°.

obtuse triangle A triangle with one obtuse angle.

octagon A polygon with eight sides.

opposite (additive inverse) The opposite, or additive inverse, of a number a is $-a$.

order of operations agreement A set of rules used to ensure consistency in the evaluation of mathematical expressions.

ordered pair A pair of numbers written in parentheses (x, y), in which the order of the numbers is important.

origin On the number line, the point associated with the real number 0; in the xy-plane, the point where the axes intersect, $(0, 0)$.

outcome A result of an experiment.

parallel lines Two or more lines in the same plane that never intersect.

parallelogram A quadrilateral with two pairs of parallel sides.

partial dividend The number formed by starting at the left end of a dividend and selecting the fewest digits that give a number that is greater than the divisor.

pentagon A polygon with five sides.

percent A ratio with a denominator of 100.

percent chance A probability written as a percent.

percent change If a quantity changes from x to y, then the percent change is $[(y - x)/x] \times 100$.

percent decrease A negative percent change.

percent increase A positive percent change.

percent problem A question asking us to find the percent, whole, or part in a percent statement.

perfect square A number with an integer square root.

perimeter The distance around an enclosed region.

period A group of digits in a whole number.

perpendicular lines Two lines in a plane that intersect to form a right (90°) angle.

pictograph A graph that uses a picture or symbol to represent a quantity visually.

place value The position of a digit in a number that is written in standard form.

plane A flat surface that continues without end.

plane geometry geometry that focuses on figures with two or fewer dimensions.

plotting Graphing points in the xy-plane.

polygon A closed plane figure determined by three or more line segments.

polynomial The sum of one or more monomials.

polynomials in one variable Polynomials that contain one variable.

positive number A number that is greater than zero.

prime factorization A number written as a product of prime numbers.

prime factorization method A method for finding the LCM that uses the prime factorizations of the given numbers.

prime number A natural number greater than 1 that has only itself and 1 as natural number factors.

principal An amount of money that is initially borrowed or invested.

principal square root The square root of a that is nonnegative, denoted $\sqrt{a}$.

probability A measure of the likelihood of an event occurring, expressed as a real number between 0 and 1. A probability of 0 indicates that an event is impossible, whereas a probability of 1 indicates that an event is certain.

product The answer to a multiplication problem.

proper fraction A fraction whose numerator is less than its denominator.

proportion A statement that two ratios are equal.

Pythagorean theorem If a right triangle has legs a and b with hypotenuse c, then $a^2 + b^2 = c^2$.

quadrants The four regions determined by the xy-plane.

quadrilateral A polygon with four sides.

quotient The answer to a division problem.

radical sign The symbol $\sqrt{\ }$.

radicand The expression under the radical sign.

radius The distance from the center of a circle to any point on the circle.

rate A ratio used to compare different kinds of quantities.

ratio A comparison of two quantities, expressed as a quotient.

rational expression A polynomial divided by a nonzero polynomial.

rational number Any number that can be expressed as the ratio of two integers p/q, where $q \neq 0$; a fraction.

real numbers All rational and irrational numbers; any number that can be written in decimal form.

reciprocal (multiplicative inverse) The reciprocal of a nonzero number a is $1/a$.

rectangle A quadrilateral in which all angles measure 90°.

rectangular coordinate system (xy-plane) The xy-plane used to plot points and graph data.

regular polygon A polygon in which all sides have equal length and all angles have equal measure.

remainder The amount left over when the quotient of two whole numbers is not a whole number.

repeat bar A bar written above a group of digits in a decimal number to indicate that the digits repeat.

repeating decimal A decimal number with digits to the right of the decimal point that continue without end in a repeating pattern.

rhombus A quadrilateral with four sides of equal length.

right angle An angle that measures 90°.

right triangle A triangle with one right angle.

rounding a number Approximating a number to a given level of accuracy.

sale price The result obtained by subtracting the discount from the original price.

sales tax A percent of the original price that is added to the original price.

scalene triangle A triangle with no sides of the same length.

scatterplot A graph of distinct points plotted in the xy-plane.

scientific notation A real number a written as $b \times 10^n$, where $1 \leq |b| < 10$ and n is an integer.

semicircle Half of a circle.

side-angle-side (SAS) Two triangles are congruent if two sides and the included angle of one triangle are congruent to two sides and the included angle of the other triangle.

side-side-side (SSS) Two triangles are congruent if three sides of one triangle are congruent to three sides of the other triangle.

sides of an angle The two rays that form an angle.

sides of a polygon The line segments that form a polygon.

signed numbers Positive numbers, negative numbers, and zero.

similar figures Two figures in which the measures of corresponding angles are equal and the measures of corresponding sides are proportional.

simple interest Interest that is based only on the original principal.

solution Each value of the variable that makes the equation true.

solution set The set of all solutions to an equation.

solving an equation Finding all of the solutions to an equation.

speed A rate that gives a distance traveled in an amount of time.

square A regular quadrilateral.

square root The number b is a square root of a number a if $b^2 = a$.

square unit A measure of area represented by a square that measures 1 unit on each side.

standard form of an equation for a line The equation given by $Ax + By = C$, where A, B, and C are fixed numbers with A and B not both 0.

standard form of a whole number A whole number written in digits with periods separated by commas.

straight angle An angle that measures 180°.

subtrahend The number b in the difference $a - b$.

sum The answer to an addition problem.

supplementary angles Two angles whose measures sum to 180°.

surface area A measure of the amount of exposed area for a given solid.

symbolic solution A solution to an equation obtained by using properties of equations; the resulting solution set is exact.

table A structure used to display information visually in a rectangular array.

term A number, a variable, or a product of numbers and variables raised to powers.

total price The sum of the original price and the sales tax.

total value (of an investment) The sum of the principal and the interest.

transversal A line that intersect two other lines in the same plane.

trapezoid A quadrilateral with one pair of parallel sides.

triangle A polygon with three sides.

undefined An expression is undefined when it involves division by zero.

unit fraction A fraction that is equivalent to 1.

unit pricing A ratio used to compare pricing.

unit rate Rates expressed with a denominator of 1.

unit ratio A ratio expressed with a denominator of 1.

unlike terms Terms that are not like terms.

variable A symbol, such as x, y, or z, used to represent any unknown quantity.

vertex of an angle The common endpoint of the two rays that form an angle.

vertex of a polygon A common endpoint of two sides of a polygon.

vertical angles Congruent pairs of angles formed by intersecting lines.

volume A measure of the amount of a substance.

weight A measure of the force on an object due to gravity.

weighted mean A process for finding the mean of a list of numbers in which each number is multiplied by the number of times it occurs in the list.

whole numbers The set of numbers $0, 1, 2, 3, 4, 5, \ldots$.

withholdings Taxes, insurance, and other deductions that are subtracted from gross pay.

word form of a whole number A whole number written in words.

x-axis The horizontal axis in the xy-plane.

x-coordinate The first number in an ordered pair.

xy-plane (rectangular coordinate system) The system used to plot points and graph data.

y-axis The vertical axis in the xy-plane.

y-coordinate The second number in an ordered pair.

zero property of addition If 0 is added to any real number a, the result is a, that is, $a + 0 = 0 + a = a$.

zero property of multiplication If any real number a is multiplied by 0, the result is 0, that is, $a \cdot 0 = 0 \cdot a = 0$.

zero properties of division If $a \neq 0$, then $0/a = 0$; for any number a, $a/0$ is undefined.

Photo Credits

Front left endsheet: **upper right,** Monkey Business Images/Shutterstock **upper left,** Stockbyte/Thinkstock **lower right,** Diego Cervo/Shutterstock **lower left,** Micha Rosenwirth/Shutterstock
Front right endsheet: **upper right,** Valua Vitaly/Shutterstock **upper left,** Diego Cervo/Shutterstock **lower right,** Natalia Bratslavsky/Shutterstock **lower left,** Sebastian Czapnik/Dreamstime **1,** Digital Vision/Thinkstock **13,** Xrrr/Dreamstime **14,** Photodisc/Thinkstock **22,** Galina Barskaya/Shutterstock **23 top,** Alaska Stock LLC/Alamy **23 bottom,** Bruce Bennett Studios/Getty Images **28,** Monkey Business Images/Shutterstock **39 top,** J.D.S./Shutterstock **39 bottom,** Stockbyte/Thinkstock **43,** JPL/NASA **44,** Horticulture/Fotolia **50,** Comstock/Thinkstock **51,** Regien Paassen/Shutterstock **56,** Paramount Pictures/Photofest **57,** Alex Segre/Alamy **65,** Faraways/Shutterstock **71,** Kenneth Sponsler/Shutterstock **76,** Gallofoto/Shutterstock **88,** Micha Rosenwirth/Shutterstock **89,** Yellowj/Shutterstock **98,** Laurin Rinder/Shutterstock **107,** Andrea Danti/Shutterstock **113,** Scoutingstock/Shutterstock **114,** Melvyn Longhurst/Alamy **123,** Herbert Kratky/Shutterstock **131,** PRNewsFoto/Microsoft Corp./Netflix, Inc./Newscom **149,** Jim Dietz/AP Images **150,** OtnaYdur/Shutterstock **158,** Schalk van Zuydam/AP Images **164,** Naluwan/Shutterstock **172,** Byron W. Moore/Shutterstock **184,** Jose Gil/Shutterstock **192,** Copyright © 2010 Toyota Motors Sales, U.S.A., Inc. **197,** Patrick Kobernus/NCTC Image Library/U.S. Fish and Wildlife Service **200,** Sdecoret/Shutterstock **201,** Valmas/Shutterstock **212,** Johnfoto18/Shutterstock **226,** Steve Cukrov/Shutterstock **238,** Goodluz/Shutterstock **239,** Evgeny Dubinchuk/Shutterstock **245,** VanHart/Shutterstock **246,** Ferenc Cegledi/Shutterstock **252,** Leksele/Shutterstock **256,** Elena Elisseeva/Shutterstock **265,** Svariophoto/Shutterstock **272,** Stefan Fierros/Shutterstock **299,** Gail Mooney-Kelly/Alamy **300,** Kira-N/Shutterstock **309,** Stuart Monk/Shutterstock **310,** Monkey Business Images/Shutterstock **319,** Levent Konuk/Shutterstock **328,** Michael Shake/Shutterstock **332,** Vlue/Shutterstock **333,** Pasieka/SPL/Getty Images **343,** Skull82/Shutterstock **348,** Monkey Business Images/Shutterstock **354,** Kochneva Tetyana/Shutterstock **359,** Donald Sawvel/Shutterstock **380,** Rob Bouwman/Shutterstock **381,** 3Dfoto/Shutterstock **389,** Natalia Bratslavsky/Shutterstock **393 left,** Reddogs/Shutterstock **393 right,** Aleksandrs Kobilanskis/Shutterstock **401,** Galushko Sergey/Shutterstock **408,** Nicolas Asfouri/AFP/Getty Images **418,** Wavebreakmedia Ltd/Shutterstock **428,** HomeStudio/Shutterstock **429,** Konstantin Sutyagin/Shutterstock **441,** Andresr/Shutterstock **442,** Susan Chiang/Shutterstock **450,** RoJo Images/Shutterstock **457,** JohnKwan/Shutterstock **463,** Bikeriderlondon/Shutterstock **471,** Tyler Olson/Shutterstock **473,** Levent Konuk/Shutterstock **480,** Image Source/Getty Images **497,** James Edward Bates/Biloxi Sun Herald/MCT/Getty Images **498,** Stockbyte/Thinkstock **512,** Diego Cervo/Shutterstock **522,** Valua Vitaly/Shutterstock **539,** Max Topchii/Shutterstock **540,** Evanto/Alamy **549,** Tan Wei Ming/Shutterstock **560,** Dmitriy Shironosov/Shutterstock **567,** Sho Shan 2/Alamy **586,** Rabbit75/Dreamstime **587,** Normadesmond/Dreamstime **605,** NHPA/Photo Researchers, Inc. **613 top,** Craig Barhorst/Shutterstock **613 bottom,** Digital Vision/Thinkstock **614 top,** Stockbyte/Thinkstock **614 bottom,** Ian Dagnall/Alamy **622,** Olga Lyubkina/Shutterstock **629,** Luciano Mortula/Shutterstock

Index